Advances in Intelligent Systems and Computing

Volume 1177

AF805282

The series "Advances in Intelligent Systems and Computing" contains publications on theory, applications, and design methods of Intelligent Systems and Intelligent Computing. Virtually all disciplines such as engineering, natural sciences, computer and information science, ICT, economics, business, e-commerce, environment, healthcare, life science are covered. The list of topics spans all the areas of modern intelligent systems and computing such as: computational intelligence, soft computing including neural networks, fuzzy systems, evolutionary computing and the fusion of these paradigms, social intelligence, ambient intelligence, computational neuroscience, artificial life, virtual worlds and society, cognitive science and systems, Perception and Vision, DNA and immune based systems, self-organizing and adaptive systems, e-Learning and teaching, human-centered and human-centric computing, recommender systems, intelligent control, robotics and mechatronics including human-machine teaming, knowledge-based paradigms, learning paradigms, machine ethics, intelligent data analysis, knowledge management, intelligent agents, intelligent decision making and support, intelligent network security, trust management, interactive entertainment, Web intelligence and multimedia.

The publications within "Advances in Intelligent Systems and Computing" are primarily proceedings of important conferences, symposia and congresses. They cover significant recent developments in the field, both of a foundational and applicable character. An important characteristic feature of the series is the short publication time and world-wide distribution. This permits a rapid and broad dissemination of research results.

**** Indexing: The books of this series are submitted to ISI Proceedings, EI-Compendex, DBLP, SCOPUS, Google Scholar and Springerlink ****

More information about this series at http://www.springer.com/series/11156

Suresh Chandra Satapathy · Yu-Dong Zhang ·
Vikrant Bhateja · Ritanjali Majhi
Editors

Intelligent Data Engineering and Analytics

Frontiers in Intelligent Computing: Theory and Applications (FICTA 2020), Volume 2

Springer

Editors
Suresh Chandra Satapathy
School of Computer Engineering
Kalinga Institute Industrial Technology
Bhubaneswar, Odisha, India

Yu-Dong Zhang
Department of Informatics
University of Leicester
Leicester, UK

Vikrant Bhateja
Department of Electronics and Communication Engineering
Shri Ramswaroop Memorial Group of Professional Colleges (SRMGPC)
Lucknow, Uttar Pradesh, India

Dr. A.P.J. Abdul Kalam Technical University
Lucknow, Uttar Pradesh, India

Ritanjali Majhi
School of Management
National Institute of Technology Karnataka
Surathkal, Karnataka, India

ISSN 2194-5357 ISSN 2194-5365 (electronic)
Advances in Intelligent Systems and Computing
ISBN 978-981-15-5678-4 ISBN 978-981-15-5679-1 (eBook)
https://doi.org/10.1007/978-981-15-5679-1

This Springer imprint is published by the registered company Springer Nature Singapore Pte Ltd.
The registered company address is: 152 Beach Road, #21-01/04 Gateway East, Singapore 189721, Singapore

Organization

Chief Patrons

Prof. K. Balaveera Reddy, Chairman, BOG, NITK Surathkal
Prof. Karanam Umamaheshwar Rao, Director, NITK Surathkal

Patrons

Prof. Ananthanarayana V. S., Deputy Director, NITK Surathkal
Prof. Aloysius H. Sequeira, Dean Faculty Welfare, NITK Surathkal
Prof. G. Ram Mohana Reddy, Professor-HAG, IT Department, NITK Surathkal
Dr. S. Pavan Kumar, Head, School of Management, NITK Surathkal

Organizing Chairs

Dr. Ritanjali Majhi, Associate Professor, School of Management, NITK Surathkal
Dr. S. Sowmya Kamath, Assistant Professor, Department of Information Technology, NITK Surathkal
Dr. Suprabha K. R., Assistant Professor, School of Management, NITK Surathkal
Dr. Geetha V., Assistant Professor, Department of Information Technology, NITK Surathkal
Dr. Rashmi Uchil, Assistant Professor, School of Management, NITK Surathkal
Dr. Biju R. Mohan, Assistant Professor and Head, Department of Information Technology, NITK Surathkal

Dr. Pradyot Ranjan Jena, Assistant Professor, School of Management, NITK Surathkal
Dr. Nagamma Patil, Assistant Professor, Department of Information Technology, NITK Surathkal

Publicity Chairs

Dr. Suprabha K. R., Assistant Professor, SOM, NITK Surathkal, India
Dr. Geetha V., Assistant Professor, Department of IT, NITK Surathkal, India
Dr. Rashmi Uchil, Assistant Professor, SOM, NITK Surathkal, India
Dr. Biju R. Mohan, Assistant Professor and Head, Department of IT, NITK Surathkal, India

Advisory Committee

Prof. Abrar A. Qureshi, Professor, University of Virginia's College at Wise, USA
Dr. Alastair W. Watson, Program Director, Faculty of Business, University of Wollongong, Dubai
Prof. Anjan K. Swain, IIM Kozhikode, India
Prof. Anurag Mittal, Department of Computer Science and Engineering, IIT Madras, India
Dr. Armin Haller, Australian National University, Canberra
Prof. Arnab K. Laha, IIM Ahmedabad, India
Prof. Ashok K. Pradhan, IIT Kharagpur, India
Prof. Athanasios V. Vasilakos, Professor, University of Western Macedonia, Greece/Athens
Prof. Atreyi Kankanhalli, NUS School of Computing, Singapore
Prof. A. H. Sequeira, Dean Faculty Welfare, NITK Surathkal, India
Prof. Carlos A. Coello Coello, Centro de Investigación y de Estudios Avanzados del Instituto
Prof. Charles Vincent,Director of Research, Buckingham University, UK
Prof. Chilukuri K. Mohan, Professor, Syracuse University, Syracuse, NY, USA
Prof. Dipankar Dasgupta, Professor, the University of Memphis, TN
Prof. Durga Toshniwal, IIT Roorkee, India
Dr. Elena Cabrio, University of Nice Sophia Antipolis, Inria, CNRS, I3S, France
Prof. Ganapati Panda, Ex Deputy director, IIT Bhubaneswar
Prof. Gerardo Beni, Professor, University of California, CA, United States
Dr. Giancarlo Giudici, Politecnico di Milano DIG School of Management, Milano, Italy
Prof. G. K. Venayagamoorthy, Professor, Clemson University, Clemson, SC, USA
Mr. Harish Kamath, Master Technologist, HP Enterprise, Bangalore

Prof. Heitor Silvério Lopes, Professor, Federal University of Technology Paraná, Brazil
Prof. Hoang Pham, Distinguished Professor, Rutgers University, Piscataway, NJ, USA
Prof. Jeng-Shyang Pan, Shandong University of Science and Technology Qingdao, China
Prof. Juan Luis Fernández Martínez, Professor, University of Oviedo, Spain
Prof. Kailash C. Patidar, Senior Professor, University of the Western Cape, South Africa
Prof. Kerry Taylor, Australian National University, Canberra
Prof. Kumkum Garg, Ex.Prof IIT Roorkee, Pro-Vice Chancellor Manipal University, Jaipur
Prof. K. Parsopoulos, Associate Professor, University of Ioannina, Greece
Prof. Leandro Dos Santos Coelho, Associate Professor, Federal University of Parana, Brazil
Prof. Lexing Xie, Professor of Computer Science, Australian National University, Canberra
Prof. Lingfeng Wang, University of Wisconsin-Milwaukee Milwaukee, WI, USA
Mr. Mahesha Nanjundaiah, Director of Engineering, HP Enterprise, Bangalore
Prof. Maurice Clerc, Independent Consultant, France Télécom, Annecy, France
Prof. M. A. Abido, King Fahd University of Petroleum and Minerals, Dhahran, Saudi Arabia
Prof. Naeem Hanoon, Multimedia University, Cyberjaya, Malaysia
Prof. Narasimha Murthy, Department of Computer Science and Automation, IISc, Bangalore
Prof. Oscar Castillo, Professor, Tijuana Institute of Technology, Mexico
Prof. Pei-Chann Chang, Professor, Yuan Ze University, Taoyuan, Taiwan
Prof. Peng Shi, Professor, University of Adelaide, Adelaide, SA, Australia
Dr. Prakash Raghavendra, Principal Member of Technical Staff, AMD India
Prof. Rafael Stubs Parpinelli, Professor, State University of Santa Catarina, Brazil
Prof. Raj Acharya, Dean and Rudy Professor of Engineering, Computer Science and Informatics, Indiana University, USA
Prof. Raghav Gowda, Professor, University of Dayton-Ohio, USA
Prof. Roderich Gross, Senior Lecturer, University of Sheffield, UK
Mr. Rudramuni, Vice President, Dell EMC, Bangalore
Prof. Saman Halgamuge, Professor, University of Melbourne, Australia
Prof. Subhadip Basu, Professor, Jadavpur University, India
Prof. Sumanth Yenduri, Professor, Kennesaw State University, USA
Prof. Sumit Kumar Jha, Department of Computer Science, University of Central Florida, USA
Prof. S. G. Ponnambalam, Professor, Subang Jaya, Malaysia
Dr. Suyash P. Awate, Department of Computer Science and Engineering, IIT Bombay
Dr. Valerio Basile, Research Fellow, University of Turin, Italy

Dr. Vineeth Balasubramanian, Department of Computer Science and Engineering, IIT Hyderabad
Dr. Vikash Ramiah, Associate Professor, Applied Finance, University of South Australia
Prof. X. Z. Gao, Docent, Aalto University School of Electrical Engineering, Finland
Prof. Ying Tan, Associate Professor, the University of Melbourne, Australia
Prof. Zong Woo Geem, Gachon University in South Korea

Technical Program Committee

Dr. Anand Kumar M., Assistant Professor, Department of IT, NITK Surathkal
Dr. Babita Majhi, Assistant Professor, Department of IT, G G University Bilashpur
Dr. Bhawana Rudra, Assistant Professor, Department of IT, NITK Surathkal
Dr. Bibhu Prasad Nayak, Associate Professor, Department of HSS, TISS Hyderabad
Dr. Bijuna C. Mohan, Assistant Professor, School of Management, NITK Surathkal
Dr. Dhishna P., Assistant Professor, School of Management, NITK Surathkal
Mr. Dinesh Naik, Assistant Professor, Department of Information Technology, NITK Surathkal
Prof. Geetha Maiya, Department of Computer Science and Engineering, MIT Manipal
Dr. Gopalakrishna B. V., Assistant Professor, School of Management, NITK Surathkal
Dr. Keshavamurthy B. N., Associate Professor, Department of CSE, NIT Goa
Dr. Kiran M., Assistant Professor, Department of Information Technology, NITK Surathkal
Prof. K. B. Kiran, Professor, School of Management, NITK Surathkal
Dr. Madhu Kumari, Assistant Professor, Department of CSE, NIT Hamirpur
Dr. Mussarrat Shaheen, Assistant Professor, IBS Hyderabad
Dr. Pilli Shubhakar, Associate Professor, Department of CSE, MNIT Jaipur
Dr. P. R. K. Gupta, Institute of Finance and International Management, Bengaluru.
Dr. Rajesh Acharya H., Assistant Professor, School of Management, NITK Surathkal
Dr. Ranjay Hazra, Assistant Professor, Department of EIE, NIT Silchar
Dr. Ravikumar Jatoth, Associate Professor, Department of ECE, NIT Warangal
Dr. Rohit Budhiraja, Assistant Professor, Department of EE, IIT Kanpur
Dr. Sandeep Kumar, Associate Professor, Department of CSE, IIT Roorkee
Dr. Savita Bhat, Assistant Professor, School of Management, NITK Surathkal
Dr. Shashikantha Koudur, Associate Professor, School of Management, NITK Surathkal
Dr. Shridhar Domanal, IBM India, Bengaluru
Dr. Sheena, Associate Professor, School of Management, NITK Surathkal
Dr. Sreejith A., Assistant Professor, School of Management, NITK Surathkal

Preface

This book is a collection of high-quality peer-reviewed research papers presented at the 8th International Conference on Frontiers in Intelligent Computing: Theory and Applications (FICTA 2020) held at National Institute of Technology, Karnataka, Surathkal, India, during 4–5 January 2020.

The idea of this conference series was conceived by a few eminent professors and researchers from premier institutions of India. The first three editions of this conference FICTA 2012, 2013, and 2014 were organized by Bhubaneswar Engineering College (BEC), Bhubaneswar, Odisha, India. The fourth edition FICTA 2015 was held at NIT, Durgapur, W.B., India. The fifth and sixth editions FICTA 2016 and FICTA 2017 were consecutively organized by KIIT University, Bhubaneswar, Odisha, India. FICTA 2018 was hosted by Duy Tan University, Da Nang City, Vietnam. All past seven editions of the FICTA conference proceedings are published in Springer AISC Series. Presently, FICTA 2020 is the eighth edition of this conference series which aims to bring together researchers, scientists, engineers, and practitioners to exchange and share their theories, methodologies, new ideas, experiences, applications in all areas of intelligent computing theories, and applications to various engineering disciplines like Computer Science, Electronics, Electrical, Mechanical, Biomedical Engineering, etc.

FICTA 2020 had received a good number of submissions from the different areas relating to computational intelligence, intelligent data engineering, data analytics, decision sciences, and associated applications in the arena of intelligent computing. These papers have undergone a rigorous peer-review process with the help of our technical program committee members (from the country as well as abroad). The review process has been very crucial with minimum 02 reviews each; and in many cases, 3–5 reviews along with due checks on similarity and content overlap as well. This conference witnessed more than 300 papers including the main track as well as special sessions. The conference featured five special sessions in various cutting-edge technologies of specialized focus which were organized and chaired by eminent professors. The total toll of papers included submissions received across the country along with 06 overseas countries. Out of this pool, only 147 papers were given acceptance and were segregated as two different volumes for publication under the proceedings. This volume consists of 72 papers from diverse areas of Intelligent Data Engineering and Analytics.

The conference featured many distinguished keynote addresses in different spheres of intelligent computing by eminent speakers like Dr. Venkat N. Gudivada, (Professor and Chair, Department of Computer Science, East Carolina University, Greenville, USA); Prof. Ganapati Panda (Professor and Former Deputy Director, Indian Institute of Technology, Bhubaneswar, Odisha, India); and Dr. Lipo Wang (School of Electrical and Electronic Engineering, Nanyang Technological University, Singapore). Last but not the least, the invited talk on "Importance of Ethics in Research Publishing" delivered by Mr. Aninda Bose (Senior Editor—Interdisciplinary Applied Sciences, Publishing Department, Springer Nature) received ample applause from the vast audience of delegates, budding researchers, faculty, and students.

We thank the advisory chairs and steering committees for rendering mentor support to the conference. An extreme note of gratitude to Prof. Suresh Chandra Satapathy (KIIT University, Bhubaneshwar, Odisha, India) for providing valuable guidelines and being an inspiration in the entire process of organizing this conference. We would also like to thank School of Management and the Department of Information Technology, NIT Karnataka, Surathkal, who jointly came forward and provided their support to organize the eighth edition of this conference series.

We take this opportunity to thank authors of all submitted papers for their hard work, adherence to the deadlines, and patience with the review process. The quality of a refereed volume depends mainly on the expertise and dedication of the reviewers. We are indebted to the technical program committee members who not only produced excellent reviews but also did these in short time frames. We would also like to thank the participants of this conference, who have participated in the conference despite all hardships.

Volume Editors

Bhubaneswar, India — Dr. Suresh Chandra Satapathy
Leicester, UK — Dr. Yu-Dong Zhang
Lucknow, India — Dr. Vikrant Bhateja
Surathkal, India — Dr. Ritanjali Majhi

About This Book

The book covers proceedings of 8th International Conference on Frontiers of Intelligent Computing: Theory and applications (FICTA 2020) that aims to bring together researchers, scientists, engineers, and practitioners to exchange their new ideas and experiences in domain of intelligent computing theories with prospective applications to various engineering disciplines. The book is divided in to two volumes: Evolution in Computational Intelligence (Volume-1) and Intelligent Data Engineering and Analytics (Volume-2).

This volume covers broad areas of Intelligent Data Engineering and Analytics. The conference papers included herein presents both theoretical as well as practical aspects of data intensive computing, data mining, big data, knowledge management, intelligent data acquisition and processing from sensors data communication networks protocols and architectures, etc. The volume will also serve as knowledge centre for students of post-graduate level in various engineering disciplines.

Contents

About the Editors

Suresh Chandra Satapathy is a Professor, School of Computer Engg, KIIT Deemed to be University, Bhubaneswar, India. His research interest includes machine learning, data mining, swarm intelligence studies and their applications to engineering. He has more than 140 publications to his credit in various reputed international journals and conference Proceedings. He has edited many volumes from Springer AISC, LNEE, SIST etc. He is a senior member of IEEE and Life Member of Computer society of India.

Prof. Dr. Yu-Dong Zhang received his Ph.D. degree from Southeast University, China, in 2010. From 2010 to 2013, he worked as post-doc and then as a research scientist at Columbia University, USA. He served as Professor at Nanjing Normal University from 2013 to 2017, and is currently a Full Professor at the University of Leicester, UK. His research interests include deep learning in communication and signal processing, and medical image processing.

Vikrant Bhateja is Associate Professor, Department of ECE in SRMGPC, Lucknow. His areas of research include digital image and video processing, computer vision, medical imaging, machine learning, pattern analysis and recognition. He has around 150 quality publications in various international journals and conference proceedings. He has edited more than 25 volumes of conference proceedings with Springer Nature (AISC, SIST, LNEE, LNNS Series). He is associate editor of IJSE and IJRSDA and presently EiC of IJNCR journal under IGI Global.

Dr. Ritanjali Majhi is an Associate Professor at the School of Management, National Institute of Technology Karnataka, Surathkal, India. She is an expert on green marketing, big data analysis, consumer decision-making, time series prediction, AI applications in management (marketing effectiveness) and marketing analytics. She has more than 15 years of research experience, including in projects funded by the Indian government. She has published about 100 research papers in various peer-reviewed international journals and at conferences.

Classification of Dry/Wet Snow Using Sentinel-2 High Spatial Resolution Optical Data

V. Nagajothi, M. Geetha Priya, Parmanand Sharma, and D. Krishnaveni

Abstract The proposed study targets to utilize satellite optical data with remote sensing techniques to classify the wet and dry snow of Himalayan glaciers. The study has been carried out for Miyar glacier, one of the largest glaciers of Miyar basin, Western Himalayas, using Sentinel-2(A and B) high-resolution, multispectral imaging data for the Hydrological year 2018–2019. To estimate the snow cover area and to classify the snow as wet/dry, optical band ratios and slicing have been adopted in the data processing algorithm. Results obtained show that the proposed algorithm is capable of mapping dry snow region, wet snow region and bed/moraine covered glacier ice for a given glacier with high spatial resolution. Dry snow and wet snow areas observed during summer (June–September) are approximately 70.11 km^2 and 5.50 km^2 on average, respectively. During winter (November–May), dry and wet snow areas observed are approximately 48.58 km^2 and 12.57 km^2 on average, respectively.

Keywords Miyar glacier · Sentinel 2 · NDSI · Dry snow · Wet snow

1 Introduction

Outside the Polar region, the Himalayas contain the largest source of freshwater, hence it is also called as "Third pole" [1]. Glacier studies in the Himalayas are very important in both economical as well as scientific view. Besides its importance towards climate change, glaciers are considered as the key indicators of Global

V. Nagajothi · M. Geetha Priya (✉)
CIIRC – Jyothy Institute of Technology, Bengaluru 560082, India
e-mail: geetha.sri82@gmail.com

P. Sharma
NCPOR – Ministry of Earth Sciences, Goa 403804, India

D. Krishnaveni
Department of ECE, Jyothy Institute of Technology, Bengaluru 560082, India

S. C. Satapathy et al. (eds.), *Intelligent Data Engineering and Analytics*, Advances in Intelligent Systems and Computing 1177,
https://doi.org/10.1007/978-981-15-5679-1_1

warming [2]. Monitoring of Himalayan glaciers is also essential for hazard assessment, effects on hydrology, including sea-level rise and water security. Understanding and monitoring of Seasonal Snow Cover Area (SCA) is important for an agricultural country like India, where meltwater from snow plays an important role in feeding the Indus, Brahmaputra and Ganga rivers [3]. Remote sensing of SCA has been used effectively to monitor replacing the conventional method [4]. Remote sensing data provide us multi-sensor and multi-temporal data which can be used to monitor glacier area, glacier length, Equilibrium Line Altitude (ELA), terminus position and accumulation/ablation rates from which mass balance can be inferred [5].

Microwave remote sensing data has significant importance in snow/glacier/ice studies since the properties of ice and water behave differently in the microwave region of the spectrum [6]. But, SAR data are difficult to interpret and process compared to optical data. Further, albedo is responsible for the snowmelt runoff process which cannot be calculated using microwave data [2]. Hence in this study, Sentinel 2 A and B from European Space Agency (ESA) and Landsat 8 from USGS have been used to classify the dry snow, wet snow and moraine covered glacier ice in the study area located in Western Himalayas.

2 Data Used and Study Area

Sentinel 2 (A and B) data is available for free download on the European Space Agency website (ESA, https://scihub.copernicus.eu/dhus/#/home) with a spatial resolution of 10 m and combined revisit period of 5 days. Sentinel 2 Level-1C data—top of atmospheric reflectances in cartographic geometry with cloud cover less than 10%—has been used for Hydrological Year (HY) October 2018–September 2019. For the months January and May, Landsat 8 C1 Level 1 data with a spatial resolution of 30 m and revisit period of 16 days has been used (Table 1) due to the unavailability of cloud free data from Sentinel 2. Similarly, Landsat-7/8 and Sentinel-2 data were unavailable for Febuary, 2019. Meteorological field data from the Indian Meteorological Department (IMD, Keylong station) has been used for validation purposes (http://www.imd.gov.in/pages/city_weather_show.php).

Miyar basin is located (32° 40′ 0″ N–33° 10′ 0″ N and 76° 30′ 0″ E–77° 0′ 0″ E) in the Lahaul and Spiti district of Himachal Pradesh (Fig. 1). The basin is one of the

Table 1 Data specifications

Satellite and sensor	Spatial resolution	Temporal resolution
Sentinel 2 A and B, MSI—Multispectral Imager	10 m	5 days (combined revisit period)
Landsat 8 OLI—Operational Land Imager	30 m	16 days

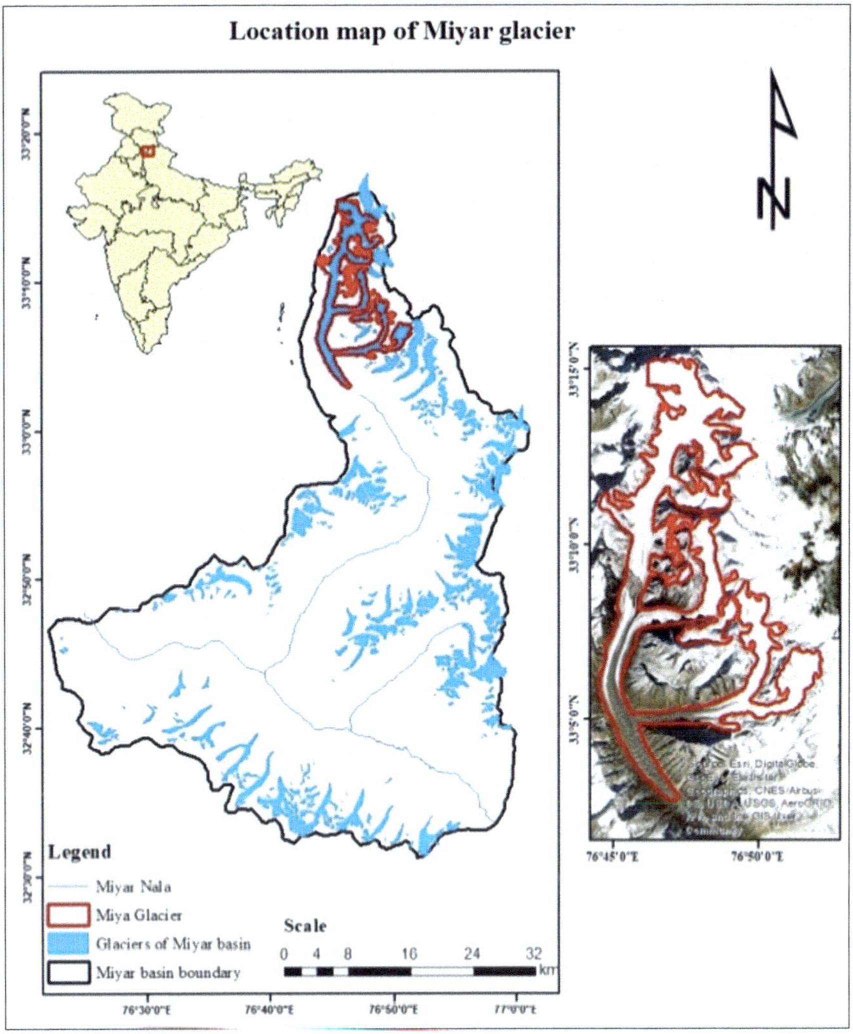

Fig. 1 Location map of the study area

important sub-basins of river Chenab, which consists of total 173 glaciers. Miyar glacier, one of the largest glaciers in the Miyar basin (76° 45′ 44″–76° 50′ 64″ E, 33° 8′ 23″–33° 15′ 53″ N) which has been least explored by the scientific community, has been selected for this study. The Miyar Nala originates from the snout at an altitude of 4200 m.a.s.l Miyar glacier and joins river Chenab at Udaipur, HP [7]. The present length of the glacier is 24 km with an area of 79.33 km^2.

3 Methodology

Sentinel 2 Level-1C data which is Top of Atmospheric (TOA) reflectance has been used for the classification of dry and wet snow (Fig. 2). For the present study, Band 3 (Green), Band 11 (SWIR) and Band 8 (NIR band) of Sentinel 2 have been used. Band 11 which is of 20 m resolution has been resampled to 10 m resolution using the nearest neighbourhood algorithm. Normalized Difference Snow Index (NDSI) [4] has been used to estimate snow cover area and NIR/SWIR ratio has been used to mask the water pixels as the water pixel has low reflectance in infrared bands. NDSI algorithm (1) is based on the spectral response of snow in Green and SWIR wavelengths.

$$\text{NDSI} = (\text{Green} - \text{SWIR}) / (\text{Green} + \text{SWIR}) \quad (1)$$

$$\text{NIR/SWIR} = (\text{NIR} - \text{SWIR}) / (\text{NIR} + \text{SWIR}) \quad (2)$$

A threshold of NDSI $\geq$ 0.4 classifies snow/ice, water pixels and snow under shadow as snow pixel which is widely used for optical satellite images [8]. NDSI has the advantage of distinguishing between clouds and snow, as the clouds have higher reflectance in the SWIR band [5]. In order to avoid misclassification of water pixels as snow, water mask has been generated using summer month data with NIR/SWIR

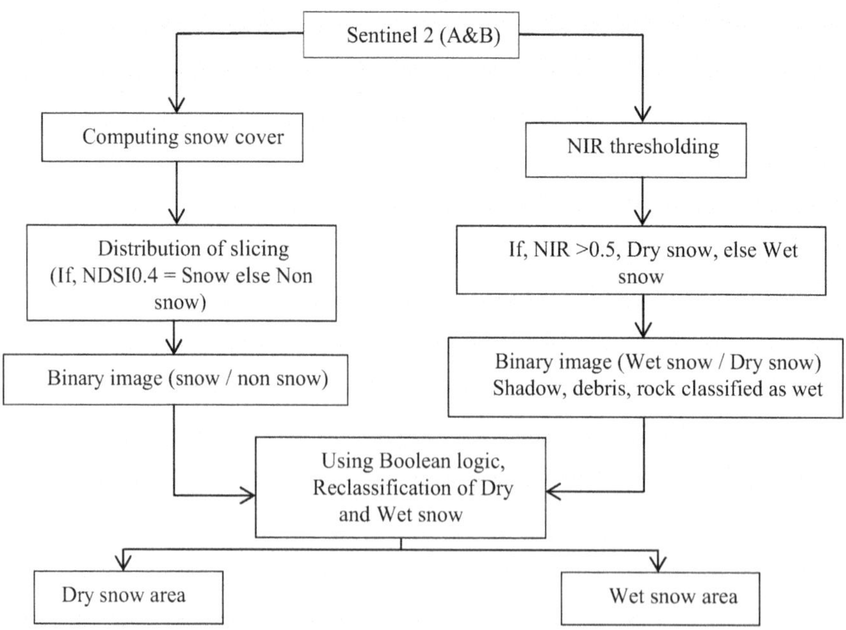

Fig. 2 Methodology

ratio (2) of threshold value 0.37 (NIR/SWIR < 0.37, water) has been adopted [9]. During summer season, snow starts melting which creates moisture surface on the top of the glaciers [10]. NIR band reflectance can be used to classify dry and wet snow with proper thresholding to discriminate water surface due to its (poor) spectral response. A threshold value of 0.5 has been adopted to map the dry ($\geq$0.5) and wet (<0.5) snow from NDSI-based snow cover map. Glacier boundary has been adopted from Randolph Glacier Inventory version 6 (RGI v6) [11]. Landsat 8 data used for the months of February and May has been converted from Digital Number (DN) to Top of Atmospheric (TOA) reflectance using appropriate equations. NDSI has been generated for every satellite image covering the study period, which has been further converted to a binary image (snow/non-snow) by a thresholding technique. Using NIR band, snow cover image has been sliced to discriminate dry and wet snow based on threshold.

4 Results

Sentinel-2 (A and B) images have been processed as discussed in the methodology section using QGIS 3.4.4, an open-source software. The following inferences can be summarized from Table 2 and Figs. 3, 4 and 5:

- More dry snow cover has been observed during October–November 2018 (rows 1 and 2 of Table 2 and Fig. 3a, b), as snowfall for the Hydrological year 2018–2019 has started in the last week of September 2018 onwards as per IMD data.
- Decrease in dry snow and increase in wet snow areas have been observed for November–December 2018 due to more number of positive degree days during

Table 2 Area for dry and wet snow Miyar glacier

Month	Dry snow area (km^2)	Wet snow area (km^2)
October 2018	69.47	8.31
November 2018	65.68	13.47
December 2018	62.98	16.04
January 2019	63.55	0.50
February 2019	Cloud cover—no data	Cloud cover—no data
March 2019	74.90	0.56
April 2019	76.46	0.63
May 2019	77.09	1.86
June 2019	67.16	4.62
July 2019	54.45	12.34
August 2019	37.34	19.39
September 2019	35.17	13.96

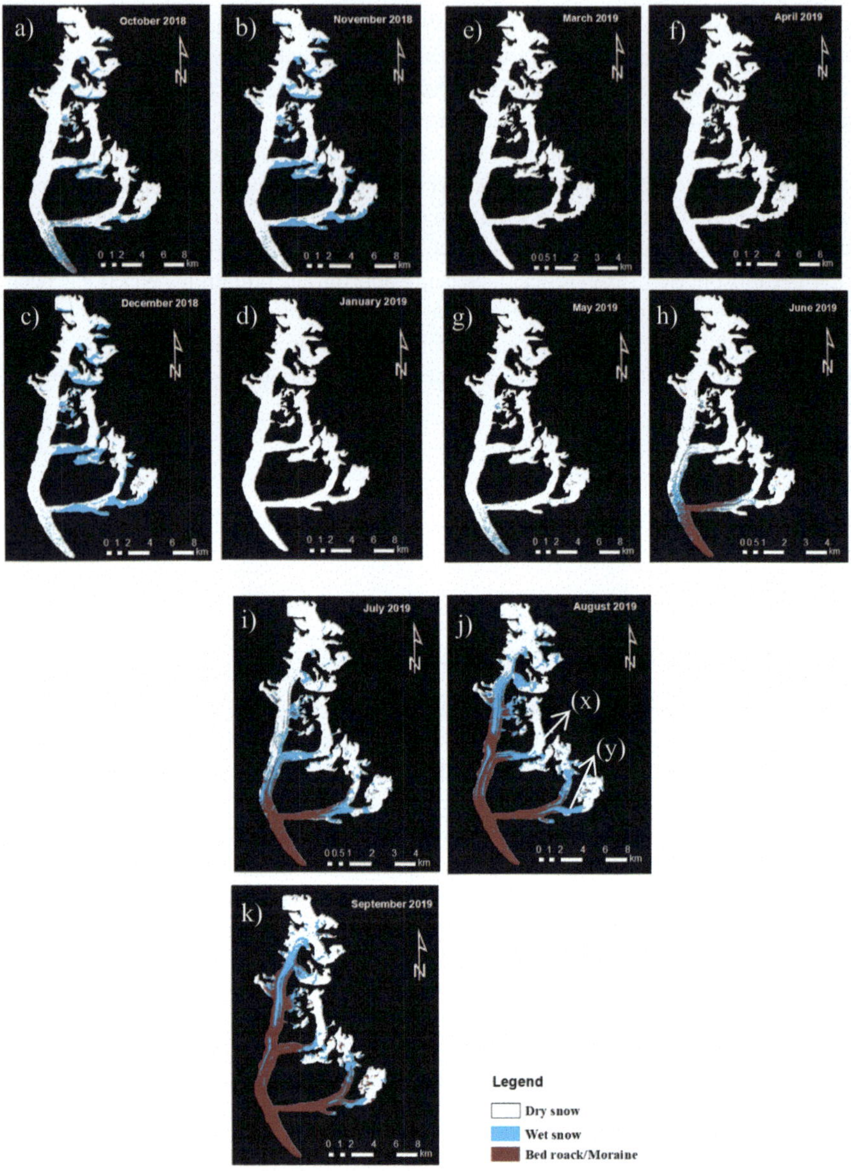

Fig. 3 a–k Spatial variation of dry and wet snow of Miyar Glacier for the HY (October 2018–September 2019)

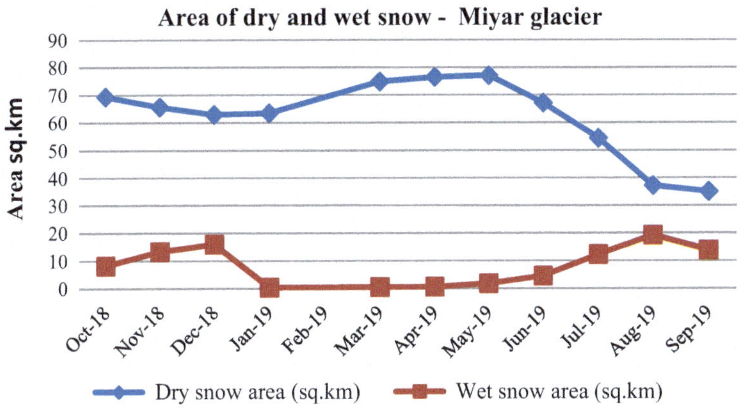

Fig. 4 Dry and wet snow areas

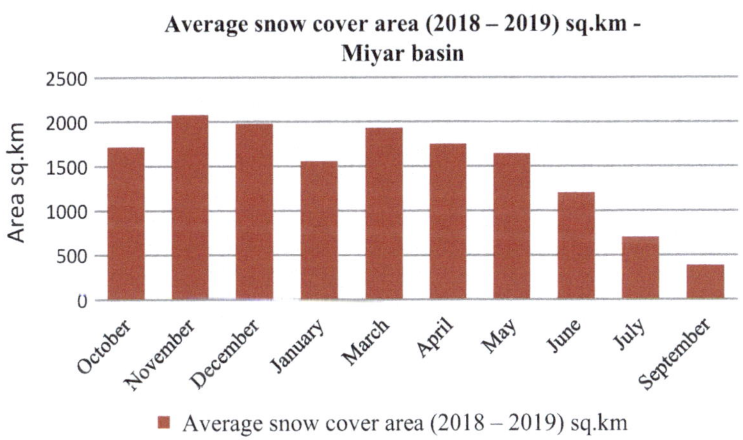

Fig. 5 Snow cover area for Miyar basin

this period as per IMD data (rows 2 and 3 of Table 2 and Fig. 3b, c). This is also evident from the snow cover area mapped using NDSI (see Fig. 5).

- With the onset of summer (May–September 2019), wet snow area has been observed to gradually increase (rows 8–12 of Table 2 and Fig. 3g–k) with maximum wet snow area observed during August corresponding to 19.39 km^2 (row 11 of Table 2 and Fig. 3j).
- Results indicate that the proposed method can clearly discriminate between wet snow (x) and the exposed glacial ice (y) without any misclassification (Fig. 3j, k). Even though the wet snow and glacial ice have the same reflectance properties, they can be distinguished with their shape and repetitive occurrence nature in optical images.

Transient snow line (wet snow fringe) shifts from lower elevation to higher elevation dynamically during the summer season (May–September) exposing ice in the ablation zone of the glacier. This process contributes to variations in dry and wet snow areas mapped during summer (see Fig. 4).

5 Conclusion

The potential of satellite images and remote sensing aspects has been used to estimate and map the dry and wet snow cover areas in Miyar glacier for the HY 2018–2019. For Miyar glacier, during summer (June–September), an average dry snow area of 70.11 km^2 (which is 88% of the total glacier area) and wet snow area of 5.50 km^2 have been observed. During winter (November–May), an average dry area of 48.58 km^2 (which is 61% of the total glacier area) and wet snow area of 12.57 km^2 approximately has been observed. It is showed that with the optical images it is possible to identify dry snow region, wet snow region, exposed glacial ice and moraine covered glacial ice. There are two main constraints observed in this work: (1) Availability of data without cloud cover at a regular interval; (2) Misclassification of mountain shadow as wet snow in NIR thresholding when calculated in the basin scale. Further, this work can be extended to estimate the snow line altitude to estimate the mass balance of the glacier.

Acknowledgements The authors acknowledge the financial support given by the ESSO-National Centre for Antarctic and Ocean Research, Ministry of Earth Sciences, under the HiCOM initiative to undertake this research. The authors gratefully acknowledge the support and cooperation given by Dr. Krishna Venkatesh, Director, CIIRC-Jyothy Institute of Technology (JIT), Bengaluru, Karnataka, and Sri Sringeri Sharada Peetham, Sringeri, Karnataka, India.

References

1. Ajai: Inventory and monitoring of snow and glaciers of the Himalaya using space data. In: Goel, P., Ravindra, R., Chattopadhyay, S. (eds). Science and Geopolitics of The White World. Springer (2017)
2. Gupta, R.P., Haritashya, U.K., Singh, P.: Mapping dry/wet snow cover in the Indian Himalayas using IRS multispectral imagery. Remote Sens. Environ. **97**(4), 458–469 (2005)
3. Bahuguna, I.M., Kulkarni, A.V., Nayak, S., Rathore, B.P., Negi, H.S., Mathur, P.: 2007 Himalayan glacier retreat using IRS 1C PAN stereo data. Int. J. Remote Sens. **28**(2), 432–437 (2007)
4. Kulkarni, A.V., Rathore, B.P., Singh, S.K., Bahuguna, I.M.: Understanding changes in the Himalayan cryosphere using remote sensing techniques. Int. J. Remote Sens. **32**(3), 601–615 (2011)
5. Geetha Priya, M., Krishnaveni, D.: An approach to measure snow depth of winter accumulation at basin scale using satellite data. Int. J. Comput. Inf. Eng. **13**(2), 70–74 (2019)

6. Kulkarni, A.V., Mathur, P., Singh, S.K., Rathore, B.P., Thakur, N.K.: Remote sensing based techniques for snow cover monitoring for the Himalayan region. In: International Symposium on Snow Monitoring and Avalanches (ISSMA-04), pp. 399–405, Manali, India (2004)
7. Patel, L.K., Sharma, P., Fathima, T.N., Thamban, M.: Geospatial observations of topographical control over the glacier retreat, Miyar basin, Western Himalaya. India Environ. Earth Sci. **77**, 19 (2018)
8. Dozier, J: Spectral signature of alpine snow cover from the Landsat Thematic Mapper. Remote Sens. Environ. (28), 9–22 (1989)
9. Nagajothi, V., Geetha Priya, M., Sharma, P.: Snow cover estimation in western Himalayas using Sentinel 2. Indian J. Ecol. **46**(1), 88–93 (2018)
10. Negi, H.S., Kulkarni, A.V., Semwal, B.S.: Study of contaminated and mixed objects snow reflectance in Indian Himalaya using spectroradiometer. Int. J. Remote Sens. **30**(2), 315–325 (2009)
11. RGI Consortium. Randolph Glacier Inventory—A Dataset of Global Glacier Outlines: Version 6.0: Technical Report, Global Land Ice Measurements from Space, Colorado, USA. Digital Media (2017)

Potential of Robust Face Recognition from Real-Time CCTV Video Stream for Biometric Attendance Using Convolutional Neural Network

Suresh Limkar, Shashank Hunashimarad, Prajwal Chinchmalatpure, Ankit Baj, and Rupali Patil

Abstract Face recognition is one of the most bothersome research issues in security systems due to various challenges like constantly changing poses, facial expressions, lighting conditions, and resolution of the image. The wellness of the recognition technique firmly depends on the accuracy of extracted features and also on the ability to deal with the low-resolution face images. The mastery to learn accurate features from raw face images makes deep convolutional neural networks (DCNNs) a suitable option for facial recognition. The DCNNs utilizes Softmax for evaluating model accuracy of a category for associate degree input image to create a forecast. However, the Softmax probabilities do not depict the real representation of model accuracy. The main aim of this paper is to maximize the accuracy of face recognition systems by minimizing false positives. The complete procedure of building a face recognition prototype is defined very well. This prototype consists of many vital steps built using most advanced methods: CNN cascade for detection of face and HOG for generating face embeddings. The primary aim of this analysis was the sensible use of those developing deep learning techniques for face recognition work, because of the reason that CNNs give almost accurate results for huge datasets. The proposed face recognition prototype can be used together with another system by making some minor changes or without making any changes as an assisting or a primary element for surveillance functions.

S. Limkar · S. Hunashimarad · P. Chinchmalatpure (✉) · A. Baj · R. Patil
Department of Computer Engineering, AISSMS IOIT, Pune-01, India
e-mail: prajwalvvc@gmail.com

S. Limkar
e-mail: sureshlimkar@gmail.com

S. Hunashimarad
e-mail: shankleo.08@gmail.com

A. Baj
e-mail: ankitbaj51@gmail.com

R. Patil
e-mail: rupalipatil14498@gmail.com

S. C. Satapathy et al. (eds.), *Intelligent Data Engineering and Analytics*, Advances in Intelligent Systems and Computing 1177,
https://doi.org/10.1007/978-981-15-5679-1_2

Keywords Facial recognition · Deep learning · Attendance monitoring system · Unconstrained face images · CNN

1 Introduction

In recent years, the use of cameras for security purposes [1] and market research [2] purposes has increased a lot. Computer vision is used for face detection. These techniques are considered applicable for Human detection [3, 4] part of the monitoring system. There are ways to detect the frontal face in the field of face detection research. Most ways use intensity pictures as feature vectors and applied math classifiers [5]. SVM [6] and sparse network of winnow (SNOW) [7] intensity-based methods focus on less potential areas which include facial features: eyes, mouth, and extensive areas of high potency as cheeks and forehead. Also, there are other methods like (refer Fig. 1) the Principle Component Analysis (PCA) method giving 60.02% accuracy, the Elastic Bunch Graph Matching (EBGM) method giving an accuracy of 65.23%, and the Local Binary Patterns (LBP) method giving an accuracy of 73.01%. However, the mentioned methods only hold the front part of the face; they can only find humans facing the cameras. There are also other methods proposed for frontal and profile face detection, which define the wavelet coefficient probability [8]. Even though the

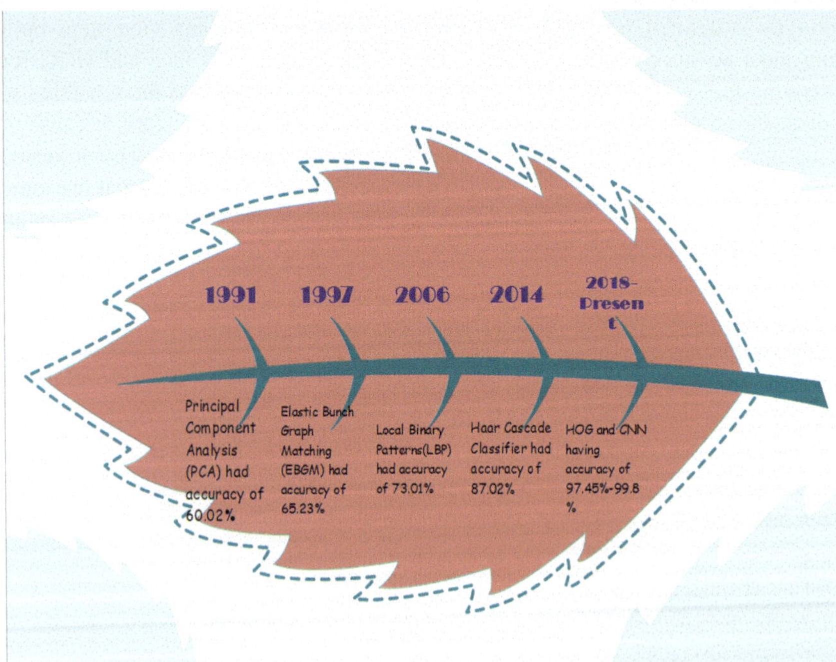

Fig. 1 Timeline diagram

method looks righteous, the result makes computing cost difficult. However, human pictures acquired by Camera are not restricted to front faces or full resemblance of a Pedestrian. Therefore, the methods mentioned are tough to try for practical use [9, 10]. The proposed system will be able to detect people who are not facing the camera and also the person will not be required to stand in front of the camera; the system will be able to detect the person as he/she walks past the camera. The proposed method employs CNN cascade for face recognition and HOG for invoking face embeddings, which gives an approximate accuracy of 97–98% and reduces the Computing Cost for Scanning and Classification of Real-time detection plan. The proposed system will also have a time slot for every face detected, and stores it in a database for monitoring purposes.

2 Literature Survey

Table 1 presents a thorough literature survey comparing our model with various recent models for face recognition.

3 Contributions

In this paper, we attempt to get the better of the constraints witnessed in the methodologies suggested above. For instance, the principal component analysis gives us an accuracy of only 68% which is not sufficient for real-world applications. The PIE and extended Yale datasets are not efficient for all the faces as they do not use CNN due to which the accuracy of detecting the image is not up to the mark. The low-resolution face recognition system cannot handle some of the face attributes like gender, age, and makeup. The high-resolution face recognition system method is only capable of recognizing datasets having High-resolution images. Other than that, it cannot compute low-resolution images. Low-Power Scalable 3-D Face Formalization Processor for CNN-based Face Recognition can only be used in mobile devices and sometimes the low-level images are not recognized. Face Detection with Different Scales Based on Faster R-CNN has low efficiency in parallel type CNN. Deep Aging Face Verification with Large Gaps has only front face pose present in the database and no different poses are available. Exploring Priors of Sparse Face Recognition on Smartphones can be used only in mobile devices. Sensor-assisted Multi-view Face Recognition System on Smart Glass gives low accuracy for different poses of images. Single Sample Face Recognition via Learning Deep Supervised Autoencoders has the image size restricted to 32 × 32 and only 20 images are chosen from CMY-PIE datasets for training. So, considering the above limitations, we propose implementing a deep learning-based face recognition module, which makes use of OpenCV, Dlib, and Convolutional Neural Networks. The HOG algorithm is used for face detection. Once an image is detected using HOG, the corresponding can

Table 1 Literature survey

Reference	Purpose	Methodology	Dataset used	Methods/Models used	Pros	Cons	Remark
Arsenovic et al. [11]	Deep learning-based face recognition system for attendance purpose in schools and institutions	CNN cascade is used for face detection and CNN for generating face embeddings	Limited dataset of 5 images	CNN model	Overall accuracy on the small dataset is 95.02% using the CNN cascade	Using the principal component analysis method (PCA) the accuracy is 68%	Suitable for small datasets
Sima et al. [12]	Face recognition in presence of space-varying motion blur comprising of arbitrarily shaped kernels	Making the blur face as convex union of modified instance and obtaining a convex set for blurred image so that the non-uniform face is detected	CMU-PIE and extended Yale dataset	MOBILAP algorithm	It can handle pose variations and has stable performance for blurred images	Significant occlusions and large changes in facial expressions cannot be handled	Efficiency is less which may affect the accuracy
Moeini et al. [13]	Face recognition under pose and expression variations	3D probabilistic facial expression recognition generic elastic model (3D PFER-GEM) is proposed to reconstruct a 3D real human face using 2D front image to detect human face	CMU-PIE and LFW dataset	PFER method	Various poses of faces can be detected	Cannot handle unconstrained face recognition that is robust to a wide range of face variations	Ability to handle all types of faces
Shiming et al. [14]	Recognize low resolution faces via selective knowledge distillation	A 2 flow of CNN is assigned to recognize faces	Low intensity images of students and teachers	CNN model	The proposed network leads to dense face recognition models with splendid productivity and skill	It cannot handle the face attributes like gender, age and makeup	Accurate for low resolution images
Ding et al. [15]	Face identification framework capable of handling the full range of pose variations within 90° of yaw	It uses PBPR (patch based partial representation) face representation scheme. Face matching is performed at patch level rather than at holistic level	LFW dataset	PBPR method	It can be applied to face images in arbitrary pose providing good accuracy	Doesn't provide good accuracy for PCA method	Very precise for occluded faces
Dong et al. [16]	A pose-invariant face-verification method, using the high-resolution information based on pore-scale facial features	PCA-SIFT is devised for extraction of compact set of distinctive pore scale facial features which help to determine human faces	High resolution images	PCA method	Computational time for matching process is reduced	The proposed method is robust to alignment errors and pose variations	Capable to handle only high resolution images

(continued)

Table 1 (continued)

Reference	Purpose	Methodology	Dataset used	Methods/Models used	Pros	Cons	Remark
Kang et al. [17]	Accurate face recognition in mobile devices using 3-D face formalization	The method followed is (Facial Feature Detection) FFD -> (Frontal Face Generation) FFG	LFPW dataset	CNN model	The proposed face formalization processor is implemented in 65 m CMOS process and shows 4.73 fps throughput which is very fast	Used in mobile devices and sometimes the low-level images are not recognized	Suitable only for mobile devices providing high accuracy
Wenqi et al. [18]	Different scales face detector (DSFD) based on Faster CNN	Difference of strategies is initiated in face detection network, inclusive of multitask learning, feature pyramid and feature concatenation	WIDER FACE dataset	Faster CNN	The proposed face detection system acquires remarkable performance on various standards of face detection	It cannot perform well on large scale face subsets	Faster CNN which consist of 3 networks and is very much efficient for small scale face subsets
Liu et al. [19]	Deep ageing face verification (DAFV)	The ageing face verification takes the synthesized aging pattern of face pair as input and is fed to CNN to detect the faces	CAFE (cross age face) dataset	CNN model	It has good accuracy for CAFE dataset	Only front face pose present in the database and no different pose are available	Lacks in performance on other poses of face
Yang et al. [20]	Exploit the prior knowledge of the training set to improve the recognition accuracy	Opti-GSRC exploits the group sparsity structure in the sparse representation classification problem to improve the recognition accuracy	Extended Yale database B	SRC (Sparse representation classification) method	Provides a good running time for finding the faces on the mobile devices	Can be used only in mobile devices	Accuracy for recognizing faces is improved in mobile devices
Weitao et al. [21]	Robust and efficient sensor-assisted face recognition system on smart glasses	Multi-view sparse representation classification (MVSRC) is used for exploiting the prolific information among multi-view face images	UCSD dataset	MVSRC method	It improves the recognition accuracy by combining multi-view face images	The accuracy is not good for different poses of the images	Improves accuracy by 15% as compared to OpenCV algorithms on smart glasses
Li et al. [22]	Ageing face recognition using two level learning	Two level learning is followed to solve the problem of face recognition	MORPH dataset	LPS (Local Pattern Selection)	This method is able to predict the images very much accurately	Some of the young faces are not detected properly	LPS feature selection is efficient than other feature detectors

be recognized from CCTV cameras. OpenCV's face recognition and Python's Dlib libraries are also used to improve the accuracy further. Dlib is a library that contains encodings of over 3 million human face images. For detecting time, the internal clock is used. We achieved an accuracy of 99.38% using the above methodology.

4 Proposed System

We propose to implement a deep learning-based face recognition module, which makes use of OpenCV [23] and Convolutional Neural Networks. The HOG algorithm is used for face detection. Once an image is detected using HOG, the corresponding can be recognized using OpenCV and Dlib library by making use of image landmarks. The image landmarks will be used to represent and localize distinctive attributes of a human face such as jaws, eyebrows, eyes, mouth, and jawline.

Dlib [24] is a library that contains encodings of over 3 million human face images. For detecting time, the internal clock is used, which computes the time following face detection. The face recognition accuracy is consistently over 99%. This number makes the system reliable even in adverse conditions such as less intensity light, images at a certain angle, low-quality images in terms of pixel density, and images with certain facial expressions. Once the input image matches with some image from the dataset, in other words, once an employee's face is recognized, we move toward capturing the login or logout time of the corresponding employee. To record the timings, we make use of a simple Python clock function. Whenever a face is recognized by the first module, a call is made to the clock function which then returns the time at that instance. Thus, by making use of these two modules, we have the recognized employee image along with the time of recognition. This time value can then be used for entering the login and log out time of the employee (Fig. 2).

When the image is captured, the face is detected using Histogram of Gradients (HOG). The face embeddings are extracted using CNN which creates a 128D vector of each image. These embeddings are then used for training the face recognition model. The model gets trained using these embeddings. Then the captured image is compared with these embeddings and it determines the recognized face.

5 Algorithm

Face detection using OpenCV and YOLO

1. Give the input image, convert to RGB
2. Loop over the facial embeddings using the bounding box created by YOLO
for encoding in encodings **do**
attempt to match the encodings with the input image
if True in **matches**:

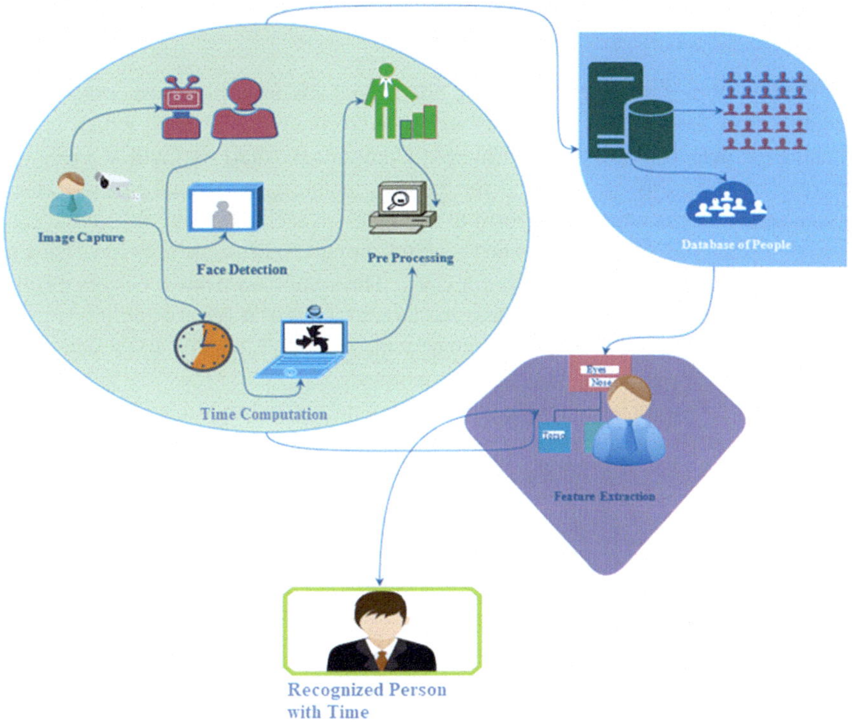

Fig. 2 Architecture diagram

3. Find the indexes of the matched faces and then initialize a dictionary to compute the total number of times each face was matched.
4. Loop over the matched indexes
for i in matched indexes **do**
decide the face with substantial percent of votes
end for
end if
end for
5. Display the output image
6. If the subject face image does not match any image from the dataset, deny the door entry and notify the reception desk.
7. Once the input image has been matched with some image from the database of employee images, mark a login entry for the recognized employee.
8. The employee will be identified by employee id and login time will be determined by returning time from an internal clock function.
9. Repeat steps 1 through 7 when an employee is leaving.

6 Results

The Results from Table 1 represent the algorithms having different accuracies for face detection. Local Binary Patterns shows an accuracy of 77.55%, which is not efficient for detection purposes. Principle Component Analysis method gives an accuracy of 80%, this can be used for datasets having good frontal images. Haar Cascade Classifier is a famous algorithm and gives an accuracy of 92.9%, and it can be used in place of CNN if there is no GPU in Computers. Multiple tests have shown the accuracy of 99.38% consistently for CNN. The system has shown correct results in low light conditions, in cases where the target face is at a certain angle from the mounted camera and also in video frames with low pixel density. Even when the faces are slightly blurred, it shows good accuracy.

From Table 2, it can be concluded that most of the algorithms are having accuracy between 85 and 90% and the error rate is also high. The majority of algorithms only recognize the frontal part of the face. The proposed system which we have used

Table 2 Comparison table

Algorithms	True positive	False positive	Accuracy (%)	Quality of dataset	Error rate (σ)	Low light condition performance	Facial angles recognized
LBPH [25]	76	34	77.55	Very low quality	31.1%	Good	Frontal part
Eigenfaces [26]	85	25	85	Low quality	25.4%	Good	Frontal part
PCA [27]	78	22	80	Low quality	29.3%	Bad	Frontal part
Haar cascade [28]	87	23	92.9	High quality	15.5%	Very Good	Side angles and frontal part
PCA-SIFT [16]	84	36	76.27	High quality	11.23%	Moderately Good	Frontal part
MOBILAP [12]	82	28	87	Low quality	14.25%	Bad	Frontal part
PFER [13]	80	20	81	Low quality	15.55%	Bad	Frontal part
Faster CNN [18]	88	22	89.4	High quality	9%	Good	Side angles and frontal part
MVSRC [21]	86	24	87.5	Low quality	17.2%	Bad	All angles recognized
Proposed system using CNN	89	11	95	High quality	0.93	Very good	All angles recognized

shows better results than other algorithms as it recognizes all facial angles and the error rate is also low with excellent performance in low light conditions.

7 Conclusion

The system being proposed here can provide comprehensive and accurate detection and recognition of human face against the given dataset for a given stream of real-time CCTV footage.

References

1. British government, CCTV initiative. http://www.crimereduction.gov.uk/cctvminisite4.htm
2. Brickstream Corp. http://www.brickstream.com
3. Murphy, T.M., Broussard, R., Schultz, R., Rakvic, R., Ngo, H.: Face detection with a Viola–Jones based hybrid network. Biometr. IET **6**(3), 200–210 (2017)
4. Hjelmas, E., Low, B.K.: Face detection: a survey. Comput. Vis. Image Underst. **83**, 236–237
5. Rowly, H., Baluja, S., Kanade, T.: Neural network-based face detection. IEEE Trans. Pattern Anal. Mach. Intell. **20**(1), 23–38
6. Kyrkou, C., Bouganis, C.-S., Theocharides, T., Polycarpou, M.M.: Embedded hardware-efficient real-time classification with cascade support vector machines. IEEE Trans. Neural Netw. Learn. Syst. **27**(1), 99–112 (2016). https://doi.org/10.1109/tnnls.2015.2428738
7. Yang, M.-H., Roth, D., Ahuja, N.: A SNoW-based face detector. Adv. Neural Inf. Process. Syst. **12**, 855–861
8. Schneiderman, H., Kanade, T.: A Statistical Method for 3D Object Detection Applied to Faces and Cars
9. Oren, M., Papageorgiou, C., Sinha, P., Osuna, E., Poggio, T.: Pedestrian detection using wavelet templates. In: Proceedings of the CVPR97, pp. 93–199
10. Mohan, A., Poapageorgiou, C., Poggio, T.: Example-based object detection in images by components. IEEE Trans. PAMI **23**(4), 349–361
11. Arsenovic, M., Sladojevic, S., Anderla, A., Stefanovic, D.: FaceTime—deep learning-based face recognition attendance system. In: 2017 IEEE 15th International Symposium on Intelligent Systems and Informatics (SISY) (2017). https://doi.org/10.1109/sisy.2017.8080587
12. Punnappurath, A., Rajagopalan, A.N., Taheri, S., Chellappa, R., Seetharaman, G.: Face recognition across non-uniform motion blur, illumination, and pose. IEEE Trans. Image Process. **24**(7), 2067–2082 (2015). https://doi.org/10.1109/tip.2015.2412379
13. Moeini, A., Moeini, H.: Real-world and rapid face recognition toward pose and expression variations via feature library matrix. IEEE Trans. Inf. Forensics Secur. **10**(5), 969–984 (2015). https://doi.org/10.1109/tifs.2015.2393553
14. Ge, S., Zhao, S., Li, C., Li, J.: Low-resolution face recognition in the wild via selective knowledge distillation. IEEE Trans. Image Process. 1–1 (2018). https://doi.org/10.1109/tip.2018.2883743
15. Ding, C., Chang, X., Tao, D.: Multi-task pose-invariant face recognition. IEEE Trans. Image Process. **24**(3), 980–993 (2015). https://doi.org/10.1109/tip.2015.2390959
16. Li, D., Zhou, H., Lam, K.-M.: High-resolution face verification using pore-scale facial features. IEEE Trans. Image Process. **24**(8), 2317–2327 (2015). https://doi.org/10.1109/tip.2015.2412374

17. Kang, S., Lee, J., Bong, K., Kim, C., Kim, Y., Yoo, H.-J.: Low-power scalable 3-D face frontalization processor for CNN-based face recognition in mobile devices. IEEE J. Emerg. Sel. Top. Circuits Syst. 1–1 (2018). https://doi.org/10.1109/jetcas.2018.2845663
18. Face detection with different scales based on faster R-CNN (2018). IEEE Trans. Cybern. 1–12. https://doi.org/10.1109/tcyb.2018.2859482
19. Liu, L., Xiong, C., Zhang, H., Niu, Z., Wang, M., Yan, S.: Deep aging face verification with large gaps. IEEE Trans. Multimedia **18**(1), 64–75 (2016). https://doi.org/10.1109/tmm.2015.2500730
20. Shen, Y., Yang, M., Wei, B., Chou, C.T., Hu, W.: Learn to recognise: exploring priors of sparse face recognition on smartphones. IEEE Trans. Mob. Comput. **16**(6), 1705–1717 (2017). https://doi.org/10.1109/tmc.2016.2593919
21. Xu, W., Shen, Y., Bergmann, N., Hu, W.: Sensor-assisted multi-view face recognition system on smart glass. IEEE Trans. Mob. Comput. **17**(1), 197–210 (2018). https://doi.org/10.1109/tmc.2017.2702634
22. Li, Z., Gong, D., Li, X., Tao, D.: Aging face recognition: a hierarchical learning model based on local patterns selection. IEEE Trans. Image Process. **25**(5), 2146–2154 (2016). https://doi.org/10.1109/tip.2016.2535284
23. Marengoni, M., Stringhini, D.: High level computer vision using OpenCV. In: 2011 24th SIBGRAPI Conference on Graphics, Patterns, and Images Tutorials (2011). https://doi.org/10.1109/sibgrapi-t.2011.11
24. Sharma, S., Shanmugasundaram, K., Ramasamy, S.K.: FAREC—CNN based efficient face recognition technique using Dlib. In: 2016 International Conference on Advanced Communication Control and Computing Technologies (ICACCCT) (2016). https://doi.org/10.1109/icaccct.2016.7831628
25. Abuzneid, M.A., Mahmood, A.: Enhanced human face recognition using LBPH descriptor, multi-KNN, and back-propagation neural network. IEEE Access **6**, 20641–20651 (2018). https://doi.org/10.1109/access.2018.2825310
26. Lei, L., Kim, S., Park, W., Kim, D., Ko, S.: Eigen directional bit-planes for robust face recognition. IEEE Trans. Consum. Electron. **60**(4), 702–709 (2014). https://doi.org/10.1109/tce.2014.7027346
27. Xiao, X., Zhou, Y.: Two-dimensional quaternion PCA and sparse PCA. IEEE Trans. Neural Netw. Learn. Syst. 1–15 (2018). https://doi.org/10.1109/tnnls.2018.2872541
28. Liu, C., Liu, C., Chang, F.: Cascaded split-level colour Haar-like features for object detection. Electron. Lett. **51**(25), 2106–2107 (2015). https://doi.org/10.1049/el.2015.2092

ATM Theft Investigation Using Convolutional Neural Network

Y. C. Satish and Bhawana Rudra

Abstract Image processing in a surveillance video has been a challenging task in research and development for several years. Crimes in Automated Teller Machine (ATM) is common nowadays, in spite of having a surveillance camera inside an ATM as it is not fully integrated to detect crime/theft. On the other hand, we have many image processing algorithms that can help us to detect the covered faces, a person wearing a helmet and some other abnormal features. This paper proposes an alert system, by extracting various features like face-covering, helmet-wearing inside an ATM system to detect theft/crime that may happen. We cannot judge theft/crime as it may happen at any time but we can alert the authorized persons to monitor the video surveillance.

Keywords Automated teller machine · Surveillance video · Image processing

1 Introduction

Criminal activities in ATM's square measure are seen frequently occurring in the news across Asian nations and worldwide [1, 2]. Not exclusively underdeveloped countries face these well-known incidents but also countries just like America, the United Kingdom, etc. also face these issues. As in the recent news, criminals are searching for new techniques to perform the action. They are not bothered of video investigation techniques, recorded video square measures used for the vocal investigation of the robbery's impact as they do not facilitate plenty of to forestall ATM robbery [3]. So the detection of the ATM robbery victimization investigation camera is an associate degree that raised a critical issue to verify a secured ATM atmosphere. With the advancement of the latest technologies, video police investigation cam-

Y. C. Satish (✉) · B. Rudra
Department of Information Technology, National Institute of Technology Karnataka Surathkal, Mangaluru 575025, India
e-mail: satishyc@hotmail.com

B. Rudra
e-mail: bhawanarudra@nitk.edu.in

S. C. Satapathy et al. (eds.), *Intelligent Data Engineering and Analytics*,
Advances in Intelligent Systems and Computing 1177,
https://doi.org/10.1007/978-981-15-5679-1_3

Table 1 Recent records of ATM theft and online fraud according to RBI

Number of cases	Year
1012	2016–2017
972	2017–2018
261	April–June 2018

eras integrated with image-processing technologies were used to discover suspicious activities in ATMs. There are various square measure image processing algorithms offered for occluded or human abnormal behavior analysis, coated face detection and black object detection; not all the qualities best in class calculations will work for ATMs. As they are entirely working on different environments (illumination, camera scan, etc.), bizarre gestures and crime pieces of equipment or accessories are to be used for the detection of the events.

Indian banks lost Rs. 109.75 crore and Rs. 168.74 crore to theft and online fraud in the financial year 2018 and the last three years, respectively, according to Reserve Bank of India. Uttar Pradesh has been consistently ranked among the riskiest states for lenders. In 2017–18, there were 85 heists at ATMs, setting back banks by Rs. 2.09 crore. West Bengal also witnessed over 100 such incidents in the past financial year [4] (Table 1).

Our proposed system allows detecting the abnormal activities when a person enters an ATM by covering his/her face with a cloth or any other mask, it will extract the features. Even if the person is wearing a helmet inside the ATM, the system extracts the features and sends the information to authorized authority. The paper consists of a brief introduction about ATM crimes followed by a literature review. Section 3 consists of the methodology, followed by Sect. 4, which consists of results and analysis and finally conclusion and future work.

2 Literature Review

Many researchers have developed various ways [5–12] to resolve the problem of detecting the person, i.e. a criminal with a helmet in traffic—in which unit of measurement is mentioned. The authors [11] have designed an algorithm to solve the problem of detecting bikers in traffic recorded videos. This algorithm divides each video frame in the videos thus keeping track of bikers, heads and helmets using a probability-based algorithm. This handles the barrier disadvantage but will not be able to process tiny variations due to illumination and noise effects. Further, it uses canny edge detection with a search window of a fixed size, thus on observe head. The authors used edge histogram-based choices, to notice motorcyclists [6, 7]. The main advantage of this technique is that it works well even for low-resolution videos because of the use of edge histograms near to the head instead of detecting the choices of the highest region. Since the sting histograms use the circular Hough transforms for classification and matching of the helmets, it leads to lots of misclassification among bikers with helmet and sometimes, helmet-like objects were collectively classified as a helmet. As a result of the helmets that were altogether totally distinct, wasn't classified as helmets.

Object detection in video investigation systems is fairly gaining quality due to its good selection of applications that involve vital processes like the investigation of abnormal events. Not only these, it can also be used for the characterization of giants in humans, count people in crowds, characteristics of the individuals, classification supported gender, detection of fall for the aged individuals, etc. Generally, the various scenes have unit of measurement results of the video on closed-circuit television unit composed of really low resolution. However, the static camera captures scenes with minimum change inside the background whereby the surface investigation has to discover the item in a larger scope. A variety of the prevailing systems depend on the human observers UN agency that would perform a period of time activity detection. This ends up in limitations similar to the problem related to co-occurring observance at intervals [9]. This wants the automation of the video investigation for analysis of human motion and has created a research attraction on the arena of pattern recognition and computer vision. The technique includes Object detection and classification. The previous is commonly distributed by processes like background subtraction, optical flow followed by spatiotemporal filtering. The first methodology is the background subtraction which is extremely normal for the detection of objects using constituent-by-constituent or block-by-block fashion. These blocks are considered to check for the excellence between the backgrounds. The present frame, whereas investigation of moving objects to boot to the present various totally different approaches embodies the mathematician mixture [13, 14]. The non-parametric background method [15, 16], temporal differencing [17, 18], deformation background model [19] and gradable background [20–23] models were used. In another attempt, associate optical flow primarily based on object detection technique was used [24–26], which uses moving object flow vectors unit of measurement utilized indefinite intervals of some time that detects the objects in motion at intervals of the image sequence. The item classification methodology is typically classified into different categories like someone wearing cap, helmet, or covering his/her face with any textile or any mask, etc.

3 Methodology

To detect and recognize the object in an image, first of all, we will find the blob image (blob image is the collection of binary data stored as a single entity in DBMS) for loaded image. This image will be forwarded to the network to identify the bounding boxes. For the detection of an object in an image, a bounding box is used to describe the target location. It is a rectangular box that can be determined using X and Y axis coordinates in the upper-left corner and the X and Y axis coordinates in the lower-right corner of the rectangle. Once we obtain this, it will be compared with the threshold value. If the value is less than the threshold value, then no object is found, so the image will be rejected. Otherwise, the coordinates of bounding boxes are calculated. Non-maximum suppression is a key post-processing step for the computer vision applications. This is used to transform a smooth response map that triggers many imprecise object window hypotheses into, ideally, a single bounding-box for each detected object. The technique was implemented on bounding boxes. If the number of bounding boxes is at least 1, then the object within the bounding box will be detected. Otherwise, the process will be aborted. To detect and classify the objects we can use convolutional neural networks like R-CNN, Mask CNN, etc.

To realize more general objects that might discover several object classes. As object detectors are made to measure a selected object category, e.g. facial recognition, a possible approach is to begin with a better work of image classification. In Image classification, unit of measurement is within the main targeted on convolutional neural network (CNN); that unit of measure is powerfully influenced by the results. VGGnet, Inception, etc. were used for classification purposes. We will give a picture as an input in image classification that provides the output a prediction of the pretrained categories or multiple classes, only just in case of multi-label classification. However, it doesn't contain any location information. The classifiers are applied for the localization of the object associated to discover along with the associate degree across the entire picture; a sliding window approach is typically used. During this phase, as the window sizes are different, the object sizes which are different will be scaled to the measurement over overlapping components. Divide these components of pictures that are to be handled carefully.

In the event that the classifier happens to acknowledge associated objects in the window, it will be named and set apart by a bounding box for the future system. The outcome unit estimates a gaggle of bounding boxes and relating class names. Be that as it may, the outcome will have associative degree outsized kind of spare overlapping predictions, with future development and by the assembly of rather loads of capable hardware. In CNN-based algorithms, the unit of measurement is to find the familiar objects and restriction of the objects. R-CNN (R stands for the region) is attempted to enhance the window approach. Before the feature extraction by the convolutional network, R-CNN will demonstrate the bouncing boxes known as region proposals. Once region proposals are retrieved, exploitation of selective search techniques are used. Some of them are support vector machine (SVM), whereas R-CNN performs a regression on region proposals with relevance determined object category to come up with tighter bounding box coordinates.

Fast R-CNN imports Region of Interest Pooling (RoIPool) that decreases the number of forwarding passes and has managed to hitch separate image features (CNN), classification (SVM) and bounding boxes modification. Faster R-CNN enhances the selective search technique by reusing CNN results for region proposals rather than running again for selective search techniques. Masked R-CNN is an associate degree supplement of faster R-CNN that summates a parallel branch for the prediction of the segmentation masks on each Region of Interest (RoI), a further extension to existing branches among the networks that produce class labels and offsets of bounding boxes. The new mask branch might be a bit fully connected network (FCN) applied to each RoI (Fig. 1).

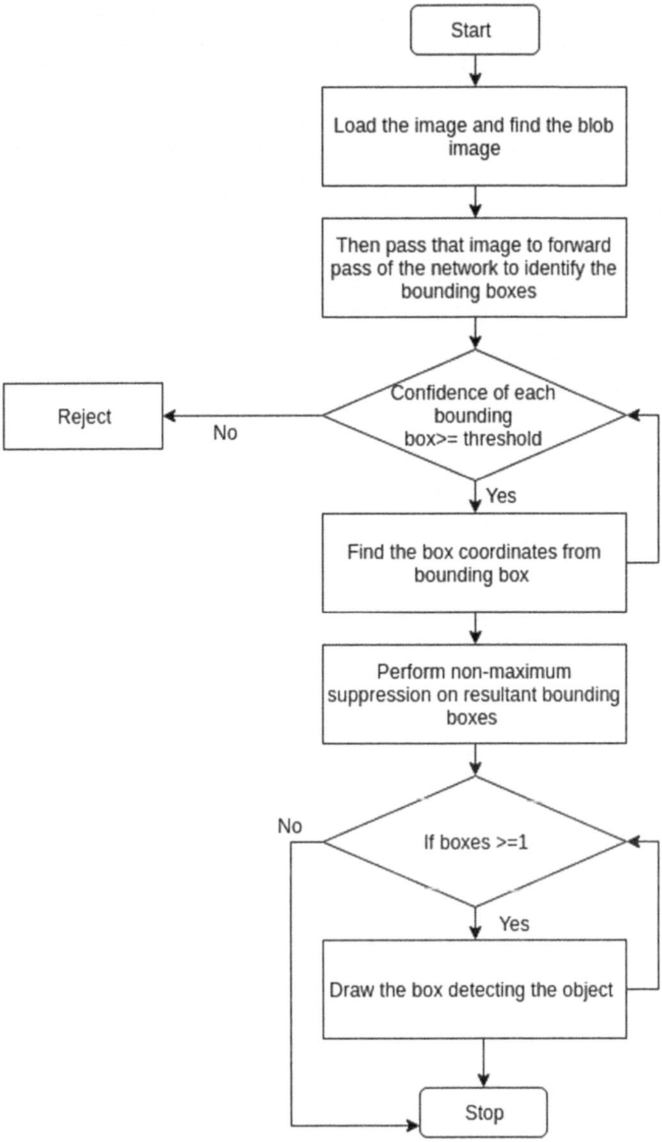

Fig. 1 Working of the system

Masked R-CNN is based on the two steps of Faster R-CNN. The first step can be a region proposal network (RPN) that uses candidate object bounding boxes or RoI. The second step of RPN consists of a deep convolutional network that takes input as an image and produces a feature map, beginning with a smaller network considering a spatial window of a feature map, in addition, cuts down the feature dimension then feeds them to a pair of fully connected layers. It will provide the planned regions bounding box coordinates; using these lines the difference can yield associate degree objectness "score for each and every box. This would be alive of enrollment to line of object class vs. background for each spatial window. The k regions unit of estimation is arranged at the consistent time supported by the reference boxes with predefined scales and aspect ratios called as anchors," representing general object shapes.

Training information for RPN unit of measurement is developed from tagged ground truth data of RoI picture. The clear tag unit of measurement is appointed to anchors with the Intersection-over-Union (IOU) that overlaps with a ground truth box with more than 0.5 throughout. This approach uses multiple anchors which may even be labeled as positive ones. The RPNs unit of measurement is trained from end to end using backpropagation and random gradient descent. The image square measure is resized to 800 pixels. From each training picture, some random RoIs are sampled that the magnitude relation of positive and negative with a 1:3 ratio, to avoid the domination of negative samples inside the data. Masked R-CNN brings about masks and bounding boxes for all available classes severally from classification, eventually, results of the classification branch accustomed to produce the selection of boxes and masks.

4 Data Processing and Data Set

We have collected helmet data set from [27]; it is a collection of images that were taken from different locations and have labels like the person wearing helmets and person not wearing helmets. For the detection of face of the feature covering and mask detection, we developed a new data set that consists of 1000 images. Data annotation of these images was performed using ImgLab [28], an open-source tool.

5 Results and Analysis

The above results are tested in the outside environment as well as inside where it will detect the person who wears a helmet as shown in the bounding boxes. It will display the probability of the objects which we use while training the model. Figure 2 shows the outside environment testing and Fig. 3 shows the inside environment test. The input image will be converted to Blob, and then it will be forwarded to the convolution network to identify the bounding boxes of an image, and finally it will be compared with the threshold value. If it is more than the threshold, we will calculate coordinates of bounding boxes. Then the non-maximum suppression

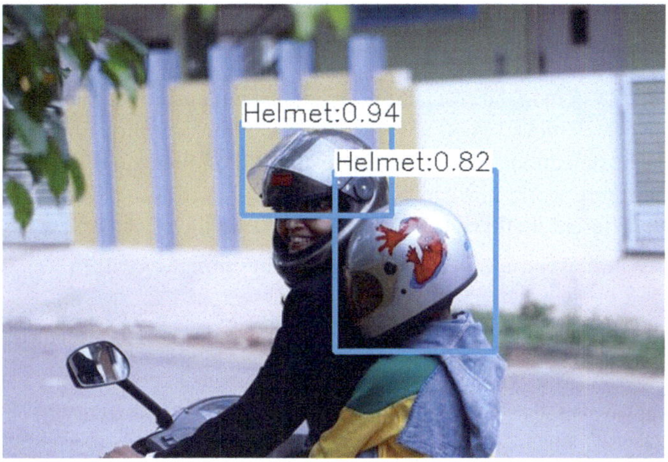

Fig. 2 Helmet detection in picture input outside the environment

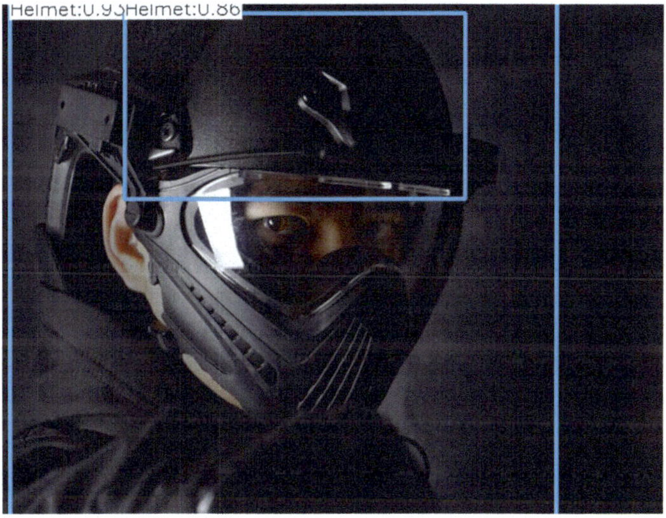

Fig. 3 Helmet detection in picture input inside the environment

technique will be performed on bounding boxes. If the number of bounding boxes is at least one, objects within the box will check with pretrained classes; otherwise, the process will be stopped.

6 Conclusion and Future Work

This is an alert system in the crucial time, it requires manual intervention. Within the environment, we achieved good accuracy for violation/non-violation, we observed some errors in detection; we can improve this by adding more data to train the network. As future work, this can be simplified and improved to a fully automated system with different theft/crime videos by analyzing their behavior.

References

1. ATM near Gorakhnath temple looted. https://timesofindia.indiatimes.com/city/varanasi/atm-near-gorakhnath-temple-looted/articleshow/70944316.cms (2019). Accessed 8 Sept 2019
2. ATM physical attacks in Europe on the increase. https://www.association-secure-transactions.eu/atm-physical-attacks-in-europe-on-the-increase/ (2019). Accessed 8 Sept 2019
3. Security cameras were not enough to stop thieves in Live Oak robbery. https://foxsanantonio.com/news/local/security-cameras-not-enough-to-stop-thieves-in-live-oak-robbery (2019). Accessed 8 Sept 2019
4. Indian banks lost Rs. 109.75 crore to theft and online fraud in FY18. https://www.moneycontrol.com/news/trends/current-affairs-trends/indian-banks-lost-rs-109-75-crore-to-theft-and-online-fraud-in-fy18-2881431.html (2019). Accessed 8 Sept 2019
5. Singh, D., Vishnu, C., Mohan, C.K.: Visual big data analytics for traffic monitoring in smart city. In: Proceedings of the IEEE Conference on Machine Learning and Application (ICMLA), Anaheim, California, 18–20 December 2016
6. Chiverton, J.: Helmet presence classification with motorcycle detection and tracking. IET Intell. Transp. Syst. (ITS) **6**(3), 259–269 (2012)
7. Silva, R., Aires, K., Santos, T., Abdala, K., Veras, R., Soares, A.: Automatic detection of motorcyclists without helmet. In: Proceedings of the Latin American Computing Conference (CLEI), Puerto Azul, Venezuela, 4–6 October 2013, pp. 1–7 (2013)
8. Silva, R.V., Aires, T., Rodrigo, V.: Helmet detection on motorcyclists using image descriptors and classifiers. In: Proceedings of the Graphics, Patterns and Images (SIBGRAPI), Rio de Janeiro, Brazil, 27–30 August 2014, pp. 141–148 (2014)
9. Rattapoom, W., Nannaphat, B., Vasan, T., Chainarong, T., Pattanawadee, P.: Machine vision techniques for motorcycle safety helmet detection. In: Proceedings of the International Conference on Image and Vision Computing New Zealand (IVCNZ), Wellington, New Zealand, 27–29 November 2013, pp. 35–40 (2013)
10. Dahiya, K., Singh, D., Mohan, C.K.: Automatic detection of bike riders without helmet using surveillance videos in real-time. In: Proceedings of the International Joint Conference on Neural Networks (IJCNN), Vancouver, Canada, 24–29 July 2016, pp. 3046–3051 (2016)
11. Chiu, C.-C., Ku, M.-Y., Chen, H.-T.: Motorcycle detection and tracking system with occlusion segmentation. In: Proceedings of the International Workshop on Image Analysis for Multimedia Interactive Services, Santorini, Greece, 6–8 June 2007, pp. 32–32 (2007)
12. Sulman, N., Sanocki, T., Goldgof, D., Kasturi, R.: How effective is human video surveillance performance? In: 19th International Conference on Pattern Recognization (ICPR 2008), pp. 1–3. IEEE, Piscataway (2008)
13. Stauffer, C., Grimson, W.: Adaptive background mixture models for real-time tracking. In: IEEE Conference on Computer Vision and Pattern Recognition (CVPR 1999), pp. 246–252. IEEE, Piscataway (1999)
14. Tian, Y.L., Feris, R.S., Liu, H., Hampapur, A., Sun, M.-T.: Robust Detection of abandoned and removed objects in complex surveillance videos. Syst. Man Cybern. Part C Appl. Rev. IEEE Trans. **41**(5), 65–576 (2011)

15. Kim, W., Kim, C.: Background subtraction for dynamic texture scenes using fuzzy color histograms. Signal Process. Lett. IEEE **19**(3), 127–13 (2012)
16. Lanza, A.: Background subtraction by non-parametric probabilistic clustering. In: 8th IEEE International Conference on Advanced Video and Signal-Based Surveillance, pp. 243–248. IEEE, Piscataway (2011)
17. Candamo, J., Shreve, M., Goldgof, D.B., Sapper, D.B., Kasturi, R.: Understanding transit scenes: a survey on human behavior-recognition algorithms. IEEE Trans. Intell. Transp. Syst. **11**(1), 206–224 (2010)
18. Cheng, F.-C., Huang, S.-C., Ruan, S.-J.: Scene analysis for object detection in advanced surveillance systems using the Laplacian distribution model. Syst. Man Cybern. Part C Trans. **41**(5), 589–598 (2011)
19. Ko, T., Soatto, S., Estrin, D.: Warping background s attraction. In: 2010 IEEE Conference on Computer Vision and Pattern Recognition (CVPR 2010), pp. 1331–1338. IEEE, Piscataway (2010)
20. Chen, S., Zhang, J., Li, Y., Zhang, J.: A hierarchical model incorporating segmented regions and pixel descriptors for video background subtraction. IEEE Trans. Ind. Inform. **8**(1), 118–127 (2012)
21. Srinivasan, K., Porkumaran, K., Sainarayanan, G.: Improved background subtraction techniques for security in video applications. In: 2009 3rd International Conference on Anti-counterfeiting, Security, and Identification in Communication, Hong Kong, pp. 114–117 (2009)
22. Bayona, A., San Miguel, J.C., Martinez, J.M.: Stationary foreground detection using background subtraction and temporal difference in video surveillance. In: 2010 IEEE International Conference on Image Processing, Hong Kong, pp. 4657–4660 (2010)
23. Razif, M.A.M., Mokji, M., Zabidi, M.M.A.: Low complexity maritime surveillance video using background subtraction on H.264. In: International Symposium on Technology Management and Emerging Technologies (ISTMET), Langkawi Island, pp. 364–368 (2015)
24. Candamo, J., Shreve, M., Goldgof, D.B., Sapper, D.B., Kasturi, R: Understanding transit scenes: a survey on human behavior- recognitional algorithms. IEEE Trans. Intell. Transp. Syst. **11**(1), 206–224 (2010)
25. Wu, X., Ou, Y., Qian, H., Xu, Y.: A detection system for human abnormal behavior. In: 2005 IEEE/RSJ International Conference on Intelligent Robots and Systems, Edmonton, Alta., pp. 1204–1208 (2005)
26. Hao, Z., Liu, M., Wang, Z., Zhan, W.: Human behavior analysis based on attention mechanism and LSTM neural network. In: 2019 IEEE 9th International Conference on Electronics Information and Emergency Communication (ICEIEC), Beijing, China, pp. 346–349 (2019)
27. Dataturks Bikers Wearing Helmet Or Not. https://dataturks.com/projects/priyaagarwal2730/Bikers%20Wearing%20Helmet%20Or%20Not (2019). Accessed 8 Sept 2019
28. Imglab. https://github.com/NaturalIntelligence/imglab (2019). Accessed 8 Sept 2019

Classification and Prediction of Rice Crop Diseases Using CNN and PNN

Suresh Limkar, Sneha Kulkarni, Prajwal Chinchmalatpure, Divya Sharma, Mithila Desai, Shivani Angadi, and Pushkar Jadhav

Abstract Rice holds a major share in India's agricultural economy. The various areas under rice cultivation in India include the jade green shaded rice cultivated in the eastern regions, dry rice fields in southern regions, etc. The country is one of the world's massive brown and white rice producers. The total yield in the year 2009 declined almost from 99.18 million tons to a total of just 89.14 million tons which affected the overall decrease in the crop yields as well as the financial outcome of the farmers. So, detecting rice diseases will help in lessening the adverse effects of the natural imbalance. Rice is one of the staple foods in India. Therefore, it becomes the main crop with the largest area under rice cultivation. As India is a tropical country, it benefits crop production as the crop needs hot and humid conditions for its efficient growth. Rice plants are grown in regions that receive heavy rainfall every year. For proper yield the crop requires an overall temperature of around 25 °C and a steady rainfall of more than 0.1mm. India being a country with extreme climatic conditions and increasing pollution cannot meet the production demand for the crops due to

S. Limkar · S. Kulkarni (✉) · P. Chinchmalatpure · M. Desai · S. Angadi · P. Jadhav
Department of Computer Engineering, AISSMS IOIT, Pune-01, India
e-mail: sneha.kulkarni13@gmail.com

S. Limkar
e-mail: sureshlimkar@gmail.com

P. Chinchmalatpure
e-mail: prajwalvvc@gmail.com

M. Desai
e-mail: mithiladesai25@gmail.com

S. Angadi
e-mail: shivaniangadi757@gmail.com

P. Jadhav
e-mail: pushkar.jadhao@gmail.com

D. Sharma
Department of Information Technology, MCC, Mumbai, India
e-mail: divzsharma19@gmail.com

S. C. Satapathy et al. (eds.), *Intelligent Data Engineering and Analytics*, Advances in Intelligent Systems and Computing 1177,
https://doi.org/10.1007/978-981-15-5679-1_4

growing diseases and abnormalities. This paper proposes a method to detect whether a rice crop is healthy or unhealthy using Convolutional neural networks, its various architectures, and probabilistic neural networks.

Keywords Convolutional neural network · Probabilistic neural network · Logical regression · Rice disease prediction · Feedforward neural network · Hyperparameter tuning

1 Introduction

Although other industries such as IT, automobiles, and trade contribute a major share in employability, no other occupation holds as much a share as agriculture holds in the terms of employability [1]. Almost a population of 2/3 earn their livelihood by working in farms and fields. As the gross domestic product value needs to be increased for a developing country like India, it is necessary to preserve these crops [2]. To do so, early prediction is required for both cultivation and monitoring purposes [3]. The main aim of this paper is to highlight this problem and to propose a system that could help or at least be a start in the problem solving [4]. The model aims to implement CNN as well as PNN [5], and gives the user the freedom to use any of the models depending upon the dataset and the accuracy expected. Physiological disorders are quite common in rice crops grown under different conditions of the soil. It is also called physiological diseases. The system focuses on the following categories of diseases. HispaRice scrapes the upper surface of leaf blades leaving only the lower epidermis. It also tunnels through the leaf tissues [6]. When damage is severe, plants become less vigorous The ultimate cause of Leaf Blast disease is by a fungus named Magnaporthe grisea. It causes an adverse effect on all the parts of the plant which lie above the ground such as leaves of the plant, nodes, small hidden portions of the neck, the node region, and also the sheath area of the leaf. This disease can affect the plant in a region where the blast spores are situated [7]. The first detection of this particular disease was recorded in the Tanjore district located in the southern region of Tamil Nadu in the year 1918. Subsequently, it spread over to other states such as Maharashtra in the year 1923 [8]. This disease includes a loss of around 70–80% in the overall grain loss. Brown Spot has been known to cause severe damage to the rice crop and is known to be a serious rice crop disease [9]. When the infection occurs in the seed, unfilled grains or spotted or discolored seeds are formed [10]. Discolored grains or unfilled grains are generated when such an infection occurs on the leaf plant. Such diseases develop in a region unfavorable to weather conditions such as high humidity and a temperature of around 16–36 °C [11]. In the years 1942–43, Bengal witnessed severe drought damage due to the occurrence of this disease [12].

2 Literature Survey

Vanitha [13] has a proposed a system for detecting three commonly occuring diseases in rice leaf plants namely Leaf brown spot, Sheath root, and Bacterial blight by using CNN. The dataset of 350 images is gathered from different sources like Google website and from various fields. Pros: Using ResNet architecture, the model is able to achieve an accuracy of 99.53%. The unhealthy plant leaves are classified into 3 classes. Cons: VGGI6 has the lowest accuracy of 96.2% as compared to ResNet. The model still lacks in efficiency and accuracy.

Shah et al. [14] have proposed a system to detect diseases occurring in rice plant diseases such as brown spot, bacterial leaf blight, and leaf smut. The paper surveys various preprocessed images and ML techniques useful for the identification of diseases in rice plants. 145 images are considered out of which 30 images belong to the healthy class, 25 rice images belong to brown spot class, 46 belong to bacterial leaf blight, and the remaining 44 belong to leaf smut class, respectively. Pros: With the help of backpropagation NN, total accuracy of 74.2% is achieved just by considering the image features. The paper provides an insight into rice disease detection using the image processing technique.

Kaur et al. [15] proposed classification on infection and scientific scenarios in various instances and whether detection of diseases is carried out. The classification includes unsupervised and supervised techniques for rice plants; self-organizing map neural network (SOM-NN) is deployed to find out brown spot diseases and rice blast diseases. 60 images of diseased plants were collected from various paddy fields from all over the country. Pros: Various classification algorithms were implemented with different datasets. The classification achieved the highest, around 97.20%; the classification accuracy of Bayes being 79.5%, SVM (88.1%), PNN (97.76%), and KNN (93.33%). Cons: Out of 60 images, only 50 images were detected accurately.

Phadikar and Sil [16] proposed a system for paddy leaf disease detection based on the images of leaves which were infected. Further, the proposed model includes the recognition of diseases based upon the damage symptoms as observed from plants. The classification algorithm implemented here is the self-organizing map (SOM) neural network. Pros: Four cases are classified namely RGB spots; paddy leaf images are classified using SOM neural network. On observation, it is found that the transformation of the image is frequent.

Akila and Deepan [17] made the use of a machine learning-based approach to predict leaf diseases using plant leaf images. Here 3 family detectors are considered which are faster region based CNN, Region-based FCNN and Single Shot Multibox Detector (SSD).The method implemented can identify various diseases which are capable of dealing any variation from plant areas. Pros: Faster Region Based CNN, Region-based FCNN and Single Shot Multibox Detector demonstrated in a single system. Cons: Images should be availabe with accurate resolution and quality or else the model won't run successfully.

Lu et al. [18] proposed a model for the identification of rice diseases with the classification methods implemented with the help of deep CNNs. The model trains

CNNs to identify 10 common diseased rice gradient-descent algorithms to train CNNs. **Pros**: Rice disease identification shows that the proposed system can correctly recognize diseases through image recognition. **Cons**: Training consumes a lot of time.

Liang et al. [19] proposed a recognition method of rice blast based on Convolutional Neural Network. LBPH along with SVM generates a lower accuracy than the use of CNN with Softmax as well as SVM. The data includes 5808 images patched out of which 2906 are positive and 2902 are negative. Pros: The evaluation results show that CNN is more effective than LBPH and Haar-WT with an accuracy percentage of 95. The proposed method generates satisfactory results but more advanced work is required to achieve more accuracy and reliability in disease detection of rice leaves. The model achieves 95% accuracy with the proposed CNN model.

3 Proposed System

This system proposed will help to predict physiological rice crop diseases depending on the structural abnormalities in rice crop leaves. The image dataset of the rice plant is taken and analyzed. After that system will classify the images using CNN and PNN model in order to predict the disease. Figure 1 shows the architecture of the proposed system.

Load Image Dataset: The first step in our proposed system involves getting images of rice crop leaves and uploading them to Google drive for Colab to access it. The dataset is loaded from the link https://www.kaggle.com/minhhuy2810/rice-diseases-image-dataset.

The image set is divided into three parts so as to allow maximum variance during training and testing.

Perform Preprocessing and Transformation Operations

In this phase, the images are resized to a standard pixel format, i.e. while passing these pixels to the convolutional layer, all the images should be in an identical format. Further, the images are normalized according to a range based on mean and standard deviation across RGB channels. After the images are preprocessed, they are further divided into training and testing phases. In our model, we have divided 80% as the training data and 20% as testing data. PyTorch [20] library of Python is used to perform image processing as well as tensor manipulations so as to normalize it and allow it to train in the CNN.

Rice Disease Prediction Using Machine Learning

The system deals with existing image data and performs analysis on that data. We are using Kaggle dataset for prediction of rice crop disease. This dataset is composed of rice leaf images. The model facilitates the use of two approaches:

Approach 1: The system uses a Convolutional Operation for feature extraction and a Feedforward Neural Network for classification. Basically, we are feeding all the normalized images to the CNN layer. Thereafter, a filter is applied to the input

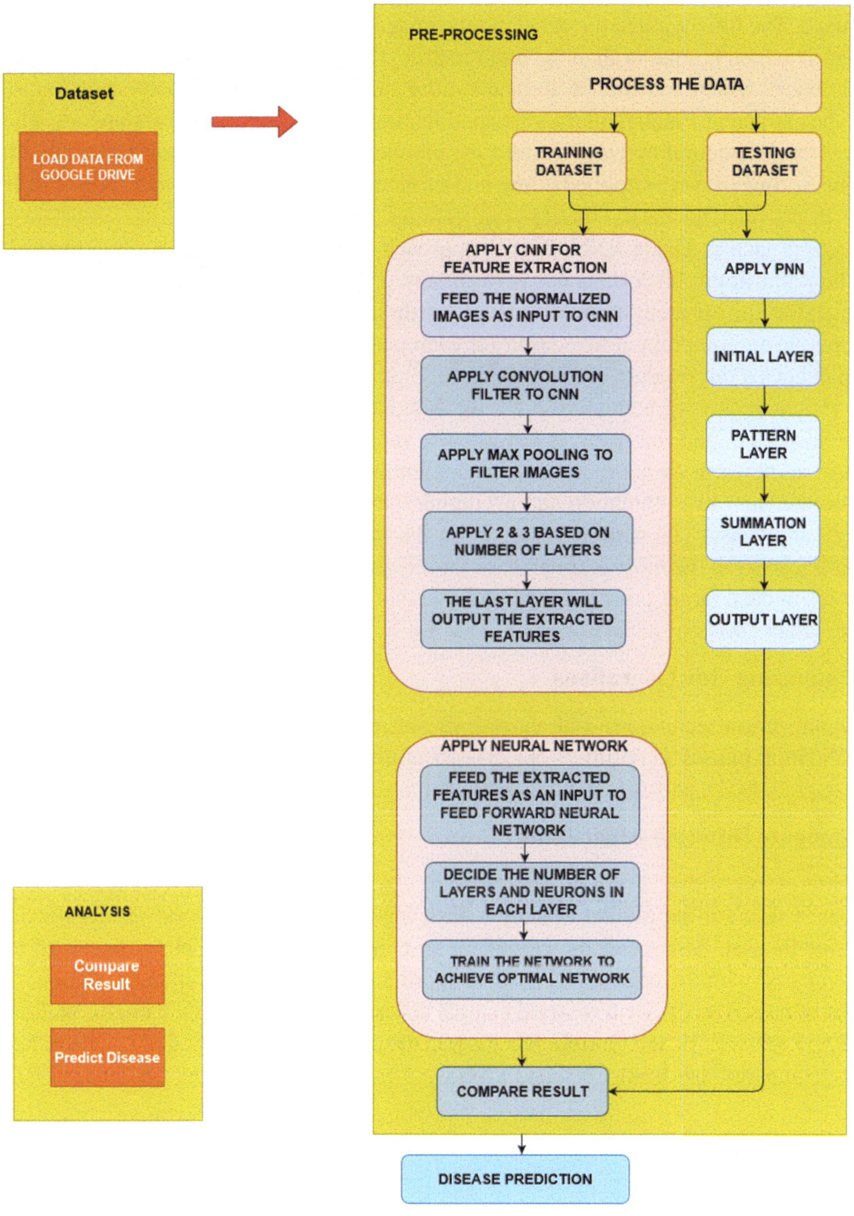

Fig. 1 System architecture

image. The filter applied moves along the image. The dot product is stored in another matrix which is smaller than the input matrix. Then we apply a max-pooling layer to the layers which are filtered. Repeat similar steps until we get the expected feature to be extracted. Finally, all the extracted features are fed to the neural network. The feedforward neural network decides the number of layers and the neurons required. The Softmax layer is applied at the end of feedforward neural network. A SoftMax layer provides means to converge our training.

Approach 2: Using PNN [21]. In approach 2, a PNN classifier is implemented. The initial Input layer does not perform any operations and further feeds the input to all the units present in the next layer. Further, the pattern layer is connected to the input layer, containing one neuron for each pattern in the training set. Each neuron calculates a dot product of the given rice sample say Y, pattern j which is stored as a weight vector wj, named $xj = Y \cdot wj$. A radial transfer function $\exp[(xj - 1)/\sigma 2]$ is calculated, and the result is fed into the summation layer. The summation layer neurons perform the computation, the maximum chances of pattern X being classified into the class by summation and averaging the output of all neurons that belong to the same class. The neuron generated in the output layer generates a binary value with respect to the most optimal class for the given example. It further compares the votes in each target present in the pattern layer and uses the highest vote to predict the target class.

Optimizing Our Operations

Optimization techniques and algorithms are used during both training as well as validation phases to optimize our classification; various algorithms like RMSprop, Adagrad, SGD, and Adam are used as well as compared.

Compare Different Algorithms

Different algorithms are applied to the dataset to train the network. A clear confusion matrix needs to be plotted to show which algorithm provides good results for the given dataset. Also, after the classification is implemented by CNN, it needs to be compared with the other CNN models in order to provide an accurate prediction. In our proposed work, we have tried using 3 different models of CNN namely, VGG16 [22], ResNet [23], and GoogLeNet [24]. In the end, results from both the approaches are compared, i.e. from CNN and PNN.

4 Algorithms

I. Logistic Regression: It is a supervised classification algorithm. The output variable y takes values that are discrete in nature with respect to the given set of features, say X. It basically predicts a value which can lie anywhere between $-\infty$ to $+\infty$. The output is considered a class variable. Therefore, the final output lies in the range of 0–1. For this, a sigmoid function is used. The required output should be in the form of 0-no, 1-yes.

$$g(x) = \frac{1}{1 + e^{-x}}$$

where $g(x)$ is the sigmoid function.

II. Convolutional Neural Network: Convolutional neural network (CNN) is a type of feedforward artificial neural network in which the connectivity pattern between its neurons resemble the organization of the human visual cortex.

(a) **Initialization**: Xavier initialization is used as the initial step. With the help of this, the activations and the gradients are controlled very efficiently.
(b) **Activation Function**: It is responsible for nonlinearly transforming the data. Rectifier linear units (ReLU) is defined as

$$f(x) = \max(0, x)$$

Results found were better than the traditional sigmoid and tangent functions. The limitation such as imposing a constant can be treated with the help of a variant called Leaky Rectifier linear unit (LReLU). In such a case, the function is defined as

$$f(x) = \max(0, x) + \mu\min(0, x)$$

where μ is the leakiness parameter. In the last FC layer, we use Softmax.

III. Probabilistic Neural Network: It is a feedforward neural network used for effective classification and recognition purposes. In this algorithm, the parent probability distribution function (PDF) of each class is calculated by Parzen window and a nonparametric function. The layers are further split into 4 layers namely

(1) Input: Small units called neurons represent a particular predictor variable. It performs Standardization by subtracting median values and dividing by a random in quartile range.
(2) Pattern Layer: It consists of one neuron per case present in the training data set. It contains the specific values for the predicted class label. Neuron present in the hidden layer calculates the Euclidean distance from its center point and eventually applies the radial basis kernel function.
(3) Summation layer: In probabilistic neural network patterns, a neuron is present for each category. The real target value is maintained by a hidden neuron and its weighted value is further passed to the pattern neuron. These pattern neurons simply add on the values to the class that they represent.
(4) Output layer: This layer inspects all the weighted votes for every category in the second layer and checks the largest vote to detect the original category.

5 Result and Discussion

The experimental results of classification are tabulated in the following table:

Neural network		Training		Testing		Validation	
Hidden layer	No. of nodes	Accuracy	Error	Accuracy	Error	Accuracy	Error
For CNN							
1	272	0.8640	0.5643	0.6890	0.9870	0.7889	0.5670
2	272 148	0.9444	0.1371	0.8670	0.7438	0.8183	0.6066
3	272 148 160	0.9100	0.280	0.7569	0.7469	0.8345	0.5961
For PNN							
1	272	0.880	0.343	0.7654	0.612	0.790	0.453
2	272 148	0.9980	0.111	0.8857	0.569	0.8239	0.562
3	272 148 160	0.9465	0.1634	0.8567	0.675	0.8674	0.498

6 Conclusion

Crop disease prediction is a popular exploration area in computer vision. The parameter on which the disease is mostly dependent is its physical structure. So, processing and using images play an important role in the system. In this paper, we give a brief review of different methodologies in the prediction of rice crop disease detection. A large collection of methods is identified for recognition of rice disease. Through results, we can conclude that PNN & CNN achieves 99.8% and 94.4% accuracy, respectively. But none of them can give 100% accuracy in the prediction. So, there is a need to develop a system which can predict various crop abnormalities with higher accuracy.

References

1. Rehman, A., Jingdong, L., Khatoon, R., Hussain, M.I.: Modern agricultural technology adoption its importance, role and usage for the improvement of agriculture. Am. Eurasian J. Agric. Environ. Sci. **16**, 284–288 (2016). https://doi.org/10.5829/idosi.aejaes.2016.16.2.12840

2. Gandhi, N., Armstrong, L.J., Petkar, O.: Predicting rice crop yield using Bayesian networks. In: 2016 International Conference on Advances in Computing, Communications and Informatics (ICACCI), Jaipur, pp. 795–799 (2016). https://doi.org/10.1109/icacci.2016.7732143
3. [Available online on 13-10-2019]. https://www.sgs.com/en/agriculture-food/seed-and-crop/crop-monitoring-and-agronomic-services/crop-monitoring
4. [Available on 13-10-2019]. https://mindbowser.com/solve-agricultural-problems-using-machine-learning
5. Yun, S., Xianfeng, W., et al.: PNN based crop disease recognition with leaf image features and meteorological data. **8**(4) (2015)
6. Hazarika, L., Deka, M., Bhuyan, M.: Oviposition behaviour of the rice hispa (2005)
7. Narmadha, R.P., Arulvadivu, G.: Detection and measurement of paddy leaf disease symptoms using image processing. In: 2017 International Conference on Computer Communication and Informatics (ICCCI), Coimbatore, pp. 1–4 (2017). https://doi.org/10.1109/iccci.2017.8117730
8. Zhang, H., Jin, Q., Chai, R., Hu, H., Zheng, K.: Monitoring rice leaves blast severity with hyperspectral reflectance. In: 2010 2nd International Conference on Information Engineering and Computer Science, Wuhan, pp. 1–4 (2010). https://doi.org/10.1109/iciecs.2010.5678125
9. Singh, R., Sunder, Agarwal, R.: Brown spot of rice: an overview. Indian Phytopathol. 201–215 (2014)
10. Liu, L., Zhou, G.: Extraction of the rice leaf disease image based on BP neural network. In: 2009 International Conference on Computational Intelligence and Software Engineering, Wuhan, pp. 1–3 (2009). https://doi.org/10.1109/cise.2009.5363225
11. Joshi, A.A., Jadhav, B.D.: Monitoring and controlling rice diseases using Image processing techniques. In: 2016 International Conference on Computing, Analytic and Security Trends (CAST), Pune, pp. 471–476 (2016). https://doi.org/10.1109/cast.2016.7915015
12. Islam, T., Sah, M., Baral, S., Roy Choudhury, R.: A faster technique on rice disease detection using image processing of affected area in agro-field. In: 2018 Second International Conference on Inventive Communication and Computational Technologies (ICICCT), Coimbatore, pp. 62–66 (2018). https://doi.org/10.1109/icicct.2018.8473322
13. Vanitha, V.: Rice Disease Detection Using Deep Learning, vol. 7, no. 5S3 (2019). ISSN: 2277-3878
14. Shah, J., Prajapati, H., Dabhi, V.: A survey on detection and classification of rice plant diseases. 1–8 (2016). https://doi.org/10.1109/icctac.2016.7567333
15. Kaur, S., Pandey, S., Goel, S.: Plants disease identification and classification through leaf images: a survey. Arch. Comput. Methods Eng. **26** (2018). https://doi.org/10.1007/s11831-018-9255-6
16. Phadikar, S., Sil, J.: Rice disease identification using pattern recognition techniques. In: 2008 11th International Conference on Computer and Information Technology, Khulna, pp. 420–423 (2008). https://doi.org/10.1109/iccitechn.2008.4803079
17. Akila, M., Deepan, P.: Detection and classification of plant leaf diseases by using deep learning algorithm. Int. J. Eng. Res. Technol. (IJERT) ICONNECT **6**(7) (2018)
18. Lu, Y., Yi, S., Zeng, N., Liu, Y., Zhang, Y.: Identification of Rice Diseases using Deep Convolutional Neural Networks (2017). https://doi.org/10.1016/j.neucom.2017.06.023
19. Liang, W., Zhang, H., Zhang, G., et al.: Rice blast disease recognition using a deep convolutional neural network. Sci. Rep. **9**, 2869 (2019). https://doi.org/10.1038/s41598-019-38966-0
20. Heghedus, C., Chakravorty, A., Rong, C.: Neural network frameworks. comparison on public transportation prediction. In: 2019 IEEE International Parallel and Distributed Processing Symposium Workshops (IPDPSW), Rio de Janeiro, Brazil, pp. 842–849 (2019). https://doi.org/10.1109/IPDPSW.2019.00138
21. https://www.cse.unr.edu/~looney/cs773b/PNNtutorial.pdf
22. https://neurohive.io/en/popular-networks/vgg16/
23. Wang, F., et al.: Residual attention network for image classification. In: 2017 IEEE Conference on Computer Vision and Pattern Recognition (CVPR), Honolulu, HI, pp. 6450–6458 (2017). https://doi.org/10.1109/cvpr.2017.683

24. Szegedy, C., et al.: Going deeper with convolutions. In: 2015 IEEE Conference on Computer Vision and Pattern Recognition (CVPR), Boston, MA, pp. 1–9 (2015). https://doi.org/10.1109/cvpr.2015.7298594
25. Singh, A., Singh, M.L.: Automated blast disease detection from paddy plant leaf—a color slicing approach. In: 2018 7th International Conference on Industrial Technology and Management (ICITM), Oxford, pp. 339–344 (2018). https://doi.org/10.1109/icitm.2018.8333972

SAGRU: A Stacked Autoencoder-Based Gated Recurrent Unit Approach to Intrusion Detection

N. G. Bhuvaneswari Amma, S. Selvakumar, and R. Leela Velusamy

Abstract The ubiquitous use of the Internet in today's technological world makes the computer systems prone to cyberattacks. This led to the emergence of Intrusion Detection System (IDS). Nowadays, IDS can be built using deep learning approaches. The issues in the existing deep learning-based IDS are the curse of dimensionality and vanishing gradient problems leading to high learning time and low accuracy. In this paper, a Stacked Autoencoder-based Gated Recurrent Unit (SAGRU) approach has been proposed to overcome these issues by extracting the relevant features by reducing the dimension of the data using Stacked Autoencoder (SA) and learning the extracted features using Gated Recurrent Unit (GRU) to construct the IDS. Experiments were conducted on NSL KDD network traffic dataset and it is evident that the proposed SAGRU approach provides promising results with low learning time and high accuracy as compared to the existing deep learning approaches.

Keywords Autoencoder · Cyberattacks · Deep learning · Gated recurrent unit · Intrusion detection

N. G. Bhuvaneswari Amma (✉) · S. Selvakumar · R. Leela Velusamy
National Institute of Technology, Tiruchirappalli 620 015, Tamil Nadu, India
e-mail: ngbhuvaneswariamma@gmail.com

S. Selvakumar
e-mail: ssk@nitt.edu; director@iiitu.ac.in

R. Leela Velusamy
e-mail: leela@nitt.edu

S. Selvakumar
Indian Institute of Information Technology, Una 177 209, Himachal Pradesh, India

S. C. Satapathy et al. (eds.), *Intelligent Data Engineering and Analytics*, Advances in Intelligent Systems and Computing 1177,
https://doi.org/10.1007/978-981-15-5679-1_5

1 Introduction

Nowadays, the Internet plays an essential role as all the businesses and customers use the Internet services to access their websites and e-mails to do various activities. The increasing usage of the Internet creates the vulnerability of cyberattacks. Therefore, securing the Internet services is needed and Intrusion Detection System (IDS) provides a defense layer against the cyberattacks [5]. The IDS has been used to detect intrusions in network traffic which could not be detected by conventional firewalls. The detection of intrusions in a network is based on the behavior of the intruder that differs from that of a legitimate network traffic. Any traffic that deviates from that of legitimate is termed as intrusion [3].

The approaches used to construct IDS can be classified into signature-based approach and anomaly-based approach [9, 12]. The anomaly-based approaches are classified into statistical, computational intelligence, data mining, and machine learning [4, 13]. In recent years, deep learning-based approaches have been proposed for building IDSs. The techniques such as Convolutional Neural Network (CNN), Autoencoder, Recurrent Neural Network (RNN), Long Short Term Memory (LSTM), and Gated Recurrent Unit (GRU) have been used in recent studies [2]. These techniques extract the features automatically and learn the features in various levels of representations. The learning process of these techniques suffers the vanishing gradient problem, i.e., small gradient value does not contribute too much for learning [10]. Further, the network generates huge features and suffers the curse of dimensionality problem [14]. These issues motivated us to propose the $SAGRU$ approach that reduces the dimension of traffic features and build an IDS without suffering the vanishing gradient problem. The contributions of this paper are listed as follows:

1. Stacked Autoencoder Feature Extraction (SAFE) to extract the features of network traffic data.
2. Gated Recurrent Unit Learning (GRUL) for learning the extracted features to build IDS.
3. SAGRU-based intrusion detection for detecting the intrusions in network traffic data.

The rest of the paper is organized as follows: Sect. 2 briefly describes related works. Section 3 introduces and describes the proposed $SAGRU$ approach. Section 4 analyzes the performance of the proposed approach. Conclusion with future directions are provided in Sect. 5.

2 Related Works

Cyberattacks are the attempts made maliciously by an attacker to breach an individual or the information system of an organization. The motive behind these attacks is that the attacker may get benefited by destroying the victim's network [3]. The most common cyberattacks include malware, phishing, Denial-of-Service attack, worms, port scans, etc. These attacks can be launched against critical infrastructures, viz., telecommunications, transportation, financial networks, etc. The attackers can disrupt the command, control, and communication of these infrastructures [4]. Further, the usage of smart devices with the Internet is increasing in our day-to-day activities which leads to multiple cyberattacks. In this circumstance, intrusion detection is needed to defend against these attacks [9]. The IDS monitors the network and alerts the administrator about abnormal behavior of the network traffic and resists to external attacks [11].

Deep learning learns the traffic data by computing the hidden relationship in the features [8]. The techniques, viz., autoencoder, CNN, RNN, LSTM, etc., can be used for network traffic classification [12]. These techniques automatically compute the correlations in network traffic features motivated to propose the *SAGRU* approach for intrusion detection.

3 Proposed SAGRU Approach

The $SAGRU$ approach consists of three modules, viz., Stacked Autoencoder Feature Extraction (SAFE), Gated Recurrent Unit Learning (GRUL), and SAGRU-based intrusion detection. The SAFE module automatically finds the correlation among features and extracts the features from the training data. The extracted features are used by GRUL which overcomes the vanishing gradient problem by remembering the relevant information and forgetting the irrelevant information. The learned $SAGRU$ model is used for detecting the intrusions in network traffic data. The block schematic of the proposed $SAGRU$ intrusion detection approach is depicted in Fig. 1.

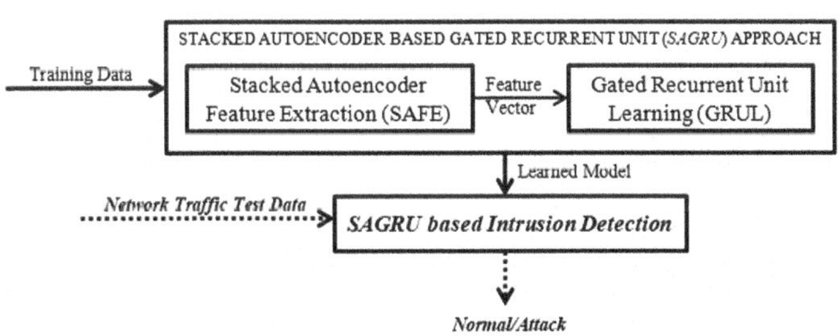

Fig. 1 Block schematic of the proposed SAGRU approach

3.1 Stacked Autoencoder Feature Extraction (SAFE)

Feature extraction plays a major role in classification tasks to improve the performance of the detection process. The proposed SAFE module captures the nonlinear relationships between the data and is capable of handling large-scale network data. The intuition behind using autoencoder for feature extraction is that the unsupervised nature of this approach is robust to noisy data.

Figure 2 depicts the architecture of the SAFE module. The autoencoder consists of encoder, discriminator, and decoder. The SA consists of more than one layer in encoder and decoder. The first layer of the encoder is the input layer and the last layer of the decoder is the output layer. Being an unsupervised approach, the autoencoder projects the input as the output. The layer which overlaps the encoder with the decoder is the discriminative or bottleneck layer [14]. The structure of the proposed SA is $x - 30 - 20 - 10 - 20 - 30 - x_r$, where x and x_r are the number of inputs and outputs of the SA. The units of the SA are represented in blue, yellow, and green for encoder, discriminator, and decoder, respectively. The discriminative layer provides the extracted features for further learning.

The traffic data is given as input to the SA which is denoted as x and the output of the SA is x_r which is similar to x. The weights, w_{H1}, w_{H2}, w_{H3}, and w_{H4}, are the encoding layer weights and w_{H1r}, w_{H2r}, w_{H3r}, and w_{H4r} are the decoding layer weights. Let $F_1, F_2, \ldots, F_x$ be the input features and the computation in the encoding layer using (1) with Rectified Linear Unit (ReLU) activation function. The intuition behind using ReLU activation function is to generate a sparse representation of a feature so as to provide separation capability. Moreover, ReLU performs faster training for large-scale network data. The decoding of the features is done by performing the reverse operations using (2) with the sigmoid activation function.

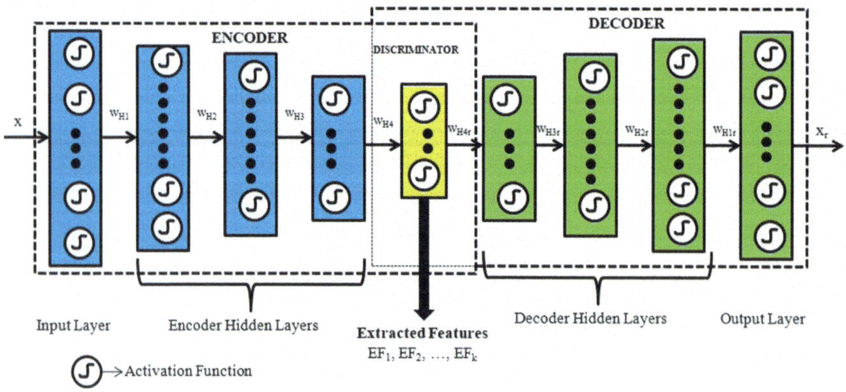

Fig. 2 Architecture of the proposed SAFE

$$E_f = f_{\text{ReLU}}\left(W_{\text{Hi}} \times F_x + b\right) \tag{1}$$

$$D_f = f_{\text{Sig}}\left(W_{\text{Hir}} \times E_f + b^1\right) \tag{2}$$

where b is the bias. The learning is performed layer-wise, i.e., first $x - 30 - x_r$ is learned, then $30 - 20 - 30$, and finally $20 - 10 - 20$ is learned. Once the data is learned, the loss occurred in learning is computed using cross entropy as follows:

$$\text{Loss}_{\text{SA}} = -\frac{1}{x}\left[\sum x_i \log\left(f_{\text{Sig}}\left(x_{\text{ri}}\right)\right) + \left(1 - x_i\right)\log\left(1 - f_{\text{Sig}}\left(x_{\text{ri}}\right)\right)\right] \tag{3}$$

Finally, the features in the discriminative layer are the extracted features which are used for fine-tuning the intrusion detection model.

3.2 Gated Recurrent Unit Learning (GRUL)

The existing deep neural networks suffer the vanishing gradient problem [7] in which the gradients of the network become zero with the usage of certain activation functions. The network that suffers this problem is hard to train. The proposed GRUL approach overcomes this problem by remembering and forgetting certain information. The extracted features are passed to the GRUL and each feature requires a separate GRU. Figure 3 depicts the architecture of proposed GRUL. The GRU consists of four phases: update gate, reset gate, current memory content, and final memory

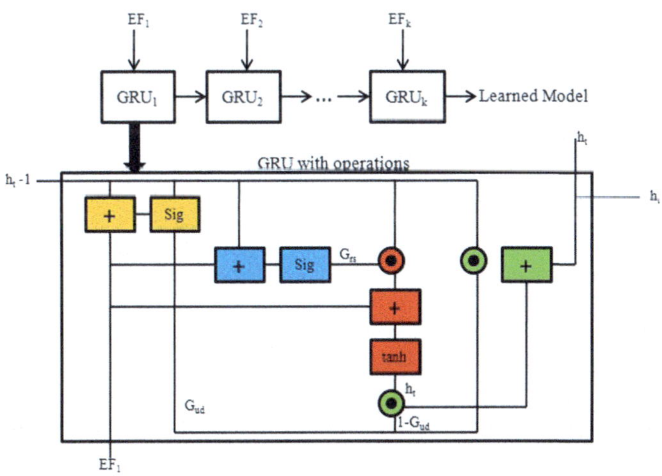

Fig. 3 Architecture of the proposed GRUL

content represented in yellow, blue, red, and green, respectively, and computed using (4), (5), (6), and (7), respectively.

Let $\mathrm{EF}_1, \mathrm{EF}_2, \ldots, \mathrm{EF}_k$ be the extracted features obtained as a result of executing the SAFE module. The update gate determines the percentage of the past information obtained in the previous steps to be passed along with the future data. This part of the GRU eliminates the information which creates the vanishing gradient problem. The reset gate decides the percentage of the past information to forget. The current memory content stores the relevant information from the past, and the final memory at the current step is passed down to the network.

$$G_{\mathrm{ud}} = f_{\mathrm{Sig}}\left(W_u \times \mathrm{EF}_k + U_u \times h_{t-1}\right) \tag{4}$$

$$G_{\mathrm{rs}} = f_{\mathrm{Sig}}\left(W_r \times \mathrm{EF}_k + U_r \times h_{t-1}\right) \tag{5}$$

$$h_t^1 = f_{\mathrm{tanh}}\left(W \times G_{\mathrm{ud}} + G_{\mathrm{rs}} \odot U \times h_{t-1}\right) \tag{6}$$

$$h_t = G_{\mathrm{ud}} \odot h_{t-1} + (1 - G_{\mathrm{ud}}) \odot h_t^1 \tag{7}$$

where $f_{\mathrm{tan}}h$ is the tangent activation function. The output of the GRUL is passed to the cross entropy computation using (8) and the learning happens.

$$\mathrm{Loss}_{\mathrm{GRU}} = -\frac{1}{k}\left[\sum_{i=1}^{k} \mathrm{tar}_i \log\left(f_{\mathrm{Sig}}\left(h_t\right)\right) + (1 - \mathrm{tar}_i)\log\left(1 - f_{\mathrm{Sig}}\left(h_t\right)\right)\right] \tag{8}$$

The learned SAGRU model which is built using SAFE and GRUL has been used for detecting the intrusions in network traffic.

3.3 SAGRU-Based Intrusion Detection

The network traffic features are extracted using the learned SAFE architecture, and the extracted features are given to the GRUL architecture. The computed output of GRUL is passed to softmax activation function, f_{sm}, in (9) for classifying the class of the network traffic.

$$f_{\mathrm{sm}}\left(\mathrm{CO}_{ij}\right) = \mathrm{Exp}\left(\mathrm{CO}_{ij}\right) / \sum_{j=0}^{k} \mathrm{Exp}\left(\mathrm{CO}_{ij}\right) \tag{9}$$

where CO_{ij} is the computed output of the SAGRU model. The output of f_{sm} is converted to a vector, and a mask operation is performed with the vector [1 1 1 1 1]. If the first element of the mask operation is 1, then the given network traffic is classified as normal or else detected as an attack.

4 Performance Analysis

The proposed SAGRU approach has been implemented in MATLAB R2018a under Windows 10 environment. The experimentation was performed using NSL KDD benchmark network traffic dataset [1]. The NSL KDD dataset consists of 41 input features categorized into basic features, content features, traffic features with the same host 2 second window, and traffic features with the same service 100 connections. The class labels are in any one of the classes such as Normal, Denial of Service (DoS), Probe, User to Root (U2R), and Remote to Local (R2L) [6]. The training dataset consists of 13449, 9234, 2289, 11, and 209 records of Normal, DoS, Probe, U2R, and R2L traffic, respectively, and the testing dataset consists of 2152, 4344, 2402, 67, and 2885 records of Normal, DoS, Probe, U2R, and R2L traffic, respectively.

The proposed approach is evaluated based on the following metrics: Precision, Recall, F-measure, False Alarm, Accuracy, and Error Rate [2]. Table 1 tabulates the performance of the proposed approach and the existing deep learning approaches, viz., RNN, LSTM, and GRU. The reason for choosing these approaches for comparison is that all these approaches are recurrent-based supervised learning approaches.

Table 1 Performance evaluation

Approach	Traffic	Precision	Recall	F-measure	False Alarm	Accuracy	Error rate
RNN	Normal	94.50	99.02	96.71	5.5	99.02	0.98
	DoS	96.21	92.45	94.29	3.79	92.45	7.55
	Probe	92.48	87.09	89.70	7.52	87.09	12.91
	U2R	32.56	83.58	46.86	67.44	83.58	16.42
	R2L	93.0	96.29	94.62	7.0	96.29	3.71
LSTM	Normal	95.53	99.35	97.40	4.47	99.35	0.65
	DoS	98.13	96.85	97.49	1.87	96.85	3.15
	Probe	95.38	93.63	94.50	4.62	93.63	6.37
	U2R	53.64	88.06	66.67	46.36	88.06	11.94
	R2L	97.55	96.60	97.07	2.45	96.60	3.40
GRU	Normal	95.74	99.21	97.44	4.26	99.21	0.79
	DoS	98.07	97.03	97.55	1.93	97.03	2.97
	Probe	95.97	93.09	94.51	4.03	93.09	6.91
	U2R	52.25	86.57	65.17	47.75	86.57	13.43
	R2L	96.98	96.85	96.91	3.02	96.85	3.15
SAGRU	Normal	96.89	99.81	98.33	3.11	99.81	0.19
	DoS	98.59	98.85	98.72	1.42	98.85	1.15
	Probe	97.81	94.92	96.34	2.19	94.92	5.08
	U2R	63.54	91.04	74.84	36.47	91.04	8.96
	R2L	98.35	97.37	97.86	1.65	97.37	2.63

The proposed SAGRU approach exhibits promising results compared to the existing approaches. All these approaches could not provide significant results with respect to U2R as the traffic patterns are similar to normal traffic. The approaches LSTM and GRU provide more or less similar results as these two approaches are similar in performance but differ in learning time.

Figure 4 shows the learning time taken by the SAGRU approach as compared to the existing approaches. It can be seen that the proposed SAGRU approach took 32 min for learning as the features were extracted using the SAFE module, and the learning was performed using the extracted features. The existing recurrent-based deep learning approaches such as RNN, LSTM, and GRU took 51, 46, and 40 min, respectively. Figure 5 depicts the results based on the dimension of features. It is

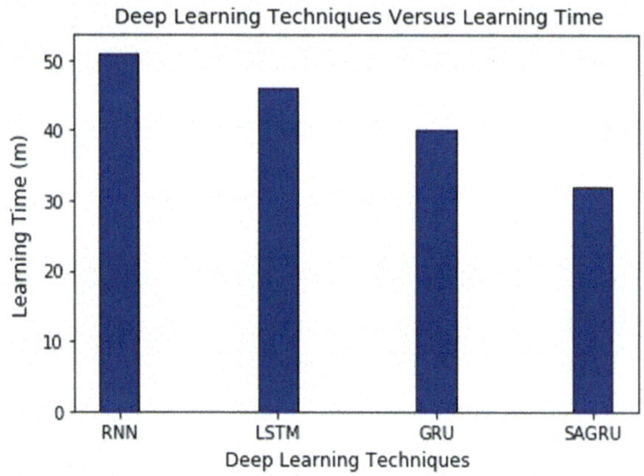

Fig. 4 Deep learning techniques versus learning time

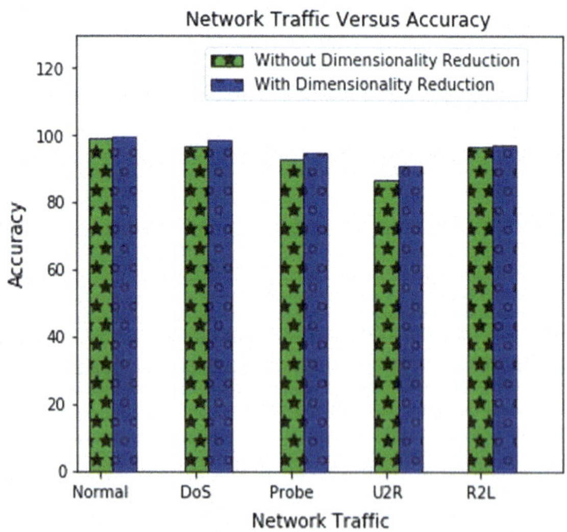

Fig. 5 Network traffic versus accuracy

observed that the SAGRU approach performs significantly better compared to GRU without dimensionality reduction as the SAGRU approach uses the SAFE module to reduce the dimension of the network traffic features.

5 Conclusion

In this paper, an anomaly-based IDS approach named $SAGRU$ is proposed to detect cyberattacks by overcoming the curse of dimensionality and vanishing gradient problems. The features were extracted using stacked autoencoder and the IDS was built using the learned GRU. The SAGRU IDS detected the network traffic and the experimentation was performed using NSL KDD network traffic dataset. The accuracy of 99.81, 98.85, 94.92, 91.04, and 97.37% for Normal, DoS, Probe, U2R, and R2L network traffic, respectively, have been obtained. Also, the proposed approach gives promising results compared to the existing RNN, LSTM, and GRU approaches. Further, it is evident that the proposed method requires less learning time compared to the existing approaches. Moreover, the reduction in the dimension of the data improves the performance of the IDS in terms of accuracy. In future, the proposed approach will be investigated with real-time network traffic data streams.

References

1. Nsl-kdd dataset. http://www.unb.ca/research/iscx/dataset/iscx-NSL-KDD-dataset.html (2009)
2. Amma, N.G.B., Subramanian, S.: Vcdeepfl: Vector convolutional deep feature learning approach for identification of known and unknown denial of service attacks. In: TENCON 2018-2018 IEEE Region 10 Conference, pp. 640–645. IEEE (2018)
3. Bhuyan, M.H., Bhattacharyya, D.K., Kalita, J.K.: Network anomaly detection: methods, systems and tools. IEEE Commun. Surv. Tutor. **16**(1), 303–336 (2013)
4. Buczak, A.L., Guven, E.: A survey of data mining and machine learning methods for cyber security intrusion detection. IEEE Commun. Surv. Tutor. **18**(2), 1153–1176 (2015)
5. Guo, C., Zhou, Y., Ping, Y., Zhang, Z., Liu, G., Yang, Y.: A distance sum-based hybrid method for intrusion detection. Appl. Intell. **40**(1), 178–188 (2014)
6. Iglesias, F., Zseby, T.: Analysis of network traffic features for anomaly detection. Mach. Learn. **101**(1–3), 59–84 (2015). https://doi.org/10.1007/s10994-014-5473-9
7. Kim, P.S., Lee, D.G., Lee, S.W.: Discriminative context learning with gated recurrent unit for group activity recognition. Pattern Recogn. **76**, 149–161 (2018)
8. LeCun, Y., Bengio, Y., Hinton, G.: Deep learning. Nature **521**(7553), 436 (2015)
9. Mishra, P., Varadharajan, V., Tupakula, U., Pilli, E.S.: A detailed investigation and analysis of using machine learning techniques for intrusion detection. IEEE Commun. Surv. Tutor. **21**(1), 686–728 (2018)
10. NG, B.A., Selvakumar, S.: Deep radial intelligence with cumulative incarnation approach for detecting denial of service attacks. Neurocomputing **340**, 294–308 (2019)
11. Rezvy, S., Petridis, M., Lasebae, A., Zebin, T.: Intrusion detection and classification with autoencoded deep neural network. In: International Conference on Security for Information Technology and Communications, pp. 142–156. Springer (2018)

12. Shone, N., Ngoc, T.N., Phai, V.D., Shi, Q.: A deep learning approach to network intrusion detection. IEEE Trans. Emerg. Top. Comput. Intell. **2**(1), 41–50 (2018)
13. Weller-Fahy, D.J., Borghetti, B.J., Sodemann, A.A.: A survey of distance and similarity measures used within network intrusion anomaly detection. IEEE Commun. Surv. Tutor. **17**(1), 70–91 (2014)
14. Yousefi-Azar, M., Varadharajan, V., Hamey, L., Tupakula, U.: Autoencoder-based feature learning for cyber security applications. In: 2017 International Joint Conference on Neural Networks (IJCNN), pp. 3854–3861. IEEE (2017)

Comparison of KNN and SVM Algorithms to Detect Clinical Mastitis in Cows Using Internet of Animal Health Things

K. Ankitha and D. H. Manjaiah

Abstract The clinical mastitis is a harmful disease in cows and many researchers working on milk parameters to detect clinical mastitis. Internet of things (IoT) is a developing era of technology where every object is connected to the Internet using sensors. Sensors are an essential unit of an IoT to collect the data for analysis. The proposed method concentrates on deploying sensors on cows to monitor the health issues and will state IoT as an Internet of Animal Health Things (IoAHT). Dairy cows are an essential unit of the Indian economy because India is a top country in milk production. Clinical mastitis affects dairy cows in the production of milk. Recent studies in the dairy industry proved the use of technologies and sensors for good growth of cows. This paper reviews a method used for detecting clinical mastitis in cows and proposes a system for the same using IoAHT. The KNN and SVM algorithms are used on the primary data set to obtain a result of the detection. In comparison to these algorithms, SVM provided better results in detecting mastitis in cows.

Keywords Mastitis · Veterinary science · Sensors · Sac

1 Introduction

The present world is evolving at a tremendous phase, where the Internet has become a basic requirement. Nowadays, people expect nearly every physical object to be connected to the Internet, which has become a reality through the sensors. This has given rise to a new technology called IoT, which has successfully attracted people from various domains—from industries to researchers, from scientists to students and teachers, etc. IoT has found its application in various domains like smart homes, smart

K. Ankitha (✉) · D. H. Manjaiah
Department of Computer Science, Mangalore University, Mangaluru, Karnataka, India
e-mail: ankithapraj@gmail.com

D. H. Manjaiah
e-mail: drmdhmu@gmail.com

S. C. Satapathy et al. (eds.), *Intelligent Data Engineering and Analytics*, Advances in Intelligent Systems and Computing 1177,
https://doi.org/10.1007/978-981-15-5679-1_6

cities, connected health, connected diaries, etc. One such upcoming IoT application can be found in connected diaries.

Earth is home to many living creatures including man. Health is an important aspect for each and every living creature. With this regard, IoT has played a major role in connected health, especially in Human Medical System (HMS), where the tasks of doctors are successfully being automated. Due to its spectacular success in HMS, Veterinary Science has also started to involve IoT for making its tasks automated.

At present, connected diaries are facing huge challenges of dealing with various kinds of diseases affecting the cows. One identified disease is clinical mastitis, which leads to less milk production in cows, and on ignorance, death is for sure. Clinical mastitis has various parameters to be considered like environmental conditions, bacteria, viruses, worms, etc. If the affected cows are not treated immediately within a specific time, it may either reduce the milk production or may lead to death. Dairy industries are one of the important economic sources of India, which, if affected by clinical mastitis, has the potential to affect the entire economy.

In better practices, for better monitoring purposes, the owners of cows are expected to use advanced technology for detecting clinical mastitis. The main idea of this paper is to propose a new methodology to find clinical mastitis by reviewing the existing technologies and overcoming their deficiencies.

Figure 1 shows the udder information which is one of the important parameters in clinical mastitis. Udder size does not vary in normal cows but in mastitis, cow's udder size increases gradually depending on the time interval.

The color of a mastitis cow udder turns to red; this is also one of the parameters to detect mastitis in cows as shown in Fig. 2.

The idea of this paper is to give a glimpse of the existing technologies to detect clinical mastitis and its drawbacks, also a new hardware is proposed to find clinical mastitis in cows.

Fig. 1 Udder variation in cows

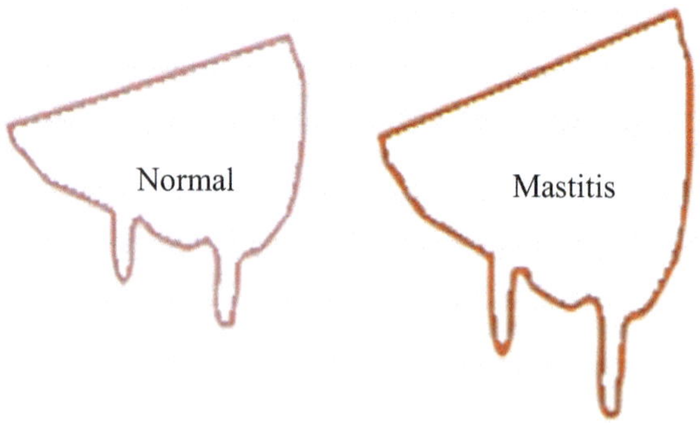

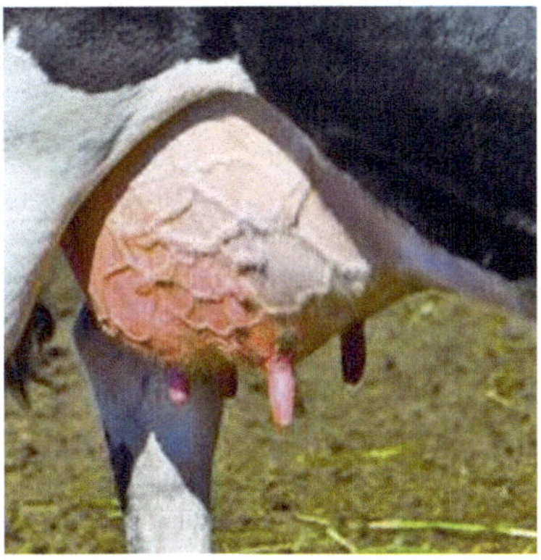

Fig. 2 Udder turns red [12]

2 Literature Study

The initial reviews started from a veterinary hospital; here, manual methods were employed to detect clinical mastitis disease. The problem of mastitis can be classified into two categories—Sub Clinical Mastitis (SCM) and clinical mastitis. In SCM, the farmer cannot identify the external symptom, whereas in the clinical mastitis latter exhibits various external symptoms like

- Swollen udder
- Increase in temperature
- Udder hardness
- Watery milk
- Variation in Somatic Cell Count (SCC) in milk
- Variation in Electric Conductivity (EC) of milk
- Variation in PH value of milk

Based on the kind of symptoms exhibited, a veterinary doctor will decide the state of mastitis using the California Mastitis Test (CMT) which is a usual indicator on the cow side to give an SCC count [1, 2].

Emma Carlén and Erling Strandberg proposed "Genetic Parameters for Clinical Mastitis, Somatic Cell Score, and Production in the First Three Lactations of Swedish", which highlighted the detection of clinical mastitis by examining genetic parameters, average SCC, and milk production in first three lactations [1]. It was found that the genetic correlation between mastitis and average SCC was higher implying low average SCC reduces the situation of mastitis. The result is proven using statistical analysis.

Caroline Viguier et al. proposed "Mastitis detection: current trends and future perspectives" which identified and discussed various methods for clinical mastitis detection. They are (i) using natural genes or characteristics differentiating a cattle with mastitis [2]. (ii) Measuring specific proteins in the milk. (iii) A nucleic acid test for pathogen detection in milk. (iv) Temperature is used as a parameter to detect mastitis in cows. Increase in temperature tells about the cow's illness. (v) Variations in the EC, SCC, and color of the milk are detected for the clinical mastitis using suitable sensors.

Indu Panchal et al. proposed "Identifying Healthy and Mastitis Sahiwal Cows Using Electro-Chemical Properties: A Connectionist Approach" which used the electrochemical property (PH, EC, SCC, and temperature of udder, milk, and skin) for classifying cows into two categories—healthy cows and mastitis-infected cows [3].

Zhifei Zhang et al. proposed "Early Mastitis Diagnosis through Topological Analysis of Biosignals from Low-Voltage Alternate Current Electrokinetics" which used Biosensors on sample milk and employed Gaussian decision tree for topological based analysis of mastitis [4]. The result proved that the proposed method takes less voltage for analysis.

J. Eric Hillerton proposed a "Detecting Mastitis Cow-Side" where electrical conductivity of milk and milk temperature were used to detect clinical mastitis [5]. Sensitivity and specificity are the parameters used to prove the result of mastitis detection.

E. Wang and S. Samarasinghe proposed "Online Detection of Mastitis in Dairy Herds Using Artificial Neural Networks" which used various properties of milk to detect mastitis [6]. The work encompassed two-stage analysis—(i) statistical data preprocessing and (ii) model development. The Multi-Layer Perception (MLP) and the Self-Organizing Map (SOM) are the classifiers used to get the result were trained to detect the presence or absence of clinical mastitis. The following set of parameters was used for their work toward mastitis detection.

- Milk pH
- Electrical conductivity (mS/cm),
- Udder temperature (°C)
- Milk temperature (°C)
- Skin temperature (°C),
- Milk SCC (100,000 cells/ml)
- Milk yield (kg) and
- Dielectric constant

For the farmer who stays away from the farm and gives charge for the workers to collect milk, it is difficult to monitor the process and cattle's health issues. So Srushti K. Sarnobat and Mali A. S. proposed a method called "Detection of Mastitis and Monitoring Milk Parameters from a Remote Location" [7]. Here the owner of a farm can monitor the milk quality and cattle health from a remote place.

Based on the survey, it is notable that IoT is still not introduced for clinical mastitis detection. At present, in veterinary science, mastitis detection is done only using milk properties.

Modern-day clinical mastitis detection is done based on the above parameters, but it is found to be inadequate. The survey says that sensors are used to detect clinical mastitis and algorithms are applied but it was found that almost all the methods use a common parameter—milk properties. Hence it is way better to rely only on milk properties to identify clinical mastitis [8].

The summary is that every animal is connected to the Internet and monitored in a remote place. The animal's illnesses data are collected dynamically. IoT for animals is basically the same as human to machine communication, except for the fact that the source of data collection will be animals instead of humans.

There are two types of sensors [9, 10],

1. Active sensors: sensors are put into the bag and fixed to the animals so that dynamically animals are monitored as they move around.
2. Passive sensors: these types of sensors are kept in one place and when the animal comes to the range, they are monitored.

The list of sensors on animals are as follows [11, 12]:

- ECG sensors: to acquire signals such as ECG (Electro Cardio Graphic), body temperature, and blood oxygen saturation.
- Motion sensors: to track animal health based on the movement.
- Environmental sensors: to monitor humidity and temperature to know the effect of it on animal health.
- RFID: to match the respective animals and to monitor activities.
- Temperature sensors: to increase the comfortable zone of animals and to find the variations of body temperature.
- Smartphone: smartphones also act as sensors to monitor remotely the health of animals.
- Heart rate sensors: it detects the heartbeat speed to monitor animal health.
- Rumination sensors: to take care of digestion related health issues, rumination sensors are used.
- Oxygen sensors: these sensors are used in fisheries to know the oxygen level.
- PH sensors: these sensors are useful to know the deficiencies in the milk or any water-related issues.
- The conclusion is that Sensors are used on animals for the following purposes:
- To monitor animals;
- To analyze the behavior of animals;
- To detect ECG data;
- To find motoric dysfunction.

Sensors are not harmful to the animals, but the way of deploying or attaching them should be known. There are a few researchers who applied the data mining technique to detect clinical mastitis and the parameters considered are from milk properties [13].

3 Methodology

The main idea behind the above survey is to propose a new methodology, which identifies clinical mastitis accurately, with less delay. Sensors will be employed to collect data, which will be trained and analyzed. The test data is collected and compared with the data set and the end result will be given to the user through a handheld device like Smartphone. The architecture of the proposed methodology is shown in Fig. 3. Initially, the sensors collect the externally visible symptoms as a part of data acquisition, which will be trained using appropriate machine learning techniques. The refined data will be stored in cloud which will be used later for further computing. The test data is compared with the training data using the same sensor, which yields the result. The final result will be provided to the owner via a handheld device (smartphone). The cows' information may be provided as SMS which alerts the cows' owners and nearby veterinary doctors. The primary data will produce an accurate result. Using milk data set, validation is done with the result produced by the data of external symptoms to prove the result.

The working procedure is

Step 1: Use a sac on the cow's udder to read the sensor values as shown in Fig. 4.
Step 2: Send the sensor data to the cloud.
Step 3: Train the data set.
Step 4: Collect test data from the sac.
Step 5: Apply algorithms on the data to detect the clinical mastitis.

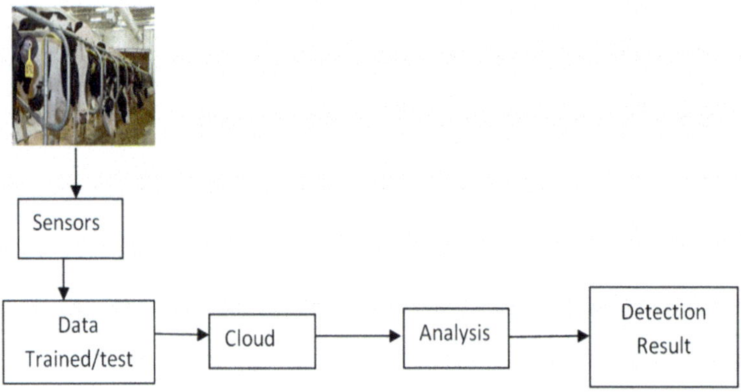

Fig. 3 Architecture of the proposed methodology

Fig. 4 A sac to detect clinical mastitis

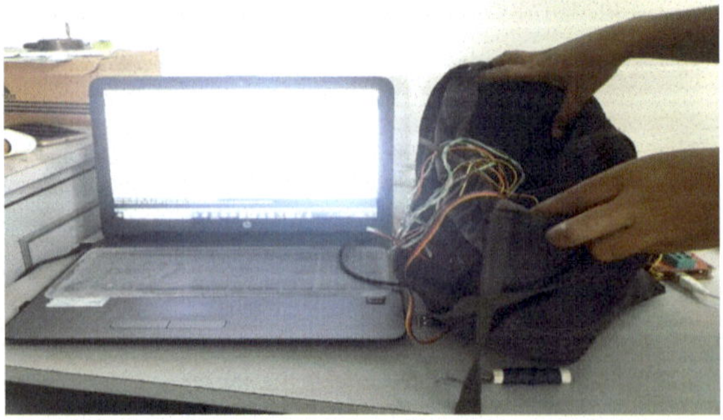

4 A Sack for Data Acquisition and Testing

As essential data is not available, data acquisition is one of the important phases in this research. A sac is designed by deploying the four flex sensors and a temperature sensor. This smart sac detects the variations in the udder size and temperature for further processing. The flex sensors are used to find the udder swellings, and temperature sensor is used to identify the temperature. Four flex sensors are required to find the health of the teat. Here we are dealing with numerical data so a combination of arduino and raspberry pi gives us the desired result. These hardware devices are put inside the sac and are to be worn to the udder before milking as shown in Fig. 4.

Cloud is used for the purpose of storage. Using the sac, the data is collected through WiFi in the dairy field and these data are stored in the cloud for further analysis. The data set attributes are eight udder size values as Upper_Value1, Upper_Value2, Upper_Value3, Upper_Value4, and Lower_Value1, Lower_Value2, Lower_Value3, Lower_Value4, and Temperature value, PH value and SCC count of cow milk. Overall, the sac includes the sensors for some purpose. These sensors will not harm the cow as they use battery power of less voltage.

5 Result

The two algorithms, K-Nearest Neighbor (KNN) and Support Vector Machine (SVM), are applied on collected primary data, and efficiency is compared. The KNN and SVM efficiencies are 73% and 86%, respectively. To conclude, SVM gives better results compared to the KNN but we cannot rely on SVM because it's a matter of a living creature and the Indian economy. The result is shown in Fig. 5, *X*-axis and *Y*-axis represent timeline and accuracy, respectively. Table 1 gives the accuracy of KNN and SVM.

The confusion matrix for the proposed system is shown below, having two classes. Class I is Mastitis and Class II is Normal.

Fig. 5 KNN and SVM comparison

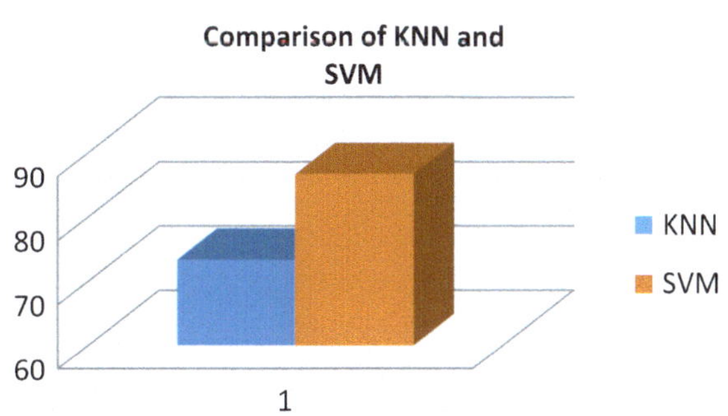

Table 1 KNN and SVM algorithm accuracy

S. No.	Algorithm	Accuracy (%)
1	KNN	73.33
2	SVM	86.66

Fig. 6 Results of SVM

```
             precision    recall  f1-score   support

          M       1.00      0.60      0.75         5
          N       0.83      1.00      0.91        10

avg / total       0.89      0.87      0.86        15
```

$$\begin{matrix} 3 & 2 \\ 0 & 10 \end{matrix}$$

The result of SVM based on the values of precession, recall, F1 score, and support are shown in Fig. 6.

The precision is calculated using formula (1),

$$\text{Precision} = \frac{\text{Mastitis Correctly Identified}}{\text{Mastitis Correctly Identified} + \text{Incorrectly Labeled as Mastitis}} \tag{1}$$

The recall is calculated using formula (2),

$$\text{Recall} = \frac{\text{Mastitis Correctly Identified}}{\text{Mastitis Correctly Identified} + \text{Incorrectly Labeled as not Mastitis}} \tag{2}$$

The F1 score is calculated using formula (3),

$$\text{f1-score} = 2 * \frac{\text{Precision} * \text{Recall}}{\text{Precision} + \text{Recall}} \tag{3}$$

The data is collected from the sensor from the fieldwork, and results are obtained using KNN and SVM. The all methods are not sufficient on the sensor data from cow udder. There are many works carried out to detect mastitis based on the milk parameters which are not sufficient to detect mastitis in cows. The advantage of the proposed system is to detect clinical mastitis more accurately compared to the existing systems, and SVM gives more accurate results on the primary data set.

6 Conclusion

The world is becoming smarter every day. To keep up with the pace and advancements, it is very essential to connect every object to the Internet. People should be aware of object communication in order to manage smart objects effectively. Everything (both living and non-living) can be connected to the Internet via sensors attached to them. Cows are an important source of economy in dairy industries, which in turn is one of the important economic sources of India. It is not accurate to consider only the milk properties to detect the clinical mastitis but, including external symptoms with milk properties will provide accurate detection of clinical mastitis. By employing IoT, clinical mastitis problem can be better tracked and assessed by cows' owners and Veterinary doctors through their smartphones.

The sac used here performs the task of a clinical mastitis detector and passes the data to the server for further analysis. The data used for analysis is a primary data so existing algorithms are applied to it. The KNN and SVM are used to find clinical mastitis which provides the desired result in detection. The future enhancement of the system is done by designing an efficient algorithm to find clinical mastitis comparing with the existing study. The argument will be concluded by saying that veterinary science required technology's assistance to find clinical mastitis accurately and on time.

Declaration We have taken permission from a competent authority to use the data as given in the paper. In case of any dispute in the future, we shall be wholly responsible.

References

1. Carlen, E., Strandberg, E.: Genetic parameters for clinical mastitis, somatic cell score, and production in the first three lactations of Swedish. J. Dairy Sci. **87**(9), 306–3071 (2004)
2. Viguier, C., Arora, S., Gilmartin, N., Welbeck, K., O'Kennedy, R.: Mastitis detection: current trends and future perspectives. Cell Press: Trends Biotechnol. **27**(8), 486–493 (2009)
3. Panchal, I., Sawhney, I.K., Sharma, A.K.: Identifying healthy and mastitis Sahiwal cows using electro-chemical properties: a connectionist approach. In: IEEE International Conference on Computing for Sustainable Global Development (INDIACom) (2015)
4. Zhang, Z., et al.: Early mastitis diagnosis through topological analysis of biosignals from low-voltage alternate current electro kinetics. In: IEEE International Conference on Engineering in Medicine and Biology Society (EMBC) (2015)
5. Eric Hillerton, J.: Detecting mastitis cow-side. In: National Mastitis Council Annual Meeting Proceedings (2000)
6. Wang, E., Samarasinghe, S.: On-Line Detection of Mastitis in Dairy Herds Using Artificial Neural Networks. Research Archive, Lincoln University (2015)
7. Sarnobat, S.K., Mali, A.S.: Detection of mastitis and monitoring milk parameters from a remote location. Int. J. Electr. Electron. Comput. Sci. Eng. **3** (2016)
8. Hogeveen, H., Kamphuis, C., Steeneveld, W., Mollenhorst, H.: Sensors and clinical mastitis—the quest for the perfect alert. MDPI-Sens. **10** (2010)
9. Hoflinger, F., et al.: Motion capture sensor to monitor movement patterns in animal models of disease. In: IEEE International Conference on Circuits and Systems (2015)
10. Kamphuis, C., Mollenhorst, H., Heesterbeek, J., Hogeveen, H.: Detection of clinical mastitis with sensor data from automatic milking systems is improved by using decision-tree induction. J. Dairy Sci. **93**, 3616–3627 (2010)
11. Jukan, A., Masip-Bruin, X., Amla, N.: Smart Computing and Sensing Technologies for Animal Welfare: A Systematic Review, pp. 1–15. National Agricultural Library (2016)

12. Udder turns to red, online available at http://explainagainplease.blogspot.com/2012/10/cow-health-mastitis-and-teat-injuries.html
13. De Mol, R.M., Ouweltjes, W., Kroeze, G.H., Hendriks, M.M.W.B.: Detection of estrus and mastitis: field performance of a model. Appl. Eng. Agric. **17**, 399–407 (2001)

Two-Way Face Scrutinizing System for Elimination of Proxy Attendances Using Deep Learning

Arvind Rathore, Ninad Patil, Shreyash Bobade, and Shilpa P. Metkar

Abstract Automation is taking over several fields ranging from home appliance automation to autonomous vehicles to industrial plant automation and various others, and also has a major impact in facilitating new cutting-edge technologies and innovations. TheInternet of Things, image processing and machine learning are evolving day by day. Many systems have completely changed due to this evolvement to achieve more accurate results. The attendance recording system is a typical example of this transition, starting from the traditional signature-based on-paper methods to fingerprint-based systems to face recognition-based systems. The major drawback of different algorithms for face recognition-based attendance system is that one person can scan his/her face by facing the camera and once the face is recognized, his/her attendance will be marked whether or not the person attends the lecture after that. In this paper, we have proposed an efficient algorithm to eliminate such proxy attendances. Furthermore, we have added IoT capabilities to our system in order to increase the ease of access to the collected attendance and to maintain transparency.

Keywords Face detection · Face recognition · HOG · CNN · Blynk

A. Rathore (✉) · N. Patil · S. Bobade · S. P. Metkar
Department of Electronics and Telecommunications, College of Engineering, Pune, India
e-mail: rathoreaa16.extc@coep.ac.in

N. Patil
e-mail: patilninada16.extc@coep.ac.in

S. Bobade
e-mail: bobdesd16.extc@coep.ac.in

S. P. Metkar
e-mail: metkars.extc@coep.ac.in

S. C. Satapathy et al. (eds.), *Intelligent Data Engineering and Analytics*, Advances in Intelligent Systems and Computing 1177,
https://doi.org/10.1007/978-981-15-5679-1_7

Abbreviations

CNN Convolutional neural networks
HOG Histogram oriented gradient
ReLU Rectified linear unit

1 Introduction

Maintaining and recording student's attendance is a difficult and time-consuming task in various educational and corporate institutions. Different institutions have different methods of recording attendance such as using traditional pen-paper-based methods or by using some other biometric methods like fingerprint or retina-based systems. But these methods consume a lot of time. There are two major traditional methods of recording attendance which are roll calls and circulating the attendance sheet. Circulating the attendance sheet from one student to another takes time as well as causes distraction. Due to such problems, some lecturers delay attendance till the end of the class, yet some students might be in a hurry to leave the class immediately, hence they might miss signing the attendance sheet. Furthermore, there are some students who never come to the class but sign attendance by proxy or false attendance [1]. In the case of roll calls, lecturers call out the names one by one to mark the attendance, but this method too results in loss of valuable lecture time for a process which can be done simultaneously in the background. Also, we do not know whether the authenticated student is responding or not. Calculation of recorded attendance is another major deed which may cause manual errors. There's always the possibility of losing the attendance sheet and thus it requires extra care and effort. These traditional methods are not environment friendly and lots of paper is wasted in the process. To overcome such troubles, we need an automated attendance management system.

Also, there are many biometric methods available and adopted recently. One of them is the fingerprint verification. In this method, first, the fingerprints of the individuals are collected and stored in the database of the fingerprint sensor. When a student places his/her fingers on the sensor, the recorded fingerprints are compared with the prints in the database. If the two fingerprints are the same, then the attendance is marked as present. But this method has some disadvantages. The students have to wait in a queue which ultimately consumes a lot of time and creates chaos. If once the finger is not kept correctly or if the fingerprint is not recognized properly, then the attendance will be marked as absent. So, this method is not 100% efficient [1].

The other biometric method adopted is eyeball detection. In this method, an eyeball sensor is used. It senses the blinking rate of the eyeball and it also senses the position of iris. In this method, first, the eyeball or iris image of each individual is stored in the database. Similarly, the obtained image of the eyeball is then compared with the eyeball in the database. If it is the same, then the attendance is marked. But practically it is not possible. As there are a large number of students in the class, eyeball detection of everyone is not possible and it may lead to chaos. To overcome such troubles, we need an automated attendance management system with proxy elimination capabilities.

1.1 Comparison with Previous Works

Face recognition-based attendance systems have two system position-based ways. One way is to implement the system inside of the classroom at a vantage point, and the second way is to implement the system at the entrance of the classroom. The former requires a good quality camera with good field angle and depth of field. This increases the cost of the overall system. This system also has the drawback of missing out on some of the persons at the back of the classroom. Also, the possibility of proxy attendances cannot be ruled out.

For the entrance-based attendance systems, there aren't many restrictions on the camera specifications. Also, the drawback of missing out on some of the persons does not arise. Most of the face recognition-based attendance systems implement the former way. Entrance-based attendance systems although implemented in some cases have the drawback of proxy attendances being marked. The proposed algorithm eliminates this drawback of proxy attendances.

2 Methodology

The steps involved in face recognition are

- Preparing a database of images of enrolled students;
- Capturing images from a camera for comparing with the database;
- HOG/CNN algorithm to detect faces;
- Encoding captured images as well as those present in the database;
- Comparing encoded images [2].

2.1 Encoding Faces Present in an Image

To recognize a detected face, the simplest way is to try and match the HOG pattern of the captured face image with the HOG pattern of the face images present in the database. The time required for recognition can be reduced by reducing the complexity present in the image. The complexity is reduced by generating and using 128 face measurements (called embeddings) instead of the HOG pattern. The face image is fed as input to the network. The network learns by itself which parts of the face are important to measure and generates 128 measurements. This process is called encoding [2].

2.2 *Face Recognition*

Once the captured image is encoded, then it can be compared with the encoded images present in the database which are in the form of vectors of 128D. K-nearest neighbor method of distance measurement is used for classification with a tolerance value of 0.6 (distance between faces to consider it a match). This algorithm uses the K-nearest neighbor of the vector of interest to find out the class to which it belongs. As per the training, the vectors of the same person will usually tend to be closer to one another. Once the nearest neighbor is found, it is said to be matched and thus the face is recognized [2].

3 Algorithm for Eliminating Proxies

Although face recognition-based attendance systems helped in reducing the time wasted in manual attendance, there were still some loopholes and errors. It was still possible to record false attendance and proxies as the camera worked on the principle of a single scan. So, a person could just simply show his face before the camera and leave, and still his attendance would be recorded. To eliminate this possibility, we have come up with an algorithm of multiple scans in sync with the timers of the Raspberry Pi running in the background.

- The LDR sensor attached to the camera will detect whether the lighting condition is enough. In case the light is insufficient, then flashlight will be turned on.
- For this case, the duration of the lecture is assumed to be 50 min.
- A database consisting of encoded images of enrolled persons is created and stored in the internal memory of Raspberry Pi.
- The camera will start capturing images, the image captured will be encoded and compared with the database of encoded images.
- If the two encoded images match, the green light will blink indicating the person's face is properly detected.
- If the image is not properly captured or if the image is not properly recognized, then the red light blinks.
- The camera will start capturing and recognizing face images at the beginning of the lecture and will continue until 15 min after the expected starting time of the lecture. It records the in-time of that person. After 15 min, the camera will stop capturing images and will turn inwards.
- The camera will again start capturing images after the lecture ends (after 50 min) and will continue up to 5 min; when the person leaves the class, the camera

captures the image and encodes it, it records the in-time and out-time of person and marks the attendance of the person in accordance with the time for which he attended the lecture.
- After 5 min, the camera is rotated outwards and is ready to record attendance for the next lecture.
- At the end of the day, the updated excel file is uploaded to the Dropbox cloud whose access is provided to respective faculty members.
- This system can be programmed according to the lecture timings which vary in different institutions (Figs. 1 and 2).

4 Blynk

Blynk cloud server is used for providing cloud access to the prototype. Blynk is an open-source IoT platform which allows us to integrate and connect any nonliving thing to the Internet by using Cloud computing. The attendance system can be rotated at the particular programmed angle by the teachers with the help of this IoT app. We have integrated a button widget on this app for this need. The excel sheet uploaded on the drive can also be accessed through this app (Figs. 3 and 4).

5 Validation

The steps are divided into 4 parts:

1. Preprocessing;
2. Initialization and 1st stage scrutinization;
3. 2nd stage scrutinization for case 1;
4. 2nd stage scrutinization for case 2.

Two cases are considered:

1. Students exiting after lecture 1;
2. Students exiting after attending multiple lectures:
    - A Python script was written to encode all the images of students present in the database.
    - Executing step 1, .py file from the command line started the timer in the background and started the camera for capturing images.
    - Low light condition was detected (with the help of LDR) [3].
    - It didn't recognize a face in the image when the face was tilted (HOG method was used) and when the face was half masked (with hand).
    - For the rest of the cases, the face was detected and recognized properly.
    - If it failed to recognize a face, then it prompted by flashing a red led; in case of success, it flashed a green led.

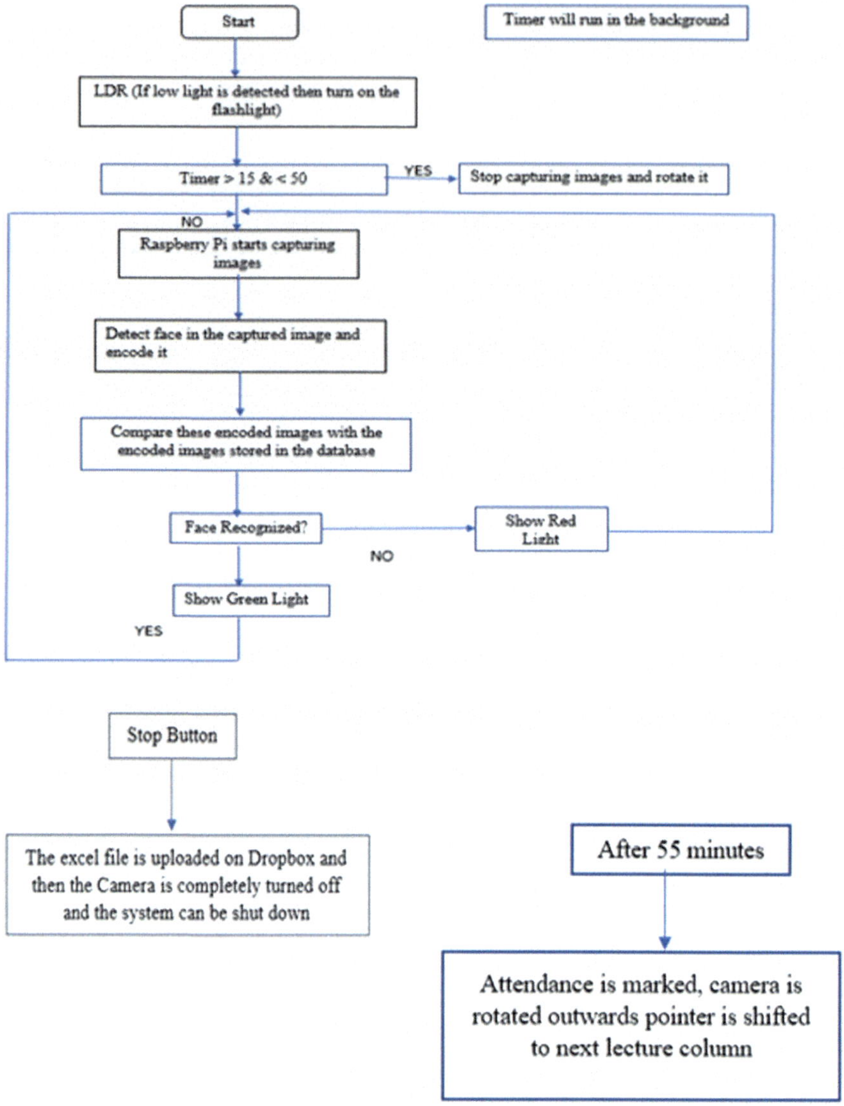

Fig. 1 Flowchart for eliminating proxies

- When the background timer reached 20 min, it disabled the camera and rotated it inwards.
- When the background timer reached 50 min (indicating the lecture was over), the camera was enabled again.
- It recognized the students exiting and marked the attendance of only those students for the 1st lecture.
- At the end of the 2nd lecture, attendance was marked for those students who were present for both the lectures.

Fig. 2 External appearance of the face recognition system

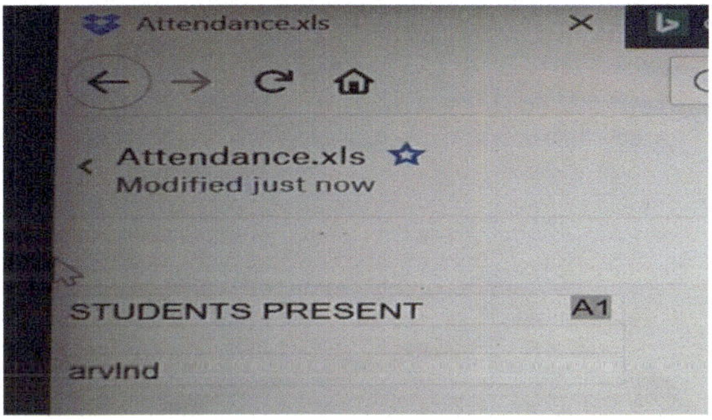

Fig. 3 Excel sheet uploaded on dropbox

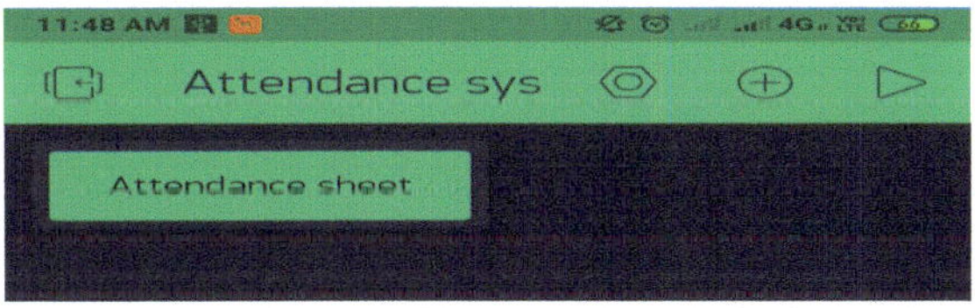

Fig. 4 Blynk cloud display

6 Conclusion

In this paper, by using technologies such as deep learning and IoT synergistically, we have implemented a model which suggests and implements a system for proxy elimination at the institute and corporate world. Blynk cloud server was used for providing cloud access and making the prototype IoT capable. This system offers a higher authority to monitor the daily activity of students or employees, along with flexibility in rescheduling events and data management. The benefits of the system will be reaped by faculties and students. The possibility of utilizing real-time face recognition for attendance will largely abate the efforts for attendance management.

Acknowledgements We would like to express our sincere gratitude to the staff of the Department of Electronics and Telecommunication, College of Engineering, Pune, for their encouragement and support. We would also like to thank Adrian Rosebrock, the founder of PyImageSearch whose content helped us in this research work.

References

1. Dharani, R., Jeevitha, S., Kavinmathi, B., Hemalatha, S., Varadharajan, E.: Automatic attendance management system using face. In: 2016 Online International Conference on Green Engineering and Technologies (IC-GET) (2016)
2. Kalenichenko, D., Philbin, J., Schroff, F.: Facenet: a unified embedding for face recognition and clustering. In: IEEE Conference on Computer Vision and Pattern Recognition, Boston (2015)
3. Pranay Kujur, K.G.: Smart interaction of object on Internet of Things. Int. J. Comput. Sci. Eng. **3**(1), 15–19, 2015

Ontology-Driven Sentiment Analysis in Indian Healthcare Sector

Abhilasha Sharma, Anmol Chandra Singh, Harsh Pandey, and Milind Srivastava

Abstract In today's world, social media platforms have emerged as one of the most prominent media for expressing opinions. Sentiment analysis for utilizing the big data socially available over the web has become one of the most researched areas. Sentimental analysis is used to get the public perception over any event, topic or subject matter by classifying data into various polarity categories like positive, neutral, negative, etc. Various computational techniques such as machine learning and deep learning are used to perform polarity analysis and to find the best classifiers. Traditional techniques tend to create feature vectors in order to quantify the data, but these feature vectors are often very large for tweets collected over a particular domain and that leads to a performance reduction. This paper proposes a combined approach which utilizes domain ontology with machine learning methods in order to reduce the size of feature vector to increase the performance for all machine learning models. The observations state that our technique provides a significant advantage over the traditional methods.

Keywords Ontology · Sentiment analysis · Machine learning · Healthcare

A. Sharma
Department of Computer Science & Engineering Delhi Technological University, Delhi-42, India
e-mail: abhi16.sharma@gmail.com

A. Chandra Singh (✉) · H. Pandey · M. Srivastava
Netaji Subhas Institute Of Technology, Delhi-78, India
e-mail: 96anmolchandra@gmail.com

H. Pandey
e-mail: hardey261996@gmail.com

M. Srivastava
e-mail: milind@ntitynetwork.com

S. C. Satapathy et al. (eds.), *Intelligent Data Engineering and Analytics*, Advances in Intelligent Systems and Computing 1177,
https://doi.org/10.1007/978-981-15-5679-1_8

1 Introduction

In the current scenario, social media has emerged as one of the leading platforms for people to raise their voice and opinion. This makes social media sites the perfect platform for extracting information to determine the mood of the public. By analysing the data about a certain topic algorithms can gauge the overall public perception. Sentiment analysis techniques quantify this data as feature vectors and then uses these feature vectors to train classifiers in order to automate the process of analysing data and placing classifying it into various sentiment polarity groups. Sentiment analysis can be used for a variety of tasks like running marketing campaigns for companies [1], early warning systems and disaster analysis [2], user content personalization [3], *etc*. Government sectors such as military, energy, public welfare, tourism, agriculture, healthcare sector, etc. can also make use of sentiment analysis as a feedback system so that they can incorporate the public opinion while formulating policies. With healthcare sectors across the world facing major issues like the rising cost of healthcare, a dearth of properly trained medical professionals, poor access to healthcare services in remote areas, lack of knowledge about the importance of personal and community hygiene, *etc*. governments need to pay more attention to this sector. The map in Fig. 1 shows the HAQ (healthcare access and quality) index across the world.

Although the Indian healthcare sector has seen noteworthy improvements in the last two decades such as improvements in infant mortality rate (IMR) and maternal mortality rate (MMR), drives for increasing sanitation, rise in medical infrastructures, increasing awareness and campaigns owing to overall increase in life expectancy, there are still many challenges facing the Indian healthcare industry.

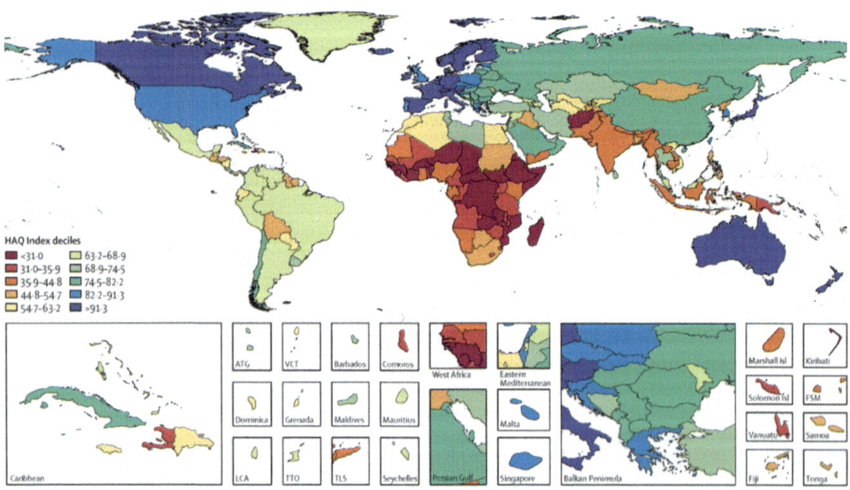

Fig. 1 Map depicting the HAQ Index for all the countries—2016 [5]

In recent years, the government has started the Digital India initiative with one of its goal being the digitization of the healthcare system in order to bring more transparency, awareness, and easy accessibility of quality healthcare to the masses. This has resulted in various online services such as the National Health Portal and the e-Hospital application. The government has also successfully launched many schemes in the healthcare sector such as Mission Indradhanush, the Affordable Medicines and Reliable Implants for Treatment (AMRIT) programme, and the Jan Aushadhi Yojana. For these schemes to be effective, the government needs to take into account the public feedback so as to alter these policies to the people's needs. The government can make use of sentiment analysis techniques and opinion mining to help improve the policies governing the healthcare sector [4].

The proposed approach aids the government in order to collect public feedback on its various schemes. Nowadays, people express their opinions on social platforms like Twitter or Facebook which makes them an excellent source for data collection. The data has been scraped from Twitter and then sentiment analysis has been applied on this data to get an overview of the public opinion. To validate the proposed approach, the Ayushman Bharat Yojana, a healthcare scheme launched by the Government of India, is chosen.

The proposed approach uses a combination of natural language processing and analysis techniques along with a domain ontology to produce more effective sentiment analysis results for the topic at hand. The model uses a domain ontology, created using the Protege tool, to assist the sentiment analysis techniques.

The rest of the paper consists of the following: Sect. 2 reports on the related research efforts and Sect. 3 lists some of the details on Ayushman Bharat Scheme. Section 4 deals with ontology creation and gives an overview of the ontology whereas Sect. 5 covers the implementation details of the proposed algorithm. Section 6 illustrates the results of the proposed approach and future work is explained in Sect. 7.

2 Literature Overview

Many techniques have been used for sentiment analysis and opinion mining tasks. More advanced ones are being created every day for optimizing the previous algorithms. Techniques which make use of a domain ontology tend to perform better because they provide the classification algorithms with the context from domain-specific knowledge.

Shein et al. [6] proposed an approach where an ontology-based approach was intended as an enhancement for existing techniques. They proposed a combination approach of POS (part Of speech) tagging, FCA (formal concept analysis)-based domain ontology and SVM (support vector machine) classifier for achieving a better result than standard models. Shein [7] also used a similar approach of combining POS tagging, creating an ontology-based on the FCA design by using the protege 2000 tool and ultimately using machine learning techniques for classification of sentiments. He used customer reviews from the IMDb database to evaluate their proposed model. Their proposed approach demonstrated an increased accuracy of classification across positive, negative and neutral sentences.

Polswat et al. [8] proposed an approach which increased the efficiency of the SentiWordNet analysis. They further proposed a method for solving the problem encountered with synonymous words by using SPARQL for accessing DBpedia to replace abbreviations with their full terms by searching the entire Wikipedia website for the abbreviations. They analysed 500 texts and the proposed algorithm yielded a precision of 97%, recall of 97% and f-measure of 94%.

Nitish et al. [9] proposed an approach in which they used a domain ontology in OWL format. They used reviews from online shopping sites for conducting analysis on various mobile phone models. They used the SNLP (Stanford Natural Language Processing) tool for POS tagging and mapped the results to the POS used in the WordNet database. Finally, they used SentiWordNet 3.0 to get sentiment scores for the words. The proposed method yielded results which agreed closely (70%) with the opinions of the market retailers.

Sam and Chatwin [10] proposed an ontology-based model for sentiment analysis of customer reviews for electronic products. They prepared two ontologies, one for the electronic products and another one for the emotions in customer reviews and then combined these two ontologies for their proposed model. The emotional ontology had different levels for each emotion and also took into account various negating and enhancing words. They used the HowNet database for bunching different emotional words into levels by calculating the semantic similarity between the words. Finally, they also added an emotional tolerance parameter to their model to make sure that their model only displayed relevant information to the user. To demonstrate their model they used 347 randomly selected customer reviews from facebook.com and the model achieved an accuracy of more than 90% with the emotional tolerance parameter set to 0.

3 Ayushman Bharat Yojana (PM Jan Arogya Yojana)

According to the World Health Statistics report of 2013 [11], India spent only 1.04% of its GDP (gross domestic product) on health expenditure. The same report stated that the out-of-pocket expenditure on healthcare in India is 61.7% in contrast to the 20.5% global average (Figs. 2 and 3).

To make healthcare services accessible to all and to help the underprivileged population of India, the government launched the AB-NHPS (Ayushman Bharat-National Health Protection Scheme) [12] on 23 September 2018. It is a national health insurance scheme covering over 10 crore poor and vulnerable families and providing coverage of up to 5 lakh rupees per family per year for secondary and tertiary care hospitalization.

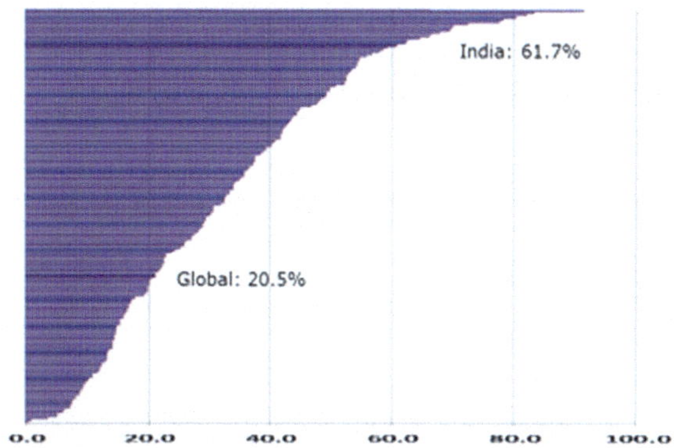

Fig. 2 OOP (out of pocket) expenditure as proportion of total healthcare expenditure [11]

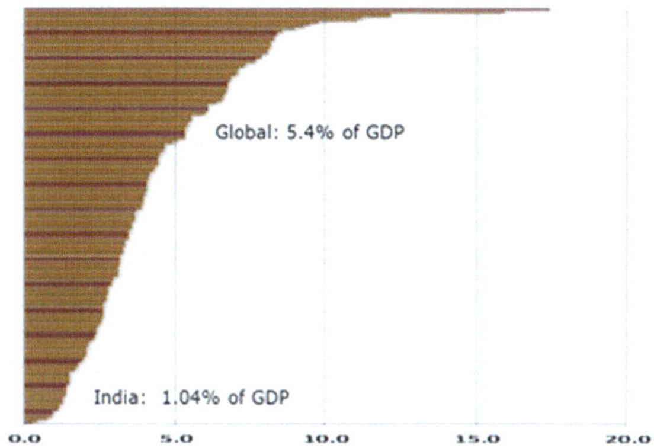

Fig. 3 Government health expenditure as proportion of GDP [11]

3.1 Key Benefits of the Scheme

The scheme offers the following benefits :

1. There will be no limit on family size and age of family members to ensure that everybody receives quality healthcare.
2. Anyone covered under the scheme will be able to take cashless benefits from any private or public enlisted hospital across India.
3. Any pre-existing health conditions will be covered once a person enrolls for the scheme.
4. A set transport allowance per hospitalization will be given to any person covered under the scheme.
5. Advanced medical treatment for cancer, cardiac surgery and other diseases will also be covered.

3.2 Beneficiaries of the Scheme

The scheme is targeted at the following groups :

1. Poor and the economically backward segment of society.
2. Beneficiaries are picked up from the SECC (Socio-Economic Caste Census) Database. These 10 crore beneficiary families comprising of 8 crore families from the rural areas and 2 crore families from urban areas of the country.

3.3 Impact of the Scheme on the Beneficiaries

This scheme is empowering the citizens of India to live their lives to the fullest. Although universal health coverage is still far away this is the first step towards that endeavour. The scheme will have a major impact on the reduction of OOP expenditure because of the following points:

1. It extends increased benefit cover to nearly 40% of the country's population.
2. It covers almost all secondary and most of the tertiary hospitalizations.
3. The government plans to provide Rs. 5 lakh to each family covered under the scheme.

4 An Overview of the Ontology

A group of concepts and categories in respect to a certain domain along with the properties and their interconnections defines an ontology. In this paper, the healthcare sector in India is chosen as the domain and a study on the new scheme, Ayushman Bharat Yojana, is conducted with the help of the ontology [13–17].

4.1 Tools and Resources Used

1. **Protégé**
   Protégé [18] is a free, open-source software for ontology creation and editing. It was developed by Stanford University for building smart systems. Protégé is used by a vast community of academic, government and corporate users in the domain of biomedicine, e-commerce and organizational modelling.
2. **Twitter**
   The data required for building the ontology is extracted from Twitter. The tweets are mainly used to recognize the modules for the ontology. Topic-specific words, trending words and slang were gathered and categorized into classes and objects.

3. **Public Data Warehouses**
   More domain-specific data is collected from online sites and the government released documents regarding the features and implementation of the policy.
4. **OntoGen**
   OntoGen [19] is used for reading the created ontology in the Python script, and getting the relation between interlinked classes and objects. It is a semi-automatic, data-driven ontology editor.

4.2 Ontology: An Overview

The created ontology comprises the classes and their individuals as listed in Table 1. The ontology is presented in Fig. 4. It comprises 25 major classes. For example, the class RelatedSlogans is showing all the slogans that were found while cleaning the tweets. Since they do not add any significant impact when treated as separate classes they all are grouped under the class—'RelatedSlogans'.

Table 1 Sample of the features and individuals in the ontology

Featured word	Children words
Agencies_Health _Service_Provider	*'health wellness counselling services _hwcs', 'indusHealth, 'eyecare', 'chc', 'bhs_babylon health services', 'geriatrics', 'phfi_public health foundation of India', 'mphrx', 'unicef_united Nation children fund', 'visulytix', 'National health system resource centre_nhsrc', 'cochrane', 'astrazeneca', 'clirNet', 'medecube', 'medgenera', 'spagasia'*
Benefits_for_netas	*'popularity', 'easy medical access', 'public support WorldWide'*
Disorders	*'obesity', 'alcoholism', 'haemophilia', 'stress', 'depression'*
Ethical reasons	*'immunization','expensive healthcare', 'help needy', 'improves_indian healthcare', 'jandhan', 'boost healthcare'*
Hospitals	*'saanvi', 'max', 'apollo', 'aiims', 'saroj'*
IT_Platform	*'eHealth', 'paperless_cashless_transactions', 'online_medium'*
Leaders	*'PM_prime minister_Narendra_Modi', 'Rajnath Singh_Home minister_HM'*
Medical_Coverage	*'fivelakh_5lakh_perfamily peryear'*
Old_Schemes	*'Rashtriya_swasthya_bima', 'senior_citizen_health_insurance_scheme _schic'*
Political_Reasons _Politics_rajneeti _rajniti	*'modicare', 'votebank'*

(continued)

Table 1 (continued)

Featured word	Children words
Related slogans	*'Screening India seven', 'swastha Bharat', 'healthy forgood', 'krimiMukt bharat', 'malaria must die', 'ayurvedic lifestyle', 'unwomen India', 'unwomen Asia', 'health4all', 'gram Swaraj Abhiyan', 'diabetes awareness', 'bjpfornation', 'Quit_tobacco_gutkha', 'Bharatin London', 'healthy lifestyle', 'buzz in India', 'doing well_doing good', 'making India healthy', 'health for all', 'mission Indra dhanush', 'end plastic pollution', 'bjp4nation'*
Surroundings	*'landfills', 'sanitation', 'waste management', 'toilets', 'manufacturing_industrial_waste', 'plastic', 'Cleanliness_hygiene', 'malnutrition_no_nutritional_food'*
Symptoms	*'bleeding', 'strokes', 'loose motion', 'high blood pressure', 'headache', 'hairfall', 'fever', 'joint muscle pain', 'coughing', 'rashes', 'swelling', 'cold', 'sneeze', 'fatigue', 'vomitting', 'pms_premenstrual syndrome', 'nausea', 'loss of appetite'*
Workers	*'gaurds', 'surgeons', 'chemists', 'pharmacists', 'doctors', 'nurses'*
Care given	*'secondary', 'primary'*
cd_communicable	*'Swine flu', 'chickenpox', 'chickenguniya', 'diarrhoe', 'malaria', 'measles', 'tuberculosis_tb', 'fungal infection', 'leprosy'*
Facilities	*'ICU_intensive care unit', 'ambulance', 'laboratory'*
Mental	*'memory', 'brain', 'mind'*
ncd _non communicable	*'cancer_tumor', 'anthrax_bacillus', 'dengue', 'hydrocephalus', 'skin infection', 'cataract', 'hypertension', 'migraine', 'heart attack', 'diabetes', 'hepatitis A', 'hepatitis B', 'hepatitis C', 'Zika virus'*
Payments	*'packagerates'*
People	*'patients', 'rural', 'smokers', 'poor', 'farmers_kisan', '100 crore'*
Physical	*'accidents', 'cuts', 'bruise', 'bam'*
Products_drugs _health care means	*'contraceptives', 'tablets_numal', 'painkillers_palliates', 'ointments', 'vaccines_arteether', 'aushadhi', 'syrups', 'physiotherapy', 'exercise_gym'*
Programmes	*'world_immunization_day', 'national_rural_livelihood_mission_nrlm', 'world_health_day', 'mcessation_programs', 'worldMalariaDay', 'village_health_nutrition_day_vhnd', 'worldliverday'*
Shortcomings _ignored	*'homeopathy', 'vedas', 'ayurvedic_ayurveda_ayur', 'herbals_herbs'*

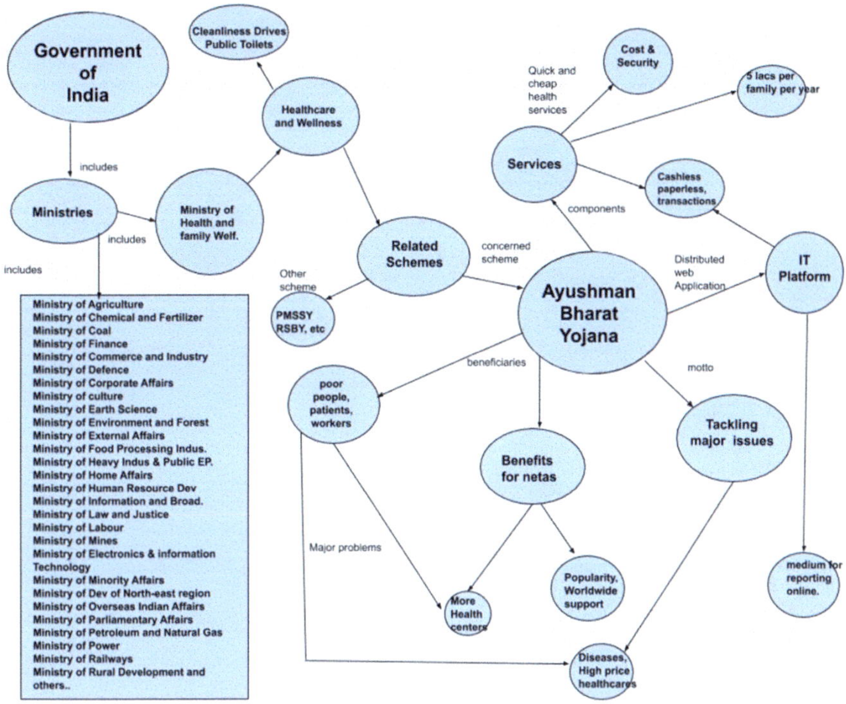

Fig. 4 Simplified version of the created ontology

5 Proposed Work

In this section, we take a look at each step of the proposed algorithm as illustrated in Fig. 5.

5.1 Data Acquisition

Initially, the relevant tweets are extracted using the GitHub repository developed by Jefferson Henrique, GetOldTweets [20]. The repository allows extraction of tweets by specifying a query term and also the range of dates for which we want the tweets. With the help of this repository, 2000 raw tweets were extracted using the hashtags ['Ayushman', 'Ayushman Bharat', 'ABY', 'Ayushman Bharat yojana'] for a duration of 6 months from 01-02-2018 to 31-07-2018. The extracted tweets were stored in a CSV format for further processing.

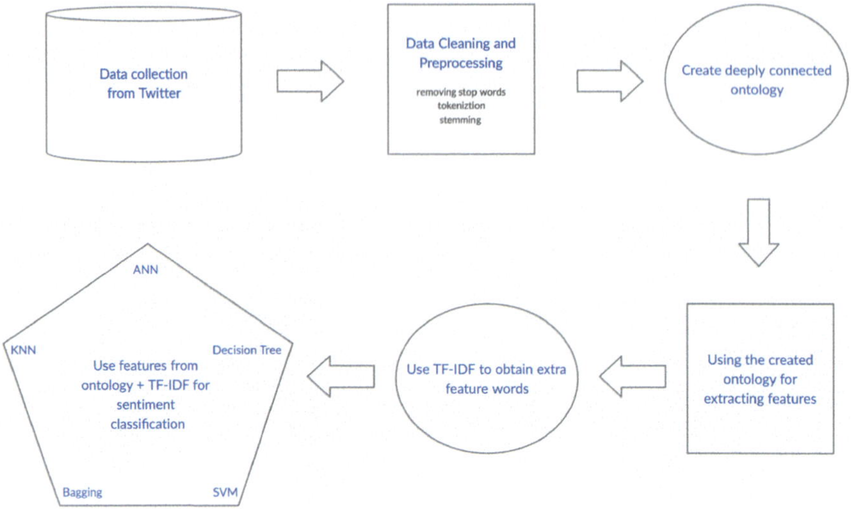

Fig. 5 Systemic flow of the proposed algorithm

5.2 Data Cleaning and Preprocessing

With the tweets stored in a CSV format, we move onto the next step of data preprocessing. Stopwords are removed using the stopwords class provided by nltk (Natural Language Toolkit) package in Python. After this, we used stemming to reduce all words to their base forms. After this step, all the collected tweets are manually tagged as positive, negative or neutral.

Table 2 reflects the weekly tweet distribution based on the polarity of the tweets and Fig. 6 visually represents how the tweet collection varied throughout the weeks.

5.3 Extracting Features

5.3.1 Proposed Algorithm

The proposed algorithm comprises categorizing the words based on their impact ratio and mapping similar meaning words under a single word. This has been described in detail as follows:

1. **Categorizing sentiment deciding words**
   Positive and negative words are categorized on three levels: basic, medium and extreme. These categorizations are done based on the impact of these words on a tweet. Neutral words are taken into a single category 'neutral words'.

Table 2 Tweets collected per week grouped by polarity

Week	Negative	Neutral	Positive
1	9	5	98
2	1	3	14
3	1	0	249
4	0	3	21
5	0	0	1
6	1	4	8
7	0	0	1
8	0	6	19
9	1	2	12
10	0	1	3
11	0	3	118
12	1	1	9
13	0	0	6
14	1	3	16
15	0	2	2
16	1	1	17
17	1	4	4
18	0	4	3
19	1	2	0
20	7	5	3
21	2	2	0
22	3	1	3
23	0	9	12
24	2	5	1
25	0	8	1
26	1	9	2
27	0	1	1

2. **Mapping Children to Parents**
Recurring words signifying the same thing are mapped to their parent word like, for example, the word govt and Government are mapped to a single word Govt to remove redundancy. Also, all the children words are mapped to their parent word instead of treating them as a separate nodes in the network. For example, all the medicine names are mapped to a class named medicine, and the occurrence of any medicine name will mark the class medicine into the final feature vector. This creation of a parent–child network has two advantages.

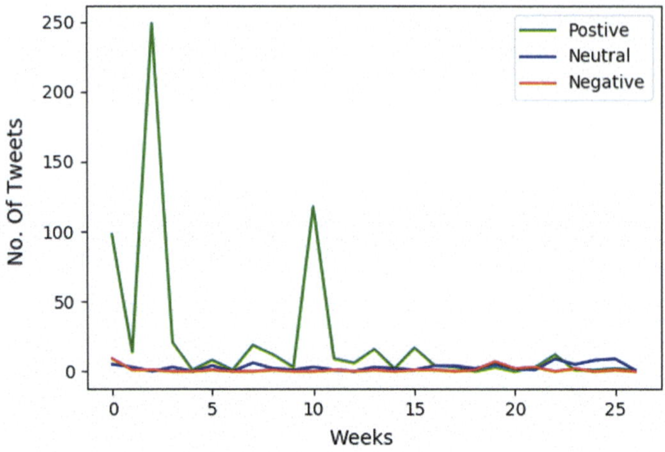

Fig. 6 Tweet collection per week

- It reduces the size of the overall feature vector column which enables the resulting model to train better, faster and find the correlation patterns more easily.
- It removes the unwanted noisy words which were increasing redundancy and decreasing the performance of the model. The leads to increased accuracy for the model as is demonstrated by the better results in comparison to the bag of words and normal ontology models.

Each tweet is represented as a binary vector of size equal to the size of the feature vector, *i.e.* the number of parent nodes or classes in ontology , where 1 represents the presence of the corresponding word of dictionary in the tweet and 0 represents its absence. This created a vector on which machine learning algorithms can be easily applied.

5.3.2 Pseudocode

Using the proposed algorithm, a deeply connected ontology is created using features from the extracted tweets. Instead of taking all the unique words found, only the words relevant to the domain were taken. Words based on current trends, popularity are also taken from online sites and data warehouses which will ensure the robustness of the created ontology. The pseudocode for the same is presented in Fig. 7.

```
GenerateFeatureMatrix(tweets):

        OntoFeatureMatrix = [ len(tweets), len(OntoClasses) ];
        ExtraSupportFeatureMatrix = [len(tweets), len(ExtraSupportFeatures)];

        For ti in tweets ( i in 1 to len(tweets)) :
              ti   = convertTweetToWordBag(ti);
              ti   = cleanTweet(ti);

              for wj in ti (j in 1 to len(ti)) :
                    If wj in OntoClasses:
                          OntoFeatureMatric[I][index_wj] = 1;
                    else if wj in OntoIndividuals:
                          ontoClassIndex = getOntoClassIndexFromIndividual(wj);
                          OntoFeatureMatrix[I][ontoClassIndex] = 1;
                    else if wj in ExtraSupportFeatures:
                          ExtraSupportFeatureMatrix[I][index_wj] = 1;

        return Concat(OntoFeatureMatrix, ExtraSupportFeatureMatrix);
```

Fig. 7 Pseudocode for the proposed algorithm

5.3.3 Extracting Additional Features Using TF-IDF Method

Term Frequency–Inverse Document Frequency (TF-IDF) is a conventional statistical weighting technique which measures how important a word is to a document.

Term frequency (TF) is the ratio of how many times a word t is present in a document d to the total number of words k in the document and it represents the frequency of that word in the document.

$$tf_{t,d} = \frac{n_{t,d}}{\sum_k n_{k,d}}$$

Inverse document frequency (IDF) is a measure of how rare a word t is in the entire corpus D and therefore the higher the IDF value the rarer the word.

$$idf_{t,D} = \log \frac{|D|}{|d : t_i \in d|}$$

TF-IDF is simply the product of the two terms calculated above

$$tfidf_{t,d,D} = tf_{t,d} \cdot idf_{t,D}$$

In addition to the feature words in the ontology, the TF-IDF method is also used to extract important words from the tweets. These features are called stand-out features. Each tweet is treated as a document and a TF-IDF score is computed for all the unique words in the collection of tweets. After this, 20 words that have the highest TF-IDF score and which are not present in the ontology are selected. The selected words and the words extracted from the ontology are merged together to form the final feature vector.

Table 3 Configurations used for training classification models

Algorithms	Configurations
SVM Classifier	*kernel = 'linear', gamma = 0.1, C = 1, random_state = 0*
Decision tree classifier	*criterion= 'entropy', random_state=0*
KNN classifier	*n_neighbors=5, metric= 'minkowski', p=2*
ANN classifier	***Layer1***: output_dim = 32, init = 'uniform', activation = 'relu' **Layer2**: output_dim = 32, init = 'uniform', activation = 'relu' **Layer3**: output_dim = 3, init = 'uniform', activation = 'softmax' **optimizer** = 'adam', **loss** = 'categorical_crossentropy' **batch_size** = 20, **nb_epoch** = 100
Ensemble learning—bagging	***Classifier***: kernel = 'linear', gamma = 0.1, C = 1, random_state = 0 **Bagging classifier**: base_estimator = classifier, n_estimators = 100, random_state = seed

5.4 *Training the Model*

The dataset is split into training and testing sets with the use of scikit-learn Python library. The 50:50, 60:40, 75:25 and 80:20 split ratios are tried for training and testing datasets, respectively, and it is concluded that the 75:25 split ratio helped the models achieve the best accuracy. Next, the scikit-learn Python library is used to configure the four types of machine learning classifiers, namely, Naïve Bayes classifier, support vector machine classifier, decision tree classifier and KNN (K nearest neighbours) classifier. The *Keras* library is used to train a ANN (artificial neural network)-based classifier. The predicted sentiments of these five classifiers are compared with the actual sentiment and then accuracy, precision and recall are calculated for all the classifiers as a measure to compare their performances. The configurations used to setup the abovementioned classifiers are shown in Table 3.

6 Observations and Results

The proposed algorithm has enabled the classifiers to perform better as compared to other traditional approaches. Decision tree classifier has shown the highest improvements, followed by SVM, ANN and Bagging while KNN classifier has shown only marginal improvements. The bagging model is a group of SVMs joined in parallel. Its configuration along with other models is presented in Table 3 (Fig. 8).

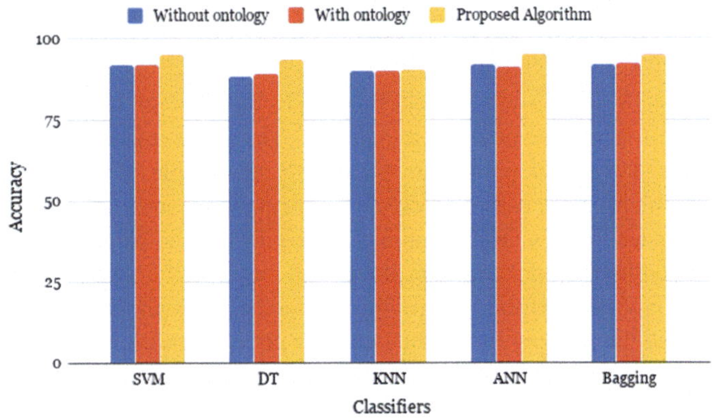

Fig. 8 Comparison between the accuracy of different classifiers

Table 4 Accuracy for different classifiers using different approaches

Models	Without ontology	With ontology	Proposed algorithm
SVM	91.9	91.9	95.3
DT	88.1	89.2	94
KNN	89.6	89.7	90.3
ANN	91.9	91.1	95.2
Bagging	91.9	92.4	95.2

Table 5 Feature vector size for different approaches

Method	Without ontology	With ontology	Proposed algorithm
Feature vector length	1069	736	72

The feature vector length has been decreased due to which misleading data is reduced and modelling accuracy has improved. Reducing feature size also forbades the decisioning on noise data. The models are trained to be robust. These are also tested on handwritten tweets and have shown promising results. The split ratio of 75:25 has performed better among all the other tried ratios, as mentioned in the training and testing sections. The reduction in feature vector length has resulted in faster training of models (Tables 4 and 5).

Stand-out features which are not included in the ontology and are supposed to be impacting the overall accuracy of the model are chosen parallelly based on their tf-idf score. It has helped in finding better correlation among features and deciding the polarity of the tweet.

This process of choosing the stand-out features will be automated so that they can be chosen dynamically depending on trend.

7 Conclusion

Social media platforms are data sources that can be used to gauge the public opinion with the help of sentiment analysis techniques. The government can make use of public opinion to guide its policies to better help the people. This paper proposes an approach which makes use of a domain ontology to improve upon standard sentiment analysis techniques. The domain ontology for Indian healthcare sector was created and used for feature extraction along with the TF-IDF method. The proposed approach not only out performs the traditional methods for all the classifiers tested but also reduced the training time. The ontology provides machine learning models with more context from the domain and therefore this leads to improved accuracy. The SVM classifier performs the best among all the classifiers with an accuracy of 95.3%. The proposed approach also drastically reduces in feature vector length by approximately 90%.

8 Future Work

Future scope includes automation of the process of tweet extraction and writing a script for updating the ontology with the newly extracted tweets. Information could also be extracted from other social networking sites so as to build a more diverse dataset. The approach of creating and maintaining the ontology could also be optimized further.

References

1. Glance, N., Hurst, M., Nigam, K., Siegler, M., Stickton, R., Tomokiyo, T.: Deriving marketing intelligence from online discussion. In: Proceedings of the Eleventh ACM SIGKDD International Conference on Knowledge Discovery in Data Mining, pp. 419–428, ACM (2011). https://doi.org/10.1145/1081870.1081919
2. Himanshu, S., Shankar, S.: Disaster analysis through tweets. In: 2015 International Conference on Advances in Computing, Communications and Informatics (ICACCI), pp. 1719–1723. (2015). https://doi.org/10.1109/ICACCI.2015.7275861
3. Ashok, M., Rajanna,, S., Joshi, P.V., Kamath, S. A personalized recommender system using machine learning based sentiment analysis over social data. In: 2016 IEEE Students' Conference on Electrical, Electronics and Computer Science (SCEECS), pp. 1-6. IEEE, March 2016
4. Kumar, A., Sharma, A.: Systematic literature review on opinion mining of big data for government intelligence. Webology **14**(2) (2017). http://www.webology.org/2017/v14n2/a156.pdf
5. GBD 2016 Healthcare Access and Quality Collaborators.: Measuring performance on the healthcare access and quality index for 195 countries and territories and selected subnational locations: a systematic analysis from the global burden of disease study 2016. Lancet **391**(10136), 2236–2271 (2018). https://doi.org/10.1016/S0140-6736(18)30994-2

6. Shein, K.P.P., Nyunt, T.T.S.: Sentiment classification based on ontology and SVM classifier. In: 2010 Second International Conference on Communication Software and Networks, pp. 169–172. Singapore (2010). https://doi.org/10.1109/ICCSN.2010.35
7. Shein, K.P.P.: Ontology based combined approach for sentiment classification, 3rd International Conference on Communications and Information Technology, CIT-09, pp 112–115, Stevens Point, Wisconsin, USA. World Scientific and Engineering Academy and Society (2009)
8. Polsawat, T., Arch-int, N., Arch-int, S., Pattanachak, A.: Sentiment analysis process for product's customer reviews using ontology-based approach. Int. Conf. Syst. Sci. Eng. (ICSSE) 1–6 (2018)
9. Nithish, R., Sabarish, S., Abirami, A.M., Askarunisa, A., Kishen, M.N.: An ontology based sentiment analysis for mobile products using tweets. In: Fifth International Conference on Advanced Computing, pp. 342–347 (2013)
10. Sam, K.M., Chatwin, C.R.: Ontology-based sentiment analysis model of customer reviews for electronic products. Proc. Int. J. e-Business, e-Management e-Learning **3**(6) (2013)
11. World Health Statistics (2013). https://www.who.int/gho/publications/world_health_statistics/2013
12. Press Information Bureau, Government of India. https://pib.gov.in/newsite/PrintRelease.aspx?relid=183624
13. Noy, N.F., McGuinness, D.L.: Ontology Development 101: A Guide to Creating Your First Ontology. Stanford University, Stanford, CA (2001)
14. Fernandez-Lope, M.: Overview of methodologies for building ontologies. In: Proceedings of the IJCAI- 99 Workshop on Ontologies and Problem-Solving Methods (KRR5), Stockholm, Sweden (1999)
15. Beck, H., Pinto, H.S.: Overview of Approach, Methodologies, Standards, and Tools for Ontologies. Agricultural Ontology Service, UNFAO (2003)
16. Kumar, A., Sharma, A.: Ontology driven social big data analytics for fog enabled sentic-social governance, Scalable Computing: Pract. Experience **20**(2) (2019). https://doi.org/10.12694/scpe.v20i2.1513
17. Kumar, A., Joshi, A.: IndiGov-O: an ontology of Indian government to empower digital governance. In: India International Conference on Information Processing (IEEE) (2016). https://doi.org/10.1109/IICIP.2016.7975373
18. Musen, M.A.: The protégé project: a look back and a look forward. AI matters, association of computing machinery specific interest group in artificial intelligence. **1**(4) (2015). https://doi.org/10.1145/2557001.25757003
19. Fortuna, B., Grobelnik, M., Mladenic, D.: OntoGen: semi-automatic ontology. In: Smith, M.J., Salvendy, G. (eds.) Human Interface and the Management of Information. Interacting in Information Environments. Human Interface. Lecture Notes in Computer Science, vol. 4558. Springer, Berlin, Heidelberg (2007). https://doi.org/10.1007/978-3-540-73354-6_34
20. Henrique, J.: GetOldTweets-python, GitHub Repository (2016). https://github.com/Jefferson-Henrique/GetOldTweets-python

Segmentation of Nuclei in Microscopy Images Across Varied Experimental Systems

Sohom Dey, Mahendra Kumar Gourisaria, Siddharth Swarup Rautray, and Manjusha Pandey

Abstract Nuclei detection in microscopy images is a major bottleneck in the discovery of new and effective drugs. Researchers need to test thousands of chemical compounds to find something of therapeutic efficacy. Nucleus being the most prominent part of a cell helps in the identification of individual cells in a sample and by analyzing the cell's reaction to various treatments the researchers can infer the underlying biological process at work. Automating the process of nuclei detection can help unlock cures faster and speedup drug discovery. In this paper, we propose a custom encoder–decoder style fully convolutional neural network architecture with residual blocks and skip connections which achieves state-of-the-art accuracy. We also use spatial transformations for data augmentation to make our model generalize better. Our proposed model is capable of segmenting nuclei effectively across a wide variety of cell types and experimental systems. Automated nuclei detection is projected to improve throughput for research in the biomedical field by saving researchers several hundred thousand hours of effort every year.

Keywords Artificial intelligence · Biomedical image processing · Computer-aided analysis · Medical expert systems · Neural networks

S. Dey (✉) · M. K. Gourisaria · S. S. Rautray · M. Pandey
Kalinga Institute of Industrial Technology, Bhubaneshwar, India
e-mail: sohom21d@gmail.com

M. K. Gourisaria
e-mail: mkgourisaria2010@gmail.com

S. S. Rautray
e-mail: siddharthfcs@kiit.ac.in

M. Pandey
e-mail: manjushafcs@kiit.ac.in

S. C. Satapathy et al. (eds.), *Intelligent Data Engineering and Analytics*, Advances in Intelligent Systems and Computing 1177,
https://doi.org/10.1007/978-981-15-5679-1_9

1 Introduction

Search for new and effective drugs requires trial of thousands of chemical compounds and observing the reactions for each to arrive at an inference. For medical analysis, batches of cells are prepared and the reaction of the cells is observed after adding different chemical compounds to each batch of cells. Preparing batches of cells and testing with different chemicals can be done on a large scale after robotic automation replaced manual labor. A major delay in the pipeline is analyzing the huge amount of cell images for various characteristics, for which we certainly need software aid. The first and the most effective approach for cell analysis is most often the detection of the nuclei. From there various properties of the cell can be calculated to find out their disease state.

Let us explain the current pipeline followed by a scientist. When the nuclei are more-or-less round and easily distinguishable from each other, a classical computational algorithm can satisfactorily segment the nuclei. But the software tends to fail if the cell images are complex and involve tissue samples, because then it becomes hard to distinguish each nucleus as they have complicated shapes and are closer to each other, sometimes even overlapping. In these cases, the scientist has to analyze each sample by eye and this costs a significant amount of time and effort. Imagine manually analyzing thousands of images to arrive at a conclusion.

An accurate software model capable of nuclei identification in medical images without any arbitration will push the boundaries of biomedical image analysis and drug discovery and shorten the time span to market a new drug. Classical image processing techniques require manual configuration, and existing models mainly specialize on specific types of cells. A single model intelligent enough to detect nuclei in different contexts and varying experimental systems would save researchers a significant amount of time and effort and speed up the analysis by a huge margin.

2 Related Works

With the recent advancements in the artificial intelligence domain, neural networks are being widely used in medical image analysis and have proven to give better results than most classical image processing algorithms. Research in the field of biomedical image segmentation has become more demanding as more powerful neural architectures and deep learning techniques are emerging every year. In this section, we discuss the recent advances in this field related to nuclei segmentation for cell analysis.

Nurzynska et al. [1] proposed a technique for searching the best parameters for color normalization for the task of segmenting the nucleus. Monte Carlo simulation was used to search for the optimal parameters for color normalization which lead to better performance in segmentation. Narotamo et al. [2] proposed a combined approach of using a fast YOLO architecture and U-Net model for detection and segmentation, respectively. The authors trained their model on 2D fluorescence microscopy images. They showed that their model is more computationally effective against mask R-CNN while sacrificing some performance. Their proposed model is nine times faster than mask R-CNN on image size of 1388 $\times$ 1040. Chen et al. [3] proposed a model for segmentation of caudate nucleus in MRI scans of brain based on a distance regularized level-set evolution. Pan et al. [4] proposed a model based on deep semantic network for segmentation of nuclei from pathological images. The authors used atrous depth-wise separable convolution layers for their model (AS-UNet) which increases the receptive field of the model. It extracts and combines features of multiple scales so that the model can perceive both small and large cells. Their model achieves promising performance. Mahbod et al. [5] proposed a U-Net architecture with two stages for segmentation of touching nuclei in sections of hematoxylin and eosin stained tissue. Semantic segmentation with U-Net was followed by the creation of a distance map with a regression U-Net model. Based on the segmentation mask and distance map a watershed algorithm is used for instance segmentation. Their model achieves a Jaccard index of 56.87%. Zeng et al. [6] proposed a U-Net-based model for nuclei segmentation which used residual blocks, and multi-scale feature and channel attention mechanism. Their model RIC-UNet achieves a Jaccard index of 0.5635 while the original U-Net achieves 0.5462 on the Cancer Genomic Atlas (TCGA) dataset. Li et al. [7] proposed a U-Net-based model which utilizes boundary and region information, which provides a huge performance boost on overlapping glioma nuclei samples. They used a classification model to predict the boundary and the distance map is predicted by a regression model. These are further used to obtain the final segmentation mask. Their proposed architecture achieves a mean IOU of 0.59 on multi-organ nuclei segmentation open dataset (MoNuSeg). Zhou et al. [8] proposed their (CIA-Net) for robust instance segmentation of nuclei. They used two separate decoders for separate tasks and a multi-level information aggregation module to capture the dependencies (spatial and texture) between the nuclei and the contour.

3 Proposed Method

3.1 Dataset Used

The BBBC038v1 dataset [9] is used for this experiment, which is accessible from Broad Bioimage Benchmark Collection [Ljosa et al., Nature Methods, 2012]. The dataset contains 670 training images with more than twenty thousand annotated nuclei. The images were gathered from various sources including biomedical professionals in hospitals and industries and researchers in various universities. The dataset has a lot of variance as the cells belong to various animals and the imaging of the treated cells has been done in different experimental systems which involves

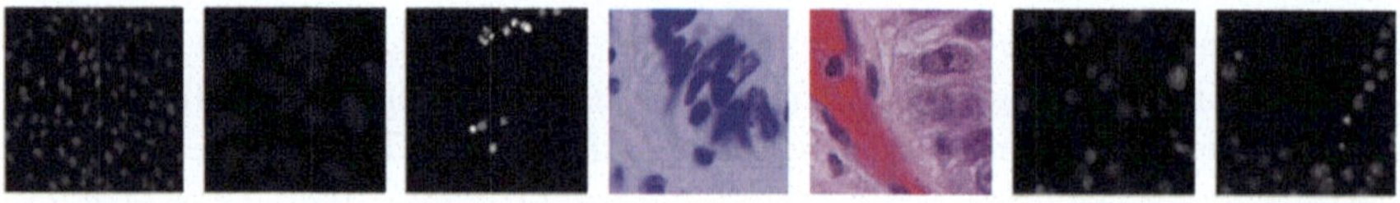

Fig. 1 Images

Fig. 2 Masks

variation in lighting conditions, microscope magnifications, and histological stains (Figs. 1 and 2).

3.2 Data Augmentation Used

Deep-learning-based approaches require a lot of input data, but it is difficult to find such huge amount of data in the medical field. The dataset we are using contains 670 images which are not sufficient for training a robust model, so we used specific data augmentation techniques to prevent our model from overfitting and make them generalize better and improve performance. In the case of medical images, spatial level transformations have already proven to give better results since they augment the data very close to real images. Especially elastic deformations and optical distortions work very well while training a segmentation network. Shift and rotation invariances also work well with microscopy images. We used a lot of heavy augmentations including horizontal flip, random contrast, random gamma, random brightness, elastic transform, grid distortion, optical distortion, shift scale rotate, etc.

3.3 Model Architecture

We used the semantic segmentation approach for our intended task of nuclei detection. Two of the most popular architectures in this domain are the mask-RCNN [10] and the FCN [11] (fully convolutional neural network)-based segmentation networks. FCN being a one-stage segmentation network is mostly preferred over two-stage networks like mask-RCNN for its simplicity and computational efficiency. The U-Net architecture [12] based on the FCN architecture has been one of the most popular architectures for medical image segmentation recently. Our model is an improvement over the U-Net architecture.

FCN-based segmentation networks replace the fully connected layers of a conventional CNN architecture with fully convolutional layers. It uses an encoder–decoder architecture to learn the segmentation mask from the input image. The encoder learns the contextual information and the decoder learns the spatial information. Skip connections help the decoder network to use the spatial information from the higher layers of the encoder network and fuses them with the upsampled features to learn the precise location of the nuclei in the images. This method gives fine-grained segmentation masks.

We use a 17 layer encoder network with residual blocks [13] which downsamples the feature map. We use convolution layers with a stride of two to downsample the images instead of using max-pooling. We only use max-pooling once at the beginning of the network. The decoder network uses transposed convolution layers to upsample the feature maps, and then concatenates features from encoder layers through skip connections, followed by residual blocks in each stage. Residual blocks allow easier optimization of deep networks while simple skip connections from encoder to decoder enable fine-grained segmentation maps to be generated using information from the previous layers of the encoder (Fig. 3).

3.4 Loss Function Used

The most commonly used loss function for segmentation models is pixel-wise cross-entropy loss which compares the class predictions for each pixel individually. Another very popular loss function used in biomedical image segmentation in soft-dice loss [14] which measures the overlap between two samples. For our task, we optimize a BCE-dice loss function which is basically binary cross-entropy added to soft-dice loss, which resulted in better performance and early convergence. Our model took only 25 epochs of training using Adam optimizer before early stopping.

$$\text{Binary cross-entropy loss} = -(y\log(p) + (1-y)\log(1-p)) \tag{1}$$

$$\text{Soft-dice loss} = \frac{2|A \cap B|}{|A| + |B|} \tag{2}$$

$$\text{BCE-dice loss} = -(y\log(p) + (1-y)\log(1-p)) + \frac{2|A \cap B|}{|A| + |B|} \tag{3}$$

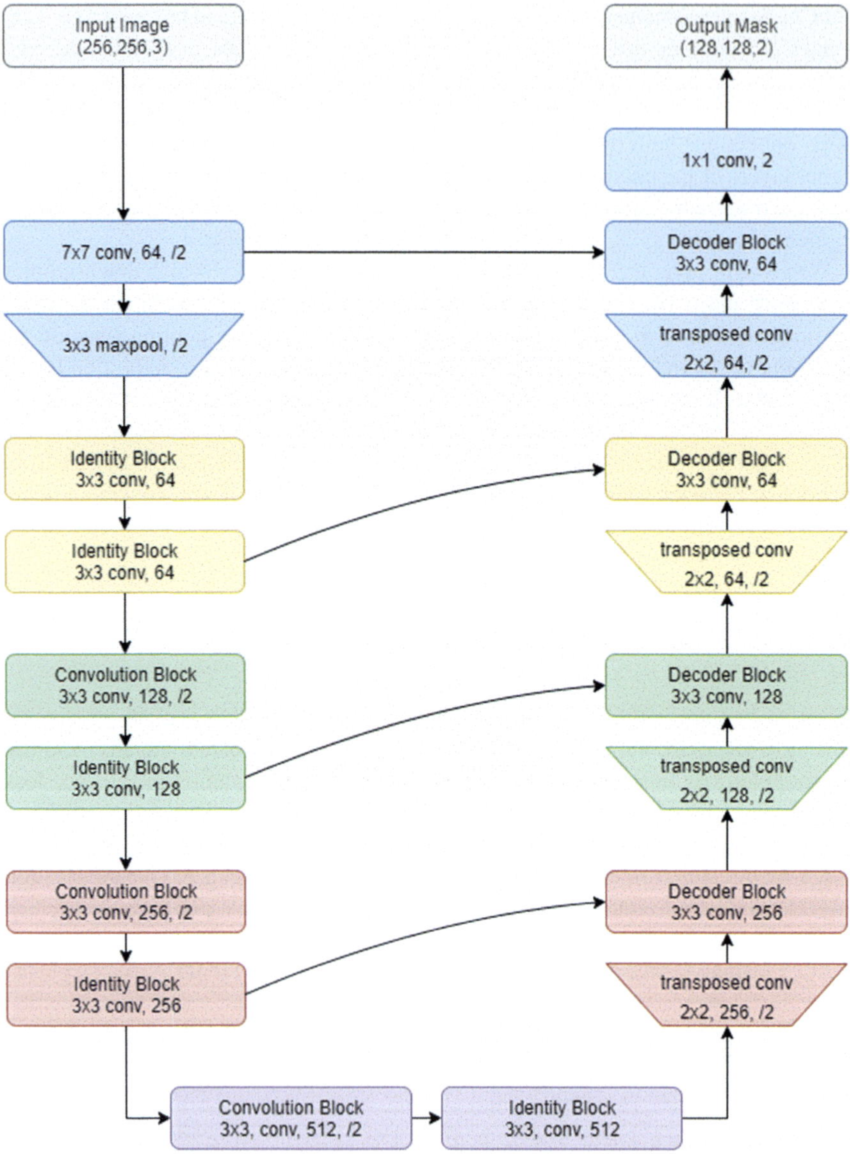

Fig. 3 Model architecture

3.5 Evaluation Metric Used

The most commonly used evaluation metrics are pixel-wise accuracy and the Jaccard index also known as the IoU. We used IoU as our primary evaluation metric which calculates the overlap between the target and prediction masks. We also choose this metric since it is closely related to dice coefficient used in the dice loss. We also calculate the precision, recall, and f1-score for a comparative evaluation of our model.

$$\text{IoU} = \frac{|A \cap B|}{|A \cup B|} \tag{4}$$

4 Experiments and Results

We resize the input images to 256 × 256 before feeding them into the network. Our network outputs masks of dimension 128 × 128. Since our model is considerably deep, we use data augmentation to prevent overfitting and thus increasing the generalizability of the model and improve overall performance. We used Adam optimizer and auto-reduced the learning rate when the learning plateaued out. Our model reached a validation IoU of 0.9486 with just 25 epochs of training before being early stopped. Using SGD optimizer gives a smooth training curve but takes 500 epochs to converge, while Adam takes 25 epochs but the initial training curve is quite abrupt. Figure 4 shows the IoU and loss function curves for the training and validation sets. Table 1 shows the results on the validation set and compares our

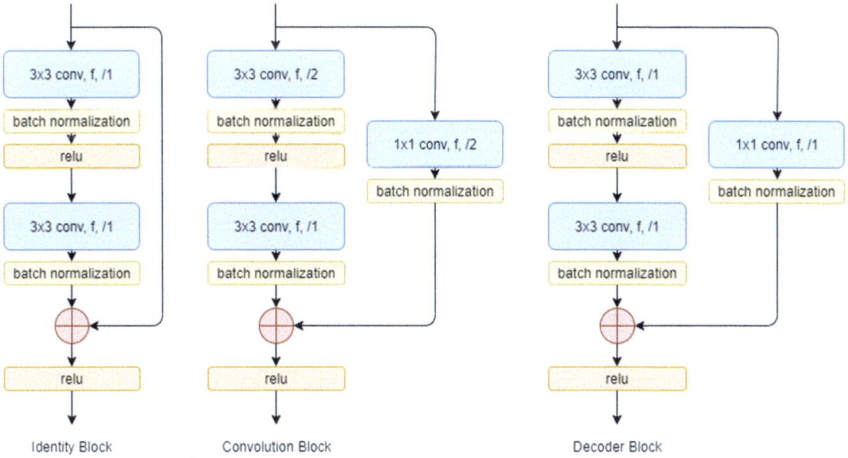

Fig. 4 Residual bottleneck blocks

Table 1 Results and comparisons

Metric	Value
Precision	0.9734
Recall	0.9738
F1-score	0.9736
IoU	0.9486
Method	IoU
U-Net	90.77
Wide U-Net	90.92
U-Net++	92.63
Our model	94.86

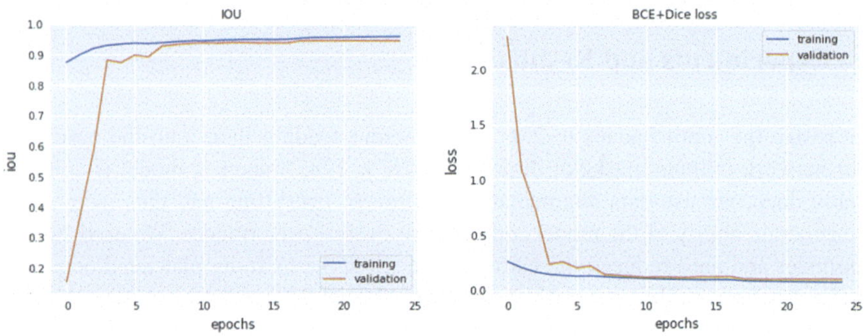

Fig. 5 Accuracy and loss curves

model with the top three state-of-the-art models for this specific task and our model performs significantly better (Fig. 5).

5 Conclusion

Medical image processing has been gaining a lot of attention recently due to the emergence of deeper and high-accuracy segmentation networks which can compete against humans and speed up biomedical research to a great extent. Nuclei detection has always been a very a crucial step for cell analysis and recently many computer-aided analysis approaches are being used for faster and more accurate medical analysis. With the inception of deep-learning-based intelligent analysis algorithms, medical industry and researchers are replacing classical computational image processing algorithms with sophisticated deep learning models. Unlike classic image processing algorithms, deep learning models do not require manual pre-processing or feature engineering, nor do they require any manual parameter tweaking. In this paper, our proposed model incorporates the latest advancements in the field of deep learning for accurate segmentation of nuclei from microscopy images of cells. Our automated nuclei detection model achieves an IoU of 0.9486 which is a significant improvement over the state-of-the-art U-Net++ network. Our model works effectively across a wide variety of types of nuclei and experimental systems. Robustness to cell types and experimental setups has been our main focus. Tackling the problem of automated nuclei detection can help to improve the rate of drug discovery and enable faster cures, thus improving overall health and quality of life of the people.

References

1. Nurzynska, K.: Optimal parameter search for colour normalization aiding cell nuclei segmentation. In: Communications in Computer and Information Science, vol. 928. Springer, Cham (2019)
2. Narotamo, H., Sanches, J.M., Silveira, M.: Segmentation of cell nuclei in fluorescence microscopy images using deep learning. In: Lecture Notes in Computer Science, vol. 11867. Springer, Cham (2019)
3. Chen, Y., Chen, G., Wang, Y., Dey, N., Sherratt, R.S., Shi, F.: A distance regularized level-set evolution model based MRI dataset segmentation of Brain's caudate nucleus. IEEE Access **7**, 124128–124140 (2019)
4. Pan, X., Li, L., Yang, D., He, Y., Liu, Z., Yang, H.: An accurate nuclei segmentation algorithm in pathological image based on deep semantic network. IEEE Access **7**, 110674–110686 (2019)
5. Mahbod, A., Schaefer, G., Ellinger, I., Ecker, R., Smedby, Ö., Wang, C.: A two-stage U-Net algorithm for segmentation of nuclei in H&E-stained tissues. In: Lecture Notes in Computer Science, vol. 11435. Springer, Cham (2019)
6. Zeng, Z., Xie, W., Zhang, Y., Lu, Y.: RIC-Unet: an improved neural network based on Unet for nuclei segmentation in histology images. IEEE Access **7**, 21420–21428 (2019)
7. Li, X., Wang, Y., Tang, Q., Fan, Z., Yu, J.: Dual U-Net for the segmentation of overlapping glioma nuclei. IEEE Access **7**, 84040–84052 (2019)
8. Zhou, Y., Onder, O.F., Dou, Q., Tsougenis, E., Chen, H., Heng, P.A.: CIA-Net: robust nuclei instance segmentation with contour-aware information aggregation. In: Lecture Notes in Computer Science, vol. 11492. Springer, Cham (2019)
9. Broad Bioimage Benchmark Collection dataset page from Broad Institute website. https://data.broadinstitute.org/bbbc/BBBC038
10. He, K., Gkioxari, G., Dollár, P., Girshick, R.: Mask R-CNN. In: IEEE International Conference on Computer Vision (ICCV) (2017)
11. Long, J., Shelhamer, E., Darrell, T.: Fully convolutional networks for semantic segmentation. In: IEEE Conference on Computer Vision and Pattern Recognition (CVPR) (2015)
12. Ronneberger, O., Fischer, P., Brox, T.: U-Net: convolutional networks for biomedical image segmentation. In: Medical Image Computing and Computer-Assisted Intervention (MICCAI). LNCS, vol. 9351, pp. 234–241. Springer (2015)
13. He, K., Zhang, X., Ren, S., Sun, J.: Deep residual learning for image recognition. In: IEEE Conference on Computer Vision and Pattern Recognition (CVPR) (2016)
14. Milletari, F., Navab, N., Ahmadi, S.A.: V-Net: fully convolutional neural networks for volumetric medical image segmentation. In: 2016 Fourth International Conference on 3D Vision (3DV), pp. 565–571. IEEE (2016)

Transitional and Parallel Approach of PSO and SGO for Solving Optimization Problems

Cherie Vartika Stephen, Snigdha Mukherjee, and Suresh Chandra Satapathy

Abstract Optimization is finding minimization or maximization of a decision variable for a given problem. In most of the engineering problems, there is a requirement of optimizing some variables or other to obtain a desired objective. Several classical techniques are existing in optimization literature. However, when the optimization problem is complex, discrete, or not derivable there is a need to look beyond classical techniques. Swarm intelligence techniques are overwhelmingly popular in current days for targeting such kind of optimization problems. Particle Swarm Optimization (PSO) and Social Group Optimization (SGO) are belonging to this category. PSO being a popular and comparatively an older algorithm to SGO, the efficiency and efficacy of PSO for function optimization are well established. In this paper, an effort is made to explore an effective alternate model in hybridizing PSO and SGO. In the proposed model, a transitional concept is used. An alternate switching between PSO and SGO is carried out after a fixed iteration. An exhaustive simulation is done on several benchmark functions and a comparative analysis is presented at the end. The results reveal that the proposed approach is a better alternative to obtain effective results in comparatively less iterations than stand-alone models.

Keywords Function optimization · Hybrid approach · PSO · SGO

C. V. Stephen · S. Mukherjee (✉) · S. C. Satapathy
Kalinga Institute of Industrial Technology, DU, Bhubaneswar, India
e-mail: snigdhabony@gmail.com

C. V. Stephen
e-mail: cherie_s20@yahoo.in

S. C. Satapathy
e-mail: Suresh.satapathyfcs@kiit.ac.in

S. C. Satapathy et al. (eds.), *Intelligent Data Engineering and Analytics*,
Advances in Intelligent Systems and Computing 1177,
https://doi.org/10.1007/978-981-15-5679-1_10

1 Introduction

The word "optimization" is very familiar and has widespread usage in our day-to-day life. It is the mathematical discipline which is concerned with finding the maximum and minimum of functions, possibly subject to constraints. Under some given circumstances this act of optimization helps in obtaining the best result. The final goal is to either minimize or maximize some parameters for obtaining optimal results. Since the required effort in any real-world situation can be expressed as a function of certain decision variables, optimization can be thought of finding the conditions that provide the maximum or minimum value of it. There are various methods available for solving optimization problems. All these optimum seeking techniques come under operations research and are called mathematical programming techniques. During this decision-making process, a number of solutions are obtained. The best solution is chosen keeping in mind several factors such as accuracy, convergence speed, robustness, etc.

In our work, we have tried to study the effectiveness of few well-known evolutionary optimization techniques such as PSO [1] and SGO [2]. We have attempted to take few benchmark functions for the simulation purpose. Later, two hybrid approaches are suggested, one is transitional approach wherein we make serial implementation of PSO and SGO, respectively, and in second a parallel approach is suggested for PSO and SGO.

1.1 Traditional Optimization Tools

Traditional optimization tools usually begin with a randomly chosen initial solution and move toward the best solution iteratively. These tools can be grouped under two categories, namely, linear search and gradient-based methods. Random search method, uni-variate method, and pattern search method belong to linear search method. Some gradient-based methods are steepest descent method, conjugate gradient method, quasi-Newton method, and others. In linear search method, the search direction is decided at each iteration randomly, whereas in gradient-based method such direction is decided by the gradient of objective function.

Drawbacks of traditional optimization tools:

- Final solution is dependent on the initially chosen random solution which is not guaranteed to be a globally optimal one. Gradient-based methods cannot tackle optimization problems involving discontinuous objective function. Moreover, the solutions of gradient-based methods may get stuck at local optimum point.
- There is no versatile optimization technique, which can be used to solve a variety of problems because a particular traditional optimization method may be suitable for solving only one type of problem.

1.2 Non-traditional Optimization Tools

The tendency of us, human beings, to follow the natural way by using some natural processes, such as biological, physical processes, etc. modeled artificially, has paved the path for solving complex optimization problems, whenever we failed to solve them using traditional optimization methods.

These tools are inspired from nature and some well know such algorithms are genetic algorithm [3], simulated annealing, ant colony optimization, cultural algorithm, particle swarm optimization, etc.

Non-traditional optimization tools were devised to overcome most of the drawbacks of the traditional optimization tools. It is more robust as one technique can solve a variety of problems.

2 Preliminaries

2.1 Function Optimization

In optimization problems, we find the largest value or the smallest value that a function can generate.

Function optimization is used in different fields such as

- Machine designing in mechanical engineering.
- Minimizing THD in multilevel inverter in electronics field.
- Optimization of upstream detention reservoir facility in civil engineering.
- Optimization in metabolic engineering and synthetic biology.
- Multi-objective optimization in chemical engineering.

2.2 Benchmark Functions

These are various types of functions that are used to evaluate, characterize, and measure the performance of optimization algorithm. It can characterize many standard optimization problems which can predict the behavior of the algorithms under various circumstances. In our work, we have chosen few such benchmark functions to simulate our proposed techniques. Some of the benchmark functions are as follows:

- Sphere function: De Jong proposed the simple and strongly convex function which converges slowly and leads to the global minimum.
  Formula: $f(x) = \sum_{i=1}^{n} x_i^2$
  Search domain: $-\infty \leq x_i \leq \infty$, $1 \leq i \leq n$
- Rastrigin function: It is obtained from sphere after adding a modulator term to it.
  Formula: $f(x) = A_n + \sum_{i=1}^{n} \left[x_i^2 - A\cos(2\pi x_i)\right]$

Search domain: $-5.12 \leq x_i \leq 5.12$

- Griewank function: It is a continuous, unimodal, and non-convex function.
  Formula: $f(x) = 1 + \sum_{i=1}^{n} \frac{x_i^2}{4000} - \prod_{i=1}^{n} \cos\left(\frac{x_i}{\sqrt{i}}\right)$
  Search domain: $x_i \in [-600, 600]$.

2.3 Evolutionary Technique

Evolutionary computation is a family of algorithms which is population-based trial and error problem solver that slowly improves an individual's adaptability to its surrounding by regulating the structure of the individual.

Swarm intelligence, as a part of evolutionary computation, iteratively produces new generation by stochastically discarding less desired solutions and thereafter small random changes are introduced. At the end, the best fitness of a function is chosen from a set of gradually evolving and increasing fitness values.

Working of Evolutionary algorithm is shown below (Flowchart 1).

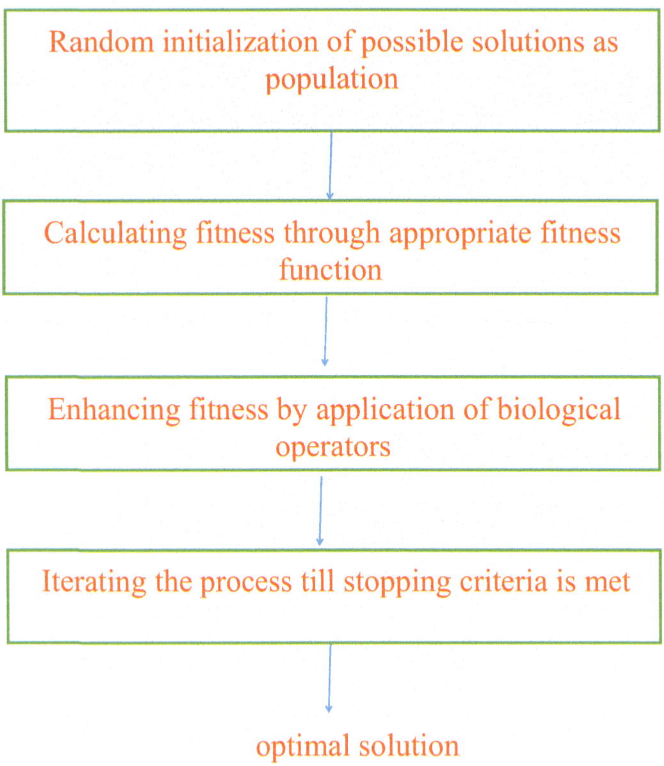

Flowchart 1 Working of evolutionary algorithm

The two evolutionary techniques we have used in our paper are PSO and SGO.

2.3.1 Particle Swarm Optimization (PSO)

In 1995, James Kennedy and Russel Eberhart designed this nature-inspired population-based evolutionary and stochastic optimization technique that solves computationally hard optimization [4] problems. This robust technique is based on the movement and intelligence of swarms, which is composed of a number of agents known as particles. This is applied successfully to a wide range of search and optimization problems. Since it is inspired from the swarms in nature, such as swarms of birds, fish, etc., in D-dimensional space where each particle, treated as a point, modifies its flying according to its own flying experience and that of other particles present in the swarm. A swarm of N particles (individuals) communicates either directly or indirectly with one another using search directions (gradients). Each particle that has been assigned a random velocity is allowed to move in the problem space to locate global optimum. During every iteration of PSO, each particle updates in position according to its previous experience and the experience of its neighbors.

A particle is composed of three vectors:

- **X-vector**: It records the current position of the particle in problem space (search).
- **P-vector (P-Best)**: It records the location of the best solution found so far by the particle.
- **V-vector**: It contains a gradient (direction) for which the particle will travel in if undisturbed (Fig. 1).

The algorithm implementation of PSO is explained in the following steps:

- We declare some initial parameters such as swarm size, maximum number of iterations, inertial weight, and acceleration coefficients $c1$ and $c2$.
- Initially, the particle position and velocity are randomly generated.
- Fitness of each particle is evaluated. Pbest and Gbest are updated accordingly.
- If Fitness $X_i >$ Fitness (Gbest), then Gbest $= X_i$.
- If Fitness $X_i >$ Fitness (Pbest), then Pbest $= X_i$.

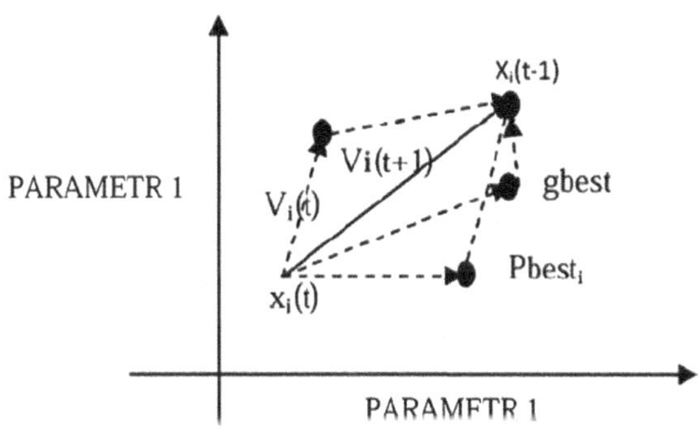

Fig. 1 Vector representation of PSO model

- Updating velocity with the following formula:
- $Vi + 1 = W * V_i + c1 * \text{rand}(0, 1) * (\text{Pbest-}X_i) + c\,2 * \text{rand}(0, 1) * (\text{Gbest-}X_i)$.
- Updation of the next particle position is simply done by adding the V-vector to the X-vector to get another X-vector –> $X_{i+1} = X_i + V_{i+1}$.
- The control falls back to step 3 until the number of iterations is satisfied.
- After termination, the final value of Gbest is the output.

2.3.2 Social Group Optimization (SGO)

This model was proposed by Satapathy et al. This is also a population-based algorithm but here each particle is a person. Its inspiration is taken from the idea of social behavior of human beings in solving a complex problem.

Each person has several behavioral traits like caring, empathy, morality, disloyalty, tolerance, politeness, fear, decency, etc., which lie in passive state in humans but need to be governed in the right direction to solve all complexities of life. However, it is often observed that these problems can also be solved when there is an influence of traits from one person to another in the society since human beings are great imitators. Group solving capabilities have been proven to be better than individual capability in exploring different traits of each person for solving a problem. Each person gains some knowledge and thus it obtains some level of capacity for solving a problem which is equivalent to "Fitness". Hence, the best person is chosen as the best solution.

This technique is divided into two phases:

- Improving phase.
- Acquiring phase.

In the improving phase, each person gains knowledge from the best person. It can be depicted as follows:

```
For i = 1 : N
    For j=1:D
        Xnew_ij=c* Xold_ij+ r* (gbest(j)−Xold_ij)
    End for
End for
where r is a random number, r ~ U(0,1)
Accept Xnew if it gives a better fitness than Xold.
```

where r is a random number, $r \sim U(0, 1)$

Accept Xnew if it gives a better fitness than Xold.

Here c is known as self-introspection parameter. Its value can be set from $0 < c < 10 < c < 1$.

In the acquiring phase, a random interaction occurs between a random person and each person in the social group for gaining or acquiring knowledge. If the random person is more knowledgeable than him/her, then it acquires something new from the random person. It can be depicted as follows:

```
(Xi's are updated values at the end of the improving phase)
For i=1:N
    Randomly select one person Xr,where i≠r
    If f(Xi)<f(Xr)
        For j=1:D
            Xnewi,j=Xoldi,j + r1* (Xi,j−Xr,j)+r2* (gbestj−Xi,j)
        End for
    Else
        For j=1:D
        Xnewi,:=Xoldi,:+r1* (Xr,:−Xi,:)+r2(gbestj−Xij)
        End for
    End If
Accept Xnew if it gives a better fitness function value.
End for
```

where $r1r1$ and $r2r2$ are two independent random sequences, $r1 \sim U(0, 1)r1 \sim U(0, 1)$ and $r2 \sim U(0, 1)r2 \sim U(0, 1)$.

The algorithm implementation of SGO is explained in the following steps:

- We declare some initial parameters such as number of people, maximum number of iterations, number of traits for each person, and self-introspection parameter c.
- Initially the population of people is randomly generated and fitness of each person is evaluated.
- Best solution as well as Gbest is identified.
- Improving phase is carried out. If the new population is better than the old population, it is accepted, else rejected.
- If accepted, the best solution and the Gbest are identified.
- Acquiring phase is carried out with the new population. If the new population is better than the old population, it is accepted, else rejected.
- If termination condition is not satisfied, the control will fall back to step 3.
- After termination, final value of solutions is obtained.

3 Proposed Approach: Hybridization PSO with SGO

From the individual studies of PSO and SGO for function optimization, we have observed that though both techniques are performing equally good and able to provide optimum solution, they have different convergence characteristics. To improve the convergence without compromising the quality of solutions, two hybridize techniques are suggested in this work. The merits of PSO and SGO are taken to form the combination in transitional and parallel way.

A. Transitional approach

In the transitional approach, we randomly initialize a population and then pass it to the PSO algorithm. The refined solution is stored and the most optimal solution is appended with the random initialized values of SGO. While taking the random values for SGO, we take one value less so that after appending the best solution from PSO, the population size remains the same. Then SGO is performed on the same population. The same routine is again followed, i.e., the best solution from SGO is appended with the remaining stored refined solution of PSO and the worst value is eliminated in order to maintain the same population size. This process is known as transitional approach. Both PSO and SGO are alternatively called in this transitional approach. The optimum value of PSO is fed to SGO and vice versa. This carries on till the termination condition is satisfied.

The transitional technique exploits the optimal solutions of both PSO and SGO, respectively. The flowchart for the transitional technique is given in Fig. 2.

B. Parallel approach

In this approach, we have passed the same randomly initialized population to both the models, namely, PSO and SGO. Both the algorithms run simultaneously for the same number of iterations. The best solution from each of the obtained refined solution is exchanged with each other (between the two techniques running simultaneously) to create a new set of population. Here, the worst value from the refined solution is eliminated in order to maintain the same population size after the best value from the other technique is added to its refined solution. This is again passed back to the initial step, where the algorithms are applied again till the termination condition is satisfied. We then obtain two sets of values from each of the algorithms, and the best of the two is the final output.

The flowchart for the parallel technique is given in Fig. 3.

4 Simulation Results and Discussion

Our simulation is done in a very systematic way using Python 3. The observation obtained after stimulation has been shown in Table 1. Firstly, we run the PSO algorithm for 50 particles and 50 iterations for dimensions 10, 20, 60, and 100 for all three functions. We repeat the same for SGO and transitional approach as well. We note down the GBest values in each case. In the parallel approach, we obtain two GBest values, one from the PSO algorithm and the other from the SGO algorithm. The better of the two values is noted down as the GBest value. We have minimized all the functions.

From graphs, Fig. 4a–c, the convergence characteristics of PSO for three different functions and four different dimensions are shown.

Similarly from graphs, Fig. 5a–c, the convergence characteristics of SGO for three

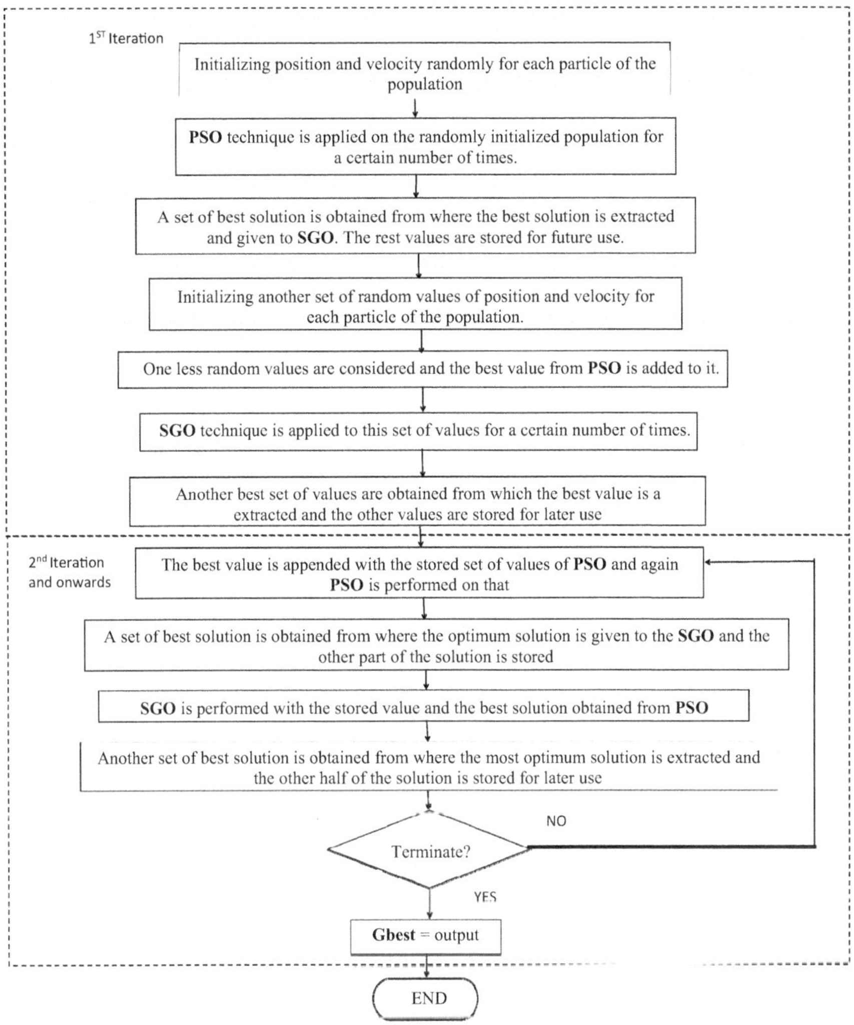

Fig. 2 Transitional approach

different functions and four different dimensions are shown.

From the above two sets of figures, it is clearly evident that SGO converges faster than PSO. However, the quality of results for both PSO and SGO is competitive as shown in Table 1.

We have shown the convergence characteristics of both transitional and parallel approach from Figs. 6a, 7, and 8c. From the figures, it can be observed clearly that our suggested approaches are able to find optimal solution faster compared to standalone PSO and SGO. This is due to the fact that SGO is having less number of user parameter to handle compared to PSO. Hence, when both techniques are combined

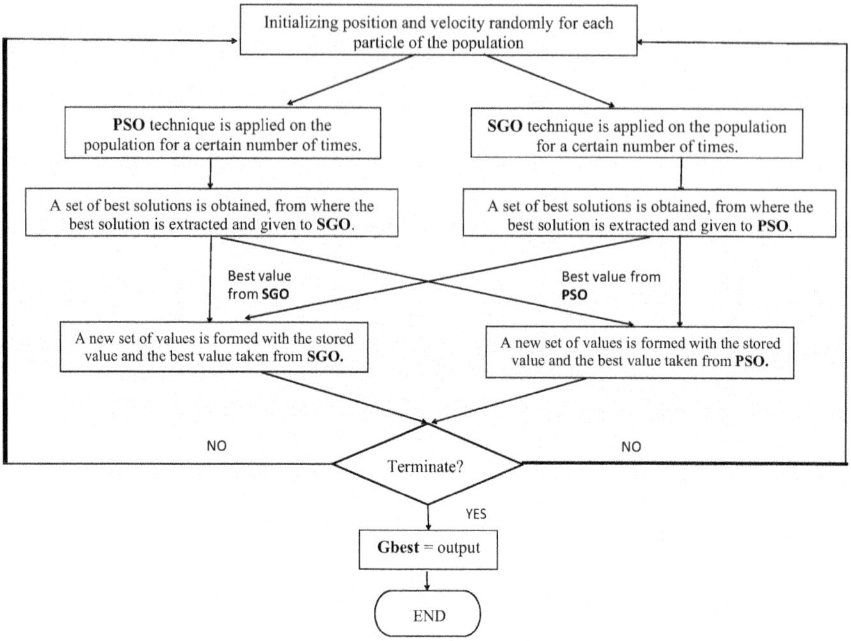

Fig. 3 Parallel approach

Table 1 Simulation results of PSO, SGO, transitional, and parallel algorithms for each function

Different approaches with dimensions		Sphere function	Rastrigin function	Griewank function
PSO	10	0.0021819640163376	6.06959112773059	0.7679597763884534
	20	0.0118905049697398	18.4981884965327	0.6866412785427738
	60	0.5538397754854892	284.8821756328698	0.9409553608295672
	100	1.0927434021122422	597.4346379750347	1.0029617532401354
SGO	10	2.1985887749826e−69	7.11472847649826e−20	2.7847645463726e−35
	20	4.1681652992716e−69	3.94529389242716e−20	3.6273627399334e−32
	60	1.6142732606372e−68	1.74859573231442e−18	0.49204044286372e−32
	100	2.6332207738009e−68	3.53242073444009e−18	9.4002044020000e−31
Transitional	10	0.0000	0.0000	0.00000
	20	0.0000	4.40835932145e−313	5.5678654355e−320
	60	0.0000	1.536849807757e−480	1.645584746546e−480
	100	0.0000	2.408609468946e−400	3.453234567546e−500
Parallel	10	0.00000	0.00000	0.00000
	20	0.00000	1.584584522225e−325	0.00000
	60	9.475838347380e−400	4.408359332145e−310	0.00000
	100	5.657754837480e−500	9.408604078946e−480	1.748348374436e−555

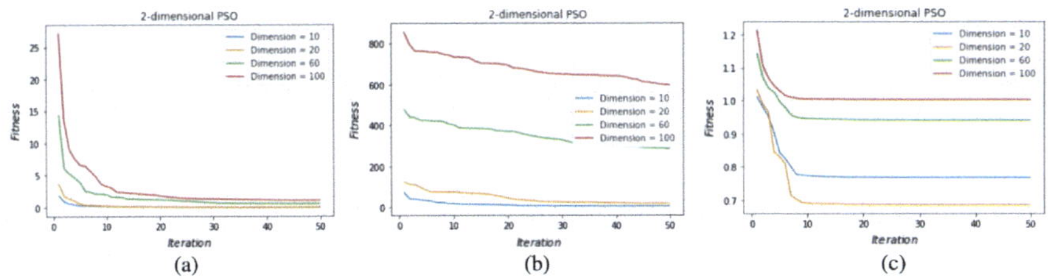

Fig. 4 Showing convergence characteristics of PSO for three functions **(a)** Sphere, **(b)** Rastrigin, and **(c)** Griewank

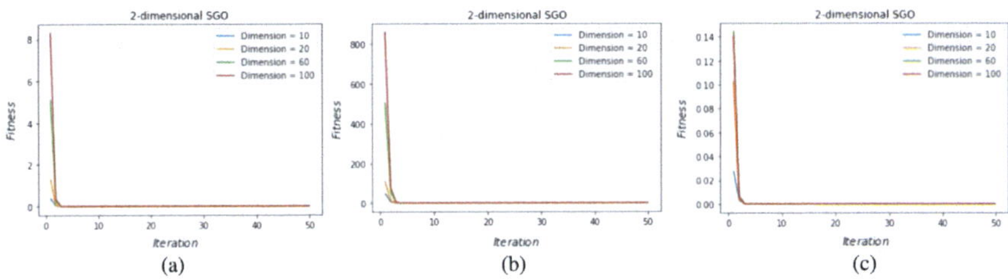

Fig. 5 Showing convergence characteristics of SGO for three functions **(a)** Sphere, **(b)** Rastrigin, and **(c)** Griewank

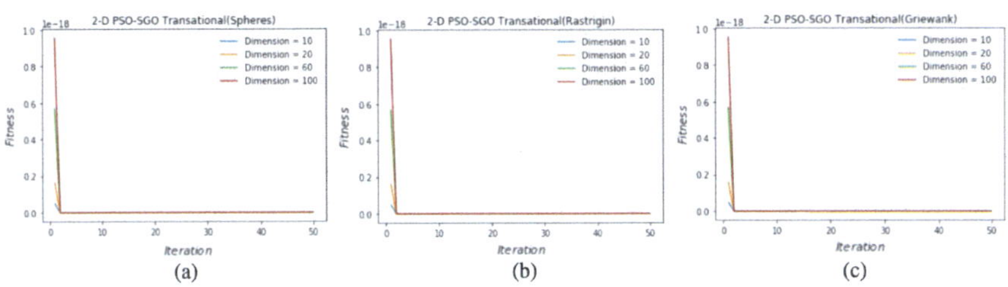

Fig. 6 Showing convergence characteristics of PSO-SGO transitional approach for three functions **(a)** Sphere, **(b)** Rastrigin, and **(c)** Griewank

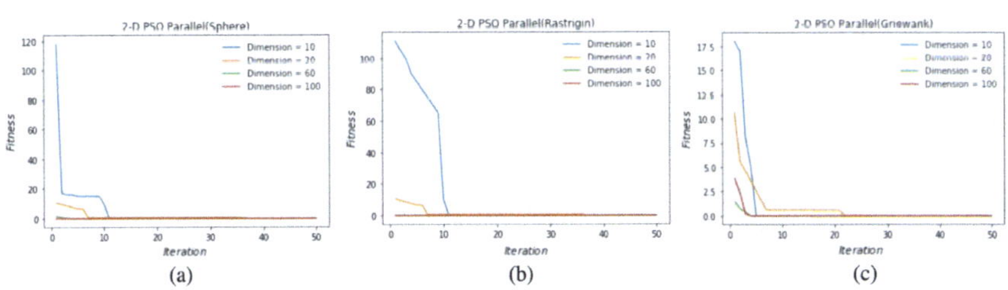

Fig. 7 Showing convergence characteristics of PSO parallel approach for three functions **(a)** Sphere, **(b)** Rastrigin, and **(c)** Griewank

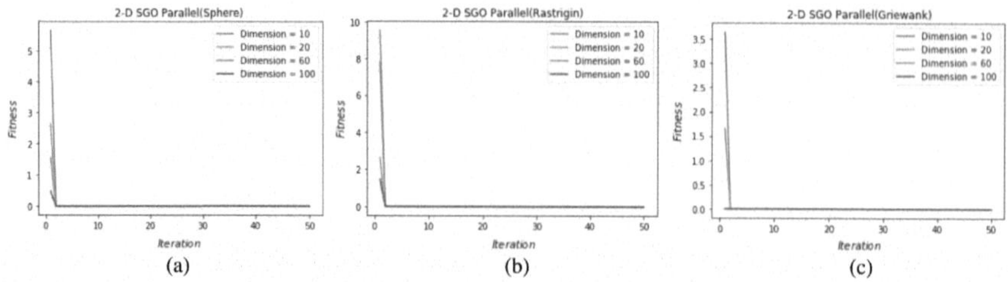

Fig. 8 Showing convergence characteristics of SGO parallel approach for three functions **(a)** Sphere, **(b)** Rastrigin, and **(c)** Griewank

not only we retain the quality solutions but also we find quality solutions at a faster rate.

5 Conclusion

We have proposed a new method of integrating PSO and SGO. In our proposed approach, PSO and SGO are integrated into a hybrid system. In the transitional approach, one algorithm runs on a fixed number of particles for a definite number of iterations, and the best solution is appended with the other set of particles on which the other algorithm is applied. In the other parallel approach, each algorithm, i.e., PSO and SGO, is applied on a fixed set of particles for a fixed number of iterations. The best solutions are then swapped with the other algorithm. The simulation results have shown the effectiveness of our approach by minimizing the functions to a very large extent. Our approach delivers a good solution in both the transitional and parallel techniques. As further research, we will like to explore how our hybrid model behaves with large datasets having large dimensions.

References

1. Kennedy, J., Eberhart, R.: Particle swarm optimization. In: Proceedings of the IEEE Congress on Evolutionary Computation, Australia, pp. 1942–1948 (1995)
2. Satapathy, S., Naik, A.: Complex Intell. Syst. **2**, 173 (2016). https://doi.org/10.1007/s40747-016-0022-8
3. Tang, K.S., et al.: Genetic algorithms and their applications. IEEE Signal Process. Mag. **13**, 22–37 (1996)
4. Suganthan, P.N., et al.: Problem definitions and evaluation criteria for the CEC 2005 special session on real-parameter optimization. Tech. Rep. KanGAL #2005005, May 2005, Nanyang Technol. Univ., Singapore, IIT Kanpur, Kanpur, India (2005)

Remote Sensing-Based Crop Identification Using Deep Learning

E. Thangadeepiga and R. A. Alagu Raja

Abstract Deep learning (DL) is a prevailing modern technique for image processing together with remote sensing (RS) data. Remote sensing data is used to obtain the object information from long distance. Remote sensing technology has been provided satellite data that can help to identify and monitor the crops in agricultural applications. This project describes crop identification from multi-spectral satellite images using deep learning algorithm. The commonly used deep learning approach in remote sensing is Convolutional Neural Network (CNN)-based approach. To achieve more accuracy the CNN algorithm is used. Dataset for this study, i.e., different types of crop images are extracted from Worldview-2 satellite data, and also images are obtained from field data collection. The number of augmented satellite images contributed for this work is 300 and the field data collection contributes 2000 images. This dataset is divided into three parts as training data, validation data, and testing data. 80% of dataset is considered as training data, whereas the validation and testing each has 10% of dataset. This dataset includes the following crop images, i.e., rice, coconut, and jasmine. This model yields 78% accuracy for satellite image dataset and it provides 83% accuracy for field data collected.

Keywords Remote sensing (RS) · Deep learning (DL) · Convolutional neural networks (CNN)

1 Introduction

This paper provides the most basic information for crop management and agricultural development, which depends on the crop types and their area developed in a region. Remote sensing technology has been used for crop identification and area estimation

E. Thangadeepiga (✉) · R. A. Alagu Raja
Thiagarajar College of Engineering, Madurai 625015, Tamil Nadu, India
e-mail: deepiga.eswar@gmail.com

R. A. Alagu Raja
e-mail: alaguraja@tce.edu

S. C. Satapathy et al. (eds.), *Intelligent Data Engineering and Analytics*, Advances in Intelligent Systems and Computing 1177,
https://doi.org/10.1007/978-981-15-5679-1_11

for many decades, ranging from aerial photography to multi-spectral satellite imaging [1]. Satellite imagery was most generally used for crop identification than aerial photography or airborne imagery due to its synoptic overview and recurring large coverage [2–4].

Deep learning is a modern technique for image processing and data analysis with promising results and great potential [5–7]. To provide more accuracy, the Convolutional Neural Networks (CNNs) model has been used from Deep Learning (DL) [8]. Image data from major satellite sensors, including Sentinel-2A and Worldview-2 satellite data, were used for crop identification and area estimation.

In this paper, Worldview-2 sample data were used as well as field collection data used for an accuracy comparison. To establish the crop area classification, the successful model for crop identification is required. The result shows that the proposed method shows quite better performance for different crop identifications that are all rice, coconut, and jasmine.

2 Literature Study

A number of methods were used to classify the croplands, and have been proposed in [2–4, 9]. An earlier comparative study on the classification of croplands using satellite images has been classified as an ENVI tool which is very well established for the detection of change [2, 10]. An earlier comparative study on classification of croplands using satellite images were classified by an ENVI tool which is very adopt to found an change detection [2, 10]. After that, the most popular and efficient approaches for crop identification are ensemble based [5–7] and deep learning (DL). These techniques are found to outperform the SVM. DL is the powerful machine learning methodology for solving a wide range of task arising in image processing [11–20]. From that CNN is introduced as an improvised version of neural network. To identifying particular object in DL, CNN is an efficient approach.

3 Methodology

3.1 Overview of Proposed Method

This section summarizes the entire process of our work. A four-level architecture is proposed for crop identification from multi-spectral satellite image and field data collection. These levels are dataset collection for input images, data augmentation, Convolutional Neural Network (CNN) algorithm, and identified image (Fig. 1).

Each block is explained in detail below. Using camera the images of various crop areas with a wide range of image variations including lighting, shadow, etc. are taken from field visits to train the CNN. For field visit dataset, a total of 2000

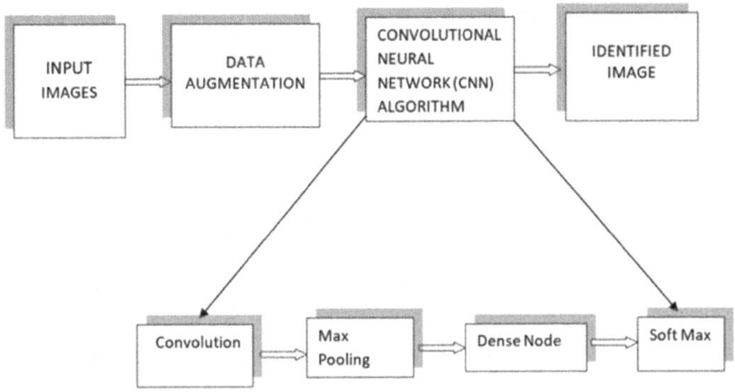

Fig. 1 Overall architecture

images were used (i.e., 1500 images with 4608 × 2128 pixel resolution for training and 300 images with 3456 × 3456 pixel resolution for validation and 200 images for testing). For the satellite images, a total of 300 images were used. The prepared training images set is fed into CNN to build CNN model for identifying the images in the validation set. When validating the CNN model through the validation image dataset, the CNN model takes the trained images and scans them to identify the test images that produce the crop-type identification report.

3.1.1 Input Images

In this project, dataset given as input, that is, the set of satellite data image samples and field collected images for the different crops such as rice, cotton, jasmine, and coconut. Field datasets are collected in various locations. Satellite data samples are collected from Worldview-2 data with the help of the earth explorer. The dataset was formed as a training dataset, validation dataset, and testing dataset. The number of augmented satellite images contributed for this work is 300 and the field data collection contributes around 2000 images. The separation of dataset has been given below (Fig. 2).

3.1.2 Data Augmentation

Image data augmentation is a method that can be used to synthetically increase the size of a training dataset by creating personalized versions of images in the dataset [21]. Training deep learning neural network models on additional data can result in more proficient models, and the augmentation techniques can create variations of the images that can progress the capability of the well models to take a broad view what they have cultured to fresh images. Data augmentation adds assessment to base data by adding information consequent from internal and external sources within a project.

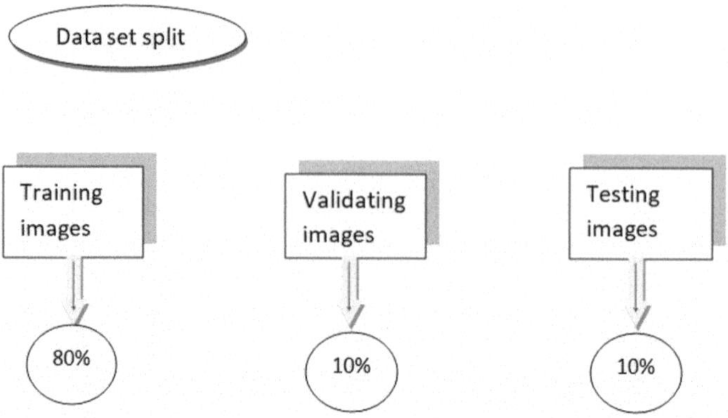

Fig. 2 Dataset split

In this paper, data augmentation methods such as rescaling, zooming, width shift, height shift, and fill mode as a reflection has been performed. Data augmentation can help reduce the manual intervention required to develop meaningful information and improve the dataset value.

3.1.3 Convolutional Neural Network (CNN) Algorithm

The general CNN architecture can be formed using multiple layers, such as input, convolution, pooling, activation, and output layers; convolution and pooling operations are conducted in the convolution and pooling layers.

A deep CNN is definite when the architecture is composed of many layers. Some other auxiliary layers, such as dropout and Batch Normalization (BN) layers, can be implemented within the aforementioned layers in accordance with the purposes of use [21–24].

AlexNet is used to perform this study. In this project, the architecture of the CNN consists of four convolutional layers such as convolution layer, max pooling, dense node, and softmax (Fig. 3).

Convolutional layer is a class of deep neural network which includes kernel size and strides. Max pooling is used to reduce the size of image. Pooling layers are used to reduce the amount of parameters and computations. It also controls the overfitting.

Dense node is a fully connected network, where single node is connected to every node in the next layer. Softmax is an activation function which is used to find probabilities for the testing data. It will help to map the non-normalized outputs of the network and predict their class.

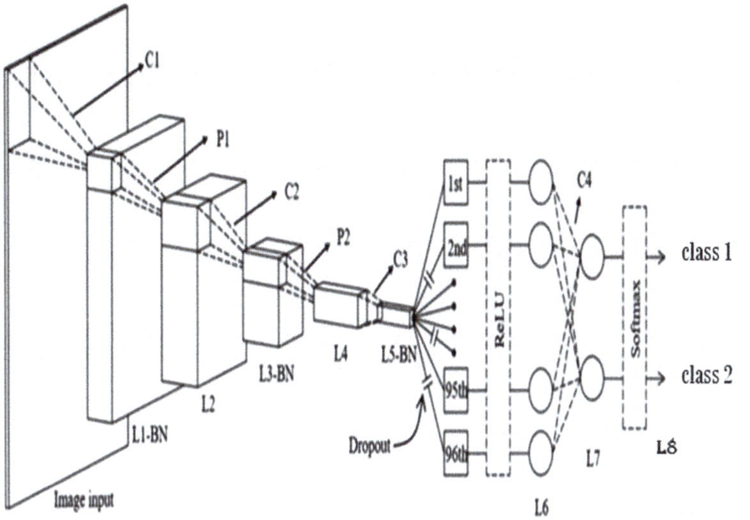

Fig. 3 Overall architecture of CNN: L#: layers corresponding to operations (L1, L3, L5, and L7: convolution layers; L2 and L4: pooling layers; L6: ReLU layer; L8: softmax layer); C#: convolution; P#: pooling; BN: batch normalization

3.2 Convolution

A convolution layer performs the following three operations throughout an input array as shown in Fig. 4. First, it performs element-by-element multiplications (i.e., dot product) between a subarray of an input array and a receptive field. The receptive field is also often called the filter or kernel. The initial weight values of a receptive field are typically randomly generated. Those of bias can be set in many ways in

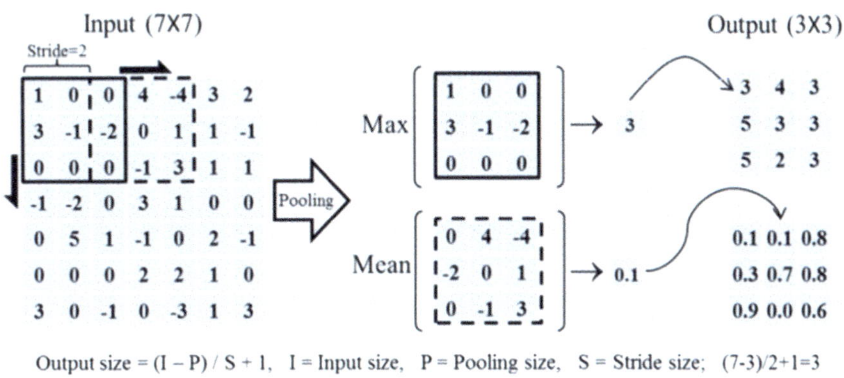

Fig. 4 Example for convolution

accordance with networks' configurations. The size of a subarray is always equal to a receptive field, but a receptive field is always smaller than the input array.

Second, the multiplied values are summed, and bias is added to the summed values. Figure 4 shows the convolution of the subarrays (solid and dashed windows) with an input array and a receptive field. One of the advantages of the convolution is that it reduces input data size, which reduces computational cost. An additional hyperparameter of the layer is the stride. The stride defines how many of the accessible field's columns and rows (pixels) slide at a time transversely the input array's width and height. A larger stride size leads to smaller amount accessible field applications and a smaller output size, which also reduces computational cost, though it may also lose features of the input data. The output size of a convolution layer is calculated by the equation shown in Fig. 4.

3.3 *Max pooling*

Another key aspect of the CNNs is a pooling layer, which reduces the spatial size of an input array. This process is often defined as downsampling. There are two different pooling options. Max pooling takes the max values from an input array's subarrays, whereas mean pooling takes the mean values. Figure 5 shows the pooling method with a stride of two, where the pooling layer output size is calculated by the equation in Fig. 5.

Owing to the stride size being larger than the convolution example in Fig. 5, the output size is further reduced to 3×3. The max pooling performance in image datasets is better than that of mean pooling. This project verified that the architecture with max pooling layers outperforms those with mean pooling layers. Thus, all the pooling layers for this study are max pooling layers.

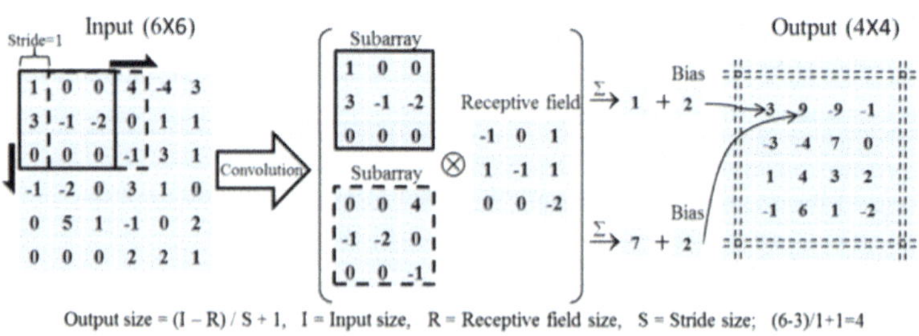

Fig. 5 Example for max pooling

3.4 *Dense Node*

Dense node is a fully connected network, where single node is connected to every node in the next layer. Fully connected layers connect every neuron in one layer to every neuron in another layer. It is in principle the same as the traditional Multi-layer Perceptron Neural Network (MLP). The flattened matrix goes through a fully connected layer to classify the images. Fully connected layers connect every neuron in one layer to every neuron in another layer. It is in principle the same as the conventional Multi-layer Perceptron Neural Network (MLP). The flattened matrix goes from first to last a fully connected layer to classify the images.

In neural networks, every neuron receives input from several numbers of locations in the previous layer. In a fully connected layer, each neuron receives input from every element of the previous layer. In a convolutional layer, neurons receive input from only a constrained subarea of the previous layer. Typically, the subarea is of a square shape (e.g., size 5 by 5). The input part of a neuron is called its receptive field. So, in a fully connected layer, the accessible field is the entire previous layer. In a convolutional layer, the receptive area is smaller than the complete previous layer.

3.5 *Softmax*

Softmax is an activation function which is used to find probabilities for the testing data [25]. It will help to map the non-normalized outputs of the network and predict their class. To classify input data, it is necessary to have a layer for predicting classes, which is usually located at the last layer of the CNN architecture. To calculate the amount of deviations between the predicted and actual classes, the softmax loss function is defined. The range of training is dependent on a mini batch size, which defines how many training samples out of the whole dataset are used. For example, if 100 images are given as the training dataset and 10 images are assigned as the mini batch size, this network updates weights 10 time, each complete update out of the whole data is called an epoch.

Whenever the epoch value increases the accuracy of a model will be increased. But the compilation time will be increased due to the increased epoch values. It will take more time to run the model. The accuracy of a model depends on dataset, epoch values, and the training data rate.

4 Results and Discussion

In this paper, the input for CNN algorithm comprises Worldview-2 satellite data and field collection data. The training and validation accuracy for field collected data is 83% which is represented in Figs. 6 and 7. We can infer training and the reliability of

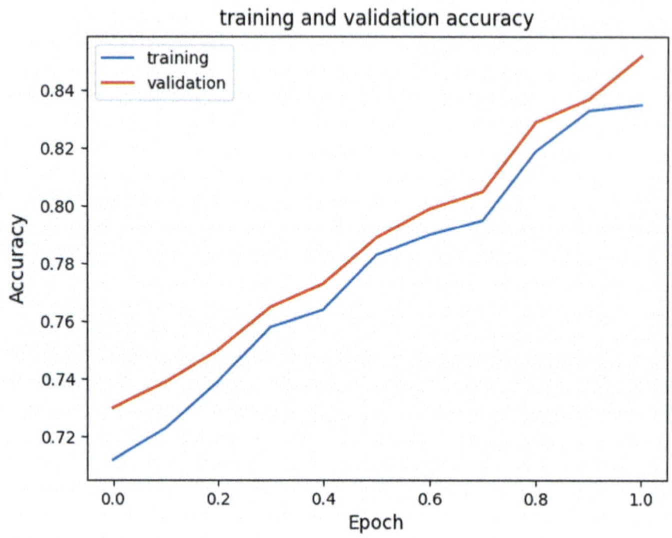

Fig. 6 Accuracy for field data collection

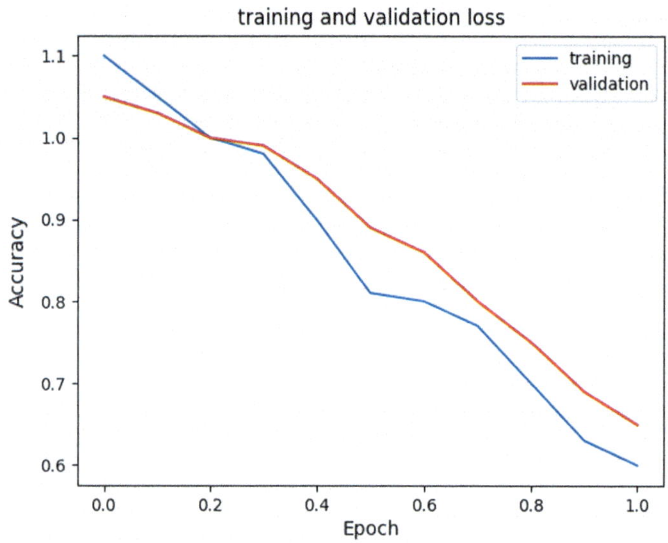

Fig. 7 Loss for field data collection

validation will be improved as the epoch values grow (Figs. 6 and 7). If any drop-off has been acquired, the validation would overfit during the training and validation. If any dropout has acquired the validation would overfit during the training and validation. The overfit should be avoided for an efficient model.

During the training process, the softmax layer assigns the probability to each image and fed into the validation images. From that the probability value of testing images was compared to trained image and identified by the softmax layer.

The tested image results are shown in Figs. 8, 9, and 10. From these figures, the resultant probability values indicate the classes. Out of the 200 images in the field collected testing dataset, 156 classes were correctly identified. Similarly, CNN model trains the satellite dataset for 300 images that include three classes, i.e., rice, jasmine, and coconut. It provides 75% accuracy for training and validation. The response of epoch will be shown in Figs. 11 and 12 and the tested images are shown in Figs. 13, 14, and 15.

The probability value of *softmax* has been used to identify the different classes in testing. Accuracy improvement depends on the training data rate, steps per epoch, and the dataset.

Fig. 8 Rice identification

Fig. 9 Jasmine identification

Fig. 10 Cotton identification

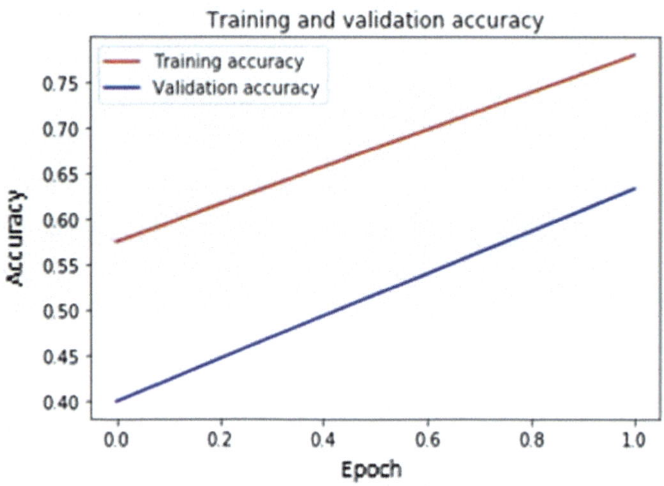

Fig. 11 Accuracy for satellite image dataset

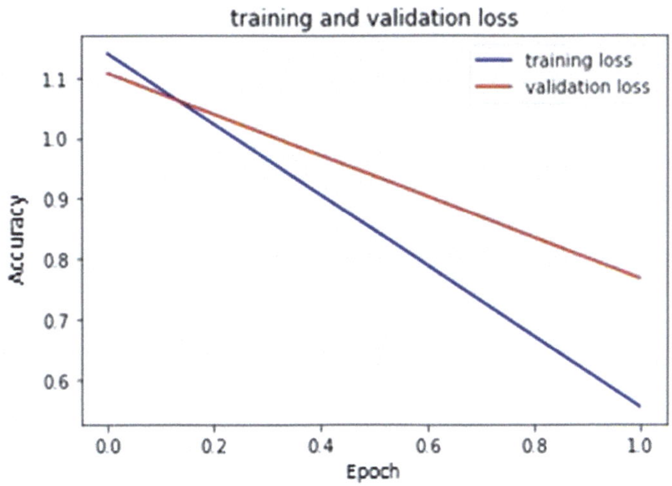

Fig. 12 Loss for satellite image dataset

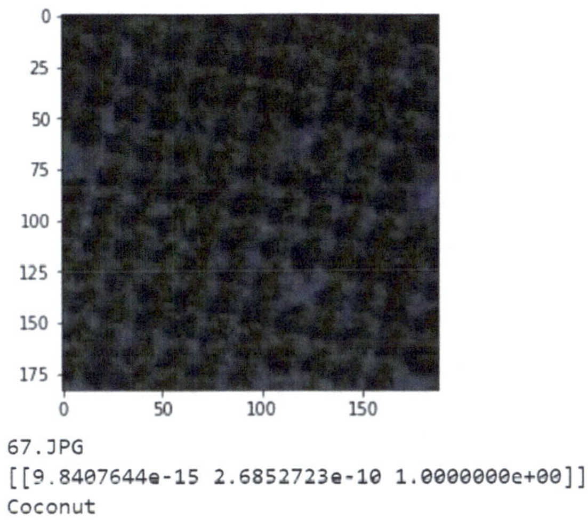

Fig. 13 Coconut identification

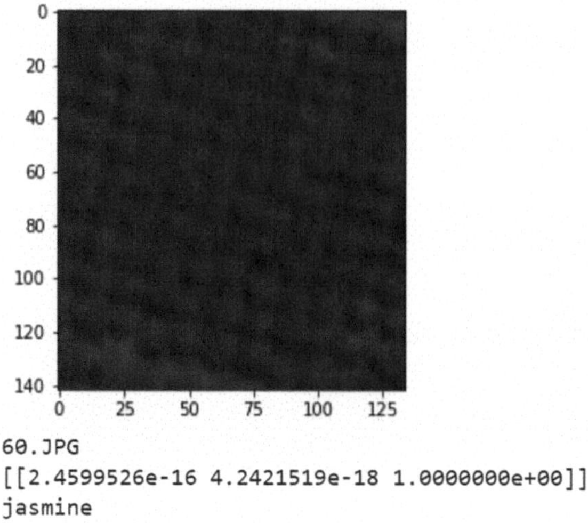

Fig. 14 Jasmine identification

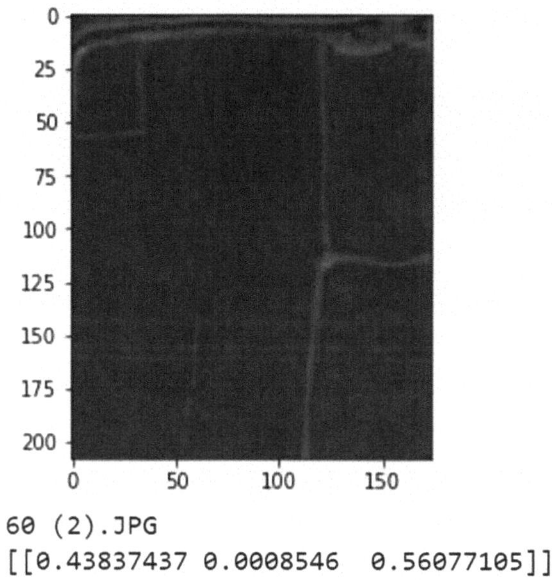

Fig. 15 Rice identification

5 Conclusion

In this project, the CNN model yields 78% validation accuracy and 63% testing accuracy for satellite image dataset. Also this model yields 83% validation accuracy and 75% testing accuracy for field collected image datasets. The accuracy can be increased by improving the factors like epoch values and training data rate. To achieve more accuracy, we can improve the size of the dataset by increasing the samples as well as different classes, which may lead to classification of the croplands.

References

1. Wójtowicz, M., Wójtowicz, A., Piekarczyk, J.: Application of remote sensing methods in agriculture. Int. J. Fac. Agric. Biol. (2016)
2. Li, Z., Long, Y., Tang, P., Tan, J., Li, Z.: Spatio-temporal changes in rice area at the northern limits of the rice cropping system in China from 1984 to 2013. J. Integr. Agric. (2017)
3. Liu, C., Chen, Z., Shao, Y., Chen, J., Tuya, H., Pan, H.: Research advances of SAR remote sensing for agriculture applications: a review. J. Integr. Agric. (2019)
4. Huang, Q., Zhang, L., Wu, W., Li, D.: MODIS-NDVI-based crop growth monitoring in China agriculture remote sensing monitoring system. In: Second IIT A International Conference on Geoscience and Remote Sensing (2010)
5. Kamilaris, A., Prenafeta-Boldú, F.X.: Deep learning in agriculture: a survey. Comput. Electron. Agric. (2018)
6. LeCun, Y., Bengio, Y., Hinton, G.: Deep Learning, 28 May 2015
7. Yalcin, H.: Phenology Recognition using Deep Learning. Visual Intelligence Laboratory, Istanbul Technical University. IEEE (2018)
8. Kussul, N., Lavreniuk, M., Skakun, S., Shelestov, A.: Deep learning classification of land cover and crop types using remote sensing data. IEEE Geosci. Remote Sens. Lett. **14**(5) (2017)
9. Zhang, J.: Multi-source remote sensing data fusion: status and trends. Int. J. Image Data Fus. (2010)
10. Qader, S.H., Dash, J., Atkinson, M., Rodriguez-Galiano, V.: Classification of vegetation type in Iraq using satellite-based phenological parameters. IEEE J. Sel. Top. Appl. Earth Observ. Remote Sens. (2016)
11. Atzberge, C.: Advances in Remote Sensing of Agriculture: Context Description, Existing Operational Monitoring Systems and Major Information Needs (2013)
12. Kussul, N., Lemoine, G.: Parcel-based crop classification in Ukraine using Landsat-8 data and Sentinel 1A data. IEEE J. Sel. Top. Appl. Earth Observ. Remote Sens.
13. Zhu, X., Zhu, W., Zhang, J., Pan, Y.: Mapping irrigated areas in china from remote sensing and statistical data. IEEE J. Sel. Top. Appl. Earth Observ. Remote Sens. **7**(11) (2014)
14. Huang, Y., Chen, Z., Yu, T., Huang, X., Gu, X.: Agricultural remote sensing big data: management and applications. J. Integr. Agric. (2018)
15. Shen, R., Huang, A., Li, B., Guo, J.: Construction of a drought monitoring model using deep learning based on multi-source remote sensing data (2019)
16. Han, M., Zhu, X., Yao, W.: Remote Sensing Image Classification Based on Neural Network Ensemble Algorithm. Elsevier (2011)
17. Kussul, N., Shelestov, A., Lavreniuk, M., Butko, I., Skakun, S.: Deep Learning Approach for Large Scale Land Cover Mapping Based on Remote Sensing Data Fusion. IEEE (2015)
18. Hu, Q., Wu, W., Song, Q., Yu, Q., Lu, M., Yang, P., Tang, H., Long, Y.: Extending the pairwise separability index for multicrop identification using time-series MODIS images. IEEE Trans. Geosci. Remote Sensi. **54**(11) (2016)

19. Jamali, S., Jönsson, P., Eklundha, L., Ardö, J., Seaquist, J.: Detecting changes in vegetation trends using time series segmentation. Remote Sens. Environ. **156** (2015)
20. Panda, S., Ames, D.P., Suranjanpanigrahi: Application of vegetation indices for agricultural crop yield prediction using neural network techniques. Remote Sens. (2010)
21. Ding, J., Chen, B., Liu, H., Huang, M.: Convolutional neural network with data augmentation for SAR target recognition. IEEE Geosci. Remote Sens. Lett. (2016)
22. Miao, F., Zheng, S., Tao, B.: Crop Weed Identification System Based on Convolutional Neural Network. IEEE (2019)
23. Huang, F.J., LeCun, Y.: Large-Scale Learning with SVM and Convolutional Nets for Generic Object Categorization (2011)
24. Zhou, Z., Li, S.: Peanut planting area change monitoring from remote sensing images based on deep learning. In: International Conference (2017)
25. Zhang, C., Woodland, P.C.: Parameterised sigmoid and ReLU hidden activation functions for DNN acoustic modelling. In: INTERSPEECH (2015)

Three-Level Hierarchical Classification Scheme: Its Application to Fractal Image Compression Technique

Utpal Nandi, Biswajit Laya, Anudyuti Ghorai, and Moirangthem Marjit Singh

Abstract Fractal-based image compression techniques are well known for its fast decoding process and resolution-independent decoded images. However, these types of techniques take more time to encode images. Domain classification strategy can greatly reduce encoding period. This paper proposed a new strategy of domain classification that groups domains in three-level hierarchical classes to speed up domain searching procedure. Then, the technique is further modified by sorting domains of each class based on frequency of matching. The results show that both the presented schemes significantly decrease the encoding duration of fractal coding and there are no effects on compression ratio and image quality.

Keywords Domain classification · Hierarchical classification · Compression ratio · Lossy compression · Loss-less compression · Encoding time

1 Introduction

Image compression [1] is special type data compression technique where digital image is encoded to reduce its size so that it takes less space in memory. Images can be compressed by either loss-less or lossy methods. In loss-less compression, image is encoded to reduce size and after decoding we get the actual image. In lossy

U. Nandi (✉) · B. Laya · A. Ghorai
Department of Computer Science, Vidyasagar University, Midnapore, West Bengal, India
e-mail: nandi.3utpal@gmail.com

B. Laya
e-mail: biswajitlaya007@gmail.com

A. Ghorai
e-mail: anudyuti@outlook.com

M. M. Singh
Department of Computer Science and Engineering, North Eastern Regional Institute of Science and Technology, Itanagar, Arunachal Pradesh, India
e-mail: marjitm@gmail.com

S. C. Satapathy et al. (eds.), *Intelligent Data Engineering and Analytics*, Advances in Intelligent Systems and Computing 1177,
https://doi.org/10.1007/978-981-15-5679-1_12

compression, image is encoded to highly reduce its size and after decoding we get nearly the actual image that is visually same with original image. Fractal image compression (FIC) [2] is a lossy method and discussed in Sect. 2. The technique is very efficient as its decoding time is low and decoded images are not depending on resolution. However, it takes more time to encode. Research works are continuing to minimize the encoding duration of the technique. Barnsley [3] initiated the idea of fractal coding for image that was automated completely by Jacquin [4, 5] depending on the idea that image similar parts can be used through self-mapping on group basis. The domain classification scheme is one of the important areas to speed up the FIC encoding. Xing et al. [6] presented a fractal-based coding using domain pools partitioned in hierarchical order. As a result, the encoding time is cut down efficiently. Then, Bhattacharya et al. [7] proposed a technique that also divided the domain pool in hierarchical order where each range is matched with similar class domains only. Jayamohon and Revathy [8, 9] proposed dynamic domain classification using B+ tree and also improved the same. Nandi and Mandal [10, 11] applied archetype classification in adaptive quad-tree-partitioning-based FIC and made some modifications to increase compression ratio. They [13, 14] also presented a classifier for adaptive quad-tree-decomposition-based FIC.

This paper proposed a new strategy of domain classification to speed up FIC encoding process (Sect. 3) that is further modified by sorting domains of each class based on frequency of matching. Section 4 analyzes the results and the conclusion is made in next section. Finally, the references are given.

2 Basic FIC Technique

The FIC technique is based on self-similarity concept of an image. In FIC, image is divided to form same size overlapping portions known as domains (DBs) as illustrated in Fig. 1. Domains are stored into domain pool. The image also partitioned into a number of non-overlapping portions called ranges (RBs). The DBs are at least twice larger than the RBs. The next step is to search the closest DB for each RB using affine transforms with operations like rotation, scaling, flip, etc. The similarity between RB and DB is measured by RMS distance. Then, affine transformation of the most similar range is stored into the new compressed file.

The encoding time of this full search scheme is very high. To reduce the searching time, classification scheme can be applied. One such scheme is proposed by Fisher [2] that classified domains based on average pixel values of four quadrants of image blocks into three major classes. These major classes are further classified based on ordering of variances of four quadrants into 24 classes. As a result, there are 72 classes and known as Fisher72 classification. The flowchart of the FIC encoding technique with quad-tree partitioning and Fisher72 classification (FICQP-Fisher72) is illustrated in Fig. 2.

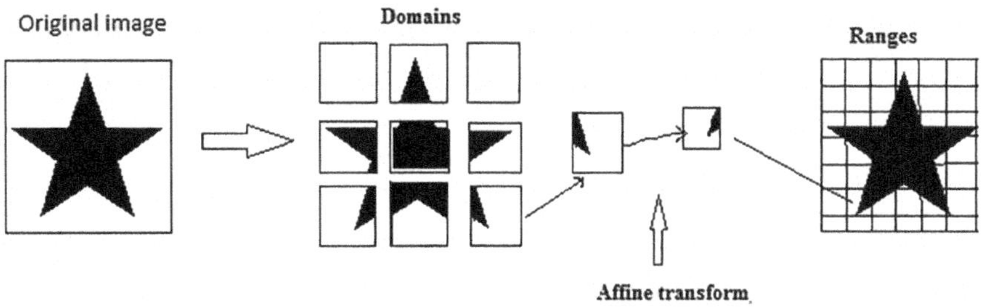

Fig. 1 The encoding process of FIC technique

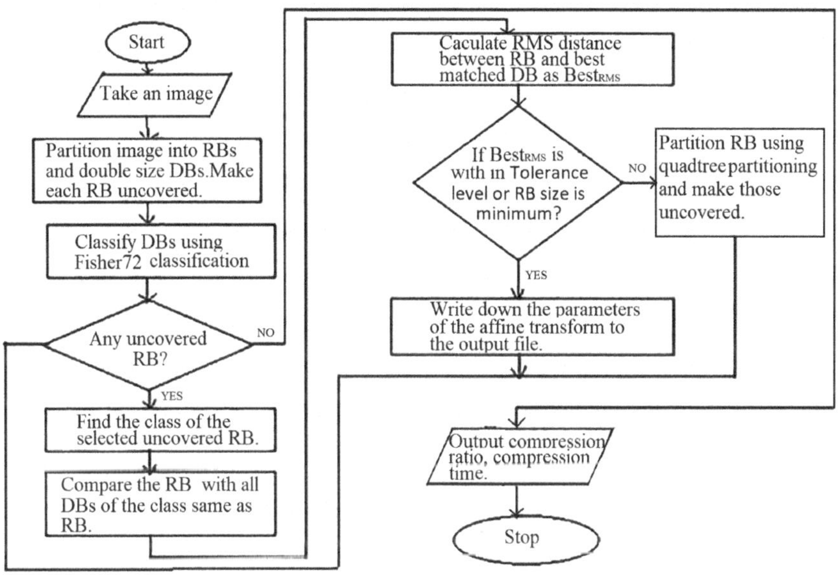

Fig. 2 The flowchart of the FIC encoding technique with quad-tree partitioning and Fisher72 classification (FICQP-Fisher72)

3 Proposed Three-Level Hierarchical Classification (3-LHC) Scheme

Take an image block, and then it is divided into four equal size quadrants as illustrated in Fig. 3. Now, the gray values of pixels of each quadrant i are added (S_i) for $0 \leq i \leq 3$ (Eq. 1) where r_{ji}, $0 \leq j \leq (n-1)$ are the pixel values of ith quadrant.

$$s_i = \sum_{j=0}^{n-1} r_j^i \quad (1)$$

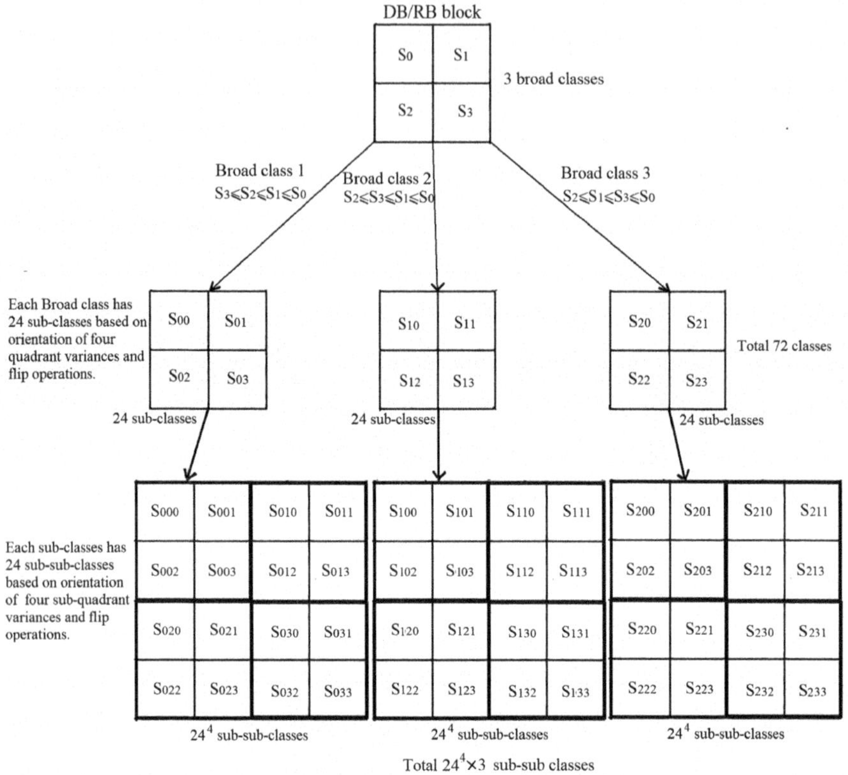

Fig. 3 The proposed three-level hierarchical classification (3-LHC) scheme

According to $S_i, 0 \leq i \leq 3$, we can classify any image block into three broad classes such as broad classes 1, 2, and 3 satisfying the conditions $S_3 \leq S_2 \leq S_1 \leq S_0$, $S_2 \leq S_3 \leq S_1 \leq S_0$, and $S_2 \leq S_1 \leq S_3 \leq S_0$, respectively. Then, the variance of each quadrant i (V_i) for $0 \leq i \leq 3$ is also obtained using Eq. 2.

$$V_i = \sum_{j=0}^{n-1} r_j^i - S_i^2 \tag{2}$$

Now, we can classify each of the broad classes depending on the orientation of variances V_i, $0 \leq i \leq 3$ and flip operation into (4! + 4!) = 24 sub-classes. There are three broad classes and each has 24 sub-classes, so there are total 243=72 sub-classes. Again, each quadrant is divided into four sub-quadrants $S_{i,j}$ for $0 \leq i \leq 3$, $0 \leq j \leq 3$ and the variance for each sub-quadrant is calculated. Depending on $V_{i,j}$ of four sub-quadrants of each quadrant, there are 24 orientations including flip operation and all four quadrants have 24^4 sub-sub-classes. As a result, each of three broad classes has 24^4 unique sub-sub-classes. Therefore, the total numbers of sub-sub-classes are 3×24^4. The FIC techniques may have different RB sizes. If the RB sizes are 2×2, 4×4 and 8×8, then the DB size should be 4×4, 8×8, and 16×16, respectively, as shown in Fig. 4. This proposed classification scheme is termed as three-level hierarchical classification (**3-LHC**) and the FIC with quad-tree partitioning and proposed three-level hierarchical classification is termed as **FICQP-3HC**.

3.1 Elementary Analysis

Consider a number of RBs and DBs in the pool are N_R and N_D, respectively. In Fisher72 classification scheme, there are 72 classes. Hence, the number of classes per DB is approximately $\frac{D_p}{72}$ in average and the number of comparisons between of DB and RB is $\frac{D_p}{72} N_R$.

In proposed classification scheme, total number of classes are 3×24^4 and number of DB and RB comparisons are $(\frac{D_p}{3 \times 24^4}) N_R$. Therefore, the number of DB and RB comparisons of the proposed scheme is exponentially reduced than Fisher72 that also reduces the encoding time of the FIC.

3.2 Proposed Modification

The proposed classifier is modified to decrease further the searching time of DB of a RB. This is done by using counters of each DB. Initially, countervalues of all DBs are assigned to zero. If a DB is selected for a RB, its counter is incremented by one. DBs of each class are sorted separately based on countervalues in descending order. This concept is implemented using max-heap tree. During searching best matching DB for a RB in a class, DBs are chosen for matching from root of the max-heap tree. The FIC quad-tree partitioning and proposed modified scheme is termed as **Modified FICQP-3HC**.

4 Results and Analysis

The experiments have been done using the standard grayscale images [15]. The compression time (in second) of proposed techniques FICQP-3LHC and modified FICQP-3LHC is compared with existing base method with Fisher's classification (FICQP-Fisher72) [2] and recent FIC with quad-tree partitioning and fast classification (FICQP-FCS) [12] and hierarchical classification (FICQP-HC) [7] techniques as given in Table 1. The average (Eq. 3) and standard deviation (Eq. 4) of the same are also calculated. The comparison of average compression time is plotted in Fig. 5a.

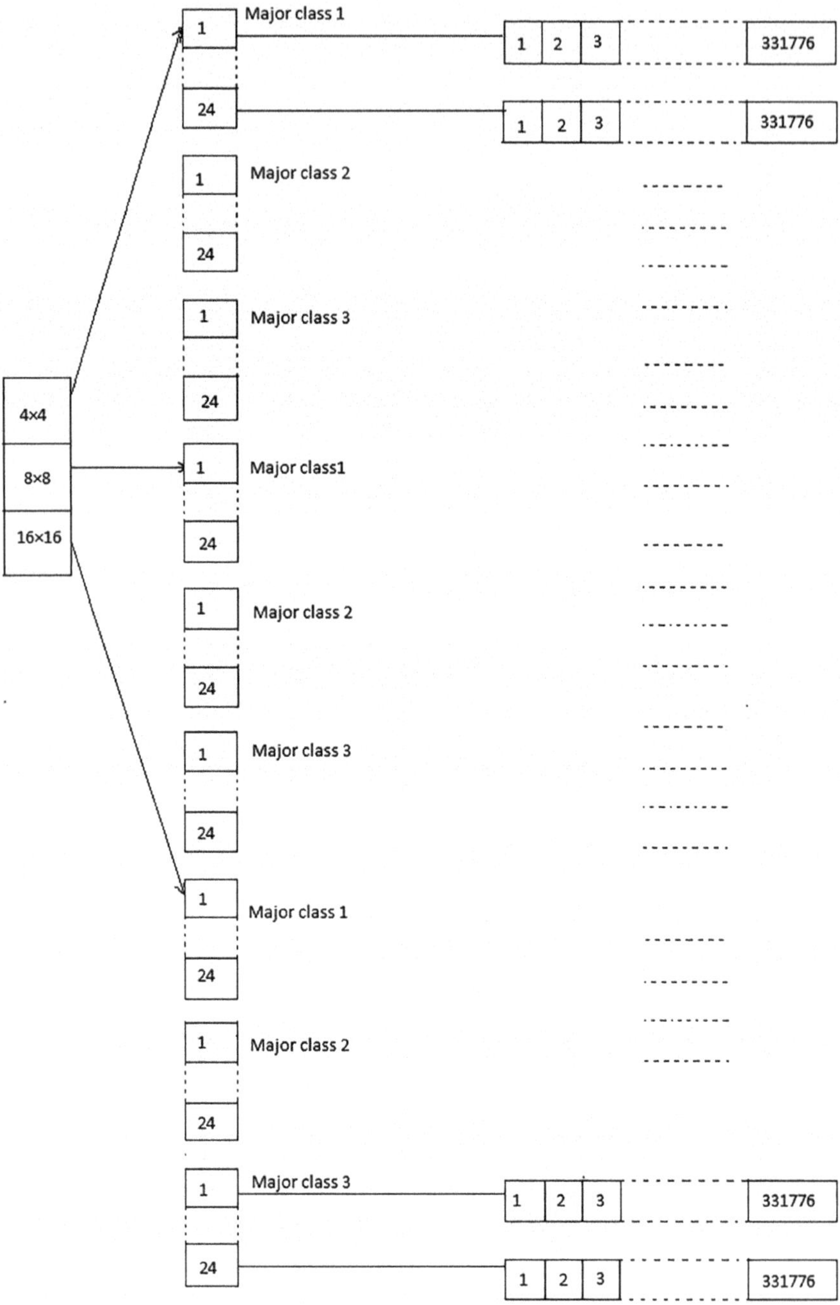

Fig. 4 The three-level hierarchical classification (3-LHC) for three different size domains

Table 1 Comparison of compression time (seconds)

Images	FICQP-Fisher72	FICQP-FCS	FICAQP-HC	Proposed FICQP-3LHC	Proposed modified FICQP-3LHC
Lena	2.27	1.16	1.37	1.09	1.03
Baboon	2.39	1.25	1.73	1.13	1.08
Cameraman	2.88	1.37	1.75	1.19	1.14
Peppers	2.39	1.18	1.08	1.10	1.07
Boats	2.06	1.29	1.16	1.11	1.10
Average	2.398	1.250	1.418	1.124	1.084
Standard deviation	0.301	0.085	0.313	0.040	0.040

The average compression time of FICQP-3LHC is significantly reduced than other techniques since it uses three-level hierarchical classification scheme. The proposed modified FICQP-3LHC further reduces the average compression time by sorting domains of each class based on frequency of selection of domains.

$$Average(\overline{x}) = \frac{1}{n}\sum_{i=0}^{n} x_i \tag{3}$$

$$Standard\ deviation = \sqrt{\frac{1}{n}\sum_{i=0}^{n} |(x_i - \overline{x})|} \tag{4}$$

The decoded image quality in terms of PSNRs (Eq. 5) and compression ratios (Eq. 6) of all the experimented methods are given in Tables 2 and 3, respectively. The graphical representations of the comparisons of average PSNR and compression ratio are depicted in Fig. 5b and c, respectively. It is observed that the PSNRs of both FICQP-3LHC and modified FICQP-3LHC are same with the base method FICQP-Fisher72. It is also noticed that compression ratios of the proposed techniques remain unaffected. Therefore, there is no change of image quality and compression ratio of the proposed techniques for speeding the compression process. These are equal with base method.

$$PSNR = 20\log_{10}^{(\frac{255}{RMS})} dB \tag{5}$$

$$Compression\ ratio = \frac{Size\ of\ file\ afther\ compression\ in\ bit}{File\ size\ in\ byte}\ bpp \tag{6}$$

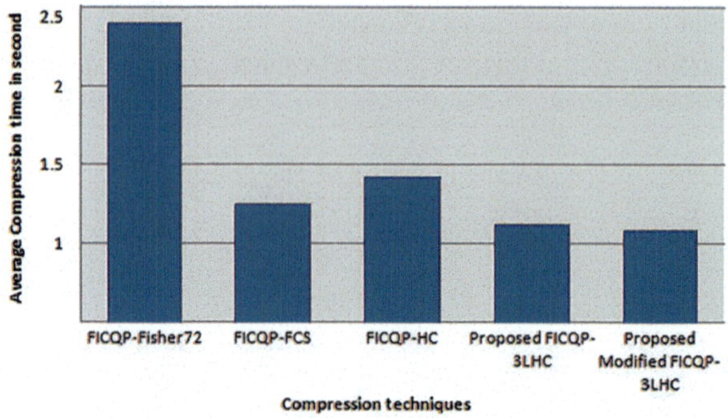

(a)

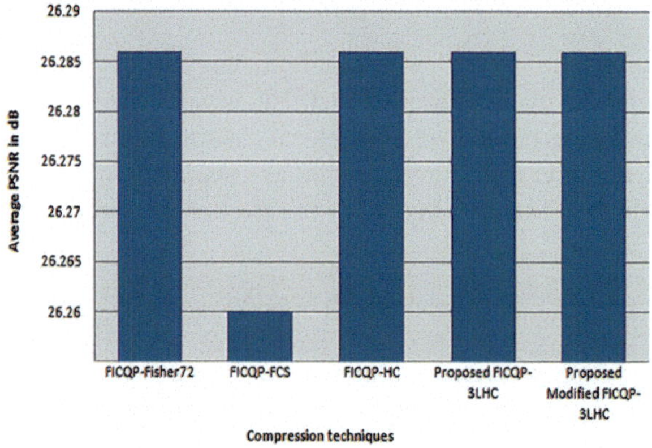

(b)

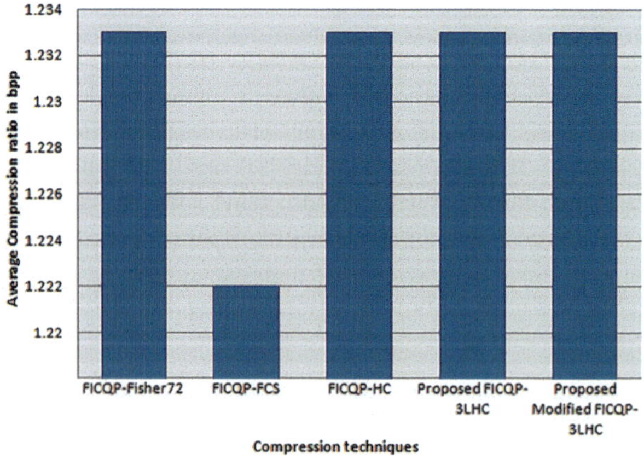

(c)

Fig. 5 The graphical representations of **a** comparison of encoding time (in second), **b** comparison of PSNR (in dB), **c** comparison of encoding time (in second) of FIC for different classification schemes

Table 2 Comparison of PSNRs of (in dB)

Images	FICQP-Fisher72	FICQP-FCS	FICQP-HC	Proposed FICQP-3LHC	Proposed modified FICQP-3LHC
Lena	22.90	28.86	28.90	28.90	28.90
Baboon	20.11	20.10	20.11	20.11	20.11
Cameraman	27.29	27.27	27.29	27.29	27.29
Peppers	29.83	29.81	29.83	29.83	29.83
Boats	25.30	25.26	25.30	25.30	25.30
Average	26.286	26.260	26.260	26.286	26.286
Standard deviation	3.857	3.851	3.857	3.857	3.857

Table 3 Comparison of compression ratio in bpp

Images	FICQP-Fisher72	FICQP-FCS	FICQP-HC	Proposed FICQP-3LHC	Proposed modified FICQP-3LHC
Lena	1.360	1.3620	1.360	1.360	1.360
Baboon	1.383	1.3660	1.383	1.383	1.383
Cameraman	1.278	1.2540	1.278	1.278	1.278
Peppers	1.026	1.0050	1.026	1.026	1.026
Boats	1.119	1.1220	1.119	1.119	1.119
Average	1.233	1.222	1.233	1.233	1.233
Standard deviation	0.155	0.157	0.155	0.155	0.155

5 Conclusion

This paper proposed a new DB/RB classifier for FIC technique to make FIC encoding more faster and also modify the same. The results yield that the FIC with both the proposed strategies greatly decrease the encoding duration. However, the ratio of compression and image quality remains unaffected. The technique has a lot of scope to improve and proposed classification scheme can be applied with other partitioning schemes also.

Acknowledgements This work is carried out by using infrastructure of the Dept. of Computer Sc., Vidyasagar University, Paschim Medinipur, West Bengal, India.

References

1. Nelson, M.: The Data Compression Book, 2nd edn. BPB Publications, India (2008)
2. Fisher, Y.: Fractal Image Compression: Theory and Application. Springer, New York (1995)
3. Barnsley, M.F.: Fractal Everywhere. Academic Press, New York (1993)
4. Jacquin, A.E.: Image coding based on a fractal theory of iterated contractive image transformations. IEEE Trans. Image Process. **1**, 18–30 (1992)
5. Jacquin, A.E.: Fractal image coding: a review. Proc. IEEE **81**(10), 1451–14654 (1993)
6. Xing, C., Ren, Y., Li, X.: A hierarchical classification matching scheme for fractal image compression. In: IEEE Congress on Image and Signal Processing (CISP08), Sanya, vol. 1, pp. 283–286. Hainan, China (2008)
7. Bhattacharya, N., Roy, S. K., Nandi, U., Banerjee, S.: Fractal image compression using hierarchical classification of sub-images. In: Proceedings of the 10th International Conference on Computer Vision Theory and Applications (VISAPP-15), pp. 46–53. Berlin, Germany (2015)
8. Jayamohan, M., Revathy, K.: Domain classification using B+ trees in fractal image compression. In: IEEE National Conference on Computing and Communication Systems (NCCCS), p. 15. Durgapur, India (2012)
9. Jayamohan, M., Revathy, K.: An improved domain classification scheme based on local fractal dimension. Indian J. Comput. Sci. Eng. (IJCSE) **3**(1), 138–145 (2012)
10. Nandi, U., Mandal, J. K.: Fractal image compression with adaptive quad-tree partitioning and archetype classification. In: IEEE International Conference on Research in Computational Intelligence and Communication Networks (ICRCICN) 2015, pp. 56–60. Kolkata, West Bengal, India (2015)
11. Nandi, U., Mandal, J.K.: Efficiency of adaptive fractal image compression with archetype classification and its modifications. Int. J. Comput. Appl. (IJCA) **38**(2–3), 156–163 (2016)
12. Nandi, U., Mandal, J.K., Santra, S., Nandi, S.: Fractal image compression with quadtree partitioning and a new fast classification strategy. In: 3rd International Conference on Computer Communication, Control and Information Technology (C3IT-2015), pp. 1–4. Hooghly, West Bengal, India (2015)
13. Nandi, U., Mandal, J. K.: A novel hierarchical classification scheme for adaptive quadtree partitioning based fractal image coding. In: 52nd Annual Convention of Computer Society of India (CSI 2017), pp. 19–21. Science City, Kolkata, West Bengal, India (2018)
14. Nandi, U.: An adaptive fractal-based image coding with hierarchical classification strategy and its modifications. Innov. Syst. Soft. Eng. **15**(1), 35–42 (2019)
15. https://doi.org/10.1007/s11334-019-00327-5

Prediction of POS Tagging for Unknown Words for Specific Hindi and Marathi Language

Kirti Chiplunkar, Meghna Kharche, Tejaswini Chaudhari, Saurabh Shaligram, and Suresh Limkar

Abstract Part of Speech (POS) tagging for Indian languages like Hindi and Marathi is generally not an investigated territory. Some of the best taggers accessible for Indian dialects utilize crossbreeds of machine learning or stochastic techniques and phonetic information. Available corpuses for Hindi and Marathi are limited. Hence, when Natural Language Processing (NLP) is applied to Hindi and Marathi sentences, desired results are not achieved. Current POS tagging techniques give UNKNOWN (UNK) POS tag for words which are not present in the corpus. This paper proposes how Hidden Markov Model (HMM)-based approach for POS tagging can be extended using Naïve Bayes theorem for prediction of UNK POS tag.

Keywords Part of speech tagging · Corpus · NLTK models · Machine learning · Viterbi algorithm · POS tag dataset · NLP for Hindi and Marathi · UNK POS tag · UNKNOWN POS tag

K. Chiplunkar (✉) · M. Kharche · T. Chaudhari · S. Limkar
Department of Computer Engineering, AISSMS Institute of Information Technology, Pune, Maharashtra, India
e-mail: chiplunkar.k.4498@gmail.com

M. Kharche
e-mail: meghnakharche1@gmail.com

T. Chaudhari
e-mail: tejaswinichaudhari29@gmail.com

S. Limkar
e-mail: sureshlimkar@gmail.com

S. Shaligram
Makers Lab, Tech Mahindra, Pune, Maharashtra, India
e-mail: saurabh.shaligram@hotmail.com

S. C. Satapathy et al. (eds.), *Intelligent Data Engineering and Analytics*, Advances in Intelligent Systems and Computing 1177,
https://doi.org/10.1007/978-981-15-5679-1_13

1 Introduction

Part of speech tagging [1–3] is a bit of programming that peruses a message in a language and extracts grammatical features out of each word, for example, thing, action word, descriptor, and so on. POS tagger forms grouping of words and joins a grammatical feature tag to each word. POS tagging is done keeping in mind the relationship of a word with its adjacent and related words in a phrase, sentence, or paragraph.

The prediction of unknown words for Hindi and Marathi language is mostly similar because of its structure of sentence, grammar, etc. They are comparative in the light of the fact that their root language is same (Sanskrit). Both Hindi and Marathi are written using Devanagari script and considered as morphologically rich languages.

1.1 POS Tagging

The knowledge of part of speech plays a vital role in NLP because it tells us how a word is used in every sentence. POS tagging is the pre-requisite for natural language processing operations like chunking, lemmatization, and building of parse trees for Named Entity Recognition (NER).

In our system, probabilistic approach is used for prediction of unknown word in Hindi and Marathi corpus. There are eight main parts of speech, viz., noun, pronoun, adjective, verb, adverb, preposition, conjunction, interjection most of it is further divided into subparts like noun is divided into proper noun, common noun, etc.

1.2 Limitations of Current POS Tagging System

In POS tagging system, the correctly tagged words are already present in the corpus (Indian corpus) which comes under Natural Language Tool Kit (NLTK) library [4]. Limitation of this system is that if the word is not present in the corpus then it is tagged with unknown "UNK" tag. Hence, the accuracy of the system degrades with increase in number of unknown words.

2 Related Work

Parts of speech tagging has drawn a lot of research interest especially for regional languages. Some of the related work is discussed below:

Deshpande and Gore [1] proposed a part of speech tagger for Marathi sentences based on hybrid methodology using vast-rule base, Marathi dictionary. However, tagger gives ambiguous output while handling derivational morphology. The 84% of accuracy was achieved by this hybrid tagger.

Mishra and Mishra [2] have developed a POS tagger for Hindi corpus. As the structure of Hindi and English languages is different, the POS tagger for English is not applicable for Hindi languages. So, that's why this system was developed. However, it needs more analysis and research work for improving accuracy.

Narayan et al. [3] proposed methodology based on artificial neural network approach for solving the problems of POS tagging. The accuracy of given methodology can be improved by various techniques that handle unknown words using ANN.

Sharma and Lehal [5] proposed a module based on hidden Markov model. A Panjabi POS tagger was developed where the bi-gram methodology is used along with hidden Markov model. This POS tagger faced difficulty in resolving the ambiguity of complex sentences and manually 90.11% of accuracy was achieved.

Singh et al. [6] used statistical method for development related to POS tagger and compared their results. Morphology complexity of Marathi is little hard. 77.38, 90.30, 91.46, and 93.82% accuracies were achieved by this proposed model.

Tian et al. [7] characterized a POS tagger based on HMM and trigram tags for Uyhur text. The proposed system is used for smoothing and parsing the data. The proposed approach provided efficient accuracy over current models.

Yuan [8] proposed a Markov family model which develops the probability of given word which depends on its own tag and previous word. Markov family model gave more accuracy over conventional HMM.

Bokaei et al. [9] solved the issues which occurred in some languages where a word can have several tokens by using HMM model. These several tokens consist of empty spaces in between them. Due to these empty spaces, user needs to specify some limitations explicitly. And this is the major drawback of this methodology. Proposed methodology had built-in tokenizer.

Ray et al. [10] characterized local word group for different regular expressions. Every language has some constraints to overcome these constraints this methodology was proposed. While grouping the problems occurred were resolved and efficient performance was achieved.

Modi et al. [11] proposed a system which yields high accuracy using a limited corpus. While analyzing it, different sub-tasks exist. Accuracy is based on rules and is corpus based. Accuracy of 91.84% is achieved by this approach.

Patil et al. [12] characterized a POS tagger specific for Marathi language. Although corpus size is relatively small compared to others, it worked efficiently compared to other taggers. Testing is performed on three datasets and time required to perform testing is increased and an accuracy of 78.82 is achieved.

Joshi and Mathur [13] proposed methodology for Hindi, with the help of HMM model. With the help of available information, proper combination of POS tag was achieved. Proposed approach achieved an accuracy of 92.13% using HMM model.

The abovementioned approaches focus POS tagging using different techniques such as Hidden Markov Model (HMM), Support Vector Machines (SVM), Artificial Neural Networks (ANN), and hybrid, rule-based, heuristic-based. Some of them focus on increasing the accuracy of existing system while others proposed entirely new approaches for POS tagging for a particular language. The problem with most of the abovementioned approaches is that they require labeled data to train their models. Also, these approaches cannot POS tag new words for which labeled data is not available, i.e., words unknown to the training corpus. In contrast, we propose an approach for prediction of POS tagging of unknown words using Naïve Bayes algorithm. Our approach is suitable for languages having limited pre-tagged data for training.

Rest of the paper is organized as follows: Sect. 3 gives our contribution, Sect. 4 discloses our proposed methodology, results and analysis is discussed in Sect. 5, Sect. 6 is the conclusion, and Sect. 7 discusses about future work.

3 Contributions

The optimal goal of the paper is to predict the POS tag of a word unknown to the trained model. This is accomplished by applying Naïve Bayes algorithm and predicting the most likely tag for the unknown word. Our technical contribution can be summarized as follows:

Presented literature survey of related work which describes the model used along with the accuracy achieved.

Presented a table containing all the parts of speech tags for NLTK's Hindi and Marathi corpus along with their meaning.

Proposed a fairly simple but effective approach to predict the POS tag of an unknown word using Naïve Bayes algorithm which is highly suited for POS tagging of languages that have a very limited training corpus.

4 Proposed Methodology

Following diagram depicts the working of our model. Raw data undergoes pre-processing. Data is split into two parts, namely, training and testing data. Training data is used to train the model and test data is used for model evaluation (Fig. 1).

4.1 Hidden Markov Model

HMM [5–7] is a probabilistic model. HMM assumes the system under consideration to be composed of unobserved/hidden states, i.e., a Markov process. It helps programs come to the most likely decision based on both the previous decision and current data. HMM is combination of tag sequence probability and word frequency measurement. HMM is widely used for two main purposes: first is assignment of proper labels to

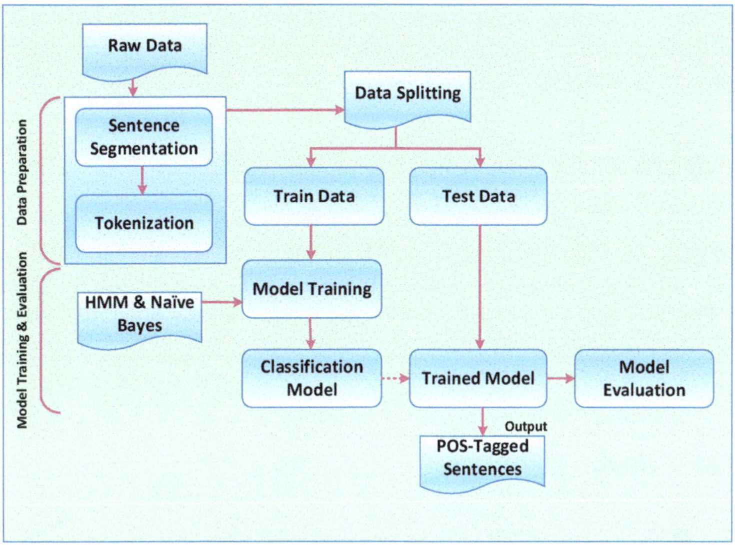

Fig. 1 Proposed model's training diagram

sequential data and second is to estimate the probability of a data sequence or label. In HMM, the observation is probabilistic function of state [14].

4.2 *Viterbi Algorithm*

Viterbi algorithm [5] is a dynamic programming algorithm. Viterbi algorithm is used to find the most probable finite sequence of hidden states also called as the Viterbi path. In our system, this finite sequence is nothing but assignment of proper parts of speech tags to the input sentence. The words are observation and the tags are states. The inputs to the Viterbi algorithm are as follows:

- A set of state (tags),
- A set of observation (words),
- Start probability of all states,
- Transition matrix, and
- Emission matrix.

The output of the Viterbi algorithm is most likely sequence of tags for the given input.

Viterbi algorithm-based model requires a training corpus. Here Hindi-tagged dataset present in Indian Corpus of NLTK is used. The model then trains on the training corpus for which frequencies, start probabilities, and transition and emission matrices are created. Formulae for calculation of start, transition and emission probability are as follows:

- Start probability of a tag $= \frac{\text{Frequency of tag}}{\text{Total Number of words}}$
- Transition probability of tag t_1 to $t_2 = \frac{\text{Frequency of transition from state } t_1 \text{ to state } t_2}{\text{Frequency of } t_1}$
- Emission probability $= \frac{\text{Frequency of } w \text{ tagged as } t}{\text{Frequency of } t}$

where t_1, t_2, t are tags and w is a word. Above calculations of start probability, transition and emission probability show the pre-processing step for Viterbi algorithm. Viterbi algorithm uses above probability values and finds the most likely sequence of tags for the given input.

4.3 *Naïve Bayes Algorithm*

Prediction of unknown (UNK) tag can be done using Naïve Bayes theorem. Naïve Bayes classifier has shown better performance in comparison with other models like logistic regression assuming that features are independent of one another. It is easy to implement, fast, and requires less training data.

Bayes' theorem in probability is stated mathematically as the following equation:

$$P\,(b \mid A) = (P\,(A \mid b)\,P(b)) \,/\, (P(A)) \tag{1}$$

By Naïve Bayes formula, prediction of unknown word tag is computed on the basis of transition and start probability. For predicting the unknown word's tag, mathematical formulae are as follows:

$$\begin{aligned} b(\text{MAP}) &= \arg\max_b [P(b \mid A)] \\ &= \arg\max_b [(P\,(a \mid b)\,P(b)) \,/\, (P(a))] \\ &= \arg\max_b [P\,(A \mid b)\,P(b)] \\ &= \arg\max_b [P\,(a_1, a_2, \ldots a_n | y) P(b)] \end{aligned} \tag{2}$$

where $A = (a_1, a_2, \ldots, a_n)$ are tags and b is unknown word's tag.

MAP is maximum of posterior which equals to most likely tag.

$$b = \arg\max_b P\left(b \prod_{i=1}^{n} P(a_i|b)\right) \tag{3}$$

By comparing the probability, the highest probability for appropriate word is considered.

Table 1 POS tag in Hindi and Marathi

SR. No.	Tags	Meaning	SR.No.	Tags	Meaning
1	NN	Noun singular	18	QFNUM	Quantifier number
2	JJ	Adjective	19	RP	Particle
3	VFM	Verb finite main	20	NEG	Negative word
4	SYM	Symbol	21	QF	Quantifier
5	NNP	Proper noun	22	JVB	Adjective in kriyamula
6	NNC	Common noun	23	NLOC	Noun location
7	INTF	Intensifier	24	VJJ	Verb non-finite adjective
8	CC	Conjunction	25	QW	Question word
9	PREP	Preposition	26	VM	Main verb
10	PRP	Pronoun	27	JJC	Adjective comparative
11	NVB	Verb past participle	28	PSP	Post position
12	VAUX	Auxiliary verb	29	NST	Spatial noun
13	PUNC	Punctuation	30	QC	Cardinal
14	NNPC	Compound proper noun	31	DEM	Demonstrative
15	VRB	Verb	32	WQ	Question word
16	VNN	Non-finite nominal	33	QO	Ordinal
17	RB	Adverb	34	RDP	Reduplication
			35	UNK	Unknown word

The following table shows the tags and their meanings present in NLTK's Indian corpus for Marathi and Hindi [14] (Table 1).

A pre-tagged dataset for Hindi present in the NLTK's Indian Corpora is used for actual training purposes. This Hindi corpus contains around 9500 words. To describe the working of our system, consider the following example:

Consider the following training corpus:

सुरज_NN उगता VB है_VAUX |_PUNC

मेरा_PRP नाम_NN सुरज_NN है_VAUX |_PUNC

हम_PRP चलते_VB है_VAUX |_PUNC

Consider the following test sentence:

सुरज सुबह उगता है ।

In the above test sentence, the word सुबह is unknown word. First sentence is passed to the Viterbi algorithm for POS tagging. Whenever an unknown word is detected it is immediately predicted using Naive Bayes. For above example, Naive Bayes assigns the “NN” tag to the unknown word सुबह. The results of Naive Bayes are further used for POS tagging the word succeeding the unknown word.

Based on the above training corpus the start probability matrix, transition matrix, and emission matrix are constructed as follows (Tables 2, 3, and 4).

Following transition diagram is drawn for better understanding (Fig. 2).

Final output is as follows:

सुरज_NN सुबह_NN उगता_VB है_VAUX ।_PUNC

POS tagging for a Marathi sentence is done in the same way as mentioned above.

Table 2 Start probability matrix

	NN	VB	PRP	VAUX	PUNC
Start probability	0.23	0.153	0.153	0.23	0.23

Table 3 Transition matrix

	NN	VB	PRP	VAUX	PUNC
NN	0.33	0.33	0	0.33	0
VB	0	0	0	1	0
PRP	0.5	0.5	0	0	0
VAUX	0	0	0	0	1
PUNC	0	0	0	0	0

Table 4 Emission matrix

	NN	VB	PRP	VAUX	PUNC
सुरज	0.667	0	0	0	0
उगता	0	0.5	0	0	0
है	0	0	0	1	0
।	0	0	0	0	1
मेरा	0	0	0.5	0	0
नाम	0.33	0	0	0	0
हम	0	0	0.5	0	0
चलते	0	0.5	0	0	0

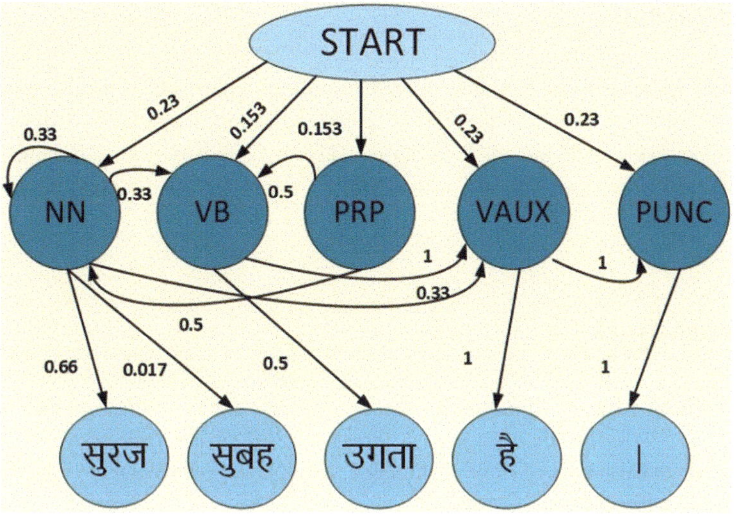

Fig. 2 Transition diagram

5 Result and Analysis

Results and analysis of our proposed system are as follows.

Accuracy of test sentences having one or two unknown words is about 90%. However, as the number of unknown words increases, the accuracy decreases up to 50%. The accuracy of our system is highly dependent on the prediction of Naïve Bayes algorithm.

Consider two cases:

Case 1: Unknown word prediction by Naïve Bayes is precise. In this case, the Viterbi algorithm gives accurate POS tags for the known words in the sentence.

Case 2: Unknown word prediction by Naïve Bayes is imprecise. In this case, the Viterbi algorithm gives incorrect POS tags to the known words succeeding the unknown word in the sentence.

It is observed that prediction of an unknown tag from previous word's tag and next word's tag gives incorrect output. But prediction of the unknown word's tag considering only the previous word's tag resulted in correct outcomes.

In general, Hindi and Marathi have morphological richness and hence very large vocabularies. This leads to the data sparseness problem. Data sparsity is usually fatal because it means that we are missing information that might be important. Large vocabularies make it impossible to have enough data to actually observe examples of all the things that people can say. There will be many phrases that won't be seen in the training data. Data sparsity can be improved by applying smoothing techniques.

6 Conclusion

In this study, the proposed system presents a unique approach to handle unknown words in a sentence. For POS tagging of Hindi and Marathi sentences, Viterbi algorithm and Naïve Bayes are used. Accuracy of the system is highly dependent on prediction of Naïve Bayes. In the further studies, size of corpus will be increased for better results.

7 Future Work

We are looking forward to manually create training data to predict unknown word POS tags for words if they appear as the first word of the sentence and predict consecutive unknown words in the sentence. Smoothing techniques can be applied to overcome data sparseness problem.

Acknowledgements We are thankful to Nikhil Malhotra, Jugnu Manhas, and Saket Apte of Maker's Lab, Tech Mahindra and Varsha Patil of AISSMS IOIT, Pune for support and help in this paper.

References

1. Deshpande, M.M., Gore, S.D.: A hybrid part-of-speech tagger for Marathi sentences. In: 2018 International Conference on Communication information and Computing Technology (ICCICT), Mumbai, pp. 1–10 (2018). https://doi.org/10.1109/iccict.2018.8325898
2. Mishra, N., Mishra, A.: Part of speech tagging for Hindi Corpus. In: 2011 International Conference on Communication Systems and Network Technologies, Katra, Jammu, pp. 554–558 (2011). https://doi.org/10.1109/csnt.2011.11
3. Narayan, R., Chakraverty, S., Singh, V.P.: Neural network based parts of speech tagger for Hindi. In: IFAC Proceedings Volumes, vol. 47, no. 1, pp. 519–524 (2014)
4. http://nltk.org/book
5. Sharma, S.K., Lehal, G.S.: Using Hidden Markov Model to improve the accuracy of Punjabi POS tagger. In: 2011 IEEE International Conference on Computer Science and Automation Engineering, Shanghai, pp. 697–701 (2011). https://doi.org/10.1109/csae.2011.5952600
6. Singh, J., Joshi, N., Mathur, I.: Development of Marathi part of speech tagger using statistical approach. In: 2013 International Conference on Advances in Computing, Communications and Informatics (ICACCI), Mysore, pp. 1554–1559 (2013). https://doi.org/10.1109/icacci.2013.66374114
7. Tian, S., Ibrahim, T., Umal, H., Yu, L.: Statistical Uyhur POS tagging with TAG predictor for unknown words. In: 2009 ISECS International Colloquium on Computing, Communication, Control, and Management, Sanya, pp. 60–62 (2009). https://doi.org/10.1109/CCCM.2009.5267823
8. Yuan, L.: Improvement for the automatic part-of-speech tagging based on hidden Markov model. In: 2010 2nd International Conference on Signal Processing Systems, Dalian, pp. V1-744–V1-747 (2010). https://doi.org/10.1109/icsps.2010.5555259

9. Bokaei, M.H., Sameti, H., Bahrani, M., Babaali, B.: Segmental HMM-based part-of-speech tagger. In: 2010 International Conference on Audio, Language and Image Processing, Shanghai, pp. 52–56 (2010). https://doi.org/10.1109/icalip.2010.5685018
10. Ray, P.R., Sudeshna, H.V., Basu, S.A.: Part of Speech Tagging and Local Word Grouping Techniques for Natural Language Parsing in Hindi. This research is funded in part by Media Lab Asia, under the auspices of the Communication Empowerment Laboratory, IIT Kharagpur (2008). oai:CiteSeerX.psu:10.1.1.114.3943
11. Modi, D., Nain, N.: Part-of-speech tagging of Hindi Corpus using rule-based method. In: Afzalpulkar, N., et al. (eds.) Proceedings of the International Conference on Recent Cognizance in Wireless Communication & Image Processing. ©Springer, India (2016). https://doi.org/10.1007/978-81-322-2638-3_28
12. Patil, H.B., Patil, A.S., Pawar, B.V.: Article: part-of-speech tagger for Marathi language using limited training corpora. In: IJCA Proceedings on National Conference on Recent Advances in Information Technology NCRAIT, no. 4, pp. 33–37 (2014)
13. Joshi, N., Mathur, I.: HMM based POS tagger for Hindi. In: Zizka, J. (ed.) CCSIT, SIPP, AISC, PDCTA-2013, pp. 341–349. ©CS & IT CSCP (2013). https://doi.org/10.5121/csit.2013.3639
14. Ekbal, A., Hasanuzzaman, Md., Bandyopadhyay, S.: Voted approach for part of speech tagging in Bengali. In: 23rd Pacific Asia Conference on Language, Information and Computation, pp. 120–129

Modified Multi-cohort Intelligence Algorithm with Panoptic Learning for Unconstrained Problems

Apoorva Shastri, Aniket Nargundkar, and Anand J. Kulkarni

Abstract In this paper, we present a new optimization algorithm referred to as Modified Multi-cohort Intelligence with Panoptic Learning (Multi-CI-PL). This proposed algorithm is a modified version of Multi-cohort Intelligence (Multi-CI), where Panoptic Learning (PL) is incorporated into Multi-CI which makes every cohort candidate learn the most from the best candidate but at same time partially learn from other candidates. A variety of well-known set of unconstrained test problems have been successfully solved by using the proposed algorithm and compared with several other evolutionary algorithms. The Multi-CI-PL approach has resulted in competent and sufficiently robust results. The associated strengths, weaknesses, and possible real-world extensions are also discussed.

Keywords Multi-cohort intelligence · Unconstrained optimization · Panoptic learning

1 Introduction and Literature Review

Several nature/bio-inspired metaheuristic techniques have been proposed so far. Fundamentally, these are the Swarm Intelligence (SI) methods and Evolutionary Algorithms (EAs). Some of the swarm intelligence methods include Ant Colony Optimization (ACO) [26], Particle Swarm Optimization (PSO) (Kennedy and Eberhart [11]), ABC [9], Bat Algorithm [30], Cuckoo Search Algorithm [20], Firefly Optimization [16], etc. Some of the important EAs are Genetic Algorithm (GA)

A. Shastri · A. Nargundkar (✉) · A. J. Kulkarni
Symbiosis Institute of Technology, Symbiosis International (Deemed University), Lavale, Pune 412115, India
e-mail: aniket.nargundkar@sitpune.edu.in

A. Shastri
e-mail: apoorva.shastri@sitpune.edu.in

A. J. Kulkarni
e-mail: anand.kulkarni@sitpune.edu.in

S. C. Satapathy et al. (eds.), *Intelligent Data Engineering and Analytics*, Advances in Intelligent Systems and Computing 1177,
https://doi.org/10.1007/978-981-15-5679-1_14

[6, 27], Evolutionary Strategies [29], Differential Evolution [28], Memetic Algorithms [17], etc. Socio-inspired optimization algorithms are the emerging class of metaheuristics which is inspired from societal behavior. In the recent past, various socio-inspired optimization algorithms are proposed and applied. These algorithms are developed based on the societal behavior. Teaching–Learning-Based Optimization (TLBO) [21], League Championship Algorithm [10], Ideology Algorithm [7] Backtracking Search Algorithm [1], Harmony Search Algorithm [4], Tabu Search Algorithm [2], Expectation Algorithm (ExA) [24, 25], etc. are some examples of such algorithms.

A new optimization approach, namely, Cohort Intelligence (CI) is based on artificial intelligence developed by Kulkarni et al. [14]. In this algorithm, candidates in a cohort interact with one another and try to follow the best behavior in order to achieve the global best solution. Further, the algorithm is applied on real-world applications such as 0–1 Knapsack problems [13], healthcare application, cross-border shipper supply chain [15], mechanical engineering applications [3, 5, 18], constrained benchmark problems and applications [23, 24, 25], and clustering problem [12]. Further, variants of (CI) are proposed by Patankar et al. [19]. The variations are based on the following strategy adopted by candidates.

It was noticed in above studies that candidates in a cohort have limited choices to learn from. Also, they quickly gather at certain location and then together search for improved solutions. Also, candidates may take significantly more time to jump out of the local minima. Very recently [22], a new variation of CI referred to as Multi-CI was proposed in which several cohorts searched the problem space at different locations and the candidates learnt certain qualities from the other cohort candidates. The Multi-CI was successfully coded and tested by solving 75 benchmark problems.

In this paper, a new learning approach referred to as Panoptic Learning (PL) is adopted to replace the current roulette wheel selection approach. The PL approach is inspired from the natural cohort learning behavior. In PL approach, candidate learns from every candidate in the cohort partially in every learning attempt as opposed to roulette wheel approach which makes every candidate learn from a single candidate. With this modified approach, every candidate learns the most from the best candidate but at the same time instead of completely ignoring the other candidates, it follows a partial behavior. Multi-CI algorithm is modified with adopting PL approach as a following mechanism. The PL-based approach is better suited to imitate the cohort learning behavior than roulette-wheel-based approach.

2 Methodology

The Multi-CI-PL algorithm implements the learning mechanisms within intra- and inter-cohorts. It focuses on the interaction among various cohorts. In Multi-CI-PL, panoptic learning approach is used for follow mechanism unlike Multi-CI where roulette wheel approach is used by the candidates.

A general minimization problem is considered which is unconstrained in nature. In Multi-CI-PL, behavior of an individual candidate is modeled as the objective function with associated set of behaviors. The process begins with initialization of learning attempts, number of cohorts and candidates in each cohort, associated sampling interval, convergence parameter, behavior variations, and sampling interval reduction factor. The further complete process steps are shown in Fig. 1.

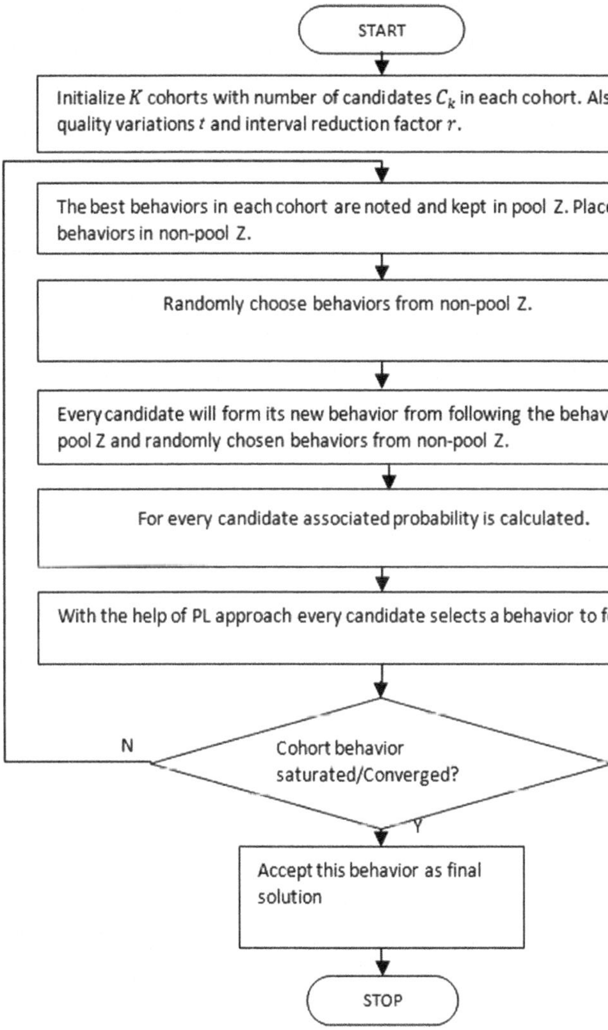

Fig. 1 Modified Multi-CI-PL flowchart

3 Results and Discussion

MATLAB R2016 is used for coding on Windows Platform with an Intel Core i3 processor with 4 GB RAM. The benchmark functions solved are taken from Civicioglu [1].

Parameters

The Multi-CI-PL factors selected for every run:

- Cohorts = 3,
- Candidates = 5, and
- Reduction factor = $0.8 < r < 0.92$.

Stopping Criteria

- Objective function value is less than 10^{-16} and
- Maximum number of learning attempts reached.

This section presents the inter-comparison of the algorithms being applied for the comparison with Modified Multi-CI-PL. The PSO algorithm is a swarm-based optimization technique in which swarm of solutions alters the positions in search space. In this paper, Multi-CI-PL is compared against Comprehensive Learning PSO (CLPSO). The CMAES [8] is mathematical-based optimization technique. The ABC algorithm works based on exploration and exploitation using scout bees and employed bees, respectively. The BSA is a metaheuristic in which genetic operators and non-uniform crossover are applied. DE is also a population-based technique similar to BSA, which adopts genetic operators. Table 1 presents the mean and best solutions along with standard deviation. The run time in seconds is also provided. It is evident from Table 1 that modified Multi-CI-PL gives best results as compared with PSO, CLPSO, ABC, DE, and SADE algorithms and also resulted in much reduced standard deviation showing robustness of algorithm. The time required for modified Multi-CI-PL is also much less as compared. Figure 2 indicates the convergence plot for best candidate across all cohorts.

4 Conclusion

In this paper, modified Multi-CI-PL methodology is proposed incorporating panoptic learning into Multi-CI. The performance of algorithm is validated by solving 15 unconstrained benchmark problems. The algorithm exhibits better results as compared with several other evolutionary algorithms, viz., PSO 2011, CMAES, ABC, JDE, CLPSO, SADE, and BSA. Multi-CI-PL outperformed in terms of standard deviation and function evaluations showing robustness of proposed algorithm. In addition, constraint handling techniques could be developed for Multi-CI-PL and constrained engineering problems could be solved in near future.

Table 1 Statistical solutions and comparison of Multi-CI-PL (Mean = mean solution; Std. Dev. = standard deviation of mean solution; Best = best solution; Runtime = mean runtime in seconds)

Sr No.	Function	Statistics	PSO20 11	CMAES	ABC	JDE	CLPSO	SADE	BSA	Multi-CI-PL
1	F9	Mean	0.00E+00	1.05E−04	6.00E−16	0.00E+00	1.94E−05	0.00E+00	0.00E+00	2.22E−16
		Std	0.00E+00	5.74E−04	3.00E−16	0.00E+00	8.47E−05	0.00E+00	0.00E+00	6.52E−16
		Best	0.00E+00	0.00E+00	1.00E−16	0.00E+00	0.00E+00	0.00E+00	0.00E+00	5.55E−17
		Runtime	1.71E+01	2.13E+00	2.17E+01	1.13E+00	3.33E+01	4.30E+00	8.29E−01	4.40E−01
2	F18	Mean	6.89E−03	1.15E−03	0.00E+00	4.82E−03	0.00E+00	2.26E−02	4.93E−04	2.62E−14
		Std	8.06E−03	3.65E−03	1.00E−16	1.33E−02	0.00E+00	2.84E−02	1.88E−03	1.32E−12
		Best	0.00E+00	0.00E+00	0.00E+00	0.00E+00	0.00E+00	0.00E+00	0.00E+00	4.44E−16
		Runtime	7.39E+01	2.65E+00	1.91E+01	6.91E+00	1.49E+01	2.59E+01	5.75E+00	2.81E−01
3	F25	Mean	0.00E+00	0.00E+00	4.00E−16	0.00E+00	4.18E−06	0.00E+00	0.00E+00	7.25E−17
		Std	0.00E+00	0.00E+00	3.00E−16	0.00E+00	1.62E−05	0.00E+00	0.00E+00	9.27E−17
		Best	0.00E+00	0.00E+00	1.00E−16	0.00E+00	0.00E+00	0.00E+00	0.00E+00	1.37E−17
		Runtime	2.54E+01	1.34E+00	1.97E+01	1.14E+00	3.16E+01	4.09E+00	8.13E−01	1.95E+00
4	F31	Mean	1.26E−04	0.00E+00	7.79E−03	2.02E−03	2.68E−04	0.00E+00	1.12E−08	1.78E−13
		Std	1.50E−04	0.00E+00	6.24E−03	7.75E−03	3.05E−04	0.00E+00	1.84E−08	4.11E−16
		Best	1.57E−08	0.00E+00	3.96E−04	0.00E+00	2.31E−06	0.00E+00	0.00E+00	1.15E−17
		Runtime	2.50E+02	1.21E+01	3.47E+01	4.87E+01	2.28E+02	2.21E+02	1.50E+02	1.49E−+00
5	F50	Mean	0.00E+00	0.00E+00	4.02E−08	0.00E+00	1.60E−10	0.00E+00	0.00E+00	1.97E−16
		Std	0.00E+00	0.00E−00	2.20E−07	0.00E+00	6.27E−10	0.00E+00	0.00E+00	2.42E−16
		Best	0.00E+00	0.00E−00	2.10E−14	0.00E+00	0.00E+00	0.00E+00	0.00E+00	4.21E−17
		Runtime	8.64E+01	1.87E+00	8.65E+01	1.41E+00	1.58E+02	4.93E+00	5.70E+00	9.80E−01

(continued)

Table 1 (continued)

Sr No.	Function	Statistics	PSO20 11	CMAES	ABC	JDE	CLPSO	SADE	BSA	Multi-CI-PL
6	F7	Mean	0.00E+00	6.22E−02	0.00E+00	0.00E+00	0.00E+00	0.00E+00	0.00E+00	1.85E−15
		Std	0.00E+00	1.35E−01	0.00E+00	0.00E+00	0.00E+00	0.00E+00	0.00E+00	1.55E−15
		Best	0.00E+00	0.00E+00	0.00E+00	0.00E+00	0.00E+00	0.00E+00	0.00E+00	2.22E−16
		Runtime	1.70E+01	6.85E+00	1.83E+00	1.14E+00	2.93E+00	4.41E+00	8.25E−01	2.63E−01
7	F47	Mean	0.00E+00	0.00E+00	5.00E−16	0.00E+00	0.00E+00	0.00E+00	0.00E+00	3.26E−16
		Std	0.00E+00	0.00E+00	0.00E+00	0.00E+00	0.00E+00	0.00E+00	0.00E+00	8.04E−16
		Best	0.00E+00	0.00E+00	3.00E−16	0.00E+00	0.00E+00	0.00E+00	0.00E+00	1.02E−16
		Runtime	5.64E+02	2.57E+00	2.42E+01	1.87E+00	1.60E+01	6.38E+00	4.31E+00	2.13E−01
8	F8	Mean	0.00E+00	7.28E−03	0.00E+00	0.00E+00	0.00E+00	0.00E+00	0.00E+00	1.36E−15
		Std	0.00E+00	3.99E−02	0.00E+00	0.00E+00	0.00E+00	0.00E+00	0.00E+00	2.16E−15
		Best	0.00E+00	0.00E+00	0.00E+00	0.00E+00	0.00E+00	0.00E+00	0.00E+00	1.11E−16
		Runtime	1.70E+01	2.17E+00	1.80E+00	1.14E+00	2.89E+00	4.42E+00	8.24E−01	2.70E−01
9	F21	Mean	3.08E−04	6.48E−03	4.42E−04	3.69E−04	3.10E−04	3.08E−04	3.08E−04	1.98E−03
		Std	0.00E+00	1.49E−02	5.68E−05	2.32E−04	5.98E−06	0.00E+00	0.00E+00	1.06E−03
		Best	3.08E−04	3.08E−04	3.23E−04	3.08E−04	3.08E−04	3.08E−04	3.08E−04	9.66E−04
		Runtime	8.45E+01	1.39E+01	2.03E+01	7.81E+00	1.56E+02	4.54E+01	1.17E+01	2.45E−01
10	F30	Mean	1.31E−05	0.00E+00	2.60E−04	1.00E−16	4.59E−02	2.73E−07	2.84E−09	1.28E−17
		Std	1.43E−06	0.00E+00	3.95E−05	2.00E−16	6.20E−02	1.79E−07	3.33E−09	1.53E−17
		Best	9.51E−06	0.00E+00	1.68E−04	0.00E+00	5.28E−04	9.44E−08	4.77E−10	2.06E−18
		Runtime	5.68E+02	1.45E+01	2.16E+02	1.94E+02	2.53E+02	3.60E+02	1.45E+02	2.20E−01

(continued)

Table 1 (continued)

Sr No.	Function	Statistics	PSO2011	CMAES	ABC	JDE	CLPSO	SADE	BSA	Multi-CI-PL
11	F32	Mean	3.55E−04	7.02E−02	2.50E−02	1.30E−03	1.96E−03	1.67E−03	2.00E−03	3.06E−02
		Std	1.41E−04	2.89E−02	7.72E−03	9.95E−04	4.34E−03	7.33E−04	9.70E−04	3.11E−02
		Best	1.01E−04	2.99E−02	9.47E−03	1.79E−04	4.21E−04	5.63E−04	6.09E−04	6.86E−03
		Runtime	2.91E+02	2.15E+00	3.50E+01	8.21E+01	1.03E+02	1.72E+02	4.82E+01	4.58E−01
12	F35	Mean	0.00E+00	4.65E−01	0.00E+00	3.89E−03	1.94E−03	6.48E−04	0.00E+00	1.78E−15
		Std	0.00E+00	9.34E−02	0.00E+00	4.84E−03	3.95E−03	2.47E−03	0.00E+00	1.84E−15
		Best	0.00E+00	9.72E−03	0.00E+00	0.00E+00	0.00E+00	0.00E+00	0.00E+00	4.44E−16
		Runtime	1.82E+01	2.40E+01	7.86E+00	4.22E+00	8.30E+00	5.90E+00	1.78E+00	6.53E−02
13	F37	Mean	0.00E+00	0.00E+00	1.46E+01	0.00E+00	6.47E+00	0.00E+00	0.00E+00	1.50E−16
		Std	0.00E+00	0.00E+00	8.71E+00	0.00E+00	8.22E+00	0.00E+00	0.00E+00	1.20E−16
		Best	0.00E+00	0.00E−00	4.04E+00	0.00E+00	1.82E−01	0.00E+00	0.00E+00	3.93E−17
		Runtime	5.43E+02	3.37E−00	1.12E+02	1.93E+01	1.79E+02	1.10E+02	5.73E+01	1.84E+00
14	F38	Mean	0.00E+00	0.00E−00	5.00E−16	0.00E+00	0.00E+00	0.00E+00	0.00E+00	1.95E−16
		Std	0.00E+00	0.00E−00	1.00E−16	0.00E+00	0.00E+00	0.00E+00	0.00E+00	5.81E−16
		Best	0.00E+00	0.00E+00	3.00E−16	0.00E+00	0.00E+00	0.00E+00	0.00E+00	2.36E−17
		Runtime	1.63E+02	2.56E+00	2.06E+01	1.49E+00	1.26E+01	5.63E+00	3.21E+00	2.97E−01
15	F44	Mean	0.00E+00	0.00E+00	4.00E−16	0.00E+00	0.00E+00	0.00E+00	0.00E+00	1.18E−17
		Std	0.00E+00	0.00E+00	1.00E−16	0.00E+00	0.00E+00	0.00E+00	0.00E+00	1.30E−16
		Best	0.00E+00	0.00E+00	3.00E−16	0.00E+00	0.00E+00	0.00E+00	0.00E+00	1.68E−20
		Runtime	1.60E+02	2.32E+00	2.19E+01	1.42E+00	1.44E+01	5.92E+00	3.30E+00	1.38E−01

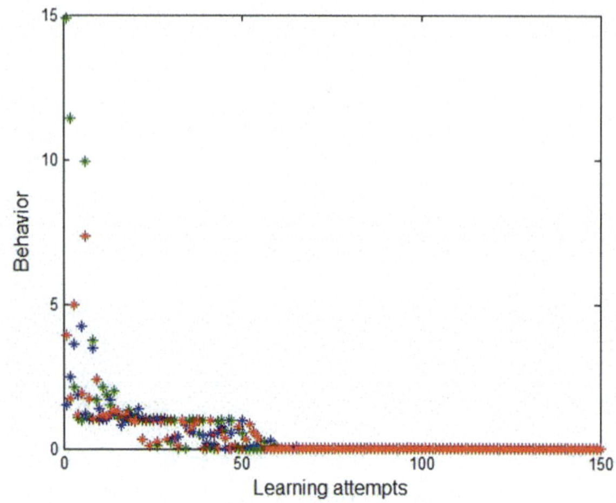

Fig. 2 Convergence plot for F7

References

1. Civicioglu, P.: Backtracking search optimization algorithm for numerical optimization problems. Appl. Math. Comput. **219**(15), 8121–8144 (2013)
2. Costa, D.: A tabu search algorithm for computing an operational timetable. Eur. J. Oper. Res. **76**(1), 98–110 (1994)
3. Dhavle, S.V., Kulkarni, A.J., Shastri, A., Kale, I.R.: Design and economic optimization of shell-and-tube heat exchanger using cohort intelligence algorithm. Neural Comput. Appl. **30**(1), 111–125 (2018)
4. Geem, Z.W.: Novel derivative of harmony search algorithm for discrete design variables. Appl. Math. Comput. **199**(1), 223–230 (2008)
5. Gulia, V., Nargundkar, A.: Optimization of process parameters of abrasive water jet machining using variations of cohort intelligence (CI). In: Applications of Artificial Intelligence Techniques in Engineering, pp. 467–474. Springer, Singapore (2019)
6. Haq, A.N., Sivakumar, K., Saravanan, R., Muthiah, V.: Tolerance design optimization of machine elements using genetic algorithm. Int. J. Adv. Manuf. Technol. **25**(3–4), 385–391 (2005)
7. Huan, T.T., Kulkarni, A.J., Kanesan, J., Huang, C.J., Abraham, A.: Ideology algorithm: a socio-inspired optimization methodology. Neural Comput. Appl. **28**(1), 845–876 (2017)
8. Igel, C., Hansen, N., Roth, S.: Covariance matrix adaptation for multi-objective optimization. Evol. Comput. **15**(1), 1–28 (2007)
9. Karaboga, D., Basturk, B.: On the performance of artificial bee colony (ABC) algorithm. Appl. Soft Comput. **8**(1), 687–697 (2008)
10. Kashan, A.H.: League championship algorithm (LCA): an algorithm for global optimization inspired by sport championships. Appl. Soft Comput. **16**, 171–200 (2014)
11. Kennedy, J., Eberhart, R.: In: Particle swarm optimization. In Proceedings of ICNN'95-International Conference on Neural Networks vol. 4, pp. 1942–1948. IEEE (1995)
12. Krishnasamy, G., Kulkarni, A.J., Paramesran, R.: A hybrid approach for data clustering based on modified cohort intelligence and K-means. Expert Syst. Appl. **41**(13), 6009–6016 (2014)
13. Kulkarni, A.J., Shabir, H.: Solving 0–1 knapsack problem using cohort intelligence algorithm. Int. J. Mach. Learn. Cybernet. **7**(3), 427–441 (2016)
14. Kulkarni, A.J., Durugkar, I.P., Kumar, M.: Cohort intelligence: a self-supervised learning behavior. In: 2013 IEEE International Conference on Systems, Man, and Cybernetics, pp. 1396–1400. IEEE (2013)

15. Kulkarni, A.J., Baki, M.F., Chaouch, B.A.: Application of the cohort-intelligence optimization method to three selected combinatorial optimization problems. Eur. J. Oper. Res. **250**(2), 427–447 (2016)
16. Łukasik, S., Żak, S.: Firefly algorithm for continuous constrained optimization tasks. In: International Conference on Computational Collective Intelligence, pp. 97–106. Springer, Berlin, Heidelberg (2009)
17. Moscato, P., Cotta, C.: A gentle introduction to memetic algorithms. In: Handbook of Metaheuristics, pp. 105–144. Springer, Boston, MA (2003)
18. Pansari, S., Mathew, A., Nargundkar, A.: An investigation of burr formation and cutting parameter optimization in micro-drilling of brass C-360 using image processing. In: Proceedings of the 2nd International Conference on Data Engineering and Communication Technology, pp. 289–302. Springer, Singapore (2019)
19. Patankar, N.S., Kulkarni, A.J.: Variations of cohort intelligence. Soft. Comput. **22**(6), 1731–1747 (2018)
20. Rajabioun, R.: Cuckoo optimization algorithm. Appl. Soft Comput. **11**(8), 5508–5518 (2011)
21. Rao, R.V., More, K.C.: Advanced optimal tolerance design of machine elements using teaching-learning-based optimization algorithm. Prod. Manuf. Res. **2**(1), 71–94 (2014)
22. Shastri A.S., Kulkarni A.J.: Multi-cohort Intelligence algorithm: an intra- and inter-group learning behavior based socio-inspired optimization methodology. Int. J. Parallel Emerg. Distrib. Syst. (2018)
23. Shastri, A.S., Jadhav, P.S., Kulkarni, A.J., Abraham, A.: Solution to constrained test problems using cohort intelligence algorithm. In: Innovations in Bio-Inspired Computing and Applications, pp. 427–435. Springer, Cham (2016)
24. Shastri, A.S., Jagetia, A., Sehgal, A., Patel, M., Kulkarni, A.J.: Expectation algorithm (ExA): a socio-inspired optimization methodology. In: Socio-cultural Inspired Metaheuristics, pp. 193–214. Springer, Singapore (2019)
25. Shastri, A.S., Thorat, E.V., Kulkarni, A.J., Jadhav, P.S.: Optimization of constrained engineering design problems using cohort intelligence method. In: Proceedings of the 2nd International Conference on Data Engineering and Communication Technology, pp. 1–11. Springer, Singapore (2019)
26. Shelokar, P.S., Siarry, P., Jayaraman, V.K., Kulkarni, B.D.: Particle swarm and ant colony algorithms hybridized for improved continuous optimization. Appl. Math. Comput. **188**(1), 129–142 (2007)
27. Singh, P.K., Jain, S.C., Jain, P.K.: Advanced optimal tolerance design of mechanical assemblies with interrelated dimension chains and process precision limits. Comput. Ind. **56**(2), 179–194 (2005)
28. Storn, R., Price, K.: Differential evolution–a simple and efficient heuristic for global optimization over continuous spaces. J. Global Optim. **11**(4), 341–359 (1997)
29. Taylor, P.D., Jonker, L.B.: Evolutionary stable strategies and game dynamics. Math. Biosci. **40**(1–2), 145–156 (1978)
30. Yang, X.S., Hossein Gandomi, A.: Bat algorithm: a novel approach for global engineering optimization. Eng. Comput. **29**(5), 464–483 (2012)

Sentiment Analysis on Movie Review Using Deep Learning RNN Method

Priya Patel, Devkishan Patel, and Chandani Naik

Abstract The usage of social media grows rapidly because of the functionality like easy to use and it will also allow user to connect with all around the globe to share the ideas. It is desired to automatically use the information which is user's interest. One of the meaningful information that is derived from the social media sites are sentiments. Sentiment analysis is used for finding relevant documents, overall sentiment, and relevant sections; quantifying the sentiment; and aggregating all sentiments to form an overview. Sentiment analysis for movie review classification is useful to analyze the information in the form of number of reviews where opinions are either positive or negative. In this paper we had applied the deep learning-based classification algorithm RNN, measured the performance of the classifier based on the pre-process of data, and obtained 94.61% accuracy. Here we had used RNN algorithm instead of machine learning algorithm because machine learning algorithm works only in single layer while RNN algorithm works on multilayer that gives you better output as compared to machine learning.

Keywords Data mining · Text mining · Natural language processing toolkit (NLTK) · Recurrent neural network (RNN)

P. Patel (✉)
Department of Computer Engineering, N. G. Polytechnic, Isroli, India
e-mail: Priya.pse@gmail.com

D. Patel
Department of Computer Engineering, Pacific School of Engineering, Palsana, India
e-mail: devkishanpatel18@gmail.com

C. Naik
Department of Computer Engineering, CGPIT, Uka Tarsadiya University, Bardoli, India
e-mail: chandni.naik@utu.ac.in

S. C. Satapathy et al. (eds.), *Intelligent Data Engineering and Analytics*, Advances in Intelligent Systems and Computing 1177,
https://doi.org/10.1007/978-981-15-5679-1_15

1 Introduction

Sentiments analysis is the field of study that analyzes people's opinions, sentiments, evaluations, appraisals, attitudes, and emotions toward entities such as products, services, organizations, individuals, issues, events, topics, and their attributes, as discussed in [1]. There are different names, and different tasks for the same are defined as sentiment analysis, opinion mining, opinion extraction, sentiment mining, subjective analysis, effect analysis, emotion analysis, review mining, and so on.

As per the authors of [1, 2] the aim of sentiment analysis or opinion mining is to automatically extract options expressed in the user-generated content. It can also be used to classify the reviews into positive, negative, or neutral. Sentiment analysis can be done at three levels as described below.

- **Sentence Level**

In sentence level, each sentence is classified into negative, positive, or neutral class. There are two types of classification known as: (1) objective and (2) subjective. Subjective sentences contain positive as well as negative opinions, while objective sentences contain no opinions, as discussed in [1] (Fig. 1).

- **Document Level**

In the document level classification, the whole document is classified into two classes, known as a positive or negative class. This level works on a single entity and contains opinion from a single holder. The whole process is composed of two steps:

a. From training dataset subjective features extract and convert ion to feature vector.
b. Classifier training data on feature vector and then classified opinions [1].

- **Aspect Level**

As per the authors of [1], it is also called as feature level which performs the finer-grained analysis. Earlier it was known as aspect level. Users often express opinions about multiple aspects in the same sentences.

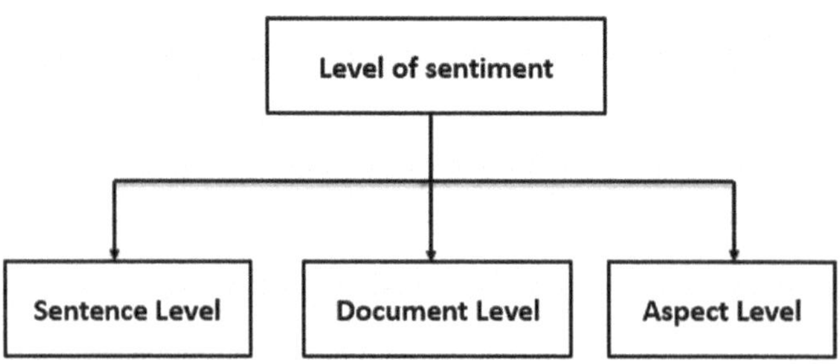

Fig. 1 Level of sentiment

Here, we had used document-level sentiment analysis classification for getting the result of proposed system and used the

movie review dataset for preprocessing. We had removed the HTML tags, punctuation, and numbers for the part of preprocessing. Here, we had also used the Word2vec and TF-IDF method for the word embedding process that is applied on deep learning LSTM method, to learn the text review on the positive and negative category.

Section 2 covers the existing works related to our proposed system, while Sect. 3 introduces our problem domain and briefly explains Word2vec with RNN. Section 4 provides observation and results that we obtained during the experiments. At last, we enclosed our work in Sect. 5 with contribution problem domain, which includes conclusions and future work.

2 Related Works

The authors of [3] enhanced the RNN language protocol, that is forward LSTM, which effectively covers all past system information and achieves better results than conventional RNN. The method proposed in [3] works more precisely than conventional RNN in multi-classification for text emotional characteristics and identifies text emotional characteristics.

The study of text sentiment analysis states that the sequential connection among words is of critical importance. The authors of [4] proposed a model that is known as recurrent neural network (RNN), which is widely suitable to develop text sequence data. There are three modules that are covered by RNN: output layer, hidden layer, and input layer.

Sentiment classification has usually been explained by linear classification methods, such as support vector machines (SVM) and logistic regression discussed in [5]. The research in [6] finds two methods, such as naive Bayes classification and maximum entropy. Deep learning method will also be applied for the sentiment analysis as discussed in [6]. For the sentiment analysis the authors proposed an approach to learn the task specially related to word and the vector as discussed in [5]. They had used an unsupervised model to learn the semantic connections between words, and a supervised module that is able to adapt the nuanced sentimental information. This model is able to achieve the similarity between the words by—semantic + sentiment model.

The authors in [7] used IMDB (Indian Movie Database). They performed preprocessing on the dataset and also performed the following tasks, like removal of hashtags, synonyms, acronyms, and so on. By using long short-term memory (LSTM) to modify version of RNN with word vector features for sentiment analysis on movie reviews, they were able to obtain 88.89% accuracy. The authors of [8] used own created movie review dataset using different sources, like BookMyShow, Netflix, Rotten Tomatoes, IMDB, and so on. After that, they applied preprocessing such as removal of HTML tags, punctuation, and numbers and also used removal of stop words. After that, they used heterogeneous features, such as SentiWordNet,

Exaggerated word shorting, negation handling, intensifier handling, emotion, term frequency-inverse document frequency (TFIDF). They used support vector machine (SVM) and naïve Bayes (NB) algorithm for detecting sentiment from movie review dataset, and proved that NB obtained higher accuracy than linear SVM.

The IMDB dataset consists of two different data, one is binary-labeled data and another one is multiclass-labeled data as discussed in [9]. They had performed skip-gram and Bag of Word (BOW) with the Word2vec model for various ML methods. In binary-labeled data, random forest, SVM, and logistic regression algorithm are used and 84.35% accuracy was obtained.

Multiclass-labeled data work on recursive neural tensor network (RNTN) algorithm and obtain 86.10% accuracy. The authors of [10] worked on document level with the IMDB dataset, TF-IDF, Bag of Word, and n-gram used different types of methods, like NB-SVM with trigram, RNN-LM, sentence vector, but combining all methods got the accurate result.

The IMDB dataset used document vector and paragraph vector and combined the approach of NB-SVM and RNN-LM with the help of BOW of n-gram, as discussed in [11]. Using NB-SVM and RNN-LM they obtained 92.10 and 92.81% accuracy, respectively, while the combination of NB-SVM with RNN-LM on unlabeled data got 93.05% accuracy. The authors of [12] worked on identifying negation scope with different domains, like movie, book, car, computer, phone on different deep learning models, like LSTM, CRF, BLSTM with task-specific word embedding because several mis-spellings word, abbreviation, and composition of word occur in dataset. They got a result of 89.38% accuracy on test data bi-LSTM), while on training data they got 89.84% accuracy.

The authors of [13] used two methods. First one is known as extending term counting and the second one is machine learning SVM. The hybrid method is used by combining both methods to obtain better result. The movie review dataset that is used was prepared by Pang and Lee from IMDB. NB algorithm with the various methods like Laplacian smoothing, handling negation, Bernoulli NB, bigram and trigram, feature selection were used.

A famous deep learning framework provided by Socher and discussed in [14] is known as the recursive neural network (RNN). He used a completely labeled parse trees to characterize the movie reviews from the *rottentomatoes.com* website. RNN with LSTM can be viewed as an enhanced model of the conventional RNN language model from the perception of language, as discussed in [15]. The benefit of using this model is that they can calculate the error of every model by pushing the text statements as the input sequence. The minor error specifies higher degree of assurance of text statement. Usually the RNN model with LSTM is more active to overwhelm the arrangements of information attenuation difficult when the text arrangement of an information is relatively long. So we had used the RNN with LSTM on the text sentiment analysis.

Our observations conclude that machine learning algorithm cannot be able to get the fruitful and accurate result, while the use of deep learning method leads to attain the better result.

3 Proposed Approach

The flow diagram of the proposed system is shown in Fig. 2. We can observe that the proposed approach has six components that are dataset of the movie review, preprocessing, feature extraction, feature selection, a classification method, and opinion classification. For an input movie review dataset is used as a text document from kaggel.

We had derived the dataset from the web only, so it might possible that the data are being affected with the noise. Hence it is important to clean the data using different preprocessing techniques, as discussed in [12].

- **Tokenization**

Tokenization means sentences are convert into pieces of word called tokens. For example "this movie was good" after the tokenization "this, movie, was, good".

- **Stop-word Removal**

The commonly used words like a, an, the, has, have, and so on which carry no meaning and that cannot help in determining the sentiment of text while analyzing should be removed from the input text. The stop-word does not emphasize on many emotions, but the main intent to remove stop-word is to compress the dataset. For example, the text "this movie was good" will be processed to "movie good".

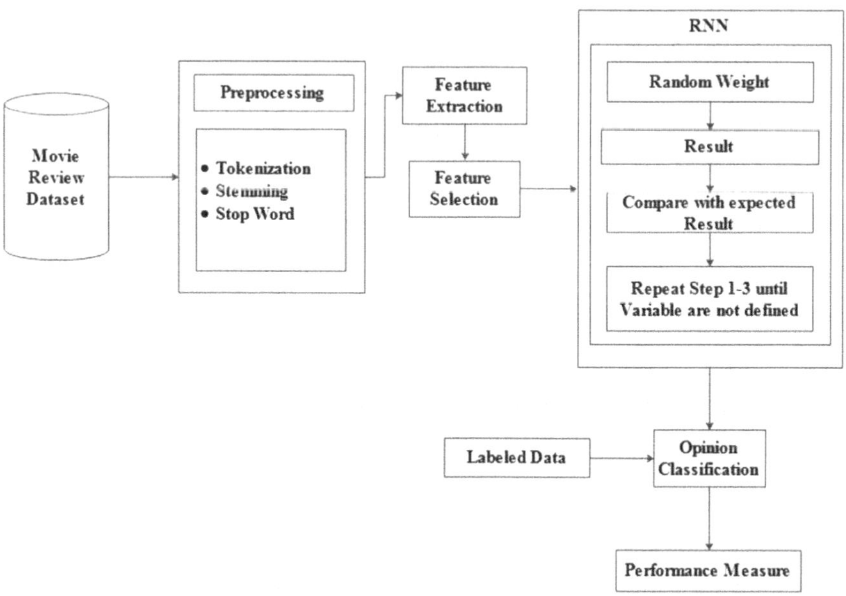

Fig. 2 Proposed flow diagram

- **Stemming**

Stemming in unnecessary character connected with the word is removed; for example, watching, watched will be processed to watch.

After the preprocessing phase, the next step is feature extraction: it analyzes the data and finds the common observable patterns that may affect the polarity of the document. In order to calculate the document polarity, it is necessary to understand that the sentiment score may be enhanced or diminished with its usage as well as their relationship with the nearby words. We used Word2vec feature vector and also used TF-IDF.

- **TF-IDF**

From TF-IDF top 10 high informative words which should be adjectives, adverbs, verbs, and nouns are selected and the score of SentiWordNet average on review was computed, but here only adjectives, adverbs, verbs, and nouns are considered.

- **Word2vec**

With the word level features it will characterize distinct words by word embedding in a constant vector-space; explicitly, tested with the Word2vec embedding. This feature works on cosine similarity. It provides an adjustable length feature sets for the document. These documents are characterized as a variable number of sentences that are symbolized as an adjustable number of fixed-length word feature vectors, as discussed in [7].

After the completion of this process apply feature selection. Best feature is selected based on Word2vec word embedding among all extracted feature set that can affect the polarity of the document. After the process of feature selection apply the RNN classification algorithm. The steps of this algorithm are as follows:

a. Input preprocesses data.
b. Model will take data and randomly initialized variables called weights.
c. Produce a predicted result and then comparing with the expected value will give us an error.
d. Propagating the error back through a similar way will change the factors.
e. Adjust variable till Steps 1–4.

A prediction is made by applying these variables to new unseen input. After the completion of this RNN model, apply opinion classification with the labeled data, which leads to the completion of all process to measure the performance of the system.

For example, we had taken a document as a dataset of movie review "This movie is good if the actor was more focused on heroin instead of villain, then this movie can be rated by more users". After tokenization and stemming, the output will be "This, movie, is, good, if, actor, was, more, focused, on, heroin, instead, of, villain, then, this, movie, can, be, rated, by, more, user".

After the removal of the punctuation and stop-word, the output will be "movie, good, actor, more, focused, heroin, instead, villain, movie, rated, more, user". After the preprocessing process is completed, the preprocessed data (tokens) are applied in Word2vec feature vector, the output of this feature will be a binary matrix. After this process the RNN model will be generated, and classify an opinion; for this particular example we get a positive opinion.

4 Implementation Tools

System description: The experiments are conducted on Python 3.6.0, using an Intel Core i3, 1.8 GHz machine with a 64-bit OS and 8 GB RAM used for testing.

Dataset description: We used IMDB movie review dataset extracted from the kaggle website, consisting of 25,000 labeled training data and 25,000 test data. Sentiment represents the positive and negative review as 1 and 0. Review represents people's opinion about the movie.

5 Performance Analysis

In this section, the performance of the proposed system is evaluated. Table 1 shows the comparative analysis of existing technique. We had tried to perform with the machine learning algorithms such as NB and SVM, but we cannot get accurate result, hence we worked on deep learning. The main advantage of using the deep learning method over machine learning is that it will work on multi-level and the error is decreased.

Table 1 shows the experiment on NB and SVM. We used TF-IDF feature vector with dimension reduction and feature selection used in NB, and obtained 80.08% accuracy. We also used Word2vec with all the given features and obtained 71.94% accuracy. We also used SVM with the same feature vector and got accuracy of 83.30% on TF-IDF and 88.02% on Word2vec.

Table 2 shows the experiment on RNN using TF-IDF feature vector and got 50.16% on 1 epoch and 87.64% accuracy on 3 epoch. TF-IDF feature vector in 3 epochs provided more accuracy because here dataset are over fit with 1 epoch.

Table 1 Comparative analysis

Approach	Feature	Accuracy (%)
Naïve Bayes	TF-IDF	80.08
	Word2vec	71.94
SVM	TF-IDF	83.30
	Word2vec	88.02

Table 2 Comparative analysis

Approach	Feature	Accuracy %	
		1 Epoch	3 Epochs
RNN	TF-IDF	50.16	87.64
	Word2vec	**94.61**	94.06

Word2vec feature is used and we obtained 94.61% accuracy on 1 epoch and 94.06% accuracy on 3 epochs. We obtain more accuracy in 1 epoch because our dataset are over fit with 3 epochs and that is the reason it gave us less accuracy with 3 epochs.

6 Conclusions and Future Work

The experiments prove that by using Word2vec with RNN we can get better accuracy even on a large amount of training dataset compared to machine learning method such as NB and SVM. As more number of people are attracted toward the digital media, this method is used to provide the efficient review related to the movie. In the future, we can work on real-time data with the use of machine learning. Also, we can work on deep learning bi-LSTM method and use different models combinations to maximize the performance.

References

1. Balaji, P., Nagaraju, O., Haritha, D.: Levels of sentiment analysis and its challenges: a literature review. In: 2017 International Conference on Big Data Analytics and Computational Intelligence (ICBDAC), pp. 436–439. IEEE (2017)
2. Bhonde, S.B., Prasad, J.R.: Sentiment analysis-methods, application and challenges. Int. J. Electron. Commun. Comput. Eng. **6**(6) (2015)
3. Li, D., Qian, J.: Text sentiment analysis based on long short-term memory. In: 2016 First IEEE International Conference on Computer Communication and the Internet (ICCCI), pp. 471–475. IEEE (2016)
4. Sepp, H. Schmidhuber, J.: long short-term memory. Neural Comput. 12–91 (1997)
5. Nair, S.K., Soni, R.: Sentiment analysis on movie reviews using recurrent neural network. (2018)
6. Bandana, R:. Sentiment analysis of movie reviews using heterogeneous features. In: 2018 2nd International Conference on Electronics, Materials Engineering and Nano-Technology (IEMENTech), pp. 1–4. IEEE (2018)
7. Pouransari, H., Ghili, S.: Deep learning for sentiment analysis of movie reviews. Tech. Rep. Stanford University (2014)
8. Mesnil, G., Mikolov, T., Ranzato, M.A., Bengio, Y.: Ensemble of generative and discriminative techniques for sentiment analysis of movie reviews. arXiv preprint arXiv:1412.5335 (2014)
9. Li, B., Liu, T., Du, X., Zhang, D., Zhao, Z.: Learning document embeddings by predicting n-grams for sentiment classification of long movie reviews. arXiv preprint arXiv:1512.08183 (2015)

10. Lazib, L., Zhao, Y., Qin, B., Liu, T.: Negation scope detection with recurrent neural networks models in review texts. In: International Conference of Young Computer Scientists, Engineers and Educators, pp. 494–508. Springer, Singapore (2016)
11. Kennedy, Alistair, Inkpen, Diana: Sentiment classification of movie reviews using contextual valence shifters. Comput. Intell. **22**(2), 110–125 (2006)
12. Pang, B., Lee, L., Vaithyanathan, S.: Thumbs up?: sentiment classification using machine learning techniques. In: Proceedings of the ACL-02 Conference on Empirical Methods In Natural Language Processing, vol. 10, pp. 79–86. Association for Computational Linguistics (2002)
13. Ahuja, R., Anand, W.: Sentiment classification of movie reviews using dual training and dual predition. In: 2017 Fourth International Conference on Image Information Processing (ICIIP), pp. 1–4. IEEE (2017)
14. Narayanan, V., Arora I, Bhatia, A.: Fast and accurate sentiment classification using an enhanced Naive Bayes model. In: International Conference on Intelligent Data Engineering and Automated Learning, pp. 194–201. Springer, Berlin, Heidelberg (2013)
15. Socher, R., Lin, C.C., Manning, C., Ng, A.Y.: Parsing natural scenes and natural language with recursive neural networks. In: Proceedings of the 28th International Conference On Machine Learning (ICML-11), pp. 129–136. (2011)

Super Sort Algorithm Using MPI and CUDA

Anaghashree, Sushmita Delcy Pereira, Rao B. Ashwath, Shwetha Rai, and N. Gopalakrishna Kini

Abstract Sorting algorithms have been a subject of research. Throughout the years various sorting algorithms have been implemented and their performance has been evaluated by comparing the space and time complexity. In this paper the super sort sorting algorithm with time complexity O(nlogn) has been implemented with MPI and CUDA. The intention is to compare the time taken by super sort algorithm when executed sequentially using C program and the time taken when implemented using CUDA and MPI.

Keywords CUDA · MPI programming · Sorting techniques · Super sort algorithm

1 Introduction

The array of n random elements is usually sorted in ascending or descending order to perform the needed operations. The sorting of elements is of two types: one uses comparison and the other doesn't. Comparison-type sorting [1] consists of bubble sorting, insertion sorting, selection sorting, merge sorting, quick sorting and so on. Non-comparison-type sorting consists of radix sorting, bucket sorting and so on [2].

Anaghashree · S. D. Pereira · R. B. Ashwath (✉) · S. Rai · N. G. Kini
Department of Computer Science and Engineering, Manipal Institute of Technology, Manipal Academy of Higher Education, Manipal 576104, Karnataka, India
e-mail: ashwath.rao.b@gmail.com

Anaghashree
e-mail: anaghashreek@gmail.com

S. D. Pereira
e-mail: sushmitapereira456@gmail.com

S. Rai
e-mail: shwetharai.cse@gmail.com

N. G. Kini
e-mail: ng.kini@manipal.edu

S. C. Satapathy et al. (eds.), *Intelligent Data Engineering and Analytics*, Advances in Intelligent Systems and Computing 1177,
https://doi.org/10.1007/978-981-15-5679-1_16

In this paper the super sort algorithm [3], a comparison sort algorithm, has been implemented using MPI and then with CUDA. A list of n elements is taken which is unsorted. Forward selection is carried out and a small sorted list is obtained. On the unsorted listed backward sort is carried out and a backward sorted list is obtained. Unsorted list with lesser elements than there were originally is left behind. This happens recursively until no elements are left in the unsorted list. Then the backward and forward sorted lists are merged to obtain the sorted list. This is done with MPI and CUDA in C programming to compare the time taken to sort.

2 Sorting Algorithm

The method discussed in this paper arranges the given components in ascending or descending order. The sorting happens in four stages as explained in the following part. When the recursive call finally returns, what is left is a fully sorted array of given input elements.

Initially, the first element in the given array is chosen as the maximum element. That element is removed from the given unsorted list of elements and added onto an empty list called the forwardSorted list. The next element is compared with this maximum element. If it happens to be greater than the present element named maximum, it is removed from the list and appended to the forwardSorted array. This element is now called maximum. If the element to compare is less than the maximum element, it is skipped, and the following element is compared. This process continues until the end of the unsorted list is reached.

In the next pass the backward selection is done where the last element of the unsorted list is called the maximum element. Like the previous forward selection, that element is removed from the unsorted list and added to another blank list called backwardSorted list. Comparison in this pass happens in the reverse direction. The element that is now the last element of the unsorted list is compared with the present maximum. In case it is greater than the maximum element, this element is called the maximum and removed from the unsorted list. If not, the maximum element remains unchanged and the previous entry is compared with the maximum. This step continues until all the elements in the array have been checked.

At the end of these two passes two fully sorted albeit smaller arrays by name forwardSorted and backwardSorted are obtained. These two sorted lists are merged by comparing the first elements of both these lists and adding them to another empty list called partialSorted1. The smaller of the two is removed from the list and added onto the new list. This continues until both the arrays are empty. The original unsorted list has been reduced in size and is taken care of in next partition step.

The partition step divides the unsorted list into two by finding the middle index. This causes the emergence of unsorted smaller lists. The super sort algorithm is recursively called on these two left and right sublists. The recursive function returns when there exists only one element in the original unsorted list. This is possible because every forward and backward selection removes some elements from the list each time, ultimately leaving only one element which is when the function returns, and a fully sorted list has been obtained.

Algorithm ***superSort(us, low, high)***:

```
//Input: Unsorted array us, leftmost position low, right-
most position high
//Output: Sorted array S with n elements in increasing
order
if size(us) < 1 then:
  return
//Initialize forwardSorted array fs and backwardSorted
array bs to NULL
fs ← NULL
bs ← NULL
//Initialize rightmost position of fs and bs arrays
fpos,bpos ← -1
//Forward pass
max ← us[low]
fs[++fpos] ← us.remove(max)
for i ← low to high-1 do
  if us[i] >= max then
    max ← us[i]
    fs[++fpos] ← max
    us.remove(max)
//Backward pass
if size(us) > 1 do
  max ← us[size(us)-1]
  bs[++bpos] ← max
  us.remove(max)
  for i ← size(us)-1 to 0 do
    if us[i] >= max then
      max ← us[i]
      bs[++bpos](max)
      us.remove(max)
partialSorted1 ← merge(fs, bs)
mid ← size(us)/2
leftSublist ← superSort(us, 0, mid)
rightSublist ← supersort(us, mid+1, size(us))
partialSorted2 ← merge(leftSublist, rightSublist)
return (merge(partialSorted1, partialSorted2))
```

```
Algorithm merge(A, B):
//Input : Arrays A and B that are sorted
//Output : mList sorted in increasing order
mList ← NULL
a ← size(A)
b ← size(B)
p,q ← 0
```

```
//Position of mList
pos ← -1
while(p+q < a+b) do
  if (p == a) then
    mList[++pos] ← B[j]
    q ← q+1
  else if (j == b) then
    mList[++pos] ← A[i]
    p ← p+1
  else if (A[p] <= B[q]) then
    mList[++pos] ← A[i]
    p ← p+1
  else if (A[p] > B[q]) then
    mList[++pos] ← B[j]
    q ← q+1
return(mList)
```

Figures 1 and 2 show how the sorting algorithm has been called on a given unsorted list. It shows the forward sort, backward sort, and partition operations on the given list.

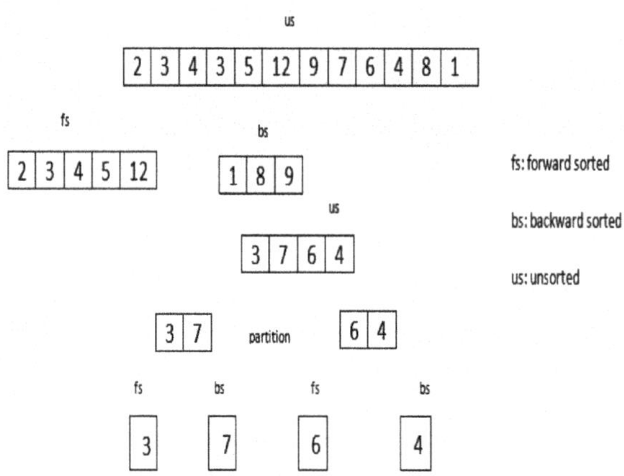

Fig. 1 Recursive calling to get intermediate lists

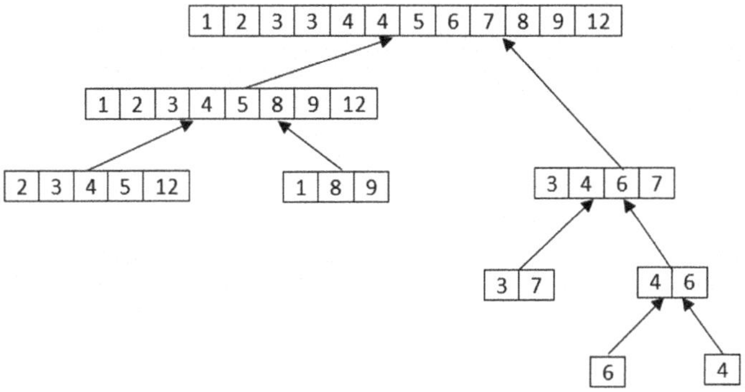

Fig. 2 Merging to get the final sorted list

3 Result

Parallel programming has been used to perform sorting and merging functions simultaneously. The input has been generated using random numbers generation for the given input size. The program was run using sequential C, MPI and CUDA and the time taken for three iterations was noted. The mean of the time taken has been recorded as shown in Table 1.

The observation is that when the input size is small sequential C programming uses less time. But as the input size increases, parallel programming with MPI proves to be better than its sequential counterpart. Also the increase in time taken with increase in input size is more in the case of sequential. With MPI, the increase is gradual.

Super sort with CUDA takes more time compared to both sequential and super sort with MPI for small input size. But when the input size taken is large, it works better than both sequential and MPI counterparts. The time taken sees very gradual increase as opposed to the drastic increase in the case of sequential program.

Figure 3 shows a graph representing the variation of time taken with respect to input sizes from 10^5 to 10^6 for sequential, MPI and CUDA.

Table 1 Time taken to sort for different input sizes by three programs

	Time(s) for different input size					
	10	100	1000	10,000	100,000	1,000,000
Sequential	0.017	0.005	0.01	0.18	2.39	6.352
MPI	0.015	0.056	0.012	0.034	0.063	0.433
CUDA	0.003	0.007	0.021	0.732	0.00034	0.003

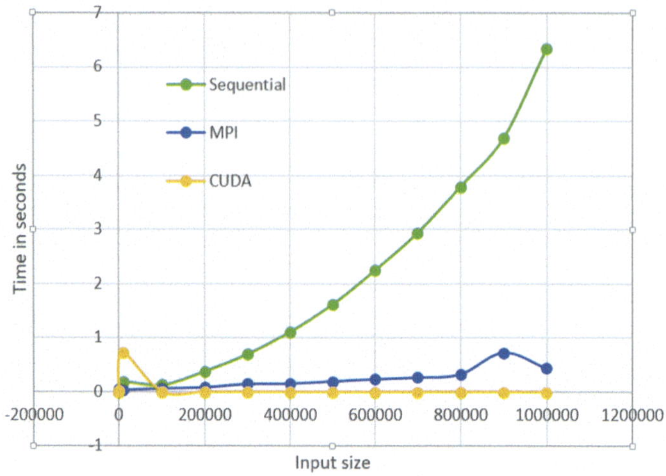

Fig. 3 Change in time for three programs with increasing input size

4 Conclusion and Future Work

The super sorting algorithm has been implemented using MPI and CUDA, and a comparative study of sequential and parallel programs has been done. It can be seen that parallel programming with MPI and CUDA works well for large input size. The future work is to explore more parallelism and thereby further reduce the time taken.

References

1. Comparison sort: (2020). https://en.wikipedia.org/wiki/Comparison_sort. Accessed 7 Dec 2019
2. Cormen, T.H., Leiserson, C.E., Rivest, R., Stein, C.: Introduction to Algorithms, 3rd edn. MIT Press, Cambridge, MA (2009)
3. Gugale, Y.: Super sort sorting algorithm. In: 3rd International Conference for Convergence in Technology (I2CT). IEEE, Pune (2018)

Significance of Network Properties of Function Words in Author Attribution

Sariga Raj, B. Kannan, and V. P. Jagathy Raj

Abstract Author identification or attribution helps in identifying the author of unknown texts and is used in plagiarism detection, identification of writers of threatening documents and resolving disputed authorship of historical documents. Stylometry and machine learning are the most popular approaches to this problem where statistical methods are employed to extract signatures of authors of known texts and these features are used to predict authorship of unknown documents. Complex network approach to feature extraction has focused on content words ignoring function words as noise. In this paper, features of function words of texts are extracted from the word co-occurrence network of texts and used for classification. The results of these experiments are found to have high accuracy. The results of the experiments using function words and content words are compared.

Keywords Author attribution · Complex networks · Natural language processing · Function words

1 Introduction

Language is the most significant tool used by man to communicate [1]. The grammar and vocabulary of the language are acquired by a man from a very young age, and through years of its usage, he tends to follow a style unique to him, intentionally or otherwise. There exist certain features or signatures in the writings that distinguish his creativity from the others [2]. Author attribution or identification is the identification

S. Raj (✉) · B. Kannan · V. P. Jagathy Raj
Cochin University of Science & Technology, Kochi, Kerala, India
e-mail: sariga@cusat.ac.in

B. Kannan
e-mail: bkannan@cusat.ac.in

V. P. Jagathy Raj
e-mail: jagathy@cusat.ac.in

S. C. Satapathy et al. (eds.), *Intelligent Data Engineering and Analytics*, Advances in Intelligent Systems and Computing 1177,
https://doi.org/10.1007/978-981-15-5679-1_17

of these writing styles and is based on understanding the author's choice of words and sentence structure [3]. The extraction of this detail from texts is computationally complex and may not be efficient [4]. Many simple style markers have been identified for the purpose with satisfactory accuracy [5, 6]. Authors of literary texts like novels, poems and plays have their characteristic styles compared to those of news, research articles and others that have limited expressiveness [6]. This research area is quite popular as there is a multitude of languages as each language has intricacies of its own. This research problem finds its application in cyber forensics, plagiarism detection, human cognition, resolving authorship conflicts among others.

Research involves searching for patterns in texts for classification and prediction. The patterns are obtained at the character, phoneme, word, phrase, sentence, paragraph and document levels. Stylometric approach focuses on measuring the features including word length, sentence length, most frequently used words, vocabulary size, word counts and frequencies of n-grams. These features prove to have considerable distinguishing properties, though are considered low on insight. Researchers have tried to involve not only single character/word features but also a combination of characters/words like n-grams to improve on the textual features with good results. In this paper, the patterns at the word-level are studied.

Words can be categorized as content words and function words. Though it is understood that content words give more information, it is found that function words are better style markers [7]. Function words have very high grammaticalized roles and frequencies that vary across the genres of text and are used by everyone unconsciously. The content words, on the other hand, have high semantic content, are less frequent and used consciously by an author. Though function words are often used as classifiers in studies, its contribution to author identification cannot be ignored. The problem is not restricted to semantics only but to the choice of words and its combination with other words. The different categories of function words are shown in Table 1[8]. Moreover, each of these categories has their roles in a text and cannot be treated alike. This study has treated the function words in different categories separately during feature extraction, unlike other studies.

Complex networks is an evolving and promising field of study with contributions in social network analysis, epidemiology, pharmacology, and network security [9, 10]. Complex network approach to the problem of text mining has also had its share of success in the last few years. This method is adopted to study the interaction

Table 1 List of function words in English with examples

Category	Example
Prepositions	of, at, in, without, between
Pronouns	he, they, anybody, it, one
Determiners	the, a that, my, more, much, either, neither
Conjunctions	and, that, when, while, although, or
Auxiliary verbs	be (is, am, are), have, got, do
Particles	No

of words with neighboring words [11]. In a word co-occurrence network (WCN), words of a text from the nodes and edges between nodes exist if they co-occur in it, as shown in Fig. 1 [12]. This approach has been used to solve many problems like word sense disambiguation [13, 14], text summarization [15], topic modeling, language clustering [16, 17] and the like. Studies in the author identification problem have shown that global and local properties of WCNs are not as efficient style markers as the stylometric approaches described above. Many of these studies have ignored the function words in the text while network creation [18, 19]. The success of using the function words as style markers in stylometric approaches motivated the authors of this paper to explore the prospects of measuring the network features of function words in the WCN for the author identification task. The main objectives of this paper are to analyze the performance of complex network features of function words as

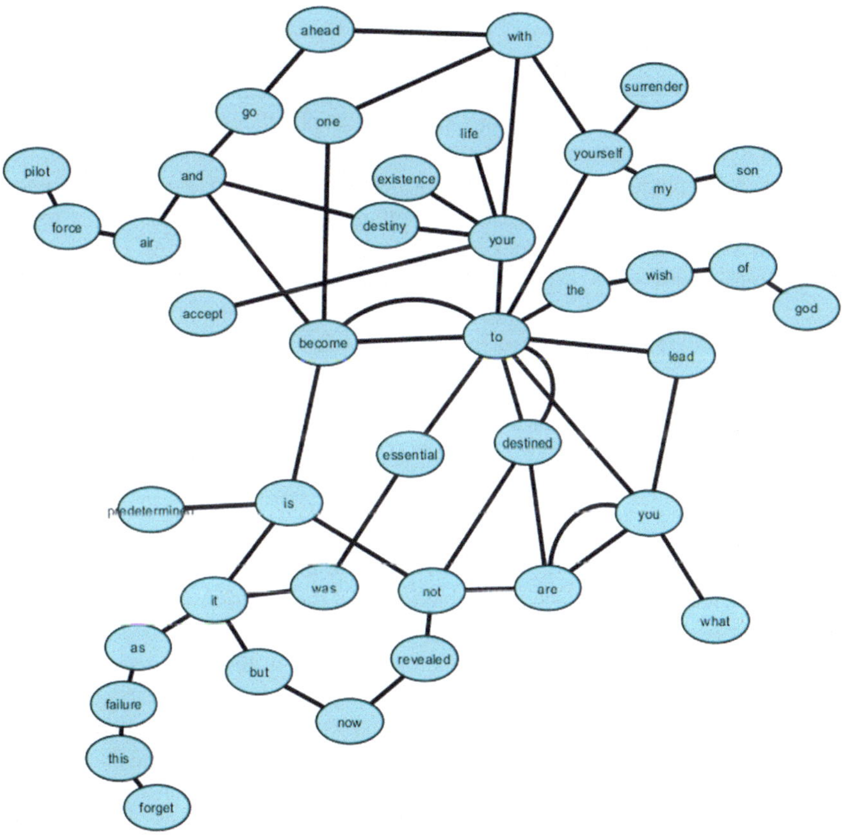

Fig. 1 WCN for the text "Accept your destiny and go ahead with your life. You are not destined to become an Air Force pilot. What you are destined to become is not revealed now but it is predetermined. Forget this failure, as it was essential to lead you to your destined path. Search, instead, for the true purpose of your existence. Become one with yourself, my son! Surrender yourself to the wish of God." Chapter 3, para 10, Wings of Fire, by Dr. APJ Abdul Kalam

style markers to the author identification problem and to compare with performances of other words in different genres of text like novels, news, and movie reviews.

The remaining of the paper explores the author identification/attribution problem first. The next section details the state-of-the-art research in the area of author identification along with complex network approach, and the following sections will describe the problem and the methodology adopted in this study. The experiments conducted and the results obtained are explained in the remaining sections.

2 Author Identification Problem

The basic problem of author identification or attribution (AA) is a multiclass classification problem and can be defined as identifying the author A_m of a text T_m from a closed set of authors $A = \{A_1, A_2, \ldots A_n\}$. The process of author identification involves two steps. The first step is to extract the features from the text and form a feature vector. The second stage involves processing the feature vectors for classification of texts according to the authors, and the most popular is the machine learning approach. Researchers have tried several feature selection methods which are mostly statistical or computational. Features are formed by quantifying text at the character, word, phrase and sentence levels. These methods have shown good results to the identification problem, but the lack of insight is always debated. As compared to human-centric approaches, these methods do not follow a deep linguistic analysis of the text, thus, restricting the approaches to their specific domains [6]. So semantic approaches were developed to extract complex textual features.

3 Related Works

All research in quantifying textual features quote the work of Mendenhall in 1887 as pioneering. Statistical approaches were used initially for author attribution (AA) [20, 21]. Mosteller and Wallace were able to break the tradition with their work on federalist papers [22]. A comprehensive account of studies in AA has been brilliantly described in the seminal work by Stamatatos [6]. Most of the approaches followed were statistical. Later with the advancement of natural language processing (NLP) more complex analyses, like lexical, syntactic and semantic analysis, were used [23–25]. Most of the methods involved had two phases of extraction of textual features and analyzing them. The analysis phase consisted of treating the problem statistically by measuring similarity [26, 27] of textual features, or the machine learning approaches were used to train and test the model with the features extracted from the text.

Text mining approaches are evolving every day with researchers discovering new ways of extracting features for AA. In stylometry, features at character, punctuations, phoneme, word, phrase, clause, sentence, paragraph and document level have been extracted [5, 6, 24], such as frequencies, word/sentence/paragraph length, burstiness, co-occurrences of characters/phonemes/words/phrases.

In the complex network approach, features are extracted from networks formed from texts. Networks or graphs of texts were created as WCNs, syntactic-dependent graphs. Local and global features or combinations have been used with machine learning approaches for AA. Lahiri and Mihalcea [28] demonstrated the application of complex network features for AA. Menon and Choi [29] brought out the significance of function words for AA. Amancio has performed many studies on complex networks of various novels for AA [30, 31] and his results form the basis of our study in this paper. A very interesting approach was to develop calligraphy of the novels-based interdependencies of paragraphs that have been used for author profiling [32]. Motifs and their influence on AA were studied in [33, 34].

Deep learning approaches have now been utilized for authorship attribution with RNN, and CNN models [35, 36]. Idiolect approach, which focuses on understanding the distinct language of the author/speaker, for AA has been used in [37]. Similarly, [38] details the study based on polarity detection of the author, to identify the author. Each of the methods explores different aspects of the writer as language, sentiments, expressiveness and personality. Every study is a step toward finding the best performance for author attribution.

4 Complex Network Approach to Author Identification Problem

A word co-occurrence network (WCN) is used to represent the text as a network of words. The edges between word nodes exist when they co-occur in the text. An illustration of the WCN is given in Fig. 1. There are other graphs like syntactic dependency graphs and semantic graphs that are used for the same application. But these require additional computations, hence are not adopted in this study.

The structural properties of these WCNs help in characterizing the network. Global and local properties are examined to find markers that differentiate one network from another. Since text can be of varying sizes, global properties like the diameter of the graph, number of nodes, number of edges may not be utilized to characterize the text. Local properties like degree, centrality measures, clustering, shortest path length give insights to the connectivity of words with other words. Feature vectors using these properties for author identification have been explored on the basis of degree, clustering coefficients, betweenness centrality, shortest path length, intermittency of words, and so on [18, 19]. But the results are not as accurate as frequency-based features.

As the saying goes "word is judged by the company it keeps", the combinations of words can also be analyzed in cliques and motifs [33]. Another approach is to find the similarity between graphs. This is on the intuition that two graphs from texts of the same author may have similar structure around highly frequent common words. Here, local properties of such nodes are compared based on similarities like cosine, Manhattan, Canberra or Jaccard to find two similar graphs [39]. In this work, machine learning algorithms such as logistic regression, SVM and random forests were utilized for the classification of text.

4.1 Methodology

The objective of this work is to prove the capability of function words as style markers in AA. Hence similar experiments were conducted on different genres of texts like novels, blogs and news. The corpora for these categories were prepared from Project Gutenberg[1] for novels, IMDB62[2] for movie reviews and C50[3] dataset for news.

The procedure has the following steps: data preprocessing, WCN creation, feature extraction, feature-aggregation or feature vector formation and classification using machine learning. Each of the corpora is first preprocessed, and the features are extracted from nodes of the WCN formed from the corpus. We have three approaches here; first, the baseline method where the stopwords are removed from the corpus and lemmatized before WCN of the remaining words are constructed (WCN-CW). In the next two methods no word is removed or lemmatized before graph construction, but feature extraction is carried out only for function words from the WCN for WCN-FW approach and all words for WCN-Complete. The feature vectors are classified using classical machine learning algorithms like logistic regression, SVM, and random forests.

Data and Preprocessing

Data for the experiments were chosen to test the performance of function words across different types of text. Three kinds having distinctive features like novels, news and movie reviews were selected. Novels of four authors were downloaded from the project Gutenberg. Novels chosen were almost of the same time period and each novel was broken down into 440 same-size chunks of 2000 words for each of the four authors to form the corpus. The list of novels selected is given in Table 2. The C50 dataset contained 100 news articles from 50 authors each. The IMDB62 dataset contained 1000 movie reviews from 62 authors. The reviews were extracted and marked for each author appropriately. Preprocessing included removal of punctuations from each sentence and words were part of speech (PoS) tagged. For WCN-CW method stopwords were removed and the remaining words were lemmatized, using the nltk package, for the experiments. Table 3 shows the details of the corpus used for experiments.

---

[1]Project Gutenberg, www.gutenberg.org/ebooks/.

[2]IMDB 62, www.imdb.com.

[3]Zhi Liu: Reuters C50, https://archive.ics.uci.edu/ml/datasets/Reuter 50 50.

Table 2 List of novels selected for processing

Author	Novel
Charles Dickens	Bleak House, The Christmas Carol, Great Expectations, Hard Times, Oliver Twist, The Pickwick Papers, A Tale of Two Cities
Mark Twain	A Tramp Aboard, A Connecticut Yankee, Double Barreled Detective Story, Huckleberry Finn, Innocents Abroad, Man Corrupted, The Prince and the Pauper, Roughing It, Adventures of Tom Sawyer
PG Wodehouse	Damsel in Distress, Adventures of Sally, A Man of Means, Prefects Uncle, Coming of Bill, Indiscretion of Archie, Jill the Reckless, Love Among Chickens, Mike, My Man Jeeves, Piccadilly Jim, Psmith in the City, Right ho Jeeves, The Clicking of Cuthbert
Thomas Hardy	A Pair of Blue Eyes, Far from the Madding Crowd, Jude the Obscure, Mayor of Casterbridge, Return of the Native, Tess of the d'Urbervilles, The Woodlanders

Table 3 Details of corpus used for experiments

Dataset	Number of authors	Average text size	Number of samples per author
Novels	4	2000	440
C50	50	400	50
IMDB62	62	340	1000

Feature Vector

Four local features of words or nodes were extracted from the WCN created [19], which were:

- Degree of the node d_i, is the frequency of the appearance of a word.
- Clustering Coefficient c_i, is equivalent to the fraction of the number of triangles among all possible triads of connected nodes and therefore ranges from 0 to 1.
- Betweenness centrality BC_i, the number of distinct shortest paths between the source node v_s, and the target node v_t, that pass through the node v_i. The sum of all the distinct paths of all nodes divided by the total number of shortest paths between the nodes is the betweenness centrality.
- Shortest path SPL_i, is the average of all shortest paths of that node to all other nodes in the network.

The aforesaid features of all nodes from the WCN created for WCN-CW and WCN-Complete were extracted and aggregated to form the feature vector. The feature vector created for WCN-FW was extracted from words of the six types of function words such as pronouns, prepositions, determiners, conjunctions, auxiliary verbs, and particles, separately.

The aggregation consisted of finding the mean, median, standard deviation, skewness and kurtosis of the four measurements. WCN-CW and WCN-Complete had five moments of four features and so had a dimension of 20, whereas the feature vector of the WCN-FW method was of 120 dimensions (6 function word types $\times$ 4 features $\times$ 5 moments). Feature vectors from different texts of three datasets were created.

Classification

The feature vectors of three categories of text were then classified using classical machine learning methods like logistic regression, SVM and random forests of scikit-learn module. Accuracy, precision, recall and F1 scores were observed for many iterations of the three approaches.

5 Results and Discussions

The experiments were conducted on three datasets using the three methods of WCN-CW, WCN-FW and WCN-Complete. All results of the classification by logistic regression (LR), SVM and random forests (RF) show that the WCN-FW method gives the best results when compared to the baseline method (WCN-CW) and WCN-Complete. Accuracy scores, average precision, recall values of the classification using these classifiers are shown in Table 4. GridSearch method of scikit-learn was applied to the classifiers, and optimal parameters for classification were obtained and tested with the test set. This proves that more frequent function words are better style markers of texts. The chances of overfitting in the case of novels could be eliminated by taking more samples. Also, experiments with text samples extracted from novels are found to be varying compared to news reports or movie reviews. Applying the best values on the parameters the accuracy, precision and recall values are given in Table 4.

The high scores of WCN-FW can be attributed to two factors: one is the high frequency of function words in the text and its treatment in feature vectors. Function

Table 4 Performance of classifiers

Dataset	Classifier	WCN-FW			WCN-CW			WCN-Complete		
		Acc.	Prec	Rec.	Acc.	Prec	Rec.	Acc.	Prec	Rec.
Novels	LR	**99.08**	0.99	0.99	56.02	0.55	0.55	50.11	0.50	0.50
	SVM	**98.57**	0.98	0.98	54.38	0.54	0.54	49.23	0.48	0.48
	RF	**99.19**	0.99	0.99	56.63	0.56	0.56	51.23	0.50	0.50
C50	LR	**98.11**	0.97	0.97	51.01	0.50	0.50	40.89	0.40	0.40
	SVM	**98.07**	0.97	0.97	50.74	0.49	0.49	40.11	0.40	0.40
	RF	**98.80**	0.98	0.98	51.13	0.50	0.50	41.09	0.41	0.41
IMDB62	LR	**98.02**	0.97	0.97	52.12	0.51	0.51	40.01	0.40	0.40
	SVM	**97.88**	0.97	0.97	50.02	0.50	0.50	39.63	0.39	0.39
	RF	**98.67**	0.98	0.98	52.48	0.52	0.52	40.18	0.41	0.41

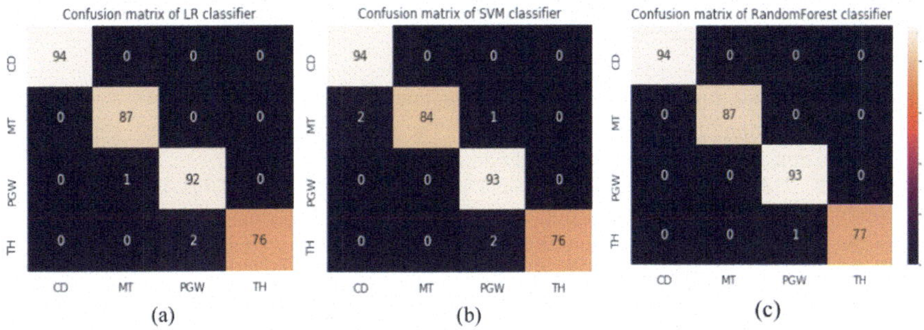

Fig. 2 Confusion matrix of the **a** LR, **b** SVM and **c** RF classifiers on novel datasets

words were overlooked while graph formation in WCN-CW, but the frequency and collocations of content words were not efficient enough. In the WCN-Complete method, feature aggregation was performed on all the words, and due to averaging of scores of all words, it did not extract author characteristic features from text. Each category of function words has different functions and cannot be treated in one class. The WCN-FW method treated the six categories separately and thereby giving high accuracy scores. Confusion matrices of the LR, SVM and random forest classifiers applied on the novel dataset is given in Fig. 2.

6 Conclusion

In this paper, different genres of the text of novels, news and movie reviews were processed to identify the authors using the complex network features. First, each text was represented as word co-occurrence network by removing the punctuations, and features degree, clustering coefficient, betweenness centrality and average path lengths of seven types of function words in the text were extracted separately. These were aggregated to a feature vector for training and testing for author classification.

Traditional methods like linear regression, SVM and random forests were used for classification. These results were compared with a baseline model that ignored stopwords in text while graph formation and another method in which features of all words were aggregated to form the feature vector. The experiments showed high accuracy, precision, recall scores for the proposed method which focused on the function word network properties. The research work signifies that function words have high discriminating properties to differentiate authorship of texts irrespective of the genre of the text.

References

1. Todorov, T., Howard, R.: Poetics of Prose. Cornell Press, New York (1977)
2. Tomori, S., Milne, J., Banjo, A., Afloyan, A.: The Morphology of Present-Day English: An Introduction. Heinemann Educational, London (1977)
3. Allan, B., Trembly, S. (eds.): The Fontana Dictionary of Modern Thoughts. Fontana, London
4. Westerhout, E.: Definition extraction using linguistic and structural features. In: Proceedings of the 1st Workshop on Definition Extraction 61–67 (2009)
5. Stamatatos, E., Fakotakis, N., Kokkinakis, G.: Computer-based authorship attribution without lexical measures. Lang. Resour. Eval. **35**, 193–214 (2001). https://doi.org/10.1023/A:1002681919510
6. Stamatatos, E.: A survey of modern authorship methods. https://doi.org/10.1080/00335634309380866
7. Kestemont, M.: Function words in authorship attribution. From Black Magic to Theory? 59–66 (2015). https://doi.org/10.3115/v1/w14-0908
8. Dang, T.N.Y., Webb, S.: Making an essential word list for beginners. In: Making and Using Word Lists for Language Learning and Testing, pp. 153–167. John Benjamins, Amsterdam (2016). https://doi.org/10.1075/z.208.15ch15
9. Estrada, E.: The structure of complex networks: theory and applications. Published to Oxford Scholarship Online (2013). https://doi.org/10.1093/acprof:oso/9780199591756.001.0001
10. Barabasi, A.-L.: Linked: how everything is connected to everything else and what it means. Plume (2003)
11. Cong, J., Liu, H.: Approaching human language with complex networks (2014). https://doi.org/10.1016/j.plrev.2014.04.004
12. Matsuo, Y., Ishizuka, M.: Flairs02. Dvi. 1–5 (2003)
13. Silva, T.C., Amancio, D.R.: Word sense disambiguation via high order of learning in complex networks. Epl. **98** (2012). https://doi.org/10.1209/0295-5075/98/58001
14. Amancio, D.R., Oliveira, O.N., Costa, L.D.F.: Unveiling the relationship between complex networks metrics and word senses. Epl. **98** (2012). https://doi.org/10.1209/0295-5075/98/18002
15. Pardo, T.A.S., Antiqueira, L., Nunes, M.D.G.V., Oliveira, O.N., Da Fontoura Costa, L.: Using complex networks for language processing: the case of summary evaluation. In: Proceedings of 2006 International Conference Communication Circuits System ICCCAS, vol 4, pp 2678–2682 (2006). https://doi.org/10.1109/ICCCAS.2006.285222
16. Aaronson, S., Aaronson, S.: Ask me anything. Quantum Comput. Since Democritus. **48**, 343–362 (2013). https://doi.org/10.1017/cbo9780511979309.023
17. Liu, J., Wang, J.: Keyword e e xthren manyicularey, as keywofdomen semantic. 129–134
18. Amancio, D.R.: A complex network approach to stylometry. PLoS ONE **10**, 1–21 (2015). https://doi.org/10.1371/journal.pone.0136076
19. Amancio, D.R., Altmann, E.G., Oliveira, O.N., Da Fontoura Costa, L.: Comparing intermittency and network measurements of words and their dependence on authorship. New J. Phys. **13** (2011). https://doi.org/10.1088/1367-2630/13/12/123024
20. Yule, G.U.: On sentence-length as a statistical characteristic of style in prose: with application to two cases of disputed authorship. Biometrika **30**, 363 (1939). https://doi.org/10.2307/2332655
21. Zipf, G.K.: Selected studies of the principle of relative frequency in language. Harvard University Press, Cambridge, MA (1932)
22. Mosteller, F., Wallace, D.: Inference in an authorship problem. J. Am. Stat. Assoc. **58**, 275–309 (1963). https://doi.org/10.2307/2283270, https://www.jstor.org/stable/2283270
23. Gorman, R.: Author identification of short texts using dependency treebanks without vocabulary 1–14 (2019)
24. NagaPrasad, S., Narsimha, V.B., Vijayapal Reddy, P., Vinaya Babu, A.: Influence of lexical, syntactic and structural features and their combination on authorship attribution for TeluguTex. Procedia Comput. Sci. **48**, 58–64(2015). https://doi.org/10.1016/j.procs.2015.04.110

25. Zhang, C., Wu, X., Niu, Z., Ding, W.: Authorship identification from unstructured texts. Knowledge-Based Syst. **66**, 99–111 (2014). https://doi.org/10.1016/j.knosys.2014.04.025
26. Adhikari, A., Subramaniyan, S.: Author identification: using text mining. Feat Eng Net Emb. SemanticScholar.Org. (2016)
27. Rexha, A., Kröll, M., Ziak, H., Kern, R.: Authorship identification of documents with high content similarity. Scientometrics **115**, 223–237 (2018). https://doi.org/10.1007/s11192-018-2661-6
28. Lahiri, S., Mihalcea, R.: Authorship attribution using word network features (2013)
29. Menon, R.K., Choi, Y.: Domain independent authorship attribution without domain adaptation (2011)
30. Akimushkin, C., Amancio, D.R., Oliveira, O.N.: On the role of words in the network structure of texts: application to authorship attribution. Phys. A Stat. Mech. Appl. **495** (2018). https://doi.org/10.1016/j.physa.2017.12.054
31. Akimushkin, C., Amancio, D.R., Oliveira, O.N.: Text authorship identified using the dynamics of word co-occurrence networks. PLoS One **12** (2017). https://doi.org/10.1371/journal.pone.0170527
32. Marinho, V.Q., de Arruda, H.F., Sinelli, T., Costa, L. da F., Amancio, D.R.: On the "calligraphy" of books. In: Proceedings of TextGraphs-11: The Workshop on Graph-Based Methods for Natural Language Processing (2017). https://doi.org/10.18653/v1/W17-2401
33. Marinho, V.Q., Hirst, G., Amancio, D.R.: Authorship attribution via network motifs identification. In: Proceedings—2016 5th Brazilian Conference on Intelligent Systems, BRACIS 2016 (2017). https://doi.org/10.1109/BRACIS.2016.071
34. Marinho, V.Q., Hirst, G., Amancio, D.R.: Labelled network subgraphs reveal stylistic subtleties in written texts. J. Complex Net. **6**, 620–638 (2018). https://doi.org/10.1093/COMNET/CNX047
35. Macke, S., Hirshman, J.: Deep sentence-level authorship attribution. CS224N Proj. 1–7 (2015). https://doi.org/10.1016/j.jpcs.2013.01.035
36. Yao, L., Liu, D.: Wallace: Author detection via recurrent neural networks. CS224N Proj. 1–7 (2015)
37. Wright, D.: Using word n-grams to identify authors and idiolects. Int. J. Corpus Linguist. **22**, 212–241 (2017). https://doi.org/10.1075/ijcl.22.2.03wri
38. Panicheva, P., Cardiff, J., Rosso, P.: Personal sense and idiolect: Combining authorship attribution and opinion analysis. In: Proceedings of 7th International Conference on Language Resources and Evaluation LR 134–1137 (2010)
39. Kocher, M., Savoy, J.: Distance measures in author profiling. Inf. Process. Manag. **53**, 1103–1119 (2017). https://doi.org/10.1016/j.ipm.2017.04.004

Performance Analysis of Periodic Defected Ground Structure for CPW-Fed Microstrip Antenna

Rajshri C. Mahajan, Vibha Vyas, and Abdulhafiz Tamboli

Abstract This research work presents a novel integration of periodic defected ground structure (PDGS) with coplanar waveguide (CPW)-fed microstrip antenna for enhancing its performance. DGS has been incorporated in microwave devices like filters and reflective surfaces for improving their performance characteristics. Additionally, the combination of DGS and the antenna is limited to offer polarization improvisation. In this paper, PDGS is used to enhance the fractional bandwidths of the antenna, which are obtained to be 43, 45, and 24% for 2.4, 5, and 7 GHz, respectively, supporting the multiband operation. The results are compared with CPW-fed microstrip antenna with woodpile electromagnetic band gap (EBG) structure-based ground surface. The comparison shows that PDGS offers better performance in fractional bandwidth and gain at all the three bands of operation of the antenna.

Keywords PDGS · Microstrip antenna · Bandwidth

1 Introduction

With the rapid development of wireless communication, the demand for the design of an antenna with high bandwidth operation recently has been increased. Modern communication systems and instruments require lightweight, small size, and low-cost antennas [1, 2]. The selection of microstrip antenna technology can fulfill these requirements. Microstrip patch is a type of antenna that offers a low profile, that is, thin and easy manufacturability. It is easy to fabricate (by using techniques like

R. C. Mahajan (✉) · V. Vyas · A. Tamboli
College of Engineering (an Autonomous Institute of the Govt. of Maharashtra), Pune (COEP), Pune 411005, Maharashtra, India
e-mail: mrc.extc@coep.ac.in

V. Vyas
e-mail: vsv.extc@coep.ac.in

A. Tamboli
e-mail: tamboliaa17.extc@coep.ac.in

S. C. Satapathy et al. (eds.), *Intelligent Data Engineering and Analytics*, Advances in Intelligent Systems and Computing 1177,
https://doi.org/10.1007/978-981-15-5679-1_18

etching), to feed, and to use in an array with moderate directivity, which provides a great advantage over traditional antennas [3, 4].

However, microstrip patch antennas are inherently having narrow bandwidth. The bandwidth enhancement is usually demanded for most of the practical applications. Several approaches have been utilized for increasing the bandwidth, like electromagnetic band gap (EBG) surfaces, metamaterials, and frequency selective surfaces (FSS) [4–6].

The defected ground structure is another technique that is integrated mostly with microwave devices and filters to specialize in their performance characteristics. The stepped impedance transmission lines are modified to obtain the bandstop, highpass, and bandpass responses. It uses dumbbell-shaped defected ground structures (DGSs), complementary split-ring resonators (CSRRs) DGSs, and inter-digitated coupling structures [7]. The reflective surfaces, in combination with a two-corner-cut square patch and a two-layer substrate with defected ground structure, are proposed for polarization conversion ratio bandwidth expansion and size reduction [8]. DGS is also developed for improving isolation between four ports in a collocated multiple-input–multiple-output (MIMO) [9].

Π-shaped and button-headed H-shaped DGS patterns are used in the filter to broaden its bandwidth and improve the rejection ratio in the low cutoff frequency range [10]. A planar lowpass filter and a fractal defected ground structure are designed to minimize the dimensions of the filter [11].

DGS is used to design a lowpass filter (LPF) that has a compact lengthwise size up to 26.3% as compared to earlier LPF [12]. DGS has also been used in combination with the antenna array for improving the isolation or for a reduction in cross-polarization. H-shaped defected ground structures are proposed to isolate a closely coupled dual-band multiple-input–multiple-output (MIMO) patch antenna that resonates at 3.7 and 4.1 GHz [13]. A back-to-back U-shaped and dumbbell-shaped DGSs are designed for the suppression of the mutual coupling between elements in a microstrip array and elimination of the scan blindness in an infinite phased array [14]. H-shaped DGS is inserted between array elements to reduce the mutual coupling and eliminate the scan blindness in a microstrip phased array design [15].

In this paper, periodic DGS (PDGS) is designed for a slotted CPW-fed wine glass-shaped microstrip antenna for improving its operating bandwidth and gain. The parametric study of the size of PDGS is carried out for better performance of the antenna. The organization of the paper is as follows: Sect. 2 describes the design of CPW-fed slotted microstrip antenna. Section 3 focuses on the parametric study of PDGS. Section 4 briefs about the fabrication of antenna and Sect. 5 presents the conclusion, followed by a list of references.

2 Design of CPW-Fed Microstrip Antenna

The design of the microstrip antenna is carried out using full-wave simulation high-frequency simulation software (HFSS 14.0). The shape microstrip is inspired by an inverted bell-shaped glass [16]. A hexagonal-shaped slot is inserted for improving the

Fig. 1 CPW-fed microstrip antenna with a hexagonal slot

radiation characteristics of the antenna. Figure 1 shows the geometrical configuration and dimensions of the proposed antenna with a hexagonal-shaped slot.

The antenna is printed on cheap and readily available FR4 (glass epoxy) substrate with thickness h = 1.6 mm, relative permittivity $\varepsilon_r = 4.4$, and loss tangent tan $\delta = 0.02$. The patch antenna has width W = 41.56 mm, top to bottom length L = 26.93 mm, ground surface width $W_g = 40.5$ mm, and length $L_g = 31$ mm. The microstrip line length $L_s = 33.56$ mm and spacing between the ground surface and microstrip antenna s = 2.56 mm. The gap between the central strip and ground surface is g = 0.7 mm. The antenna, ground plane, and CPW feed line are printed on a substrate of size 85 mm × 85 mm × 1.6 mm.

The proposed antenna resonates at three frequency bands 1.70, 4.38, and 7.36 GHz with optimum bandwidth for each band. The antenna performance is affected by electrical and geometrical parameters, and this includes the size of a slot on the ground and the ground plane length. The slot on the ground surface can produce multiple resonances providing more bandwidth. Electromagnetic coupling between the defected ground planes and a radiating patch of the antenna and slot on the ground surface can cause the increasing impedance bandwidth. In the next section, the effect of periodically placed circular defects on the ground surface of various sizes is investigated.

3 Circular-Shaped PDGS Structure

Periodic defected ground structure (PDGS) comprises periodic placement of defects on the ground surface. The size and shape of the defect is decided by the operating frequency of the microstrip antenna. In this paper, the circular-shaped defects are inserted on the ground surface as a circle in a naturally occurring shape, and it has uniform DGS coupling. As the sharp edges are absent on the circle shape, there is a reduced diffraction loss. Therefore, the transmission gain is higher compared to other shapes. Secondly, the curvature of the circular-shaped DGS exposed to the transmission is larger compared to other shapes, which in turn reduces the fringing fields, and improves the transmission gain.

Figure 2 depicts the CPW-fed microstrip antenna with a circle-shaped PDGS. The parametric study of the size of the circle is carried out. The radius of the circle is varied from 1 to 2.4 mm, and correspondingly the gap between the adjacent circles is decreased from 3.8 to 0.2 mm. Figure 3 shows the circle radius, r, and gap, g, between the two circles.

A total of 15 simulation experiments are performed for studying the effect of the size of circle-shaped defect considering the radii of circles from 1 to 2.4 mm. The results of the simulation are shown in Table 1.

The combined plot of the return loss vs. frequency of all 15 simulation experiments is shown in Fig. 4.

It is observed that for a circle with radius r = 1 mm, the multiband characteristics are achieved at the resonant frequencies 2.37, 5.00, and 7.66 GHz with return loss −33.39, −59.86, and −18.24 dB, respectively. Gain is an important parameter in the design of the wideband antenna. Figure 5 illustrates the gain patterns for the circular DGS structure of radius 1 mm. It was found that the gain of the antenna is achieved 9.5, 4, 10 dB for the frequency band of 2.4, 5.0, and 7.66 GHz, respectively.

Fig. 2 Microstrip antenna with circle-shaped PDGS

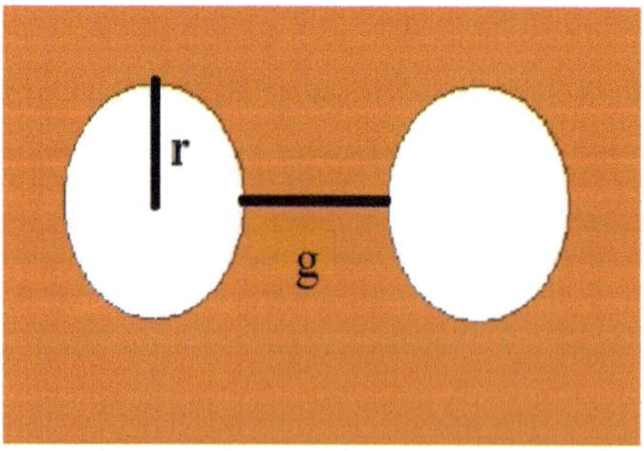

Fig. 3 Circular-shaped PDGS with gap width g between the defects

Table 1 Impedance bandwidths for various shape size of circles

S. no.	Size of circle (mm)	Frequency (GHz)	Return loss (dB)	Bandwidth (GHz)
1	1	2.3784	−33.7913	1.019
		5	−59.8692	2.2803
		7.6667	−18.2415	1.8469
2	1.1	1.9189	−18.92	0.8558
		4.9369	−68.1532	2.2703
		7.648	−18.928	1.982
3	1.2	2.2252	−21.9446	0.9
		4.8378	−35.2260	2.045
		7.6306	−17.4801	1.8018
4	1.3	2.3874	−18.1479	0.3874
		4.9369	−30.8699	2.1801
		7.6396	−19.1401	1.9279
5	1.4	1.917	−24.5	0.8198
		4.955	−32.45	2.1982
		7.638	−16.12	1.9369
6	1.5	1.855	−22.12	1.0541
		4.932	−34.11	2.279
		7.612	−20.1	1.946
7	1.6	1.921	−29.1	1.001
		4.952	−30.1	2.269
		7.65	−20.1	1.939
8	1.7	1.82	−19.91	1.055
		4.29	−28.9	2.2
		7.45	−17.2	1.92
9	1.8	2	−21.72	0.708
		4.9459	−33.98	2.2517
		7.5946	−19.272	1.95
10	1.9	1.8919	−21.33	0.9009
		4.9369	−29.60	2.1982
		7.648	−21.133	1.9817
11	2	1.851	−26.1	1.0541
		4.279	−29.0	2.288
		7.495	−19.77	1.9279
12	2.1	1.8919	−34.034	1.0811
		4.2883	−28.398	2.2612
		7.606	−20.25	1.9363

(continued)

Table 1 (continued)

S. no.	Size of circle (mm)	Frequency (GHz)	Return loss (dB)	Bandwidth (GHz)
13	2.2	1.9279	−34.44	1.4685
		4.2883	−28.39	2.225
		7.5405	−22.46	1.935
14	2.3	1.8288	−26.97	1.2973
		4.9369	−24.019	2.1351
		7.4955	−21.924	1.8288
15	2.4	1.964	−29.70	0.8018
		4.2162	−26.39	2.1427
		7.4865	−23.61	1.8559

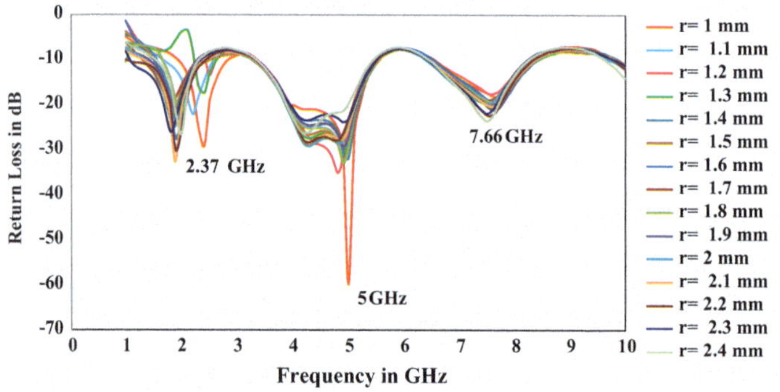

Fig. 4 Return loss vs. frequency for various sized circle shaped PDGS

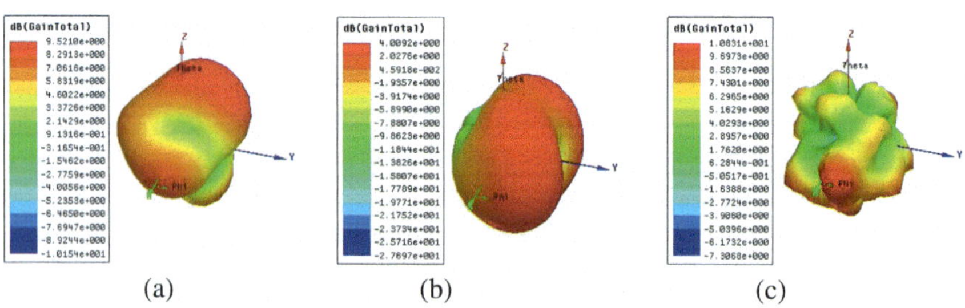

Fig. 5 3D gain of hexagonal-shaped wine glass-shaped MSA of circle radius 1 mm. **a** 2.4 GHz, **b** 5 GHz, **c** 7.66 GHz

The surface current distribution of the proposed antenna for 2.4, 5.0, and 7.6 GHz is shown in Fig. 6. It is observed that more current is concentrated on the side of the radiating patch.

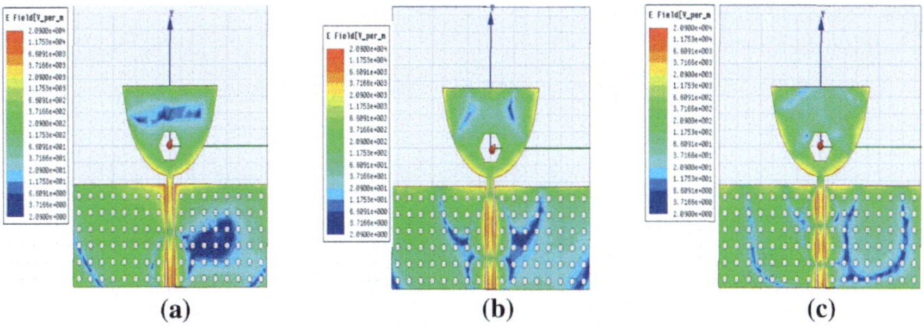

Fig. 6 Surface current distribution of wine glass-shaped MSA of circle radius 1 mm. **a** 2.4 GHz, **b** 5 GHz, and **c** 7.6 GHz

4 Fabrication and Validation of Antenna

The proposed antenna is manufactured using the EP-42AUTO PCB Prototype machine, as shown in Fig. 7. The radius of the circle-shaped PDGS is kept to be 1 mm as it is showing better performance as compared to other sizes.

Figure 8 shows the fabricated prototype of CPW-fed microstrip antenna with a circle-shaped PDGS of radius r = 1 mm as a ground plane.

The return loss measurement of the antenna is performed using the Rohde-Schwarz ZVA 8 (300 kHz to 8 GHz) vector network analyzer. Figure 9 shows the simulation setup for the S_{11} measurement.

After the measurement, it is observed that S_{11} for measured results of the fabricated antenna and simulated results of the antenna strongly agree with each other. Figure 10 shows the comparative plot of return loss for simulated and fabricated antenna.

Fig. 7 Fabrication of antenna using PCB prototype machine

Fig. 8 Fabricated prototype of the proposed antenna

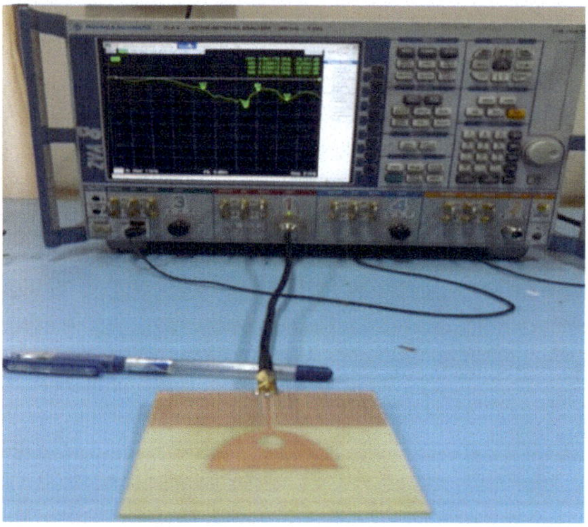

Fig. 9 Antenna measurement setup using VNA

The simulated and fabricated antenna return loss is −33.79 and −30.61 dB at 2.3784 and 2.4 GHz frequency, respectively.

For designing the wideband antenna, it requires minimum group delay. For distortion-less transmission, group delay should be less than 5 ns. The group delay for the simulated and fabricated antenna with a circle PDGS radius size of 1 mm is shown in Fig. 11. It is observed that the group delay is maintained below 5 ns for all the frequency bands of simulated and fabricated antenna.

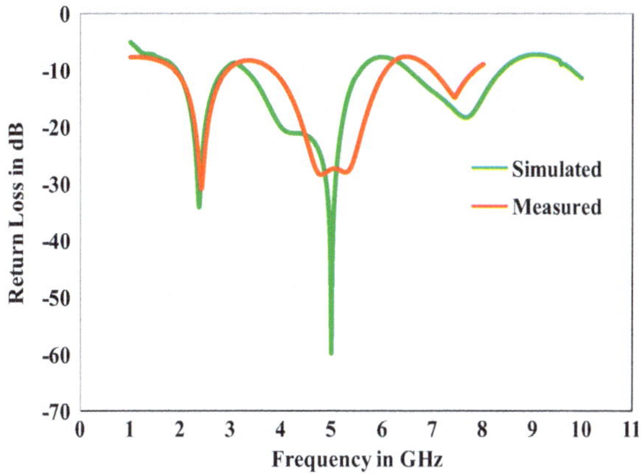

Fig. 10 Comparative plot of return loss for simulated and fabricated antenna

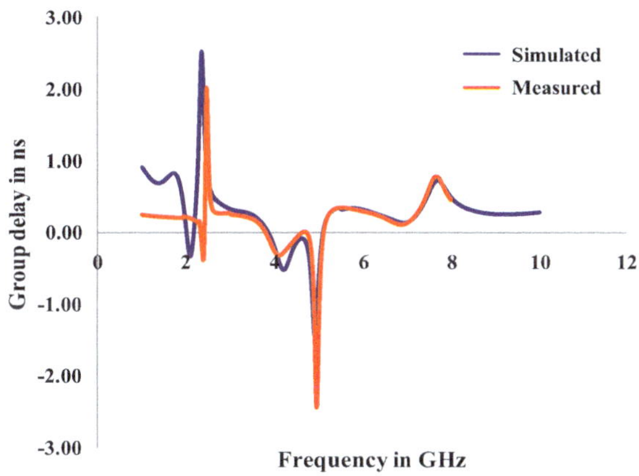

Fig. 11 Group delay plot for simulated and fabricated antenna

Here a better result for unlicensed ISM band, that is, high bandwidth and gain are achieved. So this antenna can be used in wireless applications where wide bandwidth is required.

It is hardly possible to achieve higher bandwidth and gain simultaneously for an antenna. The earlier research was mainly focused on either bandwidth or gain enhancement of the antenna [1, 5, 17]. The proposed antenna with DGS has achieved both the parameters optimally.

The performance of the proposed antenna is compared with a hexagonal slotted CPW-fed microstrip antenna with the ground surface as a woodpile EBG structure with 1 mm woodpile strip width and 1 mm gap width. The antenna resonates at 1.99, 4.94, and 7.68 GHz frequencies, and the fractional bandwidths achieved as 38, 35, and 24%, respectively. The proposed antenna offers the fractional bandwidths of the antenna which are obtained to be 43, 45, and 24% for 2.4, 5, and 7 GHz supporting the multiband operation.

5 Conclusion

The hexagonal slotted CPW-fed microstrip antenna with a circular-shaped PDGS structure of radius 1 mm is proposed. The unique shape of antenna finds useful for increasing the fractional bandwidth. The parametric study of antenna parameters depending on different-sized circle-shaped PDGS is carried out. The bandwidth of 1.091 GHz and the gain up to 10 dB are achieved using the circle-shaped DGS structure of radius 1 mm. The VSWR less than 2 is achieved for the various frequency bands. The group delay is maintained below 5 ns for all sized PDGS circles, which prove advantageous for distortion-less pulse transmission. The antenna is fabricated with 1 mm radius of the circle with 3.8 mm gap width between two circles on the ground plane. This DGS structure is validated for return loss and group delay using a vector network analyzer. The proposed antenna resonates in multiband operation at 2.37, 5, and 7.66 GHz. Also, as per the modern communication requirement, the proposed antenna is a low-cost, high-performance, compact size, comparatively high gain, and low profile.

Therefore, this antenna is useful for wireless communications especially for Wi-Fi, Bluetooth, Zigbee, wireless telephones, RFID systems for merchandise, and NFC, wireless microphones, baby monitors, garage door openers, wireless doorbells, keyless entry systems for vehicles, radio control channels for UAVs (drones), wireless surveillance systems, and wild animal tracking systems where large bandwidth is required.

References

1. Sun, L., He, M., Zhu, Y., Chen, H.: A butterfly-shaped wideband microstrip patch antennas for wireless communications. Int. J. Antennas Propag. 8. Article Id 328208 (2015)
2. Mahajan R.C., Parashar, V., Vyas, V., Sutaone, M.: Study and experimentation of defected ground surface and its implementation with transmission line. Springer Nature (SN) Appl. Sci. **1** (2019)
3. Khanna, P., Sharma, A., Shinghal, K., Kumar, A.: A defected structure shaped CPW-fed wideband microstrip antenna for wireless applications. J. Eng. Hindawi Publishing Corporation (2016)
4. Mahajan, R.C., Parashar, V., Vyas, V.: Modified unit cell analysis approach for EBG structure analysis for gap width study effect. Springer Lecture Notes in Electrical Engineering, vol. 556 (2019)
5. Azim, R., Islam, M.T., Misran, N.: Compact tapered-shape slot antenna for UWB applications. IEEE Antennas Wirel. Propag. Lett. **10**, 1190–1193 (2011)
6. Mahajan, R.C., Vyas, V., Sutaone, M.S.: Performance prediction of electromagnetic band gap structure for microstrip antenna using FDTD-PBC unit cell analysis and Taguchi's multi-objective optimization method. Elsevier Microelectro. Eng. J. **219** (2020)
7. Yuan, W., Liu, X., Lu, H., Wu, W., Yuan, N.: Flexible design method for microstrip bandstop, highpass, and bandpass filters using similar defected ground structures. IEEE Access **7**, 98453–98461 (2019)
8. Moghadam, M.S.J., Akbari, M., Samadi, F., Sebak, A.-R.: Wideband cross polarization rotation based on reflective anisotropic surfaces. IEEE Access **6**, 15919–15925 (2018)

9. Anitha, R., Sarin, V.P., Mohanan, P., Vasudevan, K.: Enhanced isolation with defected ground structure in MIMO antenna. Electron. Lett. **50**(24), 1784–1786 (2014)
10. Zeng, Z., Yao, Y., Zhuang, Y.: A wideband common-mode suppression filter with compact-defected ground structure pattern. IEEE Trans. Electromagn. Compat. **57**(5), 1277–1280 (2015)
11. Kufa, M., Raida, Z.: Lowpass filter with reduced fractal defected ground structures. Electron. Lett. **49**(3) (2013)
12. Mandal, M.K., Sanyal, S.: A novel defected ground structure for planar circuits. IEEE Microwave Wirel. Compon. Lett. **16**(2), 93–95 (2006)
13. Niu, Z., Zhang, H., Chen, Q., Zhong, T.: Isolation enhancement in closely coupled dual-band MIMO patch antennas. IEEE Antennas Wirel. Propag. Lett. **18**(8), 1686–1690 (2019)
14. Xiao, S., Tang, M.-C., Bai, Y.-Y., Gao, S., Wang, B.-Z.: Mutual coupling suppression in microstrip array using defected ground structure. IET Microwaves Antennas Propag. **5**(12), 1488–1494 (2011)
15. Hou, D.-B., Xiao, S., Wang, B.-Z., Jiang, L., Wang, J., Hong, W.: Elimination of scan blindness with compact defected ground structures in microstrip phased array. IET Microwaves, Antennas Propag. **3**(2), 269–275 (2009)
16. Mahajan, R.C., Vyas, V.: Wine glass shaped microstrip antenna with woodpile structure for wireless applications. Majlesi J. Electri. Eng. **13**(1), 37–44 (2019)
17. Zhang, L.N., Zhong, S.S., Liang, X.L., Du, C.Z.: Compact omnidirectional band notch ultra-wideband antenna. Electron. Lett. **45**(13), 659–660 (2009)

Energy Aware Task Consolidation in Fog Computing Environment

Satyabrata Rout, Sudhansu Shekhar Patra, Jnyana Ranjan Mohanty, Rabindra K. Barik, and Rakesh K. Lenka

Abstract The Internet of Things (IoT) is growing rapidly in today's world. A big challenge nowadays is the large volume of data generated between WSN and the cloud infrastructure. Fog computing is a new technology that is an extension to the cloud where processing is performed at the edge of the network, reducing latency and traffic as well. Because of its structure, it has a high demand in healthcare applications, smart homes, supply chain management, smart cities, and intelligent transportation system. Nano data centers (nDCs) are called the tiny computers at the edge of the network. Load balancing is achieved by the current fog architecture. User request allocation technique plays a vital role in fog server energy consumption. The allocation of the user request task to fog servers in a fog environment is a difficult (NP-hard) problem. This article proposes a task consolidation for energy saving by reducing the unused nDCs in a fog computing environment and maximizing CPU utilization.

Keywords Fog computing · Fog architecture · Load balancing · CPU utilization · Energy efficiency

---

S. Rout · R. K. Lenka
School of Computer Science and Engineering, IIIT Bhubaneswar, Bhubaneswar, India
e-mail: shrimansatya23@gmail.com

R. K. Lenka
e-mail: rakeshkumar@iiit-bh.ac.in

S. S. Patra (✉) · J. R. Mohanty · R. K. Barik
School of Computer Applications, KIIT Deemed to Be University, Bhubaneswar, India
e-mail: sudhanshupatra@gmail.com

J. R. Mohanty
e-mail: jnyana1@gmail.com

R. K. Barik
e-mail: rabindra.mnnit@gmail.com

S. C. Satapathy et al. (eds.), *Intelligent Data Engineering and Analytics*, Advances in Intelligent Systems and Computing 1177,
https://doi.org/10.1007/978-981-15-5679-1_19

1 Introduction

Cloud computing is a new internet computing technology that delivers pay-per-use services to the customer's request. It maintains large applications and data servers to provide customer requests or end-users with services. The main technology behind the whole process is the virtualization and central management of data-centered resources. Cloud computing manages the workloads in the data centers over the internet. Data is replicated at multiple sites of the network. The backup, data recovery is all taken care of by the cloud provider. But due to the inherent problem, some critical applications where delays cannot be tolerated cannot work efficiently in the cloud. In such applications, a low-bandwidth problem makes the transmission of data slower due to low bandwidth. Fog computing concept is introduced in such applications.

Fog computing is a technology that gives the user a digital platform close to the level of networking, processing, and storage information generation. Compared to cloud data centers, it is closer to end-users. These services are delivered closer to the user to the customer and are considered the middle-ware between the cloud and the fog servers. As much data is stored locally in a foglet, only summarized information is transmitted to the internet, thus saving bandwidth to a large extent. It reduces both latency and delay and loss of packets. Though fog computing works better as compared to the cloud in case of critical applications, we cannot replace fog by cloud; rather work together as required. The fog computing architecture shown in Fig. 1a, b depicts the fog-cloud model for IoT applications.

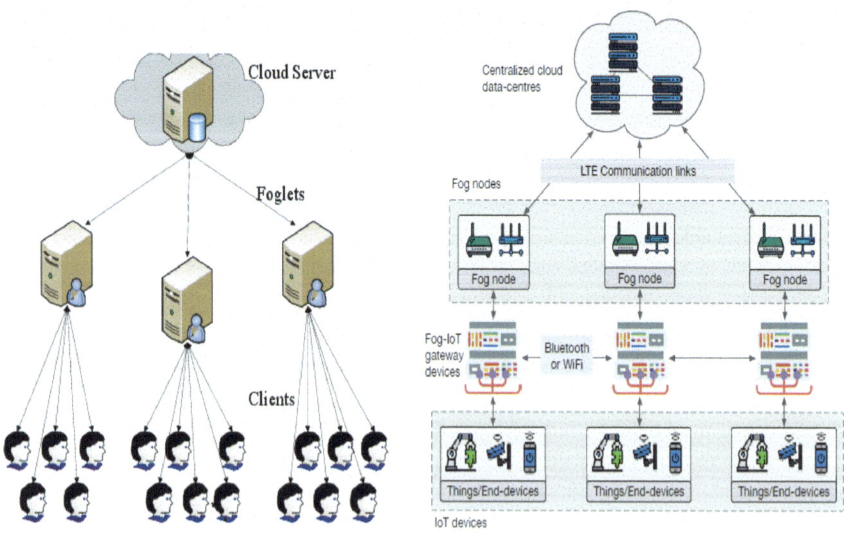

(a) Architecture of Fog Computing[20] (b) fog-cloud model for IoT applications[22]

Fig. 1 **a** Architecture of fog computing [14]. **b** Fog-cloud model for IoT applications [15]

Clients, that is end-users, send requests to the fog layer where questions are handled by the fog server and the responses are returned to the clients. Wherever there is a greater need for computing resources, they will be sent for storage to the cloud layer. When there is a workload imbalance among fog servers, the foglet resource management component balances them among fog servers. Inefficient resource management and imbalance load among the foglets degrade the QoS of the system and energy consumption of the foglets.

In the fog computing systems, energy consumption is mainly on the execution platform, cooling equipment, and air conditioning [1]. For energy, we mostly rely on fossil fuels. For the fog servers, energy saving is a vital issue for the maintenance, services, and the performance of the fog servers. As per the study in [2], the fog servers energy consumption are rendered from 38% petroleum resources, 23% natural gas products, 23% coal products, 9% by nuclear resources, and 7% by other resources. There is a need to use energy optimally as fossil fuels are the sources of non-renewable energy. A data center in the cloud or fog servers in the fog environment consumes significant amounts of energy and higher levels of carbon dioxide (CO_2) in the environment. Likewise, CO_2 emissions [3, 4] cause about 8% of global emissions, which is the main reason for global warming. The QoS can be gained by optimizing the effective utilization of computing resources. This paper concentrates on minimizing the energy utilization in the fog servers and also the makespan of the fog system as QoS constraints. By mapping all the tasks or services efficiently to the committed resources, better QoS can be achieved.

Data center resource utilization is directly dependent on task consolidation, which in turn affects the overall energy consumption in the system [1] and the cost of the system (i.e., how much energy consumption increases also the cost of the system) [5]. As the task consolidation problem is an NP-hard problem, many suboptimal solutions exist to achieve an effective technique. This paper attempts to bring down the energy consumption in a fog server and delivers the required services without compromising its capability.

The rest of the paper is arranged in the following way. Section 2 describes the works involved. In Sect. 3, the architecture for fog computing is defined. Section 4 addresses the proposed consolidation algorithm for the energy-saving function. Section 5 displays the simulation results to check the proposed system's effectiveness. Finally, Sect. 6 concludes the document.

2 Related Works

The popularity of cloud computing and the need for fog servers for mission-critical applications in computational data processing have resulted in increased demand to reduce CO_2 discharge due to substantial energy consumption in large data centers and fog servers. In the data center, CO_2 emissions are mainly due to the cooling system, electrical equipment used by data centers, and so on. In 2013, data centers in the USA are projected to consume 91 billion kWh of electricity that is equal to the output produced annually by 34 large coal-fired power plants generating 500 megawatts of electricity at an early stage, and this amount of electricity could power all New York City buildings for two years [6]. The data centers' annual use of electricity is expected to be 140 BKWh by 2022 [6, 7]. It motivates work in a common area of research, that is, research allocation techniques for fog servers or virtualized server systems [8–11]. As a multidimensional knapsack problem, Lawanyashri et al. [6] proposed a scheme to solve the service allocation problem. Integrated fog and cloud computing maximizes the cloud system's delay, load, and energy consumption. In the fog–cloud infrastructure, Barik et al. [12] applied a fixed delay workload allocation policy. We concluded that the architecture for fog computing significantly improves cloud computing and fog computing systems performance.

3 Fog Computing Architecture

Fog computing is a modern computing paradigm that enables computing to deliver new applications and services for the future of the internet directly at the edge of the network [8]. The fog nodes in fog computing are the resource providers who can provide services at the edge of the network with facilities and infrastructures. Figure 2 shows a fogNode's architecture.

Figure 3 displays the proposed fog computing architecture with three layers: a cloud layer, a fog layer, and a client-tier layer. The fog layer is the middle-tier between clients and the cloud layer.

Cloud computing paradigm has the limitation because many cloud data centers are not located nearby to the users or devices. Fog computing is an emerging technology for resolving these issues. Fog computing provides the desired data processing at the edge network that consists of fog servers.

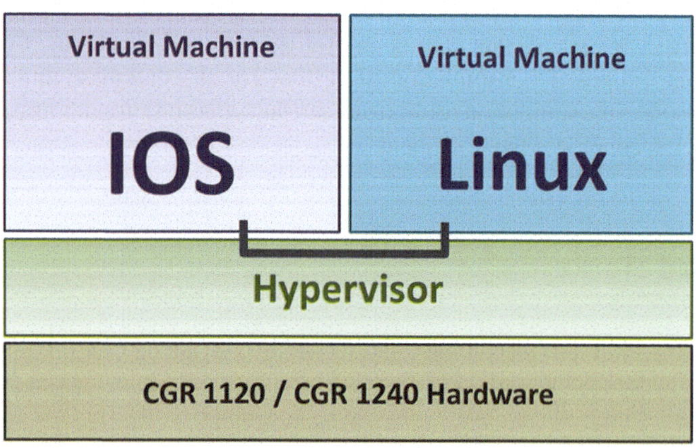

Fig. 2 A fogNode [16]

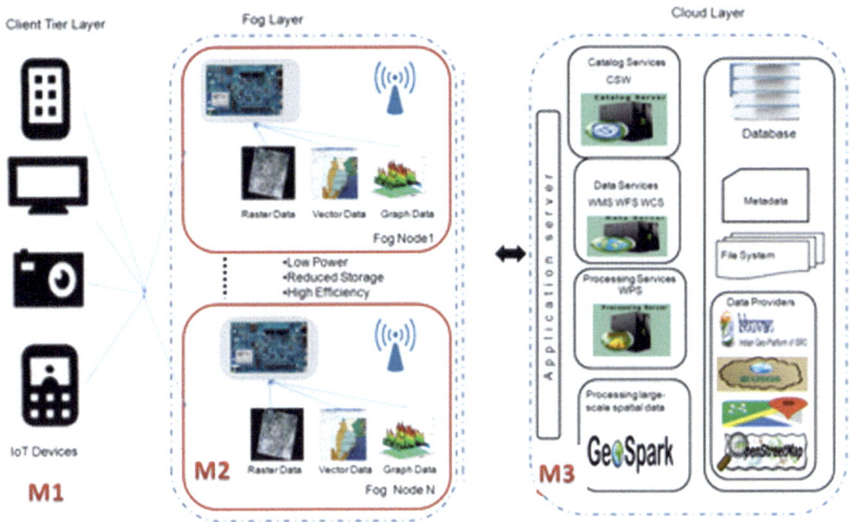

Fig. 3 Fog computing architecture with three modules that work together to perform power-efficient, high-throughput and low-latency large data processing

4 Proposed Energy-Saving Task Consolidation Algorithm

We describe fog computing for the first time and work with models of energy consumption. We also describe the task, that is, the problem of consolidation and load balancing from the client request. The client request consolidation or the task consolidation problem is the technique of assigning a set of tasks $T = \{t_0, \ldots, t_{n-1}\}$ of n client requests (services requests or services or tasks) to a resource set $R = \{r_0, \ldots, r_{m-1}\}$ of m resources by the fog nodes, with the consideration of the time constraints defined by the client requests. The problem objective is the maximization of resource utilization, which, in contrast, minimizes energy usage. The utilization Γ_i for a resource r_i can be defined at any given time.

$$\Gamma_i = \sum_{j=1}^{t} \Gamma_{i,j} \tag{1}$$

In Eq. (1) t is the number of tasks assigned to the current time whereas $\Gamma_{i,j}$ denotes the resource currently being used by the task t_j. The consumed energy EC_i of resource r_i at a time instance is defined as

$$EC_i = (l_{\max} - l_{\min}) \times \Gamma_i + l_{\min} \tag{2}$$

In Eq. (2) $l_{\max}$ is the energy usage at 100% utilization of CPU or highest load and $l_{\min}$ is the minimum energy usage of the fogNode at the extremely low load. The energy used in the VMs or fogNodes can be divided overall into six levels according to power use, the one with idle state and the other five levels of CPU use, as shown in Fig. 4.

Literature related to this work explains the significant impact of CPU utilization on the energy consumption of a process. The consumption of energy is split into two states, one working state, and the other idle one. There is a nonlinear relationship between the use of the CPU and the system's energy consumption. It is shown by a model [13] called the energy consumption model shown in Fig. 5a. The curves indicate the energy consumption on respective machines. When the CPU usage lies in between 0 and 20%, the slopes of the curves are the smallest. These states are the unutilized states of the fogNodes. When the use of the CPU drops between 20 and 50%, the consumption of energy increases marginally. The system has a restrained

$$E_i(V_i) = \begin{cases} \beta_1 \; watts/s, if \; idle \\ \beta_2 + \beta_1 \; watts/s, if \; 0\% < CPU\;Usage <= 50\% \\ 2\beta_2 + \beta_1 \; watts/s, if \; 50\% < CPU\;Usage <= 70\% \\ 3\beta_2 + \beta_1 \; watts/s, if \; 70\% < CPU\;Usage <= 80\% \\ 4\beta_2 + \beta_1 \; watts/s, if \; 80\% < CPU\;Usage <= 90\% \\ 5\beta_2 + \beta_1 \; watts/s, if \; 90\% < CPU\;Usage <= 100\% \end{cases}$$

Fig. 4 Five levels of CPU utilization [11]

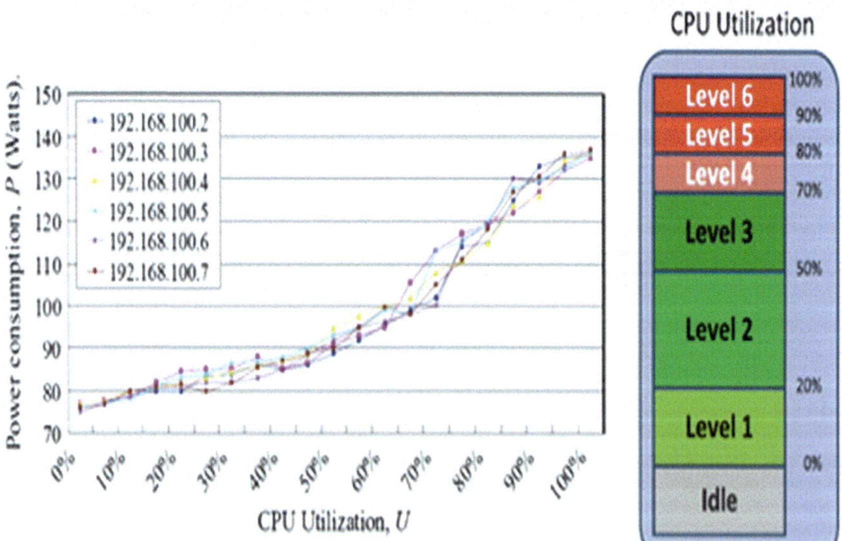

Fig. 5 How energy consumption varies with CPU utilization: **a** experimental study for various machines [13] and **b** the CPU utilization model

Table 1 Pseudo code for load balancing in fog nodes

```
Algorithm: Load Balancing in Fog_Nodes
Input: Request, FogServer (list of Requests and FogServers)
Output: Response    (List of Responses)

Function Utilization (Requests):
 for each fogNode_i € FogServer do
   if fogNode_i is not overloaded then          //fogNode can be considered for a defined threshold
     for each Request_i € Requests   do
        Response_i= fogNode_i(Request_i)         // Requesting the fogNode
           if Response_i = NORMAL  then
              return Response_i
          else
              Response_i = TransferToCloud(Request_i)  //Requesting the cloud to execute the
                                                   request by transferring the request to the cloud
          endif
     endfor
  else  // if fogNode_i is overloaded
   .  for each Request_i € Requests   do
          fogNode_j = fogNode_i((Request_i)
              Response_j= fogNode_j(Request_i)       // Requesting another fogNode
            if Response_j = NORMAL  then
              return Response_j
          else
              Response_j = TransferToCloud(Request_j)  //Requesting the cloud to execute the
                                                   request by transferring the request to the cloud
          endif
     endfor
```

increase in power usage when CPU usage ranges between 50 and 70%. At last, when the CPU usage percentage ranges between 70 and 100%, the energy usage of the machine rises significantly. Figure 5a indicates the variation in energy between energy usage and the CPU usage of a machine.

Figure 5b shows the equivalent pattern of energy consumption based on a Fig. 5a study. The fogNodes uses an idle state of the six levels in terms of energy consumption and five other levels of CPU utilization. A machine has a nonlinear relationship between CPU and the use of energy. We also anticipate that if the workload is distributed among the different CPUs the energy consumption is less. To maintain the CPU usage under a certain level, the tasks should be redistributed among the fogNodes.

The load balancing among the fogNodes is done as per the algorithm given in Table 1.

The maxUsageECTC (maximum usage energy conscious task consolidation) algorithm is defined in Table 2.

5 Simulation Results

The conduct of the proposed task consolidation algorithm was analyzed here with 1200 tasks. Multiple fogNodes from an incompatible ETC matrix carry out the tasks [3]. For 1200 tasks, Matlab 2012 code was used to evaluate the quality of the proposed heuristic. The activities arrive at the queue of the central server with an arrival level

Table 2 Pseudocode for MaxUsageECTC heuristic

Algorithm: MaxUsageECTC

```
Input: ETC Matrix (mat) having
        // The ETC matrix have the columns
        // TId, Arrival Time, CPU Usage %, Processing Time
Output : Allocation result(mat) having
        TId, MId, Task Execution Start Time, Task Execution End Time, CPU Usage %
1. Compute the MaximumArrivalTime and the MinimumArrivalTime form the input Task Matrix(mat)
2.currentTime ← MinimumArrivalTime
3.while (currentTime <= MaximumArrivalTime)
4.do
5.        CurrentTasklist ← FindTasksatArrivalTime(Task Matrix , currentTime)
6.        Create a maxHeap of the CurrentTasklist based on CPU utilization
7.      while (CurrentTasklist ≠ Ø)
8.        do
9.          task ← ExtractMax(mat)
10.         for each fogNode in fogNodelist
11.          do
12.            maxEnergyConsumed←1
13.            EC= EnergyConsumptionInclusionofTheTask(task,fogNode)
14.           //Assign the task to the fogNode in which the Increase in Energy consumption is less by the inclusion of the
              //task
15.           if (EC > maxEnergyConsumed)
16.              maxEnergyConsumed ←EC
17.             allocatedfogNode ←fogNode
18.           end if
19.        end for
20.        if (allocatedfogNode !=NULL)
21.          Assign the task to allocatedfogNode
22.          Modify the Allocation table
23.       end if
24.     end while
25. end while
```

of λ. Figures 6 and 7, respectively, show the performance of the proposed algorithms for 12 and 15 fogNodes. The energy usage in kilojoules with 15 fogNodes for various task sizes from 500 to 1500 has been shown in Fig. 8. Figure 9 shows the makespan versus the number of fogNodes in the system.

6 Conclusion

The simulations have validated the behavior of the consolidation algorithms of the heuristic task successfully. The heuristic proposed also minimizes the use of energy in a fog computing ecosystem. The algorithm for the different ETC matrix, the energy consumption, makespan, and resource utilization of the system was studied. The simulation results showed that in comparison with existing methods, the proposed heuristic algorithm performs better in different parameters such as resource utilization, energy-saving, and makespan. Since the task allocation problem in fog server is an NP-hard problem and does not exist in a polynomial-time algorithm, the heuristic algorithms have been developed, and our algorithm MaxUsageECTC is outperforming the other algorithms in many scenarios.

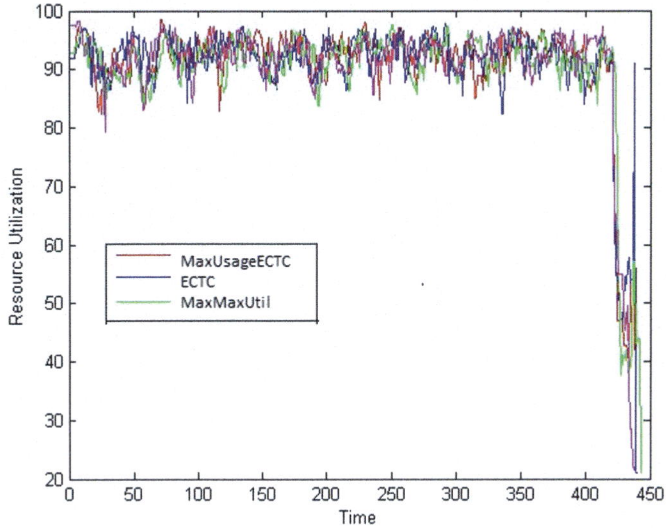

Fig. 6 Comparison of CPU utilization of 1200 tasks on 12 fogNodes

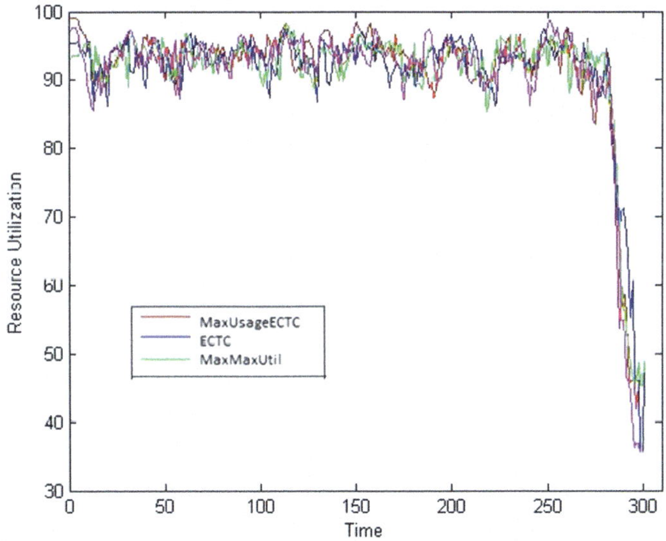

Fig. 7 Comparison of CPU utilization of 1200 tasks on 15 fogNodes

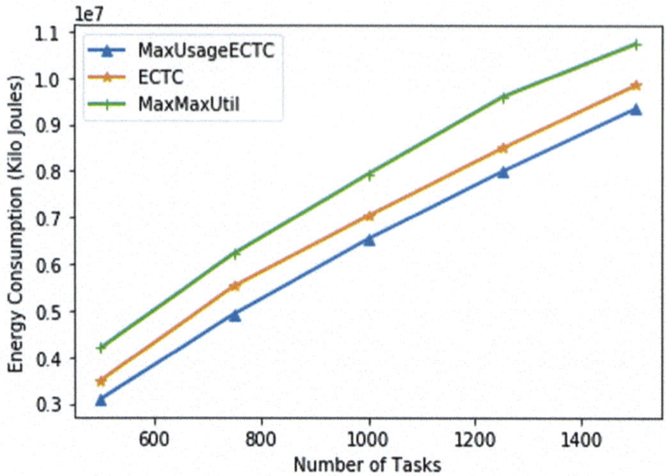

Fig. 8 Energy consumption (kilo joules) versus number of tasks on 15 fogNodes

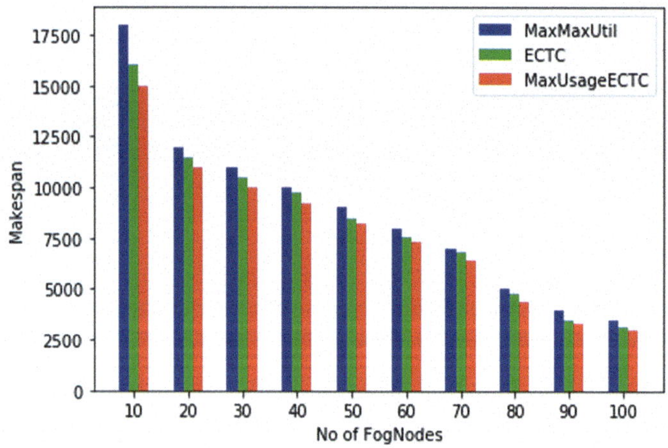

Fig. 9 Makespan versus no of fogNodes

References

1. Barik, R.K., Dubey, H., Samaddar, A.B., Gupta, R.D., Ray, P.K.: FogGIS: Fog computing for geospatial big data analytics. In: 2016 IEEE Uttar Pradesh Section International Conference on Electrical, Computer and Electronics Engineering (UPCON), pp. 613–618. IEEE (2016)
2. Dubey, H., Yang, J., Constant, N., Amiri, A.M., Yang, Q., Makodiya, K.: Fog data: enhancing telehealth big data through fog computing. In: Proceedings of the ASE Bigdata & Socialinformatics 2015, p. 14. ACM (2015)
3. Farahani, B., Firouzi, F., Chang, V., Badaroglu, M., Constant, N., Mankodiya, K.: Towards fog-driven IoT eHealth: promises and challenges of IoT in medicine and healthcare. Future Generat. Comput. Syst. **78**, 659–676 (2018)

4. Mahmoud, M.M., Rodrigues, J.J., Saleem, K., Al-Muhtadi, J., Kumar, N., Korotaev, V.: Towards energy-aware fog-enabled cloud of things for healthcare. Comput. Electr. Eng. **67**, 58–69 (2018)
5. Sun, Y., Zhang, N.: A resource-sharing model based on a repeated game in fog computing. Saudi J. Biologi. Sci. **24**(3), 687–694 (2017)
6. Lawanyashri, M., Balusamy, B., Subha, S.: Energy-aware hybrid fruitfly optimization for load balancing in cloud environments for EHR applications. Infor. Medi. Unlock. **8**, 42–50 (2017)
7. Goswami, V., Patra, S.S., Mund, G.B.: Performance analysis of cloud with queue-dependent virtual machines. In: 2012 1st International Conference on Recent Advances in Information Technology (RAIT), pp. 357–362. IEEE (2012)
8. Barik, R.K., Misra, C., Lenka, R.K., Dubey, H., Mankodiya, K.: Hybrid mist-cloud systems for large scale geospatial big data analytics and processing: opportunities and challenges. Arab. J. Geosci. **12**(2), 32 (2019)
9. Constant, N., Borthakur, D., Abtahi, M., Dubey, H., Mankodiya, K.: Fog-assisted wiot: a smart fog gateway for end-to-end analytics in wearable internet of things. arXiv:1701.08680 (2017)
10. Hu, P., Dhelim, S., Ning, H., Qiu, T.: Survey on fog computing: architecture, key technologies, applications and open issues. J. Netw. Comput. Appl. **98**, 27–42 (2017)
11. Hsu, C.-H., Chen, S.-C., Lee, C.-C., Chang, H.-Y., Lai, K.-C., Li, K.-C., Rong, C.: Energy-aware task consolidation technique for cloud computing. In: 2011 IEEE Third International Conference on Cloud Computing Technology and Science (CloudCom), pp. 115–121 (2011)
12. Barik, R.K., Dubey, H., Mankodiya, K., Sasane, S.A., Misra, C.: GeoFog4Health: a fog-based SDI framework for geospatial health big data analysis. J. Ambient Intell. Humaniz. Comput. **10**(2), 551–567 (2019)
13. Beloglazov, A.: Energy-efficient management of virtual machines in data centers for cloud computing. PhD thesis, Department of Computing and Information Systems, The University of Melbourne (2013)
14. Khattak, H.A., Arshad, H., ul Islam, S., Ahmed, G., Jabbar, S., Sharif, A.M., Khalid, S.: Utilization and load balancing in fog servers for health applications. EURASIP J. Wirel. Communi. Netw. (1), 91 (2019)
15. Adhikari, M., Mukherjee, M., Srirama, S.N.: DPTO: A deadline and priority-aware task offloading in fog computing framework leveraging multi-level feedback queueing. IEEE Inter. Things J. (2019)
16. Cisco. Iox overview. http://goo.gl/n2mfiw (2014)
17. Barik, R.K., Priyadarshini, R., Lenka, R.K., Dubey, H., Mankodiya, K.: Fog computing architecture for scalable processing of geospatial big data. Int. J. Appl. Geospat. Res. (IJAGR) **11**(1), 1–20 (2020)
18. Pooranian, Z., Shojafar, M., Naranjo, P.G.V., Chiaraviglio, L., Conti, M.: A novel distributed fog-based networked architecture to preserve energy in fog data centers. In: 2017 IEEE 14th International Conference on Mobile Ad Hoc and Sensor Systems (MASS), pp. 604–609. IEEE (2017)
19. Naranjo, P., Pooranian, Z., Shamshirband, S., Abawajy, J., Conti, M.: Fog over virtualized IoT: new opportunity for context-aware networked applications and a case study. Appl. Sci. **7**(12), 1325 (2017)
20. Mahmud, R., Kotagiri, R., Buyya, R.: Fog computing: a taxonomy, survey and future directions. In: Internet of Everything, pp. 103–130. Springer, Singapore (2018)
21. Monteiro, A., Dubey, H., Mahler, L., Yang, Q., Mankodiya, K.: Fit: a fog computing device for speech tele-treatments. In: 2016 IEEE International Conference on Smart Computing (SMARTCOMP), pp. 1–3. IEEE (2016)
22. Mishra, S.K., Puthal, D., Rodrigues, J.J., Sahoo, B., Dutkiewicz, E.: Sustainable service allocation using a metaheuristic technique in a fog server for industrial applications. IEEE Trans. Industr. Inf. **14**(10), 4497–4506 (2018)
23. Dastjerdi, A.V., Buyya, R.: Fog computing: helping the internet of things realize its potential. Computer **49**(8), 112–116 (2016)

Modelling CPU Execution Time of AES Encryption Algorithm as Employed Over a Mobile Environment

Ambili Thomas and V. Lakshmi Narasimhan

Abstract This paper presents results on modelling of AES encryption algorithm in terms of CPU execution time, considering different modelling techniques such as linear, quadratic, cubic and exponential mathematical models, each with the application of piecewise approximations. C#.net framework is used to implement this study. This study recommends quadratic piecewise approximation modelling as the most optimized model for modelling the CPU execution time of AES towards encryption of data files. The model proposed in this study can be extended to other encryption algorithms, besides taking them over a mobile cloud environment also.

Keywords Mobile computing · Mathematical modelling · Piecewise approximation

1 Introduction

Mobile environment facilitates data sharing between devices that support mobility across mobile networks. Developed and developing countries experience a tremendous growth in mobile devices' penetration and mobile technologies' usage [1]. Several studies show that the count of mobile phone subscriptions has surpassed the global population by 2018, and nearly the entire world population lives within the mobile network range [2]. Increased mobile device penetration results in significant increase in the development of mobile applications in various domains. Mobile users use numerous mobile applications in their mobile devices. Therefore, mobile devices consume substantial amount of energy to run the augmented number of mobile applications. But the mobile devices depend on the constrained energy sources to operate

A. Thomas
BOTHO University, Gaborone, Botswana
e-mail: ambili.thomas@bothouniversity.ac.bw

V. L. Narasimhan (✉)
University of Botswana, Gaborone, Botswana
e-mail: srikar1008@gmail.com

S. C. Satapathy et al. (eds.), *Intelligent Data Engineering and Analytics*, Advances in Intelligent Systems and Computing 1177,
https://doi.org/10.1007/978-981-15-5679-1_20

[3, 4]. Thus, it is important to ponder about the optimized energy consumption of mobile devices. Ubiquity of mobile phones implies that secured data transmission over the mobile environment, along with its performance, is the major area of concern. Nowadays, organizations operate their business effectively through the implementation of various mobile computing techniques. This situation demands for high security of organizations' sensitive data and optimized energy consumption of mobile devices.

A tradeoff exists between the security and the energy consumption of mobile devices. Higher security is achieved with the cryptographic algorithm having a bigger number of rounds and long encryption key sizes. Due to the higher computation complexity involved, cryptographic algorithms consume substantial amount of energy and execution time. Higher security demands higher energy consumption [4]. The execution of cryptographic algorithms to encrypt the data results in reduction of battery lifetime in mobile devices [4]. Since cryptographic algorithms are widely used to ensure security of data at rest and data in transit, it is important to examine the performance of cryptographic algorithms running within the context of energy used. Central processing unit (CPU) execution time[1], which consumes majority of the energy during execution, is used as one of the metrics to analyze cryptographic algorithms' energy consumption. The estimation of CPU execution time and energy consumption are essential [3] to be carried out in the mobile environment. Thus, an optimized energy model which supports the most possible secured data processing is essential in the mobile environment.

Considering the wide popularity, advanced encryption standard (AES) algorithm has been chosen. Montoya et al. [5] conclude AES as an optimum algorithm for mobile environment, where the battery consumption is a critical factor. An optimized model based on CPU execution time of AES algorithm has been proposed. The metric chosen for this study is the CPU execution time taken by the AES algorithm for encrypting a data file.

The objective of this study is to examine and find out the actual CPU execution time taken by the AES algorithm. This result can be used to analyze and optimize the energy consumption of the AES algorithm. The rest of the paper is organized as follows: Sect. 2 provides an overview of the related literature, while Sect. 3 describes the proposed model. Section 4 provides experimental analysis of data and, Sect. 5 compares the mathematical models. The conclusion summarizes the paper and provides the pointers for further work in this arena.

[1]Other parameters such as memory swap time, cache miss time can be included. However, encryption algorithm is usually in memory-resident state. Therefore, CPU execution time is the dominant parameter over all the other parameters to be considered.

2 Related Literature

AES algorithm is chosen for this study as this is one of the most widely used security algorithm and is suitable for the resource-constraint mobile devices. Lu and Tseng [6] have proposed an AES algorithm architecture which is suitable for the mobile devices. Toldinas et al. [4] propose an energy security tradeoff model based on cryptography which describes how the cryptographic algorithms' security and energy consumption relate. This study concludes AES as one of the most energy-efficient asymmetric algorithms among other analyzed algorithms. The AES algorithm is selected, because it is most widely used for encryption and energy efficiency. Ramesh and Suruliandi [7] have done a comparative study on the performance of cryptographic algorithms, such as AES, data encryption standard (DES) and BLOWFISH using performance metrics—execution time, memory usage and throughput. A study has been conducted by Elminaam et al. [8] to evaluate the performance of various symmetric algorithms in terms of encryption time, throughput and power consumption. Javed et al. [9] have surveyed energy consumption in mobile phones in terms of energy consumed by operating systems, applications and hardware. They conclude CPU and wireless techniques as the components of mobile phones which consume most of the energy. These facts motivated us to choose the CPU execution time of AES algorithm as the base for this study.

Dolezal et al. [10] propose mathematical models to predict energy consumption of smartphones. They analyze and evaluate the energy consumption on various hardware components, such as CPU, screen, storage and speakers. They conclude that these hardware components' energy consumption does not vary linearly, and thus suggest investigating using general mathematical models. This inspired us to consider four mathematical models, namely linear, quadratic, cubic and exponential models to develop an optimized model for this study.

Marsiglio [11] states the importance of piecewise linear approximation in the optimization domain. Fallah et al. [12] examine the performance of piecewise linear approximation (PLA) techniques in wireless sensor network (WSN) based on the energy consumption and the compression ratio. This paper uses the PLA techniques to reduce the energy consumption during data transmission. To optimize the model, we apply PLA with each of the four mathematical models so that an optimum model for measuring execution time of AES algorithm can be obtained.

Umaparvathi and Varughese [13] have studied symmetric encryption algorithms' evaluation in terms of power consumption over MANETs. This paper concludes AES as the higher performance algorithm in the battery power constrained environment. The approach encrypts input files in different hardware platforms using Java programming language, followed by a comparison based on encryption time, decryption time and throughput. They discuss the CPU clock cycle as a CPU energy consumption evaluation metric for encryption operations. As a future work, they suggest the study of encryption algorithms' battery power consumption based on CPU clock cycles. This study inspired the researcher to work with the calculation of CPU clock cycles using the assembly code of AES algorithm. With the study on various methods to measure the execution time of embedded systems, Stewart [14] considers both software-only methods and hardware-specific methods. They discuss that the software-only methods which assist to run the code and measure the execution time is much easier than the hardware-specific methods.

3 Description of Our Model

Two schemes are followed for the modelling of our problem space. With Scheme 1, CPU execution time of AES algorithm is calculated using C#.net framework. With Scheme 2, CPU execution time of AES algorithm is calculated using the assembly code of AES algorithm. Both are elaborated below;

3.1 Scheme 1

Specification of System

The specification of the laptop used for this study includes Windows 10 Pro Operating System with Intel i5 Processor. The software named Visual Studio 2015 is used to run the sample C# code of AES algorithm.

Experimental Procedure

Contents inside a data file were encrypted using the AES algorithm. The data file was encrypted 200 times within the single execution of the AES algorithm sample code and obtained 100 iterations of such execution time samples to create data set for this study. This approach is followed to ensure the data sets' consistency and accuracy through the reduction of possible cache effects [15]. TotalProcessorTime property of the C#.net framework is used to calculate the total processor time spent by AES algorithm for its execution.

The first model follows the algorithm specified in Fig. 1, in order to find out an acceptable data set for this study.

According to Kaufmann [16], the CV value less than 1 indicates less variation of the data distribution, and thus the corresponding mean value will be acceptable. The scatter graph in Fig. 2 is plotted using the acceptable data set identified in this study.

- Execute the AES encryption code in C# to encrypt the data file contents.
- Capture 100 iterations of execution time as the output of C# code.
- Obtain a data set with these 100 iterations, with X value as the CPU Execution Time and Y value as the Number of Samples.
- Calculate Mean, Standard Deviation (SD) and coefficient of variation (CV) based on the data set. [If CV value is less than 1, the Mean for the data set is acceptable].

Fig. 1 Algorithm for the acceptable data set

Fig. 2 Scatter graph of acceptable data set

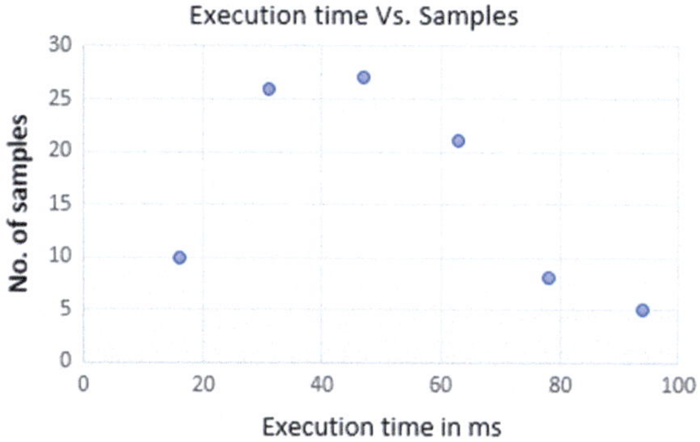

Various mathematical models, such as linear, quadratic, cubic and exponential models are created using the values of Fig. 2. Piecewise approximation is also applied with each of these models to optimize, and the model with the least root mean square error (RMSE) value is chosen as the most optimized model.

3.2 *Scheme 2*

Scheme 2 uses the CPU clock cycles used by the assembly equivalent of the AES encryption code to calculate the CPU execution time. The scheme follows the algorithm given in Fig. 3, in order to find out the CPU execution time for the AES algorithm using corresponding assembly code.

4 Experimental Analysis of Data

Experimental analysis of data based on Scheme 1 and Scheme 2 is described in the following subsections.

- Consider the AES encryption code in C# to encrypt the data file contents.
- Get the assembly equivalent code of AES encryption code.
- Calculate the total clock cycles used by the assembly code.
- Calculate the CPU execution time using the below formula;
  CPU execution time = total clock cycles/ clock rate.

Fig. 3 Algorithm for CPU execution time calculation from the assembly code

4.1 Scheme 1

Experimental analysis of data based on linear, quadratic, cubic and exponential models is described in this section. Microsoft Excel is used to create the best fit model based on the four models and to calculate the RMSE value for each of the best fit model.

Linear Model
Figure 4 depicts the best fit model created using the data set, and this linear model is given as:

$$y = -0.156x + 24.72 \tag{1}$$

The RMSE value calculated for the linear model is 7.78, which is a better fit based on the data range. The Y value (dependent variable) of the data set ranges from 5 to 27 and the RMSE value comes as an acceptable value within the data range. According to Martin [17], a lower RMSE value ensures a better fit, and hence, this linear model can be considered as a robust model.

Figure 5 depicts the linear piecewise models with RMSE values of 5.63 and 2.48.

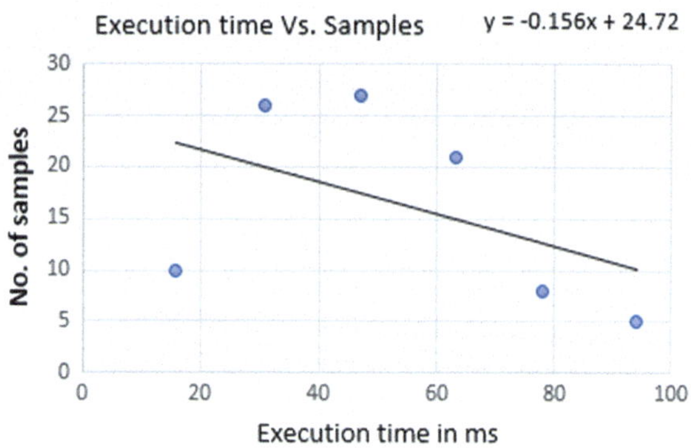

Fig. 4 Linear model of execution time versus number of samples

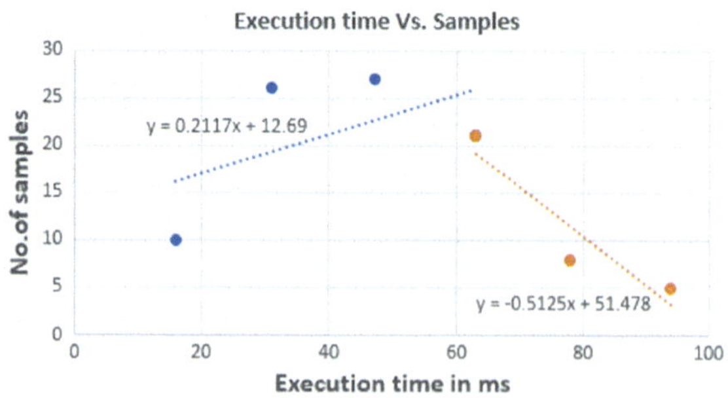

Fig. 5 Linear piecewise model of execution time versus number of samples

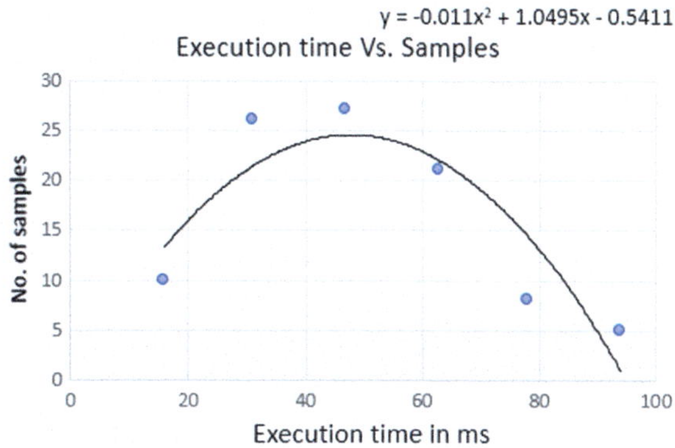

Fig. 6 Quadratic model of execution time versus number of samples

Considering the lowest RMSE value of 2.48, the model plotted using orange colour in Fig. 5 is identified as the best linear model.

Quadratic Model

Figure 6 depicts the best fit model created using the data set for the quadratic model. The quadratic model is given as

$$y = -0.011x^2 + 1.0495x - 0.5411 \tag{2}$$

The RMSE value calculated for the quadratic model is 4.03, which is a better fit based on the data range. The Y value (dependent variable) of the data set ranges from 5 to 27 and the RMSE value comes as an acceptable value within the data range.

Figure 7 depicts the quadratic piecewise models with RMSE values of 1.04 and 0.06. Considering the lowest RMSE value of 0.06, the model plotted using orange colour in Fig. 7 is identified as the best quadratic model.

Cubic Model

Figure 8 depicts the best fit model created using the data set for the cubic model. The cubic model is given as

$$y = 0.0003x^3 - 0.0625x^2 + 3.5093x - 31.615 \tag{3}$$

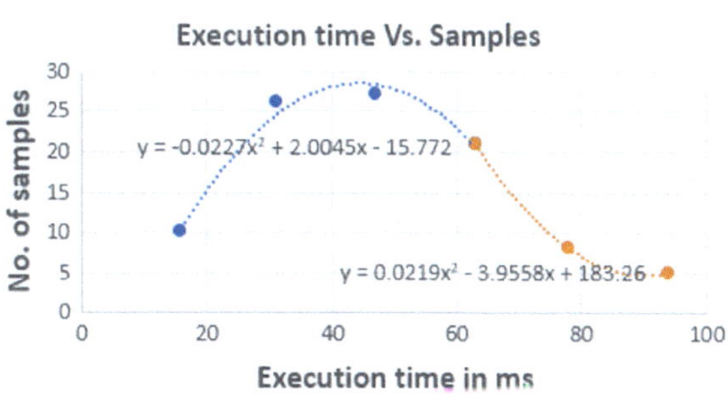

Fig. 7 Quadratic piecewise model of execution time versus number of samples

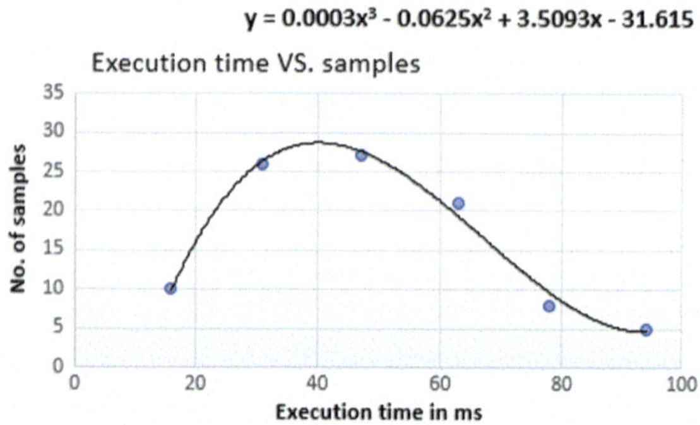

Fig. 8 Cubic model of execution time versus number of samples

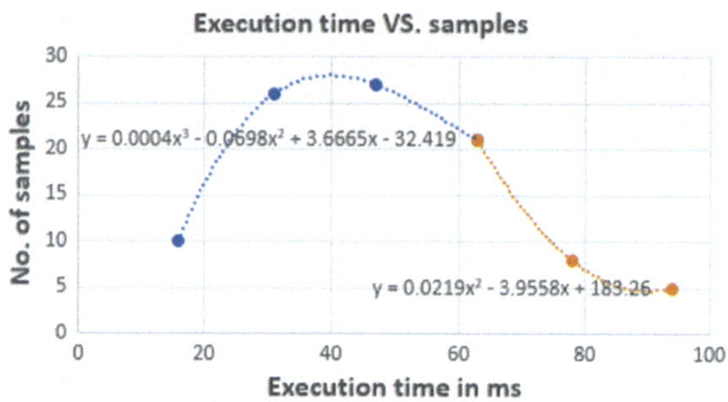

Fig. 9 Cubic piecewise model of execution time versus number of samples

The RMSE value calculated for the cubic model is 4.69, which is a better fit based on the data range. The Y value (dependent variable) of the data set ranges from 5 to 27 and the RMSE value comes as an acceptable value within the data range.

Figure 9 depicts the cubic piecewise models with RMSE values of 0.31 and 0.06. Considering the lowest RMSE value of 0.06, the model plotted using orange colour in Fig. 9 is identified as the best cubic model.

Exponential Model

Figure 10 depicts the best fit model created using the data set for the exponential model. The exponential model is given as

$$y = 27.856e^{-0.013x} \tag{4}$$

The RMSE value calculated for the exponential model is 8.62, which is a better fit based on the data range. The Y value (dependent variable) of the data set ranges from 5 to 27 and the RMSE value comes as an acceptable value within the data range.

Figure 11 depicts the exponential piecewise models with the RMSE values of 6.28 and 1.41. Considering the lowest RMSE value of 1.41, the model plotted using orange colour in Fig. 11 is identified as the best exponential model.

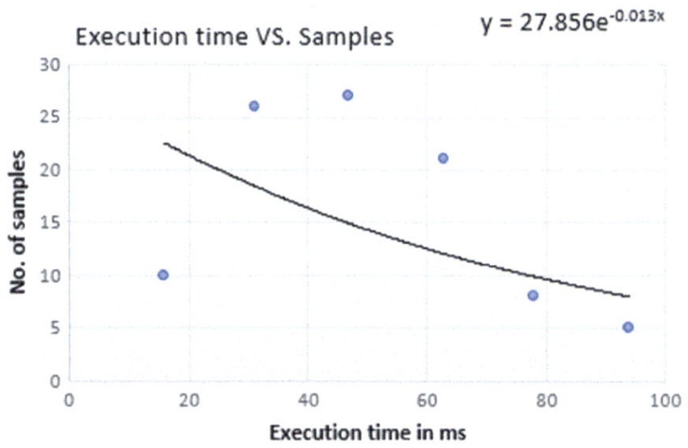

Fig. 10 Exponential model of execution time versus number of samples

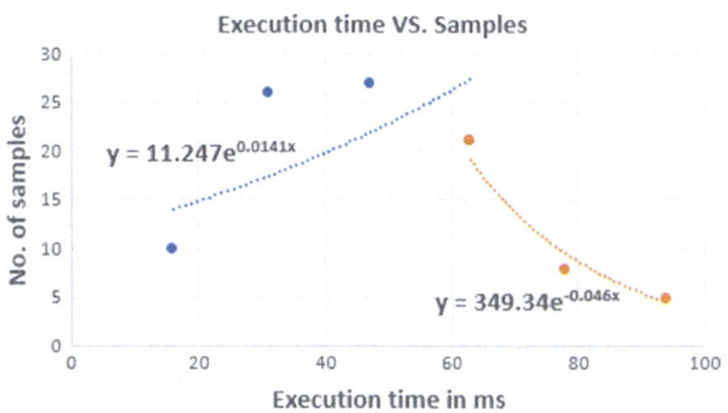

Fig. 11 Exponential piecewise model of execution time versus number of samples

4.2 Scheme 2

With Scheme 2, the algorithm depicted in Fig. 3 is used to calculate the CPU execution time for the AES encryption code using its corresponding assembly code.

5 Comparison of Models

Scheme 1 and Scheme 2 findings are compared in this section:

5.1 Scheme 1

This section performs a comparison of the four models in terms of the RMSE values. The mean value of the CPU execution time based on Scheme 1 is calculated as 46 clock cycles per millisecond. The linear, quadratic, cubic and exponential models yield RMSE value of 7.78, 4.03, 4.69 and 8.62, respectively. It is observed that the quadratic model yields the least RMSE value of 4.03.

The results after the piecewise approximation's application on the four models are:

With piecewise approximations, the linear model consists of

- $$y = 0.2117x + 12.69 \text{ where } x =< 63 \text{ with RMSE } 5.63 \quad (5)$$

- $$y = -0.5125x + 51.478 \text{ where } 63 < x =< 94 \text{ with RMSE } 2.48 \quad (6)$$

With piecewise approximations, the quadratic model consists of

- $$y = -0.0227x^2 + 2.0045x - 15.77 \text{ where } x =< 63 \text{ with RMSE } 1.04 \quad (7)$$

- $$y = 0.0219x^2 - 3.9558x + 183.26 \text{ where } 63 < x =< 94 \text{ with RMSE } 0.06 \quad (8)$$

With piecewise approximations, the cubic model consists of[2];

- $$y = 0.0004x^3 - 0.0698x^2 + 3.6665x - 32.419 \text{ where } x \leq 63 \text{ with RMSE } 0.31 \quad (9)$$

- $$y = 0.0219x^2 - 3.9558x + 183.26 \text{ where } 63 < x \leq 94 \text{ with RMSE } 0.06 \quad (10)$$

With piecewise approximations, the exponential model consists of

- $$y = 11.247e^{0.0141x} \text{ where } x \leq 63 \text{ with RMSE } 6.28 \quad (11)$$

- $$y = 349.34e^{-0.046x} \text{ where } 63 < x \leq 94 \text{ with RMSE } 1.41 \quad (12)$$

[2]This model is not considered for the comparison, as it was plotted with only three points and it yields only a quadratic equation.

Table 1 Execution times for Scheme 1 versus Scheme 2

Scheme	X (CPU execution time in milliseconds)
Scheme 1	46
Scheme 2	1.3

Quadratic model with piecewise approximation yields the least RMSE value of 0.06. After performing a comparison of the four models, it is observed that our quadratic piecewise model can be chosen as the optimized model to measure the CPU execution time of AES algorithm.

5.2 Scheme 2

Based on Scheme 2, the CPU execution time is calculated as 1.3 clock cycles per millisecond.

5.3 Comparison Between Scheme 1 and Scheme 2

To determine the relation between Scheme 1 and Scheme 2, we consider the mean value of CPU execution time obtained through Scheme 1 and the CPU execution time obtained through Scheme 2. These values are given in Table 1.

The CPU execution time calculated using Scheme 1 is higher than the CPU execution time calculated using Scheme 2. Even though Scheme 2 offers the least CPU execution time as depicted in Table 1, Scheme 2 has some drawbacks over Scheme 1, namely the tight couplings of the assembly code to the processor model and CPU clock frequency, which make the CPU execution time calculation harder to achieve [14]. Therefore, CPU execution time calculation using Scheme 2 is non-trivial. Scheme 1 is the simple and easily achievable way as software-only methods require less effort to obtain measurements [14]. It is concluded that Scheme 1 coupled with the modelling technique is considered as the best option for this study.

6 Conclusions

This paper developed an optimized model which is used to measure the CPU execution time taken by AES algorithm to encrypt a data file. Two schemes are used to carry out this study. Scheme 1 models the data set using piecewise approximation applied with four mathematical models, namely linear, quadratic, cubic and exponential models. Scheme 2 models the CPU execution time from the assembly code calculation. Even though Scheme 2 offers more accurate results, Scheme 1 is recommended as it is relatively easy to get the data set. This study presents a combined model of quadratic with piecewise approximation. With this study, quadratic model with piecewise approximation is observed as the most optimized model to measure the processor execution time of AES algorithm. Further works in this arena include the virtual hardware development using VHDL, which yield better timing analysis.

References

1. Kaliisa, R., Picard, M.: A systematic review on mobile learning in higher education: the African perspective. Turkish Online J. Educat. Technol. **16**(1) (2017)
2. Telecommunication Union, Measuring the Information Society Report Executive summary 2018, Switzerland, Geneva, ITU Publications (2018)
3. Callou, G., Maciel, P., Tavares, E., Andrade, E., Nogueira, B., Araujo, C., Cunha, P.: Energy consumption and execution time estimation of embedded system applications. Microproc. Microsyst. **35** 426–440 (2011) (2010), Elsevier
4. Toldinas, J., Damasevicius, R., Venckauskas, A., Blazauskas, T., Ceponis, J.: Energy consumption of cryptographic algorithms in mobile devices. ELEKTRONIKA IR ELEKTROTECHNIKA **20**(5) (2014). ISSN 1392–1215
5. Montoya, A.O., Munoz, M.A., Kofuji, S.T.: Performance analysis of encryption algorithms on mobile devices. In: 47th International Carnahan Conference on Security Technology. IEEE, Colombia (2013)
6. Lu, C., Tseng, S.: Integrated design of AES encrypter and decrypter. In: Proceedings IEEE International Conference on Application Specific Systems, Architectures, and Processors, USA (2002)
7. Ramesh, A., Suruliandi, A.: Performance analysis of encryption algorithms for information security. In: International Conference on Circuits, Power and Computing Technologies. IEEE, India (2013)
8. Elminaam, D.S.A., Kader, H.M.A., Hadhoud, M.M.: Tradeoffs between energy consumption and security of symmetric encryption algorithms. Int. J. Comput. Theory Eng. **1**(3), 1793–8201 (2009)
9. Javed, A., Shahid, M.A., Sharif, M., Yasmin, M.: Energy consumption in mobile phones, I. J. Comput. Netw. Informat. Secur. **12**, 18–28. Modern Education and Computer Science Press
10. Dolezal, J., Becvar, Z.: Methodology and tool for energy consumption modeling of mobile devices. In: IEEE Wireless Communications and Networking Conference Workshops, April 2014
11. Marsiglio, J.: Piecewise linear approximation. https://optimization.mccormick.northwestern.edu/index.php/Piecewise_linear_approximation. Last accessed 13 Apr 2019
12. Fallah, S.A., Arioua, M., Oualkadi, A.E., Asri, J.E.: On the performance of piecewise linear approximation techniques in WSNs. International Conference on Advanced Communication Technologies and Networking, Marrakech (2018)
13. Umaparvathi, M., Varughese, D.K.: Evaluation of symmetric encryption algorithms for MANETs. In: International Conference on Computational Intelligence and Computing Research. IEEE, India (2010)
14. Stewart, D.B.: Measuring execution time and real-time performance. In: Embedded Systems Conference, Boston (2006)
15. Pereira, R., Couto, M., Ribeiro, F., Cunha, J., Fernandes, J.P., Saraiva, J.: Energy efficiency across programming languages. In: Proceedings of Software Language Engineering, 12 pp. ACM, Canada (2017)

16. Kaufmann, J.: Reply to "What do you consider a good standard deviation?". https://www.researchgate.net/post/What_do_you_consider_a_good_standard_deviation. Last accessed 25 Apr 2019
17. Martin, K.G.: Assessing the fit of regression models. https://www.theanalysisfactor.com/assessing-the-fit-of-regression-models/. Last accessed 23 Apr 2019

Gradient-Based Feature Extraction for Early Termination and Fast Intra Prediction Mode Decision in HEVC

Yogita M. Vaidya and Shilpa P. Metkar

Abstract High-efficiency video coding is the most recent video compression standard. HEVC is designed to decrease the bit rate of video transmission without affecting the video quality. The intra prediction of HEVC features about 35 directional modes. The planar mode and DC modes are included into it. The decision about the most appropriate intra mode within the coding unit of high-efficiency video encoder is a vital component in video coding. The intra mode decision has been a crucial, computationally complex processing step and has a share of 85% in the overall video coding complexity. The optimal mode is selected by rough mode decision (RMD) process from all 35 modes and final decision of partitioning is taken through the rate distortion optimization (RDO) process. The brute force RD cost calculation process consumes a large portion of HEVC encoding complexity. This paper presents analysis of the spread of 35 directional modes over the video frame and the correlation between the homogeneous or non-homogeneous characteristics of video content and the spread of directional modes over the video frame. The proposed method is based on the sum of average gradient evaluated for each of the 35 directional modes which help to reduce the number of candidate modes for rough mode decision and RD cost calculation. The performance of the proposed algorithm is evaluated on three distinct classes of video sequences. The offline classification accuracy of the proposed scheme is measured to be 90%. The exhaustive analysis of mode decision carried out in the proposed method will be subsequently useful for training machine learning algorithm for early decision about coding unit depth and fast prediction of the appropriate intra mode. The early depth decision and reduction in the number of candidate coding units to be passed through iterative RD cost computation will drastically reduce the computation complexity and increase the encoding time of high-efficiency video encoder.

Keywords RDO · RMD · HEVC

Y. M. Vaidya (✉) · S. P. Metkar
College of Engineering, Pune, Pune, India
e-mail: ymv.extc@coep.ac.in

S. C. Satapathy et al. (eds.), *Intelligent Data Engineering and Analytics*, Advances in Intelligent Systems and Computing 1177,
https://doi.org/10.1007/978-981-15-5679-1_21

1 Introduction

High-efficiency video coding standard [1] provides approximately 50% reduction of bit rate without compromising the video quality as compared to its predecessor. This is accomplished by adopting advanced video coding techniques. The conventional block-based hybrid video coding framework with flexible quad tree subpartitioning is implemented in HEVC. The coding tree unit (CTU) is the basic block of nested quad tree structure. As shown in Fig. 1, depending upon the rate distortion cost, each CTU may consist of one or multiple CUs. Secondly, each CU may split up to four CUs on the basis of prediction mode. Finally, residual block is obtained for each PU and consequently one or more transform units (TU) are constituted.

High-efficiency encoder implements three-stage intra mode decision. Initially, the Hadamard transform is used to calculate costs of rough mode decision (RMD) to prepare the list of candidate modes. Then the three most probable modes (MPMs) that are generated from the modes of adjoining prediction units are added to the candidate mode list. Finally, the modes in the candidate mode list undergo the iterative RDO cost estimation process to select best intra prediction mode [2–4].

The RD search includes a top-down checking process and a bottom-up comparison process. The RDO search consumes largest portion of the total encoding time. In a 64 × 64 CTU, 85 probable CUs are checked. In order to check the RD cost of each CU, the encoder needs to implement pre-coding for the CU, in which possible prediction modes and transformations modes have to be encoded. Effectively the pre-coding needs to be implemented for all 85 possible CUs in the standard HEVC, consuming the maximum portion of the encoding time. However, in the final CU partition only

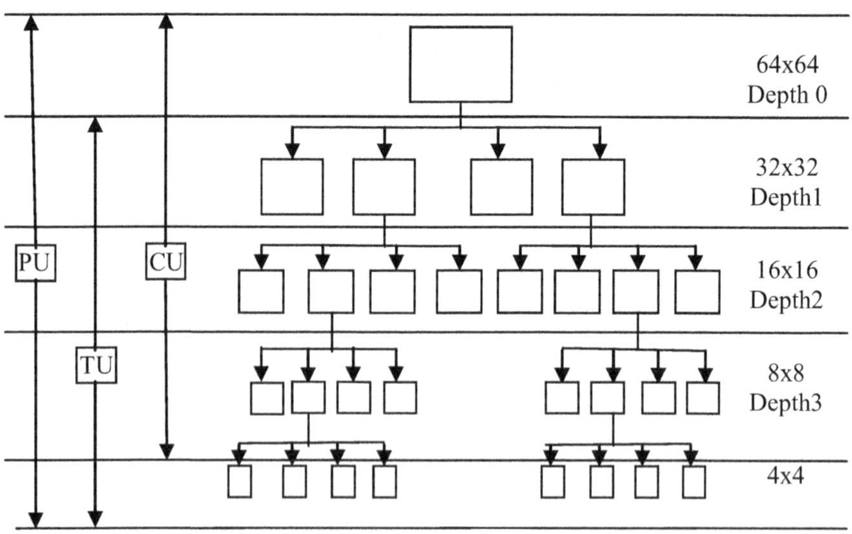

Fig. 1 Quad tree organization of HEVC

certain CUs are selected. The analysis implies that the pre-coding of maximum 84 CUs and minimum of 21 CUs may be avoided through the accurate prediction of the CU partition [5–7].

The fast intra mode decision and early CU depth decision techniques are categorized as: the techniques based on heuristics approaches, machine learning approaches and CNN-based approaches. The techniques based on heuristic approach analyze certain feature to decide the appropriate CU depth before traversing through all quad tree patterns. Later, few machine learning-based techniques have been proposed for fast intra mode decision and early depth decision. These algorithms are based on the exhaustive training through extensive data such as to formulate rules for video encoding components to bypass the execution of iterative RDO process for these components. In order to model the intra prediction and CU partition processes using machine learning approaches, it is essential to explore and extract domain features. The classical gradient operation is one of the promising techniques to determine pixel intensity variation in the coding unit. Ziang et al. [8] proposed a gradient-based technique to analyze the mode along near-horizontal and near-vertical directions in order to reduce the number of candidates for RDO and RMD.

This paper differs from the prescribed approach in the sense that the proposed method is based on the sum of average gradient evaluated for all the 33 direction modes. The angle and amplitude of the gradient is evaluated for each coding unit for all the mode directions, and the mapping approach is used to assign mode to the sum of average gradient amplitude at each pixel. The spread of histogram indicates the mode along which pixel variation is maximum in given CU size. These modes form the candidate mode list for the coding unit of the prescribed video frame. The candidate mode list helps in fast intra mode decision. The histogram is generated for coding unit partitioning of size ranging from 32×32, 16×16 and 8×8, for three distinct video sequences.

Section 2 of the paper presents description of the proposed algorithm. Section 3 presents the experimental results. Section 4 is discussion and conclusion.

2 Proposed Algorithm

The optical flow theory manifests that the direction of the gradient of a pixel represents its maximum variation. Each coding tree block is initially partitioned into coding unit, and then at each pixel the gradient is calculated using Sobel operator. The algorithm is implemented as given below.

Each pixel is convolved with 3×3 filter mask. Then the gradient vector of pixel $p_{i,j\,1}$ is calculated as given in Eqs. (1) and (2).

$$\vec{D}_{i,j} = \left\{ Dx_{i,j}, Dy_{i,j} \right\}, \text{ and}$$

$$Dx_{i,j} = P_{i+1,j-1} + 2 \times P_{i+1,j} + P_{i+1,j+1} - P_{i-1,j-1} - 2 \times P_{i-1,j} - P_{i-1,j+1} \quad (1)$$

$$Dy_{i,j} = P_{i-1,j+1} + 2 \times P_{i,j-1} + P_{i+1,j-1} - P_{i-1,j+1} - 2 \times P_{i,j+1} - P_{i+1,j+1} \quad (2)$$

where $Dx_{i,j}$ and $Dy_{i,j}$ represent the degree of difference in x and y direction, respectively. The gradient amplitude is given as

$$G_{Amp}[\underset{D_{i,j}}{\rightarrow}] = |Dx_{i,j}| + |Dy_{i,j}| \quad (3)$$

The angle of gradient is calculated by using the function given below:

$$G_{Ang}[\underset{D_{i,j}}{\rightarrow}] = \arctan\{Dy_{i,j} / Dy_{i,j}\} \quad (4)$$

The magnitudes of gradient vectors along the same angle are added and stored in the accumulator. A preloaded table look-up exhibits angle corresponding to each of the 33-mode angles. A threshold value is set to map each accumulated gradients magnitudes at nearby angle with the mode angle. The histogram plot between intra mode and "gradient amplitude sum" shows the mode distribution over the coding unit. The prominent modes with highest amplitude are considered to be the candidate modes to undergo the iterative RDO process. The analysis of mode distribution over CU partitions of typical sizes of 32 × 32, 16 × 16 and 8 × 8 help to identify the appropriate depth of CU partition. Thus the two important deductions eventually are helpful to design a machine learning approach for fast mode prediction and early termination of CU partition for high-efficiency video encoder. Reduction in the number of RD computations drastically reduces the overall complexity of the high-efficiency encoder.

3 Experimental Results

The performance of the proposed method has been evaluated for three distinct classes of video sequences. As shown in Table 1, the class A and D video sequences are characterized with complex detailing. The histogram plots in Fig. 2 indicate significant modes at each CU depth. The number of candidate modes is increasing as depth is increasing. Early termination will reduce computation complexity but will also reduce PSNR. Class B video sequences are characterized with both, camera motion and object and background change. In such cases CU size of 32 × 32 will take lesser computation complexity and less encoding time. The class E video sequences are characterized as homogeneous. The CU partition of 32 × 32 or higher will take lesser time without affecting PSNR. The classification accuracy evaluated for 50 frames per sequence is

Table 1 Test sequence

Class	Test sequence	Resolution	Characteristics description	No. of input frames
A/D	BlowingBubbles	416 × 240	No camera motion, no background change, slow motion of objects	50
B	BQTerrace	1920 × 1080	Slow camera motion, slow moving objects, slow background change	50
E	Johnny	1280 × 720	No camera motion, no background change, slow motion of objects	50

$$\%\ \text{Accuracy} = [(\text{TP} - \text{TN}) / \text{TP}\,] * 100$$

where TP is correct distribution of intra mode at the appropriate depth of partition in consistent with the content of the video frame. TN is the mode distribution not consistent with the respective depth level and the content of the video frame (Fig. 3).

4 Conclusion

Most recently, machine learning-based approaches are being used as promising technique for computational complexity reduction of high-efficiency video encoding. The performance of these approaches heavily relies on feature selection for training the machine learning algorithm. These features are mainly the hand-crafted feature. The paper proposed use of classical gradient-based technique to extract an important feature. The gradient-based feature extracted in the proposed technique will be employed in the machine learning algorithm, since it helps to optimize the number of candidate modes for iterative and most complex RDO process and predicting the appropriate CU depth. Both of these aspects will drastically reduce the computational complexity and also decrease the encoding time. The proposed approach is tested on three distinct classes of video sequences on 50 frames, each with 90% accuracy of classification. The future direction is to employ the features crafted through the exhaustive analysis presented in this paper for machine learning approach and evaluate its performance for high-efficiency video encoding process.

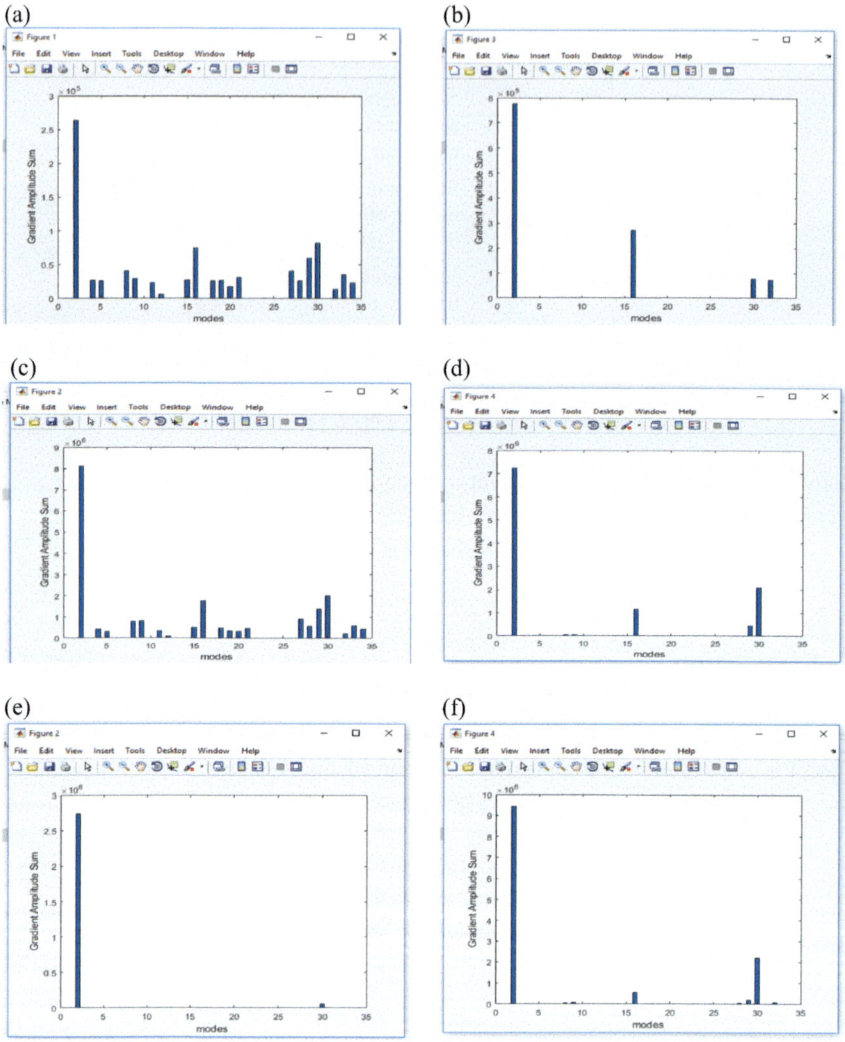

Fig. 2 **a** Mode distribution for CU size 8 × 8 for class A/D video sequence, **b** mode distribution for CU size 32 × 32 for class A/D video sequence, **c** mode distribution for CU size 8 × 8 for class B video sequence, **d** mode distribution for CU size 32 × 32 for class B video sequence, **e** mode distribution for CU size 8 × 8 for class E video sequence, **f** mode distribution for CU size 32 × 32 for class E video sequence

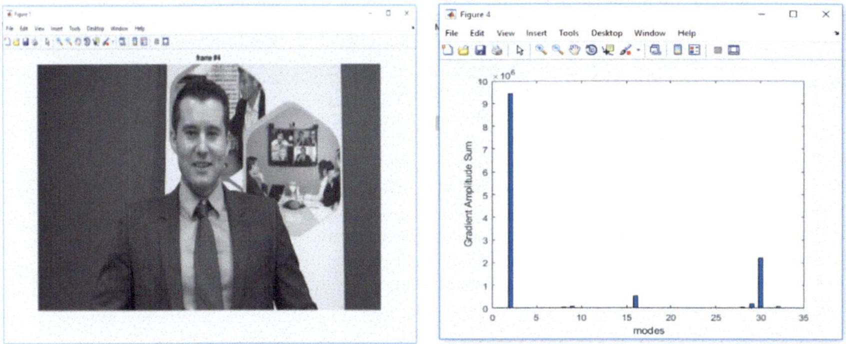

(a) (b)

Fig. 3 **a** Class E homogeneous video frame, **b** intra mode distribution

References

1. Lainema, J., Bossen, F., Han, W.-J.: Intra coding of the HEVC standard. IEEE Trans. Circuits Syst. Video Technol. **22**(12) (2012)
2. Kim, I., McCann, K., Suggimoto, K., Bross, B., Han W.-J.: High efficiency video coding (HEVC) test model 14 encoder description, document JCTVC-P1002, JCT-VC (2014)
3. Piao, Y., Min, J.H., Chen, J.: Encoder improvement of unified intra prediction, document JCTVC-C207 (2010)
4. Zhao, L., Zhang, L., Ma, S., Zhao, D.: Fast mode decision algorithm for intra prediction in HEVC. In: Proceedings of IEEE International Conference Vision Communication Image Process, pp. 1–4 (2011)
5. Jamali, M., Coulombe, S., Caron, F.: Fast HEVC intra mode decision based on edge detection and SATD costs classification. In: Proceedings of IEEE International Data Compression Conference, pp. 43–52 (2015)
6. Chen, G., Pei, Z., Sun, L., Liu, Z., Ikenaga, T.: Fast intra prediction for HEVC based on pixel gradient statistics and mode refinement. In: Proceedings of IEEE China Summit International Conference on Signal Information and Processing, pp. 514–517 (2013)
7. Kim, T.S., Sunwoo, M.H., Chung, J.G.: Hierarchical fast mode decision algorithm for intra prediction in HEVC. In: Proceedings of IEEE International Symposium Circuits and Systems, pp. 2792–2795 (2015)
8. Ziang, T., Sun, M.-T.: Fast intra mode and CU size decision for HEVC. IEEE Trans. Circuits Syst. Video Technol. **27**(8) (2017)

A Variance Model for Risk Assessment During Software Maintenance

V. Lakshmi Narasimhan

Abstract This paper presents the design of a risk management framework that facilitates large-scale software system maintenance and version controlling. The important aspects of the solution space include impact profiling (i.e., the impact of loss of a particular system or sub-system) and parametric risk modelling of the system. A variance-based model has been developed for risk assessment. The system provides number of other features including, but not limited to, report generation, alert and flash messaging, testing/testability considerations and other such records. The system also offers a limited degree of visualization capabilities in order to view risks at various layers and types.

Keywords Software risk management during maintenance · Impact profiling · Parametric modelling · Risk visualization · Variance modelling

1 Introduction

The design and development of software is not an easy exercise due to several reasons. First, unlike established disciplines like civil engineering, the software field is only relatively new. Secondly, very less historical data sets (or statistics) are available on major projects for comparison or evaluation. Thirdly, repeatability of (key) aspects of software projects and their repetition rates are at present appear to be limited, even though research in this area (e.g., component-based software engineering and product-line engineering) is showing great promise. Lastly, unlike most fields, software systems do indeed seed and shape many areas and in that process get significantly influenced themselves. As a consequence, the design and development of software systems appear to be an on-going learning exercise, at least at present with the current state-of-the-art technologies in this area. Indeed, one author [1] compares the design and development of software systems to writing a successful novel—the

V. L. Narasimhan (✉)
University of Botswana, Gaborone, Botswana
e-mail: srikar1008@gmail.com

S. C. Satapathy et al. (eds.), *Intelligent Data Engineering and Analytics*, Advances in Intelligent Systems and Computing 1177,
https://doi.org/10.1007/978-981-15-5679-1_22

more vivid and accurate the portrayal to the reality (or user satisfaction), the more successful the novel (or software) becomes! Unfortunately, the size of these "software novels" is looming bigger by the day. For example, a modern cellphone now contains 2 million lines of code (MLOC), and by 2010, it will have a size of 20 MLOC [2]. Typical motor cars will be running on software systems of size 100 MLOC by then!

Because of instabilities in the software design and development processes, the downstream issue of software system maintenance is fraught with considerable risks of various types. Further, the nature and severity of these risks usually compound during the maintenance phase and the costs to address or ameliorate or mitigate these risks also considerably increase—in some cases exponentially. The problem is very much akin to noting and trying to fix a stitch in a sweater after it has been completely knitted [2], except by then one may have to undo many stages of stitches in order to correct the original error. The cost of fixing an error can be 100 times as high as it would have been during the development stage and the associated risks in testing and verifying the correct operation of the amended software is also order of magnitude higher than during the original development of the system [3].

In this paper, we identify various types of risks during the maintenance phase through parametric attributes. For this purpose, we note the definition of the term risk as, "Risk is the net negative impact of the exercise of a vulnerability, considering both the probability and the impact of occurrence [4]". They include risks in testing (during maintenance), on-going risks to business and modelling the risk management capability. We develop a variance model for risk management, mitigation and minimization during the maintenance phase and we believe that this model serves as a step in the right direction towards addressing risk-related issues in large organizations.

The rest of the paper is organized as follows: Section 2 presents more specific details on the risks in software development, while Sect. 3 deals with factors in software risk assessment and related models. Section 4 is on the development of a variance model for evaluating software risk assessment, followed by a discussion on this model in Sect. 5. The conclusion summarizes the paper and provides further research directions.

2 Risks in Software Development and Maintenance

Software development life cycle (SDLC) has a number of steps and several types of risks occur during each stage of the SDLC. Maintenance of software is critical to the on-going success of an organization, and as software become larger and interact more with each other, maintenance becomes more profound. Indeed, a recent empirical survey by EDS, Australia, reveals that 80% of the IT budget in large organizations in Australia is on maintenance of IT systems [5]. Unfortunately, the impact of costs and risks involved in software maintenance has not permeated to financial managers, who still perceive software as replaceable or perishable commodities, such as hardware. Indeed, the Software Hall of Shame [2] showcases some 30 odd projects during the time period 1992–2005, which have been totally abandoned after spending budgets ranging from $11 million to $4 billion. These projects have been pursued in the English-speaking OECD countries and they include mission-critical projects, such as the FBI's Virtual Case File [6] project[1] and various others (see [7, 8] for more details).

Maintenance occurs in four ways: corrective maintenance, adaptive maintenance, perfective maintenance and preventative maintenance. The first two issues concern fixing of systems, while the latter two issues relate to enhancing the system. The corrective approach is intended to defect identification and removal of defects/bugs; the adaptive approach manages changes resulting from operating system, hardware or DBMS changes; the perfective approach handles changes resulting from user requests and the preventative approach deals with changes made to the software to make it more maintainable. Risks arise at all of the maintenance phases, and the two impact factors that we are interested in this paper include risks to business and risks to re-testability.

A short and quick literature survey relevant to this paper is as follows: Risks related to software systems, related frameworks [9], testability [10–13], cost-benefits [14], choice of test tools [15] and typical mistakes [16, 17] have also been extensively studied. Chillarege [18] provides a list of testing best practices, DeLano et al. [19] detail a test pattern language, while Bach [20] provides interesting pointers on the lack of functionalities of test automation tools. On non-software systems, Gits [21] provides a comprehensive literature review of maintenance concept for various systems and Dunthie et al. [22] provide risk-based approaches to maintenance. Edwards [23] provides a specific case study in the construction industry. Various IEEE and NIST Standards [24, 4] deal with software maintenance issues, analyse problems during maintenance and provide a way to go about addressing these issues.

Charette [2] lists several major factors for the failure of such projects. We have amended the original list in order to generate the list of risks that occur during the software maintenance phase (Table 1).

Several others have proposed a number of techniques to ameliorate risks in software. One of the notable ones is by Knuth, who proposed literate programming [25] as a means to improve program maintainability, readability and lucidity. Software arts [26] developed using such techniques do pose considerably reduced risks during maintenance; however, such systems are rare.

The impact of the software process chosen to design and development of systems is profoundly felt during maintenance. For instance, Cusumano et al. report [27, 28, 24] that Japanese companies have applied extensive quality control procedures that they employ for manufacturing to software development also. As a consequence, the reported median defects in Japanese IT projects are one-fourth of the corresponding

[1] Incidentally, very less data appear to be available on software projects pursued or being pursued in non-English-speaking OECD countries and other IT-service dominant countries, such as India. One of the largest IT projects, viz., the India Railway Reservation System, developed by CMC India (a Government of India organization), has not been profiled much at all in the literature. This project has been one of the major IT project success stories emanating from India in its early days (1980s).

Table 1 List of risks during software maintenance phase

• Unrealistic or unarticulated project goals • Inaccurate estimates of the needed resources for maintenance • Badly defined maintenance requests • Poor reporting of the system status under maintenance • Unmanaged risks during maintenance	• Poor communication among users and maintainers • Use of immature technology in implementation • Complexity of the system • Poor maintenance practices, processes and procedures • Stakeholder politics, and Commercial pressure

numbers in US IT projects. When measured over 100 projects, Cumano et al. [24] report that Japanese adherence to rigid software processes produced 0.02 defects per 1000 source lines of code, compared to 0.40 for corresponding US projects. Further, their rigid processes involving considerable amount of comprehensive documentation allow considerable degree of code re-use also. All these issues have a profound impact on software maintenance.

3 Factors in Software Risk Assessment During Maintenance

Risk assessment is highly related to assessment or estimation of errors in software, which are identified during repeated testing. Unfortunately, testing is expensive, limited in nature, inconclusive and sometimes impossible too [29]. Typically, test path complexity can explode super exponentially. Specifically, Deklava [30] ranks the problems in maintenance to include the following:

- Changing priorities, testing methods, performance measurement, incomplete or non-existent system documentation
- Adapting to changing business requirements, backlog size, measurement of contributions
- Low morale due to lack of recognition or respect, lack of personnel, especially experienced
- Lack of maintenance methodology, standards, procedures and tools.

Obviously, the number and types of unresolved issues and variables determine the level of risk [31] posed by the project. However, one can use high-level estimation outputs to make critical decisions on risks only if the underlying processes can be trusted. When parametric measurements of risk prove inadequate, risks can be simply stated in terms of issues that remain to be resolved yet and their perceived impact to the software system. For example, a good indication on the maintainability of a software system can be inferred by the work of Oman [32], whose key points are brought up in Table 1 (Table 2).

Table 2 Effects on maintainability of source code properties (adapted and amended from [32])

SOURCE CODE					
CONTROL STRUCTURE		INFORMATION STRUCTURE		CODE DETAIL	
SYSTEM	COMPONENT	SYSTEM	COMPONENT	SYSTEM	COMPONENT
+modularity	-complexity	-global data types	-local data types	+overall program formatting	statement formatting
-complexity	+use of structured constructs	-global data structures	-local data structures	+overall program commenting	vertical spacing
+consistency	-use of unconditional branching	+data flow consistency	-span of data	+module separation	horizontal spacing
-nesting	-nesting	+data type consistency	+data initialized	naming	+intra-module commenting
-control coupling	+cohesion	-nesting		symbol and case	
+encapsulation		-I/O complexity			
+module re-use					

+ usually makes an application more maintainable

- usually makes an application less maintainable

Our risk models are adaptations of the function point models employed in software cost and effort estimations and the work of Oman [32]. The overall system interaction model (adopted and modified from the MIT-90 model for assessing and achieving change [33]) for software maintenance is captured in Fig. 1, wherein we identify the following elements as having profound impact on the overall impact contributing to the vulnerability of an organization's IT system:

- Technology, which changes rapidly
- Maturity of skill-base, which is vital for maintenance

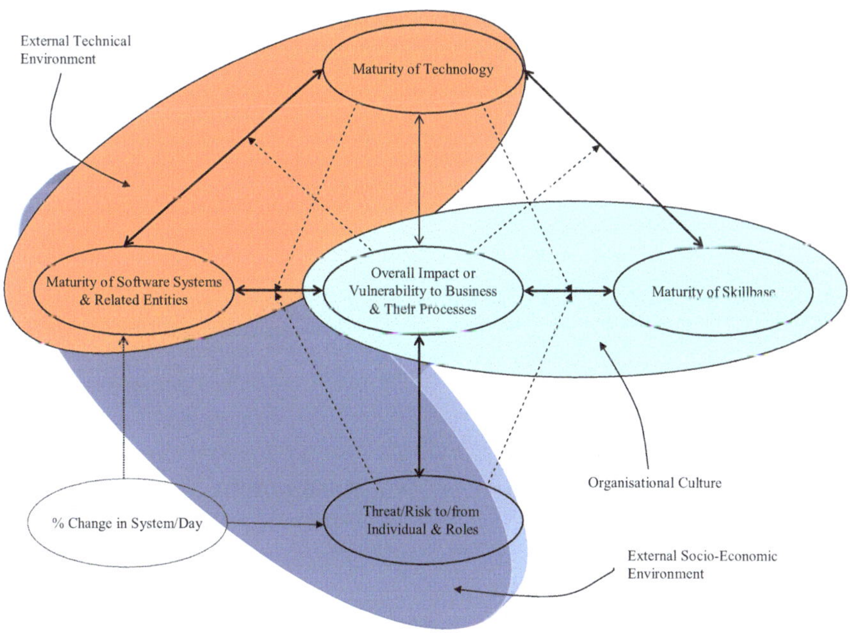

Fig. 1 Overall system interaction model for software maintenance

Table 3 Code-related risk factors (CRF)

CRF risk factors	Measurement mechanism
Size of code	For every 10 KLOC, add 1 point up to 8 points For every 1 MLOC or part of thereof, add 8 points
Number of systems affected	Actual value
Mean time to test	Actual value
Mean time to fix a bug	Actual value
Average cost per unit test	Actual value

- Maturity of software systems and related entities, which is critical to the stability of IT systems (which in turn depends on the percentage change in the system per day) and
- Threat or risk to/from individuals to the IT system. Note that this is not a malicious threat, but more along the lines that "what if a key personnel leaves the organization, then what will be the impact on IT system maintenance".

One can notice that different combinations of factors influence: (i) the organizational culture, (ii) external technical environment and (iii) external socio-economic environment. People involved in the maintenance of IT systems have to deal with these issues pragmatically and on a continual basis. The rest of the paper borrows conceptual ideas from our earlier work on asset management [34], and adapts them to software maintenance risk model. A more details on our model for software maintenance risks will be described in a forthcoming paper.

We also define the following factors as contributors to risk during software maintenance and these factors are elaborated in Tables 3, 4, 5 and 6: Code-related risk factors (CRF), process-related risk factors (PRF), practice-related risk factors (PcRF) and testing-related risk factors (TRF).

3.1 Testing-Related Risk Factors

See Table 6.

4 Variance Model for Software Risk Assessment

We define several risk factor metrics as noted in Table 7, before we develop the variance model.

We now define the following metrics as noted in Table 8. Note that each one of the functions (F#(t)) below is unique and only the key parameters from each of the risk factors from Tables 3, 4, 5 and 6 are included in the actual calculation of the

Table 4 Process-related risk factors (PRF)

PRF risk factors	Measurement mechanism
Volatility of development process	no process (=10), ad hoc, well-defined, repeatable (=1)
Perceived impact to/on organization	On a nonlinear ***multiple*** of 1 (low), 3, 7, 10, 20 (high)
Degree of clarity of system requirements	On a scale of 1 (good clarity)–10 (poor clarity)
Degree of clarity of testing requirements	On a scale of 1 (good clarity)–10 (poor clarity)
Degree of reusability requirement	On a scale of 1 (low need)–10 (high need)
Degree of availability of reusable scripts	On a scale of 1 (poor availability)–10 (good availability)
Degree of maintainability of reusable scripts	On a scale of 1 (good maintainability)–10 (poor maintainability)
Degree of comprehensiveness of the test tool & environment	Life cycle toolkit (=1), …, ad hoc *toolkit* (=10)

Table 5 Practice related risk factors (PcRF)

PcRF–risk factors	Measurement mechanism
Adherence to Standards (*a la* training time)	On a nonlinear ***multiple*** of 1 (low), 3, 7, 10, 20 (high)
Time to test	Average time to test a module—actual value
Remaining time to deploy	Actual value
(Efficacy) weighted number of testers	on a nonlinear ***multiple*** of 1 (low), 3, 7, 10, 20 (high)
Degree of (real time) performance requirement	On a scale of 1 (low) to 10 (high)
Clarity of performance requirement	On a scale of 1 (low) to 10 (high)
Degree of data/access security constraints imposed	On a scale of 1 (low) to 10 (high)
Degree of code-level migration requirement	On a scale of 1 (low) to 10 (high)
Degree of systems-level migration requirement	On a scale of 1 (low) to 10 (high)
Percentage of full load testability requirement	Actual value

Table 6 Testing related risk factors (TRF)

- Number of users that use the system, Number of databases used in the system, Number of mainframes accessible through the system, Number of refreshes per day, Number of transactions per day, Total MIPS available
- Total storage used for storing data, Number of Procedures in the total system, Number of approved code changes per day, Number of procedures changed per year, Number of Developers, Number of Testers
- Number of testing centres in the organization (there could be a "cultural difference" between the testing centres!), Number of total sub-task in testing

Table 7 Risk factor metrics and their definitions

Risk factor metrics	Definitions
Coverage	Percentage amount of the system tested after a maintenance cycle
Impact	Degree of effect of the maintenance operation from one cycle to another, which can be positive or negative
Time2Fix	The average time taken to fix a reported bug (does not include bug detection time)
Exposure	Degree of revelation of faults after a particular maintenance cycle that gets exposed to outside of the organization
Fault Likelihood	Probability of a fault occurring
Consequence	The consequence of failure of a particular system
Degree of Adherence	Degree of adherence to procedures, Standards or process and their perceived veracity and effectiveness towards software maintenance
Probability of loss of key people	The impact of the loss of key people
Remaining time	The remaining time before version release of a software

Table 8 Risk factor metrics and their measurement

Risk factor metrics	Nature of measurement
Coverage	F# (CRF-risk factors)
Impact	F# (TRF-risk factors)
Time2Fix	F# (PcRF- and TRF-risk factors)
Exposure	F# (CRF- and PcRF-risk factors)
Fault Likelihood	F# (CRF, PcRF and TRF-risk factors)
Consequence	F# (TRF-risk factors)
Degree of Adherence	F# (PRF-risk factors)
Probability of loss of key people	F# (PRF and TRF-risk factors)
Remaining time	F# (PcRF-risk factors)

metrics; the remaining parameters are ignored. The selection of the right type of parameters is based on experience, application requirement and context in which the measurement is sought. Further the function F#(●) is a measure of the variance or changes in the parameters as the entire software system goes through various stages of its maintenance cycle at any given ●.

The metrics given in Table 8 can be variably evaluated depending on which the listed risk factors are considered to dominate a particular information system. As a further extension, one can define a number of metrics for the following terms also:

- **Impact Metrics**: Integrity, availability, confidentiality, vulnerability, threat-source identification, threat-action identification.
- **Mitigation Action and Strategy Metrics**: Risk awareness, risk assumptions, risk avoidance, risk limits, risk plans, risk acknowledgement, risk transfer or outsourcing.
- **Cost-benefit Metrics**: For the following actions—analysis, assigning responsibility, priority/rating generation, cost of detection, cost for correction, targeting, process/standard generation, education & adherence to process/standards (training[2]), standards/process improvement (capability maturity), identifying and evaluating residual risks.
- **Incident Management Metrics**: Incident reporting, categorization, incident prioritization/rating, responsibility assignment, risk isolation, impact assessment.

5 Discussion of the Variance Model

We define several types of metrics for software maintenance, which include the following: risk to testability, risk to business activities and risk to business perception. These are pictorially described in Fig. 2. The empirical implications of various types of total acceptable risks (TAR) values are provided in Table 9.

Risk to Testability (R2T) = **A** = Area of triangle {Coverage, Impact, Time2Fix}
Business Perception Risk (BPR) = B = Area of triangle {Exposure, Fault Likelihood, Consequence}
Business Vulnerability Risk (BVR) = **C** = Area of triangle {Degree of Adherence, Prob. loss of key people, Remaining Time}
Total Acceptable Risk (TAR) = **A * B * C**.

Obviously, higher than normal values indicate potential to serious problems as shown in Table 9. Other factors such as risk to business can be in terms of performance, time-to-market (or market conditions) and technology alternatives. For example,

Risk2Business (●) = F3 (c1, c2, c3, c4)/F4 (d1, d2, d3, d4)
C1 = perceived impact on business (visibility, loss of life, etc.)

[2]A number of air accidents are attributed to training-related issues.

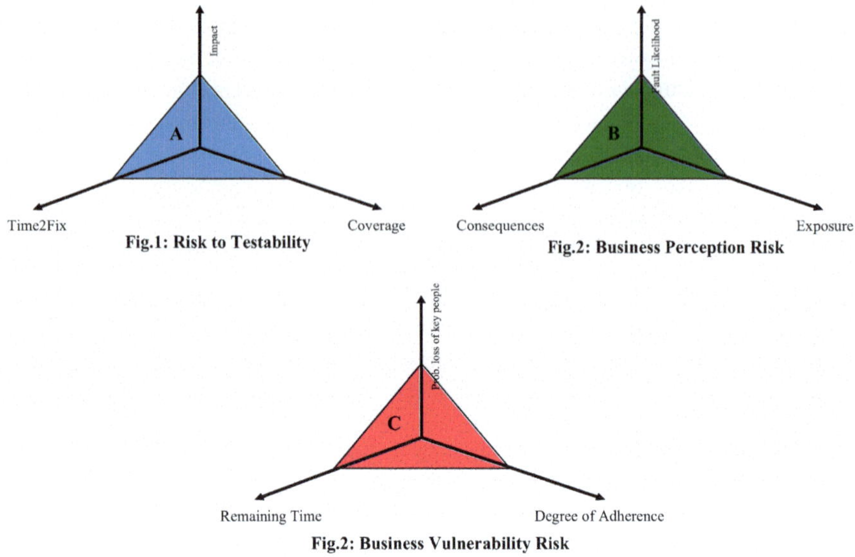

Fig. 2 Pictorial description of TAR values

Table 9 TAR values and their implications

TAR values	Implications	Recommended remedial actions
10	Fatal failures	Immediate attention with highest priority
7	Serious failures	High priority; expect consequences
4	Manageable failures	Good priority; consequences may be serious
1	Acceptable failures	Low priority, consequences may not be serious
0.1	Virtually impossible	Well tested, rare errors with minor consequences

C2 = degree of leadership support
C3 = degree of organizational culture to support the values of testing and testability.

And, the rest are explained in the forthcoming paper.

Similarly, risks can be quantified or qualified in terms of humans and systems as noted below:

Human-Related Risks (●) = F (easy of training test tool/environment, % change in policies, nature of project, recoverability from errors, # maintenance centres, etc.)
System-Related Risks (●) = F (# of sub-systems, # of controllers, # of transactions/sec, # of refreshes/day, # changes/day, degree of concurrency required, etc.).

5.1 *Development of a Risk Measurement Toolkit*

We have developed a version of our risk measurement toolkit using Excel spreadsheet, whose parameters are provided in Table 10, which itself is a modified version of the NIST Standard on Maintenance Risk Analysis Matrix [4]. We also generate some recommendations automatically and further, repositories for the creation and maintenance of a repository for each of the following items: (1) risk watch-list, (2) profile of typical errors, bugs, pitfalls and mistakes, (3) code patterns for error corrections, (4) potential impact profiles, (5) key personnel impact and (6) knowledge gained from correcting bugs, pitfalls and mistakes.

5.2 *Maturing of the Risk Model: An Aquatic Profile for Risk Maturity Management (APRMM)*

We present a new model, called the aquatic risk management (ARM) model (see Table 11), which indicates the degree of maturity of organizations in managing their risk. The APRMM model is inspired by the works of Pooch [35], Capability Maturity Model Integration of Humphry [36] and discussions with various industrial experts. At the basic step, all organizations are Calamaries, where risk management happens by accident of serendipity. As the organizations improve their risk management profiles, they move in their maturity levels from the Calamari-level, to that of whales, piranha, salmons and then sharks, one-level at time. Along with their maturity level upward movement, their capabilities also increase profoundly. At the highest level, organizations will be able to consider their entire risk management life cycle and cost their operations in terms of costs, resources, personnel and performance requirements.

6 Conclusions

The measurement of risks and process that must be placed to eliminate or mitigate the risks involves the collection of data and decision support systems. In this paper, we have identified various types of risks that occur during software maintenance. We have parameterized these attributes and developed a variance-based model for their measurement and control. We have also identified several metrics for monitoring risk during maintenance, described a toolkit and discussed an aquatic profile for risk management maturity (APRMM). We are currently working on the design and development of an intelligent agent-based framework for comprehensive risk assessment during software maintenance.

Table 10 Maintenance risk analysis matrix

Nature of risk	Priority	Recommended control actions	Planned controls	Required resources	Responsible team or persons	Start/End dates	Estimated total costs	Residual risk	Maintenance comments
•									
•									
•									

Table 11 APRMM—aquatic profile for risk maturity management model

Level	Name	Attributes	Characteristics
1	Calamari	Scavengers, nature fed, unfocussed; eat whatever available dead or alive, so long as you can eat!	Risk management happens by accident or serendipity
2	Whale	Gulp all, unfocussed irrespective of quality, no particular motivation except to fill-in	Limited risk management is performed, but just to fill the scripts, with no planning or real load based evaluation
3	Piranha	Acting as a team on easy target, predatory, fast	Automated tools are used to perform risk management, but inconsistent processes employed with limited estimate on costs
4	Salmon	Group action, movement, temperate	Near-production risk management is performed with clearly repeatable processes, control over performance requirements and costs, maintainable risk profiles
5	Shark	Group, motivation, large-scale movement across entire available space	Entire risk management life cycle is considered, well-articulated risk profiles maintained, costs, resources and performance requirements controlled

References

1. Braude, E.J.: Software Engineering: An Object-Oriented Perspective. Wiley 2001
2. Charette, R.N.: Why software fails. IEEE Spectr. **42**(9), 36–43 (2005)
3. Whittaker, J. Jorgensen, A.: Why software fails. Available at: http://www.aet-usa.com/people/aaj/WhySoftwareFails.htm
4. Stoneburner, G., Goguen A., Feringa, A.: Risk management guide for information technology systems. NIST special publication no. 800–830, US Department of Commerce
5. EDS seminar notes, EDS IT technical manager, Adelaide. Seminar held at Newcastle in April, 2005
6. Goldstein, H.: Who killed the virtual case file? IEEE Spectr. **42**(9), 18–29 (2005)
7. Armour, P.G.: To pan, two plans: a planning approach to managing risk. Comm. ACM **48**(9), 15–19 (2005)
8. “Inside Risks”, regular column. Comm. ACM (particularly 2004–05 issues are very relevant)
9. Nigle, C.: Test automation frameworks. Available at: http://safsdev.sourceforge.net/FRAMES DataDrivenTestAutomationFrameworks.htm
10. Bach, J.: Risk-based testing. Available at: http://www.stickyminds.com/sitewide.asp?ObjectId=1800&ObjectType=ART&Function=edetail
11. McMahon, K.: Risk-based testing. Available at: http://www.data-dimensions.com/testersnet/docs/riskbase.htm
12. Shafer, J.: Improving software testability. Available at: http://www.data-dimensions.com/testersnet/docs/testability.htm
13. Kaner, Falk, Nguyen.: Testing computer software, 2nd edn. John wiley (1999)
14. Kaner, C.: Quality cost analysis: benefits and risks. Available at: http://www.kaner.com/qualcost.htm

15. Fewster, M., Graham, D.: Choosing a test tool. Available at: http://www.grove.co.uk/Tool_Information/Choosing_Tools.html
16. Pettichord, B.: Seven steps to test automation success. Available at: http://www.io.com/~wazmo/papers/seven_steps.html
17. Marick, B.: Classic testing mistakes. Available at: http://www.testing.com/writings/classic/mistakes.html
18. Chillarege, R.: Software testing best practices. Available at: Software Testing Best Practices, by Ram Chillarege
19. DeLano D., Rising, L.: System test pattern language. Available at: http://www.agcs.com/supportv2/techpapers/patterns/papers/systestp.htm
20. Bach, J.: Test automation snake oil. Available at: http://www.satisfice.com/articles/test_automation_snake_oil.pdf
21. Gits, C.W.: On the maintenance concept for a technical system: II. Literature review. Maintenance Manage. Int. **6**, 181–196 (1986)
22. Duthie, J.C., Robertson M.I., Clayton A.M., Lidbury D.P.G.: Risk-based approaches to ageing and maintenance management. Nucl Eng. Design. **184**(1), 27–38(12), 1 August (1998)
23. Edwards, L.: Practical risk management in the construction industry. ISBN: 0727720643, Thomas Telford (1995)
24. Cusumano, M.A., MacCormack, A., Kemerer, C., Crandall, B.: Software development worldwide: the state of the practice. IEEE Soft. **20**(6), 28–34 (2003)
25. Knuth, D.E.: Literate programming in CSLI lecture notes, no.27.In: Centre for the Study of Language and Information. Stanford, CA (1992)
26. Bond G.W.: Software as art. Comms. ACM. **48**(8) 118–124 (2005)
27. Cusumano, M.A.: The puzzle of Japanese software. Comm. ACM **48**(9), 25–27 (2005)
28. Cusumano, M.A.: Japan's Software Factories. Oxford University Press, NY (1991)
29. Bach, J.: The challenge of "Good Enough" Software. Available at: http://www.data-dimensions.com/testersnet/docs/good.htm
30. Deklava: Delphi study of software maintenance problems. In: Proceedings of International Conference on Software Maintenance, pp. 10–17 (1992)
31. Armour, P.G.: Project portfolios: organizational management of risk. Comms. ACM. **48**(3),17–20 (2005)
32. Oman, P., Hagemeister, J.: Metrics for assessing software system maintainability. In: Proceedings Conference on Software Maintenance, pp. 337–344. IEEE CS Press, Los Alamitos Califormnia. Order No. 2980–02T (1992)
33. Model (MIT90) for assessing and achieving chang. Available at. http://www.adventengineering.com/why/mit_90/mit_90.htm
34. Lakshmi Narasimhan, V.: A risk management toolkit for integrated engineering asset maintenance. Aust. J. Mech. Eng. (AJME) (2008)
35. Pooch, P.: Application performance – a risk-based approach. In: Second test automation workshop, 1-2 Sept. Bond University, Gold Coast, Australia (2005)
36. Humphry, M.: Capability maturity model integration. http://www.sei.cmu.edu/cmm/

Cyber Attack Detection Framework for Cloud Computing

Suryakant Badde, Vikash Kumar, Kakali Chatterjee, and Ditipriya Sinha

Abstract To prevent cyber-attacks, cloud-based systems mainly depend upon different types of intrusion detection systems (IDS). Most of the approaches have high detection rate for known attacks. But in case of unknown attacks or new attacks, these intrusion detection system increases false alarm rate. Another problem is that the reduction of false alarm rate increases the computational complexities in case of genetic algorithm-based IDS and ANN-based IDS. For instance, to tackle challenges like zero-day attack, the only way is to rely upon a robust data-driven approach for security in cloud. Actually in cloud huge amount of data are processed for various activities. It is very difficult to correlate events over such huge amount of data. To improve the abilities of monitoring and fast decision-making, context management is used for correlating events and inferring contexts and evidences. In this paper, a new data-driven framework has been proposed which utilizes ontology and knowledge base to detect cyber-attack with intrusion detection system in cloud.

Keywords Cyber-attack · Cloud computing · Intrusion detection system · Context management

---

S. Badde (✉) · V. Kumar · K. Chatterjee · D. Sinha
Department of Computer Science and Engineering, National Institute of Technology, Patna, Patna 800005, Bihar, India
e-mail: suryakantb35@gmail.com

V. Kumar
e-mail: vika96snz@gmail.com

K. Chatterjee
e-mail: kakali@nitp.ac.in

D. Sinha
e-mail: ditipriya.cse@nitp.ac.in

S. C. Satapathy et al. (eds.), *Intelligent Data Engineering and Analytics*, Advances in Intelligent Systems and Computing 1177,
https://doi.org/10.1007/978-981-15-5679-1_23

1 Introduction

Cyber-terrorism is one of the most focused subjects of research in today's world. Cyber-attacks have developed and spread so fast that sometimes it is very difficult to identify these attacks. Cyber-attacks like WannaCry, Ransomware are often launched by sophisticated attackers to bypass passive, defense-based security measures [1]. To resist such attacks, the role of cyber-security mechanism is not only identifying the threats, but also has to predict the threats of tomorrow. In particular, today's world is mainly dependent on cloud computing which offers all types of public utility system in terms of scalability, flexibility, pay-per-use and so on. Cyber-attack detection in cloud includes big data analysis. For example, host log event data in cloud mainly accumulate large volumes of data. Inaccurate analysis of these data will increase false alarm rate. Hence attack estimation is one of the major challenges in cloud security.

Many attacks in cloud have been discussed in the literatures [2–5]. DoS attacks are the most common and often spoke-about threat in the cloud. Qiu et al. [2] developed a framework for recognizing denial-of-service (DoS) attacks in cloud data centers by taking advantage of virtual machine status, including CPU and networks use. They have found that when a DoS attack is launched, malicious virtual machines exhibit similar status patterns, and thus information entropy can be used to monitor the status of virtual machines to identify behaviors of attack. The classification method was developed by the authors in [3] to assess the packet behavior based on the kappa coefficient. In cloud computing, for example, when multiple virtual machines (VMs) share the same physical machine, this creates a great opportunity for cache-based side channel attack (CSCA) performance. A bloom filter (BF) detection technique [4] was built to address this issue. BF's central idea is to reduce the overhead quality and a mean calculator to predict the behavior of the cache. Alternatively, to prevent unauthorized access, the SQL injection attack detection method was introduced in [5].

The problem of attack detection can be regarded as a classification process. It can be solved using several classification techniques, such as SVM, ANN, clustering, Naïve Bayes, and so on, found in [6]. Further it can be improved by utilizing security analytics for hidden patterns. Correlating events, discovering patterns and inferring context with evidences are essential for cyber-security analytics. Some models based on probability theory [7], fuzzy logic [8] and Dempster-Shafer theory (DST) [9] are found in this area.

In order to provide an efficient detection system, we have proposed a data-driven framework which utilizes ontology and knowledge base to detect cyber-attack with intrusion detection system in cloud.

Our major contributions in this paper are:

- We design and implement a cyber-attack detection and prevention framework for cloud environment.

- The intrusion detection block of this framework has been experimented using real-time data for accuracy checking and the result shows that performance is acceptable.
- The overall performance is evaluated by applying different algorithms for better performance.

The rest of the paper is organized as follows: Sect. 2 presents cloud security attacks. Section 3 presents proposed framework. In Sect. 4, performance evaluation of proposed framework is discussed. Finally, the work is concluded in Sect. 5.

2 Cloud Security Attacks

Early detection of cyber-threat is very crucial because of the huge sensitive data stored on cloud for various purposes. Therefore, implementing correct countermeasures to prevent the risks is a challenging task. Some approaches have been proposed to detect and prevent cyber-attacks in cloud environment. Summary of cloud security requirements and security threats is discussed in Table 1.

3 Proposed Framework

The proposed framework is shown in Fig. 1 which consists of four major blocks—data capturing, feature extraction, intrusion detection and knowledge extraction block. The detail of working of the blocks in the proposed framework is explained below.

3.1 Data Capturing

In the proposed work, UNSW-NB15 dataset is used as benchmark, which consists of 47 features and 10 target classes. In this work, to analyze the proposed model, a synthetic dataset is generated with similar categories and features to validate the model. To find out the most relevant features among 47 features, information gain technique is used. It is a technique by which important features are selected which significantly contributes for decision making. If D is the total size of the given dataset and A is a feature, then information gain value for feature A is calculated as Eq. (1).

$$\text{Information Gain (A)} = \text{Entropy}\ (D) - \text{Entropy}_{\text{A}}(D) \tag{1}$$

where Entropy (D) = Expected information needed to classify a tuple in D calculated as per Eq. (2). $\text{Entropy}_{\text{A}}$ (D) = Extra needed expected information for exact classification when feature A is selected, which is calculated as per Eq. (3).

Table 1 Summary of cloud security attacks

Cloud threats	Description	Security requirement
Unauthorized access	It is possible to delete, destroy, corrupt personal sensitive data	Confidentiality and privacy
Denial of service	Lack of control over cloud infrastructure	Availability and scalability
Misuse of services	Loss of verification, product theft, heavier assault due to unexplained registration	Availability and confidentiality
Hypervisor compromised	Interfere with other user services by compromising hypervisor	Privacy and confidentiality
Insecure interface and API	Authentication and authorization unacceptable, incorrect content transmission	Confidentiality and scalability
Impersonation attack	Access the cloud's critical area, stolen user account credentials, allowing attacker jeopardize system safety	Availability, privacy, and confidentiality
Insider attack	Penetrate capital of organizations, harm property, productivity loss, affect an activity	Confidentiality and privacy
Risk profiling	Operations for internal security, security policies, breach of configuration, patching, auditing and logging	Integrity and scalability
Identity theft	An aggressor can obtain a legitimate user's identity to access user's assets and take credits or other benefits in the user's name	Privacy and availability

$$\text{Entropy } (D) = -\sum_{i=1}^{m} P_i * \log_2(P_i) \tag{2}$$

$$\text{Entropy}_{\text{A}}(D) = \sum_{j=1}^{v} \frac{|D_j|}{|D|} * \text{Entropy } \left(D_{\text{j}}\right) \tag{3}$$

where feature A has "v" distinct values and D_{j} is the number of tuples that belong to each distinct feature value of A and probability (P_{i}) defines that an arbitrary tuple in D belongs to class C_{i}. It is given by Eq. (4).

$$P_{\text{i}} = \frac{|C_i|}{|D|} \tag{4}$$

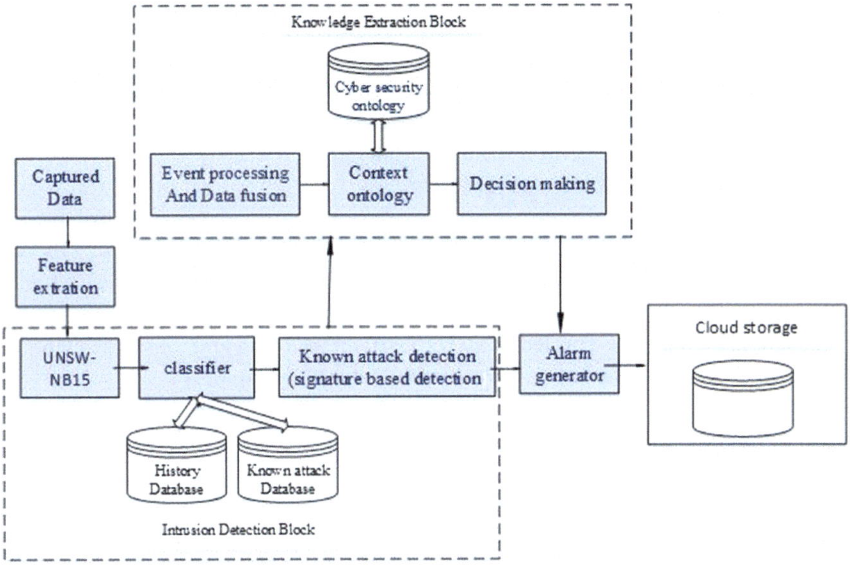

Fig. 1 Proposed framework for cyber-attack detection

Ten features are selected using the above technique to build the model. In order to create the dataset, a virtual environment is created which consists of different operating systems and other tools to generate and capture traffic. Traffic for different categories is generated in separate sessions. This virtual setup is created using several nodes, where few act as malicious users while others act as victim. This includes Kali Linux operating system along with other Ubuntu distributions. Kali Linux acts as an attacker, which performs different attacks, whereas Ubuntu systems act as victim node. In order to perform attacks, Kali Linux provides several tools and can also be done using command line. Wireshark tool is installed on all victim nodes which captures the inbound and outbound traffic on each interface. Captured data are then exported to csv file which will be used to find the features and to map the values for each feature.

3.2 *Filtering and Feature Extraction*

Data filtering is a process where the redundant instances are removed in such a way that the reduced dataset won't show the degradation in quality. This process is a critical step before the other preprocessing is applied. In the context of the proposed work, the filtering refers removing redundant instance from further phases of analysis.

Post-filtering. Set of features are extracted using combination of features. These features are same as 10 features obtained from the UNSW-NB15 dataset after applying information gain technique. Values corresponding to each feature are then calculated based on the filtered dataset. These values could be obtained by using two or more features of the filtered data. The range of the values of features could be very high to process. After applying the feature extraction and mapping of values the new form of this dataset with certain features are again checked by applying different filtering process, for example, data cleaning.

Normalization. Normalization is a process by which the values of an attribute are brought to a single scale, so that they do not lead to poor model design or model evaluation. In this work, z-score normalization is used based on the empirical studies in domain of this type of dataset.

Z-score (Zero-mean). This technique is based on the mean and standard deviation of each attribute "A" data and mathematically defined by Eq. (5)

$$V^{new=\frac{V^{old}-\bar{A}}{\sigma_A}} \tag{5}$$

where, $\bar{A}$ = mean of data A.

σ_A = standard deviation of A.

Finally, 22,000 instances are considered for the dataset, which cover all the classes.

3.3 Intrusion Detection Model Design

To design an IDS model, the generated synthetic dataset with features similar to that of UNSW-NB15 reduced features is used. This dataset is divided into training and testing part by following the ratio of 60:40. The proposed approach covers both signature and anomaly-based attacks. The training dataset is used to generate the rules which will be used to detect known attacks similar to that present in training data. These rules are maintained as knowledge base (KB) for detection. Rules follow the form

$$\boldsymbol{if}(condition_1 \cap condition_2 \cap \ldots \cap condition_n\ \boldsymbol{then}\ class = \text{"}attack\ category\text{"})$$

Each condition represents the relational expression on any attribute. For example, "proto ==http" could represent a condition for any rule. All the conditions are joined through "AND" operator. The detection of attack is done against the KB scanning through all the rules. If match found, the alert module will trigger an alert to the administrator about the malicious activity.

3.4 Knowledge Extraction Block

In this block, three modules are executed. The description of each module is given below:

Event Processing and Data Fusion. In this module the analysis of gathered information from different network events is performed. The collected evidences are used to deduce the threat level, type of attack and associated risks. Risk score is computed from vulnerability assessment defined in CRAMM [10]. In this module Dempster-Shafer evidence theory (DST), which is based on the set of evidences of an event, is used for threat estimation. This method can calculate the belief levels of individual data received from different sources so that the system will reduce the false positive and false negative of security alerts. The DST theory is based on two major functions, which are lower probability and higher probability [9]. From these functions, Shafer have attributed a belief (BEL) function shown in Eq. (6) and plausibility function (PL) shown in Eq. (7).

The belief function is defined by the formula:

$$\text{BEL (A)} = \sum_{B_{i \subset A}} m(B_i) \tag{6}$$

The plausibility function is defined by the formula:

$$\text{PL (A)} = \sum_{A \cap B_i} m(B_i) \tag{7}$$

Context Ontology. The proposed framework uses context ontology and knowledge base for correctness and uniformity of data instances. It provides a unified and formalized description to solve the problem of cloud attacks and threats. The ontology description language describes reasoning base of the IDS alert data. In the proposed framework, the ontology is build up with following security situation parameters and the inference rules using semantic web language (SWRL) is given below:

(a) **Context**: It consists of network connections, equipment and so on. In cloud environment, various client devices are sharing data. Hardware and variety of equipment are the subclasses of context to describe them. For example, the inference rule can be expressed as follows:
R1: Element(?c) ∧ hasOStype(?c, ?s) ∧ hascriticallevel(?c, High) ∧ hasvulnerability(?c, ?v) ∧ Attack(?a) ∧ hasAttackImpact(?a, High Damage) → SelectSecurityMechamism(?c, ?v, ?a)
(b) **Vulnerability**: Vulnerabilities are scanned by the attackers for future attack. In the proposed model, one of the object properties of vulnerability is "hasCVscore". CVscore is calculated for every identified vulnerabilities and leveled as high, low and medium according to the severity of the vulnerability. For example, the inference rule can be expressed as follows:

R2: Vulnerability (?v) $\wedge$ hasCVscore(?v, ?s) $\wedge$ Low(?s) $\wedge$ Attack(?a) $\wedge$ Damaged(?v, ?a) $\rightarrow$ NormalVulnerability(?v)
R3: Vulnerability (?v) $\wedge$ hasCVscore(?v, ?s) $\wedge$ Medium(?s) $\wedge$ Attack(?a) $\wedge$ Damaged (?v, ?a) $\rightarrow$ SeriousVulnerability(?v)

(c) **Attack**: The attacker performs some activity to damage the system. The alarm information is identified as attack. Attacks have an impact on assets. If denial of service attack is performed through SYN flooding method, then the inference rule can be expressed as follows:
R4: syn_flood(?x) ^ hasattackProperty(?x, ?i) ^ hasSourceIP(?a, ?c) ^ hasDestIP(?a, ?b) ^ TCP_Connect(?z) $\rightarrow$ Denial_of_Service(?d) ^ hasattack-Property(?d, ?z) ^ hasDestIP(?z, ?b) ^ hasSourceIP(?z, ?c).

(d) **Network Flow**: It helps to detect abnormal behavior of the network. In cloud environment, various client devices are sharing data. If traffic analysis shows abnormal traffic in some specific direction, then the inference rule can be expressed as follows:
R5: Netflow(?x) ^ hasProtocoltype(?x, ?i) hasSourceport(?a, ?c) ^ hasDest-port(?a, ?b) ^ hasbytes(?z) ^ AbnormalTraffic(?d) ^ hasattackProperty(?d, ?z) ^ SelectSecuri-tyMechanism(?x, ?c, ?b).

This module is mainly based on situation parameters using semantic ontology and user-defined rules to monitor the overall security behavior of the cloud. After that the decision module will be responsible for alert generation or not.

Decision Module. Normally this module is responsible for alert generation. A common threshold value of risk score is generated on the previous block. Based on this risk score an alert will generate for identification of risk cases.

4 Performance Evaluation of Proposed Framework

The proposed model is evaluated on several matrices in order to analyze every aspect of effectivity of the model. These matrices are based on the multiclass classification problem rather than binary and explained with their mathematical equations. In order to use these matrices, few parameters are needed, which are as follows:

FP_i = Represents instances with the actual class other than i^{th} class.
TP_i = Represents the correctly predicted instances of ith class.
FN_i = Total instances belonging to i^{th} class but predicted as other than i^{th} class.

- **Accuracy**: It indicates the frequency of correct classification over whole dataset and mathematically defined in Eq. (8).

$$\text{Accuracy} = \frac{\sum_{i=1}^{|C|} TP_i}{N} \tag{8}$$

- **Precision**: It shows the number of correctly predicted over the total predicted instances for each class. Mathematically, it is given by Eq. (9).

$$\text{Precision}_i = \frac{TP_i}{TP_i + FP_i} \tag{9}$$

- **Recall**: It shows the number of instances correctly classified over the total instances taken as input for that class. Mathematically, it is given by Eq. (10).

$$Recall_i = \frac{TP_i}{TP_i + FN_i} \tag{10}$$

- **F-Measure**: It shows the balance between precision and recall and calculated as in Eq. (11)

$$\text{F Measure}_i = \frac{2 * \text{Precision}_i * \text{Recall}_i}{\text{Precision}_i + \text{Recall}_i} \tag{11}$$

- **Mean F-Measure**: It is the mean of F-measures of all the classes and can be calculated by Eq. (12), where $|c|$ is the number of classes

$$MFM = \sum_{i=1}^{|c|} \frac{\text{F Measure}_i}{|c|} \tag{12}$$

- **Average Accuracy**: It is the mean value of recall over all the classes and can be calculated by Eq. (13).

$$\text{AvgAcc} = \sum_{i=1}^{|c|} \frac{\text{Recall}_i}{|c|} \tag{13}$$

- **Attack Accuracy**: It shows the efficiency of classifier in detecting attack classes only and can be calculated by Eq. (14):

$$\text{AttAcc} = \sum_{i=2}^{|c-1|} \frac{\text{Recall}_i}{|c-1|} \tag{14}$$

- **Attack Detection Rate**: It shows the rate of accuracy of model on attack categories and is calculated by Eq. (15).

$$ADR = \frac{\sum_{i=2}^{|c|} TP_i}{\sum_{i=2}^{|c|} TP_i + FP_i} \tag{15}$$

- **False Alarm Rate**: It shows the non-attack classes classified as attack and is given by Eq. (16)

$$FAR = \frac{FN_i}{TP_i + FP_i} \tag{16}$$

4.1 Result Analysis

In this section, we have analyzed the outcome of the proposed model on the above matrices. Classification of different category by the proposed model is shown by confusion matrix shown in Table 2.

Confusion matrix: It is defined by $N \times N$ matrix where the diagonal entry shows the correct classification for that particular category. In Table 2, the precision and recall is also shown, where the analysis attack shows the higher precision and recall. The overall performance of the proposed model is shown in Fig. 2, where the accuracy of the system is highest among all other parameters with FAR of 9.24.

5 Conclusion

In this paper, a data-driven rule-based framework is proposed for cyber-attack detection in cloud computing. The result shows that the accuracy of the system is highest among all other parameters with FAR of 9.24. The rule generated for the IDS engine is used for detection of known attacks and the knowledge base is used for unknown malicious activities. Hence the proposed framework can be used to detect known and unknown attacks at cloud infrastructure. It also provides a good detection rate and low false alarm for both signatures-based and anomaly-based detection. In future this framework can be used for security in IoT.

Table 2 Confusion metric obtained for the proposed model under different classes

	Normal	Backdoor	Analysis	Fuzzers	Shell code	Recon.	Exploits	DoS	Worms	Generic	Recall (%)
Normal	4568	33	0	113	6	125	25	52	1	77	90.90
Backdoor	6	1806	10	7	0	29	8	32	1	1	95.05
Analysis	20	13	1902	7	0	0	5	52	0	1	95.1
Fuzzers	99	47	5	1804	33	6	2	4	0	0	90.2
Shell code	23	0	0	54	764	119	20	2	8	10	76.4
Recon.	21	26	0	50	61	1562	0	0	0	0	90.81
Exploits	8	0	0	3	17	22	1920	30	0	0	96
DoS	5	0	0	7	82	66	9	2788	14	29	92.9
Worms	0	1	0	42	251	52	28	0	614	12	61.4
Generic	98	0	1	4	0	3	0	73	8	1813	90.65
Precision (%)	94.22	93.78	99.16	86.27	62.93	78.73	95.19	91.92	96.23	93.31	

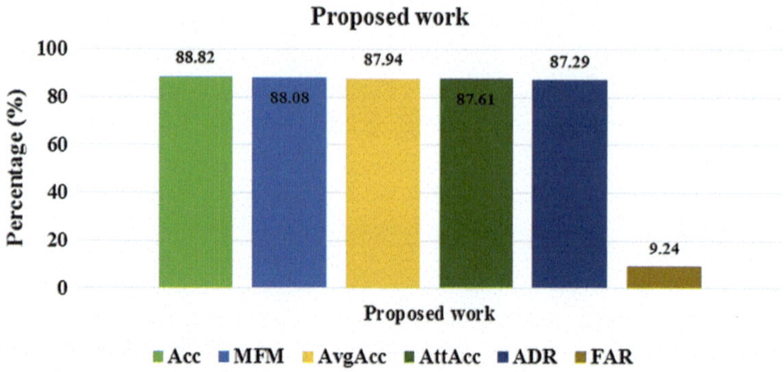

Fig. 2 Performance analysis

Acknowledgments This research was supported by Information Security Education and Awareness (ISEA) Project II funded by Ministry of Electronics and Information Technology (MeitY), Govt. of India.

References

1. Li, C. (ed.): Handbook of Research on Computational Forensics, Digital Crime, and Investigation: Methods and Solutions. IGI Global (2010)
2. Qiu, J., Wu, Q., Ding, G., Xu, Y., Feng, S.: A survey of machine learning for big data processing. EURASIP J. Adv. Signal Process. 67 (2016)
3. Shahul Kshirsagar, S.Y.: Intrusion detection systems: a survey and analysis of classification techniques. Int. J. Scienti. Res. Eng. Technol. IJRSET **3**(4), 742–747 (2014)
4. Dharmapurikar, S., Krishnamurthy, P., Sproull, T., Lockwood, J.: Deep packet inspection using parallel bloom filters. In: Proceedings of 11th Symposium on High Performance Interconnects, pp. 44–51. IEEE (2003)
5. Lee, I., Jeong, S., Yeo, S., Moon, J.: A novel method for SQL injection attack detection based on removing SQL query attribute values. Math. Comput. Model. **55**(1–2), 58–68 (2012)
6. Chen, W.-H., Hsu, S.-H., Shen, H.-P.: Application of SVM and ANN for intrusion detection. Comput. Oper. Res. **32**(10), 2617–2634 (2005)
7. Pearl, J.: Probabilistic Reasoning in Intelligent Systems: Networks of Plausible Inference. Morgan Kaufmann Publishers Inc., San Francisco, CA, USA (1988)
8. Dotcenko, S., Vladyko, A., Letenko, I.: A fuzzy logic-based information security management for software-defined networks. In: 16th International Conference on Advanced Communication Technology, pp. 167–171 (2014)
9. Shafer, G.: A mathematical theory of evidence turns 40. Int. J. Approxi. Reason. **79**, 7–25, 2016. 40 years of Research on Dempster-Shafer theory
10. Yazar, Z.: A qualitative risk analysis and management tool—CRAMM. SANS InfoSec Reading Room White Paper **11**, 12–32 (2002)

Benchmarking Semantic, Centroid, and Graph-Based Approaches for Multi-document Summarization

Anumeha Agrawal, Rosa Anil George, Selvan Sunitha Ravi, and S. Sowmya Kamath

Abstract Multi-document summarization (MDS) is a pre-programmed process to excerpt data from various documents regarding similar topics. We aim to employ three techniques for generating summaries from various document collections on the same topic. The first approach is to calculate the importance score for each sentence using features including TF-IDF matrix as well as semantic and syntax similarity. We build our algorithm to sort the sentences by importance and add it to the summary. In the second approach, we use the k-means clustering algorithm for generating the summary. The third approach makes use of the Page Ranking algorithm wherein edges of the graph are formed between sentences that are syntactically similar but are not semantically similar. All these techniques have been used to generate 100–200 word summaries for the DUC 2004 dataset. We use ROUGE scores to evaluate the system-generated summaries with respect to the manually generated summaries.

Keywords Summarization · K-means · SVM · Page rank · Gaussian mixture

---

Anumeha Agrawal, Rosa Anil George, Selvan Sunitha Ravi are equally contributed.

A. Agrawal · R. A. George · S. S. Ravi (✉) · S. S. Kamath
Department of Information Technology, National Institute of Technology Karnataka, Surathkal, Surathkal, Mangalore, India
e-mail: sunitha98selvan@gmail.com

A. Agrawal
e-mail: anumehaagrawal29@gmail.com

R. A. George
e-mail: rosageorge97@gmail.com

S. S. Kamath
e-mail: sowmyakamath@nitk.edu.in

S. C. Satapathy et al. (eds.), *Intelligent Data Engineering and Analytics*, Advances in Intelligent Systems and Computing 1177,
https://doi.org/10.1007/978-981-15-5679-1_24

1 Introduction

With the advancement of technology and data flood all over the Internet, it has become important to generate summaries of documents that are representative of the entire content of the document. Multi-document summarization is a pre-programmed process to excerpt data from various documents regarding similar topics. The summary report gives an overview of the content contained in the large collection of documents. This way a lot of time that would have been wasted in going through all the notes or articles can be saved and necessary quick summaries can be generated.

Several factors affect the multi-document summarization process, for example, speed, redundancy, and selection of paragraphs in documents. If there are thousands of articles in the document collection, the summarization will take time as all the sentences must be read, cleaned, processed, then either assigned a score or put into a cluster that it is most similar to and hence speed depends on the size of the document collection. If there are multiple similar sentences, then there is a high probability of having a lot of redundancy in the summary. Hence, setting a threshold value for dissimilarity is very crucial. Also, sentence position in an article and sentence position in document collection is significant. An important sentence placed at the bottom or end of a collection will have a high chance of not being considered. This will decrease the accuracy of the summary and is an important aspect to consider. To ensure that this does not happen, the summarization process must only begin once all sentences have been read. All these factors are crucial in the formation of valuable summaries.

The rest of this paper is organized as follows—Section 2 describes the related work in this domain. Section 3 elucidates the proposed methodology and the various algorithms that can be used to perform document summarization. Section 4 presents the results obtained and draws a comparison to the state-of-the-art results, followed by conclusions and future work.

2 Related Work

Several researchers have produced extensive work in the area of Multi-document summarization over the years. Goldstein [1] described a text extraction approach that is built on methods to summarize single documents by using implicit information about the set of documents as a whole as well as the relationships between these documents. These methods are usually domain-independent and are based mainly on fast, statistical processing. They are not based on natural language understanding or information extraction techniques. Hence, summaries lack coherence and can be fragmented, semantically unrelated, or repetitive which is undesirable.

Banerjee [2] used an integer linear programming (ILP) based multi-sentence compression to achieve abstractive summarization. His approach first identifies the most important document in the multi-document set. The sentences in the most important document are compared with sentences in other documents to generate clusters of similar sentences. They then generate K-shortest paths from the sentences in each cluster using a word-graph structure. Finally, they select sentences from the set of shortest paths generated from all the clusters employing an integer linear programming (ILP) model to produce a coherent summary. It captures the redundant information using a constructive clustering technique and it is more preferable than the baseline abstractive memorization technique. But it has several drawbacks such as the presence of phrase-level redundancies.

Erkan and Radev [3] uses the idea of weighted words in a sentence to identify important sentences. They define centrality by the presence of certain weighted words. They also came up with another method for computing sentence importance based on the concept of eigenvectors that we call LexPageRank. In their model, a sentence connectivity matrix is created based on the cosine similarity. If the cosine similarity between two sentences is greater than the threshold value, a corresponding edge is added to the connectivity matrix.

Mani et al. [7] proposed a reconstruction based approach for summarization using the distributed bag of words model. The unsupervised centroid-based document-level technique selects summary sentences to decrease the error between the documents and the summary. The sentence selection and beam search methods have also been incorporated to further improve the performance of the model. This technique was able to achieve significant gains as compared to the state-of-the-art baselines.

Bing et al. [8] proposed an abstractive framework for Multi-document summarization that adds new sentences on the basis of syntactic units such as noun and verb phrases. The sentences in documents are broken down into a set of noun phrases (NPs) and verb object phrases (VPs) which represent the key concepts and key facts, respectively. A parser is then employed to obtain a constituency tree for each of the input sentences. The new sentences are constructed through an optimization problem and each sentence containing NP and VP is considered through compatibility relation.

From the discussions, it is evident that extracting relevant content on various topics is critical since there is a rise in data overflow. Summaries can help us provide intuition and relevant knowledge on those topics. Automated summaries will reduce the time and effort taken to generate these summaries and multi-document summarization will help in capturing contexts from various sources.

In this paper, we aim to generate summaries of 100–200 words for each set of articles using four different techniques, and benchmark the proposed methods on a standard dataset. We then will compare each summary of the four techniques with the user-generated summary for that cluster which is also available online to compute the ROUGE score and then find which technique achieves the best results. Our objective is to use importance score algorithm, clustering algorithms, and a graph-based formulation to generate summaries and analyze ways to improve it.

3 Proposed Methodology

For representing the documents in a representation feature space there are several features that we have considered which include the length of the document, number of nouns, verbs, and even the sentence position in a document. The length of the document is given by the number of words in a document and is one of the features. This is important as we can normalize the other features based on this measure. The number of verbs in a sentence is considered as it helps capture the common actions in two sentences and is computed using the verb count function in NLTK [9]. Sentence Position in the respective document is taken into account as important sentences usually occur at the beginning of the document, thus helping in capturing the context well. Usually the sentences, in the beginning, introduce an idea and the sentences at the end conclude a discovery. Named Entities are another significant feature and the count of named entities is also used. For a domain-specific entity, it is difficult to label the entities. For a general entity such as location, name action, and organization we use Stanford Named Entity Recognizer (NER) [10]. We also experimented with Spacy's name tagger but got better results with Stanford's NER tagger.

The number of digits in a sentence are also counted as it is useful in finding statistical content. Sentences with a large number of digits may contain crucial information and need to be analyzed carefully. The number of adjectives in a sentence can be used to analyze and compare the degree of the problem. For example, if a sentence describes a car accident the adjectives severe and fatal are used to describe a problem at the almost same level of intensity. Thus, one of the two sentences might be chosen in the summary. The count of Uppercase words is also important as it indicates either a name, place, or something that is to be given importance.

The Term frequency-Inverse Document frequency (Tf-idf) value is calculated for the sentence. This helps in assigning the importance of term frequency in a document and reducing the importance of common words that are redundant. This feature is a statistical measure and contributes toward the importance of the word in a document in a collection of texts. The frequency of a word in a document is directly proportional to the importance of that word in the document. The term frequency is normalized by the length of the document to ensure the term frequency is consistently calculated throughout texts. After obtaining all these features, we create the feature vector and normalize them. We then proceed to use four distinct algorithms to generate the summary. This feature vector is used to calculate the importance of sentences and each feature represents some component of the sentence which contributes to summarization.

3.1 Sort by Importance Score Algorithm

All the sentences are placed in descending order of sentence importance score. This importance score is calculated from the feature vector that has been created. The total score is the weighted sum of all the features and is normalized so that the score does not blow up. The first sentence is added to the summary. Each subsequent sentence is added to the summary based on the sentence importance score. This method is very fast and is space-efficient as we use only one array instead of 100 different arrays to capture sentences of various lengths. We keep adding sentences until the total size of the summary generated exceeds the threshold length. We also compare each of the sentences in the array with incoming sentences to check their similarity. If the two sentences are similar then the incoming sentence is discarded. This stacking method helps in ensuring two things, firstly, the sentences represent a unique argument to the topic and secondly, a strict summary of a fixed number of words can be obtained.

3.2 K-Means Clustering

The K-means clustering algorithm was implemented by choosing 'k' number of sentences as the clusters such that the distance between them was maximum. We observe that with k set to 5, we obtain the best results. Cosine similarity, WordNet [11] based similarity (Lesk disambiguation), and Jaccard similarity were the distance metrics that were used in the implementation. After each iteration, we choose a cluster center such that it minimizes the average distance from the cluster center. Once all the iterations were done, we chose the cluster center of each cluster as the representative sentence for the summary followed by ranking sentences by their importance score. We use a TF-IDF transformer for the conversion of sentences into a list of vectors so that the distances between individual vectors can be calculated and this is quite intuitive as well. Then we proceed to select sentences according to their score and continue appending sentences until the threshold limit is reached.

However, K-means has some drawbacks as it is non-probabilistic in nature and considers only simple distance to distance measure. From an intuitive point of view, it is expected that the clustering center assignment for some data points is more certain than others. For example, there may be an overlap between two clusters such that we cannot confidently say which cluster a point belongs to. The k-means model has no ingrained measure of probability or uncertainty in cluster assignments.

3.3 Gaussian Mixture Method

To overcome the drawbacks of k-means clustering, we use the Gaussian Mixture model. It helps in resolving the uncertainty component in the cluster assignment by allowing a point to belong to more than one cluster. Mixture models are a better way of expressing the presentiment to consolidate information about the covariance structure as well as the centers of the latent Gaussians. Gaussian Mixture method provides an idea of the probability with which a data point belongs to a particular cluster. The parameters of each Gaussian (i.e., variance, mean, and weight) need to be estimated in order to cluster data and this takes place using the Expectation Maximization problem. The covariance type hyperparameter controls the degrees of freedom in the shape of each cluster. This helps in deciding the level of intuitiveness.

3.4 *Graph-Based Approach*

The graph G = (V, E) is formulated as given—lLet V = {SlS {sentences that need to summarized}} and E = {(U V) l sim (U, V)} < threshold. The next step is to obtain a clique. The nodes are reordered based on the importance score of the sentences and these sentences are appended in that order to generate a summary. It helps in ranking the sentences which are nodes in a graph based on the number of incoming links. If a sentence is similar to another then a new node is created and connected to the previous sentence. If a sentence is unique then a new node is used to represent this. This helps in creating cliques where each clique represents a common idea and while creating a summary we use one sentence from each of these cliques. This calculation uses the concept of Eigenvalue vectors. The Eigenvector calculation is done by the power iteration method but there is no guarantee that it will converge to a point.

4 Experimental Results and Analysis

For experimental validation, we use the DUC 2004 dataset, which consists of 50 document clusters, each with 10 news articles related to the same topic. It also contains a manually generated summary. We measure the performance of each approach using precision, recall, and F1 scores. The ROUGE score is used for measuring the accuracy of the summaries. The summaries generated by the different approaches are measured against manual summaries. Precision is the measure of exactness and takes into account the number of right instances retrieved. Recall measures the total right instances retrieved, while the F1 score is the harmonic means of the precision and recall. We measure the Rouge-L score which gives the longest matching sequence of words. We calculate the precision, recall, and F1 scores for each method using 10 different sets of documents (Fig. 1).

Table 1 describes the ROUGE-L scores obtained for Sort by Importance Score method. We obtained an average precision score of 0.45811, average recall of 0.36941, and average F1 score of 0.39149. As shown in Table 2, K-means clustering gives us average precision, recall, and F1-score of 0.39267, 0.47311, and 0.42131. K-means performs better than sort by importance in terms of recall and F1-score. Table 1 shows the ROUGE-L values for the Page Rank method. Page Rank method has an average precision of 0.42414, average recall of 0.43485, average F1-score of 0.42352. The average scores of precision, recall, and F1-scores of Gaussian mixture

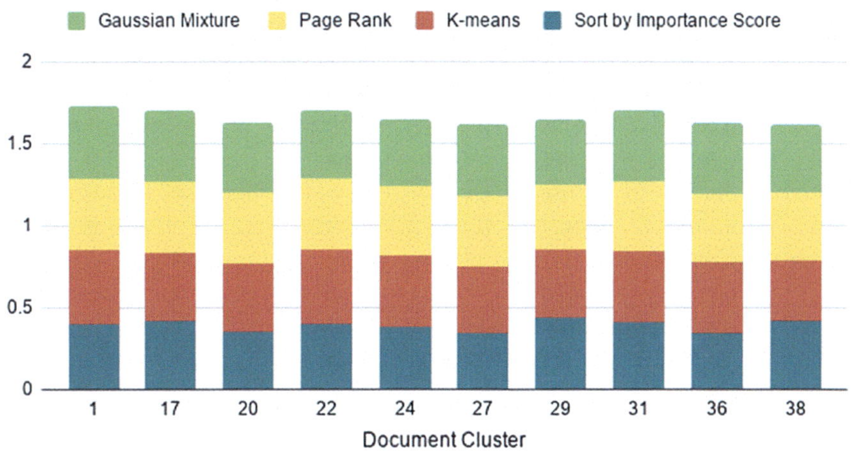

Fig. 1 F1-score performance of the proposed methods

Table 1 Results obtained by sort by importance and page ranking method

	Sort by importance			Page ranking		
Document	Precision	Recall	F1 score	Precision	Recall	F1 score
1	0.31489	0.55141	0.40086	0.41080	0.49395	0.44856
17	0.54830	0.33439	0.41543	0.40225	0.42622	0.41389
20	0.24364	0.64624	0.35387	0.36994	0.46656	0.41267
22	0.50000	0.33033	0.39783	0.40233	0.52679	0.45623
24	0.47651	0.31934	0.38240	0.44976	0.42342	0.43619
27	0.63306	0.23860	0.34657	0.42857	0.38680	0.40662
29	0.46837	0.41265	0.43875	0.38205	0.45288	0.41446
31	0.40322	0.41415	0.40861	0.46587	0.41114	0.43680
36	0.54849	0.25386	0.34709	0.36004	0.52108	0.42584
38	0.44462	0.40444	0.42358	0.25507	0.62229	0.36183

method Table 2 are 0.39476, 0.46720, and 0.42516. From this, it can be concluded that the Page Rank and Gaussian mixture algorithm perform the best and this is indicated by the F1 score.

Table 2 Results obtained by K-means and Gaussian mixture methods

	K-means			Gaussian mixture		
Document	Precision	Recall	F1 score	Precision	Recall	F1 score
1	0.41080	0.49395	0.44856	0.37083	0.53776	0.43896
17	0.40225	0.42622	0.41389	0.44035	0.44296	0.44165
20	0.36994	0.46656	0.41267	0.48437	0.39490	0.43508
22	0.40233	0.52679	0.45623	0.43390	0.42725	0.43055
24	0.44976	0.42342	0.43619	0.34917	0.53468	0.42246
27	0.42857	0.38680	0.40662	0.40870	0.44224	0.42481
29	0.38205	0.45288	0.41446	0.49874	0.32357	0.39250
31	0.46587	0.41114	0.43680	0.42768	0.41415	0.42081
36	0.36004	0.52108	0.42584	0.40112	0.43072	0.41539
38	0.25507	0.62229	0.36183	0.42651	0.40029	0.41299

5 Conclusion and Future Work

In this paper, we present four techniques to generate 100–200 word summaries from various documents on the same topic. The four approaches are sort based on the importance score, K-means clustering, Gaussian model, and a graph-based approach using the Page Ranking algorithm. From experimental results, we concluded that clustering algorithms and page ranking algorithms perform marginally better than the sort by importance score as handcrafted features cannot outperform powerful functions like Gaussian function. These algorithms can be used to generate effective summaries. The size of the summary can be altered by changing some hyper-parameters in the algorithms. These summaries are comparable to the human-generated summaries and can be used to generate quick summaries as hand generation is very cumbersome and time-consuming. This gives an unbiased opinion as well as the summary is extracted from different documents. The results show that the F1-score is comparable for each document set under each method. This guarantees the consistency of the methodology and it can be used for any document set.

References

1. Goldstein, J., Mittal, V., Carbonell, J., Kantrowitz, M.: Multi-document summarization by sentence extraction. In: NAACL-ANLP-AutoSum '00 Proceedings of the 2000 NAACL-ANLP Workshop on Automatic summarization, vol. 4 (2000)
2. Banerjee, S., Mitra, P., Sugiyama, K.: Multi-document abstractive summarization using ILP based multi-sentence compression. In: IJCAI'15 Proceedings of the 24th International Conference on Artificial Intelligence, pp. 1208–1214. AAAI Press
3. Erkan, G., Radev, D.: LexPageRank: prestige in multi-document text summarization. 365–371 (2004)

4. Kumar, A., Ahrodia, S.: Multi-document Summarization and Opinion Mining Using Stack Decoder Method and Neural Networks: Proceedings of ICDMAI 2018, vol. 2
5. Sripada, S., Gopal Kasturi, V., Parai, G.: Multi-document extraction based summarization (2019)
6. Daumé, H.III., Marcu, D.: Bayesian multi-document summarization at MSE. In: Proceedings of the Workshop on Multilingual Summarization Evaluation (MSE), Ann Arbor, MI (2005)
7. Mani, K., et al.: Multi-document summarization using distributed bag-of-words model. In: 2018 IEEE/WIC/ACM International Conference on Web Intelligence (WI). IEEE (2018)
8. Bing, L., Li P., Liao, Y., Lam, W., Guo, W., Passonneau, R.: Abstractive multi-document summarization via phrase selection and merging. In: Proceedings of the 53rd Annual Meeting of the Association for Computational Linguistics and the 7th International Joint Conference on Natural Language Processing, vol. 1, Long Papers (2015)
9. Loper, E., Bird, S.: NLTK: the natural language toolkit. arXiv preprint cs/0205028 (2002)
10. Finkel, J.R., Grenager, T., Manning, C.: Incorporating non-local information into information extraction systems by Gibbs sampling. In: Proceedings of the 43nd Annual Meeting of the Association for Computational Linguistics (ACL 2005), pp. 363–370
11. Miller, G.A.: WordNet: a lexical database for English. Commun. ACM **38**(11), 39–41 (1995)

Water Availability Prediction in Chennai City Using Machine Learning

A. P. Bhoomika

Abstract Chennai is a city located down south in India. It serves as the capital of Tamil Nadu, and the city and the surrounding area serves as a major economic center in India. In the recent past, Chennai is facing an acute water shortage. This is due to two years of inadequate monsoon seasons, and increasing urbanization that caused some encroachment on water bodies in and around the region. There are four reservoirs which are major sources of water supply to Chennai, viz., Poondi, Cholavaram, Redhills, and Chembarambakkam. This paper discusses the major causes of the water crisis while analyzing water levels of four main reservoirs and rainfall levels in reservoir regions. An attempt is made to predict water availability per person in Chennai by using machine learning, and possible measures to rescue the city from the thirst of water are discussed.

Keywords Chennai · Water scarcity · Water availability prediction · Machine learning

1 Introduction

Chennai is the capital city of the "Tamil Nadu" state in south India, located at the Coromandel Coast, of the Bay of Bengal. It is the biggest center for education, economy, and culture of south India. Chennai's population is close to 9 million and it is the 36th largest urban area by population.

During June–July 2019, the city faced an acute water shortage. This water scarcity is due to lack of monsoon rainfall for two years, i.e., in late 2017 and throughout much of 2018 [1]. This is mainly because of irrational planning and use of land, lack of rational measures for conservation of water resources and its management. Changes in rainfall patterns are also a significant reason. The city has often experienced both floods and drought, like the heavy rainfall to nil rain for 190 and odd days.

A. P. Bhoomika (✉)
ACED, Alliance University, Bengaluru, India
e-mail: kannika.bhoomi@gmail.com

S. C. Satapathy et al. (eds.), *Intelligent Data Engineering and Analytics*, Advances in Intelligent Systems and Computing 1177,
https://doi.org/10.1007/978-981-15-5679-1_25

Earlier, Chennai was a water—surplus city of the country. Decades ago, there were nearly twenty-four water bodies which include three rivers and the Buckingham canal of the British's period. But now hardly six of them can be seen.

The city's sewage polluted the rivers; since then the annual monsoon rains are the only means to replenish its reservoirs of water. Groundwater resources are being replenished by rainwater and 1,276 mm is the average rainfall of the city. Four major water supply reservoirs were completely dry due to lack of groundwater and rainwater.

The major water supply sources of Chennai are

1. Cholavaram, Poondi, Chembarambakkam, and Red Hill's are the four major water reservoirs.
2. Cauvery water from Veeranam Lake.
3. Nemelli's and Minjur's desalination plants.
4. Aquifers at Panchetty, Minjur, and Neyveli.
5. Tamaraipakkam, Poondi, and Minjur Agriculture wells.
6. CMWSSB Borewells.
7. Retteri lake.

There are four major reservoirs, in the city that supply drinking water, namely, Poondi, Cholavaram, Red Hills, and Chembarambakkam, and their combined capacity is 11,057 mcft. During June 2019, all of them reached far below the zero level and did not hold even 1% of their capacity. The city's dependency was now extremely on three mega plants of water desalination with a combined capacity of 180 mld. All these units are operating overtime to maintain efficiency not less than 80–90%.

In this paper, water levels of four major reservoirs and rainfall levels of reservoir regions for the past 15 years are analyzed to understand the water needs of the city. The effect of population growth on meeting the water needs of the city is also analyzed. Later, a machine learning technique, Support Vector Regression (SVR) is employed to predict water availability per person in the future.

2 Related Work

The authors described the demand and supply of water available from various sources in the Tamil Nadu state [2]. The study gave an insight into understanding Chennai's water supply dynamics and discussed the urban water system [3]. The authors proposed a Support Vector Regression model with optimized hyperparameters to predict the water demand accurately in a short time [4]. A model with Support Vector Regression for base prediction was discussed to forecast the water demand. Further, the base predictions were improved using the Fourier time series process [5]. The work discussed employing the Backtracking Search algorithm with Artificial Neural Networks and Gravitational Search algorithm for forecasting water demand along with studying the impact of weather on water demand [6]. The study focused on creating a model to predict water consumption using soft computing methods published between 2005 and 2015 [7]. An approach to understand water consumption behavior using a non-homogeneous Markov model was discussed [8]. The paper discussed major sources of arising challenges of water scarcity in India. The authors emphasized encouraging people to adopt traditional methods of water management [9].

3 Methodology

3.1 Dataset Description

The dataset shows the details of water availability in the four main reservoirs Poondi, Cholavaram, Redhills, and Chembarambakkam in million cubic feet (mcft) and rainfall at different reservoir regions in mm over the last 15 years. The data has been collected from the Chennai Metropolitan Water Supply and Sewage Board website [10]. The Chennai population dataset is also considered to understand the population growth and increase in the demands for water supply.

3.2 Support Vector Regression

Support Vector Regression (SVR) is a supervised machine learning technique used for the prediction of continuous values. The working principle of SVR is analogous to the Support Vector Machine. It provides an efficient prediction model by considering the presence of non-linearity in data. SVR is a non-parametric technique and its output depends on kernel functions but not on the distributions of underlying target and predictor variables. SVR also allows us to create a non-linear model without changing predictor variables. Thus, resulting model can be better interpreted.

Predictions given by SVR are not affected as long as the error (ε) is less than a particular value. This is called as principal of maximal margin. The hypermeters Cost and Epsilon of SVR are tuned to optimize the model performance. In this paper, SVR is employed to predict the availability of water in the future.

4 Experimental Results and Analysis

The datasets of water levels in all four main reservoirs and rainfall levels for the period 2004–2019 (till June 2019) is considered for experimental purpose. The data is available day wise. The experiment is conducted using the tool RStudio, and R programming language is used for implementation.

As an initial step, data pre-processing was carried out to identify missing values in the dataset and to carry out data imputation. In the next step, the exploratory analysis of the dataset is performed [11]. The exploratory analysis gave deep insight into rainfall, water levels in reservoirs, and ultimately the water scarcity.

4.1 Analysis of Water Levels of Reservoirs

The visualizations of Figs. 1 and 2 show the water levels of four different reservoirs in million cubic feet (mcft). From the plots, it can be inferred that every year the water level in all the four reservoirs follows a decremental phase during summer and a replenishment phase from October–November. Bad water scarcity was observed in 2004 in all the four reservoirs where the water level reached almost zero. There was also a bad phase in 2014–15 and there was no water availability in the two reservoirs (Poondi and Cholavaram). It is alarming that there is almost no water available in all the four major water reservoirs in recent days.

Considering water availability from four reservoirs from Fig. 3, it can be observed that there was a downfall in water levels to almost zero in 2004, 2017, and 2019. The depletion of water availability in 2017 is similar to 2019 but the water levels reached close to 0 during the end of August. But, in 2019 the water levels reached zero at the beginning of June itself.

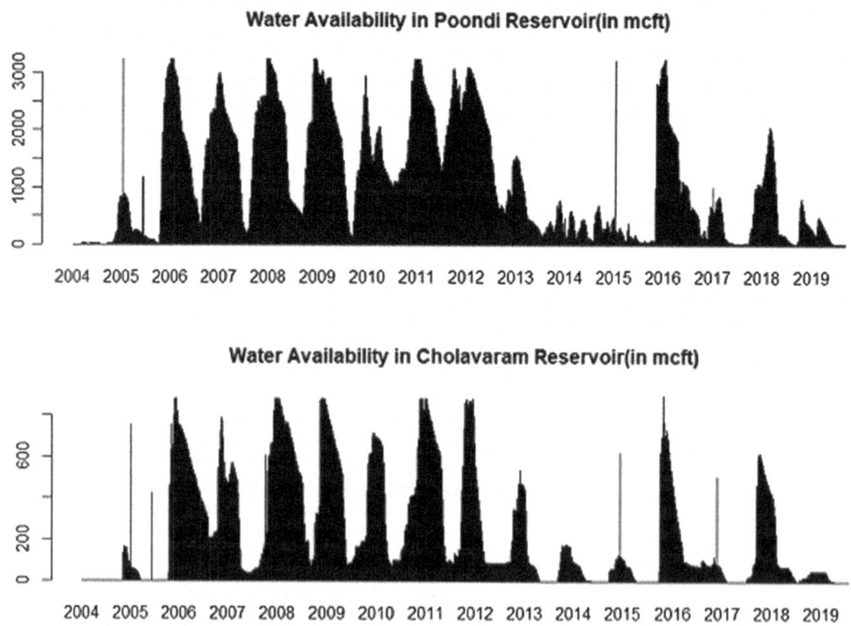

Fig. 1 Water availability in Poondi and Cholavaram reservoirs during 2004–2019

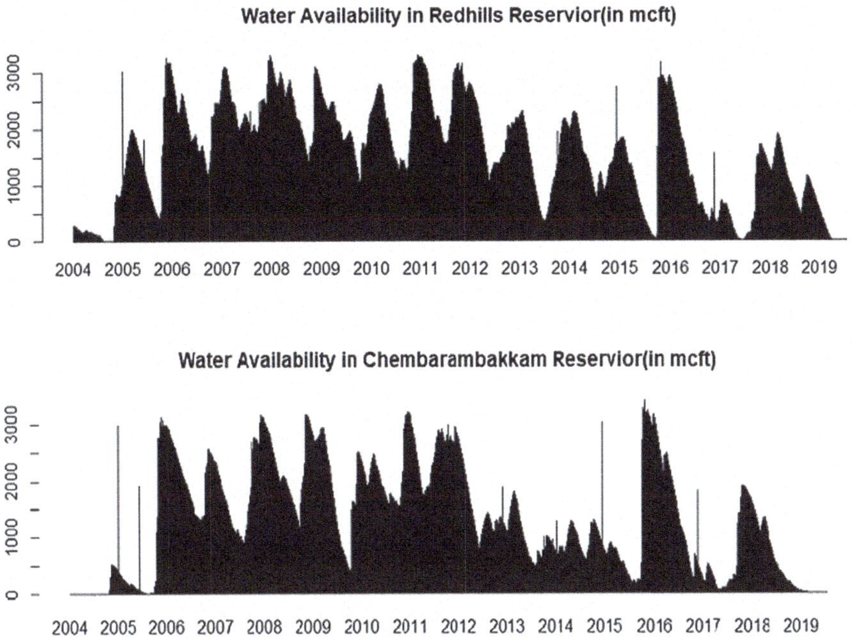

Fig. 2 Water availability in Redhills and Chembarambakkam reservoirs during 2004–2019

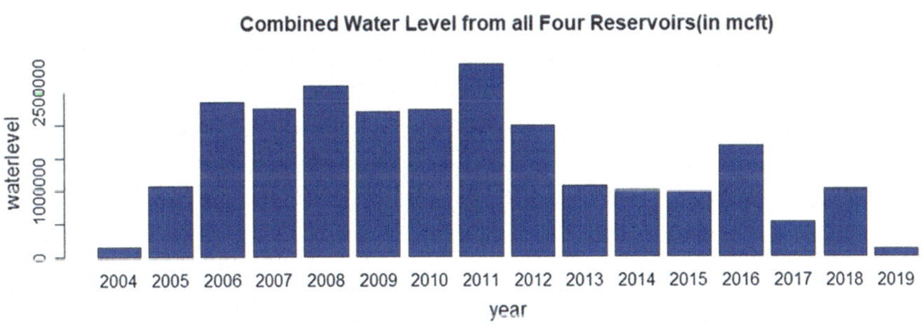

Fig. 3 Combined water levels of all four reservoirs during 2004–2019

To estimate the water shortage, the sum of water levels at the beginning of summer is compared in Fig. 4. This is because there will not be any replenishment of reservoirs till the next monsoon and the amount of water stored in the four reservoirs itself will be a clear indicator of how long can the water be managed during summer and any backup plans should be done. There is a continuous decrease in water level from 2012. But there is a spike in water levels in early 2016. This can be attributed to the severe floods the city had faced by the end of 2015.

Despite the increase in water level in early 2016, storage levels have dropped steeply than never before from 2016 to the end of 2017. It can be observed that the

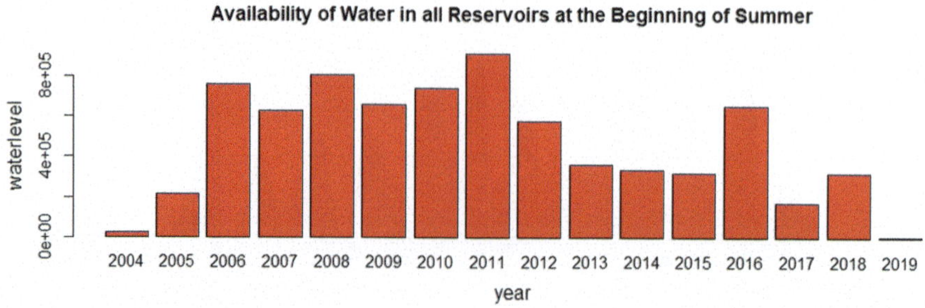

Fig. 4 Availability of water in all four reservoirs at the beginning of summer during 2004–2019

water level has almost reached zero in all the four reservoirs at the beginning of summer 2019. A similar condition was observed during 2004 as well. This situation can repeat in further years if necessary water scarcity measures are not taken ahead of time.

4.2 Analysis of Rainfall Levels of Reservoir Regions

From the analysis of water level in major reservoirs, it is clear that water levels are decreasing every year and in June 2019 there is no water in any of the major reservoirs. All reservoirs depend on rain for their replenishment.

Figures 5, 6, and 7 describe the year-wise rainfall level in four major reservoir regions and month-wise combined rainfall levels. The city gets rains in June, July, August, and September due to southwest monsoon. Major rainfall happens during October and November of every year which is due to northeast monsoon. The annual rainfall in 2018 is the lowest of all the years from 2004. During the initial years, rain from northeast monsoon is much higher than the southwest monsoon. But for the last few years, there is a reduction in rains from northeast monsoon.

4.3 Analysis of Chennai Population

The Population is also adding to the water scarcity problem at another level. Increases in the demands for water supply with the growth of the population and on the other hand decrease in rainfall and water levels of the reservoir have worsened the situation.

From Fig. 8, we can see that the population growth of Chennai from 2000 to 2019 has been exponential. Speculation and prediction also show that the population of Chennai will keep increasing according to the given data. So, as the number of consumers increase, there will be a heavy pressure on these reservoirs also to fulfill the needs of the population in the future. So, population growth will also add to the water problems on different levels.

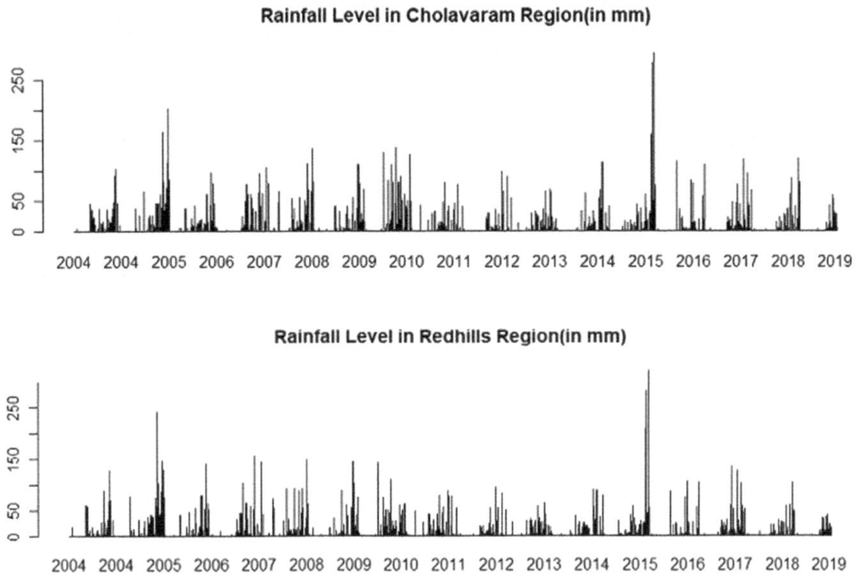

Fig. 5 Rainfall level in Cholavaram and Redhills reservoir regions during 2004–2019

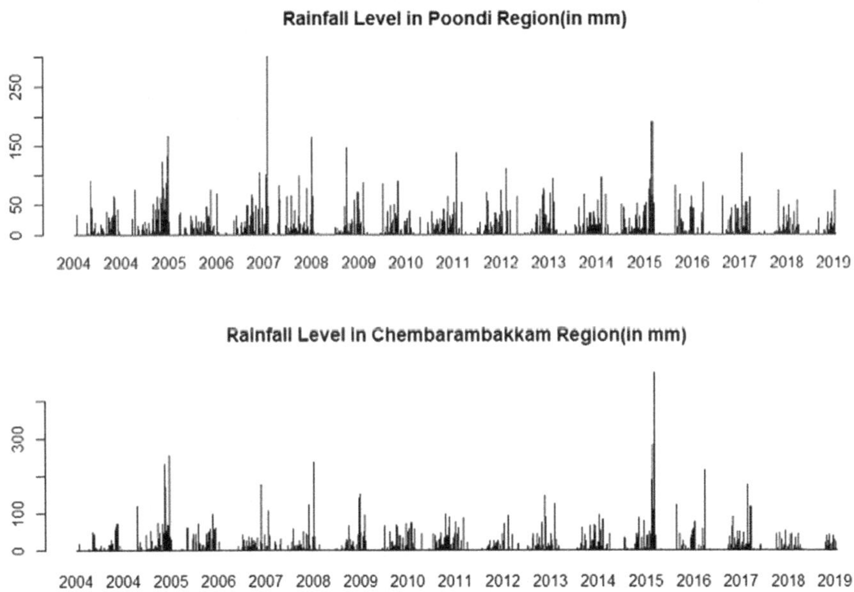

Fig. 6 Rainfall level in Poondi and Chembarambakkam reservoir regions during 2004–2019

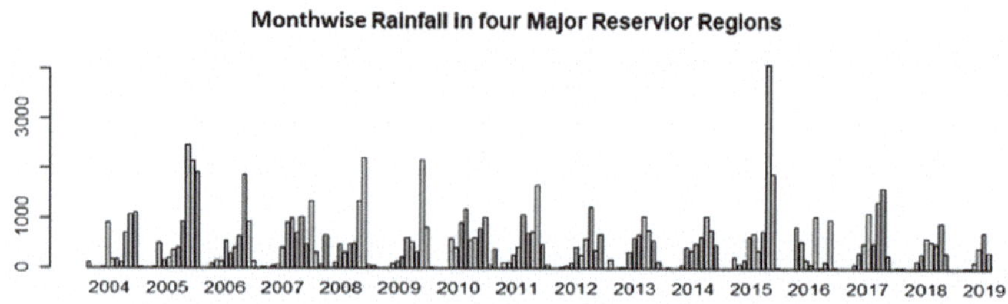

Fig. 7 Month-wise rainfall level in all four reservoir regions during 2004–2019

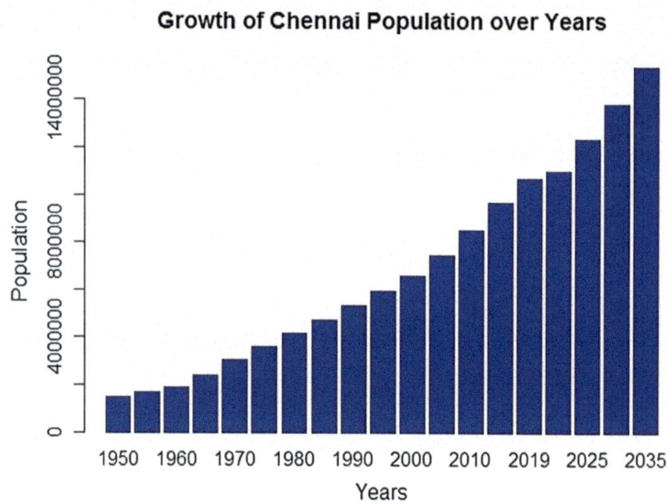

Fig. 8 Chennai population growth

4.4 *Prediction of Water Availability per Person*

Recent reports show that Chennai requires 830 MLD (Million Liters a Day) of water [12]. But during the critical days of the water crisis, Chennai Metro Water Supply and Sewage Board (CMWSSB) could supply only a 525 MLD.

To estimate the water availability in the future, an experiment is conducted to predict average water available in liters per person in Chennai city. The dataset consisted of combined water levels of four major water reservoirs, rainfall levels, population, and water availability per person for the past 15 years. Further available water in reservoirs in mcft is converted in liters as 1 mcft = 28316846.6 liters.

Support Vector Regression model is created to predict water availability. The Root Mean Square Error (RMSE) is a measure of the differences between predicted values by a model and the observed values. RMSE of the SVR model is calculated by using the difference between actual and predicted values of water availability over the test sample. RMSE is calculated by the formula,

$$\mathbf{RMSE} = \sqrt{\frac{1}{m}\sum_{j=1}^{m}(\mathbf{p}_j - \hat{\mathbf{p}}_j)^2}$$

where

$\hat{p}_1, \hat{p}_2, \ldots, \hat{p}_n$ are predicted values.

$p_1, p_2, \ldots, p_n$ are observed values.

m is the number of observations.

Mean Absolute Error (MAE) is the measure of the average magnitude of the errors in a set of predictions, without considering their direction. It is calculated as the mean of the absolute differences between predicted values by a model and actual values over the test sample, where all individual differences have equal weight. MAE is calculated by the formula,

$$\text{MAE} = \frac{1}{m}\sum_{j=1}^{m}\left|p_j - \hat{p}_j\right|$$

where

$\hat{p}_1, \hat{p}_2, \ldots, \hat{p}_n$ are predicted values.

$p_1, p_2, \ldots, p_n$ are observed values.

m is the number of observations.

The SVR model created showed a better performance as it was observed that, RMSE = 2.15, MAE = 1.79 upon 10 fold cross-validation, with epsilon = 0, Cost = 16 using radial kernel.

Finally, from Fig. 9 it can be observed that there is a decrease in available water per person from 2012. The average water available per person per day in 2019 is very low, almost 1000–1200 L. This is the lowest water availability after 2004. Observing the trends in the data, it is implied that the entire city would struggle with a thirst for water in the future if necessary water scarcity control measures are not employed.

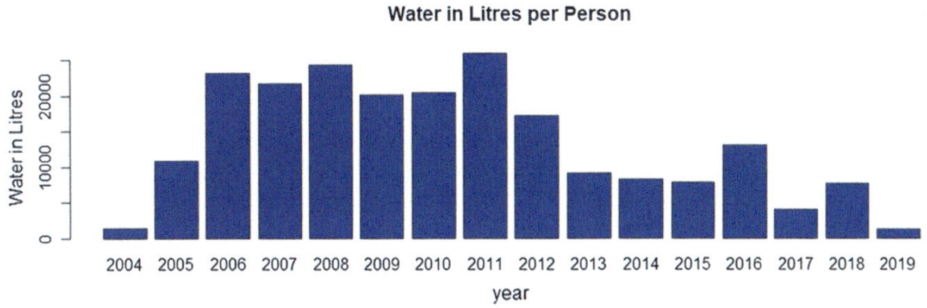

Fig. 9 Water availability on Chennai per person during 2004–2019

The major reasons for the water crisis are poor monsoons, urbanization, deforestation, heavy industrial usage of water, water pollution, poor management of existing water bodies, and several other factors. As a measure of water scarcity control, all of them have to be controlled. Rational measures must be taken to restore and manage extinct and existing water bodies, install rainwater harvesters, and promote the reuse of treated water. Necessary actions to be taken to protect reservoir catchments recharge groundwater and develop associations with stakeholders and farmers to conserve the water for the future.

5 Conclusion and Future Work

Chennai is recently facing a high water scarcity problem. It affected the lives of people badly at various levels. In this paper, an attempt is made to explore the problem in detail by analyzing the Chennai population, rainfall, and water levels of four major reservoirs of Chennai. It is alarming that rainfall is decreasing every year, as well as the water levels in reservoirs. On the other hand, population growth is leading to an increasing demand for water supply. The inability to meet this demand resulted in huge water scarcity. Analysis showed water per person in 2019 is almost 1200 liters, which is very low compared to previous years. As a step to predict water availability per person in the future, the Support Vector Regression model is created and water scarcity control measures are discussed. As future work, additional features like temperature, groundwater level can be considered to improve the predictions of the SVR model and better machine techniques may be investigated to predict water demands in the future.

References

1. Chennai Water Crisis.: https://www.indiatoday.in/india/story/how-chennai-lost-its-water-a-story-that-should-worry-you-1555096-2019-06-24
2. Angappapillai, A.B., Muthukumaran, C.K.: Demand and supply of water resource in the state of Tamilnadu: a descriptive analysis. Asia Pacific J. Market. Manage. Rev. **1**(3) (2012)
3. Bavanaa, N., Murugesanb, A., Vijayakumarb, C., Vigneshaa, T.: Water supply and demand management system: a case study of Chennai Metropolitan City, Tamil Nadu, India. Int. J. Soc. Relev. Concern (IJSRC) **3**(5), 20–33 (2015)
4. Candelieri, A. et al.: Tuning hyperparameters of an SVM-based water demand forecasting system through parallel global optimization. In: Computers and Operations Research, Elsevier. vol. 106, pp. 202–209 (2019)
5. Brentan, B.M., et al.: Hybrid regression model for near real-time urban water demand forecasting. J. Computat. Appl. Mathemat. Elsevier, vol. 309, pp. 532–541 (2017)
6. Zubaidi, S.L., Gharghan, S.K., Dooley, J., et al.: Short-term urban water demand prediction considering weather factors. In: Water Resource Management, Springer, vol. 32, pp. 4527–4542 (2018)

7. Ghalehkhondabi, I., Ardjmand, E., Young, W.A., et al.: Water demand forecasting: review of soft computing methods. In: Environmental Monitoring and Assessment, Springer, vol. 189, Article number: 313 (2017)
8. Abadi, M.L., et al.: Predictive classification of water consumption time series using non-homogeneous Markov models. In: IEEE International Conference on Data Science and Advanced Analytics (DSAA), Tokyo, pp. 323–331 (2017)
9. Kumar, R.: Emerging challenges of water scarcity in India: the way ahead. Int. J. Innov. Stud. Soc. Human. **4**(4), 6–28 (2019)
10. Chennai Metropolitan Water Supply & Sewage Board website: https://chennaimetrowater.tn.gov.in/
11. Chennai Water Management: https://www.kaggle.com/sudalairajkumar/chennai-water-management
12. Chennai Water Scarcity: https://www.downtoearth.org.in/blog/water/chennai-water-crisis-a-wake-up-call-for-indian-cities-66024

Field Extraction and Logo Recognition on Indian Bank Cheques Using Convolution Neural Networks

Gopireddy Vishnuvardhan, Vadlamani Ravi, and Amiya Ranjan Mallik

Abstract A large number of bank checks are processed manually every day, across the world. In a developing nation like India, cheques are significant instruments for achieving cashless transactions. Cheque processing is a tedious task that can be automated with advanced deep learning architectures. Cheque automation involves selecting the Regions Of Interest (ROI) and then analyzing the contents in the ROI. In this paper, we propose a novel approach to extract ROI (fields) on the cheque using a Convolutional Neural Network (CNN)-based object detection algorithms like YOLO. By virtue of employing a CNN-based model, our approach turns out to be scale, skew, and shift invariant. We achieved a mean average precision (mAP) score of 86.6% across all the fields on a publicly available database of cheques. On the extracted logo field from YOLO, we performed logo recognition using VGGnet as a feature extractor and achieved an accuracy of 99.01%.

Keywords Object detection · Convolution Neural Network · Indian bank cheques · Field extraction

G. Vishnuvardhan · V. Ravi (✉)
Center of Excellence in Analytics, Institute for Development and Research in Banking Technology, Castle Hills Road 1 Masab Tank, Hyderabad 500057, India
e-mail: rav_padma@yahoo.com

G. Vishnuvardhan
e-mail: vishnu.var.reddy@gmail.com

G. Vishnuvardhan
School of Computer and Information Sciences, University of Hyderabad, Hyderabad 500046, India

A. R. Mallik
Department of Computer Science Engineering, IIIT Bhubaneswar, Bhubaneswar 751003, India
e-mail: b116009@iiit-bh.ac.in

S. C. Satapathy et al. (eds.), *Intelligent Data Engineering and Analytics*, Advances in Intelligent Systems and Computing 1177,
https://doi.org/10.1007/978-981-15-5679-1_26

1 Introduction

With the ever-increasing cashless transactions across the world, cheque transactions are also increasing globally. These days even with the rise of digital and instant payments, people still go for cheque transactions because of security and authentication. Millions of handwritten bank cheques are processed manually by reading, entering, and then validating the data. Validation of cheques requires special attention to some fields. The date field has to be validated because Indian cheques have to be cleared before three months from the date when it is issued. The courtesy amount and legal amount has to match. Further, the signature on the cheque ensures the authentication. Cheque Automation System (CAS) is the process of automating these processes. CAS reduces the manual work and saves both time and cost for banks. CAS also provides cheque validation very precisely. Frauds like cheque tampering can be prevented. In a country like India, with the prevalence of a vast number of banks, recognizing the bank name itself is a difficult task. This also can be addressed with CAS [1].

Nowadays, with the exponential growth of data around us and the rise of powerful GPUs, training a CNN is easy. Recent advances in CNN facilitate many computer vision algorithms to automate tedious tasks across all domains. One thing we can automate for banking is the CAS. CAS is a challenging problem that can be addressed using advanced deep Convolution Neural Network (CNN) architectures. The challenging part of cheque automation is to spot the important fields (localization) followed by their recognition.

Significant fields on Indian cheques, which a CAS should take care are (i) Bank logo, (ii) Date, (iii) Payee name, (iv) Legal amount, (v) Courtesy amount, and (vi) Signature.

We now briefly survey the related works on CAS. Koerich and Ling [2] worked on Brazilian cheques and suggested to get the handwritten part by subtracting the filled cheque with a template (blank cheque) with some parameters to adjust the image. Koerich and Lee [3] proposed Hough transformations to detect horizontal lines on Brazilian cheques. Madasu and Lovell [4] handpicked features like fuzzy membership, entropy, energy, and aspect ratio to train a fuzzy neural network. Ali and Pal [5] introduced a method to detect important horizontal lines on Indian cheques from keypoint localization and Speed Up Robust Features (SURF) [6]. Bhateja et al. classified EEG/EOG signals using ANN. Raghavendra [7] localized logo from Indian cheques using blobs. Then they recognized the logo using the geometric features like centroid, eccentricity, etc. Savita [8] proposed a method to select a fixed region (top left) of a cheque and trained Artificial Neural Networks (ANN) to recognize the bank name.

All the above works involve handpicked features to localize the fields on cheques. Most of the approaches may give bad results, if the cheque is skewed or scaled. Line detection based approaches may fail to extract fields if lines are not detected due to noise. All the above techniques require human intervention to scan the cheque with utmost care.

The motivation for the present research is as follows: We attempt to solve the cheque automation process using advanced computer vision algorithms. As cheque automation depends on the localized fields of the cheque, we attempt to solve this with a robust approach. Our approach is skew invariant, can withhold 8° tilt, and can be robust to errors made while scanning the cheque.

The main contributions of this research are

- We propose a novel approach to localize important fields on a given Indian bank cheque using advanced object detection algorithm based on deep learning architecture called CNN.
- For the first time, we employed a metric, mean average precision (mAP) for measuring the performance of field extraction algorithm on cheques.
- We categorized all the fields into three class of objects and performed object detection algorithm to localize the fields.
- For the first time, name of the bank is recognized from the logo with the help of a CNN architecture.
- The techniques proposed by us for localization and recognition are very robust to the scale, shift, skew, and noise.

The rest of the paper is organized as follows: Sect. 2 presents the detail of background knowledge to understand model; Sect. 3 presents in detail our proposed model; Sect. 4 presents the dataset description and evaluation metrics; Sect. 5 presents a discussion of the results and finally Sect. 6 concludes the paper and presents future directions.

2 Background

2.1 Object Detection

With the popularity of CNN, many computer vision algorithms like object detection, face verification, semantic segmentation, and object tracking were proposed to solve some of the real-world problems. Object detection is the task of detecting and segmenting the objects of different classes in a given image. Object detection started with a traditional handpicked features like Viola–Jones algorithm [9], HOG features [10] then followed by the SIFT [11] and SURF [6] algorithms. Later, deep learning based object detection algorithms with automatic feature selection like region-based object detectors and single-shot detectors appeared in literature.

Region-based object detectors like RCNN [12] propose certain regions from the image and pass them to feature extractors to obtain features. The features are then passed to Support Vector Machines (SVM) [13] to detect objects in the proposed regions. On the other hand, single-shot detectors like You Only Look Once (YOLO) [14] can accomplish the task in a single pass.

YOLO poses object detection as a classification and regression task: Classification for the class of the object and regression for the predicting bounding box around the object. Anchor boxes are predetermined shapes of the objects and are crucial. The architecture of YOLO v3 [15] has two components: feature extractor, and box detector. Feature extractor with a 53-layer CNN extracts features and feeds them to the detector box for classification and predicting the bounding box.

3 Proposed Methodology

Given a set of cheques, we preprocessed and passed them to a trained object detection algorithm (YOLO) to segment important fields on the cheque. Then, we considered the logo field and performed bank logo classification.

3.1 Preprocessing

Scanned cheques are color images with RGB channels. We perform some preprocessing steps on the image. We observed that these simple preprocessing steps improve the overall performance.

1. First, we perform image normalization to improve the contrast of the image. This step can be skipped if the contrast of the image is good enough.
2. Threshold operation is performed on the image at value 127, which removes the background. This converts cheque to a binary image.
3. Since we deal with neural networks, the next step is to scale the pixel values in the range (0–1). So, we divide pixel values of the image in the range 0–255.

3.2 Localization

Feature extractor of YOLO extracts features from a preprocessed cheque and gives the features to a box detector which predicts rectangular bounding boxes around objects. Based on the physical properties, we group six fields on the cheque into three different classes as shown in Table 1. We performed object detection with YOLO on three classes. For a given image, YOLO returns bounding boxes across the objects along with a probability score. Preprocessing and localization images are depicted in Fig. 1. Based on the position of handwritten class boxes we further divided handwritten boxes into date and amount fields.

Table 1 Fields on a cheque and their appropriate class

Field	Class
Logo	Logo
Date	Handwritten
Payee name	Handwritten
Legal amount	Handwritten
Courtesy amount	Handwritten
Signature	Signature

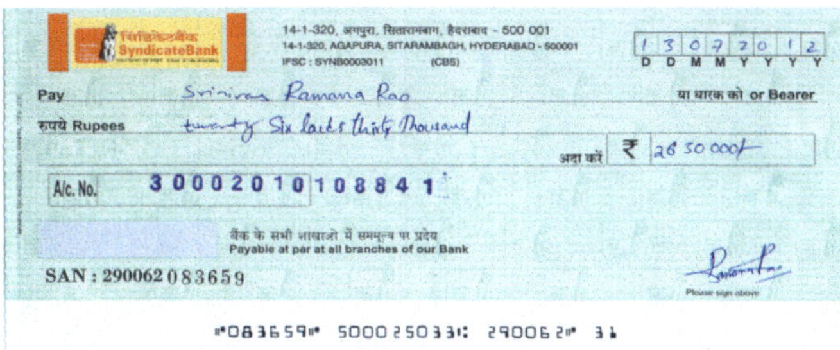

(a) Original cheque image

(b) Pre-processed image

(c) Output from YOLO

Fig. 1 Cheque image after each step of preprocessing

3.3 Bank Name and Logo Recognition

The core idea in the bank name classification is that the Euclidian distance between the feature vectors of two similar images is always minimized. A CNN (VGG) [16] without any multilayer perceptron(MLP) [17] layers can extract important features for a given image and acts as a feature extractor. In the training phase, we add feature vectors of different bank logos along with their bank name to the knowledge base. In the test phase for a given cheque, the object detection model localizes all the important fields along with a label. We consider the logo class object and get feature vectors of size (1, 1000), compute the most similar feature vector in the knowledge database, and assign that particular bank name. The overview of our proposed methodology is shown in Fig. 2.

3.4 Experimental Setting

All our experiments are carried out using tensorflow framework and python language. YOLOv3 algorithm is trained with adam optimizer (learning rate = 0.001) and a batch size of 16. We obtained eight anchor boxes after performing k-means clustering. For the bank name classification, we added a flatten layer to the pre-trained VGG16 available in tensorflow to get a 1000 dimensional vector for a given logo.

4 Dataset Description

We collected two datasets, one from a well-known commercial bank of India anonymized as XYZ bank, second from the IDRBT cheque dataset collected by IDRBT, Hyderabad [5]. Cheques from XYZ bank are grayscale images with low contrast. On the other hand, IDRBT dataset consists of color images with good contrast. We combined both datasets to get a total of 1335 bank cheques. We used a labelImg tool [18] to annotate the objects on the cheques. We split the dataset into training and test set with ratio 75%:25%. Thus, we trained the YOLO model with 1001 images and tested them with 334 images.

4.1 Evaluation Metrics

Mean Average Precision (mAP), defined in The Pascal Visual Object Classes (VOC) challenge [19], is the common metric defined to measure object detection algorithm across the literature. Therefore, we reported our localization results in terms of mAP along with the Average Precision (AP) for three classes (logo, handwritten text,

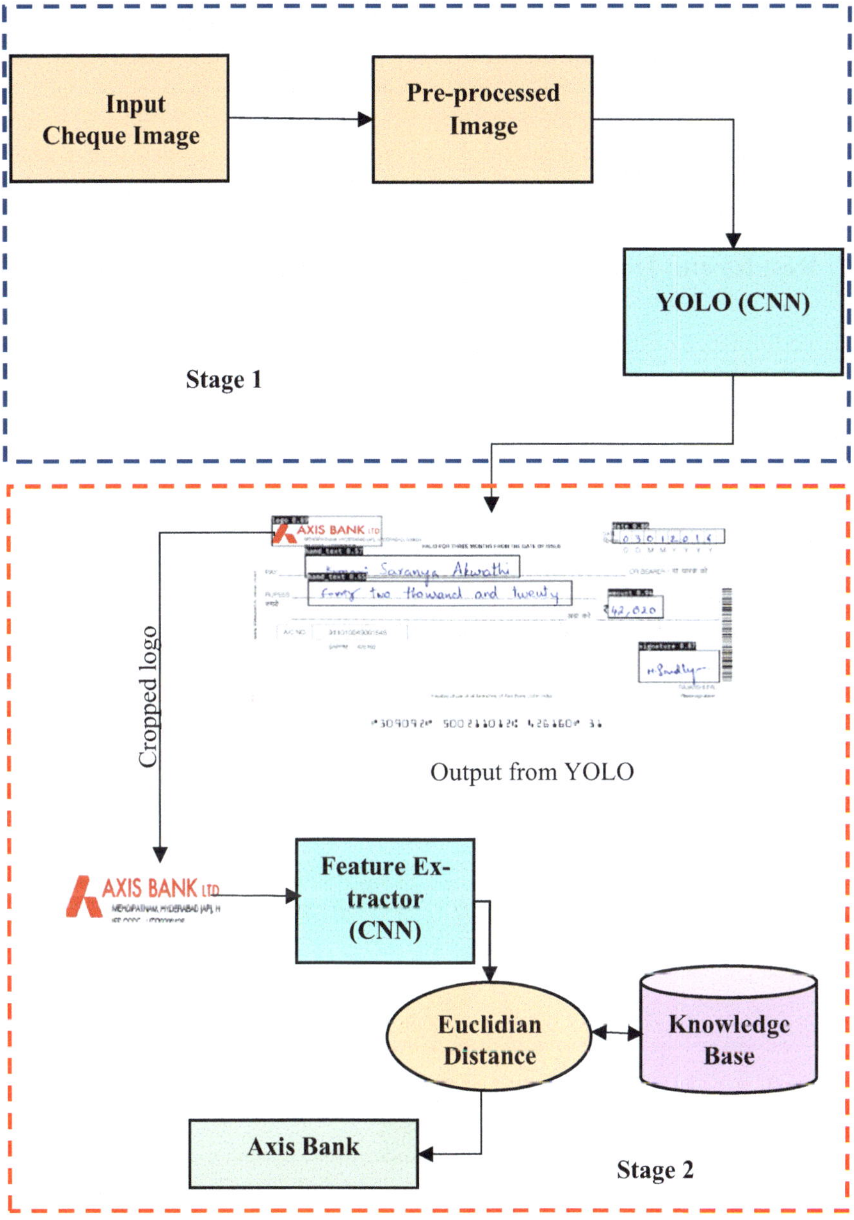

Fig. 2 Overview of the proposed methodology

signature). Mean average precision is the average precision value across all classes of objects. We define a prediction to be a true positive (TP), if the Intersection over Union (IoU) of the object is > 0.5 and the prediction label is correct.

For the bank name classification, we reported the results in terms of accuracy across each bank. We also summarized the performance with the confusion matrix.

5 Results and Discussions

The performance of the YOLO model is tested with 25% test data (334 cheques). For each class, true positive (TP) values and average precision values are depicted in Fig. 3 and 5. As signatures do not have any rectangular shape that affects the values of the IOU and mAP. Thus, TP values of signatures are high but mAP value is low compared to other classes. Because we have CNN-based object detection algorithm for localization and classification our approach is scale, skew, and shift invariant. To check skew invariance of the model, we also tested the cheque with different tilting angles as shown in Fig. 4. From the skewed images, we can say that our algorithm can withstand less than 8° tilt. This ensures that small errors while scanning cheques are negligible. We reported our localization results in terms of mAP and achieved a value of 86.6% depicted in the Fig. 5. Accuracy of the bank name recognition, turned out to be 99.1%, and confusion matrix are presented in Table 2 and Table 3, respectively.

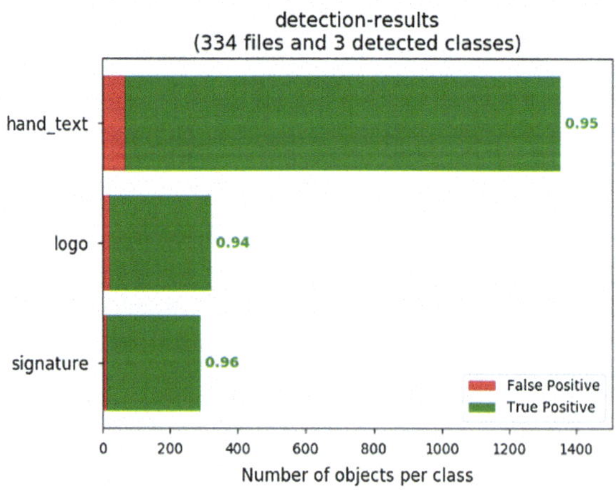

Fig. 3 Performance measure of object detection (YOLO) algorithm on test data True positive

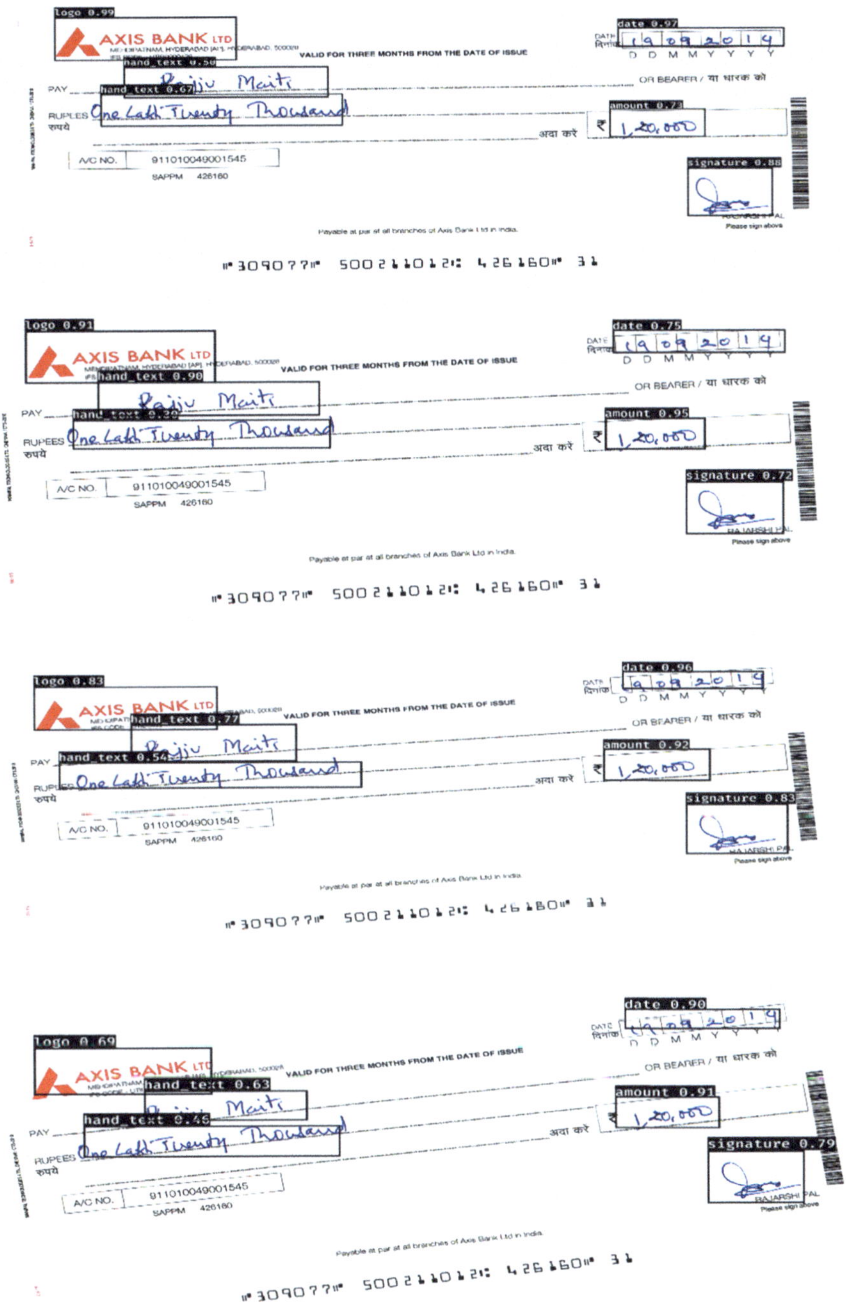

Fig. 4 YOLO output on skewed images with 1°, 3°, 5°, 8° tilt angles (from top to bottom)

Table 2 Number of cheques from each bank and accuracy of bank name classification

Bank	Number of cheques	Accuracy obtained
Axis bank	87	100
Canara bank	10	100
ICICI bank	8	87.5
Syndicate bank	7	100
Total	112	**99.1**

Table 3 Confusion matrix for the bank name recognition

Predicted					
Ground truth		Axis	Canara	ICICI	Syndicate
	Axis	87	0	0	0
	Canara	0	10	0	0
	ICICI	1	0	7	0
	Syndicate	0	0	0	7

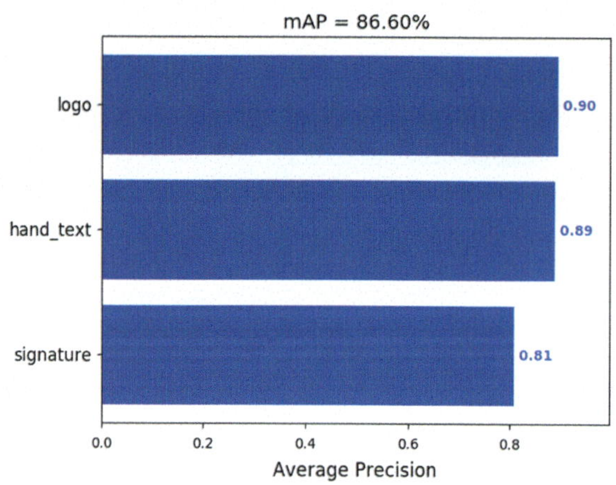

Fig. 5 Performance measure of object detection (YOLO) algorithm on test data Average Precision for each class

6 Conclusions

In this paper, we proposed a robust, skew, shift, and noise invariant approach for the localization of the fields on cheques based on CNN. We reported the localization results in mAP metric, which is first-of-its-kind and achieved a mAP of 86.6%. From the localized logo fields obtained by YOLO, we then recognized the name of the bank from Indian cheques and achieved an accuracy of 99.1%. In the future, we will work on signature verification which strengthens the security and authentication of bank cheques. We then implement a dictionary free Intelligent Character Recognition

(ICR) to recognize handwritten entries on cheques. This completes the development of smart cheque automation system.

References

1. Jayadevan, R., Kolhe, S.R., Patil, P.M., Pal, U.: Automatic processing of handwritten bank cheque images: a survey. Int. J. Doc. Anal. Recognit. **15**, 267–296 (2012). https://doi.org/10.1007/s10032-011-0170-8
2. Koerich, A., Ling, L.L.: A system for automatic extraction of the user-entered data from bankchecks. In: Proceedings—SIBGRAPI 1998: International Symposium on Computer Graphics, Image Processing, and Vision, pp. 270–277. Institute of Electrical and Electronics Engineers Inc., Rio de Janeiro (1998). https://doi.org/10.1109/SIBGRA.1998.722760
3. Koerich, A.L., Lee, L.L.: A novel approach for automatic extraction of the user entered data from bankchecks. In: Proceedings of International Workshop on Document Analysis Systems, pp. 141–144 (1998)
4. Madasu, V.K., Lovell, B.C.: Automatic segmentation and recognition of bank cheque fields. In: Proceedings of the Digital Imaging Computing: Techniques and Applications, DICTA 2005, pp. 223–228. IEEE, Adelaide (2005). https://doi.org/10.1109/DICTA.2005.1578131
5. Dansena, P., Bag, S., Pal, R.: Differentiating pen inks in handwritten bank cheques using multi-layer perceptron. In: Lecture Notes in Computer Science (including subseries Lecture Notes in Artificial Intelligence and Lecture Notes in Bioinformatics), pp. 655–663. Springer Verlag (2017). https://doi.org/10.1007/978-3-319-69900-4_83
6. Bay, H., Ess, A., Tuytelaars, T., Van Gool, L.: Speeded-up robust features (SURF). Comput. Vis. Image Underst. **110**, 346–359 (2008). https://doi.org/10.1016/j.cviu.2007.09.014
7. Raghavendra, S.P., Danti, A.: A novel recognition of Indian bank cheques based on invariant geometrical features. In: International Conference on Trends in Automation, Communication and Computing Technologies, I-TACT 2015. Institute of Electrical and Electronics Engineers Inc. (2016). https://doi.org/10.1109/ITACT.2015.7492682
8. Savita Biradar, S.S.P.: Bank cheque identification and classification using ANN. Int. J. Eng. Comput. Sci. **4** (2018)
9. Viola, P., Jones, M.J.: Robust real-time face detection. Int. J. Comput. Vis. **57**, 137–154 (2004). https://doi.org/10.1023/B:VISI.0000013087.49260.fb
10. Dalal, N., Triggs, B.: Histograms of oriented gradients for human detection
11. Lowe, D.G.: Distinctive image features from scale-invariant keypoints. Int. J. Comput. Vis. **60**, 91–110 (2004). https://doi.org/10.1023/B:VISI.0000029664.99615.94
12. Girshick, R., Donahue, J., Darrell, T., Malik, J.: Rich feature hierarchies for accurate object detection and semantic segmentation. In: Proceedings of the IEEE Computer Society Conference on Computer Vision and Pattern Recognition, pp. 580–587. IEEE Computer Society, Columbus, OH (2014). https://doi.org/10.1109/CVPR.2014.81
13. Schölkopf, B.: SVMs—a practical consequence of learning theory. IEEE Intell. Syst. Their Appl. **13**, 18–21 (1998). https://doi.org/10.1109/5254.708428
14. Redmon, J., Divvala, S., Girshick, R., Farhadi, A.: You only look once: unified, real-time object detection (2015)
15. Redmon, J., Farhadi, A.: YOLOv3: an incremental improvement (2018)
16. Simonyan, K., Zisserman, A.: Very deep convolutional networks for large-scale image recognition (2014)
17. Popescu, M.-C., Balas, V.E., Perescu-Popescu, L., Mastorakis, N.: Multilayer perceptron and neural networks. WSEAS Trans. Cir. Sys. **8**, 579–588 (2009)
18. LabelImg: A graphical image annotation tool and label object bounding boxes in images. https://github.com/tzutalin/labelImg, last accessed 2019/11/15
19. Everingham, M., Van ~ Gool, L., Williams, C.K.I., Winn, J., Zisserman, A.: The pascal visual object classes (VOC) challenge. Int. J. Comput. Vis. **88**, 303–338 (2010)

20. Bhateja, V., Gupta, A., Mishra, A., Mishra, A.: Artificial neural networks based fusion and classification of EEG/EOG signals. In: Advances in Intelligent Systems and Computing, pp. 141–148. Springer Verlag (2019). https://doi.org/10.1007/978-981-13-3338-5_14

A Genetic Algorithm Based Medical Image Watermarking for Improving Robustness and Fidelity in Wavelet Domain

Balasamy Krishnasamy, M. Balakrishnan, and Arockia Christopher

Abstract In modern medical diagnosis, protecting the medical images from vulnerability attacks gains more importance. The proposed work focuses on providing better robustness, when the image has undergone various geometric attacks combined with balancing the trade-off between imperceptibility and robustness, by introducing genetic algorithm based watermarking method combined with Discrete Wavelet Transform (DWT) and Singular Value Decomposition (SVD). False positive errors due to singular values are balanced by calculating Key Component (KC). Authentication of watermarks for patient verification is done at the watermark extraction phase by decrypting the watermark through logistic map permutation method. The Peak Signal-to-Noise Ratio (PSNR) and Normalized Cross-Correlation (NCC) values calculate multi-objective fitness values of the chromosome. The proposed method shows the better robustness and security to the medical images along with authentication to the watermark.

Keywords Genetic algorithm · Singular value decomposition · Wavelet transform · Medical image watermarking

---

B. Krishnasamy (✉) · M. Balakrishnan · A. Christopher
Dr. Mahalingam College of Engineering and Technology, Pollachi, India
e-mail: balasamyk@gmail.com

M. Balakrishnan
e-mail: balakrishnanme@gmail.com

A. Christopher
e-mail: abachristo123@gmail.com

S. C. Satapathy et al. (eds.), *Intelligent Data Engineering and Analytics*, Advances in Intelligent Systems and Computing 1177,
https://doi.org/10.1007/978-981-15-5679-1_27

1 Introduction

Recently, medical image security has become a vital issue to protect it from unauthorized access. Tele-medicine is the transmission of medical images over the Internet in the form of digital media, where they undergo attacks, which results in misdiagnosis. Digital watermarking paves the way for protecting the medical images. Many researchers propose methodologies used for watermarking medical images in the last few decades. Transform domain (DCT, DFT, and DWT) based watermarking schemes [1–4] prove to be efficient for securing watermarked data. In those methodologies, Discrete Wavelet Transform (DWT) gained more advantage for securing the data, due to its multi-resolution capability. In the existing medical image watermarking methods [5, 6], acceptable performance of watermarked image is obtained by defining the embedding rules. However, conventional watermarking methods [7–11] do not provide the intrinsic performance upper limit. Optimization algorithms are used with transform domain methods to obtain a better trade-off between imperceptibility and robustness.

Genetic Algorithms (GA) decide whether optimal watermarking parameters can be identified or in identifying suitable embedding positions for watermarking [12, 13]. However, this results in the insecure watermarking system by selecting poor embedding strength, which results in failure of maintaining trade-off between imperceptibility and robustness. In order to overcome the abovementioned issues, GA-based watermarking method is proposed for selecting the number of bits for embedding the watermark. But, in medical image watermarking, security is also as important as integrity. In order to maintain the trade-off between fidelity and robustness, multiple watermarking schemes are proposed based on the image input characteristics, one watermark for embedding strength, and another watermark for selecting the number of bits for embedding.

2 Related Work

Medicinal imaging was investigated by different specialists to extricate the data carefully from the clinical images for a better conclusion. Integer-based wavelet transformation [14–18] turns around the information covering up in the clinical images by edge-based installing strategies. The coefficients that are having littler greatness esteem than edge esteem are embedded into the LSB of the wavelet change.

Wavelet technique integrates distributed clinical images and time arrangement information into their basic constituents crosswise over scale [19, 20]. Consequently, wavelet strategies for examination and portrayal have critical effect on the study of clinical imaging [21, 22]. Due to the incredible basic scientific standards, wavelets offer energizing open doors for the structure of new multi-resolution clinical image methodologies. The essential uses of wavelets in clinical imaging include clinical image compression, reconstruction of CT scan images, wavelet denoising (for example, fluoroscopy, mammography), and examination of practical images of the cerebrum.

Soft computing technique used in watermarking systems [23] is to acquire the reversibility in clinical images and reestablish the host medical image. It likewise permits to add contortion to the clinical image to accomplish as far as possible in the limit range. Similarly, [24, 25] proposes a reversible watermarking strategy by fusing wavelet change and Genetic Algorithm(GA), where GA is utilized to choose the coefficients to embed the watermark. This technique shows better exchange off among payload and vigor of the watermarked medicinal picture.

Our proposed system deals with the abovementioned issues with the following contributions,

- In order to provide trade-off between the fidelity and robustness, GA-based approach is used by defining multiple genes, one for embedding watermark bit and another for selecting the embedding strength.
- Chromosome-based encoding strength is generated for embedding the watermark in the selected sub-bands, which results in the high security to the watermark.
- Encryption is done for selected sub-bands of the watermarks before embedded into the host image, thereby authentication is provided to the watermark at the watermark extraction process.
- False positive error, caused due to singular value extraction, is avoided by calculating the key component (KC).

3 Proposed Scheme

3.1 Proposed Multiple Watermarks Embedding in Wavelet Domain

1. The original and the watermark images are decomposed through wavelet transform, results in four sub-bands HH, HL, LH, and LL.
2. Logistic map is applied to the LH sub-band of the first watermark and HL sub-band of the second watermark, from which permuted watermark images are obtained.
3. Apply Singular Value Decomposition (SVD) to permuted sub-bands of the watermark images and calculate key component (KC) through,

$$KC_{i1} = U_{Ei1} * U_{Ei2} \tag{1}$$

$$KC_{i2} = S_{Ei1} * S_{Ei2} \tag{2}$$

4. Apply SVD to the HL and LH sub-band of the source medical image and modify the singular values of the source image with the watermark images through the key component.
5. Inverse SVD and DWT are applied; after incorporating the key component, obtain the watermarked image as represented in Fig. 1.

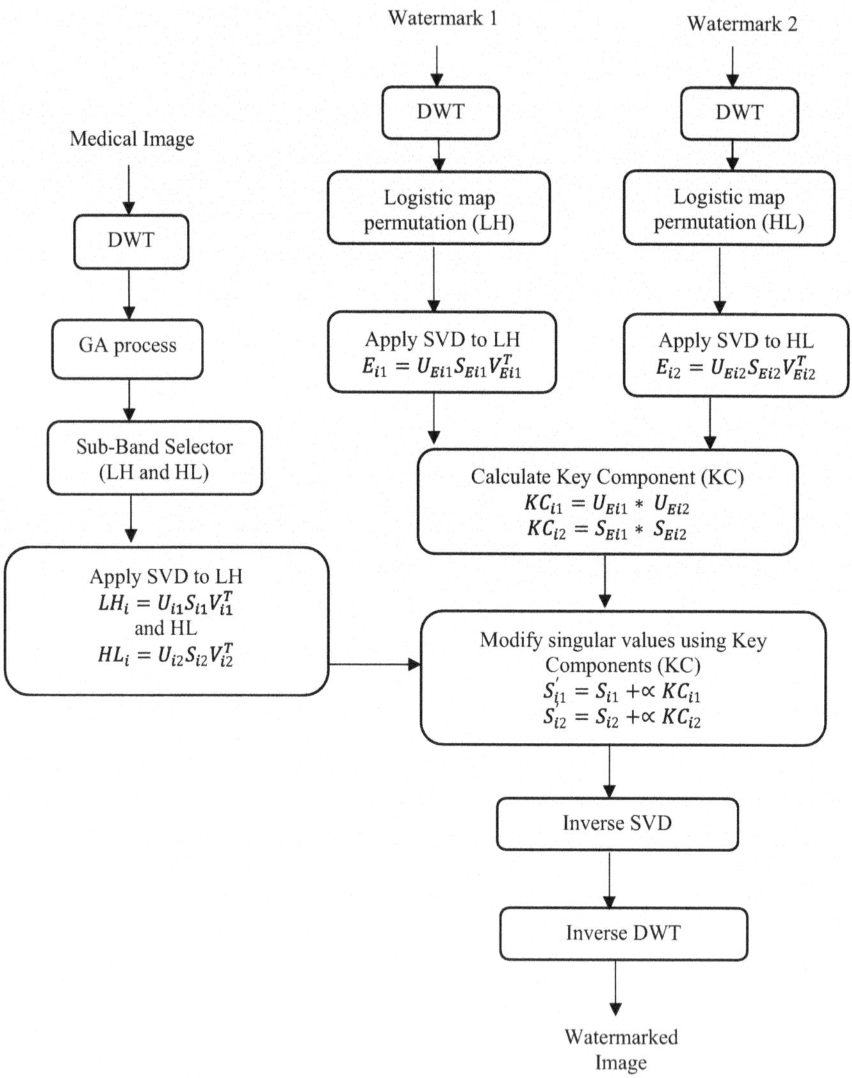

Fig. 1 Watermark embedding process

3.2 Proposed Chromosome-Based Encoding for Finding Optimal Embedding Location

General structure of GA is based on the encoding concept, which relies on finding various solutions and obtaining an optimized solution based on the fitness function. In our proposed work, chromosome genes are represented as the threshold of a given image. Wavelet threshold values were represented by real-coded chromosome, which is used for functional optimization of numerical values over binary encoding. Performance of the GA is improved in the situations when

(i) Memory required for floating point representation is less, when it is used directly.
(ii) Discretization of binary values results in no loss in precision.
(iii) Various geometric operators can be used without any condition.
(iv) Chromosome conversion to phenotypes is not needed for every function evaluation.

For embedding the watermark, HL and LH sub-bands are selected from the four sub-bands that are obtained from the 2-level wavelet decomposition and they are represented as $HL(i, j)$ and $LH(i, j)$. Due to high stability and low variations in the image pixel during watermark embedding, the approximation coefficients are selected. Watermark embedding coefficients are selected by GA process, where the coefficients will do self-modification based on the timeframe.

The Initial population is constructed with a new population set G that has a chromosome vector size as half the size of LH and HL sub-bands. Randomly fix the one's value of the chromosome vector corresponding to the watermark size. Ensure that the remaining blocks are containing value zero. Finally, chromosomes' initial sets are obtained randomly having minimum number. Each chromosome set consists of two genes, one corresponds to embedded semi-fragile watermark based on the number of bits and the second one corresponds to embedded robust watermark based on embedding strength. The number of bits used for embedding the semi-fragile watermark may vary between 1 and 8, as each coefficient of the HL_2 sub-band is an 8-bit representation. Hence, a 3-bit chromosome representation was selected to encode the number of bits. The chromosome length for representing the embedding strength is fixed as 5.

3.3 Chromosome Selection

Selecting the chromosome for watermarking is based on the roulette wheel selection method.

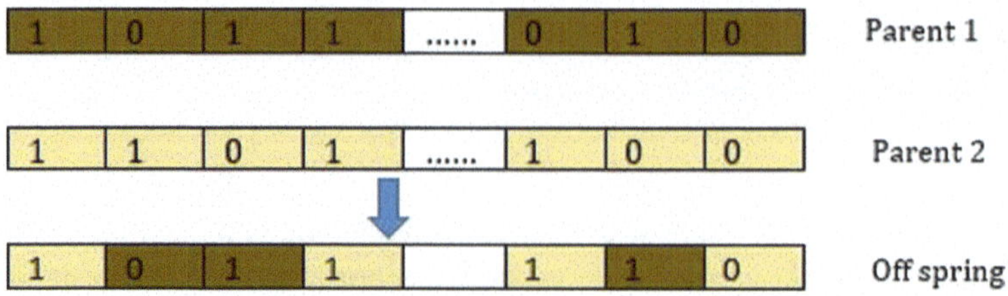

Fig. 2 Cross over for multiple gene with probability of 0.5

```
S:= random number, where 0 ≤ S < 1
    add: =0;
    For each individual i
    {
    add: = add + K(option=i);
        If S < Sum
        {
            return i;
        }
    }
```

$$\text{Where, } K(option = i) = def \frac{fitness(i)}{\sum_{j=1}^{n} fitness(j)} \tag{3}$$

3.4 Crossover and Mutation

A two-point crossover is achieved with the probability of 0.5, where improved results are obtained with n number of iterations as shown in Fig. 2.

Similarly, in our work mutation probability is applied based on Gaussian mutation function, where it shows only small changes in the binary string representation of small bit number. Initially bits between 0.01 and 0.1 were tried for mutation probability and they finally achieved best result at 0.052. The parameters taken into consideration are shown in Table 1.

4 Watermark Extraction and Genetic Algorithm Process

1. Watermarked image will undergo wavelet transformation and singular value decomposition and the key component is calculated as explained in watermark embedding process.

Table 1 Features of GA

GA features	Existing system	Proposed system
No. of generations	150	150
Size of population	50	50
Selection method	Population selection (size = 5)	Population selection (size = 5)
Crossover	One point crossover of 0.8	Two-point crossover of 0.6
Mutation	Type: Uniform Rate: 0.1	Type: Uniform Rate: 0.05
Chromosome length	3 bits	8 bits

2. Watermark image is obtained by extracting the key component and changes the singular values corresponding to it.
3. Extracted watermarks are treated with various geometric attacks, where Normalized Cross-Correlation (NCC) is calculated using Eq. (4) to measure the difference between original and watermarked images.

$$NCC = \frac{\sum_{k=1}^{w_p \times w_q} w_k w_k^{'}}{\sqrt{\sum_{k=1}^{w_p \times w_q} w_k^2 \sum_{k=1}^{w_p \times w_q} w_k^{'2}}} \quad (4)$$

4. Robustness of the watermark image is calculated through Peak Signal-to-Noise Ratio (PSNR) as represented below in using Eq. (5),

$$PSNR = 10log_{10}\left(\frac{255^2}{\left(\frac{1}{m}\right)\sum_{i=1}^{m} H_i - H_i^{'}}\right) \quad (5)$$

5. Obtain the average NCC between two watermarks as represented by using Eq. (6)

$$NCC_{avg} = avg(NCC(W_1, W_1^{'})\ NCC(W_2, W_2^{'})) \quad (6)$$

6. Imperceptibility and robustness magnitude is measured for calculating fitness function as two various representations are shown using Eq. (7) and using Eq. (8)

$$F_i = \max\left(PSNR(S, S_w) - \left(Wt * NCC_{avg}\right)\right) \quad (7)$$

$$F_i = PSNR + \left(100 \times NCC_{avg}\right) \quad (8)$$

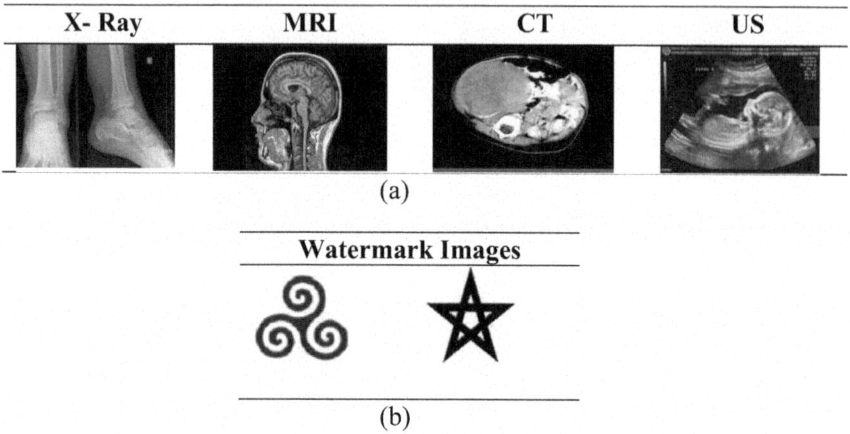

Fig. 3 **a** Medical images used for experimentation, **b** watermark images

7. Throughout the process, identify the best fitness and individual values for setting up a new population.
8. Randomly generate the new population with the features of crossover rate and mutation functions specifically on the selected individuals.
9. Repeat the steps above for n number of iterations until it reaches the predefined iteration.

5 Experimental Results and Discussions

The medical image dataset is collected from BRAT 2016 SICA medical repository. In our experiment, various gray scale medical images of 512×512 size and watermark image of 32×32 size are taken into consideration as shown in Fig. 3.

The Proposed method results are compared with various existing algorithms as shown in Table 2. The proposed method shows better results in comparison with previous algorithms and results in high robustness. The population size of GA with fitness function is shown in Fig. 4. While retrieving the attacked medical image with various noises proposed method shows high NC value as shown in Table 3. Figure 5 represents the execution time of the method with the population size.

6 Conclusion

The proposed algorithm aims at maintaining optimal balance between robustness and fidelity of watermarking algorithms. Encrypting the watermark through logistic mapping permutation before embedding into the source image will result in authenticating the extracted watermark at the receiver side. Influence of the genetic algorithm through multiple gene selection inside single chromosome based encoding,

Table 2 Comparison of the proposed method based on various parameters

Medical images	A. Al-Haj, 2015				Proposed method			
	∝	No. of bits	PSNR	NC	∝	No. of bits	PSNR	NC
MRI-Brain	0.51	3	47.62	0.9582	0.51	3	49.87	0.9961
CT-Lung	0.27	2	48.57	0.9578	0.27	2	50.71	0.9957
MRI-Chest	0.18	1	48.36	0.9473	0.18	1	50.44	0.9852
US-Abdomen	0.33	2	47.31	0.9342	0.33	2	49.39	0.9721
X-Ray-Chest	0.42	3	48.18	0.8859	0.42	3	50.41	0.9238
CT-Head	0.31	2	49.57	0.9573	0.31	2	51.75	0.9952
X-Ray-Hand	0.35	2	46.13	0.9559	0.35	2	48.31	0.9938

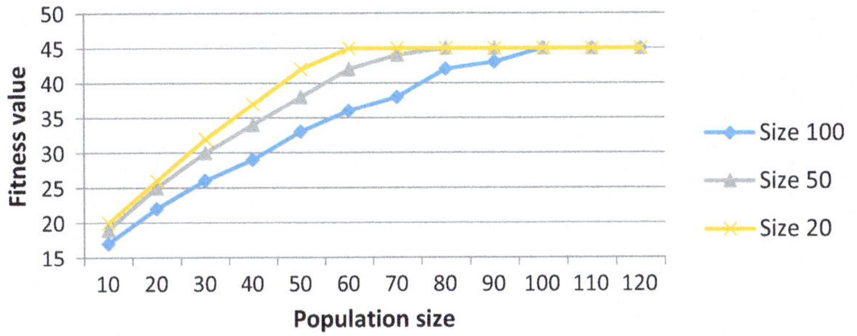

Fig. 4 Evaluation of fitness value with respect to population size

Table 3 NC values of various medical images when undergone various attacks

Medical images	Algorithm	SP	GF	CR	MF	SH	SC
MRI- Brain	Proposed method	0.979	0.990	0.993	0.989	0.991	0.9991
	SSF [16]	0.885	0.936	0.998	0.930	0.950	1.000
CT-Lung	Proposed method	0.989	0.984	0.988	0.994	0.995	0.999
	SSF [16]	0.777	0.847	0.855	0.990	0.980	0.992
MRI- Chest	Proposed method	0.970	0.973	0.949	0.976	0.988	1
	SSF [16]	0.726	0.907	0.894	0.729	0.976	1.000

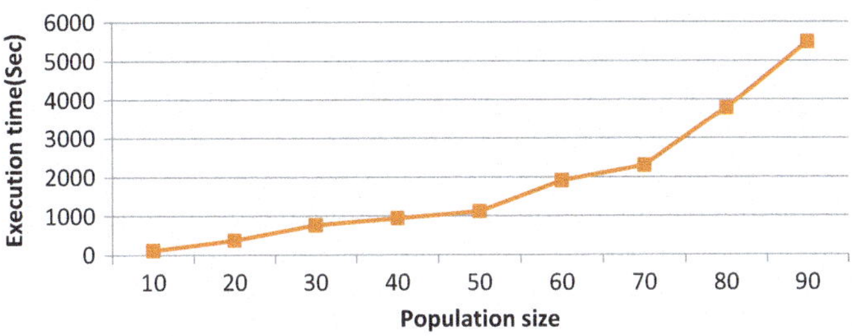

Fig. 5 Influence of population size with respect to execution time

for finding the optimal location for watermark embedding, and also analyzing the number of bits selected for watermarking, results in better trade-off between robustness and security. Our proposed algorithms achieve better robustness when using genetic algorithms and also possess good imperceptibility. Results are attained with various gray scale medical images that can be extended to color images.

References

1. Singh, A.K., Kumar, B., Dave, M., Mohan, A.: Robust and imperceptible spread-spectrum watermarking for telemedicine applications. Proc. Natl. Acad. Sci., India Sect. A: Phys. Sci. **85**(2), 295–301 (2015)
2. Ansari, I.A., Pant, M., Ahn, C.W.: Robust and false positive free watermarking in IWT domain using SVD and ABC, ng. Appl. Artif. Intell. **49**, 114–125 (2016)
3. Al-Haj, A.: Providing integrity, authenticity, and confidentiality for header and pixel data of DICOM images. J. Digit. Imaging **28**(2), 179–187 (2015)
4. Keshavarzian, R., Aghagolzadeh, A.: ROI based robust and secure image watermarking using DWT and Arnoldmap. Int. J. Electron. Commun. (AEU) **70**, 278–288 (2016)
5. Coatrieux G, Maitre, H., Sankur, B., Rolland, Y., Collorec, R.: Relevance of watermarking in medical imaging. In: Proceedings of the IEEE EMBS Conference on Information Technology Applications in Biomedicine, pp. 250–255. Arlington, USA (2000)
6. Coatrieux, G., Lecornu, L., Roux, C., Sankur, B.: A review of image watermarking applications in healthcare. In: Proceedings of IEEE-EMBC Conference, pp. 4691–4694. New York, USA (2006)
7. Zhang, H.: Compact storage of medical images with patient information. IEEE Trans. Inf Technol. Biomed. **5**(4), 320–323 (2001)
8. Giakoumaki, A., Pavlopoulos, S., Koutsouris, D.: A medical image watermarking scheme based on wavelet transform. In: Proceedings of 25th Annual International Conference of IEEE-EMBS, pp. 1541–1544. San Francisco (2004)
9. Giakoumaki, A., Pavlopoulos, S., Koutsouris, D.: Secure and efficient health data management through multiple watermarking on medical images. Med. Biol. Eng. Comput. **44**, 619–631 (2006)
10. Hong, F., Singh, H.V., Singh, S.P., Mohan, A.: Secure spread spectrum watermarking for telemedicine applications. J. Inf. Secur. **2**, 91–98 (2011)
11. Himabindu, G., Ramakrishna Murty, M., et al.: Classification of kidney lesions using bee swarm optimization. Int. J. Eng. Technol. **7**(2.33), 1046–1052 (2018)
12. Gopalakrishnan, T., Ramakrishnan, S., Balasamy, K., Murugavel, A.S.M.: Semi fragile watermarking using Gaussian mixture model for malicious image attacks. In: 2011 World Congress on Information and Communication Technologies, pp. 120–125 (2011)
13. Todmal, S., Patil, S.: Enhancing the optimal robust watermarking algorithm to high payload. Int. Arab. J. Inf. Technol. 108–117 (2013)
14. Navas, K.A., Thampy, S.A., Sasikumar, M.: ERP hiding in medical images for telemedicine. Proc. World Acad. Sci. Technol. **28**, 266–269 (2008)
15. Kannammal, A., Pavithra, K., SubhaRani, S.: Double watermarking of DICOM medical images using wavelet decomposition technique. Eur. J. Sci. Res. **70**(1), 46–55 (2012)
16. Ouhsain, M., Abdallah, E.E., Hamza, A.B.: An image watermarking scheme based on wavelet and multiple-parameter fractional Fourier transform. In: Proceedings of IEEE International Conference on Signal Processing and Communications, pp. 1375–1378. Dubai, United Arab Emirates (2007)
17. Aslantas, V., Dogan, A.L. and Ozturk, S.: DWT-SVD based image watermarking using particle swarm optimizer. In: IEEE International Conference on Multimedia and Expo, pp. 241–244 (2008)

18. Ramakrishnan, S., Gopalakrishnan, T., Balasamy, K.: SVD based robust digital watermarking for still images using wavelet transform, CCSEA 2011. CS IT **02**, 155–167 (2011)
19. Priyanka, S., Kumar, B., Dave, M., Mohan, A.: Multiple watermarking on medical images using selective DWT coefficients. J. Med. Imaging Health Inf. **5**(3), 607–614 (2015)
20. Jiansheng, M., Sukang, L., Xiaomei, T.: A digital watermarking algorithm based on DCT and DWT. In: Proceedings of International Symposium on Web Information Systems and Applications, pp. 104–107, Nanchang, P.R. China (2009)
21. Balasamy, K., et al.: An intelligent reversible watermarking system for authenticating medical images using wavelet and PSO, pp. 4431–4442. Cluster Computing, Springer (2019)
22. Hadi, A.S., Mushgil, B.M., Fadhil, H.M.: Watermarking based Fresnel transform, wavelet transform, and chaotic sequence. J. Appl. Sci. Res. **5**(10), 1463–1468 (2009)
23. Cao, C., Wang, R., Huang, M., Chen, R.: A new watermarking method based on DWT and Fresnel diffraction transforms. In: Proceedings of IEEE International Conference on Information Theory and Information Security, pp. 430–433. Beijing (2010)
24. Ramya, M.M., Murugesan, R.: Joint, image-adaptive compression and watermarking by GA-Based wavelet localization: optimal trade-off between transmission time and security. Int. J. Image Process. (IJIP) **2**, 478–487 (2012)
25. Himabindu, G., Ramakrishna Murty, M., et al.: Extraction of texture features and classification of renal masses from kidney images. Int. J. Eng. Technol. **7**(2), 1057–1063 (2018)

Developing Dialog Manager in Chatbots via Hybrid Deep Learning Architectures

Basit Ali and Vadlamani Ravi

Abstract Dialog Manager has played a great role in conversational AI so much, so that it is also called the heart of a dialog system. It has been employed in task-oriented Chatbot to learn the context of a conversation and come up with some representation which helps in executing the task. For example, booking a restaurant table, flight booking, movie tickets, etc. In this paper, a dialog manager is trained in a supervised manner in order to predict the best response given the latent state representation of the user message. The latent representation is formed by the Convolution Neural Network (CNN) and Bidirectional Long Short Term Memory network (BiLSTM) with attention. An ablation study is conducted with three different architectures. One of them achieved a state-of-the-art result in turn accuracy on babI6 dataset and dialog accuracy equivalent to the baseline model.

Keywords Word2vec (w2v) · Long Short Term Memory(LSTM) · Convolution Neural Network (CNN) · Bag of Words (BoW) · BiLSTM · One dimensional convolution neural network (1DCNN)

1 Introduction

In recent years, there is a huge spurt in the popularity of Chatbot by virtue of its availability in every web or app based service platform like an e-commerce website, banking portal, restaurant booking, and so on. All these Chatbots, independent of the domain, have a common behavior, i.e., task oriented which means every task

B. Ali
School of Computer and Information Sciences, University of Hyderabad, Hyderabad 500046, India
e-mail: alibasit78@gmail.com

B. Ali · V. Ravi (✉)
Center of Excellence in Analytics, Institute for Development and Research in Banking Technology (IDRBT) Castle Hills Road #1 Masab Tank, Hyderabad 500057, India
e-mail: rav_padma@yahoo.com

S. C. Satapathy et al. (eds.), *Intelligent Data Engineering and Analytics*, Advances in Intelligent Systems and Computing 1177,
https://doi.org/10.1007/978-981-15-5679-1_28

has a user request asking for some service and the corresponding response to fulfill that service, say "book an Italian restaurant table for two people" and the response is "where you like to have," or will give restaurant suggestions, and so forth, this process continues until the order is placed or the task is accomplished.

Past works on these Chabot's were modular-based having Natural Language Understanding (NLU), state tracking, action selection modules which lead to the dependency between modules, and individual module training is required. Due to the sequential behavior of Chatbot, Recurrent Neural Network (RNN) [15] becomes relevant in order to infer the latent representation of dialog states and to leverage this idea end-to-end dialog system is crucial. This removed the modular structure and dedicated training for each module. Several approaches related to end-to-end dialog system such as Hybrid Code networks (HCN) [18] that has domain vector knowledge assisted to learn with less training examples. It considers the average of word embedding (word2vec) [13] to represent the user input, which leads to poor representation of the sentence.

This paper proposes an end-to-end dialog system, a variation of HCN [2], based on several novel architectures to (i) learn the good representation of user input (ii) be able to capture the order of words and (iii) give weightage to relevant words. The action taken by the Chatbot is also dependent upon the past conversation that happened so far. Our proposed architecture introduces the 1DCNN layer after the input layer to create a dense vector. The other architectures applied CNN [6] and BiLSTM with attention [20] on the user input to represent sentences.

The rest of the paper is organized as follows: Sect. 2 presents the literature review; Sect. 3 presents the proposed methodology; Sect. 4 presents the dataset description; Sect. 5 presents the results, measures, and discussion. Finally, Sect. 6 concludes the paper and presents future directions.

2 Literature Review

Several works have been reported in the development of Chatbot. They are mainly divided into retrieval-based, generative-based, modular-based, and End-to-End task-oriented Chatbot.

Retrieval-based Chatbot measures the similarity between user request message and the corresponding list of responses to give the similarity score. The response with the highest similarity score is considered the correct response [2]. Each word is embedded (like word2vec, tf-idf) into a vector and the mean of vectors is taken to represent the sentences.

Another work extending this idea is Dual Encoder LSTM which learns the hidden meaning between user message and the responses. In this work, user input and responses are split into tokens, embedded into a vector and passed on to two different LSTMs to learn the hidden meaning between them [11]. However, the response output does not depend upon the previous conversation.

Generative-based Chatbot employs Seq2seq model, which learns the relation between user message and the responses. It consists of Encoder–Decoder model where encoder encodes the message into latent representation and sends as input to decoder to generate the response. However, this model cannot generate consistent responses [17]. To fix this issue, persona-based model introduces speaker model and speaker-addressee model [7]. The work in attention with intention [19] is composed of three networks, viz., encoder, intention, and decoder networks. Here, the intention network memorizes the previous turn of the conversation which gives an additional advantage to generate the responses. But still, it is prone to make a grammatical mistake.

Works are also reported in embedding to represent the sentences where word, character, context embedding are concatenated to create the best representation of user input for dialog system [5]. For extracting the slots, Named Entity Recognition Bidirectional LSTM-CNNs are used [3]. Contextual Spoken Language Understanding (SLU) [16] approach is applied to learn jointly the slots and intents classification for a particular sentence and uses RNN to learns the sequential features. These features are used in the dialog manager to represent the message. However, these features are dependent upon the SLU models.

Recently Reinforcement Learning techniques are also applied in task-oriented Chatbot in which there is no real user for training but the simulated user called an agent who does not know the goal learns the policy to accomplish the goal [9]. It is also applied to End-to-End task-oriented Chatbot that includes user simulator and neural dialog system. User input passes through a Language Understanding (LU) module and returns a semantic frame. Dialog Manager, which follows, includes a state tracker and policy learner to accumulate the semantics from each user input to predict the next action [8].

Modular-based Chatbot is composed of NLU module to extract the entities and classify the intents. Then, state tracker, which follows, maintains the conversation state, and finally, policy selection module predicts the response given the dialog state [1]. However, there is a dependency between modules in this work.

End-to-End task-oriented Chabot called as HCN combines an RNN with rule-based vector encoded as software and system action templates [18], but the rule vector is handcrafted and the user input is represented by taking the average of w2v vectors. Our proposed model is the variation of the HCN similar to Marek [12] model.

3 Proposed Methodology

In the HCN model, sentences are represented by taking the average of embedded word vector (w2v) which is concatenated with the Bag of Word (BoW) features, one-hot encoded entity vector, and the rule vector. This concatenated vector is passed as an input to the LSTM model to predict the next response. However, our proposed model ignored the rule vector and applied CNN [6] and BiLSTM with attention [20] on each sentence.

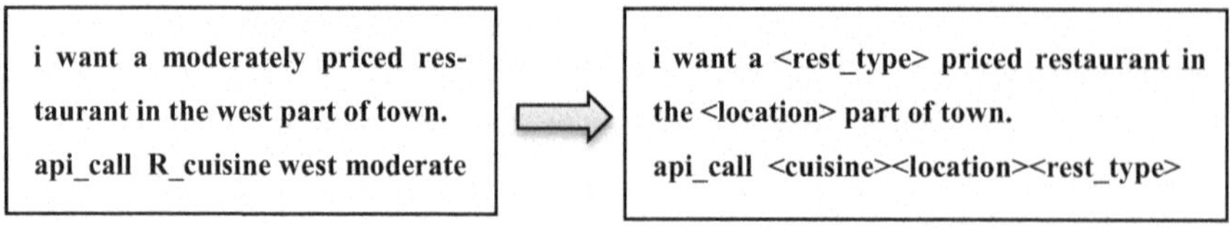

Fig. 1 Data preparation

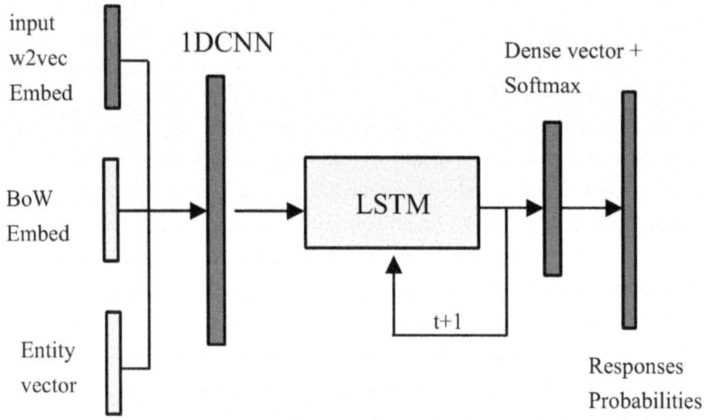

Fig. 2 W1CNNL architecture

3.1 Data Preparation

Here each data point is a user request followed by the bot response. User request and bot response sentences are parsed to extract the entities like restaurant name, price type, location, cuisine using a string matching algorithm. Finally, the extracted entities are replaced by the corresponding tags as shown below (see Fig. 1).

After pre-processing the training set, we manually built the 56 generic responses. However, there are few responses which are not present in the test dataset, those responses are placed with the largest similarity among the 56 responses.

3.2 Word2Vec Embedding + 1DCNN + LSTM (W1CNNL)

Our first proposed model is similar to HCN in which the input layer includes BoW, one-hot entity vector, utterance embedding (word2vec), i.e., average of all the word word2vec vectors. These vectors are concatenated and then passed through 1DCNN [3] to produce a dense feature vector. This dense feature vector captures the spatial features and feed as input to the LSTM [15] model (number of hidden layers 128) to predict the list of response probabilities. The output of the LSTM state vector is recursively fed back to itself at every timestamp (see Fig. 2). The LSTM present as a component in the further discussed architectures is assumed to be the same. For predicting the responses, softmax is applied on the state vectors. Categorical Cross-Entropy Loss function is used to learn the trainable parameters during backpropagation.

Pseudo Code

```
For Dialog in Dialogs:   where Dialogs = [s1, s2, ..., sn] and si is a Dialog.
 LSTM initial States are initialized
 For Sent in Dialog:   where Sent = [ x1, x2, ..., xn ] and xi is a word, i =1, 2, ..., n
  X = w2v ( Sent ) here X is the word2vec of all the words.
  Avg_w2v = avg ( X )
  Bow_emb = one hot encoding of vocabularies.
  Entity_vec = one hot encoding of entities.
  vec_1dcnn = 1DCNN_Relu ( concate ( Avg_w2v, Bow_emb, Entity_Vec ), Filters )
  S = LSTM ( vec_1dcnn, States )  where States is the LSTM initial hidden state;
States=S
  Prob_responses = Dense_Softmax ( S ) where S is the state vector at tth time stamp.
  Selected_response = Max_arg ( Prob_responses )
 End For
End For
```

3.3 Word2Vec + CNN Embedding + LSTM (WCNNL)

Our second proposed model has created a 2D matrix for each sentence which is formed by considering the word2vec vector for each word, say d dimension is the size of word vector and n is the number of words in that sentence, then n x d is the dimension of sentence matrix. However, to make all the sentences the same length, we have applied padding and convolution operation by different size filters on top of the matrix followed by the max-pooling layer, and finally achieved the flatten dense vector which is the latent vector representation of the sentence and concatenated with the rest of the feature vectors as shown below (see Fig. 3) and passed as input to the LSTM model to predict the response.

This whole architecture is the combination of CNN, LSTM model which are trained jointly. The significance of applying CNN with different size filters allows us to learn the bigram, trigram features, a pattern which in previous work of this feature engineering was not considered.

Pseudo Code

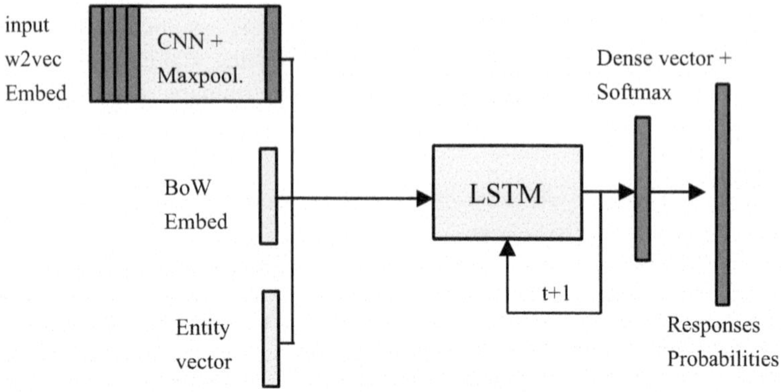

Fig. 3 WCNNL architecture

```
For Dialog in Dialogs:    where Dialogs = [s1, s2, ..., sn] and si is a Dialog.
  LSTM initial States are initialized
  For Sent in Dialog:    where Sent = [ x1, x2, ..., xn ] and xi is a word, i =1, 2, ..., n
    X_matrix = w2v ( sent ) where X_matrix is the word2vec matrix of words.
    Sent_Rep = 2DCNN_Relu_Maxpool_Flatten( X_matrix, Filters) Filters sizes of 3, 4, 5
    Final_input = concate (Sent_Rep, Bow_emb, Entity_vec )
    This Final_input is passed to the LSTM and rest are same as above implementation
  End For
End For
```

3.4 Word2Vec + BiLSTM with Attention Embedding + LSTM (WBWAL)

Our third proposed model is similar to the above model, the only difference is in place of CNN, BiLSTM with Attention is applied on the user input to produce a latent representation. Here, user input is divided into tokens, converted to word2vec vectors and each token is passed to both the forward and backward LSTM, and for all the time stamp, state vectors are added from both the LSTMs and weights are applied on those vectors to finally represents the sentence, which is the weighted sum of the state vectors [20] (see Fig. 4).

This approach is better than the previous works as the words' dependency or order of words was not handled while taking the average of word2vec embedding in the baseline model. This architecture allows us to give more attention to the relevant word in both forward and backward directions for a sentence.

Pseudo Code

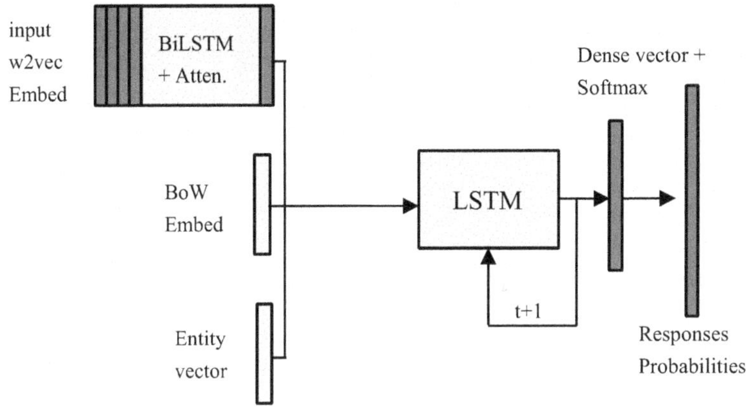

Fig. 4 WBWAL architecture

```
For Dialog in Dialogs:    where Dialogs = [s1, s2, ..., sn] and si is a Dialog.
 LSTM initial States are initialized
 For Sent in Dialog:    where Sent = [ x1, x2, ..., xn ] and xi is a word, i =1, 2, ..., n
  X_matrix = w2v ( sent ) where X_matrix is the word2vec matrix of all words
  Sent_Rep = BiLSTM_Attention ( X_matrix, Fwd_cell, Bwd_cell, Max_seq_len)
   Where Fwd_cell, Bwd_cell  - 1st LSTM, 2nd LSTM States for all the time stamp,
         Max_seq_len – max user input length
  Final_input = concate (Sent_Rep, Bow_emb, Entity_vec)
  This Final_input is passed to the LSTM and rest are same as above implementation
 End For
End For
```

4 Dataset Description

We tested the effectiveness of our models on publicly available dataset babI Task 6. It includes 1624 training dialogs, 1117 test dialogs [2]. It also has three request slots such as cuisine, price, and location. Task 6 consists of human–bot conversation but no knowledge base, so explicitly we need to extract the restaurant names and APIs from the dataset. In this dataset, pre-processing was done to extract the 56 responses.

5 Results and Discussion

We presented the results of the three proposed architectures in terms of two performance metrics, namely, turn accuracy and dialog accuracy. Turn accuracy is the

Table 1 Hyperparameters for all the architecture

Hyperparameter	W1CNNL	WCNNL	WBWAL
#hidden nodes (LSTM)	128	128	128
Learning rate (LSTM)	0.1	0.2	0.1
Activation Fn.	Relu	Relu	Relu
Optimizer	Adadelta	Adadelta	Adadelta
Dropout (BiLSTM)	–	–	0.5
Dropout (CNN)	–	0.7	–
#Filters	1	256	–
#Layers (BiLSTM)	–	–	2

number of correct responses corresponding to the user message whereas dialog accuracy measures the number of correct dialogs, i.e., all the responses must be correct in a dialog.

The hyperparameters for the three architectures are presented in Table 1. All the models are trained for 101 epochs in the case of W1CNNL, 33 epochs in the case of WCNNL, and 51 epochs in the case of WBWAL architectures, having batch size of 1. After the concatenation of the features in the input layer, it is passed through LSTM which has 128 hidden nodes in all the architectures, followed by the categorical cross-entropy with adadelta optimizer. In the case of WCNNL, dropout is applied after convolution followed by maxpool operation, whereas for WBWAL, dropout is applied after BiLSTM layers (i.e., 2 stack layers of LSTM for both forward and backward context).

Results shows (Table 2) that WCNNL outperformed the model of Marek [12] model in terms of turn accuracy by 0.87% and dialog accuracy. However, WCNNL is equivalent to the baseline model in terms of dialog accuracy. Similarly, our WBWAL model which considers the forward context and backward context of a sentence also

Table 2 Testing accuracy on babl6 task 6 dataset

Model	babI6	
	Turn Acc. (%)	Dialog Acc. (%)
Bordes and Weston (2017) [2]	41.1	0.0
Liu and Perez (2016) [10]	48.7	1.4
Eric and Manning (2017) [4]	48.0	1.5
Seo et al. (2016) [14]	51.1	–
Williams, Asadi and Zweig (2017) [18]	55.6	1.9
Marek (2019) [12]	58.9	0.5
word2vec + 1DCNN + LSTM (**W1CNNL**)	58.35	1.5
word2vec + CNN + LSTM (**WCNNL**)	**59.77**	**1.9**
word2vec + BiLSTM + LSTM (**WBWAL**)	59.28	0.0

outperforms the model of Marek by in terms of turn accuracy 0.38%, but it is worse than WCNNL. Finally, W1CNNL outperformed the Marek model in terms of dialog accuracy.

6 Conclusions

We proposed three hybrid deep learning architectures for the dialog manager to be used in Chatbot. We achieved the best result with WCNNL, which outperformed the baseline model and the model of Marek in terms of turn accuracy. The main reason behind the best performance of WCNNL and WBWAL is that we considered the bigram, trigram, order of words as the features, and applied attention mechanism to represent the sentences, unlike previous studies.

For the future work, we plan to test on different domain datasets and apply transfer learning so that we do not need to train from scratch. This can be extended by introducing a new architecture which can handle the out of domain inputs by learning the new parameters. We also plan to extend this work by introducing an entity extraction module to learn jointly with our architectures.

References

1. Bocklisch, T., Faulkner, J, Pawlowski, N., Nichol, A.: Rasa: open source language understanding and dialogue management. In: NIPS 2017 Conversational AI workshop (2017)
2. Bordes, A., Boureau, Y., Weston, J.: Learning end-to-end goal-oriented dialog. In: International Conference on Learning Representations 2017 (2017)
3. Chiu, J.P.C., Nichols, E.: Named entity recognition with bidirectional LSTM-CNN transactions of the association for computational linguistics **4**, 357–370 (2016). https://doi.org/10.1162/tacl_a_00104
4. Eric, M., Manning, C.D.: A copy-augmented sequence-to-sequence architecture gives good performance on task-oriented dialogue (2017). arXiv:1701.04024
5. Jayarao, P., Jain, C., Srivastava, A.: Exploring the importance of context and embeddings in neural NER models for task-oriented dialogue systems (2018). arXiv:1812.02370
6. Kim, Y.: Convolutional neural networks for sentence classification. In: Proceedings of the 2014 Conference on Empirical Methods in Natural Language Processing (EMNLP), pp. 1746–1751, Association for Computational Linguistics. Doha, Qatar (2014). https://doi.org/10.3115/v1/d14-1181
7. Li, J., Galley, M., Brockett, C., Spithourakis, G.P., Gao, J., Dolan, B.: A persona-based neural conversation model. In: Proceedings of the 54th Annual Meeting of the Association for Computational Linguistics, vol. 1, pp. 994–1003. Association for Computational Linguistics, Berlin, Germany (2016). https://doi.org/10.18653/v1/P16-1094
8. Li, X., Chen, Y.N., Li, L., Gao, J., Celikyilmaz, A.: End-to-end task-completion neural dialogue systems (2017). arXiv:1703.01008
9. Li, X., Chen, Y.N., Li, L., Gao, J., Celikyilmaz, A.: Investigation of language understanding impact for reinforcement learning based dialogue systems (2017). arXiv:1703.07055
10. Liu, F., Perez, J.: Gated end-to-end memory networks. In: Proceedings of the 15th Conference of the European Chapter of the Association for Computational Linguistics, vol 1, Long Papers. pp. 1–10, Association for Computational Linguistics. Valencia, Spain (2017)

11. Lowe, R., Pow, N., Serban, I., Pineau, J.: The ubuntu dialogue corpus: a large dataset for research in unstructured multi-turn dialogue systems (2015) arXiv:1506.08909
12. Marek, P.: Hybrid code networks using a convolutional neural network as an input layer achieves higher turn accuracy (2019). arXiv:1907.12162
13. Mikolov, T., Chen, K., Corrado, G., Dean, J.: Efficient estimation of word representations in vector space (2013). arXiv:1301.3781
14. Seo, M., Min, S., Farhadi, A. and Hajishirzi, H.: Query-reduction networks for question answering. In: International Conference on Learning Representations 2017 (2016)
15. Sherstinsky, A.: Fundamentals of Recurrent Neural Network (RNN) and Long Short-Term Memory (LSTM) Network (2018). arXiv:1808.03314
16. Shi, Y., Yao, K., Chen, H., Pan, Y.C., Hwang, M.Y., Peng, B.: Contextual spoken language understanding using recurrent neural networks. In: 2015 IEEE International Conference on Acoustics, Speech and Signal Processing (ICASSP), pp. 5271–5275 (2015). https://doi.org/10.1109/ICASSP.2015.7178977
17. Vinyals, O., Le, Q.: A Neural conversational model (2015). arXiv:1506.05869
18. Williams, J.D., Asadi, K., Zweig, G.: Hybrid code networks: practical and efficient end-to-end dialog control with supervised and reinforcement learning (2017). arXiv:1702.03274
19. Yao, K., Zweig, G., Peng, B.: Attention with Intention for a neural network conversation model. In: NIPS 2015 Workshop on Machine Learning for Spoken Language Understanding and Interaction (2015)
20. Zhou, P., Shi, W., Tian, J., Qi, Z., Li, B., Hao, H., Xu, B.: Attention-based bidirectional long short-term memory networks for relation classification. In: Proceedings of the 54th Annual Meeting of the Association for Computational Linguistics, vol. 2, pp. 207–212. Association for Computational Linguistics, Berlin, Germany (2016). https://doi.org/10.18653/v1/P16-2034

Experimental Analysis of Fuzzy Clustering Algorithms

Sonika Dahiya, Anushika Gosain, and Suman Mann

Abstract Fuzzy clustering is an unsupervised technique for partitioning data into fuzzy clusters. Fuzzy clustering has wide applications in various domains of science and technology. So, in this paper, we have drawn a performance comparison of five fuzzy clustering algorithms: FCM, PFCM, CFCM, IFCM, and NC. Their performance is analyzed on the bases of cluster homogeneity, clusters varying in size, shape, and density as well as when population of outliers increases. Four standard datasets: D12, D15, Dunn, and Noisy Dunn are used for this review work. This research paper will be very helpful to researchers to choose the right algorithm as per the features of their data clusters.

Keywords Clustering · Fuzzy clustering · FCM · IFCM · KFCM · NC

1 Introduction

In the digitally growing world, tremendous data is available for processing and clustering is like a very useful tool to partition the data into groups which has a high intra similarity and low inter similarity. Clustering can majorly be classified as hard clustering or soft clustering. Hard clustering gives crisp partitions whereas soft clustering has fractionally overlapping partitions. Soft clustering is a superset of hard clustering. So, it has even more wider spectrum of applications. In 1973, the first attempt was made for fuzzy clusters as ISODATA and its improved variation named Fuzzy C-Means (FCM) has been the most known fuzzy clustering algorithm. However, it

S. Dahiya (✉)
CSE, Delhi Technological University, Delhi, India
e-mail: sonika.dahiya11@gmail.com

A. Gosain · S. Mann
Maharaja Surajmal Institute of Technology, Delhi, India
e-mail: anushikagosain_123@gmail.com

S. Mann
e-mail: sumanmann@msit.in

S. C. Satapathy et al. (eds.), *Intelligent Data Engineering and Analytics*, Advances in Intelligent Systems and Computing 1177,
https://doi.org/10.1007/978-981-15-5679-1_29

fails to cluster well if data is contaminated with noise and outliers. To conquer this issue of FCM many attempts were made using the possibilistic [1] and credibility [2] concepts, which resulted in the proposal of Possibilistic Fuzzy C-Means (PFCM) and Credibilistic Fuzzy C-Means (CFCM). PFCM incorporated possibilistic membership along with fuzzy membership, which helped PFCM in dealing with noise but it fails when clusters are extremely imbalanced in size and outliers are present in the dataset. CFCM introduced a new variable named credibility, which reduced the effect of outliers in computing centroids, but sometimes assigns outliers to more than one cluster.

In 2011, based on the concept of intuitionistic fuzzy sets, fuzzy intuitionistic entropy was introduced with FCM and Intuitionistic Fuzzy C-Means (IFCM) [3] was proposed. It outperforms FCM and many variations of FCM in centroid positioning of resultant clusters. But sometimes it produces overlapping clusters.

In this paper, we have drawn a comparison on five fuzzy clustering algorithms. Although various comparisons are drawn in literature like the comparison of k-means and FCM [4], a performance analysis of various fuzzy clustering algorithms [5], survey on fuzzy clustering methods for big data [6], a review of various applications of fuzzy clustering [7], etc. But performance of fuzzy clustering algorithms depends on cluster characteristics as well as on the presence of noise and outliers. Till now no such work is done. So, the aim of this paper is to scrutinize their performance for noisy and noise-free data, with and without the presence of outliers as well as size and density varying clusters. For this analysis, we have considered four datasets: D12, D15, Dunn, and noisy Dunn dataset. D12 and D15 are standard datasets with identical symmetric 2 clusters, each of size five. Comparison of results of FCM, PFCM, CFCM, IFCM, NC on D12 and D15 helps us draw an analysis on how their performance varies with the increase of outliers. Dunn dataset is also a standard dataset, which consists of two density and size varying square-shaped clusters. Noisy Dunn dataset is noisy Dunn dataset. Comparison is drawn on Dunn and noisy Dunn dataset, with an objective to draw performance analysis of these algorithms like how their performance varies with size and density varying clusters, non-spherical clusters, and with the presence of outliers.

In the next section of this paper, a brief description of compared fuzzy clustering algorithms is given. In Section III, experimental simulation and results are presented with the help of figures, tables, and graphs. Then Section IV concludes the comparison.

2 Literature Survey

2.1 Fcm [8]

It works only when the count of clusters are known and results in optimal clusters by minimizing the following objective function (JFCM):

$$J_{\mathrm{FCM}}(U, V) = \sum_{k=1}^{c}\sum_{i=1}^{n} \mathrm{u}_{\mathrm{ki}}^{m} d_{\mathrm{ki}}^{2} \tag{1}$$

Subject to the following constraint:

$$\sum_{k=1}^{c} u_{\mathrm{ik}} = 1\, i = 1, 2, \ldots, n \tag{2}$$

For our simulation m is set to be 2. On solving optimization problem stated in Eq. no. (1), following equations are found to be used for membership and centroid updation:

$$v_k = \frac{\sum_{i=1}^{n}\left(u_{\mathrm{ik}}^{m} x_i\right)}{\sum_{i=1}^{n}\left(u_{\mathrm{ik}}^{m}\right)} \tag{3}$$

$$u_{\mathrm{ik}} = \frac{1}{\sum_{j=1}^{c}\left(\frac{d_{\mathrm{ki}}}{d_{\mathrm{ji}}}\right)^{\frac{2}{m-1}}}\ \forall\ \mathrm{k, i} \tag{4}$$

where i and j are positive integers in [1, nc] and [1, n], respectively.

However, it proves to be insufficient in the existence of noise and outliers, as it fails to identify noise and outliers. So, the centroids are inclined toward outliers.

2.2 *Possibilistic Fuzzy C-Means (PFCM) [1]*

FCM integrates possibilistic approach and fuzzy approach from PCM and FCM, respectively. Thus, PFCM has two memberships associated with each data object: (i) possibilistic membership (t) and (ii) fuzzy membership (u). It results in optimal clusters by minimizing the following objective function:

$$J_{\mathrm{PFCM}}(U, V, T) = \sum_{k=1}^{c}\sum_{i=1}^{n}\left(\mathrm{au}_{\mathrm{ki}}^{m} + \mathrm{bt}_{\mathrm{ki}}^{\eta}\right) d_{\mathrm{ki}}^{2} + \sum_{k=1}^{c} \ddot{\mathrm{Y}}_k \sum_{i=1}^{n}(1 - t_{\mathrm{ki}})^{\eta} \tag{5}$$

subject to the constraint that

$$\sum_{k=1}^{c} u_{\mathrm{ki}} = 1 \forall i \tag{6}$$

where $0 \le u_{ki} < 1$ and $0 \le t_{ki} < 1$ and m > 1, $\eta > 1$, a > 0 and b > 0.

'b' and 'a' are integer constants. They specify the relative significance of fuzzy membership and possibilistic membership in computing resultant clusters.

$$u_{\mathrm{ki}} = \frac{1}{\sum_{j=1}^{c}\left(\frac{d_{\mathrm{ki}}}{d_{\mathrm{ji}}}\right)^{\frac{2}{m-1}}} \tag{7}$$

where k and i are integers, whose values range in [1, c] and [1, n], respectively, and possibilistic membership ‘ t_{ki} ‘ is defined as

$$t_{\mathrm{ki}} = \frac{1}{1 + \left(\frac{b}{\Upsilon_k} d_{\mathrm{ki}}^2\right)^{\frac{1}{\eta-1}}} \tag{8}$$

where $1 \leq k \leq c$ and cluster center ‘v_k‘ is defined as

$$v_k = \frac{\sum_{i=1}^{n}\left(\mathrm{au}_{\mathrm{ki}}^m + \mathrm{bt}_{\mathrm{ki}}^{\eta}\right)x_i}{\sum_{i=1}^{n}\left(\mathrm{au}_{\mathrm{ki}}^m + \mathrm{bt}_{\mathrm{ki}}^{\eta}\right)} \tag{9}$$

PFCM outperforms FCM and PCM. But when clusters highly vary in their size and are contaminated with outliers, it fails to result in good clusters.

2.3 *Credibilistic Fuzzy C-Means (CFCM) [2]*

CFCM was proposed by K. K. Chintalapudi by developing a new variable, named as credibility. Credibility is defined as follows:

$$\psi_k = 1 - \frac{(1-\theta)\alpha_k}{\max_{j=1\ldots n}\left(\alpha_j\right)} \tag{10}$$

where $\alpha_k = \min_{i=1\ldots c}(d_{\mathrm{ik}})$; this is the distance of point ‘x_k’ from its nearest centroid ‘j’.

Noisiest point is assigned a credibility value equal to Φ. CFCM minimizes the following objective function:

$$J_{\mathrm{CFCM(U,\,V)}} = \sum_{j=1}^{c}\sum_{i=1}^{n} u_{\mathrm{ij}}^m d_{\mathrm{ij}}^2 \tag{11}$$

Subject to the following constraint:

$$\sum_{i=1}^{i=c} u_{\mathrm{ik}} = \psi_k, \quad \mathrm{k} = 1 \ldots \mathrm{n} \tag{12}$$

CFCM reduces the influence of outliers on resultant clusters. Thus, it improves centroid position but not the most accurate centroids. As well as outliers are assigned to more than one cluster [2, 5].

2.4 *Intuitionistic Fuzzy C-Means(IFCM) [3]*

Xu. Zeshui and Wu. Junjie based on intuitionistic fuzzy set theory proposed IFCM which is helpful in dealing with uncertain and vague data [5]. Its objective function is

$$J_{\mathrm{IFCM}} = \sum_{i=1}^{c}\sum_{k=1}^{n} u_{ik}^{*m} d_{ik}^{2} + \sum_{i=1}^{c} \eta_i^{*} e^{1-\eta_i^{*}} \tag{13}$$

with m is set to be 2, $u_{ik}^{*} = u_{ik} + \eta_{ik}$, where u_{ik}^{*} represents intuitionistic fuzzy membership and η_{ji} is hesitation degree [3] defined as

$$\eta_{ik} = 1 - u_{ik} - (1 - u_{ik}^{\acute{\alpha}})^{1/\acute{\alpha}},\ \acute{\alpha} > 0 \tag{14}$$

IFCM produces overlapping clusters and hence it becomes really difficult to assign a cluster to the points lying in the overlapping region. Also, IFCM fails to handle outliers as this algorithm and treats outliers as data objects.

2.5 *Noise Clustering [9]*

NC is a very robust clustering algorithm. It results in n + 1 clusters, where n is the required number of clusters and one cluster consists of noise and outliers. It tackles very well the major problem of FCM and computes clusters on the following objective function:

$$J(U, V) = \sum_{k=1}^{c+1}\sum_{i=1}^{N} (u_{\mathrm{ki}})^{m} (d_{\mathrm{ki}})^{2} \tag{15}$$

where 'c' is the count of good clusters. NC performs far better than FCM and PFCM. It focuses on reducing the impact of outliers on resultant clusters.

3 Experiment and Result Analysis

For simulation of results: FCM, PFCM, CFCM, IFCM, and NC are implemented using MATLAB software, version R2017a(9.2.0). Four standard datasets are considered: D12, D15, Dunn, and noisy Dunn datasets. D12 and D15 are standard dataset with two identical symmetric clusters with five data objects. D12 has additional one noise and an outlier along with 10 data points. Similarly, D15 is an extension of D12 with another three new outliers. Dunn dataset is also a standard dataset with two square-shaped clusters varying in density and size. Table 1 lists the considered datasets with their brief description.

Figure 1 and Table 2 show clustering results on D12 dataset. Coordinate of ideal centroids is (−3.34, 0) and (3.34,0). Mean Squared Error (MSE) is used to compute error in centroid positions, as shown in Table 2. Resultant clusters are represented in Fig. 1. It is observed that the performance of FCM drastically degrades with the presence of a single outlier. PFCM and CFCM perform much better than FCM but the performance is not very satisfactory. NC performs the best as it identifies the outliers and clusters these in a separate cluster.

Similarly, Fig. 2 and Table 3 show clustering results on D15 dataset. From Fig. 2 and Table 3, it is observed that when a number of outliers are quite high in dataset, FCM and PFCM completely fails to identify these outliers. However, CFCM and IFCM perform better than FCM and PFCM by identifying the right clusters and assigning the outliers to the closer cluster. NC gives best results by identifying all the outliers and provides most accurate centroids.

Figure 3 and Table 4 show clustering results on Dunn dataset, which is a density and size varying dataset. FCM and PFCM perform well for such datasets but CFCM

Table 1 Description of various datasets

S. No.	Dataset	No. of data objects	No. of noise points	No. of outliers
1	D12	10	1	1
2	D15	10	1	4
3	Dunn	117	Nil	Nil
4	Noisy Dunn	117	Nil	21

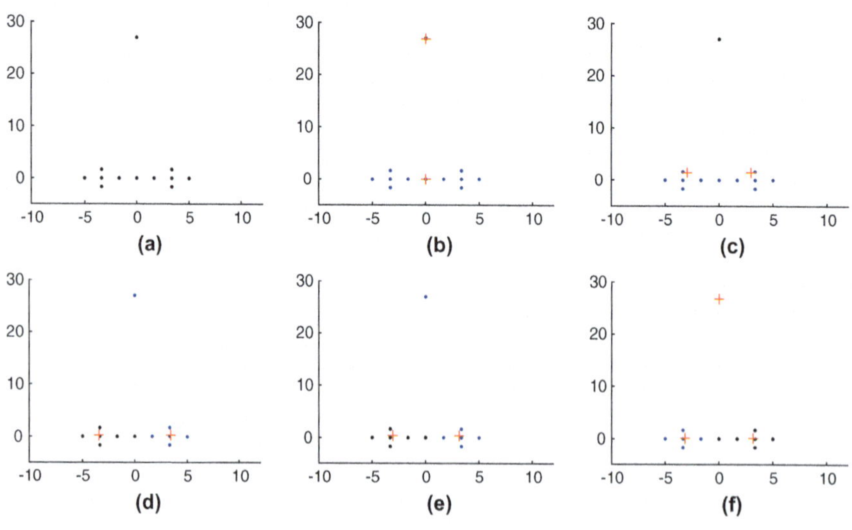

Fig. 1 Resultant clusters and dataset **a** D12 **b** FCM **c** PFCM **d** CFCM **e** IFCM **f** NC

Table 2 Clustering results' comparison on D12

Algorithm	Cluster 1		Cluster 2		Centroid error
	Cx	Cy	Cx	Cy	
FCM	0.00000	−0.00279	0.00000	26.89279	372.76674
PFCM	−2.93500	1.52135	2.93500	1.52135	2.47853
CFCM	−3.12319	0.41126	3.12321	0.41126	0.21614
NC	−3.15944	0.14883	3.15942	0.14884	0.05476
IFCM	−3.43892	0.27377	3.43897	0.27387	0.08477

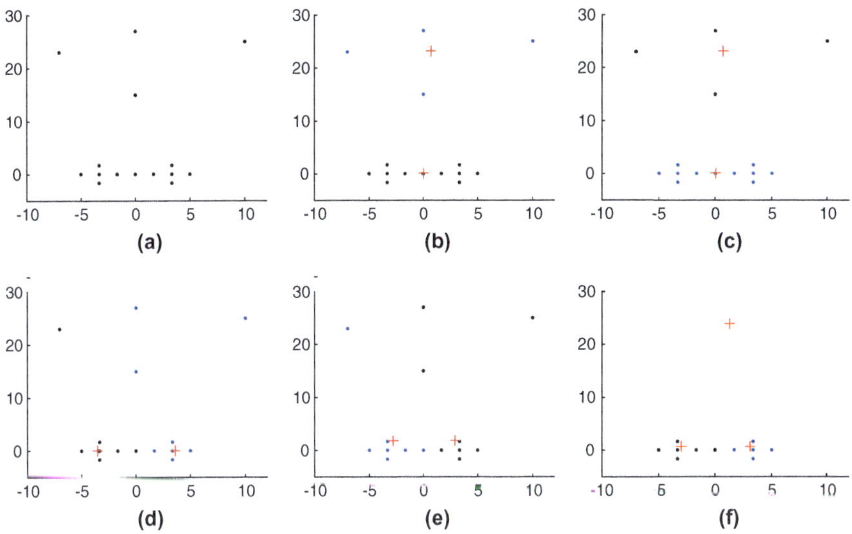

Fig. 2 Resultant clusters and dataset **a** D15 **b** FCM **c** PFCM **d** CFCM **e** IFCM **f** NC

Table 3 Clustering results' comparison on D15

Algorithm	Cluster 1		Cluster 2		Centroid error
	Cx	Cy	Cx	Cy	
FCM	0.00474	0.12274	0.67572	23.17389	277.66506
PFCM	0.00453	0.11606	0.67359	23.17087	277.59906
CFCM	−2.81821	1.82449	2.90424	1.89034	3.68215
NC	−3.56706	0.04688	3.57345	0.05262	0.05551
IFCM	−3.04603	0.77064	3.05472	0.76522	0.67362

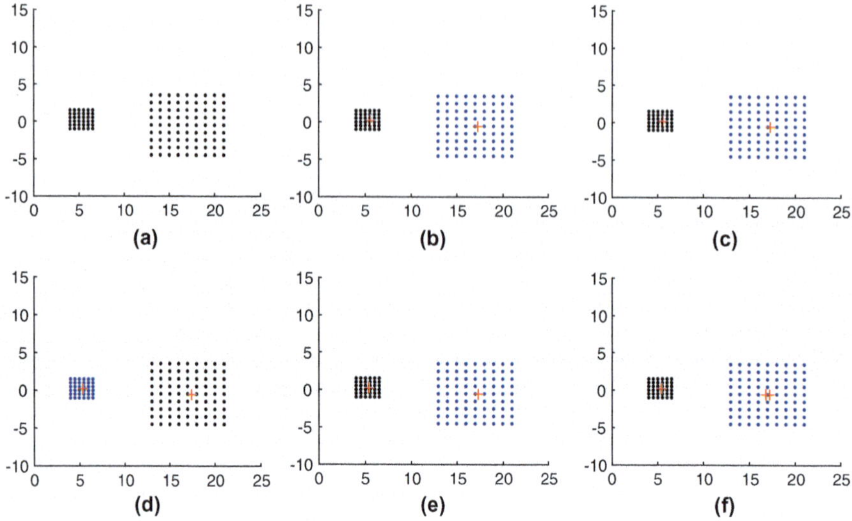

Fig. 3 Resultant clusters and dataset **a** Dunn dataset **b** FCM **c** PFCM **d** CFCM **e** IFCM **f** NC

Table 4 Clustering results' comparison on dunn dataset

Algorithm	Cluster 1		Cluster 2		Centroid error
	Cx	Cy	Cx	Cy	
FCM	5.59588	0.23119	17.30846	−0.52255	0.10782
PFCM	5.59588	0.23119	17.30846	−0.52255	0.10782
CFCM	5.42167	0.24017	17.35360	−0.52468	0.07760
NC	5.44090	0.23878	17.24442	−0.51756	0.04831
IFCM	5.40093	0.24225	17.50359	−0.53405	0.13880

and IFCM perform much better than FCM and PFCM. NC performs best among FCM, PFCM, IFCM, and CFCM.

Figure 4 and Table 5 show clustering results on Dunn dataset contaminated with noise and outliers. It is observed that the performance pattern of FCM, PFCM, CFCM, NC, and IFCM is the same as in the case of Dunn dataset. However, NC results are most robust among all.

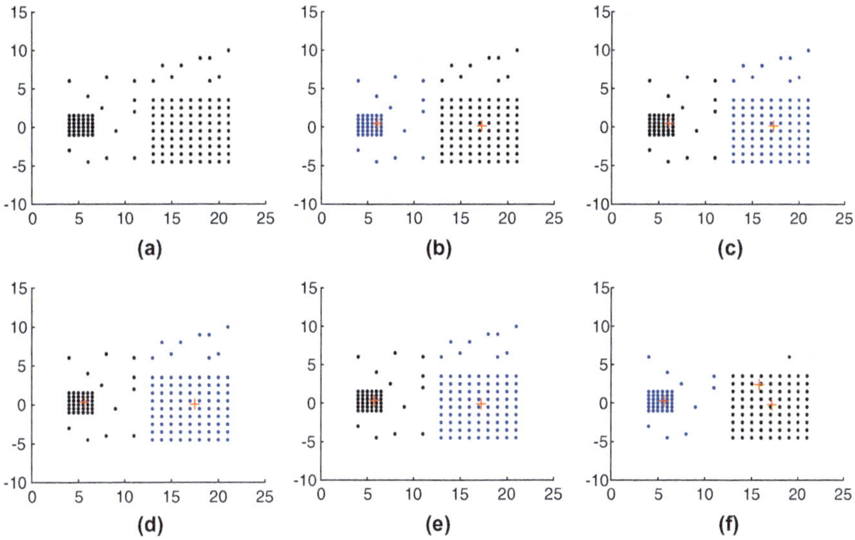

Fig. 4 Resultant clusters and dataset **a** Noisy dunn dataset **b** FCM **c** PFCM **d** CFCM **e** IFCM **f** NC

Table 5 Clustering results' comparison on dunn dataset contaminated with noise and outliers

Algorithm	Cluster 1		Cluster 2		Centroid error
	Cx	Cy	Cx	Cy	
FCM	6.11879	0.45580	17.27371	0.15904	0.65320
PFCM	6.12000	0.45658	17.27374	0.15408	0.65117
CFCM	5.81681	0.35902	17.26396	−0.07037	0.29370
NC	5.68312	0.31914	17.15851	−0.17312	0.16218
IFCM	5.67933	0.37367	17.51074	0.07384	0.39488

4 Conclusion

In this paper, we have compared FCM, IFCM, PFCM, CFCM, and NC with an objective of measuring the performance of each algorithm over different datasets. Results are analyzed on dataset with identical cluster, and dataset with clusters varying in size and density. Also, experimental results are analyzed to assess the impact on performance because of noise and outliers. It is observed that the performance of FCM and PFCM highly degrades as population of outliers increases in dataset. NC performance is most robust in the presence of outliers and it performs best among datasets with varying size and density. NC performance is followed by CFCM, IFCM, PFCM, and FCM in respective order. Thus, for the above-considered datasets, NC is the best choice for clustering but it too has its limitation in datasets with noise.

References

1. Pal, N.R., Pal, K., Keller, J.M., Bezdek, J.C.: A possibilistic fuzzy c-means clustering algorithm. IEEE Trans. Fuzzy Syst. **13**(4), 517–530 (2005)
2. Chintalapudi, K.K., Kam M.: The credibilistic fuzzy c means clustering algorithm. In: 1998 IEEE International Conference on Systems, Man, and Cybernetics, 1998, vol. 2, pp. 2034–2039. IEEE (1998)
3. Xu, Z., Junjie, W.: Intuitionistic fuzzy C-means clustering algorithms. J. Syst. Eng. Electron. **21**(4), 580–590 (2010)
4. Panda, S., Sahu, S., Jena, P. and Chattopadhyay, S.: Comparing fuzzy-C means and K-means clustering techniques: a comprehensive study. In: Advances in Computer Science, Engineering & Applications, pp. 451–460. Springer, Berlin, Heidelberg (2012)
5. Gosain, A., Dahiya, S.: Performance analysis of various fuzzy clustering algorithms: a review. Procedia Comput. Sci. **79**, 100–111 (2016)
6. Ayed, A.B., Halima, M.B., Alimi, A.M.: Survey on clustering methods: Towards fuzzy clustering for big data. In: 2014 6th International Conference of Soft Computing and Pattern Recognition (SoCPaR), pp. 331–336. IEEE (2014)
7. Li, J., Lewis, H.W.: Fuzzy clustering algorithms—review of the applications. In: 2016 IEEE International Conference on Smart Cloud (SmartCloud), pp. 282–288. IEEE (2016)
8. Bezdek, J.C., Ehrlich, R., Full, W.: FCM: the fuzzy c-means clustering algorithm. Comput. Geosci. **10**, 191–203 (1984)
9. Keller, A.: Fuzzy clustering with outliers. In: Fuzzy Information Processing Society, 2000. NAFIPS. 19th International Conference of the North American, pp. 143–147. IEEE (2000)

A Regularization-Based Feature Scoring Criterion on Candidate Genetic Marker Selection of Sporadic Motor Neuron Disease

S. Karthik and M. Sudha

Abstract Sporadic Motor Neuron Diseases (sMND) are a group of neurodegenerative conditions. It causes severe damage to the nerves in the brain and spine, makes it lose the function over time. The progression of this disease has a strong relationship with the genetics of an affected individual. Analyzing the gene expressions of sMND affected cases unveil the diagnostic genetic markers of the condition. But, higher dimensionality of the data affects the predictive performance due to the presence of vague, imprecise features. To address these issues, an effective hybrid feature selection technique called Correlation-based Feature Selection-L2 Regularization (CBFS-L2) is proposed to identify the candidate genes of sMND by eliminating inconsistent, redundant features. The proposed CBFS-L2 model revealed 26 significant Single Nucleotide Polymorphism (SNP) gene biomarkers of sMND. The performance of the identified subset is evaluated with four state-of-the-art supervised machine learning classifiers. The proposed feature selection technique attained a high accuracy of 94.31% on sMND dataset, outperformed benchmarked results, and other feature selection techniques.

Keywords Computational genomics · Dimensionality reduction · Molecular diagnostics · Regularization · Sporadic motor neuron disease

1 Introduction

Computational diagnostic system supports accurate decision-making when the process is highly critical. The advent of high-performance gene sequencing technologies transformed the treatment strategies into another dimension. Gene therapy is becoming more popular among developed countries. The reason for developing

S. Karthik · M. Sudha (✉)
School of Information Technology and Engineering, Vellore Institute of Technology, Vellore, India
e-mail: msudha@vit.ac.in

S. Karthik
e-mail: skarthik@vit.ac.in

S. C. Satapathy et al. (eds.), *Intelligent Data Engineering and Analytics*, Advances in Intelligent Systems and Computing 1177,
https://doi.org/10.1007/978-981-15-5679-1_30

diseases is mapped and analyzed at the genomic level. DNA, otherwise deoxyribonucleic acid, makes a complete genome in a cell. Changes that occurred in these base pairs of DNA are called Single Nucleotide Polymorphism (SNP). These modifications are responsible for the changes in character, appearance, and behavior of the species. Also, it can increase the risk of diabetes, neurological complications, psychiatric disorders, and other diseases in humans, since it has the potential to affect the phenotypes directly.

According to recent studies, SNPs were helpful in finding the adverse reactions and responsiveness of drugs in human metabolism [1]. So, analyzing SNP data could be helpful in identifying new pathways for better disease diagnosis. Most of the SNP datasets have only a few samples but contains millions of SNP features. Also, genetic data are highly heterogeneous. Machine learning algorithms are more effective in handling complex data. It can be applied to both feature selection and classification phases of SNP-related disease diagnosis techniques. Feature Selection is a crucial task in medical data analysis. In specific, SNP data is high in its dimension in nature. So, it contains irrelevant features such as redundant SNPs, missing values, and noisy data [2]. In order to improve the model performance, it is important to eliminate the inconsistent data and selection of an optimal SNP subset.

In this work, a hybrid fusion of filter-embedded method is proposed to select the most discriminative SNP features from two cancer datasets. Correlation-Based Feature Selection (CBFS) is a filter-based technique and L2 Regularization or Ridge Regression is an embedded feature selection model that combines with CBFS. Experimental results show better results on the proposed model benchmarked with the state-of-the-art methods.

2 Background Study

Recently, many scientific studies intend to develop "hybrid" models by combining different feature selection strategies such as filter, wrapper, and embedded methods. These models have shown improved performance on many critical applications. A regularization-based SNP biomarker identification model is constructed for colorectal cancer diagnosis and prediction. Lasso and Elastic Net penalization regression techniques were used in the pipeline of the system. Two novel gene biomarkers are found in this work which has shown promising results in predicting the disease [3].

Patients with breast cancer are likely treated with aromatase inhibitors and those are in a high chance of developing arthralgia. A Novel Analytic Algorithm is framed to identify the genes with a high-risk factor of the above condition. This model generates a subset of 70 SNPs, where 57 are highly correlated and have a strong association with each other [4]. Swarm-based SNP–SNP interaction detection framework is deployed for identifying breast cancer related genes. This model calculates the maximum deviation between the normal and abnormal SNPs using the Chaotic Particle Swarm Optimization algorithm. Seven novel gene interactions are pointed out and it acts as the predictors [5].

Table 1 Dataset description

Dataset	Accession ID	Case	Control	SNP count
sMND	GSE15826	52	36	909622

A similar study on SNP interaction adopts a two-stage machine learning approach that binds Multivariate Adaptive Regression Splines (MARS) with a random forest ensemble algorithm. RF identifies best SNP predictors, whereas MARS identifies interaction patterns. 100 candidate SNP biomarkers are highlighted in this work as they are having strong genetic associations [6].

3 Materials and Methods

This section presents the processes involved in constructing the proposed pipeline of the sMND biomarker selection and prediction model. The initial phase discusses the data pre-processing methods followed to eliminate redundant information from the SNP data. Further, the proposed hybrid CBFS-L2 feature selection method is projected in algorithmic representation. The subset performance evaluation process is briefed in the classification section.

3.1 Dataset Information

The SNP dataset is accessed from the Gene Expression Omnibus repository to conduct this study. The accession number of sMND is GSE15826 [7]. The details of the dataset are briefed in Table 1.

3.2 Data Pre-processing

The SNP values AA, AB, BB, and NC in the dataset are encoded to 11, 01, 10, and 00 by direct replacement method as AA = 11, BB = 10, AB = 01, NC = 00. Then the redundant features with the same entries replicated more than once are eliminated. Features with NC entry have discarded the genotypes with a minimum of 30% of contribution in each category (i.e., AA, BB, and AB) are retained.

3.3 Proposed CBFS-L2 Feature Selection Technique

The pre-processed data is then forwarded into the feature selection process to identify the optimal SNP subset. A hybrid filter-embedded model is proposed in this work. CBFS-L2 Regularization is combined together to achieve optimal performance. The algorithm of the CBFS-L2 method is given below.

Proposed CBFS-L2 Algorithm

Input: X Case and Control Samples with SNP Features and Labels

Output: C Vector of Coefficients as Feature Subset

Set $S = \emptyset$, $\rho = \mathrm{Corr}(X)$

for $0 \leq m < \mathrm{len}(\rho)$ do

$Z_m = 0$

for $0 \leq n < \mathrm{len}(\rho_m)$ do

$l_m = \mathrm{abs}\,(\rho_m n)$, $ax_m = ax_m + l_m$

end

$Z_m = V_m \,/\, ax_m$

end

$S = \mathrm{sort}\,(Z_m, \mathrm{desc})$

Set $D = S$, $F = \emptyset$ and $P(w) = \sum_{j=1}^{D} w_j^2$

Set $w = (X^T X + \alpha I)\text{-}1\ X^T Y$, where $\alpha = 0.05$

for $0 \leq D < \mathrm{len}(Z_m)$ do

$L{:}\frac{1}{2}\,||Y - X_w||_2^2 + \lambda \sum_{j=1}^{D} |W_j|^2 \Rightarrow \min_w \in \mathbb{R}^D$; n, x = L

end

$\overline{x} = \sum_{i=1}^{\mathrm{len}\,(n)} x_i$, $\sigma = \sqrt{\frac{1}{x-1}(x_1 - \overline{x})^2 + \ldots + (x_n - \overline{x})^2}$

$S.E = \frac{\sigma}{\sqrt{n}}$ and $T = (S.E * f_c)$, where f_c are the coefficients of the features

if(x_i >T)

C = F, otherwise Eliminate F

the end, return C

The threshold is measured from standard error, and the value calculated is 0.263 for sporadic motor neuron disease. SNPs having the coefficient value lesser than the threshold value are labeled as redundant features and are eliminated further. Table 2 represents the count of features selected in each phase. The architecture of the proposed framework is depicted in Fig. 1.

Table 2 Number of SNP features identified in different stages

Dataset	Total SNP's	Pre-process	CBFS	CBFS-L2
sMND Dataset	909622	80559	59	26

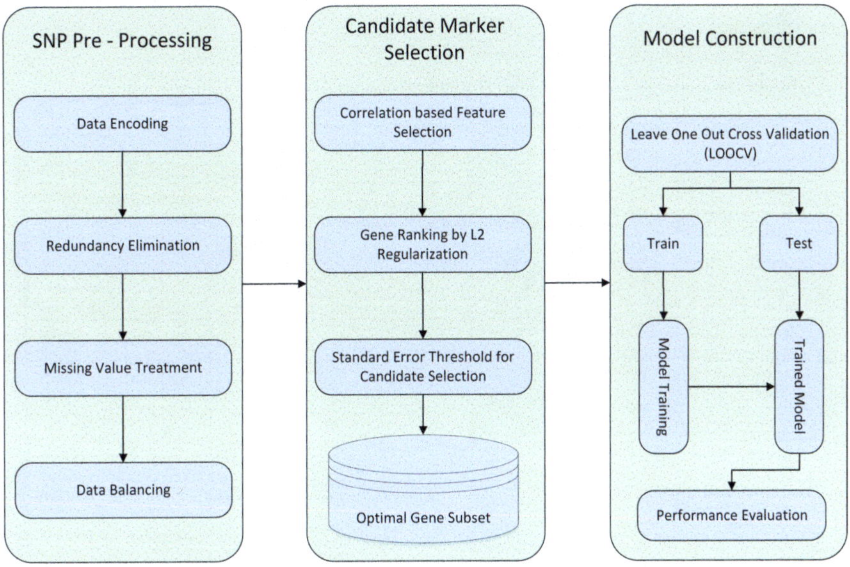

Fig. 1 Proposed CBFS-L2 framework

3.4 *Classification*

Machine learning algorithms are more prominent in computational genomics and biomedical researches [8–10]. These algorithms tend to learn complex patterns from heterogeneous data sources. This knowledge can be applied to a new set of data to perform to a prediction on future possible events. According to the "No free lunch" theorem, it is a single ML algorithm which is not expected to perform well on different applications. So, many ML algorithms are developed, each serves on its own purpose. The challenge here comes on choosing a suitable algorithm for the identified problem. In this work, four different ML algorithms LDA, SVM, NB, and k-NN are used to evaluate the performance of the model.

3.5 *Heat Map Analysis*

In gene expression data, the up-regulated and down-regulated genes are identified based on the appearance of the color in the heat map. In the plotted heat map from Fig. 2, blue color represents up-regulation; yellow indicates the down-regulation, and green implies the absence of regulatory activity.

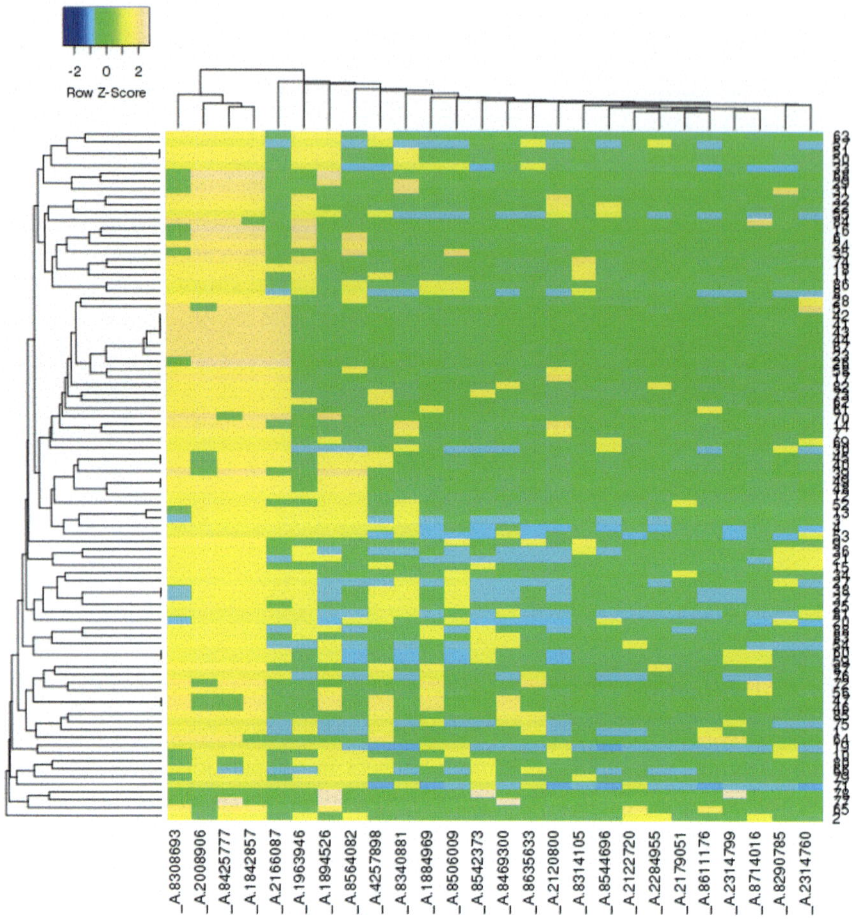

Fig. 2 Heat map of sMND with the features selected by CBFS-L2

4 Results

This experimental work is implemented in Python with Anaconda Distribution. Four supervised machine learning algorithms such as LDA, SVM, NB, and k-NN were employed to evaluate the performance of the proposed feature selection method. For model, validation of Leave one out Cross-Validation (LOOCV) method is followed. A confusion matrix is an important evaluation metric for classification of models that can be constructed with these four individual parameters TP, TN, FP, FN where each represents True Positive, True Negative, False Positive, and False Negative, respectively. The formula to calculate the score of the performance metrics were given as equations below.

$$\mathrm{Acc} = \frac{(\mathrm{TP} + \mathrm{TN})}{(\mathrm{TP} + \mathrm{TN} + \mathrm{FP} + \mathrm{FN})} \tag{1}$$

$$\mathrm{F-Score} = \frac{(2*(\mathrm{Re}*\mathrm{Pre}))}{(\mathrm{Re} + \mathrm{Pre})} \tag{2}$$

Mathews Correlation Coefficient (MCC) measure is calculated to measure the quality of binary classification models. The formula for MCC is given below.

$$\mathrm{MCC} = \frac{(\mathrm{TP}*\mathrm{TN} - \mathrm{FP}*\mathrm{FN})}{\sqrt{(\mathrm{TP} + \mathrm{FP})(\mathrm{FN} + \mathrm{TP})(\mathrm{FP} + \mathrm{TN})(\mathrm{TN} + \mathrm{FN})}} \tag{3}$$

The error rate is the calculation of the wrongly predicted instances. It can also be calculated by 1-Accuracy.

$$\mathrm{ErrorRate} = \frac{(\mathrm{FP} + \mathrm{FN})}{(\mathrm{TP} + \mathrm{TN} + \mathrm{FP} + \mathrm{FN})} \tag{4}$$

In Fig. 3, precision, recall, and AUC score are plotted as bar graphs for all the four classifiers. This plot shows the significance of each classifier on both the dataset.

The results obtained from the proposed model are benchmarked with different classifiers and feature selection models, projected in Tables 3 and 4, respectively. CBFS-L2 with NB evaluator achieved 94.31% accuracy on the sMND dataset, outperformed other feature selection methods, and benchmarked learning algorithms. As evidence, this proposed system will be further enhanced to diagnose complex neurodegenerative [11, 12] and psychiatric disorders [13] with cutting-edge mathematical models.

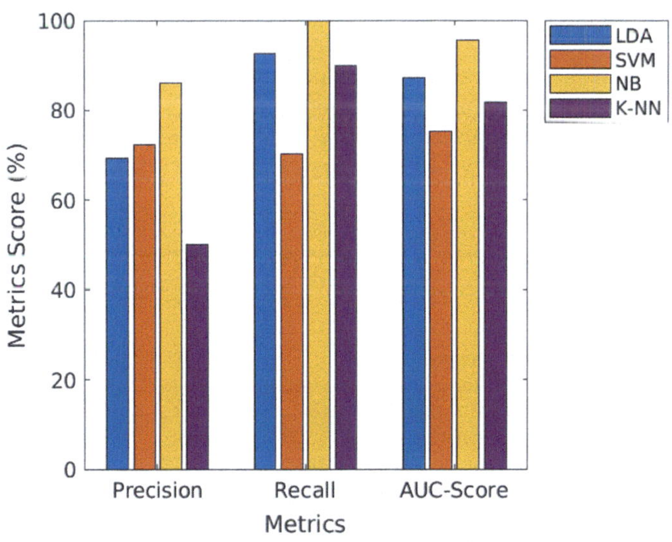

Fig. 3 Performance of different classifiers validated on various metrics

Table 3 Performance comparison of sMND dataset with 4 classifiers

sMND	Accuracy (%)	F-Score (%)	MCC (%)	Error rate (%)
LDA	85.22	88.49	69.93	14.77
SVM	76.13	79.61	88.63	23.86
NB	94.31	95.41	86.69	5.68
k-NN	77.27	83.33	54.14	22.72

Table 4 Accuracy comparison on sMND dataset with existing methods

Accuracy (%)	LDA	SVM	NB	k-NN
CBFS	82.64	72.56	89.67	76.52
PSO	79.97	74.61	88.90	71.04
CMIM-RFE	81.35	72.88	92.14	74.91
Proposed	85.22	76.13	**94.31**	77.27

5 Conclusion

The proposed model identified highly informative, distinct SNPs that effectively discriminate the unhealthy and healthy samples. The performance of the proposed CBFS-L2 model attains 94.31% accuracy with the NB classifier, higher than the benchmarked algorithm, where the rest of the model score is 85.22, 76.13, 77.27% on LDA, SVM, and k-NN, respectively. In addition to that the same algorithms were employed to evaluate the performance of the feature subset generated by CBFS, PSO, and CMIM-RFE methods. Among them, the CBFS-L2-NB pipeline outperformed the other fused models. DNA variants provide more useful patterns from the genetic alterations and mutations occurred in a genome that helps to identify the prognostic disease markers. These genetic variants act as a key factor in categorizing the patients having similar gene patterns to provide more personalized treatment near in future.

References

1. Shastry, B.S., SNPs: impact on gene function and phenotype. In: Single Nucleotide Polymorphisms, pp. 3–22. Humana Press, Totowa, NJ (2009)
2. Li, J., Cheng, K., Wang, S., Morstatter, F., Trevino, R.P., Tang, J., Liu, H.: Feature selection: a data perspective. ACM Comput. Surv. (CSUR) **50**(6), 94 (2018)
3. Barat, A., Smeets, D., Moran, B., Das, S., Betge, J., Murphy, V., Ebert, M.P.: A machine-learning approach for the identification of highly predictive germline SNPs as biomarkers for response to bevacizumab in metastatic colorectal cancer using Elastic Net and Lasso (2018)
4. Reinbolt, R.E., Sonis, S., Timmers, C.D., Fernández-Martínez, J.L., Cernea, A., de Andrés-Galiana, E.J., Lustberg, M.B.: Genomic risk prediction of aromatase inhibitor-related arthralgia in patients with breast cancer using a novel machine-learning algorithm. Cancer Med. **7**(1), 240–253 (2018)

5. Chuang, L.Y., Chang, H.W., Lin, M.C., Yang, C.H.: Chaotic particle swarm optimization for detecting SNP–SNP interactions for CXCL12-related genes in breast cancer prevention. Eur. J. Cancer Prev. **21**(4), 336–342 (2012)
6. Lin, H.Y., Ann Chen, Y., Tsai, Y.Y., Qu, X., Tseng, T.S., Park, J.Y.: TRM: a powerful two-stage machine learning approach for identifying SNP-SNP interactions. Ann. Hum. Genet. **76**(1), 53–62 (2012)
7. Pamphlett, R., Morahan, J.M.: Copy number imbalances in blood and hair in monozygotic twins discordant for amyotrophic lateral sclerosis. J. Clin. Neurosci. **18**(9), 1231–1234 (2011)
8. Sudha, M.: Evolutionary and neural computing based decision support system for disease diagnosis from clinical data sets in medical practice. J. Med. Syst. **41**(11), 178 (2017)
9. Bhateja, V., Tiwari, A., Gautam, A.: Classification of mammograms using sigmoidal transformation and SVM. In: Smart Computing and Informatics, pp. 193–199. Springer, Singapore (2018)
10. Dey, N., Bhateja, V., & Hassanien, A. E. (2016). Medical Imaging in Clinical Applications. Springer International Publishing **10**, 973–978
11. S. Karthik, Sudha M.: A survey on machine learning approaches in gene expression classification in modelling computational diagnostic system for complex diseases. Int. J. Eng. Adv. Technol. **8**(2) (2018)
12. Karthik, S., Sudha M.: Diagnostic gene biomarker selection for alzheimer's classification using machine learning. Int. J. Innov. Technol. Explor. Eng. **8**(12) (2019)
13. Sekaran, K., Sudha, M.: Prediction of lipopolysaccharides simulation responsiveness on gene expression profiles of major depression disorder affected cases using machine learning. Int. J. Sci. Technol. Res. **8**(11), 21–24, 23 Nov 2019

A Study for ANN Model for Spam Classification

Shreyasi Sinha, Isha Ghosh, and Suresh Chandra Satapathy

Abstract The classical way of detecting spam emails based on the signature is not very effective in recent days due to the huge uses of emails in various activities. The online recommendation and push emails make the spam detection job very complex and tedious. Machine learning happens to be a widely used approach for automated email spam detection. Among various machine learning algorithms, Artificial Neural Network (ANN) is gaining popularity due to its powerful approximation and generalization characteristic. The effectiveness of the email spam classifier is heavily dependent on the learning capability of ANN. In our work, we have developed a BP and a BP+M model to do the spam classification and find the accuracy of classification. We have compared the two models so that we can conclude that the BP+M model gives the same or better result than the BP model using fewer epochs. Though state-of-the-art and classical learning algorithms like backpropagation (BP) and backpropagation with momentum (BP+M) are very popular and well researched, it is understood that often it gets trapped in local optima. In our future work, we can use recent optimization techniques like SGO which can elevate the results and can eradicate the drawbacks of BP and BP+M model. After thorough simulations and results analysis, we conclude that backpropagation + momentum optimized ANN provides superior classification results than BP optimized ANN.

Keywords Spam classification · Artificial Neural Network · Backpropagation · Momentum factor

S. Sinha (✉) · I. Ghosh · S. C. Satapathy
Kalinga Institute of Industrial Technology, Deemed to be University, Bhubaneswar, India
e-mail: shreyasi22knp@gmail.com

I. Ghosh
e-mail: ishaghosh1819@gmail.com

S. C. Satapathy
e-mail: sureshsatapathy@gmail.com

S. C. Satapathy et al. (eds.), *Intelligent Data Engineering and Analytics*, Advances in Intelligent Systems and Computing 1177,
https://doi.org/10.1007/978-981-15-5679-1_31

1 Introduction

Spam emails are those emails that we have not asked or requested for, i.e., these are unbidden commercial mails that agglomerate our inbox. Spam mails are the emails sent to a large number of people and consist mostly of advertisements. These spam mails are sent in large amounts to a purchased (or stolen) mailing list that consists of our email addresses. It will be wrong to classify all spam mails are either "advertisements" or "commercials". Spam mails can also be mails related to politics, financial scam emails, emails that are sent to spread malware, and false charity request mails. Spam mails must be checked and deleted by the recipient. But in this process, some legitimate emails like mails of charity appeal or an invite for the recipient or a newsletter which are actually unbidden mails but not spam mails may also get deleted. So, in this case, we need to know whether the unbidden mails are actually spam or not. Therefore, many approaches have been put forward to differentiate spam mails and block them.

Erosheva and Fienberg, in their paper, have described and applied Bayesian approach to classification and soft clustering by using membership model [1]. Cortez, Lopes and Sousa, et al., have used a Hybrid which combines Collaborative Filtering(CF) as well as Content Based Filtering (CBF) called Symbiotic data mining [2]. To train multi-layer perceptron [3], Yang and Elfayoumy have used a genetic algorithm and then for spam filtering they have used the trained MLP. The filtering system achieves an accuracy of 89% to detect legitimate mails and 94% accuracy to detect spam mails. In this paper, classification using Artificial Neural Network (ANN) is implied to detect spam mails. Classification deals with finding new observation as to what belong to which set of categories. Through classification, we predict the class of given data points classes that can also be referred to as categories, targets, or labels. For example, spam email detection comes under the classification problems. Here there are only 2 classes: one spam and other not spam, hence it is also known as a binary classification problem. Here, the different spam emails and non-spam emails are used as training data. Then the classifier is being trained accurately, and it is used to detect spam mail. Classification is important for clearly identifying the data, studying, and observing them. It is a way of differentiating the type of data, making predictions about the observation of the same type and classifying the relationship between different data points. Knowing classification helps us to predict the result of characteristics of data, based on the observation of another presence in the same class.

Different methods available for classification are

- Linear classifiers: Logistic Regression, Naive Bayes
- Nearest neighbor
- Support vector machine
- Decision Trees
- Boosted Trees
- Random Forest
- Neural Network.

Here, I have used Neural Network for classification. A neural network is a powerful machine learning algorithm and primarily used for classification problems. Spam detection is the most common classification problem. Neural network translates all real-world data, be it text or time series, images, and sound into patterns that are recognized by them. The pattern they recognize are numerical, contained in vectors. A neural network is useful for spam classification as it helps in clustering and classifying of data set. We can also think of them as the classification layer above the data we store. In spam detection, based on similarities among different inputs, they are used to group the unlabeled data and classify the data having labeled data that we can train on. The outcomes will be a label that could be applied to the data. In spam detection, these outcomes are spam and not spam.

Artificial Neural Network (ANN) consists of a set of connected input and output network in which each connection is associated with weights. There is one input layer, one output layer, and one or more intermediate layer. Learning of Neural Network is carried out by adjusting weight that is associated with each connection. Performance of Neural Network is improved by iteratively updating the weights.

In this paper, to train neural network I have used two approaches, i.e., backpropagation and backpropagation along with momentum. Firstly backpropagation is trained on email data set which is in csv format and then neural network is used to classify the spam email and regular email. But on using backpropagation algorithm, from different simulations we discover that it takes a long time to converge. Another approach is using Backpropagation with momentum (BPM) which gives compatible result in less number of epochs and is able to converge faster. Both these approaches are measured by accuracy. But through comparisons, we infer that both these approaches are not able to give optimal results or they are trapped in local minima. The present paper discusses email spam detection by BP and BPM and then analyzing that whether we are able to achieve optimal results through these approaches.

2 Preliminaries

2.1 Spam

Spam email, aka trash email, is an email sent without clear-cut consent from the receiver. Spam emails usually try to sell obsolete goods. This is the demerit of email marketing. The spam email has been becoming a more advanced phenomenon in terms to go overboard and the technical solutions for dodging limitations.

The idea of sending spam emails is to basically make a profit. The main reason behind this is that send a heaping amount of email to all the receivers globally. Though the percentage of users who take the desired action is very less even if one person is replying over so many spam emails which takes fewer efforts than doing the promotions or marketing manually is worth it. In most cases, whatever spam is received is mostly concealed under the mask of something which is appealing to the eyes of the users and offers something of users' interest.

Getting spam emails is not rare, and this is because of the fact that there are many ways for spammers to gather emails online. There are some spammers that will go to any extent to reap emails that may be from companies that sell a list of emails because through these only the spammers get access to your email accounts. So it does not matter whatever method they are putting into use to accumulate your emails. They are sending out spam that may not satisfy or match the user's requirements.

It is necessary to get rid of the spam because they acquire much of your inbox space and it can consume your time if you wish to clear them out of your inbox area. These emails may be harmful as it may carry viruses or malware that can badly affect your computer and can pose a threat to the security of your system and your private data which you do not want to share. So it is essential to use spam filters in order to avoid all these hindrances.

2.2 ANN

ANN stands for Artificial Neural Networks. It is a computational system deduced from the processing methods and the learning capacity of a human brain. It is basically a representation of a human brain and how the human brain works. The main aim of the Artificial Neural Network is that as the name suggests is the network of neurons that processes whatever you are interpreting and tries to learn it subsequently. A human brain consists of billions of neurons. Neuron is basically a basic unit of the human brain. A basic unit denotes the smallest indivisible unit and in the case of the human brain its a neuron. The sensory organs present in our body like mouth, tongue, ears, eyes, skin senses the environment and sends a signal to the brain. These signals are received by the neurons. The neurons interpret and process the signals and generate an appropriate output to take appropriate action at a given instance. So, when we try to achieve this functionality artificially then it falls under an artificial neural network. Below is a diagram of a node which basically is a replica of the neuron and describes the functionality. A node is divided into two major parts. First being the summation part and second being the function part. As your brain consists of millions of neurons, the network will also consist of multiple nodes to generate the output. At each node, there will be a signal and each signal will be assigned its respective weights (like for x1- > w1 and x2- > w2 and so on as shown in the figure below). Then this all will pass through the summation part which will calculate the weighted sum. Subsequently, this weighted sum is entered into the function part which is basically the transfer function. The main work of the transfer function is that if we are providing any input to the function then an appropriate output or the designated action will be generated. So in a way this activation function generates or defines a particular output for a given node based on the given input that is being provided. The output being generated is defined by the function (Fig. 1).

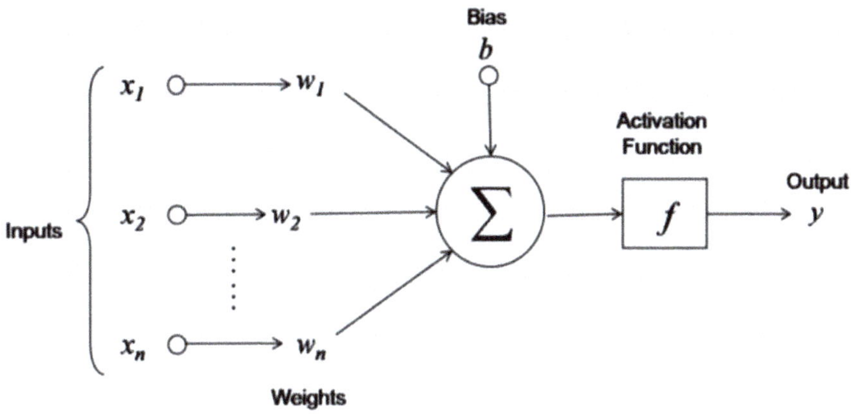

Fig. 1 Structure of ANN

2.3 *Back Propagation*

Back Propagation is used in the feed-forward network. This algorithm uses a technique called gradient descent or delta rule to search for a minimum value of the error function in weight space. In this algorithm, we first calculate the error, i.e., how far is output of our model from the actual one and then we check whether this error is minimum or not. If error is huge, then parameters that include weights and bias are updated. After updating, we check the error again. We need to repeat this process until our error becomes minimum. Once our error is minimized, we can feed the inputs to our model and produce the output.

Consider the graph given below.

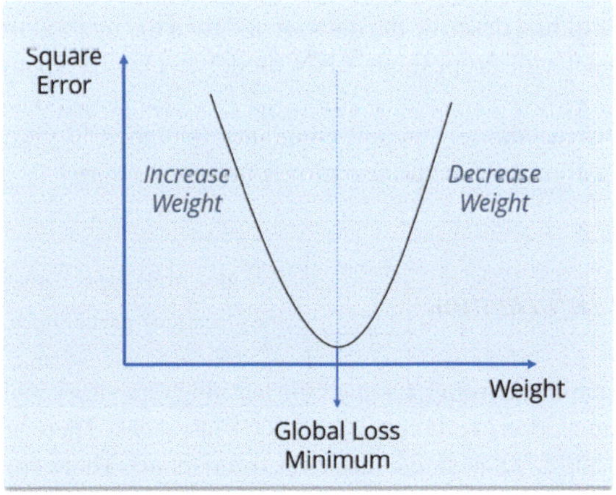

Here we need to reach the "Global Loss Minimum". This is backpropagation

2.4 Backpropagation + Momentum

As we know that backpropagation uses a technique called gradient descent. The introduction of momentum in this algorithm causes attenuation of oscillations in the gradient descent.

Given a network with n different weights W_k, then using backpropagation with momentum, the i-th correction for weight W_k is given by

$$\Delta w_k(i) = -\frac{\alpha \partial E}{\partial w_k} + \mu \, \Delta W_k(i-1), \ \frac{\partial E}{\partial w_k} \tag{1}$$

Equation (1) is the variation of the loss with respect to W_k.

where α is the learning rate, and μ is the momentum term. If α term is smaller than μ term in the next iteration, then from the previous iteration ΔW will have greater influence on the weight than the current one.

Therefore, the basic concept behind using momentum is that previous changes in the weight would influence the current current direction of movement in weight space. Momentum pushes our output toward global optimum, i.e., momentum changes the path that we take for the optimum. For example, we have decided to move around an objective function then the simplest approach is the steepest gradient but fluctuations can cause a big problem. This problem can be solved by adding momentum.

Note: High moment should always be escorted by low learning rate, otherwise we will overshoot global optimum.

3 Methods

In this section, I will first describe the data set and then the method for preprocessing of data. After that I will propose an ANN model to classify spam and non-spam emails.

Then I will discuss the results and simulation obtained by training our model using backpropagation and backpropagation with momentum.

3.1 Data set Information

The data set is a spambase from UCI machine learning repository which is created by Dua and karra Taniskidon [4]. It consists of 4601 instances. There are 1813 positive instances (spam) and 2788 negative instances (non-spam). There are 57 attributes or features and 1 label in each instance as shown in Table 1.

Table 1 Consists of information about dataset

Attributes number	Name	Type	Description
1–48	word_freq_word	Continues real [0, 100] attributes	Percentage of word in the email
49–54	char_freq_CHAR	Continues real [0, 100] attributes	Percentage of characters in email
55	capital_run_length_average	Continuous real [1, ...] attribute	Average length of uninterrupted sequences of capital letters
56	capital_run_length_longest	Continuous integer [1, ...] attribute	Length of longest uninterrupted sequence of capital letters
57	capital_run_length_total	Continuous integer [1, ...] attribute	Total number of capital letters in the email

3.2 *Data Preprocessing*

In general, normalization can improve convergence speed of the gradient descent and the accuracy of the model. In this paper, input data is first standardized to have a mean of 0 and standard deviation of 1. After that, the standardized data is normalized to range [0, 1]. Let x in attribute A is normalized to x^{new} then we have Formula (2) to calculate x^{new}.

$$x^{new} = (x^{new} - \text{mean})/(\text{std} + 1 * e - 8) \quad (2)$$

Here mean is the mean value of input data and std is the standard deviation.

3.3 *Proposed Multi-layered ANN Model*

These models have been chosen randomly. Basically two models that vary in their modular structure has been chosen as one contains one hidden layer and on the other hand the second models consist of two hidden layers. This selection can help us to find better result and help us to see much clear difference in the final result and can help us to draw unbiased conclusions.

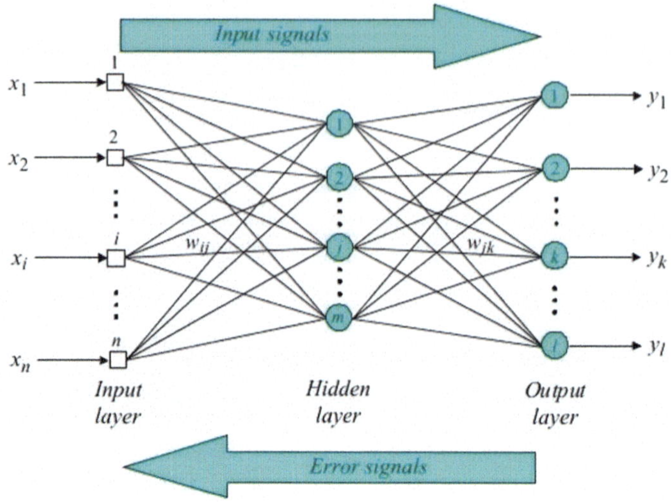

Fig. 2 ANN classifier structure, including one input layer, one hidden layer and two output layers

4 1st Model

It consists of input layer which has 57 neurons, which is equal to the number of features and only one hidden layer containing 12 neurons. The structure is shown in Fig. 2.

5 2nd Model

It consists of 57 inputs and 2 hidden layers. First hidden layer has 12 neurons and second hidden layer has 3 neurons. The structure is shown in Fig. 3.

6 Result and Discussion

6.1 Training Methods

Spam classification is done using ANN and for training the neural network I have used two algorithms BP and BP−M which are both gradient methods.

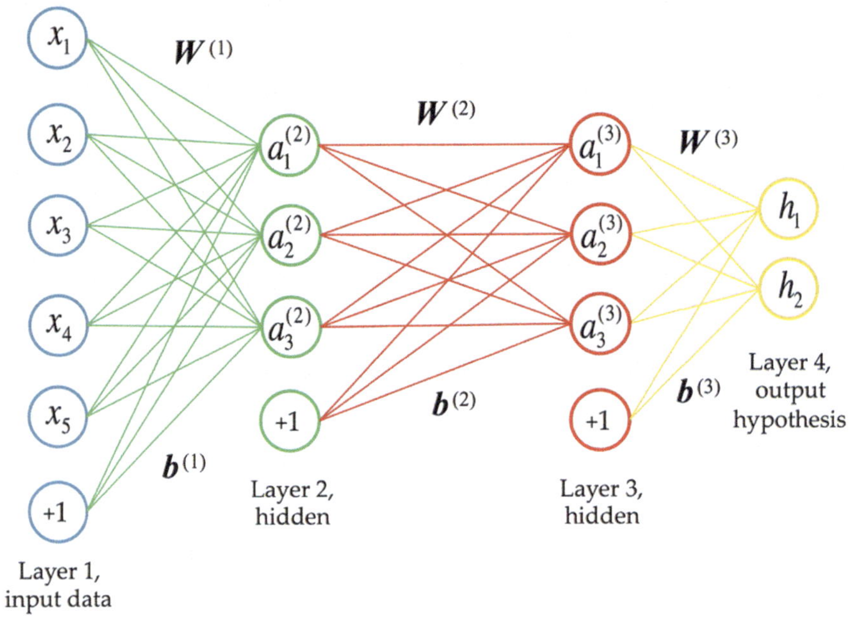

Fig. 3 ANN classifier structure, including one input layer, 2 hidden layers and one output layer

6.2 *Experiments and Results*

Experiment 1. Here, training of ANN model is done using BP. The motivation for this experiment is to find average accuracy in a specified number of epochs.

First Model. Here we have performed the experiment with a model consisting of 57 inputs, 12 neurons in the hidden layer and 1 neuron in the output layer. The training algo (BP) is applied on this model up to 300 epochs to get the average accuracy (Table 2).

Table 2 Observation table for first model using backpropagation

ANN model	Training algo	Number of epochs	Simulations	% correct classification (%)	Average % of accuracy (%)
57-12-1	BP	500	1	93.50	93.10
57-12-1	BP	500	2	94.60	
57-12-1	BP	500	3	91.70	
57-12-1	BP	500	4	92.40	
57-12-1	BP	500	5	93.30	

Table 3 Observation table for second model using backpropagation

ANN model	Training Algo	Number of epochs	Simulation	% correct classification (%)	Average % of accuracy (%)
57-12-3-1	BP	500	1	94.80	94.48
57-12-3-1	BP	500	2	93.50	
57-12-3-1	BP	500	3	93.90	
57-12-3-1	BP	500	4	94.80	
57-12-3-1	**BP**	**500**	**5**	**95.40**	

Second Model. Here we have performed the experiment with a model consisting of 57 inputs, 12 neurons in the first hidden layer, 3 neurons in the second hidden layer and 1 neuron in the output layer. The training algo(BP+M) is applied on this model up to 500 epochs to get the average accuracy (Table 3).

Experiment 2. Here, training of ANN model is done using BP+M. The motivation for this experiment is to find average accuracy in the fewer number of epochs.

First Model. Here we have performed the experiment with a model consisting of 57 inputs, 12 neurons in the hidden layer and 1 neuron in the output layer. The training algo (BP+M) is applied on this model up to 300 epochs to get the average accuracy (Table 4).

Second Model. Here we have performed the experiment with a model consisting of 57 inputs, 12 neurons in the first hidden layer, 3 neurons in the second hidden layer and 1 neuron in the output layer. The training algo (BP+M) is applied on this model up to 300 epochs to get the average accuracy (Table 5).

Inference. Here, from both the experiment we infer that BP−M gives compatible results in less number of epochs, i.e., 95.38% in just 300 epochs whereas BP gives 94.48% in 500 epochs. We get compatible results in less number of epochs because when momentum is applied in BP, it pushes or increases the steps we take toward minimum by trying to jump from a local minimum. Therefore, it speeds up the convergence toward minimum, and hence requiring lesser epochs.

Table 4 Observation table for first model using backpropagation along with momentum

ANN model	Training Algo	Number of epochs	Simulation	% correct classification (%)	Average % of accuracy (%)
57-12-1	BP+M	300	1	93.91	93.35
57-12-1	BP+M	300	2	93.91	
57-12-1	BP+M	300	3	91.95	
57-12-1	BP+M	300	4	93.47	
57-12-1	BP+M	300	5	93.50	

Table 5 Observation table for second model using backpropagation along with momentum

ANN model	Training algo	Number of epochs	Simulation	% correct classification (%)	Average % of accuracy (%)
57-12-3-1	BP−M	300	1	94.78	95.38
57-12-3-1	BP−M	300	2	94.56	
57-12-3-1	BP−M	300	3	95.65	
57-12-3-1	BP−M	300	4	95.86	
57-12-3-1	BP−M	**300**	**5**	**96.08**	

6.3 *Discussion*

For each simulation we get the output for each model and each training algorithm.

BP+M gives compatible results but in fewer numbers of epochs.

BP and BP+M are not able to give an optimal result or in other words, they are trapped in the local minima. This is due to BP and BP+M are gradient methods (Fig. 4).

Firstly, the data set was loaded using the panda's library. Then using the Keras library in anaconda, we created the suitable environment and the backpropagation code for training the data that was applied to a 57-12-3-1 model where there are 57 inputs, 12 neurons in the first hidden layer, 3 neurons in the second hidden layer, and 1 node in output layer and was run up to 500 epochs, and then the testing code up to 100 epochs which gave us an average accuracy (taken up to 5 tries) of 94.48%. Similarly, the same data was fed to the backpropagation +momentum model which contained

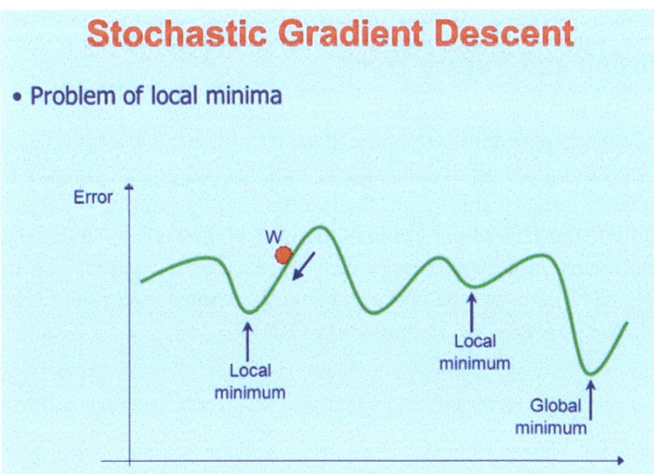

Fig. 4 Graph showing the position of local minimum with respect to global minimum

a momentum factor initialized to 1 and a learning rate to 1 because this combination gave the best accuracy of 93.69% in a 57-12-1 ANN model. After feeding the data to this model the code was trained up to 300 epochs and then was tested and we got an accuracy of 95.38%.

Similarly, another model (57-12-1) was designed which consisted of 57 inputs, 12 nodes in the first hidden layer and 1 node in the output layer. For the backpropagation model, we got an accuracy of 93.10% for 500 epochs. For the backpropagation + momentum model, we got an accuracy of 91.82% for 200 epochs and accuracy of 93.35% for 300 epochs.

6.4 Shortcomings of BP Over BP−M

The backpropagation algorithm is generally slow because to achieve stable learning it requires small learning rates whereas BP−M is usually faster as it needs higher learning rate while maintaining stability. But then also BP−M is slow for many practical applications. Another shortcoming is that, depending on the initial starting conditions it is possible for the network solution to get trapped in one of the local minima as gradient descent method is performed on the error surface. The local minimum may be good or bad depending on how close the local minimum is to the global minimum as well as how low the error is needed. And also BP will not always find the accurate weights for optimal solution. We may have to reinitialize the network as well as re-train number of times to achieve or we can say, guarantee the best solution. Although BP−M gives compatible results in less number of epochs but both BP and BP−M are trapped in local minimum as they both are gradient methods.

7 Conclusion and Future Work

From the above experiment we infer that BP+M, i.e., backpropagation with momentum gives compatible results in less number of epochs. BP and BP+M are not able to give optimal result or in other words they are trapped in local minima. This is due to BP and BP+M are gradient method. Hence, we wish to explore some popular evolutionary optimisation approach. In place of BP and BP+M, techniques like PSO, DE, SGO, etc. can be applied. From the above experiment we infer that BP+M is able to converge faster compared to BP.

References

1. Erosheva, E.A., Fienberg, S.E.: Bayesian Mixed Membership Models for Soft Clustering and Classification. Studies in Classification, Data Analysis, and Knowledge Organization Classification —the Ubiquitous Challenge, pp. 11–26 (2005)
2. Cortez, P., Lopes, C., Sousa, P., Rocha, M., Rio, M.: Symbiotic data mining for personalized spam filtering. In: 2009 IEEE/WIC/ACM International Joint Conference on Web Intelligence and Intelligent Agent Technology (2009)
3. Yang, Y., Elfayoumy, S.: Anti-spam filtering using neural networks and bayesian classifiers. In: 2007 International Symposium on Computational Intelligence in Robotics and Automation (2007)
4. Dua, D., Karra Taniskidou, E.: "Spambase", UCI Machine Learning Repository: Spambase Data Set. https://archive.ics.uci.edu/ml/datasets/spambase. Last accessed: 29 Apr 2018
5. Ioffe, S., Szegedy, C.: Batch normalization: accelerating deep network training by reducing internal covariate shift (2015)

Automated Synthesis of Memristor Crossbars Using Deep Neural Networks

Dwaipayan Chakraborty, Andy Michel, Jodh S. Pannu, Sunny Raj, Suresh Chandra Satapathy, Steven L. Fernandes, and Sumit K. Jha

Abstract We present a machine learning based approach for automatically synthesizing a memristor crossbar design from the specification of a Boolean formula. In particular, our approach employs deep neural networks to explore the design space of crossbar circuits and conjecture the design of an approximately correct crossbar. Then, we employ simulated annealing to obtain the correct crossbar design from the approximately correct design. Our experimental investigations show that the deep learning system is able to prune the search space to less than 0.0000011% of the original search space with high probability; thereby, making it easier for the simulated annealing algorithm to identify a correct crossbar design. We automatically design an adder, subtractor, comparator, and parity circuit using this combination of deep learning and simulated annealing, and demonstrate their correctness using circuit simulations. We also compare our approach to vanilla simulated annealing without the deep learning component, and show that our approach needs only 6.08% to

D. Chakraborty
Oak Ridge National Laboratory, Oak Ridge, TN 37830, USA
e-mail: chakrabortyd@ornl.gov

A. Michel · J. S. Pannu · S. Raj · S. L. Fernandes (✉) · S. K. Jha
University of Central Florida, Orlando, FL 32816, USA
e-mail: steven@cs.ucf.edu

A. Michel
e-mail: andymichel@cs.ucf.edu

J. S. Pannu
e-mail: jodh@cs.ucf.edu

S. Raj
e-mail: sraj@cs.ucf.edu

S. K. Jha
e-mail: jha@cs.ucf.edu

S. C. Satapathy
Kalinga Institute of Industrial Technology, Odisha 751024, India
e-mail: sureshsatapathy@ieee.org

S. C. Satapathy et al. (eds.), *Intelligent Data Engineering and Analytics*, Advances in Intelligent Systems and Computing 1177,
https://doi.org/10.1007/978-981-15-5679-1_32

69.22% of the number of circuit simulation queries required by simulated annealing alone.

Keywords Memristors · Crossbars · Deep learning

1 Introduction

A new wave of artificial intelligence is rapidly changing the landscape of computer science—from computer vision [1] and natural language processing [2] to control systems [3] and robotics [4]. Driven by deep learning algorithms [5] running on compute-class graphics processing units (GPUs) with thousands of processor cores, these learning systems have outperformed human beings in a number of tasks and games that are considered intellectually challenging [6].

While some recent efforts [7–10] have focused on employing simulated annealing and other classical AI-based search methods to automatically design memristor crossbars for implementing Boolean functions, little work has been done on applying deep neural networks for transforming specifications of Boolean computations onto designs of memristor circuits. There is a huge gap between the classical AI methods employed in memristor circuit design and the modern deep learning methods being rapidly deployed in other settings, such as AlphaGo [6] and self-driving cars [11]. Overview of our proposed approach is shown in Fig. 1.

In this paper, we make the following contributions:

1. We demonstrate how deep neural networks can be used to automatically prune the search space of memristor circuits to a small fraction of the original space of all possible circuits. In our experiments, we are able to prune the search space to less than 0.0000011% of the original search space with high probability.
2. We show how a variant of simulated annealing can be used to effectively search the pruned design space, and automatically generate the correct design for a

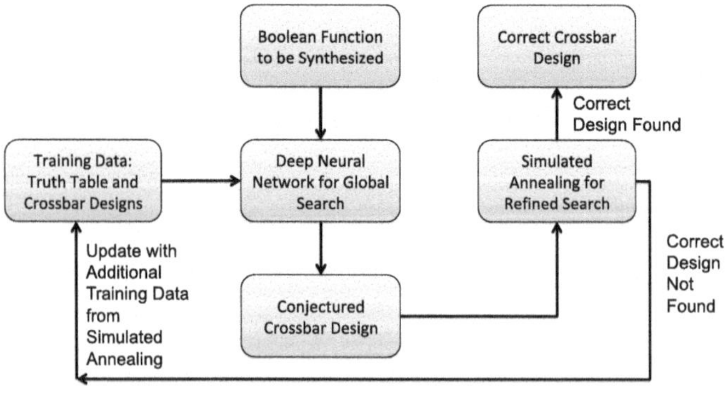

Fig. 1 Overview of our proposal approach

given Boolean formula specification. A pictorial representation of our deep neural network architecture is given in Fig. 2.

3. We establish the correctness of our approach by automatically synthesizing the design of the most significant bit of an adder, subtractor, comparator, and parity circuit, and analyzing its behavior using circuit simulations.

2 Related Work

Over the last decade, a suite of memristor-based logic design methodologies have been proposed [12–18]. The authors of [19] have proposed an end-to-end VLIW architecture based on RRAM switches. An overview of existing memristive logic families is presented in [20]. Significant efforts have also been applied toward implementing neuromorphic computing on memristor crossbars [21–24]. Such approaches often rely on integration with CMOS devices [25–27].

However, our effort is directed toward the use of deep learning for synthesizing memristor crossbars that implement a desired logical formula. Flow-based computing enables the use of data stored in non-volatile memristor crossbars to implement Boolean formulae.

In this approach, the data on which a Boolean logical computation to be performed is loaded on the crossbar in a manner such that the flow of current via sneak paths through the crossbar reaches an output nanowire from an input nanowire if and only if the Boolean formula evaluates to *True*.

Several approaches including those based on reduced ordered Boolean decision diagrams [28] and free Boolean decision diagrams [29] have been used to design flow-based crossbar computing circuits. Automated synthesis [30] via satisfiability modulo theory (SMT) and AI-based search algorithms [7] such as simulated annealing have been used to create designs of memristor crossbars for implementing Boolean formulae.

However, to the best of our knowledge, deep neural networks have not been used to aid in the design of nanoscale memristor crossbars for implementing Boolean computations.

3 Our Approach

First, we employ message passing interface (MPI) based parallel computing to generate random memristor crossbar designs and the Boolean formula is computed by these flow-based computing designs. Since it is easier to compute the Boolean formula implemented by a given flow-based memristor crossbar design (compared to the inverse problem), we can generate hundreds of such designs per hour as training data. The ability to generate massive training dataset automatically facilitates the next step involving deep learning.

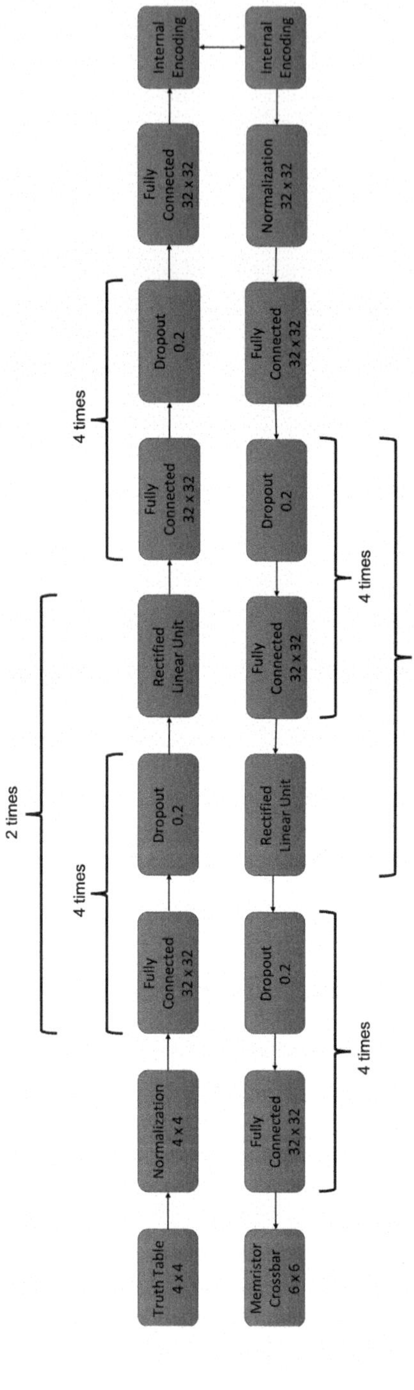

Fig. 2 Overall architecture of our deep neural network mapping truth tables to crossbars

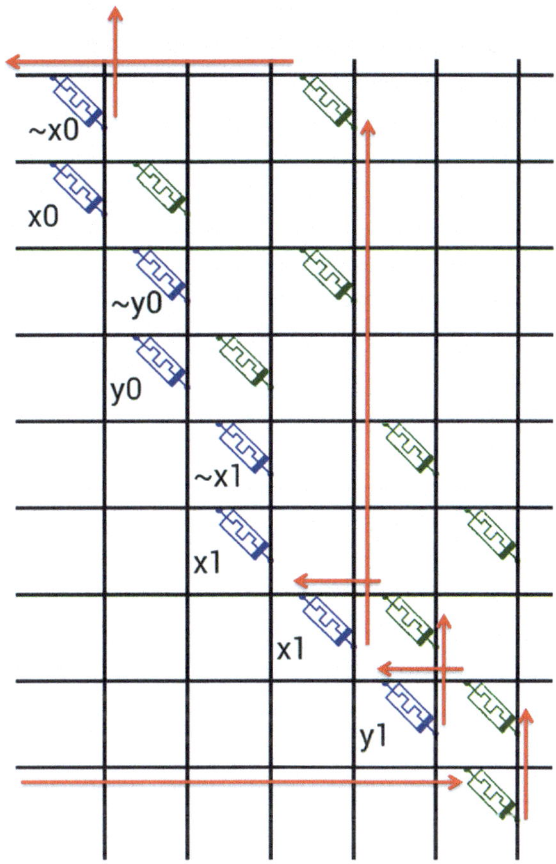

Fig. 3 Memristor crossbar design for computing the carry bit of a 2-bit adder, as presented in [7]

Figure 3 shows how they carry bit of a two-bit adder that can be computed by a crossbar of 9 rows and 6 columns. In particular, the flow of current through the crossbar is shown using red arrows for the input $x_1 = 1, y_1 = 1, x_0 = 0$, and $y_0 - 0/1$.

Second, we train a deep neural network involving an encoder–decoder pair to map a Boolean formula to its flow-based memristor crossbar design. The input to our neural network is a pictorial representation of the truth table of the Boolean formula to be implemented. Then, a 28-layered neural network consisting of fully-connected, drop-out, and rectified linear units (ReLU) layers encodes the Boolean formula into a linear encoding. Subsequently, another 28-layered neural network consisting of fully-connected, ReLU and drop-out layers decodes the linear encoding into a crossbar design.

Third, we employ the memristor crossbar design generated by the neural network as the starting point for a variant of simulated annealing. The crossbar designed by the deep neural network is structurally close to the correct crossbar design. For example, a memristor in a *random* crossbar design involving a Boolean formula with 4 bits (i.e.,

4 positive literals, 4 negative literals, *True* and *False*, or 10 values) is 5 substitutions away from the correct memristor value on average. However, a memristor in the crossbar designed by the deep neural network may only be 3 substitutions away from the correct value of the memristor.

Hence, our simulated annealing algorithm can exploit this information by assigning a value to each memristor using a two-pronged approach: (i) probabilistic assignment of a memristor value in an interval with its mean being the value predicted by the deep neural network and the end-points of the interval being the mean square error of the predicted values by the deep neural network; and (ii) probabilistic assignment of a random memristor value.

Either our variant of simulated annealing obtains the correct design or it explores the neighborhood of the design produced by the deep neural network and fails to produce a correct design. Then, we update these crossbar designs and the corresponding Boolean formula is being computed to the training set of deep neural networks. Then, the deep neural network is queried again to produce the crossbar design corresponding to the specified Boolean formula. The additional information produced by simulated annealing aids the deep learning algorithm to conjecture a better crossbar design.

4 Experimental Results

4.1 Performance Comparisons

We first compare our approach based on conjecturing a crossbar via a deep neural network and then employing a variant of simulated annealing on it with the performance of a vanilla simulated annealing approach. Our results are shown in Table 1 on the first page. A simulated annealing based search for the correct crossbar design starting at all memristors initialized with a random value queries 163, 185 designs to produce a correct 2-bit adder. On the other hand, our approach first uses a neural network to conjecture a crossbar design corresponding to the truth table of a 2-bit adder and then employs simulated annealing based search starting at this imagined crossbar design.

Table 1 Comparison between our deep learning based conjecture generation approach and simulated annealing

Example	# Design evaluations required		
	Simulated annealing	Conjecture via deep learning	Query reduction (%)
Adder	163,185	9,910	6.08
Subtractor	120,874	45,633	37.76
Comparator	7,930	5,489	69.22
Parity	575,698	378,026	65.67

In our search, we choose memristor values that are within 3 values of the imagined crossbar design with 50% probability and choose any other value with the remaining 50% probability. The search seeks to identify crossbar designs that are closer to the ones conjectured based on deep learning. This approach produces a correct crossbar design with as few as 9, 910 queries to the design simulator—this corresponds to only 6.08 % of the original number of queries made by simulated annealing alone.

Our deep learning based system for conjecturing the crossbar design of a truth table coupled to a simulated annealing algorithm also performs well on designs of subtractors, comparators, and parity circuits requiring only 37.76%, 69.22%, and 65.67% of the original number of queries to the design simulator, respectively. These designs are verified using circuit simulations in Sect. 4.3.

4.2 Evaluation of Deep Learning

We employed two NVIDIA Tesla V100 GPUs to train a 56-layered autoencoder network with 5,000 pairs of crossbar designs and truth tables, the losses for which are presented in Fig. 4. Our deep learning system predicts a crossbar design of 6 rows and 6 columns from a truth table with 16 entries. Our experiments show that the crossbar designed by our deep neural network is structurally close to the correct crossbar design. For example, a memristor in a *random* crossbar design involving a Boolean formula with 4 bits (i.e., 4 positive literals, 4 negative literals, *True* and

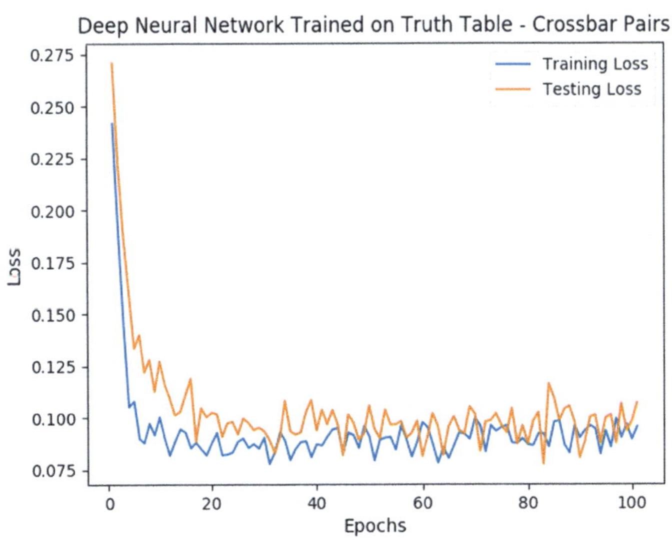

Fig. 4 The loss of the deep neural network as a function of the number of training epochs. Both the training and the testing losses, computed as the mean squared error, become smaller indicating that the model is not overfitted

False, or 10 values) is 5 away from the correct memristor value on average. However, a memristor in the crossbar designed by the deep neural network is only 3 units away from the correct value of the memristor.

Our examples involve crossbars with 6 rows and 6 columns with each memristor having 10 possible values; hence, a random search involves 10^{36} possible designs. However, a search involving errors of 3 units with high probability leads to only 6^{36} possible designs—this corresponds to less than 0.0000011% of the original search space. Even a 15–20% change in the ratio of the turned-on to turn-off resistance of the memristors has a little perceptible impact on the output produced by our automatically synthesized designs. Hence, these designs are robust to noise in the resistance of the memristor. Table 1 shows the performance comparison between simulated annealing and our approach combining deep learning with annealing.

4.3 Designs and Circuit Simulations

The flow-based memristor crossbar circuits generated by our deep learning based conjecture generation coupled to a simulated annealing approach are illustrated in Fig. 5. Each design was verified using circuit simulations for all possible input values with R_{ON}/R_{OFF} ratio of 10^4. For the sake of brevity, we are omitting the truth tables for subtractor and parity. Figure 6 shows the impact of memristor drift on the correctness of our automatically synthesized flow-based memristor crossbar computing designs.

5 Conclusion and Future Work

This paper presents a new machine learning approach to the automated synthesis of memristor crossbars. Our approach leverages deep neural networks for conjecturing a crossbar design given a truth table for a Boolean formula, and then employs a variant of simulated annealing to synthesize the correct memristor crossbar design. Our variant of simulated annealing approach seeded by the design conjectured via a deep neural network needs as few as 6.08–69.22% of the number of samples explored. We verify the synthesized designs using circuit simulations and demonstrate that the designs remain correct even when the ratios of the turned-off to turned-on resistance drift by as much as 15–20%.

Our work is a preliminary effort at understanding the use of deep learning systems running on multiple GPUs to synthesize memristor crossbars from Boolean specifications. See Table 2 for the inputs provided to our 2-bit adder and observed outputs. Table 3 shows the observed truth table for a 2-bit comparator. Hence, several directions for future research remain open. First, we have determined the architecture of a deep neural network that can learn the mapping from a truth table to a flow-based crossbar computing design. One needs to investigate if there are other neural network architectures that achieve similar or better performance. Second, our search

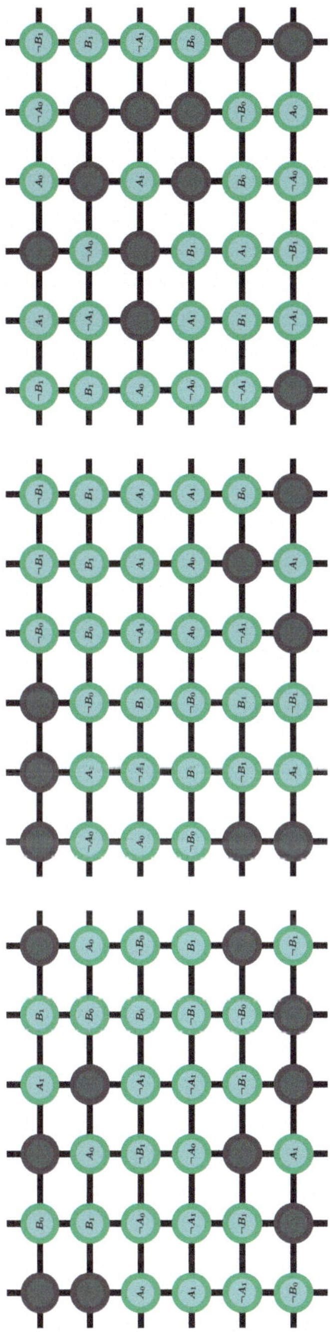

(a) Second sum bit of 2-bit addition

(b) 2-bit comparator

(c) 4-bit odd parity checking

Fig. 5 Crossbar designs synthesized using our deep learning based conjecture generation and simulated annealing

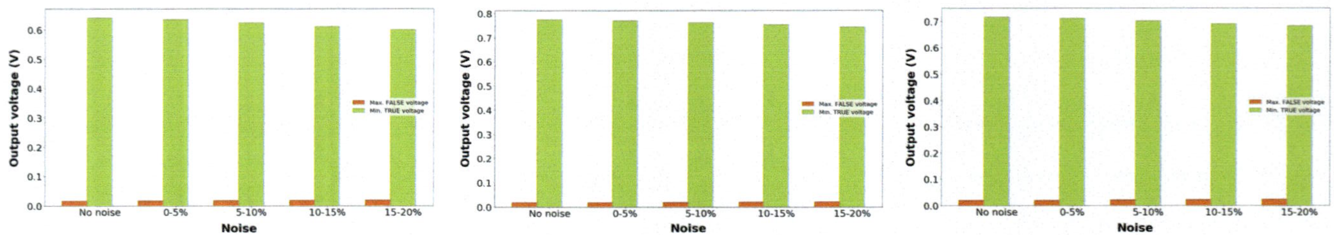

(a) 2-bit binary addition (b) 2-bit binary comparison (c) 4-bit odd parity checking

Fig. 6 Variation of output voltages with R_{ON}/R_{OFF} variance Variation of output voltages with R_{ON}/R_{OFF} variance

Table 2 Second sum bit of 2-bit addition

A_1	A_0	B_1	B_0	S_1	Output voltage (V)
0	0	0	0	0	0.00596
0	0	0	1	0	0.00989
0	0	1	0	1	0.7146
0	0	1	1	0	0.00586
0	1	0	0	0	0.00596
0	1	0	1	1	0.6821
0	1	1	0	1	0.7147
0	1	1	1	0	0.00596
1	0	0	0	1	0.7357
1	0	0	1	1	0.7412
1	0	1	0	0	0.0157
1	0	1	1	0	0.0162
1	1	0	0	1	0.6394
1	1	0	1	0	0.0157
1	1	1	0	0	0.0157
1	1	1	1	1	0.7721

Table 3 Second sum bit of 2-bit addition

A_1	A_0	B_1	B_0	S_1	Output voltage (V)
0	0	0	0	0	0.01764
0	0	0	1	0	0.00972
0	0	1	0	0	0.00596
0	0	1	1	0	0.00299
0	1	0	0	1	0.7738
0	1	0	1	0	0.00989
0	1	1	0	0	0.00596
0	1	1	1	0	0.00299
1	0	0	0	1	0.8905
1	0	0	1	1	0.885
1	0	1	0	0	0.00989
1	0	1	1	0	0.00596
1	1	0	0	1	0.8999
1	1	0	1	1	0.8905
1	1	1	0	1	0.7797
1	1	1	1	0	0.00596

specification to the deep neural network accepts a truth table as an input. The size of a truth table can be exponential in the number of variables. One needs to investigate if graph-based symbolic representations like decision diagrams can be used to represent the input to the deep learning system.

Based on our initial success with deep neural networks for the design of memristor crossbar circuits, we anticipate that deep learning approaches will serve as important and orthogonal methods complementary to the robust set of tools and algorithms are already deployed in the computer-aided design of circuits.

References

1. Szegedy, C., Liu, W., Jia, Y., Sermanet, P., Reed, S., Anguelov, D., Erhan, D., Vanhoucke, V., Rabinovich, A.: Going deeper with convolutions. In: Proceedings of the IEEE Conference on Computer Vision and Pattern Recognition, pp. 1–9 (2015)
2. Blunsom, P., Cho, K., Dyer, C., Schütze, H.: From characters to understanding natural language (c2nlu): robust end-to-end deep learning for NLP (dagstuhl seminar 17042). In: Dagstuhl Reports, vol. 7, no. 1. Schloss Dagstuhl-Leibniz-Zentrum fuer Informatik (2017)
3. Andersson, O., Wzorek, M., Doherty, P.: Deep learning quadcopter control via risk-aware active learning. In: AAAI, pp. 3812–3818 (2017)
4. Sünderhauf, N., Brock, O., Scheirer, W., Hadsell, R., Fox, D., Leitner, J., Upcroft, B., Abbeel, P., Burgard, W., Milford, M., et al.: The limits and potentials of deep learning for robotics. Int. J. Robot. Res. **37**(4–5), 405–420 (2018)
5. LeCun, Y., Bengio, Y., Hinton, G.: Deep learning. Nature **521**(7553), 436 (2015)
6. Gibney, E.: Google ai algorithm masters ancient game of go. Nat. News **529**(7587), 445 (2016)
7. Chakraborty, D., Jha, S.K.: Automated synthesis of compact crossbars for sneak-path based in-memory computing. In: Design, Automation & Test in Europe Conference & Exhibition (DATE), pp. 770–775. IEEE (2017)
8. Thangkhiew, P.L., Zulehner, A., Wille, R., Datta, K., Sengupta, I.: An efficient memristor crossbar architecture for mapping Boolean functions using binary decision diagrams (BDD). Integration (2019). http://www.sciencedirect.com/science/article/pii/S0167926019301646
9. Xie, L.: Mosaic: an automated synthesis flow for Boolean logic based on memristor crossbar. In: Proceedings of the 24th Asia and South Pacific Design Automation Conference, ser. ASPDAC 19. New York, NY, USA: Association for Computing Machinery, pp. 432–437 (2019). https://doi.org/10.1145/3287624.3287702
10. Pannu, J.S., Raj, S., Fernandes, S.L., Jha, S.K., Chakraborty, D., Rafiq, S., Cady, N.: Data-driven approximate edge detection using flow-based computing on memristor crossbars. In: 2019 IEEE Albany Nanotechnology Symposium (ANS), Nov 2019, pp. 1–6
11. Falcini, F., Lami, G., Costanza, A.M.: Deep learning in automotive software. IEEE Softw. **3**, 56–63 (2017)
12. Kvatinsky, S., Belousov, D., Liman, S., Satat, G., Wald, N., Friedman, E.G., Kolodny, A., Weiser, U.C.: Magic memristor-aided logic. IEEE Transa. Circuits Syst. II: Express Briefs **61**(11), 895–899 (2014)
13. Kvatinsky, S., Satat, G., Wald, N., Friedman, E.G., Kolodny, A., Weiser, U.C.: Memristor-based material implication (imply) logic: design principles and methodologies. IEEE Trans. Very Large Scale Integr. (VLSI) Syst. **22**(10), 2054–2066 (2014)
14. Lehtonen, E., Laiho, M.: Stateful implication logic with memristors. In: Proceedings of the 2009 IEEE/ACM International Symposium on Nanoscale Architectures, ser. NANOARCH '09. Washington, DC, USA: IEEE Computer Society, pp. 33–36. https://doi.org/10.1109/NANOARCH.2009.5226356 (2009)

15. Prezioso, M., Riminucci, A., Graziosi, P., Bergenti, I., Rakshit, R., Cecchini, R., Vianelli, A., Borgatti, F., Haag, N., Willis, M., Drew, A.J., Gillin, W.P., Dediu, V.A.: A single-device universal logic gate based on a magnetically enhanced memristor. Adv. Mater. **25**(4), 534–538. https://doi.org/10.1002/adma.201202031
16. Haj-Ali, A., Ben-Hur, R., Wald, N., Kvatinsky, S.: Efficient algorithms for in-memory fixed point multiplication using magic. In: IEEE International Symposium on Circuits and Systems (ISCAS), pp. 1–5. IEEE (2018)
17. Shirinzadeh, S., Drechsler, R., Logic synthesis for in-memory computing using resistive memories. In: IEEE Computer Society Annual Symposium on VLSI (ISVLSI), pp. 375–380. IEEE (2018)
18. Vatwani, T., Dutt, A., Bhattacharjee, D., Chattopadhyay, A.: Floating point multiplication mapping on reram based in-memory computing architecture. In: 2018 31st International Conference on VLSI Design and 2018 17th International Conference on Embedded Systems (VLSID), pp. 439–444. IEEE (2018)
19. Bhattacharjee, D., Devadoss, R., Chattopadhyay, A., Revamp: Reram based VLIW architecture for in-memory computing. In: Design, Automation & Test in Europe Conference & Exhibition (DATE), pp. 782–787. IEEE (2017)
20. Reuben, J., Ben-Hur, R., Wald, N., Talati, N., Ali, A.H., Gaillardon, P.-E., Kvatinsky, S.: Memristive logic: a framework for evaluation and comparison. In: 27th International Symposium on Power and Timing Modeling, Optimization and Simulation (PATMOS), pp. 1–8. IEEE (2017)
21. Jo, S.H., Chang, T., Ebong, I., Bhadviya, B.B., Mazumder, P., Lu, W.: Nanoscale memristor device as synapse in neuromorphic systems. Nano Lett. **10**(4), 1297–1301 (2010)
22. Hu, M., Li, H., Chen, Y., Wu, Q., Rose, G.S., Linderman, R.W.: Memristor crossbar-based neuromorphic computing system: a case study. IEEE Trans. Neural Netw. Learn. Syst. **25**(10), 1864–1878 (2014)
23. Prezioso, M., Merrikh-Bayat, F., Hoskins, B., Adam, G., Likharev, K.K., Strukov, D.B.: Training and operation of an integrated neuromorphic network based on metal-oxide memristors. Nature **521**(7550), 61 (2015)
24. Schuller, I.K., Stevens, R., Pino, R., Pechan, M.: Neuromorphic computing-from materials research to systems architecture roundtable. Technical Report, USDOE Office of Science (SC)(United States) (2015)
25. Chu, M., Kim, B., Park, S., Hwang, H., Jeon, M., Lee, B.H., Lee, B.-G.: Neuromorphic hardware system for visual pattern recognition with memristor array and cmos neuron. IEEE Trans. Ind. Electron. **62**(4), 2410–2419 (2015)
26. Kim, K.-H., Gaba, S., Wheeler, D., Cruz-Albrecht, J.M., Hussain, T., Srinivasa, N., Lu, W.: A functional hybrid memristor crossbar-array/cmos system for data storage and neuromorphic applications. Nano Lett. **12**(1), 389–395 (2011)
27. Serrano-Gotarredona, T., Prodromakis, T., Linares-Barranco, B.: A proposal for hybrid memristor-cmos spiking neuromorphic learning systems. IEEE Circuits Syst. Mag. **13**(2), 74–88 (2013)
28. Chakraborti, S., Chowdhary, P.V., Datta, K., Sengupta, I.: BDD based synthesis of Boolean functions using memristors. In: 2014 9th International Design & Test Symposium (IDT), vol. 00, Dec. 2014, pp. 136–141. https://doi.org/10.1109/IDT.2014.7038601
29. Hassen, A.U., Chakraborty, D., Jha, S.K.: Free binary decision diagram-based synthesis of compact crossbars for in-memory computing. IEEE Trans. Circuits Syst. II: Express Briefs **65**(5), 622–626 (2018)
30. Alamgir, Z., Beckmann, K., Cady, N., Velasquez, A., Jha, S.K.: Flow-based computing on nanoscale crossbars: design and implementation of full adders. In: IEEE International Symposium on Circuits and Systems (ISCAS), pp. 1870–1873. IEEE (2016)

Training Time Reduction in Transfer Learning for a Similar Dataset Using Deep Learning

Ekansh Gayakwad, J. Prabhu, R. Vijay Anand, and M. Sandeep Kumar

Abstract Training deep neural networks take a lot of time and computation. In this paper, we have discussed how we can reduce the training time for deep learning models if we have already trained a model for a similar dataset. The basic logic here is that for similar dataset the features stored in the deep neural net are similar and the only difference comes for the classification layers of deep neural so instead of training the whole net we just train the last layers for classifying the data and use the trained model weights on the rest of the layers; this method saves a lot of time.

Keywords Deep learning · Machine learning · Natural language · Transfer learning model

1 Introduction

Transfer learning is a machine learning method where a model developed for a task is reused as the starting point for a model on a second task [1–3]. It is a popular approach in deep learning where pre-trained models are used as the starting point on computer vision and natural language processing tasks given the vast compute and time resources required to develop neural network models on these problems and from the huge jumps in the skill that they provide on related problems [4, 5].

E. Gayakwad · J. Prabhu (✉) · R. V. Anand · M. S. Kumar
Vellore Institute of TechnologyVellore Institute of Technology, Vellore 632014, Tamil Nadu, India
e-mail: j.prabhu@vit.ac.in

E. Gayakwad
e-mail: ekansh.gayakwad2016@vitstudent.ac.in

R. V. Anand
e-mail: vijayanand.r@vit.ac.in

M. S. Kumar
e-mail: sandeepkumarm322@gmail.com

S. C. Satapathy et al. (eds.), *Intelligent Data Engineering and Analytics*, Advances in Intelligent Systems and Computing 1177,
https://doi.org/10.1007/978-981-15-5679-1_33

Transfer learning is good for similar dataset as the features are similar, just the classification part is different, but transfer learning can also be applied for the data which is not similar, for example, a good melanoma cancer predictor was built over inception v3 [6, 7] which is trained for ImageNet dataset which does not have an image similar to that of derma cancer [8].

Transfer learning is also very beneficial in Natural language processing as the textual data cannot be easily classified, mostly expert knowledge is required to create large labeled datasets. However, here we are focusing on transfer learning using image datasets.

2 Deep Learning

Transfer learning is an important machine learning method for solving the fundamental problem of inadequate training information. It helps to transfer the information from the source domain to the target domain by removing the presumption of the learning data and test data. This will have a great positive impact on many fields that are challenging to improve due to inadequate training [9].

The learning process of transfer learning is described elaborately in Fig. 1.

(a) **Transfer learning**

With learning task T_t based on D_t and we can get assistance from D_a for the learning task T_s.

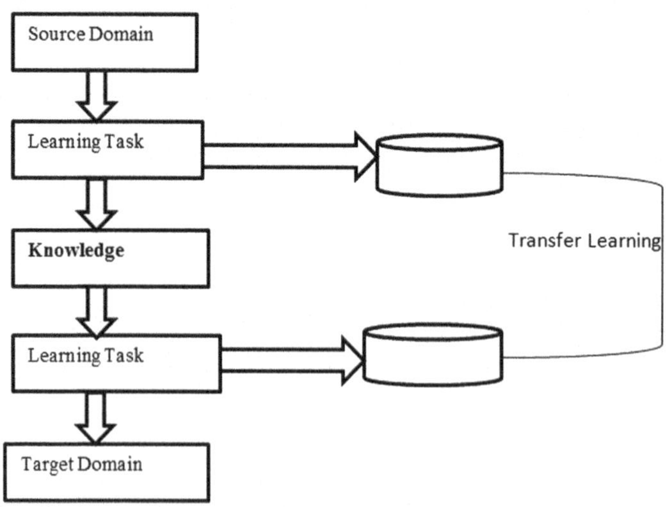

Fig. 1 Learning process of transfer learning

Transfer learning task work toward optimization to enhance the performance of predictive functions $f_Y(.)$ for learning task T_t by exploring and passing latent information from D_s and T_s.

(b) **Deep transfer learning**

According to the transfer learning task, it is defined based on $\{D_s, T_s, D_s, T_t, f_Y(.)\}$ where $f_Y(.)$ is a non-linear function that illustrates a deep neural network.

3 Major Categories of Deep Learning

Deep transfer learning investigates how deep neural networks might use expertise from several other areas. Since deep neural networks are becoming popular in various fields, it has already implied with a substantial amount of deep transfer learning methods that categorize and illustrate themselves as quite important. The premise of a technique used in deep transfer learning is categorized into four, namely instance-based deep transfer learning, mapping–based deep transfer learning, adversarial-based deep transfer learning, and network-based deep transfer learning which have been described in Table 1.

(a) **Advantage of transfer learning**

There are many benefits over time and energy savings in the use of transfer learning. A key advantage of your problem domain is the availability of an appropriate marked training set. If there is insufficient training information, an existing model (from a related problem domain) ith additional training can be used to support the new problem domain.

In feature transfer, a deep learning model provides extraction and classification functionality with a smaller topology of the neural network. Usually, output commonly varies in b/w two problems, depending on the problem domain. For this purpose, for the new problem domain, the classification layer is usually substituted and rebuilt. It needs substantially few resources to train and test while using the pre-trained extraction functions of a pipeline.

Table 1 Categories of deep learning

Approaches	Description
Instances-based	Use the appropriate weight for instance in the source domain
Mapping-based	Mapping instances for improved similarities from two domains into a data space
Network-based	Reuse the pre-trained part of the network in the origin domain
Adversarial-based	Utilize adversarial technology to identify the transferable features that are suitable for two fields

(b) **Challenges of transfer learning**

The concepts behind transfer learning are not new, and it has the potential to reduce the research needed to build complex neural networks in deep learning. Negative transition is one of the earliest problems found in transfer learning. Negative transfer leads to a decrease in the reliability of a deep learning model after retraining.

It can be influenced by strong problematic model dissimilarity or the model's inability to compensate for both the dataset of the new field (beside the new data itself). It has evolved to strategize experimentally recognizing similarity of problematic domains in order to know and better understand about an opportunity for negative transfer to learn both the transfer of feasibility among domains.

4 Model Architectures

We have created TWO different neural nets to classify the dog breeds: the first one is a basic convolutional neural network and the second one is transfer learning on the exception model [10] The description of the deep convolutional neural net, the first layer is a 2d convolutional layer with 16 filters and a kernel size of 3 × 3 the input shape of the network is (299,299,3) which is the size of the image. The first layer uses relu as the activation, also every convolutional layer used in this network uses padding [8]. The second layer is a 2d convolutional layer with 32 filters and a kernel size of 3 × 3 with relu activation which is followed by a max-pooling layer that has a 2 × 2 kernel, and the stride of the max pool filter is 2. The third layer is a dropout with a probability parameter equal to 0.4 just to reduce the chances for overfitting. The next layer is the convolutional layer with 64 filters and 3 × 3 kernel and relu activator, the next layers are the same as the previous one but with 128 and 256 filters. The next layer is the max-pooling layer with a kernel size of 2 × 2 and the stride is 2. Again, a dropout layer is introduced to reduce the risk of overfitting.

To flatten the outputs a Global Average pooling layer is introduced which makes all the feature linear followed by a dense layer of 128 nodes and relu activator. The output layer or the last layer is a dense layer with 120 nodes and has a softmax activator to get the probability of the classes (breeds). The total trainable parameters of the model are equal to 442,661.

The description of the second neural net: the second neural network is a transfer learning model built over the exception model as the dataset is similar to the dataset on which exception was trained (i.e., image-net) [11]. The dog breed data used as the primary dataset is a subset of image-net. For the model, we have used exactly the same layers and the same weights of exception but with a little change. The last layer of exception model is a dense layer with 1000 nodes, and to give the probability for the classes in image-net we removed this last layer and added a dense layer with 120 nodes and softmax activation which is the output layer for the model and the model can now predict the dog breed dataset.

The total number of trainable parameters for the model for all layers is 21052832 and the total number of trainable parameter of actual exception models are 22855952. The model for the data dog dataset has 1803120 parameters less than the actual inception model. But for this model, we only have to train the last layer that is the dense layer with 120 nodes, therefore, the total number of trainable parameters becomes equal to 245880.

(a) **Dataset**

The dataset is Stanford dog dataset; it consists of 757 MB and 20580 images of dogs with 120 breeds. Dataset was built over the ImageNet dataset. The size of the images in the dataset was converted into (299, 299, and 3) for making a standard input for the neural network. All the pixels in the dataset were normalized by dividing it with 255. For working on the classifiers the dataset was split into training testing and cross-validation data. The training data was a total of 60% of the data and the testing data and cross-validation data both were 20% of the data which makes the total number of images in training data equal to 12348 and the number of images in testing and cross-validation data equal to 4116 each. For both the models same training, testing, and cross-validation data were used.

(b) **Training the models**

The first model was trained normally, for the compilation of the model the optimizer used was rms prop and the loss was categorical cross-entropy. The model was trained up to 100 epochs on training data using the cross-validation data to save the best weights the model yielded the testing accuracy of 28.63%. The time taken for each epoch is 88 s.

The second model is a transfer learning application on a similar dataset and we only have to train the last layer to get the output so there is no need for updating the weights for all the layers of the model. So we froze all the layers so that their weights do not get updated except the last dense layer with 120 nodes. After freezing the layers and training the model the time taken by a model to run one epoch was 12790 s which is a very long time. To reduce the training time we used the method discussed and updated the weights the time taken to save the output of layer preceding to the last dense layer with 120 nodes which was 21318 s (for all of the data including the testing data). Then a new model was created with input the same as the output of the layer preceding the last dense layer followed by the dense layer of 120 nodes. Then the new model was trained and the time per epoch was 13 s and in about 20 epochs the model yielded the testing accuracy of 84.45%.

(c) **Methodology used to reduce the training time**

The exception model is a pretty big non-sequential model and shows some great results over other models trained on image-net. The model takes weeks of time to train and we can use this model to use transfer learning for a similar or dissimilar dataset, in our case we have a similar dataset.

So we only have to manipulate some last layers of the mode to train our dataset [11–13].

We take the exception net and remove its last layer which is a fully connected dense layer of 1000 nodes with softmax activation; this layer was built for image-net dataset, after removing the last layer we add a fully connected dense layer with total number of nodes equal to number of classes of our target dataset, i.e., Stanford dog dataset which is equal to 120.

The activation of the last layer will again be softmax as we want the probability of the dog breed. The task is to train the model with Stanford dog dataset, but the main problem is that we do now want to update the weights of the entire network and also the network takes a lot of time to train. The only layer that has to train is the last dense layer with 120 nodes. So we create a new model and for that model we will generate the input dataset. Our exception model was twitched and the last layer was made for 120 nodes. We take the output of the layer prior to the dense layer, global average layer, and batch normal layer [14, 15] which is separable 2d convolution layer and the output of the layer is of dimension (10, 10, 2048), the reason behind getting the output of this layer is that we do not have to fit our data over the layers preceding to these layers again and again as every time we will get the same output as we are not updating the weights of these layers, so we save a lot of time in every epoch. Now we store the output of the separable 2d convolution layer.

We create a new model which will be the actual training model which will use the output of separable 2d convolution layer as input and will train the last dense layer of 120 nodes, and the new model has an input dimension of (10, 10, 2048) followed by a batch normal layer than by a global average pooling layer followed by the dense layer of 120 nodes and softmax activation. When training the model the input is the output of the separable 2d convolution layer and the output is the probability of dog breeds (Fig. 2).

The training plot of the model shows that the model achieved an accuracy of around 85% in the first 2 epochs then the model started overfitting (Fig. 3).

Summary of the newly created model with separable convolution layer output as input for the model, this model updates the transfer learning weights.

```
_________________________________________________________________
Layer (type)                 Output Shape              Param #
=================================================================
global_average_pooling2d_3 ( (None, 2048)              0
_________________________________________________________________
dense_4 (Dense)              (None, 133)               272517
=================================================================
Total params: 272,517
Trainable params: 272,517
Non-trainable params: 0
_________________________________________________________________
```

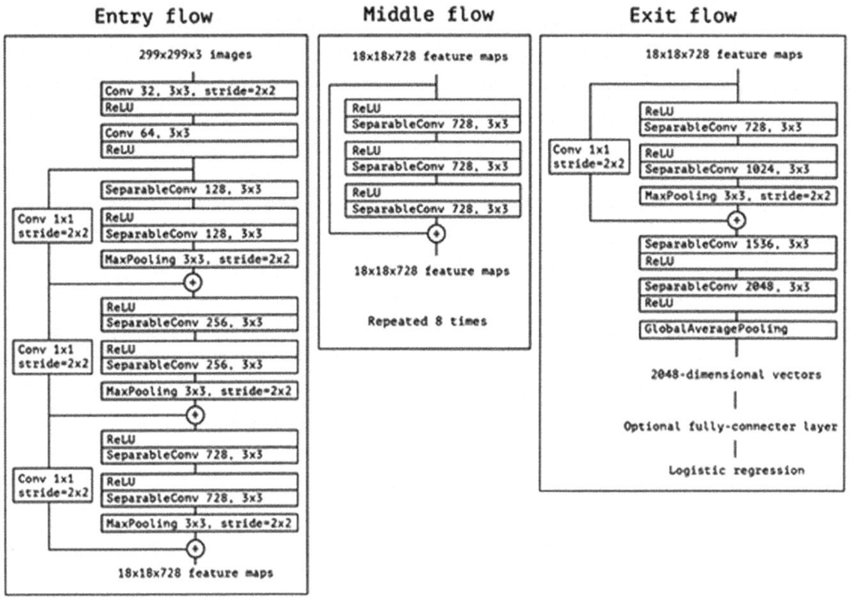

Fig. 2 Layers of the exception model

(d) **Results**

The first model which is the deep neural net gained an accuracy of 56% for the Stanford dog dataset. The time of training for the model in each epoch was approximately 25 s and the model was trained for 200 epochs but the best weights were saved around the 75th epoch as the model started overfitting the training data after that.

For the second model when the output of separable convolution layers was not stored and a new model was not created the time to train the model for one epoch was approx 12790 s but after saving the output of last separable convolution of the exception model and creating a new model the training time per epoch was approx. 13 s.

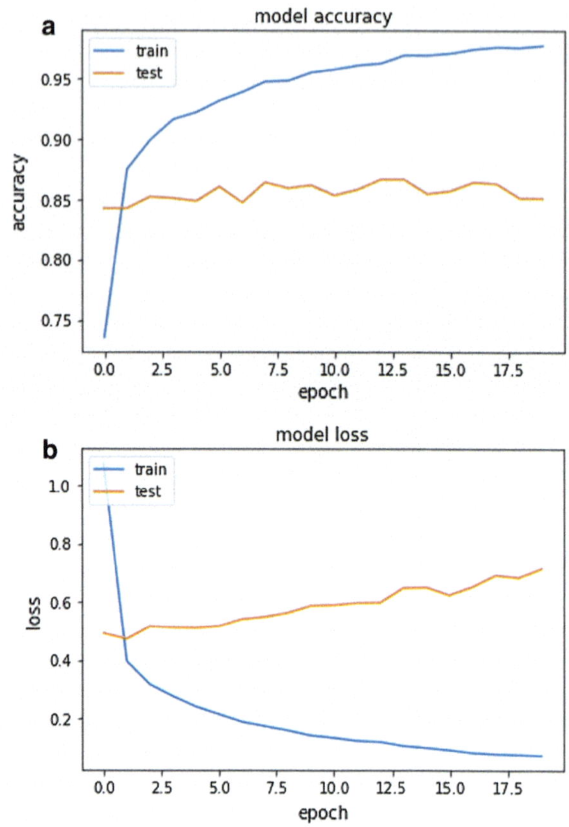

Fig. 3 **a** Model accuracy, **b** Model loss

5 Conclusion

The time taken by the transfer learning model to train is very short when we use the saved output of the last separable convolution layer and use it as input to the new model which classifies the dog data. As we are not updating the whole network weight but the weight of the last layers it is obvious not to pass data through the layers which weights will not be updated again and again so we save the output of the layer up to which the weight will not get updated.

References

1. Lu, J., Behbood, V., Hao, P., Zuo, H., Xue, S., Zhang, G.: Transfer learning using computational intelligence: a survey. Knowl. -Based Syst. **80**, 14–23 (2015)
2. Tan, C., Sun, F., Kong, T., Zhang, W., Yang, C., Liu, C.: A survey on deep transfer learning. In: International Conference on Artificial Neural Networks, pp. 270–279. Springer, Cham

(2018)

3. Sandeep, K.M., Prabhu, J.: Recent development in big data analytics: research perspective. In: Applications of Security, Mobile, Analytic, and Cloud (SMAC) Technologies for Effective Information Processing and Management. IGI Global, pp. 233–257 (2018)
4. Pan, S.J., Yang, Q.: A survey on transfer learning. IEEE Trans. Knowl. Data Eng. **22**(10), 1345–1359 (2009)
5. MR, P.K.: Role of sentiment classification in sentiment analysis: a survey. Ann. Libr. Inf. Stud. (ALIS) **65**(3), 196–209 (2018)
6. Torrey, L., Shavlik, J.: Transfer learning. In: Handbook of Research on Machine Learning Applications and Trends: Algorithms, Methods, and Techniques, pp. 242–264. IGI Global (2010)
7. Szegedy, C., Vanhoucke, V., Ioffe, S., Shlens, J., Wojna, Z.: Rethinking the inception architecture for computer vision. In: Proceedings of the IEEE Conference on Computer Vision and Pattern Recognition, pp. 2818–2826 (2016)
8. Surya, K., Gayakwad, E., Nallakaruppan, M.K.: Deep learning for short answer scoring. Int. J. Recent. Technol. Eng. (IJRTE) **7**(6) (2019)
9. Esteva, A., Kuprel, B., Novoa, R.A., Ko, J., Swetter, S.M., Blau, H.M., Thrun, S.: Dermatologist-level classification of skin cancer with deep neural networks. Nature **542**(7639), 115 (2017)
10. Chollet, F.: Xception: deep learning with depthwise separable convolutions. In: Proceedings of the IEEE Conference on Computer Vision and Pattern Recognition, pp. 1251–1258 (2017)
11. Xia, X., Xu, C., Nan, B.: Inception-v3 for flower classification. In: 2017 2nd International Conference on Image, Vision and Computing (ICIVC), pp. 783–787. IEEE (2017)
12. Ioffe, S., Szegedy, C.: Batch normalization: accelerating deep network training by reducing internal covariate shift (2015). arXiv:1502.03167
13. Kornblith, S., Shlens, J., Le, Q.V.: Do better imagenet models transfer better? In: Proceedings of the IEEE Conference on Computer Vision and Pattern Recognition, pp. 2661–2671 (2019)
14. Mun, S., Shon, S., Kim, W., Han, D. K., Ko, H.: Deep neural network based learning and transferring mid-level audio features for acoustic scene classification. In: 2017 IEEE International Conference on Acoustics, Speech and Signal Processing (ICASSP), pp. 796–800. IEEE (2017)
15. Mun, S., Shon, S., Kim, W., Ko, H.: Deep neural network bottleneck features for acoustic event recognition. In: Interspeech, pp. 2954–2957 (2016)

A Novel Model Object Oriented Approach to the Software Design

Rahul Yadav, Vikrant Singh, and J. Prabhu

Abstract This paper discusses the problems concerning several object-oriented approach developed by researchers, academies, designers and developers which has led to the use of various object-oriented methods for the software system development. The technique and approach used in various object-oriented designs lack in process model and do not include mechanisms necessary for user requirements, specifications, understanding ability and better identification with end user during software development process. These aspects are very important in software system design, where the user interaction with software is very high and significant. The software systems developed as per the user's requirements without proper approach lead to unsustainable, robust and of no use to the end user. Therefore, it is important for the designers and developers to make proper design model and then start the implementation process. This paper explains the existing object-oriented models and the problems faced by designers in design and implementation process and it also proposes a new technique. The proposed technique will provide a better object-oriented approach to different level of designers in software design.

Keywords Object-oriented business engineering · Software model · Responsibility-driven design · Object-oriented software engineering

R. Yadav · V. Singh · J. Prabhu (✉)
School of Information Technology and Engineering, Vellore Institute of Technology, Vellore 632014, Tamil Nadu, India
e-mail: jprabhuit@gmail.com

R. Yadav
e-mail: rahulveron@gmail.com

V. Singh
e-mail: vikrantpbh@gmail.com

S. C. Satapathy et al. (eds.), *Intelligent Data Engineering and Analytics*, Advances in Intelligent Systems and Computing 1177,
https://doi.org/10.1007/978-981-15-5679-1_34

1 Introduction

Previous papers have stated that user interface implementation is given more time than the application code implementation [1]. It is a design in the technical field which provides an interaction space between man-made machines and humans to provide an easier, efficient and user-friendly way to operate machines with lesser input to achieve the required output. Easy to use interface but includes complex development. User Interface involves analysis, design and implementation in development but several reasons explain its complexity in implementation, but the main reason is the difficulty in understanding the tasks performed by the user with the system and the features of different users handling the system [2]. Object-oriented user interface design involves mapping of objects and creating a relationship among them on the basis of analysis and design. Designer creates the design structure on the basis of constraints, user requirements which is not an easy process [3]. A new designer needs to learn techniques to process models, methods, principles to simply the object-oriented design. The model-based representation of an interface will make a relationship among different interface sections making it more explicit and can help to synthesize user requirements and needs more specifically and can over complex issues in object-oriented user interface approach. OOD is directly mapped to object-oriented programming language in order to increase its maintainability which will help in software modification to avoid and correct common faults, reusability of the design and to adapt to the changed environment [4]. This paper discusses part of object-oriented model-based approach for designing user interface by reviewing already discovered object-oriented process models and will discuss which model and approach are best suitable for user and designer on the basis of designer's level of experience and user requirements and demands on OOD (object-oriented design).

2 Related Work

Software development consists of three main components: system analysis, system design and finally the implementation part. Design process includes activities that are used to make a process model on the basis of user requirements, availability and specifications. Development of design model is based on the combination of judgment and intuition, principles, heuristics and process iteration which results in final design specifications [5]. Many OOD process models have been proposed and proven as a good approach to object-oriented user interface design. These models focus on basic components of object-oriented design development for designers with different levels of experience.

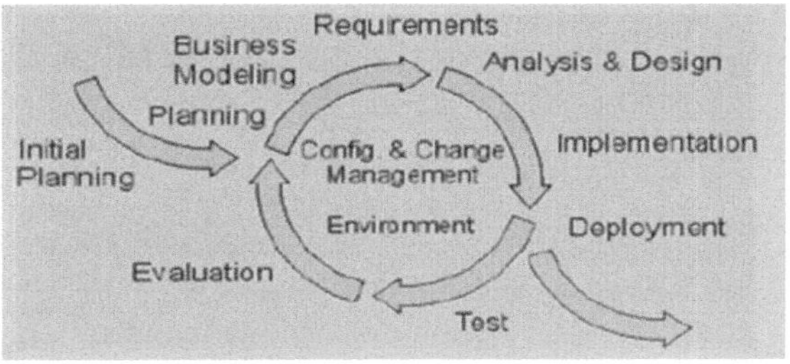

Fig. 1 Object-oriented analysis design

2.1 *Booch Methodology (OOA/OOD)*

Grady Booch [6] speaks about the importance of a person more than a process. Rational Unified Process (RUP) is a popular and effective software development process based on the idea of iterative development where iterations are time boxed and each iteration consists of analysis, design, requirements and implementation. RUP consists of four phases: inception, elaboration, construction and transition. Modelling Disciplines of RUP includes as shown in Fig. 1. OOAD is widely used by designers due to its effective management in software complexity and focus on data analysis rather than structured analysis. But due to limited functionality within objects and more design specific makes it costly and is also criticized for its large set of symbols.

Diagrams that OOAD is as follows:

Class diagram, Object diagram, State transition diagram, Module diagram, Process diagram, Interaction diagram.

Booch Methodology stated two methods:

- Macro development process—responsible for technical management of the system.
- Micro development process—responsible for day to day activities in the identification of classes and objects.

2.2 *Rumbaugh's Object Modelling Technique Methodology (OMT)*

OMT is widely used in object-oriented design approach. It consists of analysis phase, design phase and implementation phase, which are used for developing the object model and this model is used to develop object-oriented software. It is the most accepted technique used by developers as it eliminates the process of transforming one model to another. OMT involves a functional and intuitive approach which makes it demanding in various domains like transportation, telecommunication and compilers. The applications using OMT support full stack development [7]. Figure 2 explains the workflow.

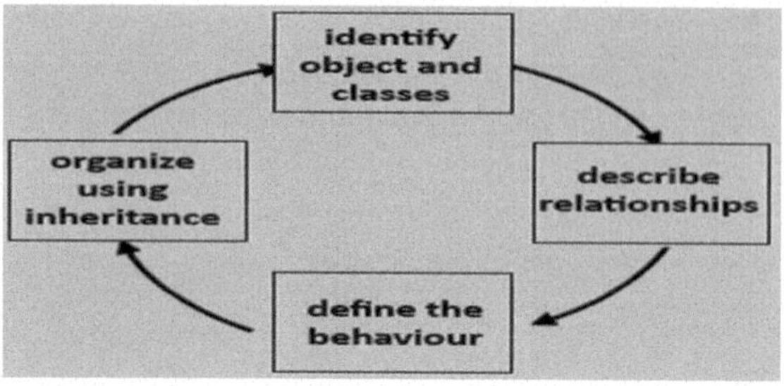

Fig. 2 Object modelling technique

OMT is divided into three models which are Object model, Functional model, Dynamic model.

OMT consists of four phases which are

- Analysis phase—consists of object, dynamic, and functional model.
- System design—it is a structure of basic architecture of the system.
- Object design—it is a design document which consists of object, dynamic, and functional model.
- Implementation phase—involves reusability of code.

2.3 *Jacobson Methodology (OOBE and OOSE)*

OOSE [8] is a popular design technique used to design software in object-oriented programming. OOSE design methodology includes use case in software design development and is of same kind as Unified Modelling Language such as Booch [6] and OMT [7]. OOSE is also called Objector and the system development based on it is a process of industrialized development. It includes various testing models as shown in Fig. 3.

Use cases involve

- Functional and non-functional requirements analysis through scenarios.
- Informal text with no clear flow of events.
- Simple and clear reading of text.
- Formal styling using pseudocode.
- Allow view which includes
  - Understanding system requirements
  - Interaction between user and system
  - Express user's goal and responsibility of the system.

Object-Oriented Business Engineering (OOBE) includes

Analysis phases define the system as

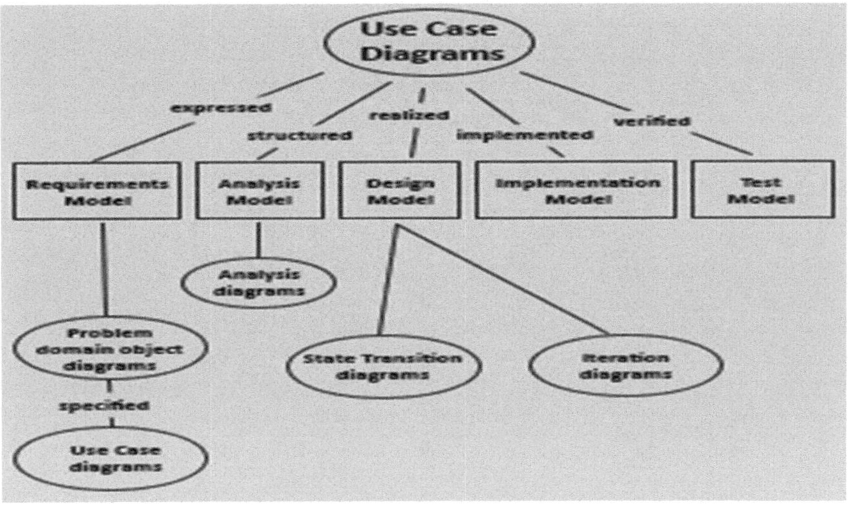

Fig. 3 OOSE

- Problem-domain object model
- Requirements model
- Analysis model.

Design phase and Implementation:

- It consists of design modelling and system implementation.

Testing phase:

- It consists of unit testing, integration testing, and system testing.

2.4 *RDD Methodology*

The Responsibility-Driven is defined by WirfsBrock [9]. RDD improves encapsulation using the client–server model where client and server are instances of classes. It emphasizes object behaviour and relationships with other objects where responsibilities are assigned to classes of objects during object-oriented design. Figure 4 explains the process model phases:

- Exploratory phase—includes class identification with similar objects then class collaboration with other classes and finally providing responsibilities to class objects.
- Analysis phase—includes analysis of hierarchies and subsystem and finally creating a protocol for the design.

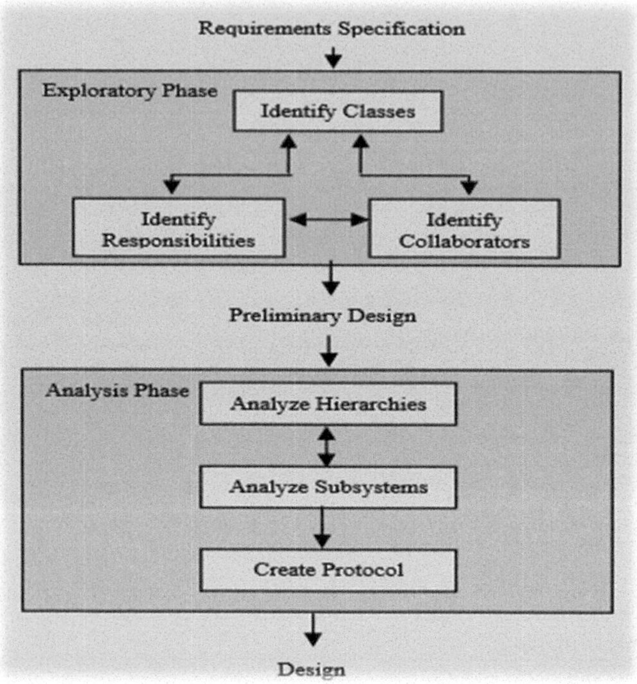

Fig. 4 Responsibility-driven design

2.5 Coad–Yourdon Methodology

The methodology is specialized in system analysis based upon a technique called "SOSAS", where each term helps in making up the analysis [10]. The terms are defined as follows:

- Subjects—are data flow diagrams for objects.
- Objects—identify class hierarchies.
- Structures—are of two types, classification structure handles the connection between related classes and the composition structure handles all other connections among classes.
- Attributes
- Services—identify methods or behaviours for each class.

Coad and Yourdon define four domain components: Problem, Human, Task, Data.

2.6 Shlaer–Mellor Methodology

This method is specialized for software system design and also works on system analysis [11].

It includes three models:

- Process model—includes data flow diagram.
- State model—documents different states and changes that occur between the objects.
- Information data model—it contains variables, objects, and all relationships between the objects.
- Comparison of already discovered model is shown in Fig. 5.

3 Design Approach

Suppapitnarm & Ahmed [12] reviewed how different designers approach their design task. Their study explained the problems, efforts, and time taken to understand the knowledge of how to approach their design process development. In object-oriented user interface design, different approach and mechanism are used to produce different outcomes which differ in the level of designer experience depending on how much similar the outcome is on the basis of user demands, requirements, and specifications. A good designer has a better understanding to user needs. Designers who are new or inexperienced in non-object-oriented approach or object-oriented

Elements	RDD	Booch (OOA/OOD)	OOSE& OOBE	Rumbaugh's (OMT)	Coad Yourdon	Shlaer-Mellor
Class	*	*	*	*	*	*
Attribute	*	*	*	*	*	*
Method	*	*	*	*	*	*
Collaboration	*		*	*	*	*
Abstraction	*		*	*	*	*
Relationship	*	*	*	*	*	*
Visibility	*			*		
Interface		*				
Subsystem			*			*
Information		*				*
Hiding						
Polymorphism		*			*	*

Fig. 5 Elements of process model

approach with no mechanism (models) need expert guidance in order to produce better design practise [5].

3.1 Problems Faced by Designers in Object-Oriented Design

This section discusses about already existing problems faced by the different level of designers and their object-oriented approach in designing field. Object-oriented designing is considered one of the famous and commonly used design approaches but several studies [13–16] show that especially student designers among different level of designers do not understand the complexity of object-oriented design approach.

Ryan [17] has explained an empirical study of various design disciplines and explained the difference and comparison between expertise and new designers on the basis of the years of experience, design approach, and other factors. The study explains the difference in level of designers on the basis of intensive analysis, experimental approach and various other factors that are involved. Object-oriented approach due to its complexity is difficult for new designers with some level of experience but the expertise will take years of experience.

Various prototypes explain the selection of design patterns that are used in object-oriented design implementation which will help in software reuse and improves the software development productivity [18]. Designers approach to design patterns used in object-oriented approach affects the software productivity if design used is not suitable thus explaining the complexity of object-oriented design. The study of Sim, Wright [19] explained the problems faced by the designers who are students in understanding the concepts of object-oriented approach and its process modelling. Students also faced difficulties in learning its analysis and design components and the implementation procedure. Recent study [15] proved that object-oriented design approach for new designers is difficult especially for college students who opted computer science engineering and still find difficult to make OOA models and simply fail to design the simple software system and their object-oriented approach is more of procedural based. Or-Bach [20] study also explains the difficulty faced by students who are learning object-oriented design concepts. Dig, Johnson [21] study also explains the change needed in an object-oriented design development, a lot of modification and changes are required to improve the development of the software design. Their keen focus is to take software development engineering from manual development to semi-automated development of software.

4 Existing Proposed Models

Din and Idris [5] proposed their idea of process model by combining the above explained methodologies which improved the design process and reduced the complexity allowing new designers to avoid common design faults and approach better the object-oriented design approach.

Their model is a hybrid of discovered models including all the elements required by the designers in software development. Designers on the basis of their work complexity level can club discovered process models for better object-oriented design development. This model workflow is better than using the existing process models. But designers who are college students need to learn all the elements, relationship, process modelling, initially, and then after that object-oriented design can be done which leads to common mistakes, miscommunication between designer and developer in software development implementation that cause code and design redo which is a waste of time. These mistakes are done by designers due to lower level experience and manual approach towards design implementation.

5 Proposed Model

5.1 Proposed Model Elements Description

Class—A class is a collection of similar objects which will act as an interface and sub-system where user will navigate while using the system. Every class has its own attributes which will be extended as a new class based upon the different classes relationship. Attributes representation is in the form of an interface design or sitemap so developer is able to understand what designer wants in the design implementation on the basis of user requirements and availability.

Attributes—An attribute is a component of a class which refers to the data belonging to a class, i.e., interface which will be extended further as a new class making the system architecture easier to understand. Each attribute of a class will become a new class and it will hence contain its own attributes hence the architecture and flow of system will be implemented very effectively.

Collaboration Different classes can collaborate with each other and implement their responsibilities where responsibilities are in the form of method and these methods execution done by developer is based on the interface design. This process is used for interaction between classes for message passing depending upon the relationship between them. Relationship like aggregation, dependency, and inheritance.

Encapsulation—On the basis of modules present in an interface of a class necessary information hiding and data protection will be done which is important in analysis, design, and programming. Concept of polymorphism is used to hide implementation not required by the end user. Polymorphism can be of any form and it is helpful for code reusability.

5.2 Proposed Model Workflow

Basic object-oriented software development involves Design process which is followed by Coding process with further process to go. Main problem arises when developer could not understand what user wants. All the proposed design models focus on Classes, Objects, and other features of object-oriented programming. All these diagrams do not depict a clear picture of how the interface of a system should look like and how the navigation between components is suitable to user. A class diagram only tells about the classes and its attributes and functions but does not clearly tell which function leads to which component of the system. This sometimes arises confusion for developer and he/she might have to recode and it will waste a lot of time and resources.

Presenting a model which is based upon class diagram covers most of the properties of object-oriented programming. In this model, we have two components—Class and its attributes which include all the concepts and elements of existing process models. Class is the interface where user sees and developer codes accordingly. Every class attribute leads to a new class hence an attribute which is a function becomes a new class, i.e., an interface for user to interact with. Also the model can make use of already existing models methodology like class collaboration can be done which includes aggregation, dependency, and inheritance. In our model, each class interface can navigate to other depending upon the class relationship. If a class is a part of another class, i.e., association, dependency is where one class manipulate with the object of other class and in inheritance the child class exhibits the behaviour of parent class and can use its properties. Further, class encapsulation is done by developer on the basis of relationship among different interface component of a class and necessary data hiding is done as per the requirements and availability. Below mentioned diagram (Fig. 6) represent our proposed model—Attribute Class Model.

Below mentioned image in (Fig. 7) presents an example for better understanding of proposed model.

Here library is a system which has several attributes like books, magazines, etc. Each attribute leads to a class of its own which is another subsystem. Like book which is an attribute of one class has its separate system which can be divided into Academic and Non-Academic which can further have their own attributes and these are represented in the form of user interface making it easy for the developer to understand the design workflow and no common mistakes are done and complexity is reduced. As a developer, he/she can see that landing page should contain navigation to different attributes, i.e., functions which should result into a new interface.

The error and design fault can be easily detected and analysed as each module consisting of various attributes of a class has separate implementation and also these modules can be reused in other object-oriented design. Here the workflow is similar to combination of other existing models which involves class and attributes identification, then class collaboration, and relationship to communicate and interact through message passing and class encapsulation which hides necessary information and

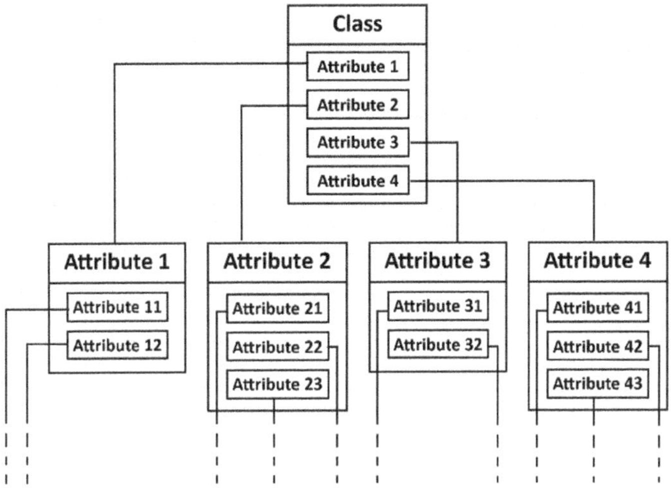

Fig. 6 Attribute class model

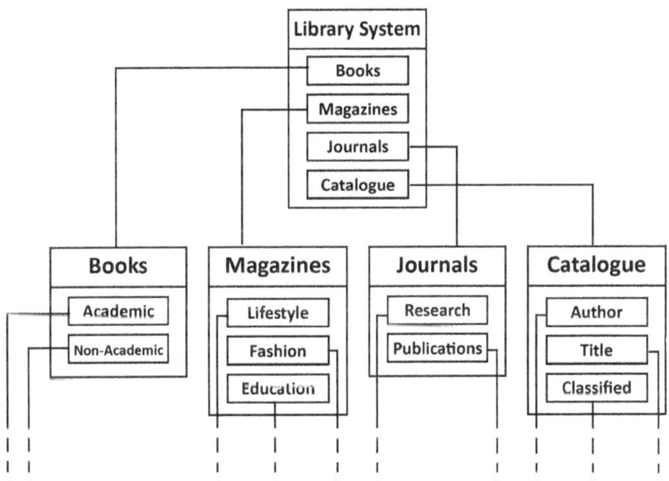

Fig. 7 Library system example

what should be visible. But the workflow representation is done in the form of user interface basically sitemaps which includes all the necessary elements and the output is that proper guidance system is created for all level of designers and side by side process model elements understanding and design implementation can be done.

6 Conclusion

The communication between designer and developer saves a lot of money, time and resources. Hence the proper design of system should be done and developer should easily understand the user requirements from it. Designing a user interface is not an easy task especially now-a-days UI implementation that is more user-friendly is a competitive job and thus designers with lower level face difficulties leading to faulty implementation and poor UI design. A good process model helps in better design implementation process and here we propose our Attribute Class Model. This is hybrid and modified version for new designers covering all object-oriented design components. With the help of our proposed model, all the requirement of user is cleared, basics of Object-Oriented programming is covered and the developer easily understand the navigation among the components of the design system and he/she can easily develop a system which satisfies user needs and saves resources and time.

References

1. Myers, B.A., Rosson, M.B.: Survey on user interface programming. In: Proceedings of the SIGCHI Conference on Human Factors in Computing Systems, pp. 195–202. ACM (1992)
2. Gould, J.D., Lewis, C.: Designing for usability: key principles and what designers think. Commun. ACM **28**(3), 300–311 (1985)
3. Biddle, R.: A lightweight case tool for learning OO design. In: Proceedings of Oopsla 2000 Educators Symposium, pp. 78–83 (2000)
4. Lewis, T.L., Pérez-Quiñones, M.A., Rosson, M.B.: A comprehensive analysis of object-oriented design: towards a measure of assessing design ability. In: 34th Annual Frontiers in Education, 2004. FIE 2004, pp. S3H–16. IEEE (2004)
5. Din, J., Idris, S.: Object-oriented design process model. Int. J. Comput. Sci. Netw. Secur. **9**(10), 71–79 (2009)
6. Booch, G.: Object-oriented analysis and design with applications. In: The Benjamin/Cummings Publishing Company, Inc (1994)
7. Rumbaugh, J., Blaha, M., Premerlani, W., Eddy, F., Lorensen, W.E.: Object-Oriented Modeling and Design, vol. 199, no. 1. Prentice-hall, Englewood Cliffs, NJ (1991)
8. Jacobson, I.: Object-oriented software engineering: a use case driven approach. Pearson Education India (1993)
9. Wirfs-Brock, R.J., Johnson, R.E.: Surveying current research in object-oriented design. Commun. ACM **33**(9), 104–124 (1990)
10. Coad, P., Yourdon, E., Coad, P.: Object-Oriented Analysis, vol. 2. Yourdon press, Englewood Cliffs, NJ (1991)
11. Shlaer, S.: The shlaer-mellor method. In: Project Technology White Paper (1996)
12. Suppapitnarm, A., Ahmed, S.: E-learning from knowledge and experience capture in design. In: The First National Conference of Electronic Business. N/A (2002)
13. Garner, S., Haden, P., Robins, A.: My program is correct but it doesn't run: a preliminary investigation of novice programmers' problems. In: Proceedings of the 7th Australasian Conference on Computing Education, vol. 42, pp. 173–180. Australian Computer Society, Inc. (2005)
14. Robins, A., Haden, P., Garner, S.: Problem distributions in a CS1 course. In: Proceedings of the 8th australasian conference on computing education, vol. 52, pp. 165–173. Australian Computer Society, Inc. (2006)

15. Eckerdal, A., McCartney, R., Moström, J.E., Ratcliffe, M., Zander, C.: Can graduating students design software systems? In: SIGCSE'06, pp. 403–407. ACM (2006)
16. Simon, B., Hanks, B.: First-year students' impressions of pair programming in CS1. J. Educ. Resour. Comput. (JERIC) **7**(4), 5 (2008)
17. Ryan, C.: A Methodology for the Empirical Study of Object-Oriented Designers. RMIT University (2002)
18. Moynihan, G.P., Suki, A., Fonseca, D.J.: An expert system for the selection of software design patterns. Expert Syst. **23**(1), 39–52 (2006)
19. Sim, E.R., Wright, G.: The difficulties of learning object-oriented analysis and design: an exploratory study. J. Comput. Inf. Syst. **42**(2), 95–100 (2002)
20. Or-Bach, R., Lavy, I.: Cognitive activities of abstraction in object orientation: an empirical study. ACM SIGCSE Bull. **36**(2), 82–86 (2004)
21. Dig, D., Johnson, R., Marinov, D., Bailey, B., Batory, D.: COPE: vision for a change-oriented programming environment. In: Proceedings of the 38th International Conference on Software Engineering Companion, pp. 773–776. ACM (2016)

Optimal Energy Distribution in Smart Grid

T. Aditya Sai Srinivas, Somula Ramasubbareddy, Adya Sharma, and K. Govinda

Abstract Almost nothing in today's world runs without power. Right from the air conditioner and water heaters to phone charging and electricity. Energy has its own role to play in everything that happens. But with increasing number of homes, the power consumption increases. The number of electricity sources does not follow the same rate of increase. So it becomes very difficult to supply all sections of a place with power simultaneously. Some areas will have to face blackout, whereas other places will have proper supply. And the electricity department needs revenue every month. This paper provides an optimal solution to provide electricity to a city divided into different sections or areas, when only a limited amount of energy units are available or generated. Assuming that each grid has its own power consumption and as per that and the revenue, the approach used in the proposed work is 0/1 knapsack problem to provide energy in a smart grid such that almost all of the power is used and maximum revenue is generated. This can be done through various methods like dynamic programming, greedy approach, brute force, backtracking, etc. We also find out that which approach will give the best solution with least time complexity.

Keywords Generation · Weight · Consumption · Revenue · Area

1 Introduction

Knapsack problem provides an algorithm in which a bag of a specific capacity is given, and n different elements each with different capacities are to be fit in the bag in such a way that maximum elements get fit in the bag and minimum space is left empty. Also, each element is a whole and cannot be divided further into smaller elements. So either we can select a particular object or we cannot. That's why it is

---

T. Aditya Sai Srinivas (✉) · A. Sharma · K. Govinda
Scope School, Vit University, Vellore, Tamil Nadu, India
e-mail: taditya1033@gmail.com

S. Ramasubbareddy
Information Technology, Vnrvjiet, Hyderabad, Telangana, India

S. C. Satapathy et al. (eds.), *Intelligent Data Engineering and Analytics*, Advances in Intelligent Systems and Computing 1177,
https://doi.org/10.1007/978-981-15-5679-1_36

Table 1 Different areas and power consumption under the grid

Section name	Area	Power units
grid1	450	150
grid2	500	180
grid3	1250	400
grid4	490	170
grid5	500	180
grid6	550	200

called a 0/1 knapsack problem in which '0' means that we are not considering the object and '1' means we are considering. A total of 2^n combinations are formed, and time taken can be exponential for normal programming. Knapsack helps in providing a simpler method with a lesser time complexity. So, we should fill the bag such that we get the maximum profit (value), and the weights of all the objects that are taken should be less than or equal to the maximum capacity of the bag (i.e., maximum weight capacity). In mathematical terms,

Maximum **(Σpixi) and Σwixi <= W.**

where

Pi The profit of that particular object.
Wi The weight of that particular object.
Xi 1 or 0 (i.e., object is taken or left behind, respectively).
W Maximum weight capacity of the sack.

Let us consider the electricity department. It's not an easy task for distributing electricity because there is lot of sections in a city, and all of them have different power requirements. Let the maximum power units available for distribution be 1000 (Table 1).

We can solve this problem using different approaches such as brute force, dynamic programming, backtracking, greedy, etc.

2 Related Works

In the last 10 years, a vast literature on the topic of Smart City has been produced to define strategies, contents, and objectives. Simultaneously, multiple contributions have been devoted to "measuring" the level of smartness of the different cities, to define their strategies (Battarra, et al., 2015) [1]. When we need to manage energy in a smart grid, then it is formulated typically as an optimization problem that cannot be considered linear. In literature, a proposal has been made for various centralized methods, such as mixed integer programming [2, 3], particle swarm optimization [4], neural networks [5], etc. In smart grid, since multi-terminal DC systems are being used, decentralized control for energy distribution has become feasible [6]. The concept of energy distribution might be connected to energy trading, which in turn is conducted by the coordination of energy flow like vehicle-to-grid (V2G) services [7]. So, the distributed optimization algorithms that have already been recognized [8, 9] might be used for finding optimal path in a distributed way; but these existing ones have not been specified to generation and distribution of energy [10–15].

Apart from the abovementioned application, there are other applications as well. Suppose we need to go on a journey and we are allowed to carry only 23 kg worth items. We have n different indivisible important items such as clothes, phone, charger, edibles, etc. all of different weights. We need to decide which item to take and which item to leave behind and 0/1 knapsack helps here to choose. The spaceships being used in the mars one project will be using the same method to maximize the values of goods which they want to carry [15–20].

If a teacher wants to set a question paper for 100 marks and there are multiple chapters all with different weightage of marks, then 0/1 knapsack can be used to set the question paper by choosing the most optimum set of questions. A question can be kept or removed as per the requirement and the paper can be set in such a way that maximum number of chapters get covered [21–23].

One application of this algorithm is download managers (e.g., Internet Download Manager). The data is fragmented into tiny parts. According to the maximum amount of data that can be recovered at a time, the server makes use of this algorithm and combines the small fragments in order to utilize the full-size limit. It is one of the multiple algorithms that allows managers to use apart from compressing, encrypting, etc.

If a building is supplied with a fixed number of electricity power units which is to be distributed among the different flats each with a certain power consumption unit different or same. Then 0/1 knapsack can be used to find the proper permutation in which maximum flats get the required amount of electricity and minimum number of flats are left out. In this way, there will be power cut for a specific period of time for only particular areas. After that tenure, power cut will be shifted to other areas and then the remaining areas will get the power supply.

In worst cases, suppose we decide that an approximation is not good enough, i.e., we want the best possible solution. Such solution is called optimal. Although 0/1 knapsack problem is based on NP-C (Non-deterministic Polynomial Completeness), it provides optimal solution in a reasonable amount of time when using some particular algorithms for worst cases. The applications of knapsack can be seen in investment decisions, debris collection, job scheduling, project selection, capital budgeting, resource allocation, cargo packing, and other fields.

The main goal is to present the comparative study of the approaches to find the performance of the different algorithms used to solve the 0/1 knapsack problem, based on the time complexity of each algorithm. To compare the performance of dynamic programming, greedy approach and brute force are used to solve 0/1 knapsack problem.

In the problem that we use in this project, we assume that we are designing a code for a person working in the electricity department. Every day, the person is allowed to supply only a limited amount of power units. The city is divided into grids. Each grid has specific power consumption and has a certain revenue. The main objective is to distribute the electricity in the entire city such that maximum units of electricity are distributed, least are left out, and maximum revenue is also collected.

3 Proposed Methods

3.1 Dynamic Programming

The dynamic programming solution for a bag packing scenario lists out a solution for the entire space with all kinds of combinations that can be put into use to pack a bag. On one hand, greedy approach gives the most optimal algorithm, and on the other hand dynamic programming is able to find the most optimal global solution.

Dynamic programming makes use of memorization in order to store previously computed operation results and then returns the cached result on reoccurrence of the operation. It ensures that the previous combination is remembered. When compared to re-computation of the answer, this takes lesser time.

Dynamic programming can also solve all of the smaller sub-divisions of the problem and instead of solving overlapping problems repeatedly, it stores the result in a table. This table is further used to derive a solution for the original problem. Normally, bottom-up approach is used.

Algorithm

We start by creating a matrix that will represent all the subsets of the items—which is basically the solution space—where rows represent items and columns represent the remaining weight capacity of the bag.

Next, we loop through this created matrix and using each combination of item at each stage we decide the worth that can be obtained.

At last, the completed matrix is examined to decide which items should be added to the bag so that a maximum possible worth of the bag is obtained.

From the above codes, we can find out the time complexity.

Dynamic Programming is:

1 + 1 + n*W*(1 + 1 + 1 + 1+1 + 1) + n

2 + 6*n*W + n

O(nW)

Suppose we have three items with value = [60,100,120] and weight = [10, 20, 24] and total capacity 50.

The formula to fill in the cells is V[i,w] = max{V[i−1,w],V[i-1,w−w(i)] + P[i]} (Table 2).

Now, to start with, we pick one item which gives us two possibilities—either to choose it or to remove it.

Table 2 Values of different parameters

Value	Weight	Number	0	10	20	30	40	50
0	0	0	0	0	0	0	0	0
60	10	1	0	60	60	60	60	60
100	20	2	0	60	100	160	160	160
120	30	3	0	60	100	120	180	220

```
                  K(3,50)
                 /       \
                /         \
          K(2,40)            K(2,50)
          /     \            /    \
         /       \          /      \
    K(1,20)  K(1,40)  K(1,30)     K(1,50)
    /    \    /    \    /     \     /   \
   /      \  /      \  /       \   /     \
K(0,0) K(0,20)K(0,20) K(0,40)K(0,10)K(0,30)K(0,30)K(0,30)
```

Every time we select an item and we either take it or do not. Hence, the most optimal solution is found.

3.2 Greedy Approach

In this approach, we first find the value by weight ratio(val/wt) of each item and arrange the items in decreasing order of their ratio.

Then, we start with the highest worth item (item with the highest value to weight ratio). Next, we start filling the bag until we cannot fit any more items in it. If there is anything remaining item that can fit, we try to fit that [24–33].

This approach does not improve upon the solution it returns. The only thing it does is adding the next highest density item into the bag.

Time complexity for greedy approach is O(nlogn) (Table 3).

So the item is with the greatest density, then the item with next highest, and so on. So the first item is taken first. So now remaining W = 40. Next, the second highest density is 5, so the second item is selected. New W = 20. But the third item cannot be selected completely because enough space is not there so it will be rejected.

Table 3 Values of different parameters

Number	1	2	3
Value	60	100	120
Weight	10	20	30
Density (V/W)	6	5	4

Hence, maximum weight that can be fit into it is 30 with value 160.

3.3 *Brute Force*

This is a simple approach to find the solution to a problem, which is normally directly based on the definition of the concept involved and the problem statement.

Algorithm

- Let us assume that there are n items. This creates 2n different choices of items for the knapsack.
- Any item can have two possibilities: it can be selected or not selected.
- For this, a bit string is created which contains only 0 and 1's.
- If at cell index, the bit value is one, it means that the respective item is selected and if it is zero then it is not selected.

Time complexity for brute force is **O(n*2n)**.

4 Result Analysis

Enter the maximum units of electricity available to supply: 1000 (Tables 4, 5 and Figs. 1, 2).

Table 4 Power consumption in different areas

S. No	Grid name	Power consumption (Kwt/hr)	Area (Sq. Feet)
1	Area1	100	300
2	Area2	140	420
3	Area3	150	750
4	Area4	70	210
5	Area5	250	1000
6	Area6	300	1500
7	Area7	200	800
8	Area8	130	390
9	Area9	170	510
10	Area10	300	1500
11	Area11	80	120
12	Area12	190	500

Table 5 The revenue generation

S. No.	Grid name	Power consumption	Revenue (in rs)
1	Area1	300	1500
2	Area2	300	1500
3	Area3	250	1000
4	Area4	150	150

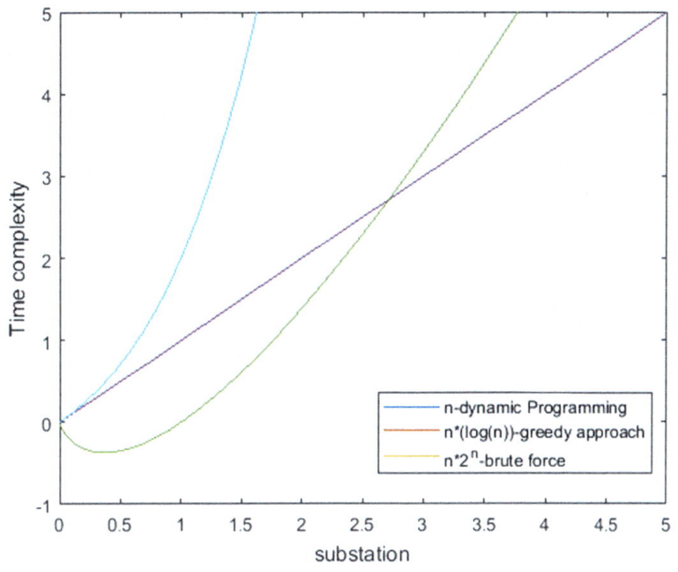

Fig. 1 Time complexity versus substation(n)

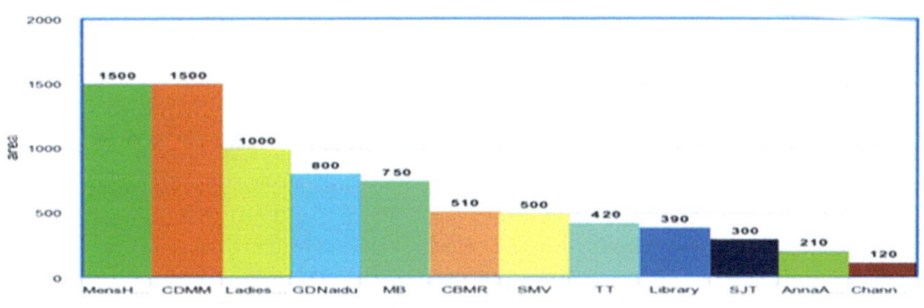

Fig. 2 Power consumption versus area

5 Conclusions

Optimization plays an important role in many engineering domains. The energy distribution in smart grid is the key factor for optimization; from the above results, it is clear that the dynamic programming has least time and also when compared with other approaches, we can see that dynamic programming provides the optimal solution. Hence, we can say that out of the presented approaches, dynamic programming, greedy and brute force, dynamic programming are the best to solve 0/1 knapsack problem.

References

1. Gyamera, L., Atripatri, I.: Energy efficiency of smart cities: an analysis of the literature (2017)
2. Choi, S., Park, S., Kang, D.-J., Han, S.-J., Kim, H.-M.: A microgrid energy management system for inducing optimal demand response. In: IEEE International Conference on Smart Grid Communications (SmartGridComm), pp. 19–24. Brussels, Belgium (2011)
3. Cecati, C., Citro, C., Siano, P.: Combined operations of renewable energy systems and responsive demand in a smart grid. IEEE Trans. Sustain. Energy **2**(4), 468–476 (2011)
4. Pourmousavi, S., Nehrir, M., Colson, C., Wang, C.: Real-time energy management of a stand-alone hybrid wind-microturbine energy system using particle swarm optimization. IEEE Trans. Sustain. Energy **1**(3), 193–201 (2010)
5. Siano, P., Cecati, C., Yu, H., Kolbusz, J.: Real time operation of smart grids via FCN networks and optimal power flow. IEEE Trans. Ind. Informat. **8**(4), 944–952 (2012)
6. Gavriluta, C., Candela, J.I., Citro, C., Rocabert, J., Luna, A., Rodri guez, P.: Decentralized primary control of MTDC networks with energy storage and distributed generation. IEEE Trans. Ind. Appl. **50**(6), 4122–4131 (2014)
7. Al-Awami, A.T., Sortomme, E.: Coordinating vehicle-to-grid services with energy trading. IEEE Trans. Smart Grid **3**(1), 453–462 (2012)
8. Johansson, B.: On Distributed Optimization in Networked Systems. Ph.D Thesis, Royal Institute of Technology (KTH) (2008)
9. Nedic, A., Ozdaglar, A., Parrilo, P.A.: Constrained consensus and optimization in multi-agent networks. IEEE Trans. Autom. Control **55**(4), 922–938 (2010)
10. Basu, S., Kannayaram, G., Ramasubbareddy, S., Venkatasubbaiah, C.: Improved genetic algorithm for monitoring of virtual machines in cloud environment. In: Smart Intelligent Computing and Applications, pp. 319–326. Springer, Singapore
11. Somula, R., Sasikala, R.: Round robin with load degree: An algorithm for optimal cloudlet discovery in mobile cloud computing. Scalable Comput.: Pract. Exp. **19**(1), 39–52 (2018)
12. Somula, R., Anilkumar, C., Venkatesh, B., Karrothu, A., Kumar, C.P., Sasikala, R.: Cloudlet services for healthcare applications in mobile cloud computing. In: Proceedings of the 2nd International Conference on Data Engineering and Communication Technology, pp. 535–543. Springer, Singapore (2019)
13. Somula, R.S., Sasikala, R.: A survey on mobile cloud computing: mobile computing + cloud computing (MCC = MC + CC). Scalable Comput.: Pract. Exp. **19**(4), 309–337 (2018)
14. Somula, R., Sasikala, R.: A load and distance aware cloudlet selection strategy in multi-cloudlet environment. Int. J. Grid High Perform. Comput. (IJGHPC) **11**(2), 85–102 (2019)
15. Somula, R., Sasikala, R.: A honey bee inspired cloudlet selection for resource allocation. In: Smart Intelligent Computing and Applications, pp. 335–343. Springer, Singapore (2019)
16. Nalluri, S., Ramasubbareddy, S., Kannayaram, G.: Weather prediction using clustering strategies in machine learning. J. Comput. Theor. Nanosci. **16**(5–6), 1977–1981 (2019)
17. Sahoo, K.S., Tiwary, M., Mishra, P., Reddy, S.R.S., Balusamy, B., Gandomi, A.H.: Improving end-users utility in software-defined wide area network systems. IEEE Trans. Netw. Serv. Manag. (2019)
18. Sahoo, K.S., Tiwary, M., Sahoo, B., Mishra, B.K., RamaSubbaReddy, S., Luhach, A.K.: RTSM: response time optimisation during switch migration in software-defined wide area network. IET Wirel. Sens. Syst. IET Wirel. Sens. Syst. (2019)

19. Somula, R., Kumar, K.D., Aravindharamanan, S., Govinda, K.: Twitter sentiment analysis based on US presidential election 2016. In: Smart Intelligent Computing and Applications, pp. 363–373. Springer, Singapore (2016)
20. Sai, K.B.K., Subbareddy, S.R., Luhach, A.K.: IOT based air quality monitoring system using MQ135 and MQ7 with machine learning analysis. Scalable Comput.: Pract. Exp. **20**(4), 599–606 (2019)
21. Somula, R., Narayana, Y., Nalluri, S., Chunduru, A., Sree, K.V.: POUPR: properly utilizing user-provided recourses for energy saving in mobile cloud computing. In: Proceedings of the 2nd International Conference on Data Engineering and Communication Technology, pp. 585–595. Springer, Singapore (2019)
22. Vaishali, R., Sasikala, R., Ramasubbareddy, S., Remya, S., Nalluri, S.: Genetic algorithm based feature selection and MOE fuzzy classification algorithm on Pima Indians diabetes dataset. In: 2017 International Conference on Computing Networking and Informatics (ICCNI), pp. 1–5. IEEE (2017)
23. Somula, R., Sasikala, R.: A research review on energy consumption of different frameworks in mobile cloud computing. In: Innovations in Computer Science and Engineering, pp. 129–142. Springer, Singapore (2019)
24. Saraswathi, R.V., Nalluri, S., Ramasubbareddy, S., Govinda, K., Swetha, E.: Brilliant corp yield prediction utilizing internet of things. In: Data Engineering and Communication Technology, pp. 893–902. Springer, Singapore (2020)
25. Kumar, I.P., Sambangi, S., Somukoa, R., Nalluri, S., Govinda, K.: Server security in cloud computing using block-chaining technique. In: Data Engineering and Communication Technology, pp. 913–920. Springer, Singapore (2020)
26. Kumar, I.P., Gopal, V.H., Ramasubbareddy, S., Nalluri, S., Govinda, K.: Dominant color palette extraction by k-means clustering algorithm and reconstruction of image. In: Data Engineering and Communication Technology, pp. 921–929. Springer, Singapore (2020)
27. Nalluri, S., Saraswathi, R. V., Ramasubbareddy, S., Govinda, K., Swetha, E.: Chronic heart disease prediction using data mining techniques. In: Data Engineering and Communication Technology, pp. 903–912. Springer, Singapore (2020)
28. Krishna, A.V., Ramasubbareddy, S., Govinda, K.: Task scheduling based on hybrid algorithm for cloud computing. In: International Conference on Intelligent Computing and Smart Communication 2019, pp. 415–421. Springer, Singapore (2020)
29. Srinivas, T.A.S., Ramasubbareddy, S., Govinda, K., Manivannan, S.S.: Web image authentication using embedding invisible watermarking. In: International Conference on Intelligent Computing and Smart Communication 2019, pp. 207–218. Springer, Singapore (2020)
30. Krishna, A.V., Ramasubbareddy, S., Govinda, K.: A unified platform for crisis mapping using web enabled crowd sourcing powered by knowledge management. In: International Conference on Intelligent Computing and Smart Communication 2019, pp. 195–205. Springer, Singapore (2020)
31. Kalyani, D., Ramasubbareddy, S., Govinda, K., Kumar, V.: Location-based proactive handoff mechanism in mobile ad hoc network. In: International Conference on Intelligent Computing and Smart Communication 2019, pp. 85–94. Springer, Singapore (2020)
32. Bhukya, K.A., Ramasubbareddy, S., Govinda, K., Srinivas, T.A.S.: Adaptive mechanism for smart street lighting system. In: Smart Intelligent Computing and Applications, pp. 69–76. Springer, Singapore (2020)
33. Srinivas, T.A.S., Somula, R., Govinda, K.: Privacy and security in aadhaar. In: Smart Intelligent Computing and Applications, pp. 405–410. Springer, Singapore (2020)

Robust Automation Testing Tool for GUI Applications in Agile World—Faster to Market

Madhu Dande and Somula Ramasubbareddy

Abstract In this digital world, technology changes exponentially to increase the speed, efficiency, and accuracy. To achieve these features, we need good programming language, high-end hardware configurations, permutation, and combinations of scenarios based on testing. Applications are developed to make more interactive and reduce the complexity, reduce the transaction response time, and without failure at the end users. For any graphical user interface application, they need to be tested either by Manual/Automation Testing tools. Robust Automation Testing (RAT) tool is built on the Hybrid Automation Framework which is easy to learn and reduces the automation scripting time/coding, while execution increases the permutation and combination of the test scenarios without changing the test steps. There is no dependency on the test data and maintenance-free. RAT tool is for testing the application from creating the manual/automation test scripts, generating the test data, executing the automation scripts, and generating the customized reports. RAT tool shows that the performance is increased the accuracy of validation by 97%, no cost to the tool. Manual tester is enough to complete the automation script execution, and frequency of execution is increased and reduces the maintenance of the scripts to less than 10% cost as well resource cost reduced to 38%.

Keywords GUI automation testing framework · Robust · Efficient and effective automation · Agile-based and script-less automation and hybrid framework · Script-less agile-based and script-less automation and hybrid framework · Customized test execution results

M. Dande
Site School, Vit University, Vellore, Tamilnadu, India

S. Ramasubbareddy (✉)
Information Technology, VNRVJIET, Hyderabad, Telangana, India
e-mail: svramasubbareddy1219@gmail.com

S. C. Satapathy et al. (eds.), *Intelligent Data Engineering and Analytics*, Advances in Intelligent Systems and Computing 1177,
https://doi.org/10.1007/978-981-15-5679-1_37

1 Introduction

Most of the applications were developed in the late 1970s and 80s that were either client/server or non-GUI based. These applications were mostly developed in COBOL [1–3] and VB language [4]. In the Internet generation world started in the year of 1990 [5], most of the applications developed in web-based languages and made easy to end users in the recent Internet world. Current world is running towards high-end technology and hardware/software to develop their applications [6, 7].

Entire world is using the software (applications) to do business either online or in-Shoppe. To continue their business in this competitive world, they need to change the requirements to make more flexible for customers in this agile world to continue the business and increase their growth [8, 9].

Any software application is developed to validate whether the requirement to be validated by the testing team. Testing plays an important role in providing quality product to the end user. Testing fundamentals to be strong to drive the testing and need to understand the types of testing. In this paper, concentrating on black-box testing is mainly on functionality of the application [10–12].

2 Background

Manual testing process will take time while executing the test cases and cannot be increased the frequency of execution without increasing the workforce. To overcome this problem, they need to automate those test cases using the tools. Each application has a different technology, business purpose, and size of the business [13].

Application is analyzed by the test architect to provide the feasibility. Based on that, high-level design document with architectures/methodologies for automation will be created and shared with the testing team for the further process [14].

One needs basis types of frameworks used to complete the automation, which are data-driven testing, model-driven testing, library-driven testing, keyword-driven testing, action-driven testing, Excel-driven testing, and hybrid-driven testing. In this Agile world, IT is moving toward either TDD or BDD at high level to increase the efficiency of the automation [15, 16].

However, nowadays the organizations have become aware of the shortcoming of these tools. The main burden with these already established tools is need of maintenance. On a regular basis, GUI application needs a change in the front end, so functionality of the application will be in system under test (SUT) changes [17]. This change leads to the changes in the automaton test scripts, which is A huge maintenance of the automation scripts [18].

Another problem is learning the features of these tools with the release of each version for using its features and writing the complex scripts, which require skilled resources. This includes tools such as Win Runner, UFT, Selenium, SOAPUI, and QARun based on their own scripting language. To use these tools, we need to hire the resources who are well expertise and have good experience with these tools. As well, license cost is high and spending on maintenance.

Organization of the paper with brief details of each section

- Section I contains the introduction toward the development of software and importance of testing with the elevation of automation testing tools.
- Literature review is covered under the background in section II.
- In Section III, Discussion on RAT architecture, data flow of RAT tool, and essential steps of flowchart and its procedures are explained.
- Section IV explains RAT execution methodology, summary results, and discussion metrics.
- Section V contains the limitation.

3 Design and Development of Rat Architecture

This paper describes how the Robust Automation Testing (RAT) tool has designed and developed procedure explained in layered structure to reduce the maintenance of automation scripts which is explained in Fig. 1. RAT tool is developed using Visual

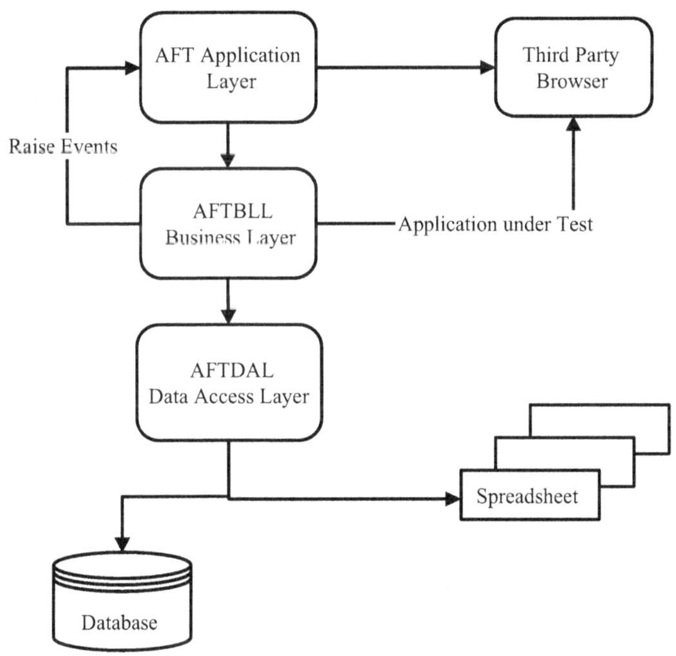

Fig. 1 High-level RAT architecture

Studio 2012 and C#. NET language. RAT followed three-layered architecture, i.e., for database, business logic, and graphical user Interface (Application layer).

- AFTDAL—Database access layer contains classed for SQLDataAccess and ExcelDataAccess.
- AFTBLL—Business logic layer contains all the business classed including test case, test data, action, page component, and query.
- AFT—Application layer will help the tester to fill the required details to complete the automation of the application.

RAT tool uses different types of frameworks are combined together known as hybrid automation framework, which has implemented [17, 19, 20]. This tool internally extracts application components for both Windows and Web-based and creates a unique ID for those extracted components. RAT tool GUI layer is directly accessible to the testers. Web application is accessible in the container of web browser, i.e., csExWB (Microsoft customized) browser which loads the web pages to be tested. This layer calls the business classes to load the test cases and execute the corresponding actions on the browser. This layer captures the events raised in the business layer during the execution of the actions and shows the progress of the results to the users in the UI screen of the RAT tool.

3.1 Rat Data Flow Diagrams

Since this tool is action keyword—data-driven framework and does not require any prior knowledge of writing test case/scripts—it reduces the burden of hiring highly skilled resources for writing the scripts and modifying the test data. With the help of Subject Matter Expert (SME), the testing entire application can be completed by recording the systematic process from entering to validation of the complete detailed scenario.

- RAT tool designed to facilitate the user with the following features.
- Powerful framework for organizing test automation, execution around keywords, test report generation, and test data generation.
- Highly productive approach of writing test case sheet from English like language organized as user actions, objects, and values.
- Functionality to generate component/page control list [21–27].
- Generates test results consisting of test summary, detailed reporting, and with screenshots.
- Features such as error logging, status viewer of every test case actions.

Application-related components would create in the spreadsheet with unique ID for each component [28, 29]. Similarly, create a sheet with test steps with parameters either entering or validating but the values will be updated in the separate sheet which will be made easy to create multiple test conditions for the same scenario. These test cases are either stored in Excel or database (Fig. 2).

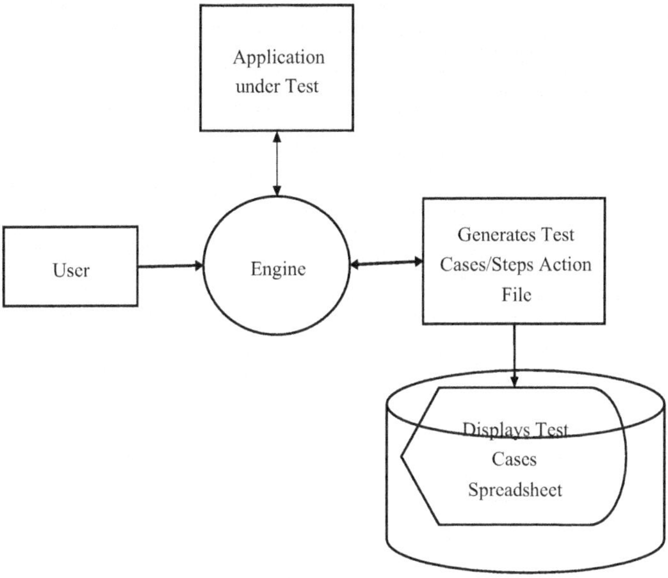

Fig. 2 Data flow of RAT-scenario-1

Record/playout test scenarios and execution results updated in new instance of the spreadsheet that includes logs and screenshots of the execution steps in the local system or in test management tool by updating the test cases. In the final stage of the execution, it will display a message whether it needs to create manual test cases with OK and cancel buttons. If you click on OK button, it will automatically create manual test cases in the standard and simple English language (Figs. 3 and 4).

Fig. 3 Data flow of RAT-scenario-2

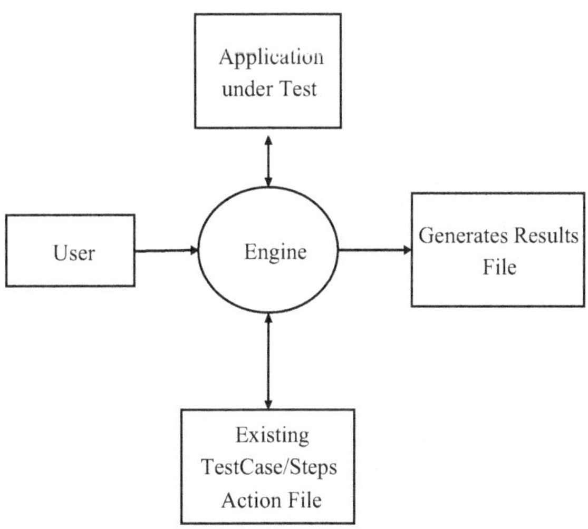

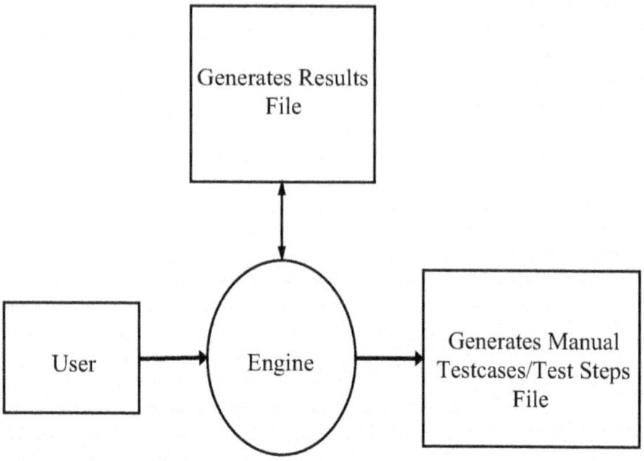

Fig. 4 Data flow of RAT-scenario-3

RAT fully supports the execution of manually created test case and recorded test cases as well for execution. Test cases will have action input file and creation of the manual test cases based on the execution results.

3.2 *Rat Flowchart and Its Steps*

Steps to execute RAT

1. Launch RAT.exe application from the desktop.
2. Select the Old/New Spreadsheet with test steps and select the application type either Web-based or Windows.
3. Click on New Script to record and select the Spreadsheet with template.
4. Select the Application as Web-based.
5. Enter the URL and valid login credentials.
6. Either New recording or using the existing test cases for enhancement business functionality scenarios.
7. For new functionality, Click on New and click on record the scenario by systematic process.
8. Save the complete scenario by giving the valid name and SME will update with Test data in the test case sheet (Spreadsheet).
9. While execution, Click on Execute button to execute the test cases from spreadsheet which is loaded into the tool.
10. Create the Test results with metrics and represented in graphical view in the specific location.
11. Results are stored in the C:\\Drive\Application Name\MMDDYY folder with Appname and Time concatenated which will be stored in the spreadsheet.
12. Similarly, Screenshots/log files stored in the same location for future validation
13. Generate the customized Report (Fig. 5).

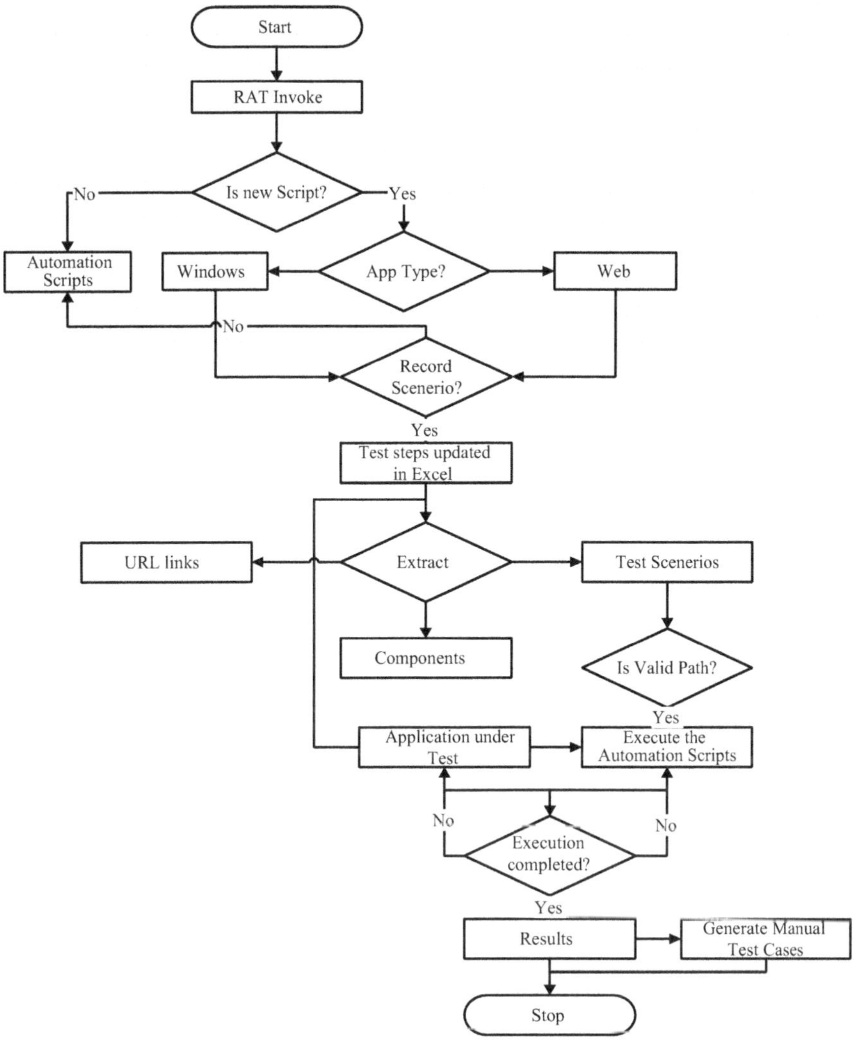

Fig. 5 Flowchart of RAT

4 Rat Input File Structure

Users need to pass the Excel file which contains all the details of test cases and actions/steps, and test data for each test case. Along with these, Excel spreadsheet contains the required components and queries for database transactions. Now to make it simple we have segregated the Excel file into multiple sheets each for each kind of data to remove the dependency.

Standard Spreadsheet details as follows: Test Cases, Actions, Group Actions, Test Data, Queries, Page Components, Page Links, and Action Results.

Test Cases sheet contains a header row and describes list of test cases to be executed as well when to stop the execution. Execute field to mark whether to execute the test case or not and Result field to update the execution status (Figs. 6 and 7).

Actions and Group Actions sheet contains the Test steps with keywords and assigning the test data as parameters. Group the test steps, and create the reusable test case in the group action sheet.

TestCase	Description	Execute	Result	StopOnFailure
TC0	SHAFT Login test case	Y	Pass	N
TC1	Broker Information test case	Y	Pass	N
TC2	Broker Contract test case	Y	Pass	N
TC3	Broker History test case	Y	Pass	N
TC4	Broker Telephone and Address	Y	Pass	N
TC5	All State test cases	Y	Pass	N
TC6	Century Club test case	Y	Pass	N
TC7	Check Information test case	Y	Pass	N
TC8	Commission test case	Y	Pass	N
TC9	County History test case	Y	Pass	N
TC10	County Service Fees test case	Y	Pass	N
TC11	Daily Business test case	Y	Pass	N
TC12	District Recognition	Y	Pass	N
TC13_Removec Fall Checkup Test Case		Y	Fail	Y

Fig. 6 Test case spreadsheet

TestData	TestCase	Execute	TestConditionType	Passed	Failed	Done	NotExecuted	InProgress	OutputPath	Result	E1	E2	E3	E4	E5	E6
TD0	TC0	Y	Valid	Pass					C:\GUIAFT\Date Time\Filenamewithtime	Pass	You are currently logged on to:					
TD1	TC1	Y	Valid	Pass					C:\GUIAFT\Date Time\Filenamewithtime	Pass						
TD2	TC2	Y	Valid	Pass					C:\GUIAFT\Date Time\Filenamewithtime	Pass						
TD3	TC3	Y	Valid	Pass					C:\GUIAFT\Date Time\Filenamewithtime	Pass	Kevin, Walter					
TD4	TC4	Y	Valid	Pass					C:\GUIAFT\Date Time\Filenamewithtime	Pass	5701	173	38			
TD5	TC5	Y	Valid	Pass					C:\GUIAFT\Date Time\Filenamewithtime	Pass						
TD6	TC6	Y	Valid	Pass					C:\GUIAFT\Date Time\Filenamewithtime	Pass	4026	2952	27	61	23	20
TD7	TC7	Y	Valid	Pass					C:\GUIAFT\Date Time\Filenamewithtime	Pass	04704 Kevin, Walter					
TD8	TC8	Y	Valid	Pass					C:\GUIAFT\Date Time\Filenamewithtime	Pass	Totals for: 04704 Kevin, Walter	Detail for:				
TD9	TC9	Y	Valid	Pass					C:\GUIAFT\Date Time\Filenamewithtime	Pass	Active Agents for County Code	All Agents				
TD10	TC10	Y	Valid	Pass					C:\GUIAFT\Date Time\Filenamewithtime	Pass	County Service Fees					
TD11	TC11	Y	Valid	Pass					C:\GUIAFT\Date Time\Filenamewithtime	Pass	Daily Paid-For Business - Applications Information	Daily New				
TD12	TC12	Y	Valid	Pass					C:\GUIAFT\Date Time\Filenamewithtime	Pass	20	59				
TD13	TC13	Y	Valid		Fail				C:\GUIAFT\Date Time\Filenamewithtime	Fail	District Best Of The Best - Combined Submitted	District	District	District	State Best	
TD14	TC14	Y	Valid				NotExecuted				First Year Collected Premium	First Year				
TD15	TC15	Y	Valid				NotExecuted				Policy Count Persistency	Policy	Policy	Policy	75%	
TD16	TC16	Y	Valid				NotExecuted				Agents and Managers Production - Personal Data					
TD17	TC17	Y	Valid				NotExecuted				Production Club	Agency	No Data			
TD18	TC18	Y	Valid				NotExecuted									
TD19	TC19	Y	Valid				NotExecuted				Round Table Contest Request	Round	04704	Round	Round Table	14441
TD20	TC20	Y	Valid				NotExecuted				Summer Sizzle	Click	Totals for:	Totals	As of	As of
TD21	TC21	Y	Valid				NotExecuted				20	20	20	20		
TD22	TC22	Y	Valid				NotExecuted				Winner's Challenge	Details for				
TD23	TC23	Y	Valid				NotExecuted				Record Month Reports	Monthly All	Monthly All	Monthly	Record	Record
TD24	TC24	Y	Valid				NotExecuted				Robert, Smith	Flexible	Non-	Inforce	Error:	Error
TD25	TC25	Y	Valid				NotExecuted				BULLETIN BOARD	test2	Complian	SFB	Winter	Winter
TD26	TC26	Y	Valid				NotExecuted				Advertising & Promos	Advertisin	Delivery	Delivery	No	Delivery
TD27	TC27	Y	Valid				NotExecuted				Claims	New	There are	Combina	Adding	Memos

Fig. 7 Test data spreadsheet

Actions are nothing but steps defined as a keyword, i.e., Open_URL, ClickButtonBy_ID, Select Dropdown By_Value, Type_Text, etc. These keywords represent the events that any user performs on a web page.

The Group actions are combinations of Actions, which performed frequently, Such as GA_Login which is a combination of Open_URL, Type_Text, and ClickButtonBy_ID actions. So instead of writing all these actions every time, we can write GA_Login group action in the Group Action Sheet.

Once added, the actions or group actions add the component ref name (defined in page component sheet) on which the actions will be performed, such as ClickButtonBy_ID action can be performed on a button. If any step has this action, then under component column we need to mention the ref name of the component from the Page Components sheet.

Similarly, Input Parameter and Expected Value columns will have data used, while execution of the action and data is validated against the expected value.

For example, if we are using a Type_Text action, then we need to pass component ref name of the textbox and the input parameter value to be entered in the textbox.

Page Components (Page Object Model) sheet will have all the components along with the page name, type of the component, value of the component if exists, and a unique Component_Ref_Name which has to be used in the Action and Group Action sheet. The structure of the Component_Ref_Name depends on page name, type, and id of the control. The benefit of this is to identify a particular control on a particular page. It will help us in maintaining the input sheet easily if any of the page structure and control name changes after any release [30–34].

Queries sheet contains all the queries with input parameters marked as '@P' and a unique query name. This query name has to be used in the input parameter column in action sheet for the database-driven actions, e.g., V_Table_Data, V_Export_Data.

While using these Queries, we have to pass the parameter values as "~" separated after the name of the query.

4.1 Rat Test Data Sheet

Test Datasheet contains multiple combinations of data used for a particular test case. Each test data execute all the actions of that respective test case. The test datasheet contains columns for Input Parameters, i.e., P1, P2, P3….. Column for Expected Values E1, E2, E3….

Along with these columns as well contains the columns like [Actual Values] and [Result], which updates after the execution of all the actions of that particular test case.

ActionResults contains the data of columns, which used for creating manual test cases. Need basis updated with each row after execution of each action with data, e.g., Input_Parameter, Actual Value, ExpectedValue, and Result.

Copy the Input file (Result file) with updated Result columns and ActionResults sheet.

StepNo	TestCase	TestData	Action	Componen	InputParamete	ExpectedValues	ActualValues	Comments	ResultStatus
TC0_TD0_	TC0	TD0	Open_	URL	http://[redacted]/servlet/pagesen			Action exe	Done
TC0_TD0_	TC0	TD0	Type_	PageServer_	[redacted]			Action exe	Done
TC0_TD0_	TC0	TD0	Type_	PageServer_	[redacted]			Action exe	Done
TC0_TD0_	TC0	TD0	Select	PageServer_	Default Company			Action exe	Done
TC0_TD0_	TC0	TD0	Click_	PageServer_	W5			5 Seconds	Done
TC0_TD0_	TC0	TD0	V_Tex	PageServer_	SPAN_MIR-USE	[redacted]	[redacted]	Verificatio	Passed
TC0_TD0_	TC0	TD0	Click_	PAGESERV	W5			5 Seconds	Done
TC0_TD0_	TC0	TD0	Click_	LinkByCapti	Traditional Life\Term, W5			5 Seconds	Done
TC0_TD0_	TC0	TD0	Type_	INDEX.HTM	777445334, W5			5 Seconds	Done
TC0_TD0_	TC0	TD0	Click_	INDEX.HTM	W5			5 Seconds	Done
TC0_TD0_	TC0	TD0	V_PageText			NO RECORDS MATCHED YOU		Page does	Failed
TC0_TD0_	TC0	TD0	Click_	LinkByCapti	Traditional Life\Term, W5			Skipped du	Condition_Not_Matched
TC0_TD0_	TC0	TD0	Type_	INDEX.HTM	NM68			Skipped du	Condition_Not_Matched
TC0_TD0_	TC0	TD0	Type_	INDEX.HTM	NML68			Skipped du	Condition_Not_Matched
TC0_TD0_	TC0	TD0	Click_	INDEX.HTM	W5			Skipped du	Condition_Not_Matched
TC0_TD0_	TC0	TD0	V_PageText			NO RECORDS MATCHED YOU		Skipped du	Condition_Not_Matched
TC0_TD0_	TC0	TD0	Type_	INDEX.HTM	NM68			Skipped du	Condition_Not_Matched
TC0_TD0_	TC0	TD0	Type_	INDEX.HTM	NML68			Skipped du	Condition_Not_Matched
TC0_TD0_	TC0	TD0	Type_	INDEX.HTM	01/01/1971			Skipped du	Condition_Not_Matched
TC0 TD0	TC0	TD0	Select	INDEX.HTM	Arkansas			Skipped du	Condition Not Matched

Fig. 8 Execution results in spreadsheet

Execute Output file or log file saved in the respective folder called [Test Results\MMDDYYY_HHmmss].

A folder [Screenshots] is with all the screenshots in a subfolder with name of the corresponding test case and test data [35–38].

4.2 Manual Test Cases Creation

Once the execution completes, the user can create the manual test cases by selecting the result file as mentioned (Fig. 8).

4.3 Execution Results and Discussion

Execution Status and Error Logging

Status/Output Viewer is the progress viewer window that shows the actions and their status, i.e., Started, In-Progress, Passed, or Failed. If fails, it shows the reason for the failure as close as possible.

Create a folder structure in C drive for the first time while execution. C:\Project Name\AFT\Data time folder will have two more folders, i.e., Logs and Results (Fig. 9).

Along with this, the tabbed window shows the error viewer with the errors that the Robust Automation Testing system encounters while executing the actions without stopping the execution (Fig. 10).

These errors logged in a Log file inside the [Logs] folder, which consist of detailed step-by-step execution results; it will be easy to debug the error/issues which is encountered while execution.

Output | Errors |

TestCase	TestData	Action	Status	Message
TC0	TD0	GA_Signin > Open_URL	Done	Action executed successfully.
TC0	TD0	GA_Signin > V_Title	Passed	Verification done successfully.
TC0	TD0	GA_Signin > Type_Text	Done	Action executed successfully.
TC0	TD0	GA_Signin > Type_Text	Done	Action executed successfully.
TC0	*TD0*	*GA_Signin > Click_ButtonById*	*In Process...*	

Fig. 9 Execution results in spreadsheet

Output Errors |

Class Name	Method Name	Line Number	Description
WebAction	ClickLinkByUrl	0	Element [Remarks.aspx] does not exist on the page.
WebAction	VText	0	Element [AgentHistory_Label_AgentName] does not exist on the page.

Fig. 10 Execution results in spreadsheet

Open_URL		http://sfbagentportaltest/Signon.aspx			14-May-2013 11:32:56:372	14-May-2013 11:33:03:091	Action executed
V_Title			SFBLI Agent Web Site Sign-On	SFBLI Agent Web Site Sign-C	14-May-2013 11:33:03:242	14-May-2013 11:33:03:472	Verification don
Type_Text	Login_Text_UserID	lct0088			14-May-2013 11:33:03:592	14-May-2013 11:33:04:273	Action executed
Type_Text	Login_Text_Pwd	apr12345			14-May-2013 11:33:04:513	14-May-2013 11:33:04:654	Action executed
Click_ButtonById	Login_Button_SignOn				14-May-2013 11:33:05:014	14-May-2013 11:33:06:276	Action executed
Click_LinkByUrl		http://sfbagentportaltest/MenuItem/80			14-May-2013 11:33:06:386	14-May-2013 11:33:19:465	Action executed
Click_LinkByUrl		http://sfbagentportaltest/ACIS_Web/DailyBusiness/DailyBusiness.aspx?Status=			14-May-2013 11:33:19:575	14-May-2013 11:33:20:727	Action executed
V_PageText			Daily Paid-For Business - Applic	Daily Paid-For Business - App	14-May-2013 11:33:21:368	14-May-2013 11:33:28:749	Verification don
Select_DropDownByValue	DailyBusinessPaid_SELECT_As	04/30/2013			14-May-2013 11:33:28:869	14-May-2013 11:33:32:995	Action executed
Select_DropDownByValue	DailyBusinessPaid_SELECT_Bu	PAID			14-May-2013 11:33:33:105	14-May-2013 11:33:33:205	Action executed
Select_DropDownByValue	DailyBusinessPaid_SELECT_Da	Applications			14-May-2013 11:33:33:335	14-May-2013 11:33:33:576	Action executed

Fig. 11 Execution results

Execution Summary Report

RAT execution results are in RAW data from the ActionResult sheet with our framework engine able to generate the manual test cases in the standard format, which will be in easy and readable format to the testing team (Fig. 11).

Manual Test steps created by RAT understand the structure of the input Excel sheets, which creates the test steps based on the functional execution steps in the automation testing (Fig. 12).

In this paper, performed comprehensive method called normalized mean data imputation technique for data imputation is presented. After imputation, this methodology has been tested on benchmark datasets with the percentage of MVs varied from 48.39% to 2.29%. The proposed method is imputed plausible data value in the original dataset and evaluated the classifier accuracy with ETrees, variance scores, and AUC curve values that are computed. In addition, we observed that after imputation some of the outliers are also eliminated in a dataset from our approach. Our experiment results are shown that the proposed imputation method accuracy is better than the other traditional mean, median, and mode imputation methods (Figs. 13 and 14).

TestData	StepNo	Description	Input Value	Expected Value	Actual Value	Result
TD0	StepNo 1	Navigate to URL.	http://[redacted]ervlet/pageserver?ProcessClassID=SignOn&PageReturn=index&LocaleID=E			Done
	StepNo 2	Entered the value in the Login textbox	[redacted]			Done
	StepNo 3	Entered the value in the Pwd textbox	[redacted]			Done
	StepNo 4	Selected text from Company dropdown list	Default Company			Done
	StepNo 5	Clicked on control.	WS			Done
	StepNo 6	Verified the text of the page label.		[redacted]	[redacted]	Passed
	StepNo 7	Clicked on control.	WS			Done
	StepNo 8	Clicked on link.	[redacted] WS			Done
	StepNo 9	Entered the value in the SSN textbox	[redacted] WS			Done
	StepNo 10	Clicked on control.	WS			Done

Fig. 12 Execution results in spreadsheet

Test_Case	Test_Case_Execution_Date	Test_Scenario_Name	No_of_Test_Steps	Passed	Failed	Skipped	In_Progress	ExecutionTime	Execution_Percentage	Completion_Percen
Daily Business test case	05/14/2013 11:38	TD11	44	40	4	0	0	00:01:29:090	100	90
Total	05/14/2013 11:38		44	40	4	0	0	00:01:29:090	100	90

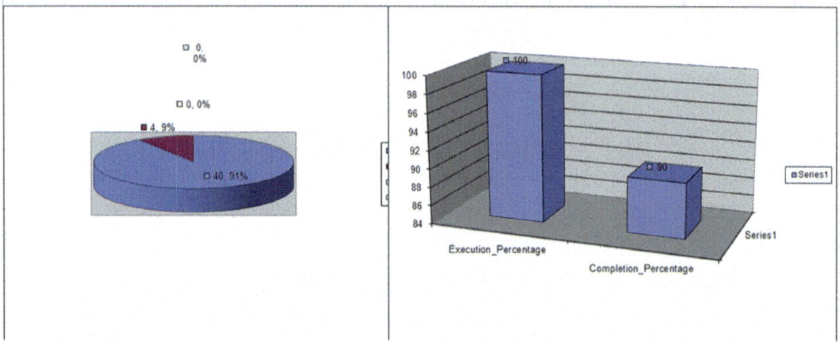

Fig. 13 Pie chart execution automation execution summary report will provide the summary metrics and graphs based on the data. Even create the customized reports and send them to the respective owners in the email (Figs. 13 and 14). Results in spreadsheet

Test_Case	Test_Case_Execution_Date	Test_Scenario_Name	No_of_Test_Steps	Passed	Failed	Skipped	In_Progress	ExecutionTime	Execution_Percentage	Completion_Percen
Daily Business test case	05/14/2013 11:38	TD11	44	40	4	0	0	00:01:29:090	100	90
Total	05/14/2013 11:38		44	40	4	0	0	00:01:29:090	100	90

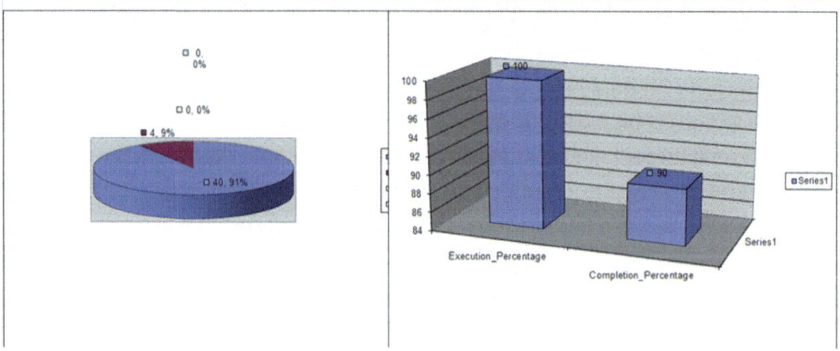

Fig. 14 Execution results in spreadsheet

5 Rat Benefits

- Easy to maintain and modify as per the enhancements.
- Easy to create test cases and actions in the Excel sheet.
- Once test cases created and reviewed the scenario, then it is ready to execute with different test data with conditions tested.
- Easy to use. No programming skills required.
- Specific test case executed on need basis.
- If there is any property of an object, it differs between the versions identified.
- Virtual objects identification done through the indexing, location, and unique ID.
- Maximize the test conditions.
- Reduce the dependency between the test cases and test data.
- Manual intervention reduced.
- Customized report generated with metrics and graphs.
- No tool license is needed.

6 Limitations

- Control ids of each page of the website are generated in advance by using any of the available tools (Daemon WebUI Utility) [21].
- Manual intervention is required to write the input test case and action file.
- Create SQL query for data verifications from database manually.
- This version does not support Nosql database and record and play feature for generating test case actions file.
- Intelligence is not implemented using any AI frameworks.

7 Conclusion

RAT tool enables in any IT organization to increase the automation coverage, easy to execute, and frequency to the execution to be increased, and there is no dependency on Test data, Code/Script-less code to automate the application to validate the business functionality.

Total time to automate the scenarios/test cases has reduced to 57% when compared with the existing automation process. Manual test cases' creation is done by using RAT tool automatically. Execution is done in sequential steps but executed in multiple systems. Efficient and accuracy of the test step validation achieved 97% without missing any validation. Total workforce effort and total cost of the testing effort have reduced by 62%. RAT is used as minimal performance testing tool for the components and images, which provides response time of the application. RAT is a perfect functional testing tool to the agile world, faster to market, and reduces the testing effort drastically.

References

1. Sammet, J.E.: Programming languages: history and future. IBM Corporation Commun. ACM **15**(7), 601–610 (1972)
2. Sammet, J.E.: Programming Languages: History and Fundamentals. Prentice-Hall, Inc. (1969). ISBN:0137299885. http://www.internetnews.com/asp-news/article.php/936061/EDS+Enhances+MetaVance+Software.htm
3. Shaw, R.S.: A study of the relationships among learning styles, participation types, and performance in programming language learning supported by online forums. Comput. Educ. **58**(1), 111–120 (2012)
4. Berners-Lee, T., Hendler, J., Lassila, O.: The semantic web—a new form of web content that is meaningful to computers will unleash a revolution of new possibilities. Sci. Am. Feature Art. Semant Web (2001)
5. Schaller, R.R.: Moore's law: past, present and future. IEEE Spectr. **34**(6), 52–59 (1997)
6. Messerschmitt, D.G., Szyperski, C.: Industrial and Economic Properties of Software Technology, Processes, and Value. Microsoft Corporation (2000)
7. Chapman, R.L., Soosay, C., Kandampully, J.: Innovation in logistic services and the new business model: a conceptual framework Manag. Serv. Qual. Int. J. ISSN: 0960-4529-2002
8. Edwards, S.: A Framework for practical, Automated Black-Box Testing of Component Based software. Virgina Tech University, Wiley (2001)
9. Patton, R.: Software Testing, pp. 53–56, Sams Publishing (2006)
10. Pettichord, B., Kaner, C., Bach, J.M.: Lessons Learned in Software Testing: a Context-Driven Approach. Wiley (2001)
11. Hoffman, D.: Test automation architectures: planning for test automation. Software Quality Methods, LLC (1999)
12. Polo, M., Reales, P., Piattini, M., Ebert, C.: Test automation. In: IEEE Software, vol. 30(1), pp. 84–89 (Jan–Feb 2013)
13. Vieira, M., Leduc, J., Hasling, B., Subramanyan, R., Kazmeier, J.: Automation of GUI Testing Using a Model-driven Approach AST'06. Shanghai, China (23 May 2006)
14. Palani, N.: Software Automation Testing Secrets Revealed. Educreation Publishing (2016)
15. Kagan, D., Saba, K., Dishon, N., Tel-Aviv, Himmelreich, E., Modiin.: Framework for Automated Testing of Enterprise Computer Systems. US 7.620, 856 B2 USPTO (2009)
16. Noller, J.A., Mason, R.: Automated Software Testing Framework. US 7, 694, 181 B2USPTO (2010)
17. Basu, S., Kannayaram, G., Ramasubbareddy, S., Venkatasubbaiah, C.: Improved genetic algorithm for monitoring of virtual machines in cloud environment. In: Smart Intelligent Computing and Applications, pp. 319–326. Springer, Singapore (2019)
18. Parker, H.M, Kepple, L.R, Newton, Sklar, L.R, Laroche, D.C.: Automated Guinterface Testing. US 5, 781, 720 USPTO (1998)
19. Somula, R., Sasikala, R.: Round robin with load degree: an algorithm for optimal cloudlet discovery in mobile cloud computing. Scalable Comput. Pract. Experience **19**(1), 39–52 (2018)
20. Somula, R., Anilkumar, C., Venkatesh, B., Karrothu, A., Kumar, C.P., Sasikala, R.: Cloudlet services for healthcare applications in mobile cloud computing. In: Proceedings of the 2nd International Conference on Data Engineering and Communication Technology, pp. 535–543. Springer, Singapore (2019)
21. Somula, R., Sasikala, R.: A honey bee inspired cloudlet selection for resource allocation. In: Smart Intelligent Computing and Applications, pp. 335–343. Springer, Singapore (2019)
22. Nalluri, S., Ramasubbareddy, S., Kannayaram, G.: Weather prediction using clustering strategies in machine learning. J. Comput. Theor. Nanosci. **16**(5–6), 1977–1981 (2019)
23. Sahoo, K.S., Tiwary, M., Mishra, P., Reddy, S.R.S., Balusamy, B., Gandomi, A.H.: Improving end-users utility in software-defined wide area network systems. IEEE Trans. Netw. Serv. Manag. (2019)

24. Sahoo, K.S., Tiwary, M., Sahoo, B., Mishra, B.K., RamaSubbaReddy, S., Luhach, A.K.: RTSM: response time optimisation during switch migration in software-defined wide area network. IET Wirel. Sens. Syst. (2019)
25. Somula, R., Kumar, K.D., Aravindharamanan, S., Govinda, K.: Twitter sentiment analysis based on us presidential election 2016. In: Smart Intelligent Computing and Applications, pp. 363–373. Springer, Singapore (2020)
26. Sai, K.B.K., Subbareddy, S.R., Luhach, A.K.: IOT based air quality monitoring system using MQ135 and MQ7 with machine learning analysis. Scalable Comput. Pract. Experience **20**(4), 599–606 (2019)
27. Somula, R., Narayana, Y., Nalluri, S., Chunduru, A., Sree, K.V.: POUPR: properly utilizing user-provided recourses for energy saving in mobile cloud computing. In: Proceedings of the 2nd International Conference on Data Engineering and Communication Technology, pp. 585–595. Springer, Singapore (2019)
28. Somula, R.S., Sasikala, R.: A survey on mobile cloud computing: mobile computing + cloud computing (MCC = MC + CC). Scalable Comput. Pract. Experience **19**(4), 309–337 (2018)
29. Somula, R., Sasikala, R.: A load and distance aware cloudlet selection strategy in multi-cloudlet environment. Int. J. Grid. High Perform. Comput. (IJGHPC) **11**(2), 85–102 (2019)
30. Vaishali, R., Sasikala, R., Ramasubbareddy, S., Remya, S., Nalluri, S.: Genetic algorithm based feature selection and MOE Fuzzy classification algorithm on Pima Indians Diabetes dataset. In: 2017 International Conference on Computing Networking and Informatics (ICCNI), pp. 1–5). IEE (2017, October)
31. Somula, R., Sasikala, R.: A research review on energy consumption of different frameworks in mobile cloud computing. In: Innovations in Computer Science and Engineering, pp. 129–142. Springer, Singapore, (2019)
32. Kumar, I.P., Sambangi, S., Somukoa, R., Nalluri, S., Govinda, K.: Server security in cloud computing using block-chaining technique. In: Data Engineering and Communication Technology, pp. 913–920. Springer, Singapore (2020)
33. Kumar, I.P., Gopal, V.H., Ramasubbareddy, S., Nalluri, S., Govinda, K.: Dominant color palette extraction by k-means clustering algorithm and reconstruction of image. In: Data Engineering and Communication Technology, pp. 921–929. Springer, Singapore (2020)
34. Nalluri, S., Saraswathi, R.V., Ramasubbareddy, S., Govinda, K., Swetha, E. Chronic heart disease prediction using data mining techniques. In: Data Engineering and Communication Technology, pp. 903–912. Springer, Singapore (2020)
35. Krishna, A.V., Ramasubbareddy, S., Govinda, K.: Task scheduling based on hybrid algorithm for cloud computing. In: International Conference on Intelligent Computing and Smart Communication 2019, pp. 415–421. Springer, Singapore (2020)
36. Srinivas, T.A.S., Ramasubbareddy, S., Govinda, K., Manivannan, S.S.: Web image authentication using embedding invisible watermarking. In: International Conference on Intelligent Computing and Smart Communication 2019, pp. 207–218. Springer, Singapore (2020)
37. Krishna, A.V., Ramasubbareddy, S., Govinda, K.: A unified platform for crisis mapping using web enabled crowdsourcing powered by knowledge management. In: International Conference on Intelligent Computing and Smart Communication 2019, pp. 195–205. Springer, Singapore (2020)
38. Saraswathi, R.V., Nalluri, S., Ramasubbareddy, S., Govinda, K., Swetha, E.: Brilliant corp yield prediction utilizing internet of things. In: Data Engineering and Communication Technology, pp. 893–902. Springer, Singapore (2020)

Storage Optimization Using File Compression Techniques for Big Data

T. Aditya Sai Srinivas, Somula Ramasubbareddy, K. Govinda, and C. S. Pavan Kumar

Abstract The world is surrounded by technology. There are lots of devices everywhere around us. It is impossible to imagine our lives without technology, as we have got dependent on it for most of our work. One of the primary functions for which we use technology or computers especially is to store and transfer data from a host system or network to another one having similar credentials. The restriction in the capacity of computers means that there's restriction on amount of data which can be stored or has to transport. So, in order to tackle this problem, computer scientists came up with data compression algorithms. A file compression system's objective is to build an efficient software which can help to reduce the size of user files to smaller bytes so that it can easily be transferred over a slower Internet connection and it takes less space on the disk. Data compression or the diminishing of rate of bit includes encoding data utilizing less number of bits as compared to the first portrayal. Compression can be of two writes lossless and lossy. The first one decreases bits by recognizing and disposing of measurable excesses, and due to this reason, no data is lost or every info is retained. The latter type lessens record estimate by expelling pointless or less vital data. This paper proposed a file compression system for big data as system utility software, and the users would also be able to use it on the desktop and lossless compression takes place in this work.

Keywords Data · Lossy · Lossless · Compression · Huffman

T. Aditya Sai Srinivas · K. Govinda · C. S. Pavan Kumar (✉)
SCOPE School, VIT University, Vellore, Tamil Nadu, India
e-mail: pavan540.mic@gmail.com

S. Ramasubbareddy
Information Technology, VNRVJIET, Hyderabad, Telangana, India

S. C. Satapathy et al. (eds.), *Intelligent Data Engineering and Analytics*, Advances in Intelligent Systems and Computing 1177,
https://doi.org/10.1007/978-981-15-5679-1_38

1 Introduction

Compression is the plan of tending to or watching out for data in a shorter shape, rather than its rise or uncompressed plot. By the day's end, utilizing this system, the level of a specific record can be lessened. This is strikingly tremendous while arranging, securing, or exchanging a titanic measure of data as a report, which needs social affairs of purposes of intrigue. On the off chance that the checks used to scramble works truly, there ought to be a tremendous unconventionality between the vital chronicle and the stuffed record. Right when information weight is utilized as a touch of an information transmission application, speed is the crucial target. Speed of information is measured through the count of bits that are sent and time taken by encoder to convert the plaintext into ciphertext and decoder who converts the encoded text into plain text. The level of information is considered as the basic requirement in an information totaling application. Weights acquire two ways of allocation, lossy and lossless. In lossless allocation, some techniques are used to change the key information from the smashed record without any loss. This is the reason for data not being changed during the weight and decompression plots. The weight checks in which the decompression system copies the central message is called the reversible compression. Packing satisfying pictures, substance and picture safeguarding, pc executable report is possible through lossless weight approaches, whereas lossy data leads to depreciation in the original data and is called irreversible data. It is named as irreversible as the standard data cannot be reproduced once it was lost. An uncertain re-attempting occurs due to decompression framework.

The confinement in the limit of PCs implies that there's a limitation on the measure of information which can be put away or exchanged. Numerous times it happens that we have to exchange a record over web to some other customer, or just need to store a major document on the capacity gadget; however, as of now said because of those confinements, it winds up being troublesome. Henceforth, we need to concoct a document/file compression framework for big data with a specific end goal to eradicate or lessen those challenges.

As far as compression proportion is concerned, **CMIX** is the apex method or algorithm; however, the main issue with it is that it requires a PC with 32 GB of memory to run it, and after that likewise it takes 4 days to pack or decompress 1 GB of content information. It utilizes word reference preprocessing and PAQ style setting blending. The preprocessor replaces words with 1–3 bit images from a lexicon and does other handling, for example, supplanting capitalized letters with an uncommon image and the relating lower case image.

Microsoft Point-to-Point Compression is spilling information procedure in light of an execution of Lempel–Ziv utilizing a sliding window cushion.

Shannon coding, which as the name suggests, is named after its maker, Claude Shannon, and is one of the techniques of lossless information pressure method for building a prefix code, in light of an arrangement of images and their probabilities (evaluated or estimated). It is problematic as it doesn't accomplish the most reduced conceivable expected codeword length.

PackBits is a quick, basic lossless pressure plot for run-length encoding of information. Run-length encoding (RLE) is an extremely basic type of lossless information pressure in which "runs" of information (i.e., successions in which similar information esteem happens in numerous continuous information components) are put away as a solitary information esteem and tally, instead of as the first run. It isn't valuable with documents that don't have numerous keeps running as it could extraordinarily expand the record estimate.

HTTP Compression is a method that can be incorporated with web servers and web customers to enhance exchange speed and transmission capacity usage. HTTP information is compacted before it is sent from the server, and agreeable programs will report what techniques are upheld to the server before downloading the right arrangement. Programs that don't bolster agreeable pressure strategy will download uncompressed information.

These are only some of the many algorithms present for file compression systems, but the most popular one is Huffman's algorithm.

2 Related Works

The course toward diminishing the measure or size of a data record is as often as possible proposed as data weight. When we examine transmission ponder, it is called "source coding".

It is useful in light of the way that it decreases the points of confinement which we require to hold and give various sorts of information. Computational resources are used as a piece of this method and, for most by a wide margin of the part, in the reversal of the framework, named as decompression. It is in peril to a space–time tradeoff. Let us assume a situation where an expensive gear is required for a video to make it appear as it is being compressed. It is also useful in making the decision about how the video should be organized before fully watching the video and its decompression. The blueprint of reducing report gauge diagrams joins trade-offs among various parts, including the level of weight, the measure of bowing showed (while using lossy data weight), and as far as possible which are basically remembering the true objective to decrease the estimations of the required data.

The process of representing information in a neatly packed form is known as compression. It has been one of the critical enabling technologies. Different formats of information can be achieved through different compression algorithms. These algorithms include both lossy and lossless compressions. This paper focuses on different algorithms on data compression algorithms. Text data is useful for experimental results. Statistical and dictionary-based compression techniques are useful for comparing lossless algorithms. Shannon–Fano coding, Huffman coding, adaptive Huffman coding, run-length encoding, and arithmetic coding are the algorithms used from statistical coding techniques [1].

Source alphabet $\sum$ represents the compression algorithms that are performed on the text. 8-bit ASCII codes consist of this source alphabet $\sum$. This alphabet may be the symbols that contain words of English, strings, and alphanumeric and non-alphanumeric characters. Better compression can be accomplished through taking an advantage of longer range correlation among words [2].

The transformation of information between two end parties needs to be secured. Security can be achieved through plaintext or binary data. Information can be converted in unreadable format so that its access to unauthorized parties is avoided. This format changing process is possible by making use of schemes. The field used for securing the information between two end users is cryptography. Cryptography abstracts the original data to avoid unauthorized persons from accessing the secret data [3].

Almost every computer application needed to be able to perform data compression. This is possible by using different data compression algorithms for different data formats. And again different approaches are used for a single data type for converting the data. This paper deals with lossless data and compares its performance. The performance of compression on test data is done by selecting a set of algorithms to measure the performance. This paper also includes different experimental compressions based on lossless data compression. Finally, conclude with the best algorithm to perform the compression on text data [4–8].

Huffman's algorithm allocates less number of bits or shorter code words for most every now and again utilized characters or words in a file (according to the factual information available), and this will spare a great deal of storage room.

Let's comprehend this better with the assistance of a case—assume we need to allot 26 extraordinary codes to the English letters in order and need to store an English novel (message just) in terms of these codes. Presently, we will require less memory on the off chance that we relegate short length codes to most oftentimes happening characters. It depends on the comparable guideline as in the case of representing information in Morse code, we don't utilize a similar number of spots and dashes for each letter of the letter set. Actually, 'E', the most incessant letter, is spoken by a solitary dot, while every single other letters are spoken by a mix of dashes and dots. This is on account of as E happens all the more oftentimes; it will be better in the event that we speak to it by some littler, less tedious code [9–13].

We can watch comparable ideas on the way that the postal and STD codes for imperative urban communities are typically shorter (as they are utilized all the time). This is an extremely essential idea in information hypothesis. Subsequently, by the above idea, Huffman's coding is the most proficient one, as it is fit for accomplishing high compression proportions, without trading off preparing efficiencies as on account of some different calculations like CMIX. One of the extra highlights of Huffman's algorithm is that it can likewise be coupled or joined with some different calculations, to shape new ones. For instance, it can be joined with LZ77 encoding algorithm to shape DEFLATE algorithm. This DEFLATE algorithm is utilized as a part of more well-known applications like WinZip [14–18].

2.1 Algorithm

There are essentially two sections in Huffman's coding. They are as follows:

(i.) Character would be taken as input and from them prepare Huffman's tree.
(ii.) Visiting the tree generated in the previous step to this step, and in the process giving out variable-length binary codes to all the nodes.

2.2 Steps

Array of one of kind characters alongside their recurrence frequencies is the input and the yield which would be obtained as an output would be the ever important Huffman's tree.

1. Make a leaf hub for every one of a kind character and assemble a min heap of all leaf hubs (Min heap is utilized as a priority queue. The estimation of recurrence field is utilized to think about two hubs in min pile. At first, the slightest incessant character is at root).
2. Take out two hubs or nodes with the minimum recurrence from the min heap.
3. Make another interior hub/node with recurrence equivalent to the addition of the two hubs frequencies. Make the primary separated hub as its left child and the other extricated hub as its right child. Add this to the original graph.
4. Redo steps 2 and 3 until the point that only a solitary node is remaining.

Traverse or cross the tree framed beginning from the root. Keep up a helper array. While moving to left child, compose 0 to the cluster. While moving to the right child, compose 1 to the exhibit. Array is printed whenever a leaf node is experienced.

Subsequent to making the Huffman's tree and doling out the variable-length codes (binary) to every one of the letter sets alongside space, in view of their frequencies in English language (data accessible on the Internet), we simply need to peruse the first content record or in simple language a "file", letter by letter and to yield the particular parallel code.

In the event that we need to decompress the document, then again, we simply need to peruse that record a little bit at a time and move along the Huffman tree until the point when we discover a letter, and soon thereafter we move back to the foundation of the tree and further keep preparing the bits of that record which we need to decompress [19–25].

3 Experiments and Results

Similar kind of tests was done on some other files also, albeit different sizes. The following compression ratios were achieved for those [26–30].

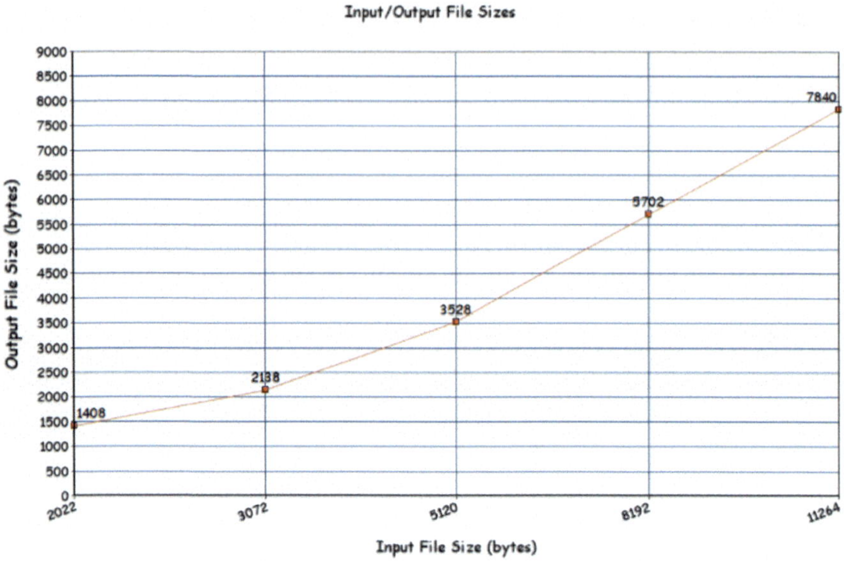

Fig. 1 Compression ration

Input file size (in bytes)	Output file size (in bytes)	Compression ratio (in %)
2022	1408	69.68
3072	2138	69.59
5120	3528	68.90
8192	5702	69.61
11264	7840	69.60

In the below graph, the X-axis contains the "input file size", and Y-axis contains the "output file size".

Hence, from the above information, it is clear that the curve or the straight line (in this case) that we obtain would be the compression fraction, i.e., the slope of the graph. If we multiply the compression fraction by 100, then we would get the compression ratio in percentage (Fig. 1).

4 Conclusions

The file compression system was successfully developed using Huffman's algorithm, which is a lossless file compression algorithm. It assigns variable-length codes to various letters, hence assuring higher compression ratios (approximately 65–70%). The systems work on all kinds of files. It can be used wherever you want to compress your files either just to store them in less space for big data, or to send them over a network of low bandwidth, etc.

References

1. Shanmugasundaram, S., Lourdusamy, R.: A comparative study of text compression algorithms. Int. J. Wisdom Based Comput. **1**(3), 68–76 (2011)
2. Horspool, R.N., Cormack, G.V.: Constructing word-based text compression algorithms. In: Data Compression Conference, pp. 62–71 (1992)
3. Sangwan, N.: Text encryption with Huffman compression. Int. J. Comput. Appl. **54**(6) (2012)
4. Kodituwakku, S.R., Amarasinghe, U.S.: Comparison of lossless data compression algorithms for text data. Indian J. Comput. Sci. Eng. **1**(4), 416–425 (2010)
5. Basu, S., Kannayaram, G., Ramasubbareddy, S., Venkatasubbaiah, C.: Improved genetic algorithm for monitoring of virtual machines in cloud environment. In: Smart Intelligent Computing and Applications, pp. 319–326. Springer, Singapore (2019)
6. Somula, R., Sasikala, R.: Round robin with load degree: an algorithm for optimal cloudlet discovery in mobile cloud computing. Scalable Comput. Pract. Exp. **19**(1), 39–52 (2018)
7. Somula, R., Anilkumar, C., Venkatesh, B., Karrothu, A., Kumar, C.P., Sasikala, R.: Cloudlet services for healthcare applications in mobile cloud computing. In: Proceedings of the 2nd International Conference on Data Engineering and Communication Technology, pp. 535–543. Springer, Singapore (2019)
8. Somula, R.S., Sasikala, R.: A survey on mobile cloud computing: mobile computing + cloud computing (MCC = MC + CC). Scalable Comput. Pract. Exp. **19**(4), 309–337 (2018)
9. Somula, R., Sasikala, R.: A load and distance aware cloudlet selection strategy in multi-cloudlet environment. Int. J. Grid High Perform. Comput. (IJGHPC) **11**(2), 85–102 (2019)
10. Somula, R., Sasikala, R.: A honey bee inspired cloudlet selection for resource allocation. In: Smart Intelligent Computing and Applications, pp. 335–343. Springer, Singapore (2019)
11. Nalluri, S., Ramasubbareddy, S., Kannayaram, G.: Weather prediction using clustering strategies in machine learning. J. Comput. Theor. Nanosci. **16**(5–6), 1977–1981 (2019)
12. Sahoo, K.S., Tiwary, M., Mishra, P., Reddy, S.R.S., Balusamy, B., Gandomi, A.H.: Improving end-users utility in software-defined wide area network systems. IEEE Trans. Netw. Serv. Manag. (2019)
13. Sahoo, K.S., Tiwary, M., Sahoo, B., Mishra, B.K., RamaSubbaReddy, S., Luhach, A.K.: RTSM: response time optimisation during switch migration in software-defined wide area network. IET Wirel. Sens. Syst. (2019)
14. Somula, R., Kumar, K.D., Aravindharamanan, S., Govinda, K.: Twitter sentiment analysis based on US presidential election 2016. In: Smart Intelligent Computing and Applications, pp. 363–373. Springer, Singapore (2020)
15. Sai, K.B.K., Subbareddy, S.R., Luhach, A.K.: IOT based air quality monitoring system using MQ135 and MQ7 with machine learning analysis. Scalable Comput. Pract. Exp. **20**(4), 599–606 (2019)
16. Somula, R., Narayana, Y., Nalluri, S., Chunduru, A., Sree, K.V.: POUPR: properly utilizing user-provided recourses for energy saving in mobile cloud computing. In: Proceedings of the 2nd International Conference on Data Engineering and Communication Technology, pp. 585–595. Springer, Singapore (2019)
17. Vaishali, R., Sasikala, R., Ramasubbareddy, S., Remya, S., Nalluri, S.: Genetic algorithm based feature selection and MOE fuzzy classification algorithm on Pima Indians Diabetes dataset. In: 2017 International Conference on Computing Networking and Informatics (ICCNI), pp. 1–5. IEEE (2017)
18. Somula, R., Sasikala, R.: A research review on energy consumption of different frameworks in mobile cloud computing. In: Innovations in Computer Science and Engineering, pp. 129–142. Springer, Singapore (2019)
19. Rao, N.P., Kannayaram, G., Ramasubbareddy, S., Swetha, E., Srinivas, A.S.: Software fault management using scheduling algorithms. J. Comput. Theor. Nanosci. **16**(5–6), 2124–2127 (2019)

20. Pramod Reddy, A., Ramasubbareddy, S., Kannayaram, G.: Parallel processed multi-lingual optical character recognition application. J. Comput. Theor. Nanosci. **16**(5–6), 2091–2095 (2019)
21. Kumar, I.P., Sambangi, S., Somukoa, R., Nalluri, S., Govinda, K.: Server security in cloud computing using block-chaining technique. In: Data Engineering and Communication Technology, pp. 913–920. Springer, Singapore (2020)
22. Kumar, I.P., Gopal, V.H., Ramasubbareddy, S., Nalluri, S., Govinda, K.: Dominant color palette extraction by K-means clustering algorithm and reconstruction of image. In: Data Engineering and Communication Technology, pp. 921–929. Springer, Singapore (2020)
23. Nalluri, S., Saraswathi, R.V., Ramasubbareddy, S., Govinda, K., Swetha, E.: Chronic heart disease prediction using data mining techniques. In: Data Engineering and Communication Technology, pp. 903–912. Springer, Singapore (2020)
24. Krishna, A.V., Ramasubbareddy, S., Govinda, K.: Task scheduling based on hybrid algorithm for cloud computing. In: International Conference on Intelligent Computing and Smart Communication 2019, pp. 415–421. Springer, Singapore (2020)
25. Srinivas, T.A.S., Ramasubbareddy, S., Govinda, K., Manivannan, S.S.: Web image authentication using embedding invisible watermarking. In: International Conference on Intelligent Computing and Smart Communication 2019, pp. 207–218. Springer, Singapore (2020)
26. Krishna, A.V., Ramasubbareddy, S., Govinda, K.: A unified platform for crisis mapping using web enabled crowdsourcing powered by knowledge management. In: International Conference on Intelligent Computing and Smart Communication 2019, pp. 195–205. Springer, Singapore (2020)
27. Saraswathi, R.V., Nalluri, S., Ramasubbareddy, S., Govinda, K., Swetha, E.: Brilliant corp yield prediction utilizing internet of things. In: Data Engineering and Communication Technology, pp. 893–902. Springer, Singapore (2020)
28. Kalyani, D., Ramasubbareddy, S., Govinda, K., Kumar, V.: Location-based proactive handoff mechanism in mobile ad hoc network. In: International Conference on Intelligent Computing and Smart Communication 2019, pp. 85–94. Springer, Singapore (2020)
29. Bhukya, K.A., Ramasubbareddy, S., Govinda, K., Srinivas, T.A.S.: Adaptive mechanism for smart street lighting system. In: Smart Intelligent Computing and Applications, pp. 69–76. Springer, Singapore (2020)
30. Srinivas, T.A.S., Somula, R., Govinda, K.: Privacy and security in Aadhaar. In: Smart Intelligent Computing and Applications, pp. 405–410. Springer, Singapore (2020)

Statistical Granular Framework Towards Dealing Inconsistent Scenarios for Parkinson's Disease Classification Big Data

D. Saidulu and R. Sasikala

Abstract While the medicinal and healthcare services sector is being changed by the competence to record gigantic measures of data about individual patients, the tremendous volume of information being gathered is outlandish for people to dissect/analyze. Over the past years, numerous techniques have been proposed so as to manage inconsistent data frameworks. Statistical applied ML facilitates an approach to consequently discover examples and reasoning about information. How one can transform raw data into valuable information that can empower healthcare professionals to make inventive automated clinical choices. The prior forecast and the location of disease cells can be profitable in curing the ailment in medical/healthcare appliances. This paper presents a novel statistical granular framework that deals with inconsistent instances, knowledge discovery, and further performs classification-based disease prediction. The experimental simulation is carried out on Parkinson's disease classification dataset. The experimental results and comparative analysis with some significant existing approaches prove the novelty and optimality of our proposed prototype.

Keywords Healthcare sector · Big Data · Machine learning · Inconsistent system · Medical applications · Knowledge discovery · Supervised learning

---

D. Saidulu
School of Computer Science and Engineering, VIT University, Vellore 632014, Tamilnadu, India
e-mail: fly2.sai@gmail.com

Department of Information Technology, Guru Nanak Institutions Technical Campus, Hyderabad, Telangana, India

R. Sasikala (✉)
Department of Computational Intelligence, School of Computer Science and Engineering, VIT University, Vellore 632014, India
e-mail: sasikala.ra@vit.ac.in

S. C. Satapathy et al. (eds.), *Intelligent Data Engineering and Analytics*, Advances in Intelligent Systems and Computing 1177,
https://doi.org/10.1007/978-981-15-5679-1_39

1 Introduction

The paradigm of Big Data isn't a recently evolved domain, be that as it may, the demeanor in which it is portrayed is continually emerging. Different cravings at describing Big Data primarily portray it as an assortment of knowledge granular factors whose size, streaming capability, type, as well as unpredictability, conjecture one to look for, receive and constitute new hardware-oriented and programming prototypes so as to adequately store, break down, process, and analyze the information components [1–3]. The healthcare sector is a prime case of how the three V's of information, velocity of the procreation of data, variation, and volume [4], are an intrinsic part of the knowledge set it yields.

1.1 Contribution Highlights

- A novel statistical granular framework is proposed that deals with inconsistent instances, knowledge discovery, and further performs classification-based disease prediction.
- We performed experiments on Parkinson's disease classification dataset. We discussed the stepwise algorithmic procedures and obtained results. We also performed the comparative analysis with significant other state-of-the-art approach.

1.2 Related Work

Wang [7] summarized different variety of methods and pathways for the Big Data analytics boom in particular to healthcare sector. Xing [8] discusses the strategies and prime principles of distributed machine learning on Big Data. Peek et al. [9], in their investigation talked about a portion of the Hadoop-based Big Data handling strategies like Oozie and Pig, Spark. Authors in [10] uncover that incorporating Big Data analysis into the medicinal healthcare sector can give the response to several significant inquiries in this sector. Discoveries from the investigation of Wang et al. [11] show that the advantages of Big Data analysis are upgraded its viability, proficiency, and enhancement of the specific clinical tasks. One of the investigations by Lee et al. [12] states that the unmistakable purposes behind the absence of clinical coordination of Big Data innovation are the deficiency of proof of functional advantages of Big Data paradigm in the healthcare sector. F. Rahman et al. proposed a novel and practically viable approach [13] of building a Big Data framework that can be adapted to diverse healthcare scenarios with particular compatible use. Vanathi et al. [14] presented a vigorous architectural schema for Big Data cascade computing.

1.3 Organization of the Paper

Section 2 discusses some significant preliminaries. Our proposed framework is given in Sect. 3. Experiments results' discussion is provided in Sect. 4. Finally, conclusion is given in Sect. 5.

2 Significant Preliminaries

2.1 Dimensionality and Heterogeneity of Data

Dimensionality in machine learning refers to how many features are present in dataset. For instance, medicinal services information is prominent for having tremendous measures of factors. In perfect system conditions, this information could be represented in a spreadsheet, with one section corresponding to each measurement [15–17].

2.2 Inconsistency Measure

Sometimes input data might be of inconsistent natured, i.e., some particular scenarios may rivalry nature with each other. Conflicting type cases have the ditto attribute (variable) values yet divergent decision values. Of the considerable number of points of view about what irregularities involve and how we can deal with them, one that escapes our consideration is that irregularities can fill in as powerful improvements to learning since they frequently help uncover the insufficiencies, gaps, lacks, or limit conditions in an operator's critical thinking knowledge.

3 Adopted Procedure

3.1 Detailed Algorithmic Steps

Procedure I: Dealing with Inconsistency -
BEGIN PROCEDURE

1. *i/p:* The tabular representation of attribute space where, $\{I_1, I_2, \cdots I_m\}$: represents m number of instances; $\{V_1, V_2, \cdots V_n\}$: represents total n number of vectors, act as conditional attributes; L: represents labeled decision vector.
2. $G(V_i)$: represents granular-set for i^{th} conditional attribute vector.
   G(L): represents granular-set for labeled decision vector.
   $\mathbb{L} \leftarrow\!\!\!\!\!\vdash G(L)$
3. $\forall \mathcal{S} = \{V_1, V_2, \cdots V_n\}$ in approximation universal space for particular concept $\mathbb{L}$, compute -
   - $\underline{\mathcal{S}}\mathbb{L} = \{x \in U | [x] \subseteq \mathbb{L}\}$
   - $\overline{\mathcal{S}}\mathbb{L} = \{x \in U | [x] \cap \mathbb{L} \neq \emptyset\}$

   where, $[x]$ is equivalence class
4. Compute, $\mathcal{R}_I = \overline{\mathcal{S}}\mathbb{L}$ - $\underline{\mathcal{S}}\mathbb{L}$
5. IF ($\mathcal{R}_I$ == NULL)
   {
       No inconsistent region
   }
   ELSE
   {
       $|\mathcal{R}_I| \leftarrow$ measure of inconsistency region
   }

END PROCEDURE

Procedure II: Classification based Disease Prediction -

BEGIN PROCEDURE

1. Every data object is plotted as a point in n-dimensional space (n: number of feature variables). Value of each feature will be mapped as values of particular co-ordinate. Classification is being performed by finding the hyperplane that differentiates the classes well.
2. Derive relaxed loss function as -
   $\theta_i = \{1 - f_{y_i}(x_i) + \frac{1}{k-1}\sum_{m \neq y} f_m(x_i)\}_+$
3. With the obtained bound in step 2, the unbiased primal problem is -
   $MIN_{w_m \in H, \theta \in R^l}(\frac{1}{2}\sum_{m=1}^{k} w_m^T w_m + C\sum_{i=1}^{l}\theta_i)$
   with dual optimization constraint,
   $w_{y_i}^T\theta(x_i) - \frac{1}{k-1}\sum_{m \neq y_i} w_m^T\theta(x_{i-1})\phi(x_i) \geq (1 - \theta_i);$
   $\theta_i \geq 0; i = 1 \cdots l$
   [marginal constraints are diminished to l]
4. Add dual set of variables, get the lagrangian -
   $L(w, \theta, \alpha, \lambda) = \frac{1}{2}\sum_{m=1}^{k} w_m^T w_m + C\sum_{i=1}^{l}\theta_i - \sum_{i=1}^{l}\lambda_i\theta_i - \sum_{i=1}^{l}\alpha_i(w_{y_i}^T\phi(x_i) - \frac{1}{k-1}\sum_{m \neq y_i} w_m^T\phi(x_i) - 1 + \theta_i)$
5. Differentiate the above lagrangian,
   $\frac{\partial L}{\partial w_m} = 0 \Longleftrightarrow w_m = \sum_{i:y_i=m}\alpha_i\phi(x_i) - \frac{1}{k-1}\sum_{i:y_i \neq m}\alpha_i\phi(x_i),$
   $\frac{\partial L}{\partial \theta} = 0 \Longleftrightarrow C_e = \lambda + \alpha,$
   constraints: $\alpha \geq 0; \lambda \geq 0$
6. Eliminate w_m, θ, λ and obtain -
   $MINIMIZE_{\alpha \in R^l}[\frac{1}{2}\alpha^T G\alpha - e^T\alpha]$
   subjected to interval: $\alpha \in [0, C_e]$
7. $l \times l$ Hessian matrix G possess its element entries -
   $G_{i,j} = \frac{K}{K-1}K_{i,j};\ if\ y_i = y_j$
   $G_{i,j} = \frac{-K}{(K-1)^2}K_{i,j};\ if\ y_i \neq y_j$
   here, the selected kernel method (either RBF or Poly case) value $\Gamma(x_i, x_j) \cong \phi(x_i)^T\phi(x_j)$ for $K_{i,j}$
8. Next, consider V be an $l \times K$ matrix with entries as -
   $V_{i,j} = 1, if y_i = j$
   $V_{i,j} = \frac{-1}{K-1}, if y_i \neq j$
9. We have $G = K \odot VV^T$ (Hadamard Product). Here, kernel matrix K & VV^T are both +ve semi-definite, so same for their Hadamard Product.
10. Optimize computation by avoiding the division operation in kernel computation, for this - assume, $\alpha = \frac{K}{(K-1)^2}\alpha$. Rewrite -
    $MINIMIZE_{\overline{\alpha} \in R^l}[\frac{1}{2}\overline{\alpha}^T\overline{G\alpha} - e^T\overline{\alpha}]$
    subjected to constraint $0 \leq \overline{\alpha} \leq \overline{C}$, where, $\overline{C} - \frac{K}{(K-1)^2.C}$
    $\overline{G}_{i,j} = (K-1)K_{i,j};\ if\ y_i = y_j$
    $\overline{G}_{i,j} = -K_{i,j};\ if\ y_i \neq y_j$
11. From step 5, obtain decision fn as -
    $\Gamma_m(x) = \sum_{i:y_i=m}\alpha_i^*\Gamma(x_i, x) - \frac{1}{k-1}\sum_{i,y_i \neq m}\alpha_i^*\Gamma(x_i, x)$
12. After simplification, final decision fn -
    $arg\ max_m\ f_m^*(x) = argmax_m\sum_{i:y_i=m}\alpha_i^*\Gamma(x_i, x)$

END PROCEDURE

3.2 Novelty Analysis of Adopted Framework

The proposed framework is a novel statistical granular framework that deals with inconsistent instances inside data (through procedure-I), carry out knowledge discovery, and further performs classification-based disease prediction (through procedure-II). Procedure-II is computeed efficiently that reasonably outperform as compared to some other methods existing in literature. It efficiently reduces the size of resultant dual problem from ($l \times K$) to l (l: number of samples; K: number of classes) by admitting more relaxed classification error bounds. This strategy with kernel methods—*RBF* & *Poly*—will result in competitive categorization and prediction accuracy.

4 Experimental Results Discussion

4.1 Setup, Simulation Environment, and Dataset Details

Our experiments were simulated using TensorFlow v' 1.2.1, with installed Python 3.7. Experiment instances were carried out on workstation having OS as Ubuntu 16.04.2, inbuilt 64 GB RAM, Intel Xeon processor having 12 cores with 2.0 GHz clock speed, in addition to NVIDIA GeForce GTX 1080 GPU which possesses 12 GB of global memory. The dataset utilized [18, 19] in this investigation was assembled from 188 patients with Parkinson's Disease (PD).

Data Set essence as: Multivariate; *Instances are:* 756; *Variable Characteristics:* Integer, Real; *Attributes count:* 754; *Missing Values:* N/A.

4.2 Obtained Results

The experiments are performed using the adopted framework. Table 1 shows the classification-based disease prediction accuracy, Matthew's correlation coefficient (MCC), and F1-score obtained. Figure 1a, b shows the graphical representation of the obtained results.

Table 1 Statistical performance results

Performance parameter	Value
Accuracy	0.89
F1-Score	0.87
Matthew's correlation coefficient (MCC)	0.61

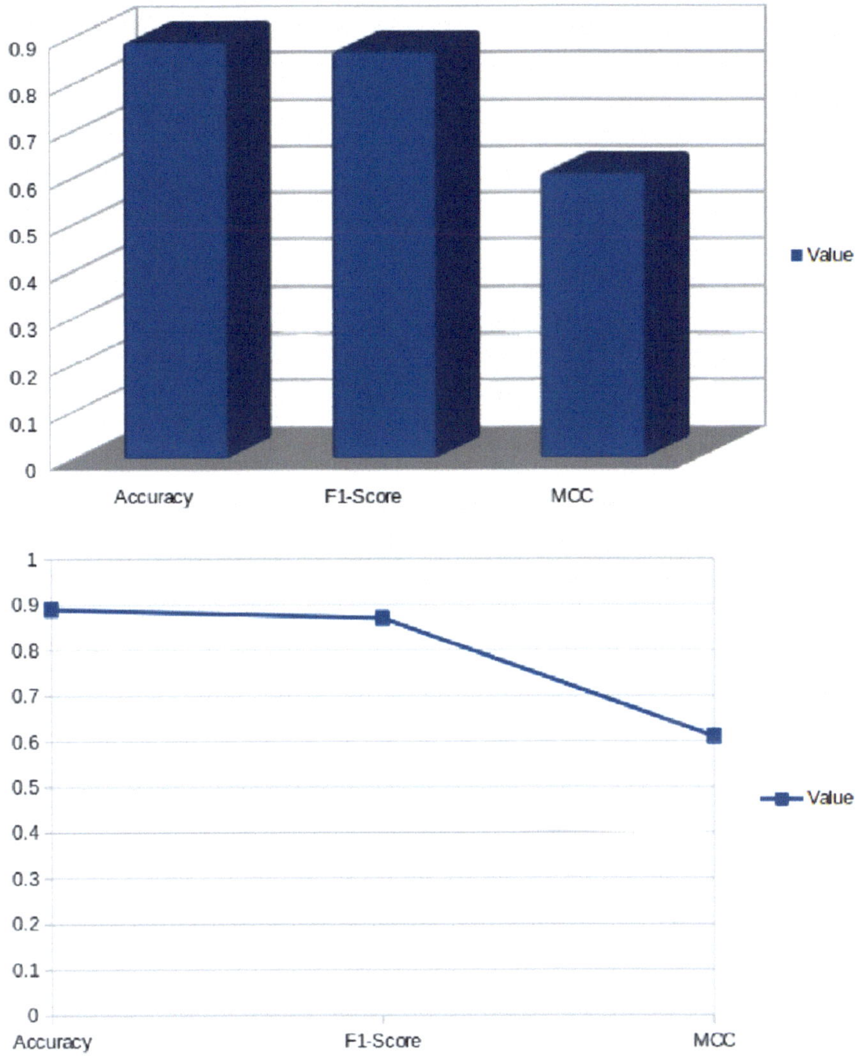

Fig. 1 **a** Obtained results Graph-1. **b** Obtained results Graph-2

4.3 Comparisons

Here, the comparisons are performed with significant existing approaches, i.e., Naive Bayes, Random Forest, SVM (RBF kernel), and SVM (Linear kernel). The comparative analysis is done with respect to some statistical performance parameters such as disease prediction accuracy, Matthew's correlation coefficient (MCC), and F1-score. The representations are given in Table 2 and corresponding graphs (Fig. 2a and b).

Table 2 Comparative analysis

Method	Accuracy	F1-Score	MCC
Naive Bayes	0.83	0.83	0.54
Random Forest	0.85	0.84	0.57
SVM (RBF)	0.86	0.84	0.59
SVM (Linear)	0.83	0.82	0.52
Our Adopted Framework	0.89	0.87	0.61

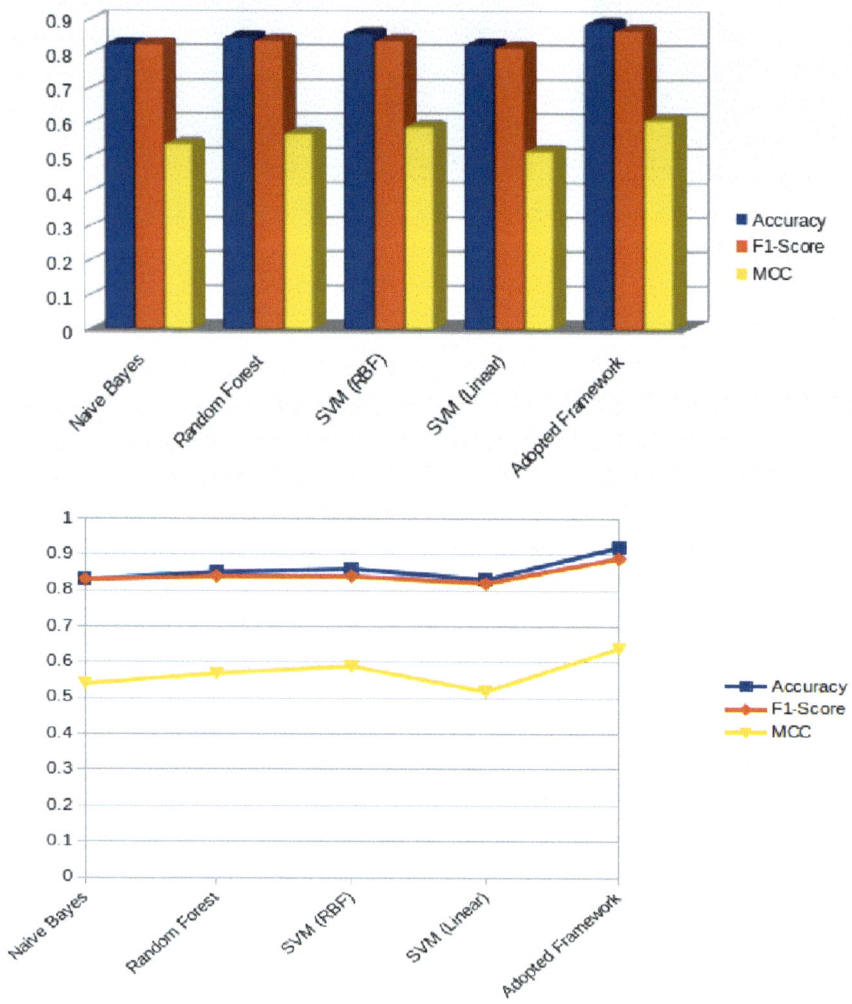

Fig. 2 **a** Comparisons Graph-2. **b** Comparisons Graph-2

The comparative analysis results show that the results obtained by our adopted framework outperform over other existing methods.

5 Conclusion

Today, to process the gigantic measured unstructured, ceaseless and ambiguous information by computing machines is a difficult exercise. In this study, we designed a framework which can process large-sized and inconsistent data efficiently and can optimally predict the mapping of unknown data instances. Comparisons are also performed with significant existing approaches to prove the novelty of proposed strategy.

References

1. McAfee, A., Brynjolfsson, E., Davenport, T.H., Patil, D.J., Barton, D.: Big data: the management revolution. Harvard Bus. Rev. **90**(10):60–68 (2012)
2. Lynch, C.: Big data: how do your data grow? Nature **455**(7209), 28–29 (2008)
3. Jacobs, A.: The pathologies of big data. Commun. ACM **52**(8), 36–44 (2009)
4. Zikopoulos, P., Eaton, C., et al.: Understanding big data: analytics for enterprise class hadoop and streaming data. McGraw-Hill Osborne Media (2011)
5. Rosler, O., Suendermann, D.: A first step towards eye state prediction using EEG. In: Proceedings of the AIHLS (2013)
6. Rajesh K.K., Sabarinathan, V., Kumar, S., Sugumaran V.: Eye state prediction using EEG signal and C4.5 decision tree algorithm, Int. J. Appl. Eng. Res. **10**(68) (2015). ISSN 0973-4562
7. Wang, Y., Hajli, N.: Exploring the path to big data analytics success in healthcare. J. Bus. Res. **70**, 287–299 (2017)
8. Xing, E.P., Ho, Q., Xie, P., Wei, D.: Strategies and principles of distributed machine learning on big data. Engineering **2**, 179–195 (2016)
9. Peek, N., Holmes, J., Sun, J.: Technical challenges for big data in biomedicine and health: data sources, infrastructure, and analytics. IMIA Yearb. **9**, 42–47 (2014)
10. Sukumar, S.R., Natarajan, R., Ferrell, R.K.: Quality of big data in health care. Int. J. Health Care Qual. Assur. **28**, 621–634 (2015)
11. Wang, Y., Hajli, N.: Exploring the path to big data analytics success in healthcare. J. Bus. Res. **70**, 287–299 (2017)
12. Cox, M., Ellsworth, D.: Application-controlled demand paging for out-of-core vi- sualization. Proc. Vis. **97**, 235–244 (1997)
13. Rahman, F., Slepian, M., Mitra, A.: A novel big-data processing framwork for healthcare applications: big-data-healthcare-in-a-box. In: IEEE International Conference on Big Data (Big Data) (2016). https://doi.org/10.1109/BigData.2016.7841018.
14. Vanathi, R., Khadir, A.S.A.: A robust architectural framework for big data stream computing in personal healthcare real time analytics. In: World Congress on Computing and Communication Technologies (WCCCT) (2017). https://doi.org/10.1109/WCCCT.2016.32
15. Wang, L., Alexander, C.A.: Machine learning in big data. Int. J. Math., Eng. Manag. Sci. **1**(2), 52–61 (2016)
16. L'Heureux, A., Grolinger, K., Elyaman, H.F., Capretz, A.M.: Machine learning with big data: challenges and approaches. IEEE Access (2017)

17. Qiu, J., Wu, Q., Ding, G., Xu, Y., Feng, S.: A survey of machine learning for big data processing. EURASIP J. Adv. Signal Process. (2016)
18. Sakar, C.O., Serbes, G., Gunduz, A., Tunc, H.C., Nizam, H., Sakar, B.E., Tutuncu, M., Aydin, T., Isenkul, M.E., Apaydin, H.: A comparative analysis of speech signal processing algorithms for Parkinson's disease classification and the use of the tunable Q-factor wavelet transform. Appl. Soft Comput. J. **74**(2019), 255–263 (2018)
19. https://archive.ics.uci.edu/ml/datasets/Parkinson%27s+Disease+Classification

Estimation of Sediment Load Using Adaptive Neuro-Fuzzy Inference System at Indus River Basin, India

Nihar Ranjan Mohanta, Paresh Biswal, Senapati Suman Kumari, Sandeep Samantaray, and Abinash Sahoo

Abstract Assessment of suspended sediments carried by streams and rivers is vital for planning and management of water resources structures and estimation of various hydrological parameters. More recently, soft computing techniques have been used in hydrological and environmental modeling. Adaptive Neuro-Fuzzy Inference System (ANFIS) is employed here to estimate sediment load at Indus River basin, India. Three different scenarios are considered to predict sediment load using ANFIS. Scenario one includes precipitation, temperature, and humidity as model input, but in case of scenario two, another one constraint infiltration loss is added with scenario one. Inclusion of evapotranspiration loss with scenario two forms scenario three that gives prominent value of performance. Mean square error (MAE) and coefficient determination (R^2) are applied here to evaluate efficiency of model. Six different membership functions Pi, Trap, Tri, Gauss, Gauss2, and Gbell are applied for model development. In case if Gbell functions, scenario three shows best value of efficacy with R^2 value 0.9811 and 0.9622 for training and testing phases, which is superior as compared to other two scenarios.

Keywords River basin · Sediment load · ANFIS · Evapotranspiration loss · Gbell function

---

N. R. Mohanta · P. Biswal · S. S. Kumari
Department of Civil Engineering, GIET University, Gunupur, Odisha, India
e-mail: niharthenew@gmail.com

P. Biswal
e-mail: biswalparesh3@gmail.com

S. S. Kumari
e-mail: senapatisuman02@gmail.com

S. Samantaray (✉) · A. Sahoo
Department of Civil Engineering, NIT Silchar, Silchar, Assam, India
e-mail: Samantaraysandeep963@gmail.com

A. Sahoo
e-mail: bablusahoo1992@gmail.com

S. C. Satapathy et al. (eds.), *Intelligent Data Engineering and Analytics*, Advances in Intelligent Systems and Computing 1177,
https://doi.org/10.1007/978-981-15-5679-1_40

1 Introduction

In the past decade, the necessity to accurately model suspended sediment has quickly increased in planning and management of water resources engineering. Recently, machine learning models to predict SSL of rivers have increasingly grown popularity amid investigators since advancement in computer models. ANN and ANFIS are two eminent models to predict hydraulic and hydrological proceedings. Numerous researches have been carried out on the applicability of ANNs in very important subject of hydrological modeling, for instance, to predict sediment load, design rainfall–runoff model, predict flow discharge, groundwater level, etc.

Buyukyildiz and Kumcu [1] investigated potential of Support Vector Machine (SVM), ANNs, and ANFIS for estimating SSL of Ispir Bridge gauge location of River Coruh. Samantaray and Ghose [2], [3] used black-box models and different ANN techniques to simulate and estimate SSL at Salebhata gauging station, Bolangir, Odisha. Sahoo et al. [4] compared prediction performances of BPNN and ANFIS approaches for flood susceptibility mapping at Basantpur watershed, Odisha, India. Rajaee et al. [5] considered ANNs, ANFIS, MLR, and predictable sediment rating curve models for modeling time series SSL in rivers. Ghose and Samantaray [6, 7] used regression and ANN models for predicting and developing flow and sediment prediction models for different tributaries of Mahanadi River basin during monsoon period. Azamathulla et al. [8] urbanized ANFIS, regression model, and gene expression programming (GEP) techniques for predicting SSL in Muda, Langat, and Kurau Rivers, Malaysia. Yekta et al. [9] used ANFIS and ANN to predict the SSL as a function of water discharge data, and results obtained were compared with rating curve method. Adnan et al. [10] proposed a dynamic evolving neural fuzzy inference system (DENFIS) as a substitute means for estimating SSL on basis of previous values of streamflow and sediment at Guangyuan and Beibei, China. Vafakhah [11] used ANN, ANFIS, cokriging, and normal kriging utilizing precipitation and streamflow data to forecast SSL of Kojor forest catchment close to Caspian Sea. Olyaie et al. [12] contrasted accuracy of ANNs, ANFIS, coupled wavelet ANN, and conventional SRC approaches to estimate daily SSL in two gauging stations in the USA. Samantaray et al. [13] applied RNN, SVM, and ANFIS for studying precipitation forecasting of Bolangir district, Odisha, India. Nivesh et al. [14] developed ANFIS, MLR, and SRC models for estimating SSL from Vamsadhara River basin, Odisha. Samantaray and Sahoo [15– 17] applied various machine learning algorithms and techniques for prediction and estimation of various hydrological parameters. The objective of this research is to explore sediment load via ANFIS.

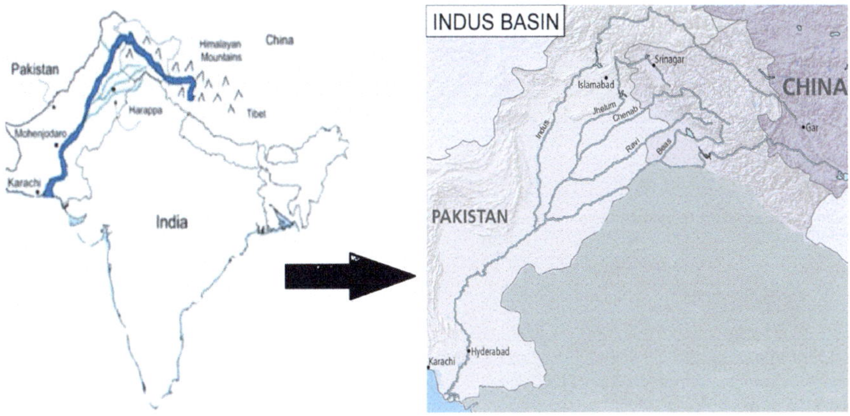

Fig. 1 Proposed research area

2 Study Area and Data

Indus is one among the longest flowing rivers of Asia. It has an overall drainage area of more than 1,165,000 km^2 having an annual flow approximately estimating at 243 km^3. It originates from Tibetan Plateau in neighborhood of Lake Manasarovar and discharges into the Arabian Sea. Length of the river is 3,180 km with 23°59′40″N 67°25′51″E coordinates. Indus plays a significant source of water for Pakistan and its financial system development. Particularly, it is the bread bin of Punjab territory that serves as country's major manufacture of agriculture goods (Fig. 1).

3 Methodology

3.1 ANFIS

ANFIS is a soft computing technique where specified input–output dataset is articulated in an FIS [18]. It is a type of FIS applied to the structure of adaptive networks. The FIS employs a nonlinear map from its input to output space. Effectiveness of fuzzy inference system (FIS) depends on the estimated parameters. Simulation of neuro-fuzzy relates to a function of applying various machine learning methods fabricated in NN literature to FIS [19]. This process is achieved by fuzzification of input using membership functions (MF), where curved relation records input value within an interval of [0–1]. Fuzzy membership constraints are optimized either by utilizing a backpropagation (BP) function or by the combination of both BP and least square techniques (Fig. 2).

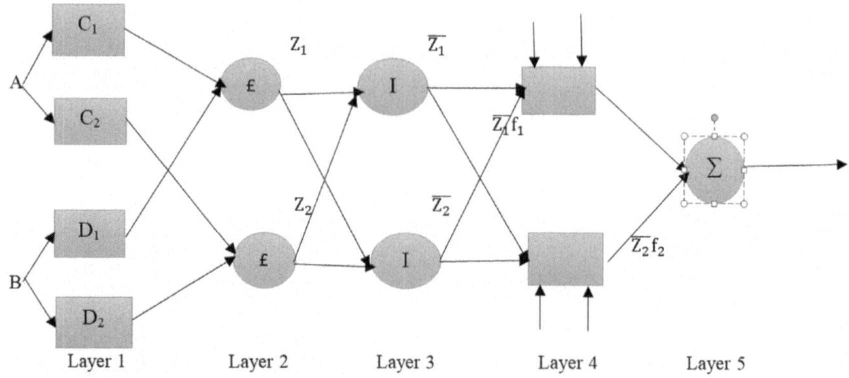

Fig. 2 Architecture of ANFIS

3.2 Data Set

Various climatic constraints like precipitation, temperature, infiltration loss, humidity, evapotranspiration losses, and sediment load of 30-year monsoons data (2018) are collected from IMD Delhi. From the entire data set, 70% of data are used for training purposes while 20% of data are considered for testing and rest 10% of data are considered for validation purposes. Before utilizing data, all the data sets are normalized to develop the consistency of input. Normalization of data was completed in accordance with equation referred here. Every data was scaled from range 0 to 1.

$$Km = (Kj - Kmin) / (Kmax - Kmin) \tag{1}$$

where Km = normalization.

Kj = actual value.

Kmax and Kmin = maximum and minimum measurement values.

Normalization eradicates random consequences of resemblance amid items and raises the answer data rate to input signal.

4 Results and Discussion

Table 1 illustrates relative potential of all scenarios deemed in the present study utilizing ANFIS. Six different MFs are regarded for ANFIS for finding finest model which can proficiently help in predicting sediment load in the proposed area. Three scenarios are considered here to develop model consistency. For scenario, one precipitation, temperature, and humidity are considered for input parameter to develop model. Results show that Gbell membership function gives best value of performance with R^2 0.8928 and 0.8635 for training and testing phases. Similarly, for

Table 1 Comparative performance of ANFIS under different scenarios

Scenario	Function	MAE		R^2	
		Training	Testing	Training	Testing
Precipitation Temperature Humidity	Pi	0.009536	0.173208	0.8487	0.8009
	Trap	0.012841	0.224851	0.8576	0.8187
	Tri	0.022638	0.361476	0.8682	0.8226
	Gauss	0.035187	0.419432	0.8714	0.8412
	Gauss2	0.043329	0.558641	0.8849	0.8526
	Gbell	0.055743	0.617854	0.8928	0.8635
Precipitation Temperature Humidity Infiltration loss	Pi	0.022543	0.031546	0.8775	0.8385
	Trap	0.035438	0.047529	0.8814	0.8412
	Tri	0.041876	0.058732	0.8997	0.8537
	Gauss	0.059466	0.069143	0.9058	0.8698
	Gauss2	0.068421	0.073427	0.9129	0.8835
	Gbell	0.077396	0.088421	0.9247	0.899
Precipitation Temperature Infiltration loss Humidity Evapotranspiration losses	Pi	0.034428	0.047114	0.9366	0.9109
	Trap	0.044164	0.058321	0.9428	0.9221
	Tri	0.050052	0.068435	0.9564	0.9316
	Gauss	0.064386	0.073998	0.9681	0.9402
	Gauss2	0.077538	0.081675	0.9786	0.9512
	Gbell	0.087485	0.099834	0.9811	0.9622

scenario two (humidity is added with the previous scenario) Gbell function shows the best value of performance with R^2 value 0.9247 and 0.899. For scenario three, precipitation, temperature, infiltration loss, humidity, and evapotranspiration losses are employed for model input. Same as previous one, Gbell shows the paramount value of efficiency with R^2 value 0.9811 and 0.9622. For fairly comparing goal of model and constraints for predicting sediment load, precipitation, temperature, and infiltration loss are taken the same for the entire scenario.

Actual verses predicted sediment loads for three different scenarios of the proposed watershed are shown in Fig. 3.

5 Conclusion

Present study evaluates the potential of ANFIS model in sediment load prediction considering different performance standards. Three scenarios were urbanized for studying the effect of evapotranspiration and infiltration losses to estimate sediment yield. Scenario 3 gives better outcomes than other scenarios because of the addition

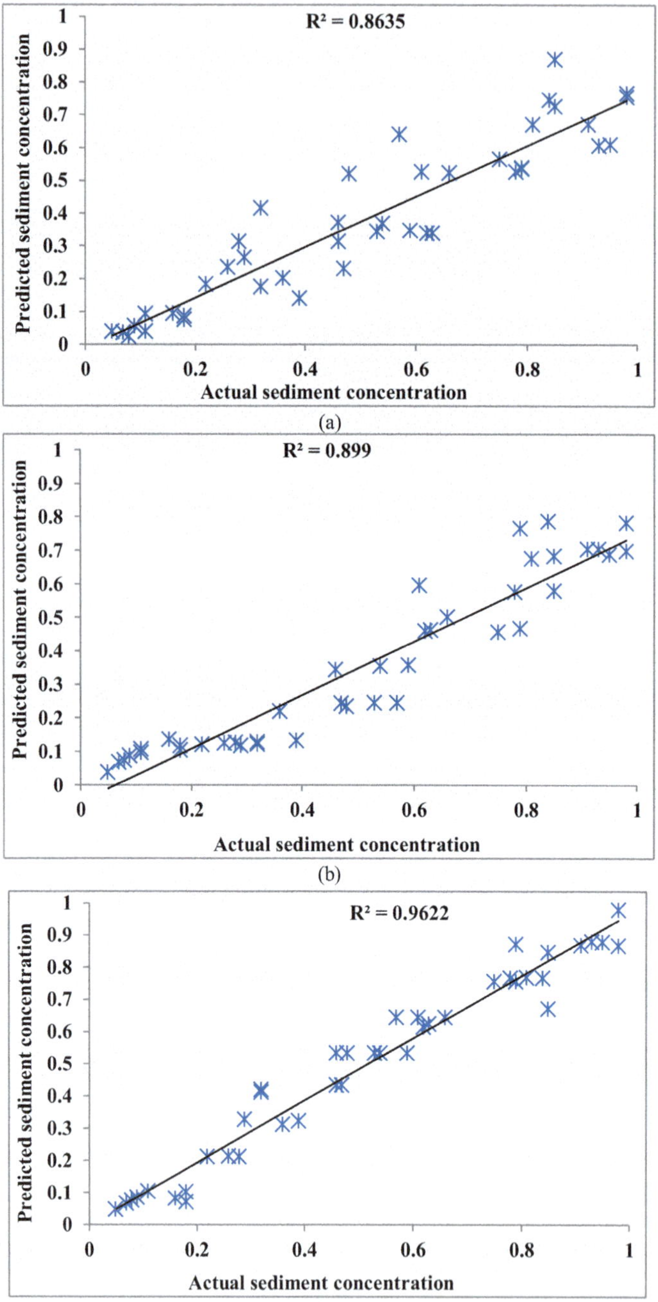

Fig. 3 Actual verses predicted sediment concentration of **a** scenario one, **b** scenario two, **c** scenario three

of losses due to evapotranspiration. Obtained outcomes show that inclusions of evapotranspiration to rainfall, temperature, and humidity are important aspects to predict sediment yield. From the present study, it can be found that Gbell function develops potential of ANFIS model by a considerable amount, and hence gives better performance than other membership functions. Results show that scenario 3 of ANFIS model produces best value of R^2 for both training and testing phases. The proposed model can also be utilized for other catchments where the sediment load data is not available for future research purposes.

References

1. Buyukyildiz, M., Kumcu, S.Y.: An estimation of the suspended sediment load using adaptive network based fuzzy inference system, support vector machine and artificial neural network models. Water Resour. Manag. **31**(4), 1343–1359 (2017)
2. Samantaray, S., Ghose, D.K.: Evaluation of suspended sediment concentration using descent neural networks. Procedia Comput. Sci. **132**, 1824–1831 (2018a)
3. Samantaray, S., Ghose, D.K.: Evaluation of suspended sediment concentration using descent neural networks. Procedia comput Sci. **132**, 1824–1831 (2018b)
4. Sahoo, A., Samantaray, S., Bankuru, S., Ghose, D.K.: Prediction of flood using adaptive neuro-fuzzy inference systems: a case study. In: Smart Intelligent Computing and Applications, pp. 733–739. Springer, Singapore (2020)
5. Rajaee, T., Mirbagheri, S.A., Zounemat-Kermani, M., Nourani, V.: Daily suspended sediment concentration simulation using ANN and neuro-fuzzy models. Sci. Total Environ. **407**(17), 4916–4927 (2009)
6. Ghose, D.K., Samantaray, S.: Modelling sediment concentration using back propagation neural network and regression coupled with genetic algorithm. Procedia Comput. Sci. **125**, 85–92 (2018)
7. Ghose, D.K., Samantaray, S.: Sedimentation process and its assessment through integrated sensor networks and machine learning process. In: Computational Intelligence in Sensor Networks, pp. 473–488. Springer, Berlin, Heidelberg (2019)
8. Azamathulla, H.M., Cuan, Y.C., Ghani, A.A., Chang, C.K.: Suspended sediment load prediction of river systems: GEP approach. Arab. J. Geosci. **6**(9), 3469–3480 (2013)
9. Yekta, A.H.A., Marsooli, R., Soltani, F.: Suspended sediment estimation of Ekbatan reservoir sub basin using adaptive neuro-fuzzy inference systems (ANFIS), artificial neural networks (ANN), and sediment rating curves (SRC). In: Dittrich, Koll, Aberle, Geisenhainer (eds.) River Flow, pp. 807–813 (2010)
10. Adnan, R.M., Liang, Z., El-Shafie, A., Zounemat-Kermani, M., Kisi, O.: Prediction of suspended sediment load using data-driven models. Water **11**(10), 2060 (2019)
11. Vafakhah, M.: Comparison of cokriging and adaptive neuro-fuzzy inference system models for suspended sediment load forecasting. Arab. J. Geosci. **6**(8), 3003–3018 (2013)
12. Olyaie, E., Banejad, H., Chau, K.W., Melesse, A.M.: A comparison of various artificial intelligence approaches performance for estimating suspended sediment load of river systems: a case study in United States. Environ. Monit. Assess. **187**(4), 189 (2015)
13. Samantaray, S., Sahoo, A., Ghose, D.K.: Assessment of runoff via precipitation using neural networks: watershed modelling for developing environment in arid region. Pertan. J. Sci. Technol. **27**(4), 2245–2263 (2019)
14. Nivesh, S., Kumar, P.: River suspended sediment load prediction using neuro-fuzzy and statistical models: Vamsadhara river basin, India. World **2**, 1 (2018)

15. Samantaray, S., Sahoo, A.: Estimation of runoff through BPNN and SVM in Agalpur watershed. In: Frontiers in Intelligent Computing: Theory and Applications, pp. 268–275. Springer, Singapore (2020)
16. Samantaray, S., Sahoo, A.: Appraisal of runoff through BPNN, RNN, and RBFN in Tentulikhunti watershed: a case study. In: Frontiers in Intelligent Computing: Theory and Applications, pp. 258–267. Springer, Singapore (2020)
17. Samantaray, S., Sahoo, A.: Assessment of sediment concentration through RBNN and SVM-FFA in Arid watershed, India. In: Smart Intelligent Computing and Applications, pp. 701–709. Springer, Singapore (2020)
18. Jang, J.S.R.: ANFIS adaptive–network-based-fuzzy inference systems. IEEE Trans. Syst. Man Cybern. **23**(3), 665–685 (1993)
19. Brown, M., Harris, C.: Neuro-fuzzy Adaptive Modelling and Control. Prentice-Hall, Upper Saddle River, New Jersey (1994)

Efficiency of River Flow Prediction in River Using Wavelet-CANFIS: A Case Study

Nihar Ranjan Mohanta, Niharika Patel, Kamaldeep Beck, Sandeep Samantaray, and Abinash Sahoo

Abstract Application of coactive neuro-fuzzy inference system (CANFIS) and wavelet coactive neuro-fuzzy inference system (WCANFIS) models to predict river flow time series is investigated in the present study. Monthly river flow time series for a period of 1989–2011 of Ganga River, India were used. To obtain the best input–output mapping, different input combinations of antecedent monthly river flow and a time index were evaluated. Both model outcomes were contrasted using mean absolute error (MAE) and coefficient of determination (R^2). Assessment of models signifies that WCANFIS model predicts more accurately than CANFIS model for monthly river flow time series. In addition, outcomes revealed that inclusion of surface runoff and evapotranspiration loss parameters to input of models enhances the accuracy of prediction more appreciably.

Keywords CANFIS · Wavelet-CANFIS · River · Flow discharge · India

N. R. Mohanta · N. Patel · K. Beck
Department of Civil Engineering, GIET University, Gunupur, Odisha, India
e-mail: niharthenew@gmail.com

N. Patel
e-mail: niharikadream@gmail.com

K. Beck
e-mail: kamalbeck.789@gmail.com

S. Samantaray (✉) · A. Sahoo
Department of Civil Engineering, NIT Silchar, Silchar, Assam, India
e-mail: Samantaraysandeep963@gmail.com

A. Sahoo
e-mail: bablusahoo1992@gmail.com

S. C. Satapathy et al. (eds.), *Intelligent Data Engineering and Analytics*, Advances in Intelligent Systems and Computing 1177,
https://doi.org/10.1007/978-981-15-5679-1_41

1 Introduction

Amid various activities linked with how to plan and operate different constituents of a water resource system requires prediction of the occurrence of future proceedings. Most significant course in lifecycle of water is the location where precipitation takes place and resulting in runoff flow. Flow becomes a very vital parameter for numerous actions like design of structures relating to flood safety purposes for built-up localities and farming land and to assess the quantity of water that might be extorted from a stream for supplying water for different uses or for irrigation. Because accurateness to estimate flow is extremely essential, few models that deal with meteorological, hydrological, and geologic parameters should be enhanced. Hence, managing water and functioning of water-related structures successfully will be achievable.

Shoaib and Shamseldin [1] explored potential of WCANFIS hybrid model for simulating rainfall–runoff model alteration in Baihe watershed, China and examined a suitable setting assortment of NF rainfall–runoff process on basis of wavelets. Rathod and Singh [2] investigated and assessed usefulness of CANFIS models for recreating rainfall from a catchment, and accuracy of models is accessed on the basis of root mean square error (MSE), coefficient of efficiency (CE), and correlation coefficient (r) for the period of June to September in Nagpur, Maharashtra, India. Malik and Kumar [3] compared the potential of CANFIS, MLP, and multiple nonlinear and linear regressions for simulating daily discharge at Tekra site on Pranhita River basin, India. Abghari and Ahmadi [4] surveyed various sorts of mother wavelet as activated functions rather than usually utilized sigmoid for finding the principle contrasts in the consequences of day-by-day skillet dissipation forecast in the Lar synoptic station by using wavelet theory and multilayer perceptron (MLP) network. Gholami and Khaleghi [5] utilized CANFIS for simulating groundwater quality. In addition, geographic information system (GIS) was utilized as preprocessor and postprocessor system to exhibit spatial variety of groundwater quality. Heydari and Talaee [6] inspected capability of CANFIS to estimate course through trapezoidal and rectangular rockfill dams and outcomes demonstrated that precise stream forecasts can be accomplished using CANFIS with Takagi–Sugeno–Kang (TSK) fuzzy model and Bell membership function. Various neural network techniques are utilized for prediction or evaluation of climatic indices on monthly and yearly basis in gauged watershed in India [7–13]. Malik and Kumar [14] utilized CANFIS, MLP, MLR, and MNLR, and sediment rating curve (SRC) methods to simulate daily suspended sediment concentration at Tekra gauging site on Pranhita River, Andhra Pradesh, India. Memarian et al. [15] evaluated the ability of CANFIS to forecast drought in Birjand, Iran by combining global climatic signals with precipitation and delayed values of standardized rainfall indicator. Tajdari and Chavoshi [16] developed radial overcut prescient models utilizing multiple regression analysis, ANN, and co-dynamic neuro-FIS to forecast the radial overcut during electro-compound penetration with vacuum extraction of electrolyte. The objective of this research is to predict flow discharge in Ganga River basin, India.

2 Study Area

Ganga is a trans-frontier river stream of Asia flowing in the course of India and Bangladesh located within coordinates 30°59′N 78°55′E. Ganga emerges from the western Himalayas in Uttarakhand, India and streams through southern and eastern parts of India in the course of Gangetic Plain in India and Bangladesh, ultimately draining into the Bay of Bengal. Major stem of this river starts at convergence of Bhagirathi and Alaknanda streams in Devprayag in Garhwal town, Uttarakhand. Ganges length is often to some extent more than 2,500 km long with basin area 1,080,000 km^2. Maximum and minimum discharges of Ganges are 70,000 m^3/s and 2,000 m^3/s, respectively, with an average discharge of 16,648 m^3/s. Over 95% of upper plain of river Ganges has been besmirched or transformed into farming or town areas (Fig. 1).

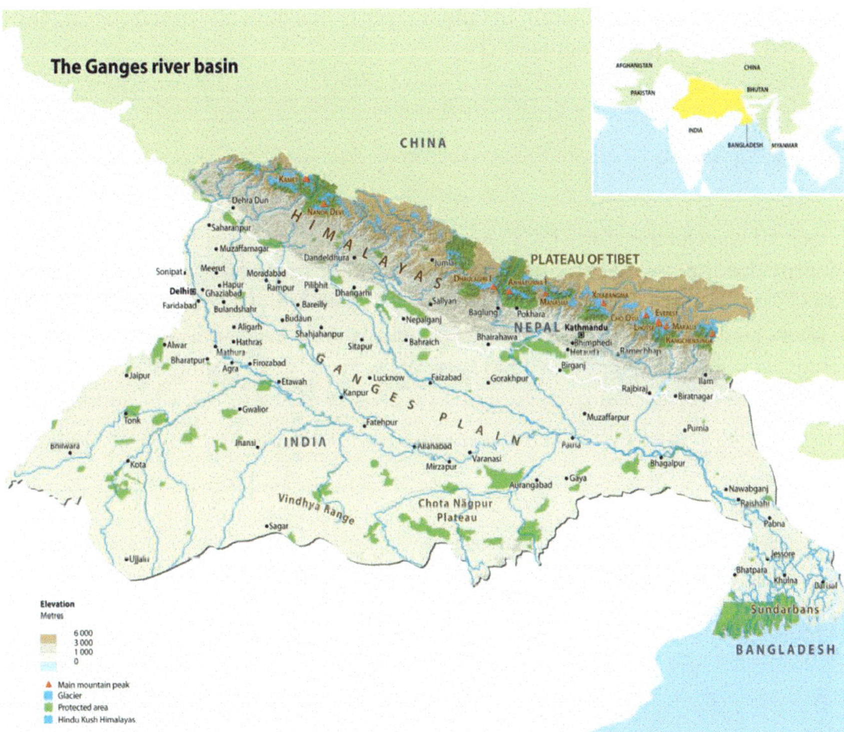

Fig. 1 Proposed river basin

3 Methodology

3.1 CANFIS

Combination of ANN and fuzzy rules results in the formation of NF architecture and can be utilized for total estimation of all kinds of nonlinear functions. Major component of a CANFIS network constitutes of fuzzy neuron that pertains to membership function (MF) to inputs. Bell and Gaussian functions are frequently utilized as MFs. This network embraces a normalizing axon as well which normalizes output amid 0 and 1. Second major constituent of this structure is a modular network which pertains functional rules to inputs. Modular network quantity matches output quantity, and dispensation rudiment amount is equivalent to the amount of MFs. Amid the majority of FIS general kind of fuzzy structure that has the capability to place in an adaptive structure is Sugeno FIS and its outcome is on the basis of linear regression equation. It is notable that transfer function in output layer is linear. In NF structure, CANFIS is utilized as feed-forward network (Fig. 2).

3.2 WCANFIS

Concept of wavelet transform (WT) was introduced for representing time series data. WT provides time-scale depiction of non-stationary time series data and all of its associations. Fourier Transform (FT) is utilized as an initial tip for introducing WT. FT alters signal from time field into a frequency field with time information loss in frequency field. There are two kinds of WTs, namely, incessant WT and distinct WT. Various WTs are categorized based on distinct characteristics of support area and declining instances number. Support area of WT is connected with wavelet span

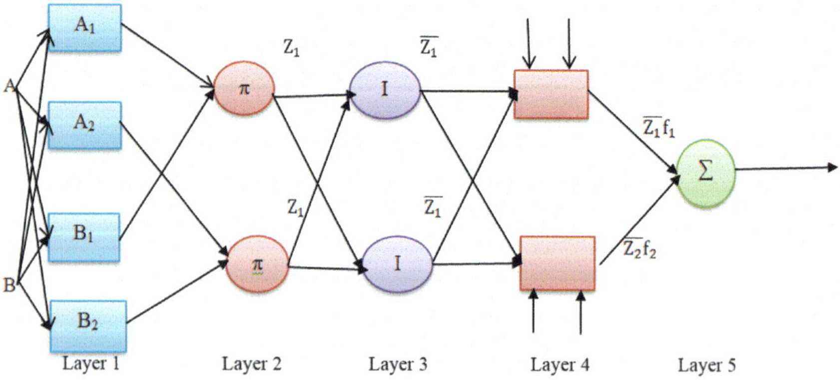

Fig. 2 Architecture of CANFIS

extent. Localized assets and information contented in a signal are mostly exaggerated by span length of a wavelet. The present study uses CANFIS model integrating with DWT to develop hybrid wavelet-CANFIS model. CANFIS model can be urbanized inclusive and exclusive of WT.

3.3 Model Formulation

The monthly average (2004–2018) river flow, precipitation, temperature, and seepage losses data are considered as model evaluation. 126 data are regarded for training network; straddling over 2004–2013 for the model design and 2014–2018 data sets are employed for testing. Primarily data sets are standardized for falling inside range 0–1. Subsequent after being standardized 70% of chronological input data are utilized for training and 30% utilized for testing. In this study, networks are trained with various membership functions on the requirement basis of developing and designing the model.

3.4 Model Performance

In determining the performance results of certain network models, two statistical constraints are utilized, i.e., mean absolute error (MAE) and coefficient of determination (R^2).

MAE can be evaluated utilizing equation mentioned below:

$$\mathrm{MAE} = \frac{1}{m}\sum_{i=1}^{m}\left|Z_{\mathrm{pre}} - Z_{\mathrm{act}}\right| \tag{1}$$

where Z_{act} is the actual output and Z_{pre} is the predicted output. R^2 represents square of correlation amid actual and predicted results. R^2 estimates variance inferred by model and ranges from 0 to 1.

4 Results and Discussions

Performance evaluations of various indicators for five different functions, that is, Pi, Tri, Trap, Gbell, and Gauss are described in Table 1. Three scenarios P (scenario one), P-T (scenario two), and P-T-L (scenario three) are accessible for model efficiency at the proposed river basin. Scenario one (P) gives the best value of performance for CANFIS and wavelet-CANFIS with MAE value of 0.050012, 0.151849 and 0.077213, 0.178456. While considering scenario two, (P-T) for CANFIS

Table 1 Performance of model using CANFIS and wavelet-CANFIS

Scenario	Function	CANFIS				Wavelet-CANFIS			
		MAE		R^2		MAE		R^2	
		Training	Testing	Training	Testing	Training	Testing	Training	Testing
P	Pi	0.021116	0.100009	0.8315	0.7812	0.040097	0.100087	0.8669	0.8149
	Tri	0.029745	0.100054	0.8574	0.8083	0.048723	0.101673	0.8891	0.8359
	Trap	0.038478	0.132587	0.8843	0.8148	0.056348	0.155675	0.9138	0.8458
	Gbell	**0.050012**	**0.151849**	**0.9001**	**0.8702**	**0.077213**	**0.178456**	**0.9384**	**0.9017**
	Gauss	0.043643	0.146278	0.8995	0.8368	0.066785	0.164879	0.9245	0.8619
P-T	Pi	0.030914	0.110087	0.8589	0.7927	0.051123	0.142671	0.8809	0.8264
	Tri	0.031467	0.119376	0.8738	0.8137	0.053897	0.133597	0.9028	0.8469
	Trap	0.042765	0.138791	0.8915	0.8309	0.066751	0.155176	0.9226	0.8684
	Gbell	**0.069748**	**0.178548**	**0.9138**	**0.8895**	**0.081784**	**0.192741**	**0.9429**	**0.9138**
	Gauss	0.050265	0.170006	0.9075	0.8525	0.078542	0.193325	0.9341	0.8878
P-T-L	Pi	0.036487	0.126538	0.8806	0.8018	0.055342	0.148537	0.9184	0.8317
	Tri	0.046785	0.142674	0.8982	0.8254	0.069874	0.162743	0.9257	0.8558
	Trap	0.050737	0.155785	0.9089	0.8423	0.076657	0.178509	0.9332	0.8749
	Gbell	**0.074768**	**0.172876**	**0.9263**	**0.8974**	**0.091734**	**0.192004**	**0.9552**	**0.9273**
	Gauss	0.069254	0.163961	0.9117	0.8699	0.084276	0.180036	0.9338	0.8997

and Wavelet-CANFIS best value of MAE are 0.069748, 0.178548 and 0.081784, 0.192741. Similarly, scenario three (P-T-L) shows paramount value of MAE of 0.074768, 0.172876 and 0.091734, 0.192004 for CANFIS and Wavelet-CANFIS, respectively.

It can be viewed from Table 1 that every scenario comes inside tolerance limit of error. It is established that insertion of losses in scenarios 1 and 2 develops model effectiveness. For the entire scenario, Gbell function shows the best value of performance for both training and testing phases. In the case of CANFIS (scenario 3), best values of R^2 are 0.9263 and 0.8974 for training and testing phases. Similarly, in the case of Wavelet-CANFIS, paramount values of R^2 are 0.9552 and 0.9273. Performance graph in context to R^2 is presented in Fig. 3.

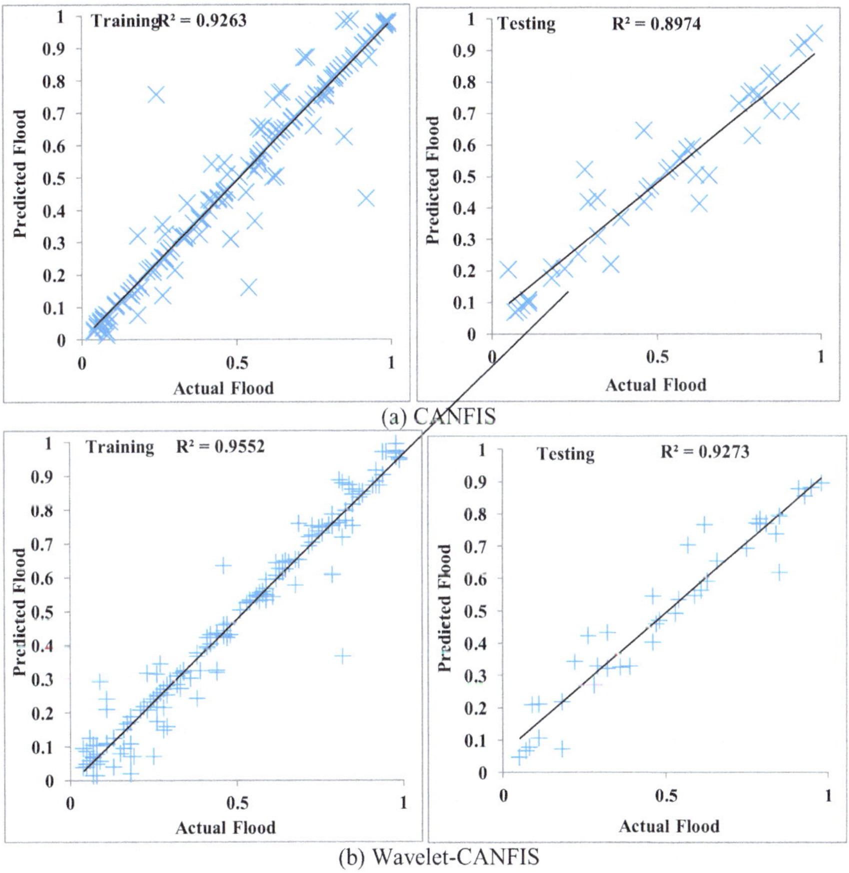

Fig. 3 Actual versus predicted flood using **a** CANFIS and **b** wavelet-CANFIS

5 Conclusions

Present study evaluates the performance of CANFIS and Wavelet-CANFIS models to predict flow discharge using different performance criteria along with six numbers of membership functions. Various scenarios were urbanized for studying effect of rainfall, infiltration losses, and evapotranspiration losses to estimate flow discharge. Scenario (4 and 5) gives better results as compared to others because of insertion of surface runoff and evapotranspiration losses. Outcomes depicted that the addition of these selected parameters to rainfall, temperature, and humidity play important aspects to predict flow discharge. Even though both CANFIS and Wavelet-CANFIS can predict flow with soaring accurateness, the present study concludes that Wavelet-CANFIS helps in improving model performance by a considerable amount and hence presents better results than CANFIS. Both numerical and empirical methods may assist in this investigation, so that a better efficiently working model can help in predicting flow discharge more accurately. Major benefit of utilizing Wavelet-CANFIS is that it incorporates both with NN and fuzzy logic principles.

References

1. Shoaib, M., Shamseldin, A.Y.: Hybrid wavelet neuro-fuzzy approach for rainfall-runoff modeling. J. Comput. Civil Eng. **30**(1) (2016)
2. Rathod, T., Singh, V.: Rainfall prediction using co-active neuro fuzzy inference system for Umargaon watershed Nagpur India. J. Pharmacogn. Phytochem. **7**(5), 658–662 (2018)
3. Malik, A., Kumar, A.: Comparison of soft-computing and statistical techniques in simulating daily river flow: a case study in India. J. Soil Water Conserv. **17**(2), 192–199 (2018)
4. Abghari, H., Ahmadi, H.: Prediction of daily pan evaporation using wavelet neural networks. Water Resour. Manag. **26**(12), 3639–3652 (2012)
5. Gholami, V., Khaleghi, M.R.: A method of groundwater quality assessment based on fuzzy network-CANFIS and geographic information system (GIS). Appl. Water Sci. **7**(7), 3633–3647 (2016)
6. Heydari, M., Talaee, P.H.: Prediction of flow through rockfill dams using a neuro-fuzzy computing technique. Int. J. Appl. Math. Comput. Sci. **2**(3), 515–528 (2011)
7. Ghose, D.K., Samantaray, S.: Integrated sensor networking for estimating groundwater potential in scanty rainfall region: challenges and evaluation. Computational Intelligence in Sensor Networks. Studies in Computational Intelligence, vol. 776, pp. 335–352 (2019)
8. Samantaray, S., Sahoo, A.: Appraisal of runoff through BPNN, RNN, and RBFN in Tentulikhunti watershed: a case study. In: Satapathy, S., Bhateja, V., Nguyen, B., Nguyen, N., Le, D.N. (eds.) Frontiers in Intelligent Computing: Theory and Applications. Advances in Intelligent Systems and Computing, vol. 1014. Springer, Singapore (2020)
9. Samantaray, S., Sahoo, A.: Estimation of runoff through BPNN and SVM in Agalpur watershed. In: Satapathy, S., Bhateja, V., Nguyen, B., Nguyen, N., Le, D.N. (eds.) Frontiers in Intelligent Computing: Theory and Applications. Advances in Intelligent Systems and Computing, vol. 1014. Springer, Singapore (2020)
10. Samantaray, S., Sahoo, A.: Assessment of sediment concentration through RBNN and SVM-FFA in Arid watershed, India. In: Satapathy, S., Bhateja, V., Mohanty, J., Udgata, S. (eds.) Smart Intelligent Computing and Applications. Smart Innovation, Systems and Technologies, vol. 159. Springer, Singapore (2020)

11. Samantaray, S., Ghose, D.K.: Sediment assessment for a watershed in arid region via neural networks. Sādhanā **44**(10), 219 (2019)
12. Samantaray, S., Sahoo, A., Ghose, D.K.: Assessment of runoff via precipitation using neural networks: watershed modelling for developing environment in arid region. Pertan. J. Sci. Technol. **27**(4), 2245–2263 (2019)
13. Das, U.K., Samantaray, S., Ghose, D.K., Roy, P.: Estimation of aquifer potential using BPNN, RBFN, RNN, and ANFIS. In: Smart Intelligent Computing and Applications. Smart Innovation, Systems and Technologies, vol. 105, pp. 569–576 (2019)
14. Malik, A., Kumar, A.: Daily suspended sediment concentration simulation using hydrological data of Pranhita River Basin, India. Comput. Electron. Agric. **138**, 20–28 (2017)
15. Memarian, H., Bilondi, M.P.: Drought prediction using co-active neuro-fuzzy inference system, validation, and uncertainty analysis (case study: Birjand, Iran). Theor. Appl. Climatol. **125**(3–4), 541–554 (2015)
16. Tajdari, M., Chavoshi, S.Z.: Prediction and analysis of radial overcut in holes drilled by electrochemical machining process. Central Eur. J. Eng. **3**(3), 466–474 (2013)

Customer Support Chatbot Using Machine Learning

R. Madana Mohana, Nagarjuna Pitty, and P. Lalitha Surya Kumari

Abstract In customer support, chatbot by using machine learning customer can converse by a chatbot and acquire the query intent information. With the enhancement of globalization and industrialization, it becomes a problem for enterprises to interact with the customer and listen to their difficulties to a big extent. Chatbots make ease the pain that the industries nowadays facing. The aim of this chatbot is to support and reply to the client by giving him/her the relevant intent depending on the query request from the customers.

Keywords Chatbot · Query · Machine learning · Natural language processing · Artificial intelligence

1 Introduction

1.1 Chatbot

A chatbot could be a piece of software that conducts an oral communication via sensory system or textual strategies. Such programs are usually designed to convincingly simulate; however, a person would behave as a usual associate, though as of 2019, they are in need of having the ability to pass the Turing test. Chatbots generally

R. Madana Mohana (✉)
Department of Computer Science and Engineering, Bharat Institute of Engineering and Technology, Ibrahimpatnam, Hyderabad 501510, Telangana, India
e-mail: madanmohanr@biet.ac.in

N. Pitty
Indian Institute of Science, Bangalore, India
e-mail: nagarjuna@iisc.ac.in

P. Lalitha Surya Kumari
Department of Computer Science and Engineering, Koneru Lakshmaiah Education Foundation, Deemed to be University, Hyderabad 500075, Telangana, India
e-mail: vlalithanagesh@gmail.com

S. C. Satapathy et al. (eds.), *Intelligent Data Engineering and Analytics*, Advances in Intelligent Systems and Computing 1177,
https://doi.org/10.1007/978-981-15-5679-1_42

measured in dialog systems for sensible functions together with customer service or information acquisition. Some chatbots use NLP systems; however, several easier ones scan for keyword within the input, so pull a reply with the foremost matching keywords, or the foremost similar pattern, from the dataset [1]. There are typically three types of chatbots. They are:

- Rule-based chatbots,
- Retrieval-based chatbots, and
- Self-learning chatbots.

1.2 Natural Language Processing

Natural language processing is a related field of artificial intelligence which can be used to process the normal language data such as audio, text, video, and image. Natural language processing acted as a tool for computer to know and examine the real-time data in human language. Natural language processing application areas are information extraction, machine translation, question answering, and text summarization [2].

The essence of natural language processing lies in making computers perceive the natural language. It is not a simple task. Computers will know the structured variety of information resembling the tables within the data and also spreadsheets; however, texts, voices, and human languages form the class of knowledge from unstructured data, and it gets troublesome for the computer to know it, and there arises the necessity for language process. There's huge amount of language information out there in varied forms, and it might get terribly simple if computers will perceive and method that information. We are able to prepare the models in accordance with expected output in numerous ways by training the data. Various challenges floating out there like understanding that correct named entity recognition, meaning of the sentence, and correct prediction of varied elements of speech, conference resolution [3].

1.3 Machine Learning (ML)

ML is a process to learn from knowledge based on some tasks and performance measures. A variety of ML areas are statistics, Bayesian methods, information theory, philosophy, computational complexity theory, psychology and neurobiology, artificial intelligence, and control theory. Some of the applications of machine learning are learning to drive an autonomous vehicle, learning to recognize spoken words, etc. [4].

2 Related Work

AIM and Facebook chatbots are the most popular in scientific discourse; many chat applications have been set up since their advent to chat with users. The classic example is, which played the role of a psychotherapist in 1966, followed by Parry, which was developed in 1972 given in [5, 6].

Many chatbots have been developed based on different platforms and concepts. Use of conversational agents [7] is growing, but there are many issues related to their functionality. My chatbot is very important when it comes to the customer service scenario. It can be accessed through both laptops and desktops, providing user interfaces for resolving the queries of customers with which our chatbot is linked. Most of the chatbots follow three implementation mechanisms. The initial ever chatbot uses a rule-based system; a set of questions and a set of answers were given in the program and the bot answers the users based on if and else statements. The next version of chatbots used retrieval-based [8] system where a dataset [3] is given in the form of paragraphs or intents, and NLP is used to understand the questions of the customer. The third type is known to be self-learning bot which learns from the user's questions and applies AI and ML techniques and highly sophisticated algorithms to give the appropriate answers to the users. Some of the examples of self-learning bots are Siri [1], Alexa [1], Cortana, Natasha, and Watson. The main disadvantage of rule-based model is that they are very outdated and are not suitable for the customer service work, and there are high chances of not getting the answer to the query you want. Self-learning bots are not widely used for the customer service scenario since they are very sophisticated; they need expertise involvement like data scientists and analysts, which takes years to develop them. Considering all the above reasons is why only big organizations such as Google, Amazon, Apple, Adobe, IBM, etc. are using them. We need a customer service chatbot which can be used by small businesses and also medium-scale establishments.

3 Proposed Methodology

Figure 1 describes the entire process of implementation of chatbot proposed approach. It also shows the input, output, and processing pathways of the application and also the flow of direction of input, output, and processing directions of the application.

The proposed idea/approach consists of the following steps:

Step-1: *Customer Query/Request*: Customer types the phrase in the chatbox.

Step-2: *Chatbot*: It packs the data and responds to the customer and the phrase sent to ML-NLP engine (ML-NLP).

Step-3: *Machine Learning NLP engine* (*ML-NLP*): Extracted user intent and entities sent back to chatbot.

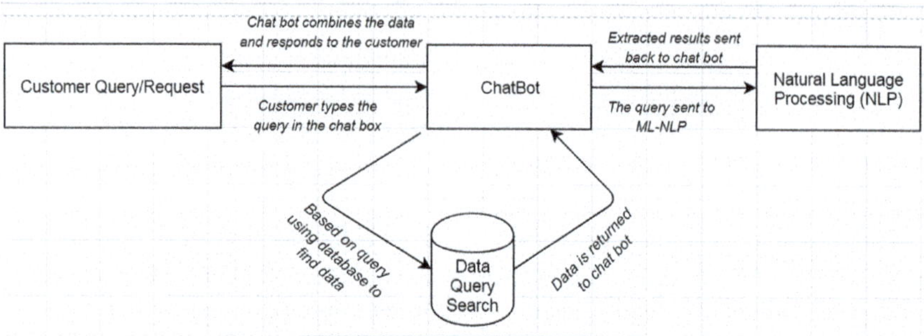

Fig. 1 Proposed idea/approach

Step-4: *Data Query Search Engine*: Chatbot based on intent call upon services using entity information to find data from database. And data is returned to the chatbot.

The use case for the idea/approach of chatbot is shown in Fig. 2.

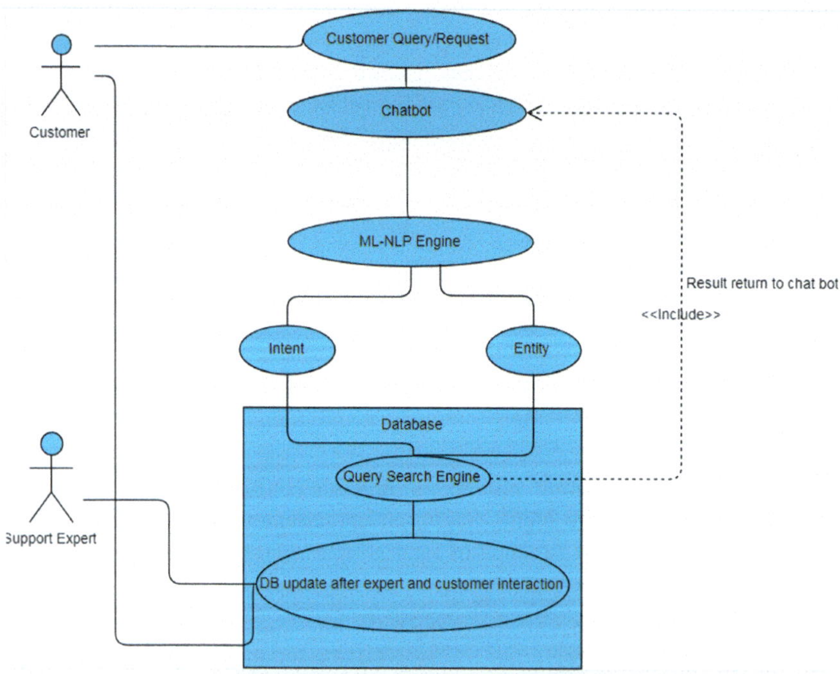

Fig. 2 Use case for the proposed idea/approach

4 Prototype

4.1 Implementation of Chatbot Using ML in Python

(i) Natural Language Processing [6]	**(ii) Machine Learning algorithm** [6]
def text_process(mess): *nopunc = [char for char in mess if char not in string.punctuation]* *print(nopunc)* *nopunc = ''.join(nopunc)* *return [word for word in nopunc.split() if word.lower() not in stopwords.words('english')]*	*rf = RandomForestClassifier(n_estimators = 100,max_depth = 3)* *rf.fit(x_train,y_train)* *pre = mnb.predict(x_test)* *acc = metrics.accuracy_score(y_test,pre)* *print("Score:",acc)*

The main dependencies/show stopper is datasets for creating a functioning database.

The prototype for chatbox text entry is shown in Fig. 3.

The prototype for query test is shown in Fig. 4.

The prototype for support expert is shown in Fig. 5.

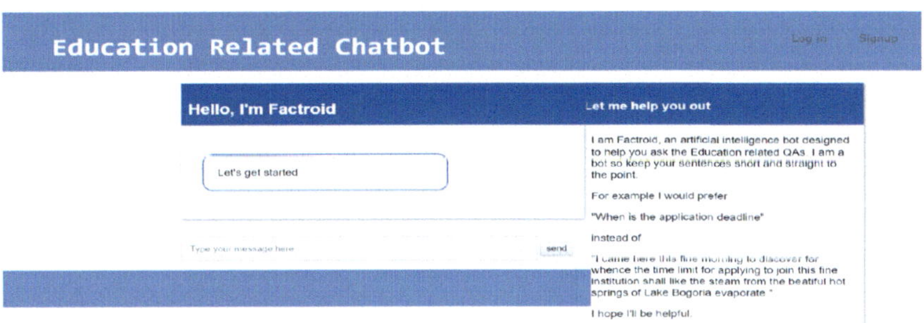

Fig. 3 Prototype for chatbox text entry

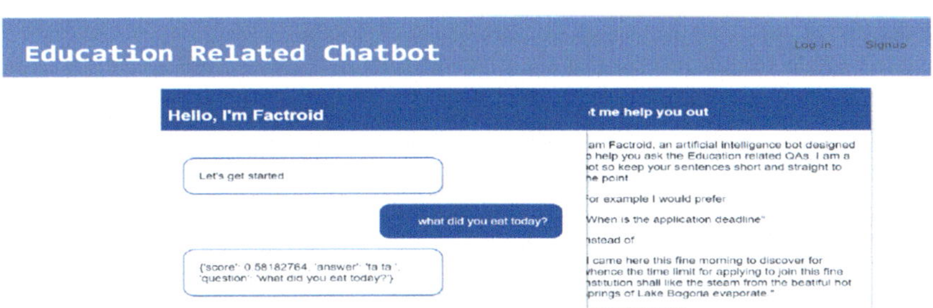

Fig. 4 Prototype for query test

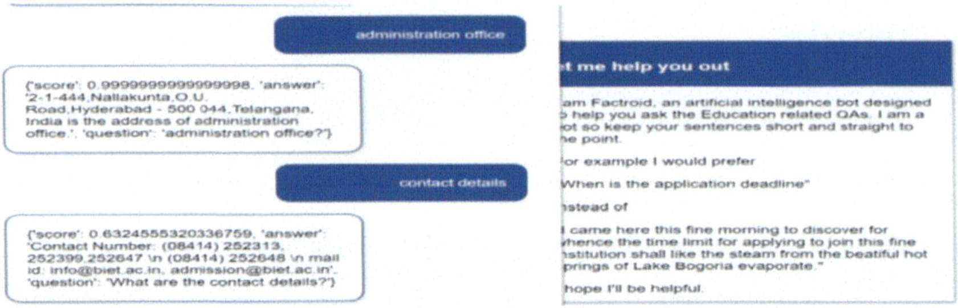

Fig. 5 Prototype for support expert

5 Conclusion

The contribution is the development of customer support chatbot by using ML and NLP in python. There are many chatbots; both rule-based and self-learning are in the market, and unfortunately nothing are used in the customer service scene. Rule-based chatbots are mostly rigid and do not have the capability to know the exact words typed by the customer. When we want to urgently know any information about any organization, we can directly contact a designated chatbot and know those details quickly without talking to a person or mailing the organization. Small details like opening or closing hours or contact information of an organization can be easily found with the help of customer support chatbot. In future, we are thinking of using voice data set to increase the communication between the user and the chatbot through audio chatting. Some more ways to achieve good communication are to add a chatbot to our college website which can help the newly joining students. Voice intake for asking queries and image facilities will be focused on which enhances the entire customer service scene.

References

1. Nimavat, K., Chempanaria, T.: Chatbots: an overview types, architecture, tools and future possibilities
2. Bird, S.: NLTK: the natural language toolkit, pp. 69–72 (2006)
3. Yordanov, V.: Introduction to NLP for text. https://towardsdatascience.com/introduction-to-natural-language-processing-for-text-df845750fb63
4. Mitchell, T.M.: Machine Learning, 1st edn. McGraw Hill Education (2017)
5. Weizenbaum, J.: ELIZA–a computer program for the study of natural language communication between man and machine. Commun. ACM **9**(1), 36–45 (1966)
6. Selvi, V., Saranya, S., Chidida, K., Abarna R.: Chatbot and bullyfree chat. In: International Conference on Systems Computation Automation and Networking (2019)
7. Keikha, M., Park, J.H., Croft, W.B., Sanderson, M.: Retrieving passages and finding answers. In: Proceedings of Australasian Document Computing Symposium, pp. 81–84 (2014)
8. Nguyen, T., et al.: MS MARCO: a human generated machine reading comprehension dataset. In: Proceedings of 30th Conference on Neural Information Processing Systems Workshop, p. 10 (2016)

9. Bernstein, M.S., Teevan, J., Dumais, S., Liebling, D., Horvitz, E.: Direct answers for search queries in the long tail. In: Proceedings of SIGCHI Conference on Human Factors Computing Systems, pp. 237–246 (2012)
10. Haller, E., Rebedea, T.: Designing a chat-bot that simulates an historical figure. IEEE Conference Publications (2013)
11. Kolomiyets, O., Moens, M.-F.: A survey on question answering technology from an information retrieval perspective. Inf. Sci. **181**(24), 5412–5434 (2011)
12. Molnár, G., Szűts, Z.: The role of chatbots in formal education. In: IEEE 16th International Symposium on Intelligent Systems and Informatics, SISY 2018, Subotica, Serbia (2018)

Prediction of Diabetes Using Internet of Things (IoT) and Decision Trees: SLDPS

Viswanatha Reddy Allugunti, C. Kishor Kumar Reddy, N. M. Elango, and P. R. Anisha

Abstract Diabetes is one of the most feared diseases currently faced by humanity. The disease is due to a poor reaction of the body to insulin: it is an important hormone in our body that converts sugar into energy that is necessary for the proper functioning of a normal life. Diabetic disease has serious complications on our body because it increases the risk of developing kidney disease, heart disease, retinal disease, nerve damage, and blood vessels. In this article, we have proposed a decision tree model: SLDPS (Diabetes Prediction System with Supervised Learning). The data set is collected via IoT sensors. The classification accuracy obtained with this model was improved to 94.63% after the rebalancing of the data set and shows a potential relative to other classification models in the literature.

Keywords Accuracy · Decision tree · Diabetes · Error rate · IoT · Kaggle

1 Introduction

Diabetes is one of the most deadly, debilitating, and costly illnesses seen today in many countries, and the disease continues to grow at an alarming rate. Women tend to be most affected by diabetes with 9.6 million women with diabetes. This represented 8.8% of the total female adult population of 18 years and older in 2003, almost double

---

V. R. Allugunti
VIT University, Vellore, India

C. Kishor Kumar Reddy (✉)
Stanley College of Engineering & Technology for Women, Hyderabad, India
e-mail: kishoar23@gmail.com

N. M. Elango
School of Information Technology and Engineering, VIT University, Vellore, India

P. R. Anisha
Department of CSE, Stanley College of Engineering & Technology for Women, Hyderabad, India

S. C. Satapathy et al. (eds.), *Intelligent Data Engineering and Analytics*, Advances in Intelligent Systems and Computing 1177,
https://doi.org/10.1007/978-981-15-5679-1_43

the percentage in 1995 (4.7%). Women from racial and ethnic minority groups have the highest prevalence rates, with rates that are two to four times higher than those of the white population. With an increasing number of minority populations, the number of women in these diagnosed groups will increase considerably in the coming years. By 2050, the expected number of people with diabetes will increase from 17 million to 29 million [1]. Diabetes is a metabolic disorder in which people who suffer from it either have a shortage of insulin or a reduced ability to use their insulin. Insulin is a hormone produced by the pancreas that converts glucose into energy at the cellular level. Uncontrolled diabetes, consistently high blood glucose levels (>200 mg/dL), result in complications of micro- and macro-vascular diseases, such as blindness, lower limb amputations, end-stage renal disease, and coronary heart disease and stroke. Diabetes affects around one in ten people, but the chances increase to one in five if the age group is 65 or older [2, 3].

Diabetes mellitus is a chronic and progressive metabolic disorder. According to the World Health Organization, around one million people worldwide have diabetes. The number of diabetic patients is expected to increase by more than 100% in 2030. The common symptoms of diabetes are characterized by insufficient production of insulin by the pancreas and ineffective use of insulin produced by the pancreas or hyperglycaemia. Causes such as obesity, high blood pressure, high cholesterol, a high-fat diet, and a sedentary lifestyle are common factors that contribute to the prevalence of diabetes. The development of renal failure, blindness, kidney disease, and coronary artery disease are types of serious lesions due to improper management and late diagnosis of diabetes [4]. Although there is no treatment for diabetes, the blood glucose levels of diabetic patients can be controlled by established treatments, adequate nutrition, and regular exercise.

Signs or symptoms of diabetes are frequent urination, increased thirst, increased hunger, fatigue/sleepiness, weight loss, blurred vision, mood swings, confusion and concentration problems, and frequent infections/insufficient healing [1, 5]. Type 1 diabetes: In type 1 diabetes, beta cells in the pancreas are injured or attacked by the body's immune system (autoimmunity). As a result of this attack, the beta cells die and are therefore unable to produce the amount of insulin needed to allow glucose to enter the cells, resulting in high blood sugar (hyperglycaemia). Type 1 diabetes affects approximately 5–10% of people with diabetes and usually people younger than 30 but can occur at any age. The signs and symptoms appear quickly and are usually intense in nature. Because type 1 diabetes is caused by a shortage of insulin, it is necessary to replace what the body cannot produce itself. According to the latest heart disease and stroke statistics in the American Heart Association, about 8 million people aged 18 and over in the United States have type 2 diabetes and don't know it. Often type 1 diabetes does not remain diagnosed until the symptoms worsen and hospitalization is required. Left untreated, diabetes can lead to many health complications. That is why, it is so important to know the warning signs and to regularly consult a healthcare provider for routine screenings. Computer-assisted diagnostics is a fast-growing field of dynamic research in the medical industry. Recent researchers from machine learning promise to improve the accuracy of disease perception and diagnosis [6–9].

In this study, we proposed a decision tree model: SLDPS (Diabetes Prediction System with Supervised Learning). The data set is collected via the IoT diabetes sensors. Initially, the algorithm is competent with 75% of the facts and is examined in more detail with 25% of the records. Here, the measurement of the entropy attribute selection is taken to identify the best part point. The classification accuracy obtained with this model was improved to 94.63% after the rebalancing of the data set and shows a potential relative to other classification models in the literature [10, 11].

The rest of the article is organized as follows: Sect. 2 presents the algorithm for predicting guided learning for diabetics, Sect. 3 illustrates the results obtained and the comparison with existing approaches, and Sect. 4 concludes the document with references.

2 Proposed SLDPS

1. Read the training data set in ascending order.
2. Evaluate partial points according to the interval range.
3. Calculate the characteristic attribute using formula 1:

$$\text{Attribute Entropy} = \sum_{j=1}^{N} P_j \left[-\sum_{i=1}^{M} P_i \log_2 P_i \right] \quad (1)$$

4. Calculate the entropy class with formula 2:

$$\text{Class Entropy} = -\sum_{i=1}^{M} P_i \log_2 P_i \quad (2)$$

5. Calculate entropy of formula 3:

$$\text{Entropy} = \text{Class Entropy} - \text{Attribute Entropy} \quad (3)$$

6. The maximum entropy value is chosen as the best part point and becomes the basic node using formula 4:

$$\text{Best Division Point} = \text{Maximum (Entropy)} \quad (4)$$

3 Results and Discussion

For the experiment, a set of data with 15,000 realities and eight qualities was compiled using the IoT diabetes sensors. In the beginning, the ranking of the standards is taught with 75% of the recordings and more tried with 25% of the realities. In the proposed set of principles, the elements of the distribution are assessed on the basis of the provisional assortment rather than any exchange in the class name. In order to choose the five-star separation point, the reality of the degree of determination of the brand is continued. The layout of the principles is coded with Net Beans IDE and realized in an Intel i3 processor, 4 GB RAM.

The precision of the proposed model, the Supervised Learning Diabetes Prediction System (SLDPS), presented in Table 1, is correlated with current techniques: random forest, pockets, decision tree, artificial neural networks, amplification, gullible Bayes, and carrier vector machines separately. The proposed model gave an accuracy of 94.63%, higher than that of the contrasting and earlier methods. The graphical representation of the accuracy assessment is illustrated in Fig. 1. In Table 1, RF stands for random forest, B is bagging, DT is decision tree, ANN is artificial neural network, BO is boosting, NB is Naïve Bayes, DPA is diabetes prediction algorithm, ADPA is advanced diabetes prediction algorithm, and SLDPS is supervised learning diabetics prediction system.

The comparison of the error rate of the proposed SLDPS model in Table 2 is examined with the current approaches: random forest, pockets, decision tree, artificial neural networks, amplification, innocence. Bayes, and support vector machines separately. The proposed model gave a rate of 5.37%, better compared to earlier systems. The diagram of the error percentage comparison is illustrated in Fig. 2.

In Table 2, RF stands for random forest, B is bagging, DT is decision tree, ANN is artificial neural network, BO is boosting, NB is Naïve Bayes, DPA is diabetes prediction algorithm, ADPA is advanced diabetes prediction algorithm, and SLDPS is supervised learning diabetics prediction system.

Table 1 Comparison of accuracy with existing approaches

Model name	Accuracy (%)
RF	85.55
B	85.33
DT	85.09
ANN	84.53
BO	84.09
NB	81.01
SVM	87.6
DPA	93.8
ADPA	94.23
SLDPS	94.63

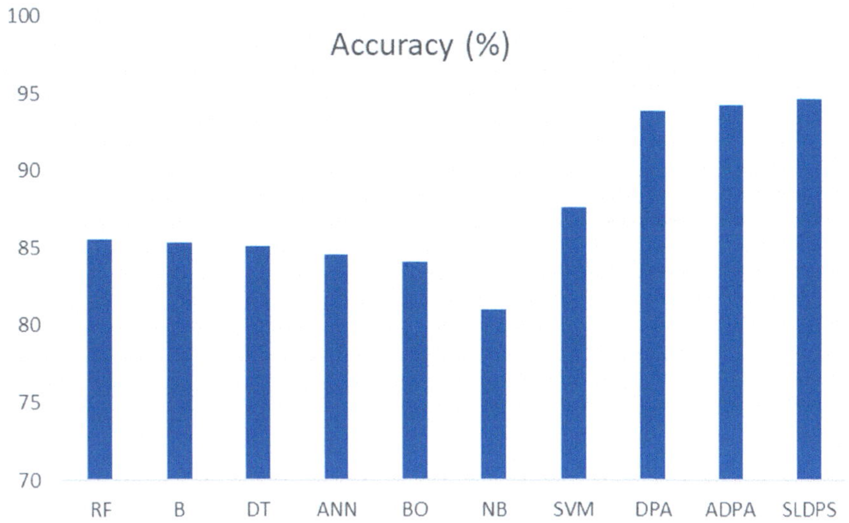

Fig. 1 Comparison of accuracy with existing approaches

Table 2 Comparison of error rate with existing approaches

Model name	Error rate (%)
RF	14.44
B	14.66
DT	14.91
ANN	15.46
BO	15.90
NB	18.99
SVM	12.4
DPA	6.2
ADPA	5.77
SLDPS	5.37

In addition, the proposed SLDPS is compared with the simple logistics algorithms of decision stump, Hoeffding tree, Naive Bayes, and simple using the data collected by the IoT diabetes sensors in terms of accuracy. The results are shown in Table 3 and Fig. 3. Here, we used the WEKA tool to find the accuracy of existing algorithms. In Table 3, DS stands for decision stump, HT is Hoeffding tree, NB is Naïve Bayes, SL is simple logistics, DPS is a prediction algorithm for diabetics, ADPS is an advanced prediction algorithm for diabetics, and SLDPS is a guidance system for assisted learning for diabetics.

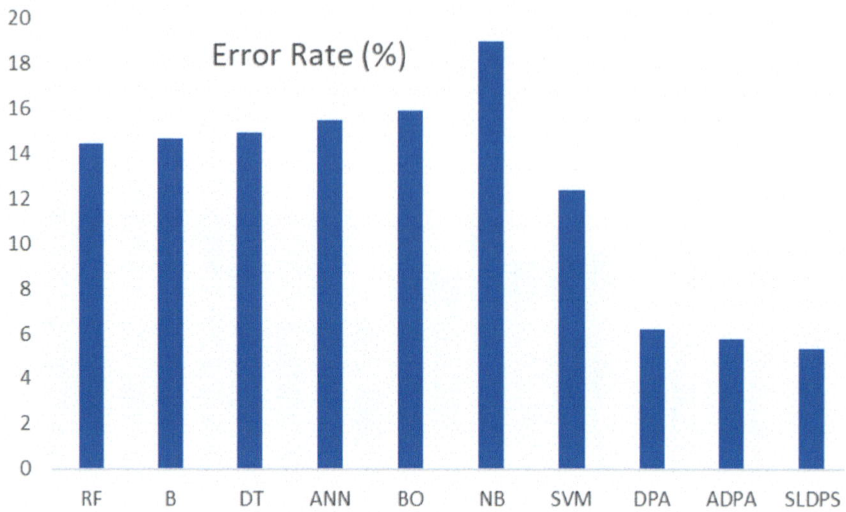

Fig. 2 Comparison of error rate with existing approaches

Table 3 Comparison of accuracy with other algorithms using Weka

Model name	Accuracy (%)
DS	78
HT	87.36
NB	79.36
SL	79.14
DPA	93.8
ADPA	94.23
SLDPS	94.63

In addition, the proposed SLDPS is compared with the simple logistic algorithms of decision stump, Hoeffding tree, Naive Bayes, and simple using the data collected by the IoT diabetes sensors in terms of error rates. The results are shown in Table 4 and Fig. 4. Here, we have used the WEKA tool to find the error rate of existing algorithms. In Table 4, DS stands for decision stump, HT is Hoeffding tree, NB is Naïve Bayes, SL is simple logistics, DPS is a prediction algorithm for diabetics, ADPS is an advanced prediction algorithm for diabetics, and SLDPS is a guidance system for assisted learning for diabetics.

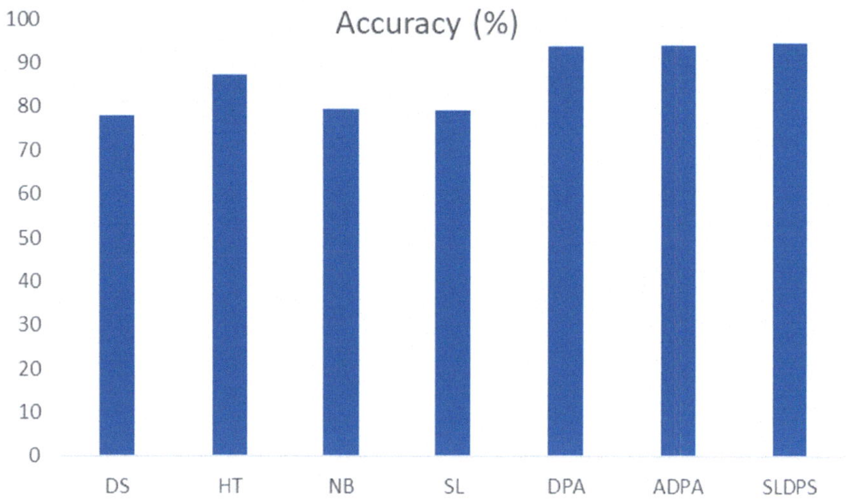

Fig. 3 Comparison of accuracy with existing approaches using WEKA

Table 4 Comparison of error rate with other algorithms using Weka

Model name	Error rate (%)
DS	22
HT	12.64
NB	20.64
SL	20.85
DPA	6.2
ADPA	5.77
SLDPS	5.37

4 Conclusions

Because expert systems and tools for machine learning have improved considerably, more and more application areas have been invaded and the medical field is not exempt. Making medical decisions can sometimes be very embarrassing. The classification systems used to make medical decisions receive medical data that they examine in a more complete form but faster. In this study, we have proposed a system based on the decision tree: SLDPS. The data set is collected via IoT sensors. The accuracy of the classification obtained with this model is improved to 94.63% after the rebalancing of the data set and shows a potential relative to other classification models in the literature.

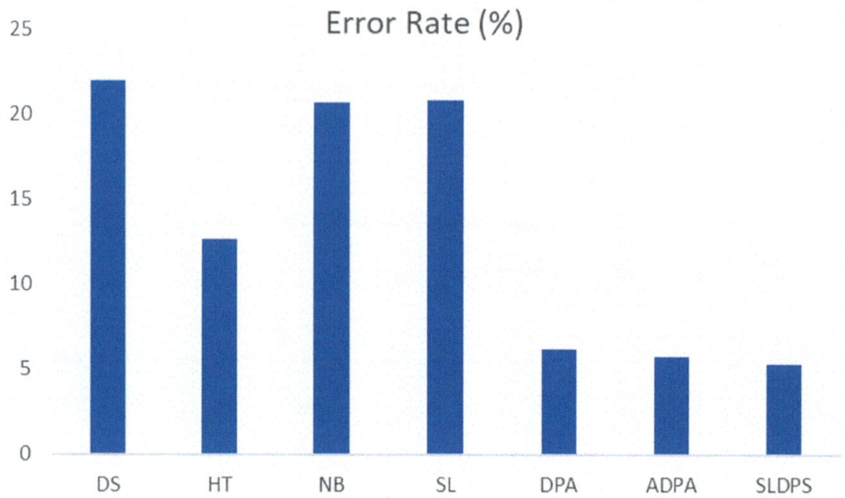

Fig. 4 Comparison of error rate with existing approaches using WEKA

References

1. Akolekar, R., Syngelaki, A., Sarquis, R., Zvanca, M., Nicolaides, K.H.: Prediction of early, intermediate and late pre-eclampsia from maternal factors, biophysical and biochemical markers at 11–13 weeks. Prenatal Diagn. **31**(1), 66–74 (2011)
2. Alssema, M., Vistisen, D., Heymans, M.W., Nijpels, G., Glümer, C., Zimmet, P.Z., Shaw, J.E., et al.: The evaluation of screening and early detection strategies for type 2 diabetes and impaired glucose tolerance (DETECT-2) update of the Finnish diabetes risk score for prediction of incident type 2 diabetes. Diabetologia **54**(5), 1004–1012 (2011)
3. Farran, B., Channanath, A.M., Behbehani, K., Thanaraj, T.A.: Predictive models to assess risk of type 2 diabetes, hypertension and comorbidity: machine-learning algorithms and validation using national health data from Kuwait—a cohort study. BMJ Open **3**(5), e002457 (2013)
4. Faust, O., Acharya, R., Ng, E.Y.-K., Ng, K.-H., Suri, J.S.: Algorithms for the automated detection of diabetic retinopathy using digital fundus images: a review. J. Med. Syst. **36**(1), 145–157 (2012)
5. Huang, G.-B., Zhou, H., Ding, X., Zhang, R.: Extreme learning machine for regression and multiclass classification. IEEE Trans. Syst. Man Cybern. Part B Cybern. **42**(2), 513–529 (2012)
6. Jensen, M.H., Mahmoudi, Z., Christensen, T.F., Tarnow, L., Seto, E., Johansen, M.D., Hejlesen, O.K.: Evaluation of an algorithm for retrospective hypoglycemia detection using professional continuous glucose monitoring data. J. Diabetes Sci. Technol. **8**(1), 117–122 (2014)
7. Kalaiselvi, C., Nasira, G.M.: Classification and prediction of heart disease from diabetes patients using hybrid particle swarm optimization and library support vector machine algorithm
8. Karthikeyan, T., Vembandasamy, K.: A novel algorithm to diagnosis type II diabetes mellitus based on association rule mining using MPSO-LSSVM with outlier detection method. Indian J. Sci. Technol. **8**(S8), 310–320 (2015)
9. Karthikeyan, T., Vembandasamy, K.: A refined continuous ant colony optimization based FP-growth association rule technique on type 2 diabetes. Int. Rev. Comput. Softw. (IRECOS) **9**(8), 1476–1483 (2014)

10. Kuo, R.J., Lin, S.Y., Shih, C.W.: Mining association rules through integration of clustering analysis and ant colony system for health insurance database in Taiwan. Expert Syst. Appl. **33**(3), 794–808 (2007)
11. Nahar, J., Imam, T., Tickle, K.S., Chen, Y.-P.P.: Association rule mining to detect factors which contribute to heart disease in males and females. Expert Syst. Appl. **40**(4), 1086–1093 (2013)

Review Paper on Fourth Industrial Revolution and Its Impact on Humans

D. Srija Harshika

Abstract The fourth industrial revolution, a term instituted by Klaus Schwab, organizer and official executive of the World Financial Gathering, depicts an existence where people move between computerized areas and disconnected reality with the utilization of associated innovation to empower and deal with their lives (Mill operator 2015, 3). The principal mechanical upheaval transformed us and economy from an agrarian and handiwork economy to one ruled by industry and machine fabricating. Oil and power encouraged large-scale manufacturing in the second mechanical insurgency. In the third modern unrest, data innovation was utilized to mechanize creation. Albeit each mechanical unrest is regularly viewed as a different occasion, together they can be better comprehended as a progression of occasions heaps of the past unrest and prompting further developed types of creation. Another technological development zone as of late has been analytics. Financial organizations track and gather a wide range of information on customers, for example, what customers purchase, how they get it, and when they do their shopping. Mobile phones are another key player in enormous information since they can likewise follow shopping information, just as information on media utilization and even your area for the duration of the day. This article examines the significant highlights of the four mechanical insurgencies, the chances of the fourth modern transformation, and the difficulties of the fourth industrial unrest. With so much information accessible, what job will it have in the up and coming fourth industrial transformation?

Keywords Fourth industrial revolution · Analytics · Analysis · Data science · Data analytics

D. Srija Harshika (✉)
Cyient Ltd., Madhapur, Hyderabad, India
e-mail: srijaharshika.d@gmail.com

S. C. Satapathy et al. (eds.), *Intelligent Data Engineering and Analytics*, Advances in Intelligent Systems and Computing 1177,
https://doi.org/10.1007/978-981-15-5679-1_44

1 Introduction

1.1 History Behind

It was in the Swiss mountains that the world was first introduced to the phrase the "**Fourth Industrial Revolution**," and it's been a topic of discussion among academics, politicians, and business leaders ever since [1].

But having heard of it here and there and many a times, ever wondered what exactly it refers to? The term "**Fourth Industrial Revolution**" was coined by the founder of the World Economic Forum, a former professor named Klaus Schwab, in his book titled "**The Fourth Industrial Revolution**" to predominantly describe an era marked by the ingrain of technologies like Artificial Intelligence, Autonomous vehicles or the Internet of Things that are rapidly becoming an essential part of our day-to-day lives, and more sooner becoming the necessities of human bodies. Think of the voice-activated virtual assistants like the Alexa's and the Google Assistants of the world, Face ID recognition on our phones, healthcare sensors on our fitness bands, and many more.

Schwab first presented his vision of the Fourth Industrial Revolution at the World Economic Forum's annual meeting in Davos, Switzerland in 2016. However, to understand his vision further in detail, he looks back in history to the **First Industrial Revolution** [2], which was started in Great Britain around 1760 and spread to Europe and North America through the early 1800s. It was powered by a major invention, the Steam Engine, resulting in the new manufacturing processes, the creation of factories, and a booming textiles industry.

From the late 1800s, the **Second Industrial Revolution** [3] was marked by mass production and addition of more new industries like steel, oil, and electricity. The light bulb, the telephone, and internal combustion engine were a few of the major inventions of this era.

The **Third Industrial Revolution** [4] sometimes referred as the Digital Revolution occurred in the second half of the twentieth century. During this, in just a few decades, we saw the invention of the semiconductor, the personal computer, and the Internet (Fig. 1).

1.2 The Differentiator

Experts say the main differentiator lies in the technology that is merging progressively with human's lives and that technological change is happening faster than ever. Consider this: It took 75 years for 100 million users to adopt the telephone, but Instagram signed up 100 million users in just 2 years, and Pokémon Go caught up with that number in just 1 month [5].

3D printing is just another example of this fast-paced technology in the Fourth Industrial Revolution. The industry has gone from a business idea to a big business

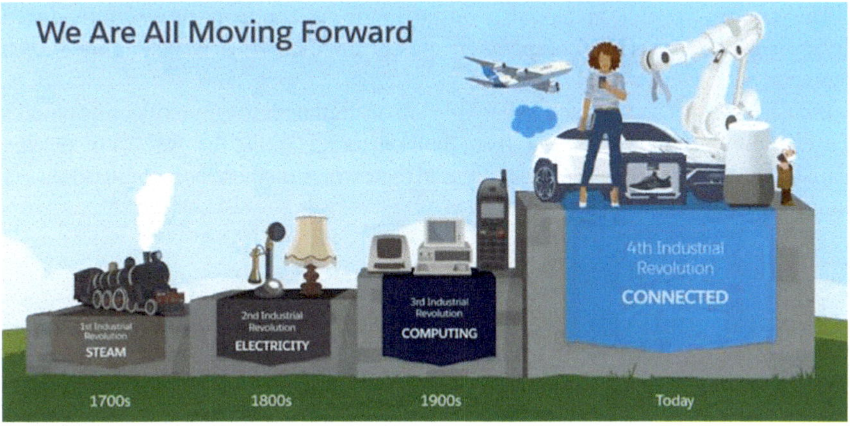

Fig. 1 Evolution of industrial revolution throughout the time

opportunity, with 3D printer shipments expected to increase from just under 200,000 in 2015 to 2.4 million in 2020. Today, hip replacements are being done from a 3D-printed bone or an arm replacement using a 3D-printed bionic arm. Talk about blurring the line between humans and technology, right?

Technologies like 3D printing or AI have been accelerating upward since the early 2000s [6–8]. Organizations are embracing these next-gen technologies to make their businesses more efficient, like how they embraced the steam engine during the First Industrial Revolution. Research shows innovators, investors, and shareholders benefit the most from these innovations.

But having said all that, there is also a great amount of risk involved in this super-fast Technological Fourth Industrial Revolution which is churning out inequality at a larger scale and driving organizations out of business for their inability to cope up with the technological trends and huge market demands. There are still many organizations and companies, and governments, who are struggling to keep up with the fast pace of this technological change, along with the huge chunks of manpower at all levels who are forcefully being driven to learn these in order to survive. Can it get anything worse than that? The World Economic Forum says most leaders do not have confidence that their organizations are ready for the changes associated with the Fourth Industrial Revolution.

Another study has found billionaires have driven almost 80% of the 40 main breakthrough innovations over the last 40 years. Well that's the actual problem since the richest one percent of households already owns nearly half of the world's entire wealth. And that is why the famous saying holds true for this economy that the "winner-takes-it-all", where the high-skilled workers are rewarded with high pay, and the remaining rest are left out in the race. Must have heard about this recently around you right with the news of layoffs coming out almost every week in every other part of the world. Studies confirm technologies like AI will eliminate many jobs and create demand for new skills that many do not have.

This current revolution has also raised immense concerns on an individual's privacy since every other company in mostly all industries is becoming a tech company.

Industries from food to retail to banking are on digital platforms, collecting chunks of user experience data every day from their customers along the process of serving them. Users across the globe have expressed their worry on these companies knowing too much about their private digital lives.

2 Analysis and Analytics: Same or Different?

It is often believed that analysis and analytics share the same meaning and thus are used interchangeably. Technically, this isn't right. There is in fact a distinct difference between the two, and the reason for one often being used instead of the other is the lack of a transparent understanding of both [9].

First, let's understand analysis. Consider this: We have a huge data set containing data of various types. Instead of tackling the entire dataset and running the risk of becoming overwhelmed, we separated this data set into easier to digest chunks and study them individually and examine how they relate to other parts. That is analysis in a nutshell.

One important thing to remember, however, is that we perform analysis on events that have already occurred in the past such as using an analysis to explain how a story ended the way it did or how there was a decrease in the cells last summer. All this means that we do analysis to explain how and or why something happened.

Analytics generally refers to the future instead of explaining past events. It explores potential future sequences. Analytics is essentially the application of logical and computational reasoning to the component parts obtained in an analysis and in doing this we are looking for patterns in exploring what we can do with them in the future.

Analytics branches into two main areas: Qualitative analytics and quantitative analytics. Analytics used our intuition and experience in conjunction with the analysis to plan our next business move and quantitative analytics [10–12]. This is then done by applying formulas and algorithms to numbers gathered from the analysis.

Here are some examples:

Say an owner of an online clothing store is ahead of the competition and have a great understanding of what are his customers' needs and wants [13]. He has performed a very detailed analysis from women's clothing articles and feels sure about which fashion trends to follow. He may use this intuition to decide on which styles of clothing to start selling. This would be qualitative analytics, but he might not know when to introduce the new collection. In that case, relying on past sales data and user experience data could predict the best month to do so. This is quantitative analytics.

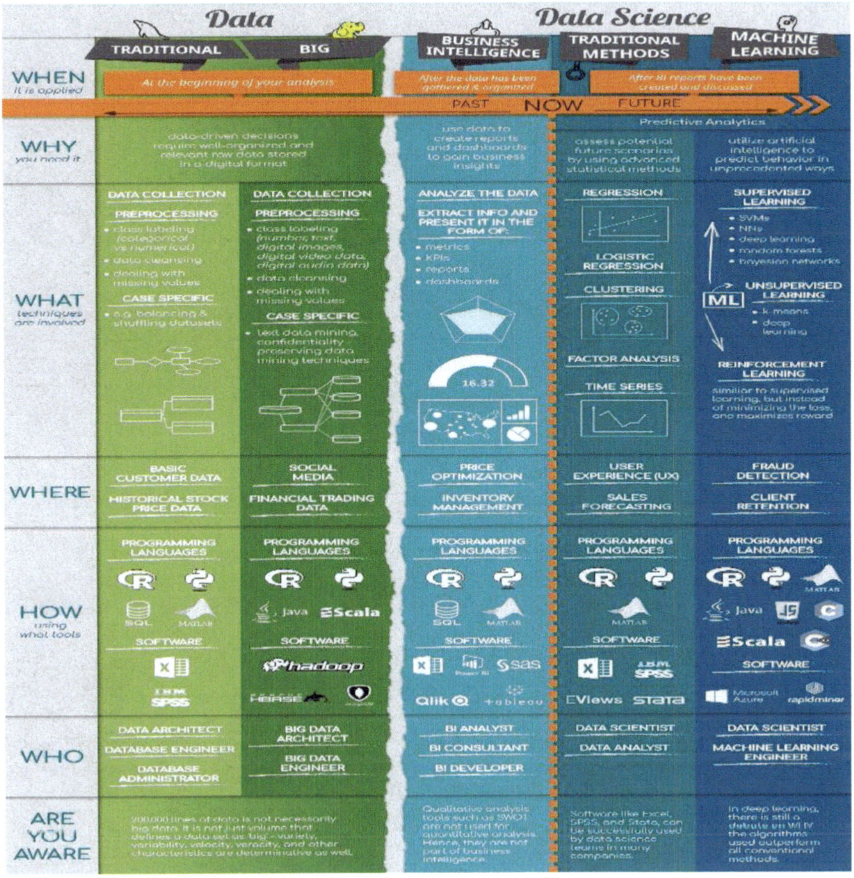

Fig. 2 Differences between data and data science with multiple parameters

2.1 *Data Versus Data Science [14]*

See Fig. 2.

3 Conclusion

The Internet of Things (IoT), Artificial Intelligence (AI), Augmented Reality, the rundown goes on. These are viewed as significant components of the Fourth Industrial Transformation that hazily spots the lines between physical, computerized, and organic. Integral to this lies data and analytics which is essential for the enormous data captured meanwhile.

4 Declaration

"I have taken permission from competent authorities to use the images/data as given in the paper. In case of any dispute in the future, we shall be wholly responsible.".

References

1. https://www.udemy.com/course/the-business-intelligence-analyst-course-2018/learn/lecture/10117282#overview
2. https://www.forbes.com/sites/theyec/2019/11/07/data-science-the-fourth-industrial-revolution-and-the-future-of-entrepreneurship/#6c5570801a6d
3. https://www.weforum.org/reports/data-science-in-the-new-economy-a-new-race-for-talent-in-the-fourth-industrial-revolution
4. https://www.salesforce.com/blog/2018/12/what-is-the-fourth-industrial-revolution-4IR.html
5. https://trailhead.salesforce.com/en/content/learn/modules/learn-about-the-fourth-industrial-revolution/meet-the-three-industrial-revolutions
6. https://en.wikipedia.org/wiki/Technological_revolution
7. https://en.wikipedia.org/wiki/Industry_4.0
8. https://www.researchgate.net/publication/325616277_Big_Data_Analytics_for_Decision_Making_in_the_4th_Industrial_Revolution
9. https://www.researchgate.net/publication/324451812_Nowhere_to_Hide_Artificial_Intelligence_and_Privacy_in_the_Fourth_Industrial_Revolution
10. https://www.britannica.com/topic/The-Fourth-Industrial-Revolution-2119734
11. https://arstechnica.com/information-technology/2019/06/the-revolution-will-be-roboticized-how-ai-is-driving-industry-4-0/
12. https://towardsdatascience.com/the-non-technical-guide-to-artificial-intelligence-e9e5da1a15c5
13. https://katecarruthers.com/2018/03/13/data-is-the-new-oil/
14. https://www.datanami.com/2019/04/25/big-data-challenges-of-industry-4-0/

Edge Detection Canny Algorithm Using Adaptive Threshold Technique

R. N. Ojashwini, R. Gangadhar Reddy, R. N. Rani, and B. Pruthvija

Abstract Detection of edge is most basic operations which is needed in processing of objects in image processing identification. Hence, edge detection is the most likely operation for the processing of image in real-time applications with optimized results which is accurate, and architecture with less complexity results in less latency. Hence, edge detection with adaptive threshold technique plays a vital role in present-day edge detection techniques. The computation is carried with threshold values which are automatically adopted according to the image specification which helps to reduce the memory and computations along with decision-making will take less time. Hence, delay gets reduced with improved detection performance along with increased efficiency. The proposed architecture is implemented using Xilinx system generator tool on Spartan6 ATLYS board.

Keywords Canny edge algorithm · Adaptive threshold technique · System generator · Parallel processing

1 Introduction

Detection of edges is the set of mathematical calculations with different methods that are adopted according to the specification of the image; the value of the pixel has more discontinuity at the edges as compared with the remaining part of the image. Hence, only the detection of edges needs different mathematical calculations that are adopted to get efficient results. The contour of the image is helpful to identify the image as an object for edge detection. Edge detection will give the outline of

R. N. Ojashwini (✉) · R. Gangadhar Reddy
Department of ECE, Raja Rajeswari College of Engineering, Bangalore, India
e-mail: ojashwini21@gmail.com

R. N. Rani
Department of ECE, R V College of Engineering, Bangalore, India

B. Pruthvija
Department of ECE, BMS College of Engineering, Bangalore, India

S. C. Satapathy et al. (eds.), *Intelligent Data Engineering and Analytics*, Advances in Intelligent Systems and Computing 1177,
https://doi.org/10.1007/978-981-15-5679-1_45

the image which is pre-processed according to the specified image calculations with adaptive threshold technique [1–3].

The edge detection techniques will have the different orientations: For the particular image sequences, different adaptive thresholds are adopted for the image depending on

1. Depth variations.
2. Variations in the image sequences according to the orientation.
3. Irregularities according to the properties of the material.
4. Illumination according to the image sequences.

One of the standard detection algorithms for the detecting of the edges is the Canny algorithm where the threshold values are physically given, where calculations are done according to the image specifications according to the size, color, and orientation of the image sequences. In the original, canny detection is based on the frame-level statistics, that is, complete image is considered as single frame and the threshold value depends on the specifications of the image according to the mathematical calculations of the Canny algorithm. Hence, computations have to take place for each image sequence; hence, Canny algorithm has more computations and has more complexity with higher latency which results in less efficient and less computational performance. The another technique has been introduced for the edge detection process which is block-based statistics, that is, each image sequence will divide into block and each block gets processed, and detection of the will take place and each block has processed and pipelined to get the edge of the processed image [4–6].

2 System Architecture

Tee proposed diagram of adaptive threshold Canny edge algorithm is shown in Fig. 1.

Pre-Processing
This is the first step in adaptive threshold technique of canny edge algorithm is to resize the input image into suitable size (256 * 256) and the resized image get converted into a gray level for the purpose of hardware optimization.

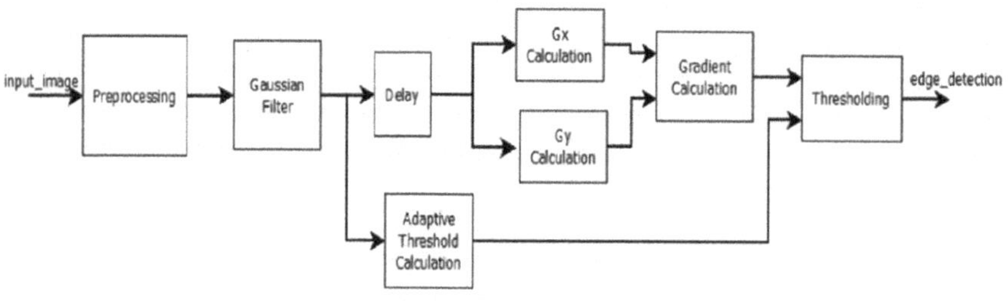

Fig. 1 Proposed diagram of adaptive threshold Canny edge algorithm

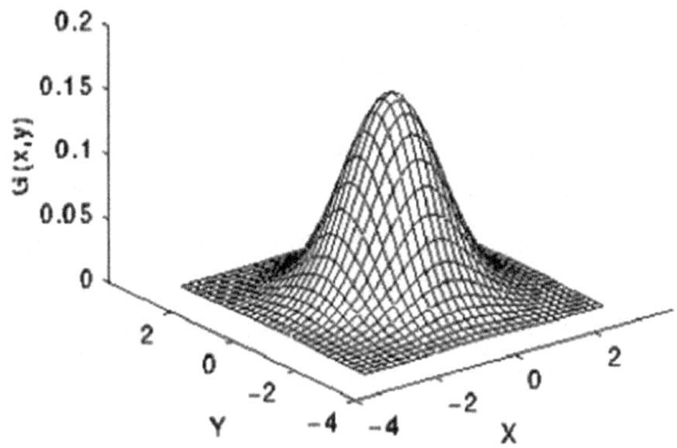

Fig. 2 Gaussian graphical representation

Gaussian Filter

It is one of the two-dimensional operators applied for convolution which is adopted to reduce the noise in the input image. The matrix represents the kernel of the system that is in Gaussian shape with bell-shaped representation.

The matrix is represented as below:

$$\text{Gaussian Filter} = \frac{1}{16}\begin{bmatrix} 1 & 2 & 1 \\ 2 & 4 & 2 \\ 1 & 2 & 1 \end{bmatrix} * \begin{bmatrix} d_0 & d_1 & d_2 \\ d_3 & d_4 & d_5 \\ d_6 & d_7 & d_8 \end{bmatrix} \quad (1)$$

where d_0 to d_8 are the 3×3 image sub-matrix pixel values. The given image is consolidated into blocks, and each consolidated block gets divided into 3 * 3 matrix along with the standard deviation value; increase in the standard deviation value the blurring of image is more; hence, with the determined standard deviation value along with the Gaussian filter coefficient the convolution will take place which results in the smoothing of the image. The representation of the image is as shown in Fig. 2.

Gaussian kernel 2D definition is as follows:

$$G(x, y) = \frac{1}{2\pi\sigma^2} e^{\frac{-(x^2+y^2)}{2\sigma^2}} \quad (2)$$

where σ is the standard deviation.

The moving window architecture to implement 3×3 image sub-matrix is shown in Fig. 3. In the moving window architecture or in 3×3 pixel generation block, nine shift registers and two FIFO structures are used. The architecture of the shift register is as shown in Fig. 4. To access 3×3 pixel, the above architecture used shift register as part of this architecture. Here, in the shift register if the clock is high the data will move to the output variable or else the input variable is assigned with the previous data. The architecture of the FIFO is shown in Fig. 5. FIFO is the part of the 3×3 pixel generation block. After convolution with the Gaussian kernel noiseless image is found.

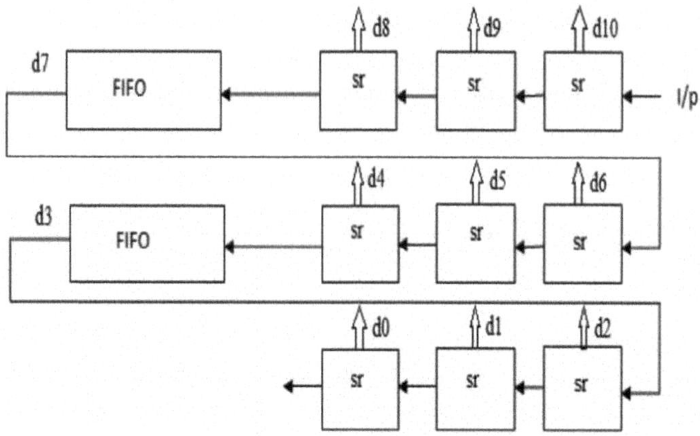

Fig. 3 Moving window architecture (3 × 3 pixel generation)

Fig. 4 Shift register block diagram

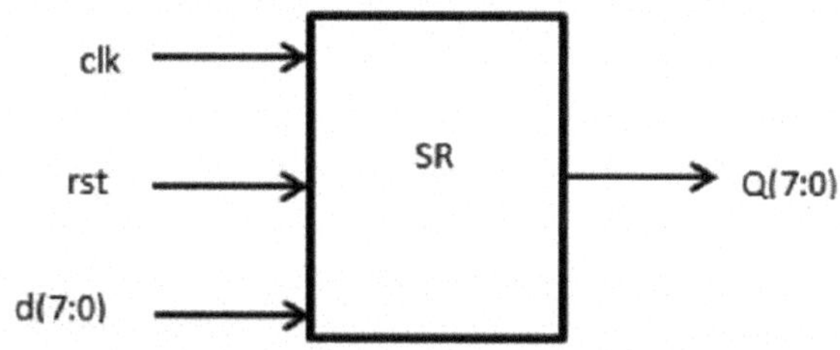

Fig. 5 FIFO block diagram

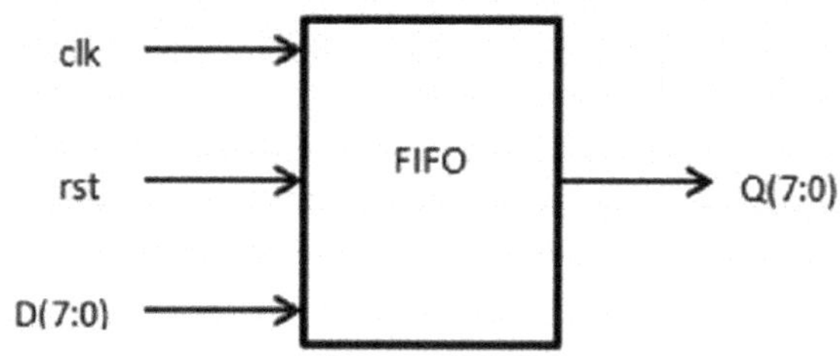

Finding Gradients

The modified Canny operator uses two 3 × 3 kernel matrix; one is horizontal gradient and one is vertical gradient. They are as follows:

$$G_x = \begin{bmatrix} -\frac{1}{4} & 0 & \frac{1}{4} \\ -1 & 0 & 1 \\ -\frac{1}{4} & 0 & \frac{1}{4} \end{bmatrix}, \quad G_x = \begin{bmatrix} \frac{1}{4} & 0 & \frac{1}{4} \\ 0 & 0 & 0 \\ -\frac{1}{4} & -1 & -\frac{1}{4} \end{bmatrix} \tag{3}$$

The convolution of the image is according to horizontal and vertical directions according to its gradient. And the magnitude is calculated from the below:

$$\text{Gradient } (G) = |G_x| + |G_y| \tag{4}$$

From the moving window, architecture pixel values are taken and the convolution of the image is according to horizontal and vertical directions according to its gradient. The hardware structure is implemented by using only shifters and adders/subtractors.

Adaptive Threshold

This is the step where Canny algorithm get differentiated from the adaptive threshold technique for the detection of the edge. In the original, canny threshold values are manually given; as a result the computations will get increase but in the adaptive threshold technique the threshold values are automatically adjusted according to the image specifications. To calculate adaptive thresholding value, the equation is as follows:

$$S = \sum_{i=1}^{N} \frac{(A_i)^2}{8N} \tag{5}$$

where S is the summation, N is the dimension of the input image ($N = 256 \times 256$), and $A1, A2, \ldots AN$ are the intensity values of the image pixel. Three modes of suppression needed for the final edge detection which as low in magnitude, medium-dark edges, and the sure-shot edges.

Non-Maximum Suppression

The non-maximum values are removed in this process in the given image based on the threshold values. It is used to suppress the low in magnitude edges.

There are many steps:

1. Let us consider the single pixel value θ and round the gradient direction into corresponding nearest value to the 90^θ according to its eight connected neighbor values that is the remaining direction pixel value get suppressed because the lesser in the threshold value.
2. Here, the comparison of the current and corresponding edges is according to its gradient direction. If the direction of the gradient is north ($\theta = 90^\theta$). Hence, the comparison will take place in both north and south directions according to its threshold values.
3. The edge gradient direction considered as

$$\text{Del}+ = (1, 0)(1, 1)(0, 1)(-1, 1)$$

$$\text{Del}- = (-1, 0)(-1, -1)(0, -1)(1, 1)$$

Let us consider for the each pixel value (i, j):

4. The direction of the gradient is normal to the edge

$$D = (\text{Dir}(i, j) + \frac{\pi}{8} \bmod \frac{\pi}{4}$$

5. If the magnitude is smaller than any one of its neighbor, along the gradient direction d, then In(i,j) = θ; otherwise, In (i,j) = magnitude(i,j)
6. If magnitude(i,j) < magnitude(i,j) + Del + (d) then In(i,j)=0 else if magnitude(i,j) < magnitude(i,j) + Del -(d) then In(i,j) = 0 which results in thinned edge image or else In = magnitude(i,j).

Double Thresholding
This mode is used to suppress the shot edges. The thresholding is taken place between the background and foreground threshold values. Low threshold corresponds to the background = 0.66* mean value of the pixel. High threshold corresponds to the foreground = 1.33* mean value of the pixel. Edge pixels weaker than the low threshold are suppressed, and edge pixels between the two thresholds are marked as weak.

Edge Tracking by Hysteresis
This mode of suppression is mainly used for medium intensity images. Here the comparison of the each pixel with the corresponding eight pixel values in all corresponding directions. According to the threshold values by comparing the larger value pixel will remain and the remaining lower pixel values get suppressed.

3 System Design

System generator is instrument created from Xilinx that enable usage of Simulink design environment for FPGA plan. Designs are captured as block set in Simulink modeling environment, and all FPGA implementation steps are performed without human intervention to produce FPGA programming files. More than 80 DSP blocks are delivered in Xilinx DSP block set for Simulink such as registers, multipliers, and adders. It also offers combination platform for design of FPGA that allow RTL and Simulink components to come organized in single simulation and employment environment.

Image Processing
When image pre-processing is done using Matlab, it delivers inputs to FPGA as vectors which is appropriate for bitstream gathering by system generator. Following functions are done for image processing as shown in Fig. 6.

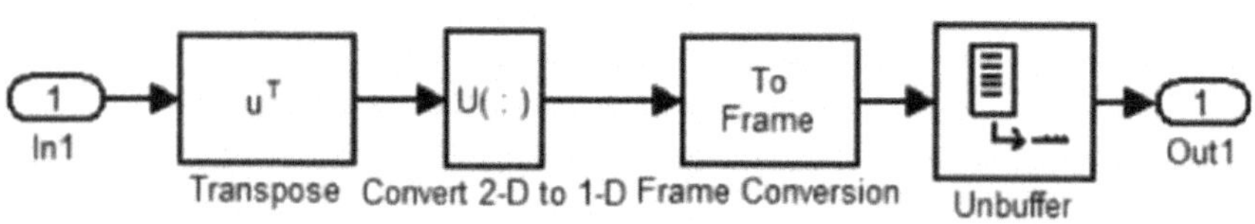

Fig. 6 Image pre-processing

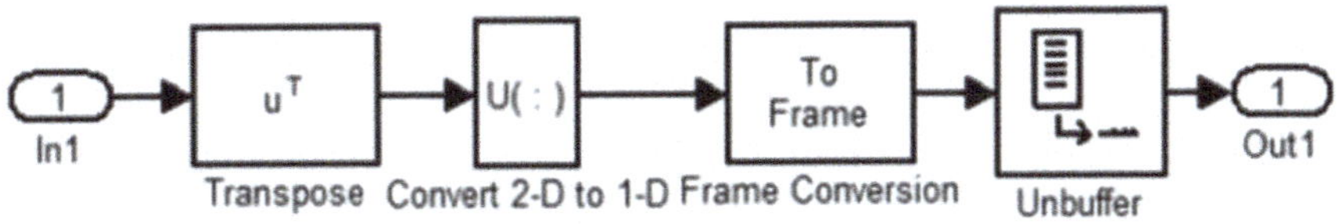

Fig. 7 Image post-processing

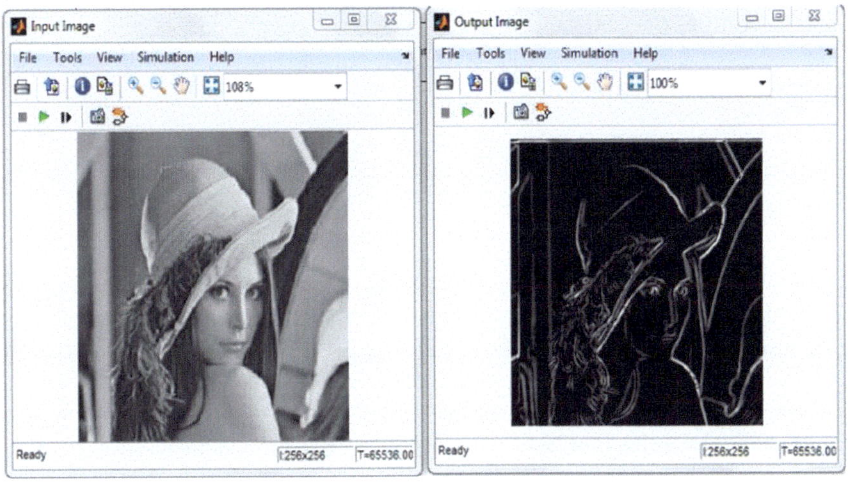

Fig. 8 Edge detected by adaptive threshold technique

Pre-Processing Procedure

1. Data form alteration: it alters image to unnamed integer setup.
2. Buffer: changes scalar illustrations to frame output. It's done at low sampling level.
3. 2D to 1D converter: it alters one-dimensional image to two-dimensional image matrix.

Post-Processing Procedure Post image processing is performed as shown in Fig. 7.

1. Data form alteration: it alters image to unnamed integer setup.
2. Buffer: changes scalar illustrations to frame output. It is done at low sampling level.
3. 1D to 2D converter: it alters one-dimensional image to two-dimensional image matrix.

The image input and output for software model is shown in Fig. 8. Here comparison of the each threshold value will take place, and the remaining pixels of image get suppressed according to its adopted threshold value and the larger threshold value will remain as an edge.

Table 1 Availability of devices and its utilization summary, a synthesis report

Logic utilization	Used	Available	Utilization (%)
No. of slice registers	533	69,120	0
No. of slice LUTs	5478	69,120	7
No. of fully used LUT-FF pairs	373	5638	6
No. of bonded IOBs	28	640	4
No. of BUFG/BUFGC TRLs	4	32	12
No. of DSP48Es	1	64	1

4 Simulation Results

Results analysis explains about the detection of efficient edge analysis by adaptive threshold technique, where the implementation of parallel processing, which results in the less computations because the each frame gets divided into blocks as a result delay will be less due to parallel programming and the memory utilization will be less. The synthesis results of the proposed block are shown in Table 1. It shows the number of device utilization takes place according to its availability.

5 Conclusion

The paper describes the "Implementation of Canny Edge Detection using adaptive technique" that used to detect the edge of any image as a complete image with dividing it into blocks. The proposed block-level Canny Edge detector has overcome the limitation of existing edge detection algorithms by reducing the delay and area. The design of block-level Canny edge detector is coded in VHDL language. The simulation and synthesis of the design is carried out using Xilinx ISE tool. The proposed method takes less area, and less computational time result of this decreases latency and increases throughput. In future, it can be possible to propose dynamic-based edge detection algorithm which can adapt for different variations of lighting conditions in image and also can be extended to video processing in detection of real-time edges required for broadcasting.

References

1. Derichee, R.: Using canny criteria to derive a recursively implemented optimal edge detector. Int. J. Comput. Vis. **1**(2), 167–187 (1987)
2. Torres, L., Robert, M., Bourennaneae, E., Paindavoineae, M.: Implementation of a recursive real time edge detector using retiming technique. In: International Conference on Very Large Scale Integration, pp. 811–816 (2017)
3. Lorcaaa, F.G., Kessalaa, L., Demignyaa, D.: Efficient ASIC and FPGA implementation of IIR filters for real time edge detection. IEEE Int. Conf. Image Process. **2**, 406–409 (2015)
4. Raao, D.V., Venkatesanan, M.: An efficient reconfigurable architecture and implementation of edge detection algorithm using handle-C. In: Proceedings of the International Conference on Information Technology: Coding and Computing, vol. 2, pp. 843–847 (2004)
5. Gentsos, C., Sotiropoulou, C., Nikolaidis, S., Vassiliadis, N.: Realtime canny edge detection parallel implementation for FPGAs. In: Proceedings of the International Conference on Electronics, Circuits and Systems, Rio de Janeiro, Brazil, pp. 499–502 (2010)
6. Heon, W., Yuan, K.: An improved canny edge detector and its realization on FPGA. In: Proceedings of the World Congress on Intelligent Control and Automation, pp. 6561–6564 (2008)

Fashion Express—All-Time Memory App

V. Sai Deepa Reddy, G. Sanjana, and G. Shreya

Abstract This paper is written with the aim to reduce the pain of those people, who own enough clothes, that they forget about the red top they wore a couple of times because it entered the black hole of the closet. No one knows, what's in there. Alexa is one such person who has that problem, she's got other problems too, such as having to wear ten different outfits to decide on one, that's some serious commitment. And during the time of AI, ML, DS, automation, etc., we having to go through that pain is not needed cause the booming technology can solve all our problems. Like how going to a grocery store manually turned into some clicks away, I want to see how I look in all the possible outfits by just clicks and not manually having to change outfits so many times. This can be easily done by the 3D software (3D printer as the upcoming technology) and an app that stores all our outfits and segregates it. The app would be made to work for android phones because according to analytics, Android is rising with its worldwide market share with 71.61% and iOS has 19.5% but varies differently for different regions. Finally, it should be accessible for everyone, therefore, in the near future, it will work on all these major mobile OS iOS, Android and Windows.

Keywords Virtual closet · 3D trial room · 3DLOOK · App · Android operating system

V. Sai Deepa Reddy (✉) · G. Sanjana · G. Shreya
Department of Computer Science, Stanley College of Engineering and Technology for Women, Hyderabad, India
e-mail: saideepa_v@outlook.com

G. Sanjana
e-mail: ganjisanjana2002@gmail.com

G. Shreya
e-mail: ganjishreya2002@gmail.com

S. C. Satapathy et al. (eds.), *Intelligent Data Engineering and Analytics*, Advances in Intelligent Systems and Computing 1177,
https://doi.org/10.1007/978-981-15-5679-1_46

1 Introduction

My idea requires 3D software to come to life. 3D modelling is the process of developing a mathematical representation of any surface of an object (either inanimate or living) in three dimensions via specialized software. The product is called a 3D model. 3D models are used in a wide variety of fields. The medical industry uses detailed models of organs; movie industry uses them as characters and objects for animated and real life motion pictures; video game industry uses them as assets for computer and video games; science sector uses them as highly detailed models of chemical compounds; architect uses them to demonstrate proposed buildings and landscape; engineering community uses them as designs of new devices; earth science community has started to construct 3D geological models. 3D models can also be the basis for physical devices that are built with 3D printers or CNC machines. When we are trying to model everything in 3D, it's obvious that we tried to model mannequins in 3D and dressed it up. And I want a 3D mannequin (which is me, scanned by 3D software) on my phone, so that I can mix and match and try many outfits and choose one easily and not spend an hour in front of the mirror trying to rotate my neck 360° and change a hundred times.

The software I am using to bring my idea to life is 3DLOOK. It's a body data platform which is a cost-saving idea for business retailers and fashion industry because customers have a hard time doing online shopping because they are not sure of the right size, and customers cannot relate how they would look wearing it, by looking at an xs size model wearing the dress. This 3DLOOK software helps solve all those problems cause of what it can do (Fig. 1).

Human Body Measuring: Time-saving and cost-reducing measurement software that helps clients to quickly and accurately measure their customers.

3D Model Generation: 3D body model generation software that powers VIRTUAL DRESSING for product designs and development, and this is the application I am using to recreate a customized digital closet app so that everyone can have their own virtual closet.

Fig. 1 All three images portray the uses of 3D eplained below

Size and Fit recommendation: Size and fit recommendation software which reduces the guesswork of finding the right size for the customers helping to reduce returns [1].

All this can be done by just taking two photos and entering some basic details like age, gender and height.

My aim is to have this accessible to everyone and anywhere, that is possible because nowadays everyone owns a PDA (personal digital assistant) which are now called smartphones. Hence, building a mobile app that will run on mobile phones operating system, android. The essential step for developing an app is making of the mobile user interface design. Mobile UI considers constraints, context, screen, input and mobility as outlines of the design. Making an app, using which the user can interact with it to find the solution to their problem and the app being able to do that in the most efficient way possible making their professional or personal life easier.

2 Literature Survey

In light of that, researchers are working hard to incorporate IoT to many systems of our daily life, including smart closet, to bring more user satisfaction by reducing the workload of the users. An analyst analyzed the consumers' attitude towards smart wardrobe using a Technology Acceptance Model (TAM), which shows the influential factors that attract users to this advanced concept. Considering the hassle of users to manage clothing.

By reading through the previous research done or the economically failed projects, I think what's lacking mostly is the strategic marketing for the idea. Anything new in the market will be shown hesitance from the customers to use it, therefore, we need to prove them that it's useful by showcasing its applications and motivate major stores to advertise their clothes in that manner. Customers who online shop can be attracted by giving them a chance to try out their wish list clothes on the 3D model. 3D apparel Design Software brings the power of 3D to designers who work with passion, and they would like to put their time and effort in their art, not spend time doing errands all over the world to try to picture out their imagination in real like D software will change the game and make their process of making art beautiful and soulful. And according to [5], Lyst's year in fashion report, the three fashion services that are changing the ways customers shop are resale, retail and the rise of virtual and augmented reality. In May, Nike launched its Nike Fit mobile scanning app that scans user's feet and recommends its best size in a range of their own branded footwear. The sportswear giant said that it had spent the past year developing a solution after learning that more than 60% of people wear shoes in the wrong size.

Currently, millions of apps are available in different online stores to smartphone users. The most successful mobile applications have been downloaded over a billon times and each day new applications are launched to the mobile market, making it extremely attractive both for companies and independent developers to invest their time and money. Such demand has often led mobile software developers to adapt established software development methodologies or submit new proposals that fit the constraints related to mobile software development. The mobile software development particularities are diverse, but surely include short and frequent development cycles, frequent technological changes like platform, operating systems, sensors, etc., limited documentation, specific requirements and resources of the development team and the client, among others. In addition, all these possible factors are prone to constant innovation. Nowadays as science and technology regarding hardware and software applications are improving or moving forward at a faster rate, expectations for the UI have increased a lot. As Android OS is used by most of the population, in mainly Asian countries, Android SDK is attracting more attention. Unfortunately, these days apps have become extremely business orientated and hence user interface is not very pleasant because of too many pop-up ads.

Operating system is the software part of electronics. Better the OS better will be the user's experience, as the time taken to run applications, opening and closing files is reduced and user can tackle many more tasks at the same time with better OS. People have different kinds of needs and hence the operating system they choose may vary, but all of them would want a fashion express app to function at high speed and beautifully on all their electronics [2–6].

3 Fashion Express Model

APP is a customizable app, because everyone has got their own unique style and lifestyle by which they live in, hence the needs for each will be varying to a great extent.

App Sign In-1: First step is to make an account, either by using google or Facebook accounts or by going through the sign in process.

Inquires for building the basic structure of App-2: After the first step there will be a series of questions that are to be answered to build the basic structure of the app.

INQUIRY:1 Draw a basic structure of your closet.

Name the sections if your closet is segregated in a certain manner (colour/occasion/category) or can be numbered or both numbered and named (Figs. 2 and 3).

If your closet looks like the first picture, then digitally pictured like the second pictures, with the text written on them as the title for that section

Having all the items segregated will help during shopping to know how many ways the new item can be matched with your already existing's clothes

INQUIRY:2 Frequent occasions you get ready for are?

1. Casual/college wear
2. Date
3. Ethnic
4. Workout

Fig. 2 Assuming user's general closet would appear like the image above

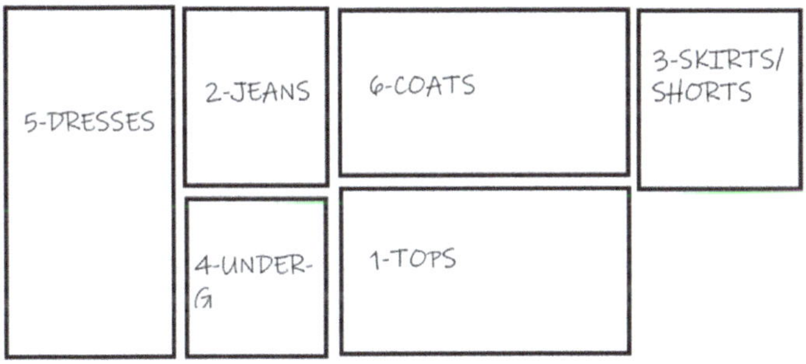

Fig. 3 The user should portray their closet (Fig. 2) in the app, like the image above

5. Formal.

User can enter any number of occasions.

Each occasion entered by the user will have a library of its own containing all the clothes of their closet which are preferred for that particular occasion.

In the add item option (explained below) user needs to check the box beside the occasion, to add that particular item to the library (of the occasion).

INQUIRY:3 Any other category by which you would want to segregate your virtual closet.

Examples: colour, looks (sleek, extra, etc.)

Fig. 4 To generate a 3D model, user needs to upload images standing in postures like the images above

- After answering all the questions, the user will be directed to the last step before where they need to take two photos which look alike (Fig. 4).

After taking the pictures, software in the app will create a 3D MODEL of the user. That will be the 3D model on which we put different outfits, save the pics, then compare and decide the look for the occasion.

- Now the app is ready for use. The main screen of the app looks like the picture below (Fig. 5).

Explanation of all icons on the main screen of the App:

Bottom Icons: The rectangular shape icons are representing the user's closet, which the user-specified about in the first question. When the user clicks those icons, it will show all the items for which the user selected the location as tops/jeans, etc.

Circular Icons: The icons represent the occasions specified by the user in the second question. When the user clicks those icons, all the items for which the user checked the box for that occasion are shown. In other words, all items that the user thought were appropriate for that particular occasion are shown. And when we click them, it is worn by the virtual 3D Model.

Add Item Icon: This icon is used to add items into the virtual closet. The user needs to add the photo and select the location of the item in the closet and check the

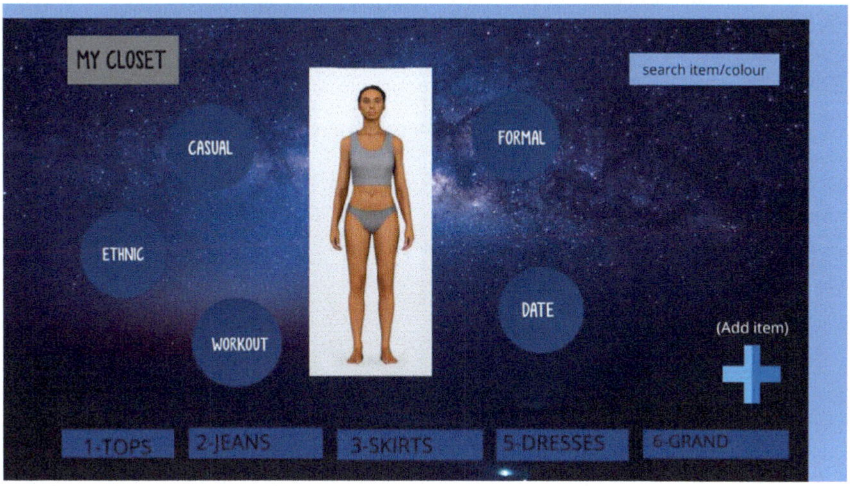

Fig. 5 Main screen of the Fashion Express app

circles for the occasion, the user thinks the item is suitable for. The user can check more than one circle, i.e. they can choose one item to be appropriate for more than one occasion (Fig. 6).

Mix and Match: Option to save the pictures and compare which outfit looks better.

From the gallery of your looks, the user can just select the pictures, and all of them are aligned side by side on the screen or the user (Fig. 7).

Fig. 6 Appearance of the add item option

Fig. 7 All the selected outfits are aligned like the images above for user to compare and choose

4 Situations

Where this App's presence would matter:

- Designers in the fashion industry won't have to wait to see how models would look on the runway.

- For runways, choosing which outfit would look better on which model would become very easy, because instead of telling the model to change outfits 10 times, they can digitally look at the 3D Model wearing all the outfits and choose easily.
- Celebrities have a try on session before with their stylist before events. The time taken for them to travel to the stylist place and change the outfit and the energy consumed of the person during the process can be replaced by five-minute discussion with the stylist on phone, both looking at the 3D model.
- Even websites instead of using models, they can use an accurate 3D Model for advertising the clothes and save money.
- College students can mix and match outfits and see which jean looks best on a particular shirt. No one has got enough time in the morning to do that, hence the app would come in handy.
- By having all your clothes organized on your phone, users easy access helps with the shopping.

5 Conclusion

This is the time or era, where everything and anything can be done just by sitting on your sofa. All your work can be done on the net using the laptop. When you can get food to eat through swiggy/zomato/uber eats, etc., order your favourite dress from amazon, get the needed makeup products from nykaa app and order a cab, using ola/uber, (all done, with a few clicks) why is there a need to get up to try on the clothes, even that should be done on our electronics with a few clicks.

References

1. 3D software. https://3dlook.me/virtual-dressing/
2. Many of my questions were answered by the help of https://www.quora.com/
3. General information regarding its use www.google.com
4. Marketexpert24.com. https://www.marketexpert24.com/2019/11/20/3d-apparel-design-software-market-emerging-trends-and-prospects-2026-with-leading-vendors-clo-efi-optitex browzwear-g2-tommy-hilfiger-3dlook/
5. Lyst's year of fashion. https://www.lyst.com/year-in-fashion-2019/. https://www.lyst.com/year-in-fashion-2018/
6. Regarding mobile application development market. https://yourstory.com/mystory/market-research-for-mobile-application-development
7. LizethChandi. https://www.researchgate.net/publication/318019805_Mobile_application_development_process_A_practical_experience
8. Comparative study of Google Android, Apple iOS and Microsoft Windows Phone mobile operating systems. https://ieeexplore.ieee.org/document/7980403
9. Mobile operating systems. https://www.webopedia.com/DidYouKnow/Hardware_Software/mobile-operating-systems-mobile-os-explained.html

Local Production of Sustainable Electricity from Domestic Wet Waste in India

P. Sahithi Reddy, M. Goda Sreya, and R. Nithya Reddy

Abstract India's issue with solid waste management couldn't be any more evident. This is because of the country's inability to keep up with the waste it has generated due to rapid urbanization, industrialization, and population explosion; implementation of an effective waste management system hasn't been fruitful. Indian domestic waste is found to be comparatively moister in nature and of lower calorific value. Hence, thermal technologies of management fail. A preeminent technology for Indian solid waste management would be its conversion into biomethane, which also happens to be an eco-friendly option. With utmost concern of this fact, a sustainable solution regarding the Public sector has been created in the effort of benefiting society at a community level. With an effort to bring back bio methanation with the perks of waste management (similar to ancient biogas stoves); this model aims at properly disposing the household wet waste and generating methane-based electricity from it to fuel a local park indefinitely.

Keywords Bio methanation · Food waste · Wet waste · Sustainable electricity · Waste management

1 Introduction

An Indian household's domestic comprehensive wet waste would constitute of kitchen waste including food waste of all kinds, cooked and uncooked, including eggshells and bones; Flower and house-plant waste; Garden sweeping or yard waste

---

P. Sahithi Reddy (✉) · M. Goda Sreya · R. Nithya Reddy
Computer Science Department of Engineering, Stanley College of Engineering and Technology for Women, Chapel Road, Abids, Hyderabad 500001, Telangana, India
e-mail: sahithihs3@gmail.com

M. Goda Sreya
e-mail: shreyashanu8@gmail.com

R. Nithya Reddy
e-mail: nithya_reddy1@hotmail.com

S. C. Satapathy et al. (eds.), *Intelligent Data Engineering and Analytics*, Advances in Intelligent Systems and Computing 1177,
https://doi.org/10.1007/978-981-15-5679-1_47

Primary Research Results Generated Across 500 Households				
Average Waste Generated per day per Household (in kgs)				
	Wet	Dry	Hazardous	Total
High Income Households	0.902	0.378	0.216	1.496
Middle Income Households	0.887	0.235	0.200	1.322
Final Averages	0.894	0.306	0.208	1.409

Fig. 1 Survey conducted concerning household waste generation

consisting of green/dry leaves, and other sanitary wastes and kitchen waste is abundant making the overall dry waste more moist, nutrient and organic matter rich, and easily biodegradable. Kitchen waste will also remain abundant for a long time due to the home-cooking culture of India. On an average, an Indian household produces 1 kg wet waste per day, and the volume of waste is projected to increase from 64 to 72 million tons at present to 125 million tons by 2031 (Fig. 1).

A huge chunk of this untreated waste from Indian cities lies for months and years at dumpsites where land was originally allocated for developing landfills for safe disposal of only the residual waste. Waste also accumulates in local areas in neighborhoods. The ill effects this waste cause include Health hazards of toxic waste such as the spread of infectious diseases; unattended waste lying around attracts flies, rats, and other creatures that, in turn, spread disease and also serves as a breeding ground for mosquitoes. The wet waste decomposes and releases an unpleasant odor, causes contamination of groundwater and harm to living creatures in that locality. Thus, we require proper means of not only disposal but also waste management to prevent degradation of the condition of the environment, avoid pollution, and make use of potent resources hidden in the trash. Current methods of management include incineration and landfill construction. Indian domestic waste is found to be moister in nature and of comparatively lower calorific value. Hence, thermal technologies (i.e. incineration) of management fail economically. Landfills in this country aren't executed properly and are a colossal disaster as they are a source of environmental pollution themselves. The landfills at few places are namesake, kept as large dump grounds.

A preeminent technology of Indian solid waste management would be its conversion into biomethane because Indian solid waste is moist and nutrient rich, and this process is a much eco-friendlier option than any other alternative, it is cost-effective and produces organic fertilizer. It is time to bring back the production of biogas as in olden days- system of cow dung stoves (this traditional method only vanishing due to LPG gas' easy availability and portability). The Biomethanation process used here is capturing of methane generated during anaerobic degradation of food waste/domestic wet waste and usage of it in the production of electricity.

The model we produced involves collecting the wet waste in each household of a colony daily (approximately 100–150 houses) assuming that the waste has been segregated into dry and wet. The waste collected is turned into a slurry using water. This slurry is fed into the inlet of the underground anaerobic digestion pit. Within the digestion pit, methane-60% and carbon dioxide-40% is produced. A pressure pump extracts the methane produced and directs it into a sizable gas turbine, where the methane is converted into electricity (alternating current) which is then stored in high capacity batteries. This electricity will successfully power a handful of led street lights and a few charging points within the park proving as social service. The tank will be periodically cleaned out and the decomposed matter obtained from the pit is to be used as fertilizer for the vegetation in the park.

This model not only provides a method to dispose of garbage, but also manage it with fruitful returns at a local level. This model is also low maintenance, minimalist, and environment friendly. The biggest boon of this model is the society-coming-together aspect of it. The gifts it bestows upon society include employment to the family gathering the waste, the locality's dispense with pollution and hence disease and irritation. The model is a local solution, the community need not rely on higher organizations, this system is sustainable as domestic households will not run out of wet waste, fertilizer is produced that can be used not only in the park but in personal gardens also, minimization of transporting the waste, hence reducing the carbon footprint of the model and the community comes together to do collective good which is a lead towards prosperity.

The rest of the paper is organized as follows: Section 2 gives the literature survey taken up for the present research. Section 3 gives the framework of the model and its methodology. Section 4 gives the results. Section 5 gives enhancements planned out in view of the future. Section 6 gives the concluding statements followed by the references.

2 Literature Survey

The generation of waste especially wet/organic waste results in environmental pollution problems if not well managed. 70% of wet waste ends up in landfills and incineration factories. As the land has many finite resources, space is limited, the current models are inefficient to confront this section of landfills, a waste-to-energy biofuel technology has been developed [1]. Biofuels can provide a clean, easily controlled source of renewable energy from organic waste materials replacing firewood or fossil fuels which are becoming more expensive as demand is rising above supply. The waste-to-energy conversion technology from municipal solid waste is the biochemical conversion method (anaerobic digestion). Anaerobic digestion can be used to treat organic farm, industrial, and domestic waste [2]. Understanding the properties of the landfill waste generated in the USA and feasibility of power generation using food waste and its benefits [3]. Assessment of economic factors, such as the financial returns of an anaerobic digester and analysis of biomethane potential test (BMP) [4]. Further understanding of bio-methane chemical production [5, 7]. Analysis of waste-to-energy alternatives for a low-capacity power plant and suitable characteristics of the gas turbine. A comparative analysis demonstrated that the cycle with gasification from solid waste has proved to be technically more appealing than the hybrid cycle integrated with incineration because of its greater efficiency and considering the initially defined guidelines for electricity generation [6]. Understanding the legal views of these sustainable energy policies [8]. Implementation of biomethanation technology in Solapur, a small town in Maharashtra, India, inspired many environmental enthusiasts. The technology used in Solapur is "Thermophilic anaerobic digestion biomethanation" in which organic material is decomposed anaerobically (absence of oxygen conditions) at elevated temperature. Biogas from the plant is converted into electricity and simultaneously compost is also produced. Every day Solapur's plant generates 3 MW of electricity and the compost is packed and sold.

3 Framework of the Intended Plan

Biomethanation is a process in which organic material in general waste is converted into biogas (methane and carbon dioxide) by microbes under anaerobic (absence of oxygen) conditions. In the biomethane production system, the function of the digester requires periodic attention and daily looking after. The relative abundance of methane on earth makes it an attractive fuel (Fig. 2).

It is noted that approximately 100–150 kg of wet waste is produced every day in a locality, each household contributing about 1 kg. This waste is collected and taken to the park. Here the pair wet waste and water termed as the "feedstock" is prepared. This feedstock is fed into the concrete digester pit through the inlet, and the anaerobic digestion of feedstock takes place. This digester pit is an underground

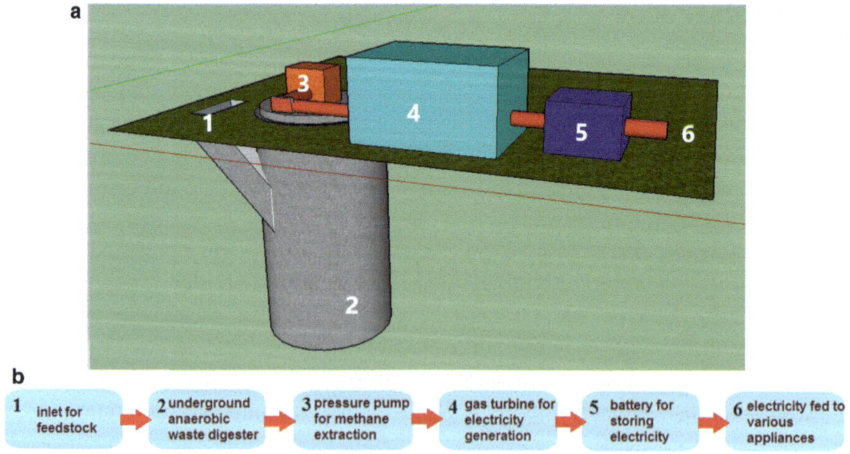

Fig. 2 **a** A simple diagrammatic explanation of the model. **b** A simple explanation of the model using flow

Fig. 3 Model of an anaerobic digester

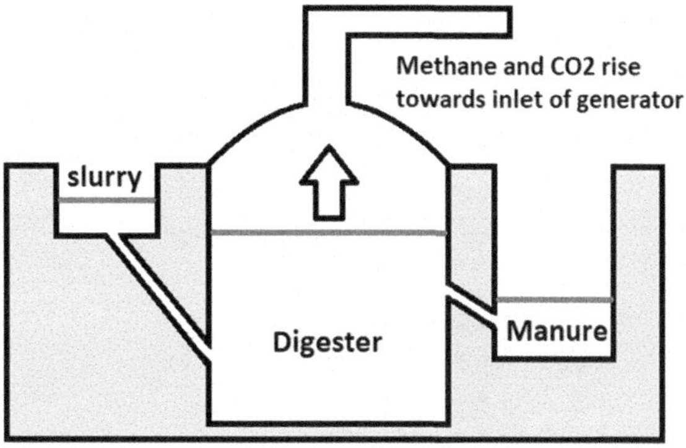

airtight concrete pit (similar to that of a drainage pit), targeted at enhancing the result of the anaerobic digestion of feedstock. The concrete digester pit is 20-m-deep and 3-m-wide in diameter. The dimensions of the pit ensure a great rate of production of methane. A high rate system biogas digester is used where methane forming bacteria are trapped in the digester to enhance the biogas production efficiency. It takes 21 days to generate one cycle of biogas, this biogas generated from the digester concrete pit contains 60% methane, 40% carbon dioxide. Several factors affect the anaerobic digestion process, hence altering the amount of methane produced; variation in feedstock will cause degradation at different rates and produces different amounts of methane. Some of these factors are season, temperature, and human lifestyle. According to the proposed model, the yield of methane produced in the digester is 40–60 cubic meters per day. The waste matter is to be cleaned out regularly, the debris serves as organic fertilizer for the vegetation in the park (Fig. 3).

The digester being located underground creates pressure within the chamber. The A pressure pump is connected to the outlet of the digester where the methane is then directed towards a gas turbine. Methane is drawn by the gas turbine and it is converted into electricity. The gas turbine mainly consists of three parts 1. The compressor, 2. The combustion chamber, and 3. The turbine. The compressor draws air into the engine, pressurizes it, and sends it to the combustion chamber with a high speed. The acceleration of the air increases the pressure and reduces the volume of the air. The compressed air is mixed with fuel injectors. The fuel-air mixtures ignite under constant pressure and the hot combustion products, i.e., gasses are directed through the turbine where it expands rapidly and imparts the rotation of the shaft. This rotation of shaft drives the compressor to draw in and compress more air for making the process continuous. And the remaining shaft power is used to drive a generator that produces electricity. This model can produce 80–120 kW electricity from wet waste every day. The electricity generated is then stored in a battery that is then used by multiple appliances of the park, according to required utilities (Fig. 4).

Fig. 4 A local park with the plant, street lights, public charging ports and led ad boards

4 Results

The statistics of this model are as such

- Amount of wet waste generated in one household per day—1 kg
- Amount of wet waste generated in one household per day—100–150 kg
- Amount of methane produced in the digester per day—40–60 cubic meters
- Amount of electricity produced by the plant per day—80–120 kW
- The produced electricity can be used to power LED street lights, mobile charging stations, water pumps, LED poster ads, electric vehicle charging stations, etc.
- With the 80–120 kW of electricity about 200 standard LED street lights can be powered and fully charge 20 phones (Figs. 5 and 6).

5 Enhancements

The addition of sensors would make the garbage disposal system "smarter". This model attacks the issue of wet waste disposal. A legitimate enhancement would be to tackle the mixed waste issue. The fact is that the waste produced in many localities is not efficiently disposed of with the current techniques.

This model's enhancements use sensors that are capable of separating different components of the waste in the garbage bin which is to be placed in the colony.

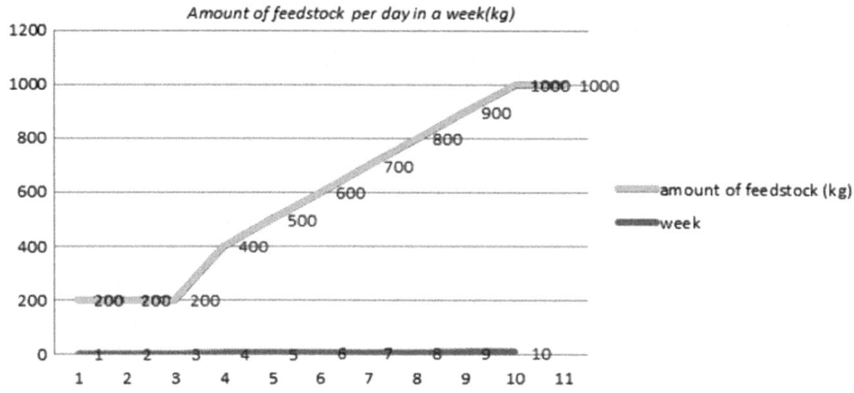

Fig. 5 The amount of feedstock used per week [2]

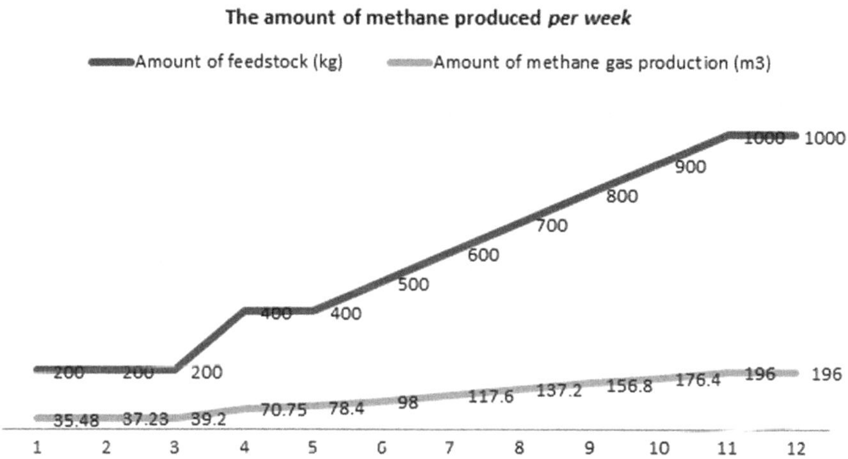

Fig. 6 The amount of methane produced per week [2]

It is known that the waste is basically two types: Biodegradable: paper, wood, sawdust and Nonbiodegradable: plastic, glass, rubber, metal. This model separates nonbiodegradable material from biodegradable material using sensors. The polymer type, i.e., nonbiodegradable material is separated using chemical sensor and metal type nonbiodegradable waste is separated using a proximity sensor. It is noted that 250–400 kg of mixed waste is produced in a locality. The mixed waste is put on a conveyer belt, as the belt moves the dry waste matter is separated as mentioned above and are directed towards different chambers. The biodegradable waste is converted into compost and electricity. The waste is sent into concrete biogas digester through a conveyer belt. The technology used is "Thermophilic anaerobic digestion biomethanation" in which feedstock is decomposed anaerobically at 50–55 °C. Advantages of this technology—Because of higher operating temperature in this process the operating loading rate is slightly higher. As the temperature is high, no pathogens are present in the final output. It takes 21 days for waste to convert into biogas. Biogas mainly contains 60–65% methane in this process and 40% carbon dioxide. A pressure pump is connected to the outlet of the digester where the methane is then directed towards a gas turbine. Methane is drawn by the gas turbine and it is converted into electricity. It is observed that 350–450 kW of electricity is generated. The compost is collected from the digester which acts as an organic fertilizer.

6 Conclusion

As Indian waste consists of more moisture and less calorific value, this model is the best way one can manage the waste in India. It is a better way for recycling the food waste generated by household chores. This method also instills a sense of community among the people. As we are producing electricity by solid waste, we reduce the consumption of fossil fuels such as coal which is used for generating energy. It is an eco-friendly way for solid waste management. Energy generation from waste releases less harmful gasses into the environment, whereas the decomposition of waste in landfills releases methane, a greenhouse gas, into the environment. Energy from waste facilitates us by reducing the cost of transportation of waste to landfills, while at the same time it produces energy which has some monetary value. Through this, we are reducing the waste going to the landfills, which ultimately reduces the need for huge landscapes for dumping the waste. Using waste to generate electricity can help reduce fluctuations in price. Also, there are no wide fluctuations and shortages in availability.

References

1. Handen, E., Diaz Padilla, M., Rears, H., Rodgers, L.: Food waste to bio-products. repository.upenn.edu. Accessed 18 Apr 2017
2. Othuman Mydin, M.A., Nik Abllah, N.F., Ghazali, N.: Development of environmentally friendly mini biogas to generate electricity by means of food waste. J. Mater. Environ. Sci. **5**(4), 1218–1223 (2014)
3. Park, M., Deginal, P., Mandac, M., Hughes, A., Chiyak, E.: Decreasing food waste deposited into landfill. digitalcommons.kent.edu. Accessed 05 Apr 2018
4. Zeynali, R., Khojastehpour, M., Ebrahimi-Nik, M.: Farm biogas plants, a sustainable waste to energy and bio-fertilizer opportunity for Iran. J. Clean. Prod. **253**, 119876 (2020)
5. Wiley, P.E., Campbell, J., McKuin, B.: Water environment research, production of biodiesel and biogas from algae: a review of process train options. Water Environ. Res. **83**, 326–338 (2011)
6. Ferreira, E.T. de F., Balestieri, J.A.P.: Comparative analysis of waste-to-energy alternatives for a low-capacity power plant in Brazil. Waste Manag. Res. **36**(3), 247–258. First Published 27 Jan 2018
7. Sialve, B., Bernet, N., Bernard, O.: Anaerobic digestion of microalgae as a necessary step to make microalgal biodiesel sustainable. Biotechnol. Adv. **27**(4), 409–416 (2009). Elsevier

8. Alexiou, A., Berardino, D., Alexiou, G.E., Kalyuzhny, S.V., Angelidaki, I.: The global role of anaerobic digestion through various governmental waste and energy sustainability policies. In: Anaerobic Digestion: 10th World Congress, 29th August–2 September 2004, Montreal, Proceedings, Montreal: NRC & IWA, vol. 4. pp. 2526–2530

GPS Tracking and Level Analysis of River Water Flow

Pasham Akshatha Sai, Tandra Hyde Celestia, and Kasturi Nischitha

Abstract The global challenge that people face in the present situation is Water Resources Management (WRM). This paper supports the creation of an application for the better analysis of the water resources in the country without any faulty information misleading the records. This was thought so that the people can themselves know the situation of the water availability in their area and use water precisely. The Central Water Commission (CWC) can now keep a constant record of the water availability by itself without any intermediary. It is important for us to evaluate the water flow in different areas.

Keywords IOT · Feature extraction · Multi-resolution satellite image · Global positioning system (GPS) · IBM Cloudant geospatial

1 Introduction

Satellite imagery is used for mapping the natural resources like water bodies and forests. The monitoring and sustainable management of these natural resources is imperative at regular intervals. The global carbon cycle and the climate variations are dependent on the water bodies which are analyzed by mapping from the satellite imagery. It provides us the assessment of the water degradation and the conservation measures to be taken by the spatiotemporal domain. The satellite data provides the visual interpretation of the water bodies of different measures. The satellites in large numbers are orbited around the earth enabling imagery of the surface, which

P. A. Sai (✉) · T. H. Celestia · K. Nischitha
Stanley College of Engineering and Technology for Women, Hyderabad, Telangana, India
e-mail: akshathasai14@gmail.com

T. H. Celestia
e-mail: hyde.celestia7@gmail.com

K. Nischitha
e-mail: kasturinischitha06@gmail.com

S. C. Satapathy et al. (eds.), *Intelligent Data Engineering and Analytics*, Advances in Intelligent Systems and Computing 1177,
https://doi.org/10.1007/978-981-15-5679-1_48

also helps in frequent targeting of a particular region repeatedly with the satellite instruments.

The observed satellite remote sensing visuals of the water body has a particular importance from the past few years.

- Nowadays problems related to water scarcity are given worldwide importance, that is, need of access to existing water resources and water resource change.
- As water getting increased in hills and mountains, by appropriate data given by satellite we could approach and misfortune the floods.

The IBM Cloudant database can be used to enhance the application by using geospatial operations. It also integrates with the existing Geographic Information System (GIS), which is an application used to analyze the spatial data of different sizes from multiple locations and users.

This focuses on creating an app to monitor the water flow of a river in a particular area.

2 Water Flow-GPS Tracking and Level Analysis

The main motive here is to create an app that shows the river flow in a particular area when typed and the water level. It is all about the management of water supply throughout the scale, right from small societies to the entire urban infrastructure. Smart water management aspires to work as a technology, where the flow of the water across the state or country is demonstrated by the satellite sensor.

Satellite Imagery:

Satellite imaging or remote sensing is the scanning of the earth by satellite to obtain information. It is useful because different surfaces and objects can be identified by the way they react to radiation. Satellite remote sensing is used in accommodation of the information retrieved from the non accessible regions at any condition. It also helps in monitoring nearly all components of water balance in a particular area (Fig. 1).

Working of Satellite Sensor:

1. Satellite sensors are used to measure infrared radiations. They give information as to how much heat is emitted from an object at the earth's surface. At a ground station, the multispectral satellite sensor data are collected and stored in the digital form of magnetic tapes for processing.
2. The other observation system used by remote sensing is False Color Composite (FCC), which is commonly used to compare true colors due to the absence of pure blue color as further scattering is dominant in the wavelength of blue. As infrared is an absorption band, the water bodies look darker if it is deep or clear (Fig. 2).

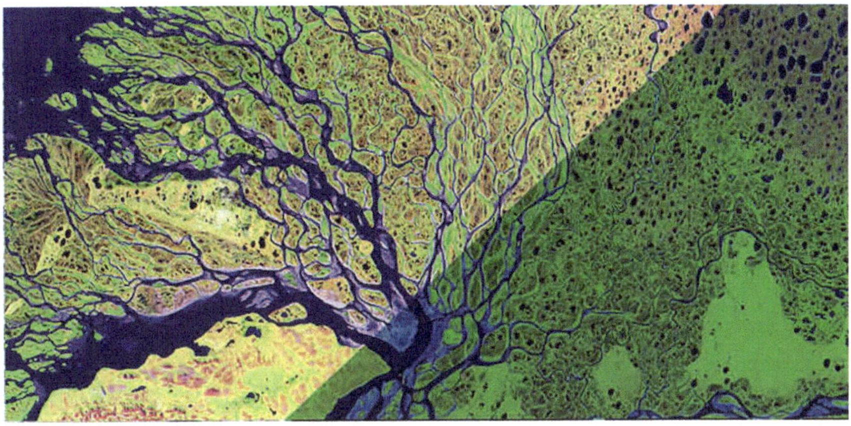

Fig. 1 Satellite image through remote sensing

This data is stored to be later retrieved using IBM Cloudant Geospatial, which combines the advanced geospatial queries of geographic information systems using visualizations powered by mapbox.

GPS Tracking System

Global Positioning System uses the satellite to send information to the receivers on the ground. It also helps in tracking the flow of water across the state and also in a particular area. A GPS tracking system of water can analyze both real-time and historic navigation data on any aspect of function. This data also is retrieved by IBM Cloudant Geospatial (Fig. 3).

Ultra sonic devices

1. As the sensor network can be flexibly expanded and shrunk according to the requirements of setup, it can also be used for analyzing the level of water fluctuating in the streamline.
2. The ultrasonic meter can also be used which allows us to calculate the velocity and volume of the flow of water.

These should be installed where the water streamline enters a particular area so as to keep a record of the water flow. These records will be sent to the IBM Cloud which transfers it to the application using IoT Cloud Connectivity.

Working of the Application

All the above data will be sent to the application template. The main feature of this application is—when a particular area around a river is entered, the satellite image and the water rate details appear. Also, the river flow can be tracked by the GPS facility to know its stream (Fig. 4).

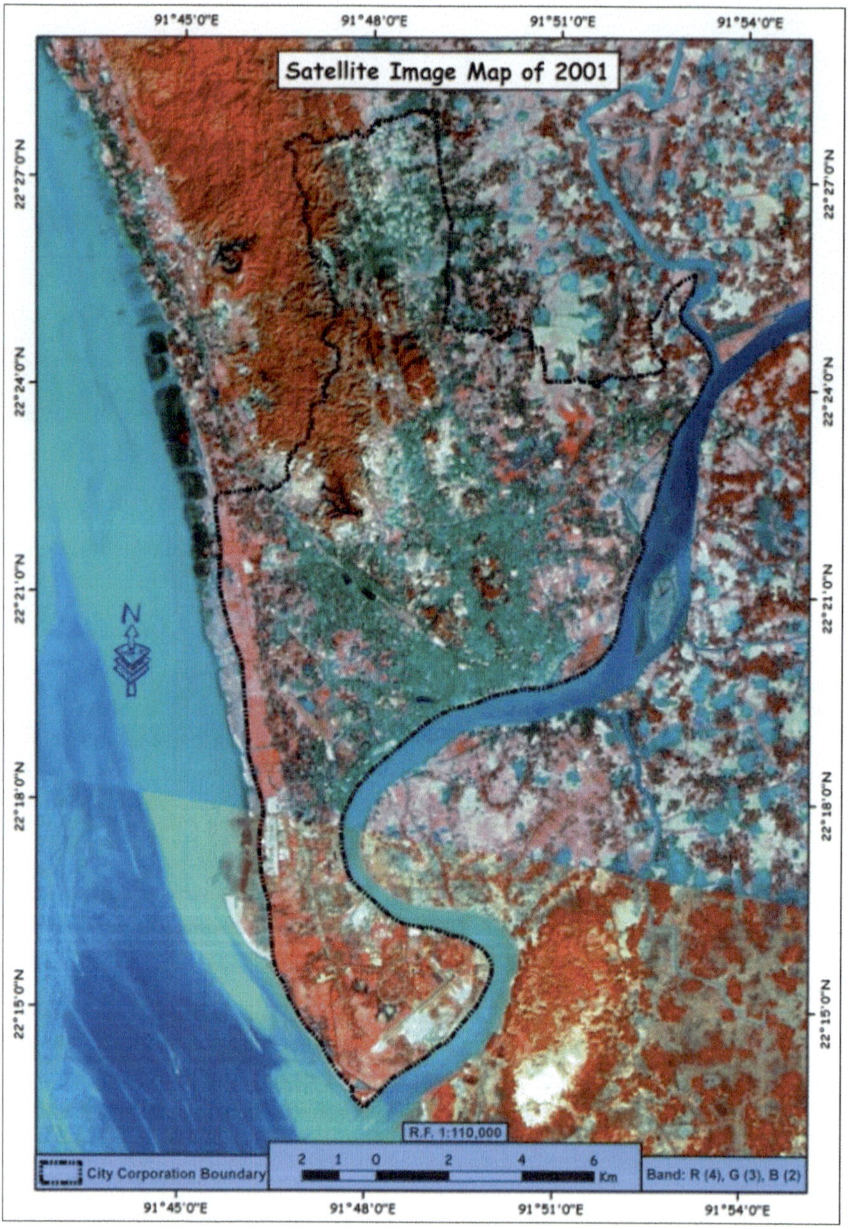

Fig. 2 Satellite FCC image

Fig. 3 GPS image by IBM Cloudant

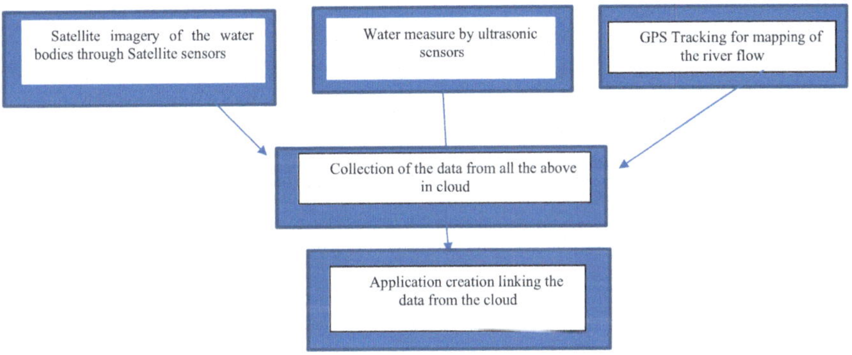

Fig. 4 Flowchart of the work done by the app

3 Conclusion

This idea will help us to know the amount of water flowing in different regions. Through this analysis, we can evaluate the amount of water being available for the residents of a particular area. The people can also monitor the availability of water and should use it accordingly. If the water level at some places is less then it can be easier to notify the government. The distribution of a common water resource for states must be done accordingly so that every state gets its share as decided by the CWC. This can also be helpful to monitor the river water flow; whether it is consistent or not. This can also decrease to some extent the problem of facing droughts; where the water level is really low. Similarly, we can also avoid the problem of river water submerging nearby places whenever the water level exceeds the normal measure [1].

For identifying various land-use classes on satellite imagery and enhanced products and identifying changes in time sequences in land-use patterns, the remote sensing GIS technique is used [2]. A new model that can identify the water body and collect data by the criteria of NDWI < −0.1 or NDVI2.0 was created based on the EOS/MOSDIS model [3]. Even from a great distances, the measurements can be calculated (hundreds or even thousands of kilometers in case of satellite sensors). Vast areas on the ground can also be covered easily with the help of remote sensing. Observing a target repeatedly (each day or several times per day) is also possible with satellite instruments. To provide frequent imagery of the earth's surface, many observation satellites have orbited, and are orbiting our planet. Most of the satellites can provide important data useful for detecting soil erosion, although less number of satellites have actually been used for this purpose. For water body extraction study, spaceborne sensors are used. The sensors can be categorized as the ones measuring the reflection of sunlight in the infrared and visible part of the electromagnetic spectrum and thermal infrared radiance, and of those actively transmitting microwave pulses and recording the signals which are received (imaging radars). In water body extraction research optical satellite systems have most frequently been applied. The Visible and Near-infrared (from 0.4 to 1.3 μm), the Shortwave infrared (between 1.3 and 3.0 μm), the Thermal infrared (from 3.0 to 15.0 μm), and the Long-wavelength infrared (from 7 to 14 μm) are the parts of the electromagnetic spectrum that these sensors include [4]. For collecting waterbody data from flood affected areas, the decision tree and programming technique is used. And for extracting water features form satellite images, a semi-automated change detection approach is used [5]. For extracting water resource information from IKONOS and other high resolution satellite images, an automatic extraction method is used [6]. Pixel by pixel classification and object-oriented image analysis, for categorizing of water bodies and different land covers in a satellite image were the two approaches proposed by the authors [7]. A mathematical morphological analysis technique for recognizing the water bodies from satellite images was also proposed by the authors. For the removal of the differences in atmospheric elements between images, a chromaticity analysis is suggested[8]. A classification algorithm [9] (using the average intracluster distance within the Bayesian algorithm) of remote sensing satellite image [10, 11]; is used which is sometimes the combination of supervised and unsupervised classification [12]. To outline the damage of the flood in 1993, in the Midwest of St. Luis, USA, the data fusion method and edge detection algorithm [13] were used [14]. To estimate and note the changing of water quality using historical land use data for a watershed, a remote sensing and Geographical Information System (GIS), was used in England [15].

Declaration

We have taken permission from competent authorities to use the images/data as given in the paper. In case of any dispute in the future, we shall be wholly responsible.

References

1. Chunxi, X., Jixian, Z., Guoman, H., Zheng, Z., Jiaoa, W.: Water body information extraction from high resolution airborne synthetic aperture radar image with technique of imaging in different directions and object-oriented
2. Prakash, A., Gupta, R.P.: Land-use mapping and change detection in a coal mining area—a case study in the Jharia coalfield, India. Int. J. Remote Sens. **19**(3), 391–410 (1998)
3. Nath, R.K., Deb, S.K.: Water-body area extraction from high resolution satellite images—an introduction, review and comparison
4. Armenik, C., Savopol, F.: Image processing and GIS tools for feature and change extraction. In: Proceeding of the ISPRS Congress Geo-Imagery Bridging Continents, Istanbul, Turkey, 12–13 July 2004, pp. 611–616
5. Yang, C., Yang, C., He, R., Wang, S.: Extracting water-body from Beijing-1 micro-satellite image based on knowledge discovery. In: The Proceeding of the IEEE International Geoscience & Remote Sensing Symposium, Boston, Massachusetts, U.S.A, 6–11 July 2008
6. Mouchot, M.-C., Alfoldi, T., De Lisle, D., Mccullough, G.: Monitoring the water bodies of the Mackenzie Delta by remote sensing methods. ARCTIC **44**(SUPP. 1), 21–28 (1991)
7. Van de, T., De Genst, W., Canters, F., Stephens, N., Wolf, E., Binard, M.: Extraction of land use/land cover-related information from very high resolution data in urban and suburban areas. In: Proceeding of the 23rd EARSeL Annual Symposium on June 3, 2003
8. Abbasi, H.U., Baluch, M.A., Soomro, A.S.: Impact assessment on Mancher lake of water scarcity through remote sensing based study. In: Proceeding of GIS, Saudi Arabia
9. da Rocha Gracioso, A.C.N., da Silva, F.F., Paris, A.C., de Freitas Góes, R.: Gabor filter applied in supervised classification of remote sensing images. In: Symposium Proceeding of the SIBGRAPI 2005
10. Jeon, Y.-J., Choi, J.-G., Kim, J.-I.: A study on supervised classification of remote sensing satellite image by bayesian algorithm using average fuzzy intracluster distance. In: Klette, R., Žunić, J. (eds.) IWCIA 2004, vol. 3322, pp. 597–606. LNCS (2004)
11. Alecu, C., Oancea, S., Bryant, E.: Multi-resolution analysis of MODIS and ASTER satellite data for water classification. In: Proceedings of the SPIE, the International Society for Optical Engineering, San Jose CA, ETATS-UNIS 2006
12. Fuller, L.M., Morgan, T.R., Aichele, S.S.: Wetland delineation with IKONOS high resolution satellite imagery, Fort Custer Training Center, Battle Creek, Michigan, 2005. Scientific Investigations Report 2006–5051
13. Cayula, J.-F., Cornillon, P.: Edge detection algorithm for SST algorithm. J. Atmos. Ocean. Technol. **9**, 67–80 (1992)
14. Petrie, G.M., Wukelic, G.E., Kimball, C.S., Steinmau, K.L., Beaver, D.E.: Responsiveness of satellite remote sensing and image processing technologies for monitoring and evaluating 1993 Mississippi River flood development using ERS-1 SAR, LANDAST, and SPOT digital data. In: Proceeding of the ASPRS/ACSM, Reno, NV (1994)
15. Mattikalli, N.M., Richards, K.S.: Estimation of surface water quality changes in response to land use change: application of the export coefficient model using remote sensing and geographical information system. J. Environ. Manag. **48**, 263–282 (1996)

Ensuring Data Privacy Using Machine Learning for Responsible Data Science

Millena Debaprada Jena, Sunil Samanta Singhar, Bhabendu Kumar Mohanta, and Somula Ramasubbareddy

Abstract With the advancement use of computers extensively the use of data has also grown to big data level. Nowadays data is collected without any specific purpose, every activity of a machine or a human being is recorded, If needed in the future then the data will be analyzed. But here the question of trust arises as the data will go through many phases for the analysis by different parties. The data may contain some sensitive or private information which can be misutilized by the organizations involved in the analysis stages. So it is needed for the hour to consider the data privacy issues very seriously. Different types of methods have been proposed in this paper to ensure data privacy and also different machine learning algorithms have been discussed which have been used to design the proposed methods to ensure data privacy.

Keywords Data · Privacy issues · Machine learning · Privacy · Cryptography · AI

1 Introduction

The fight for future markets and greater pieces of share in all sectors is going full speed ahead. The world's most powerful organizations are in a relentless race to grow better mechanized frameworks. and thus support man-made brainpower innovation

M. D. Jena
Vellore Institute of Technology, Chennai 600127, India
e-mail: millenadebaprada.2018@vitstudent.ac.in

S. S. Singhar (✉) · B. K. Mohanta
International Institute of Information Technology, Bhubaneswar 751003, India
e-mail: c119004@iiit-bh.ac.in

B. K. Mohanta
e-mail: c116004@iiit-bh.ac.in

S. Ramasubbareddy
Department of IT, VNRVJIET, Hyderabad 500090, India
e-mail: svramasubbareddy1219@gmail.com

S. C. Satapathy et al. (eds.), *Intelligent Data Engineering and Analytics*, Advances in Intelligent Systems and Computing 1177,
https://doi.org/10.1007/978-981-15-5679-1_49

to take them in front of their rivals. By the end of 2020, AI is relied upon to turn over 21 billion dollars around the world. In any case, further improvement of Machine Learning and Artificial Intelligence innovations is by all accounts obstructed by a significant deterrent: Data Security and Data Privacy [1].

Our pursuit of inquiries, browsing history, online transactions, the videos we watch, and the pictures we post in social media are nevertheless scarcely any sorts of data that are being gathered and put away on a consistent schedule. This variety of information occurs inside our cell phones and PCs, in the city, and even in our own workplaces and homes [1]. Such private information is being utilized for an assortment of AI applications. Even some ML applications require people's private information for analysis purposes [1].

Such private information is transferred to concentrated areas in a clear message for ML calculations and to manufacture models from them [1]. The issue isn't restricted to the dangers related to having this private information presented to an insider in these organizations, or even outcast risk if the organizations holding this sensitive information were hacked. What's more, it is conceivable to gather additional data about the private information regardless of whether the information was anonymized, or the information itself [2].

2 Motivation

The essential goal of data science is to create experiences discovering designs, patterns about the world utilizing an assortment of systems including Big Data, Machine learning, Probability Statistics, Data Mining, Data Visualization, and so forth [3]. There are numerous examples of overcoming the adversity of organizations that became quicker by utilizing a viably information-driven component to make choices. This drove the utilization of data science more which is putting a gigantic effect on society with the worry of irresponsible information use [4]. This has pushed for Responsible Data Science which will yield helpful bits of knowledge yet don't damage the security/privacy of the individuals.

3 Current Scenario

As AI and ML-based business openings grow and become increasingly inescapable, protection and security experts in varying backgrounds are probably going to confront predicaments of where to attract line terms of social criteria and holding individual/private information and being straightforward about its utilization. It was never a dark or white thing in any case, however, now there will probably be a lot more shades of gray [5, 6].

Despite the fact that AI/ML pushes the difficulties to an easy level, but worries about security and privacy of data have been over owing since the beginning of huge data use. Also, even before AI returned to its current standard. Databases containing information about individuals have different sections, typically from security and privacy angles can be of one of the accompanying types:

(a) Personally Identifiable Information (PII) | these are sections that can practically and straightforwardly connect to or recognize an individual [6] (e.g., Adhaar number, telephone number, email address, etc.).
(b) Quasi-Identifiers (QI) | these are sections that may not be helpful without anyone else's input yet can be joined with other Quasi-Identifiers, inquiry results and some outside data to recognize an individual (e.g., PIN code, age, sex, etc.).
(c) Sensitive Columns | these are traits that don't have a place with the over two classes yet comprise information about the individual that should be secured for different reasons (e.g., pay, HIV funding, Bank account subtleties, live geo-area, etc.).
(d) Non-touchy Columns | these are the rest of the characteristics that don't fall into (a), (b) or (c) mentioned above (e.g., nation, college, etc.).

Obviously, within the sight of QIs, simply expelling PII sections from a dataset isn't sufficient for security assurance. For example, if essential statistic information (which qualifies as QI) is available in a dataset, at that point it very well may be joined with other open information sources | for example, a voter enrollment list | to recognize the people with absolute exactness. This methodology was utilized by scientists a couple of years prior when they found that the"anonymized" dataset shared for the Netflix Prize challenge could be undermined by the partner with certain information from another open information source | viz. film evaluations by clients on IMDB [7]. In that circumstance, the specialists utilized only a couple of information indicates accessible freely from IMDB and Netflix dataset and uncovered the whole movies viewing the historical backdrop of people (something that is viewed as crucial and ensured under US security guidelines).

Responsible Data Science necessitates that the advancement of experiences doesn't damage people's protection. There is obviously, an extremely direct approach to guarantee total security: don't gather the information. Since this methodology nullifies the very point of information-driven basic approach of data science. In large, the information assurance and protection approaches proposed in this paper have tried to accomplish an offset with information utility. Three wide approaches have developed, such as:

1. Access Control. The access control approach is based on respect to the reason that entry to information requires knowing the character, as well as the job of the individual looking for access to the information, and an unmistakable comprehension of what information should be gotten to access control is ordinarily accomplished by a mix of a strategy determination and appropriate innovative and different intends to implement the method.
2. Data Anonymization. The data anonymization-based methodology means to alter the information to anticipate the ID of people. De-recognizable proof methods have been recommended that encode delicate identifiers [8]. To lessen the danger

of redistinguishing proof of people in the information, while supporting restricted investigations, methods that change "semi"-identifiers (a lot of fields that can act as an identifier in the mix) by means of speculation and concealment have been proposed. In any case, these systems don't accompany formal security ensures.
3. Privacy-protecting Data Sharing. This methodology depends on utilizing Secure Multiparty Computation (SMC) to remotely get to wellsprings of private information with characterized and controlled protection ensures [6].

4 Vision for Future

Before information can be investigated adequately, the information researcher needs to create trust in the information.

Not simply whether the "right information" (e.g., significant, fair) is utilized for the current examination, yet in addition I whether the "information is right" (e.g., precise enough for the investigation).

It is significant that "trust" is a flighty and emotional thought I one may well confide in the information for one examination and not confide in it for another. Further, it may not be sufficient that the researcher believes in the data collector , each with various thoughts of trust should likewise confide in the information [6]. Now and again the significant inquiry shouldn't be "Do we believe in the information?" yet "Do we believe in it more than the other option?" or "Which information would it be advisable for us to believe?"

The issues (and their goals) of trust in the nature of private information for responsible data science as driven by bigger and clashing cultural powers. Two essential issues can be featured.

1. To begin with, interest in the progressively advanced investigation of private information will develop hugely and from all quarters. Organizations remain to benefit there own; governments remain to convey more to their residents at a lower cost, and both will endeavor to influence popular sentiment [7]. At times, their aims will line up with the interests of clients and residents.
2. The second force, i.e., increased demand for privacy and protection of individual rights through legislation, regulation, and societal norms will be unleashed and activated [4]. Importantly, different people feel differently about confidentiality, and privacy at the individual level must increasingly be defined and implemented.

5 Privacy Preserving Machine Learning (PPM)

Numerous security improving methods focused on enabling various input gatherings to cooperatively prepare ML models without miss utilizing their private information in its unique structure [9, 10]. This was basically performed by using cryptographic methodologies, or differentially private information discharge (annoyance systems).

Differential protection is particularly successful in anticipating enrollment surmising assaults [11].

6 Cryptographic Approaches

Cryptographic protocols should be used to perform ML training/testing on encrypted data when a defnite ML software requires data from a few parties. In many of these techniques, achieving better effectiveness involves data proprietors making contributions to the computation servers with their encrypted information, thus mitigating the problem to an impervious two/three-party computing setting [11]. In addition to improved efficiency, these techniques now have the benefit that input parties are not required to remain online [10]. Most of these processes deal with the case of horizontally partitioned data: each proprietor of records has accumulated the same set of elements for separate objects of information. Focus is a case in which every man or woman who wants an ML mannequin educated for him/her can post a few characteristic vectors extracted from their own photographs. The same set of aspects are obtained in each of these cases by using each holder of the evidence [13].

The most widely used cryptographic techniques for achieving PPML are homomorphic encryption, garbled circuits, secret sharing.

1. Homomorphic Encryption: Fully homomorphic encryption empowers the calculation on encoded information, with tasks, for example, expansion and increase that can be utilized as a reason for progressively complex discretionary capacities. Because of the significant expense related with much of the time bootstrapping the figure content (reviving the figure content as a result of the aggregated clamor), added sufficient amount of error to homomorphic encryption to be utilized in PPML approaches. Such conspires just empower expansion tasks on scrambled information and augmentation by a plain message. A well-known model among them is Paillier cryptosystem [11].
2. Garbled Circuits: Garbled circuit is a cryptographic protocol that enables secure two-party computing in which two suspicious parties can jointly evaluate a function over their private inputs without the intervention of a trusted third party. The role must be defined as a Boolean circuit in the garbled circuit protocol. Assuming that a two-party agreement with Alice and Bob needs to acquire the product of a resource processed on their private data sources, Alice can turn capacity over to a skewed circuit and send this circuit along with its confused information. Bob acquires Alice's skewed modification of his input without picking up anything about Bob's private information (e.g., using oblivious transfer). Bob would now be able to use his confused input to the jumbled circuit to achieve the product of the required power (and be able to impart it to Alice alternatively). Some PPML approaches homomorphic integrated content encryption with Garbled circuits [10, 12].

3. Secret Sharing: Secret Sharing refers to cryptographic methods of taking a secret, splitting it into multiple shares, and sharing the shares to multiple parties, so that the secret can be recovered only when the parties combine their respective shares. In particular, the holder of a secret, sometimes called the dealer, creates and shares of a secret and defines a threshold 't' for the number of shares required to reconstruct the secret. The dealer then distributes the shares in such a way that they are controlled by various parties [13].

7 Perturbation Approaches

Perturbation theory comprises mathematical methods for finding an approximate solution to a problem, by starting from the exact solution of a related, simpler problem. Perturbation approachuses in Differential privacy strategies in PPML. Differential Privacy (DP) strategies oppose enrollment deduction assaults by adding arbitrary clamor to the information, to emphases in a specific calculation, or to the calculation yield. While most DP approaches accept confided aggregator of the information, neighborhood differential security enables each information gathering to include the commotion locally requiring non-confided in server. At last, dimensional decrease bothers the information by anticipating it to a lower dimensional hyperplane to avoid remaking the first information, as well as to limit the surmising of delicate data.

1. Differential Privacy (DP): Differential security is a framework for freely sharing data about a dataset by portraying the examples of gatherings inside the dataset while retaining data about people in the dataset. Another approach to portray differential protection is as an imperative on the calculations used to distribute total data about a factual database that restricts the exposure of private data of records whose data is in the database. For instance, differentially private calculations are utilized by some administration forces to distribute statistic data or other measurable totals while guaranteeing classification of review reactions, and by organizations to gather data about client conduct while controlling what is noticeable even to deeper investigators.
2. Local Differential Privacy: When the info parties need more data to prepare a ML model, it may be smarter to use moves toward that depend on neighborhood differential security (LDP). With LDP, each information gathering would annoy their information, and just discharge this dark perspective on the information [9, 12]. An old and surely understood rendition of neighborhood security is a randomized reaction, which gave conceivable deniability to respondents to touchy questions. For instance, a respondent would flip a reasonable coin: (a) on the off chance that "tails", the respondent answers honestly, and (b) in the event that "heads", at that point flip a subsequent coin, and react "Yes" if heads, and "No" if tails. This rendition of randomized response (RR) is differentially private [14, 15].

3. Dimensionality Reduction (DR): It bothers the information by anticipating it to a lower dimensional hyperplane. Such change is lossy, and it was recommended by Liu [10], that it would improve the protection, since recovering the accurate unique information from a diminished measurement adaptation would not be conceivable (the potential arrangements are unending as the quantity of conditions is not exactly the quantity of questions) [10]. Henceforth, Liu proposed to utilize an arbitrary grid to diminish the components of the info of the information. Since an irregular framework may diminish the utility, different methodologies utilized both unaided and managed DR procedures, for example, Principal Component Analysis (PCA), Detrended Correspondence Analysis (DCA), and Multidimensional Scaling (MDS). These methodologies attempt to locate the best projection network for utility purposes while depending on the diminished dimensionality perspective to improve the protection.

8 Conclusion

The problem addressed in this article I how can a data scientist establish the appropriate trust among users that their private data can not be misused? We have described several ML algorithms discussed and explained how they can be used to protect data privacy. Present case scenarios need to be successfully addressed in order to answer this key question. As our vision for the future shows, we understand that vast social issues in a heartbeat will alter privacy or trust in data landscape quality.

References

1. Srivastava, Divesh, Scannapieco, Monica, Redman, Thomas C.: Ensuring high-quality private data for responsible data science: vision and challenges. J. Data. Info. Q. (JDIQ) **11**(1), 1 (2019)
2. Singh, S., Prabhakar, S.: Ensuring correctness over untrusted private database. In: Proceedings of the 11th International Conference on Extending Database Technology: Advances in Database Technology, pp. 476–486. ACM (March 2008)
3. Woodall, P.M.: The data repurposing challenge: new pressures from data analytics. (2017)
4. Chen, D.L., Jess, E.: Can machine learning help predict the outcome of asylum adjudications?. In: Proceedings of the 16th Edition of the International Conference on Articial Intelligence and Law. ACM (2017)
5. Andrews, V.: Analyzing awareness on data privacy. In: Proceedings of the 2019 ACM Southeast Conference. ACM, (2019)
6. Liu, X., et al.: "Preserving patient privacy when sharing same-disease data. J. Data Info. Q. (JDIQ) **7**(4), 17 (2016)
7. Bishop, C.M.: Pattern recognition and machine learning. In: Springer Science Business Media, (2006)
8. Smith, M., et al.: Big data privacy issues in public social media. In: 2012 6th IEEE International Conference on Digital Ecosystems and Technologies (DEST). IEEE, (2012)
9. Chen, D., Hong, Z.: Data security and privacy protection issues in cloud computing. In: 2012 International Conference on Computer Science and Electronics Engineering, vol. 1. IEEE, (2012)

10. Buczak, Anna L., Guven, Erhan: A survey of data mining and machine learn-ing methods for cyber security intrusion detection. IEEE Commun. Surv. Tutorials **18**(2), 1153–1176 (2015)
11. Liu, Kun, Kargupta, Hillol, Ryan, Jessica: Random projection-based multi-plicative data perturbation for privacy preserving distributed data mining. IEEE Trans. Knowl. Data Eng. **18**(1), 92–106 (2005)
12. Mohanta, B.K., Panda, S.S., Jena, D.: An overview of smart contract and use cases in blockchain tech-nology. In: 2018 9th International Conference on Computing, Communication and Networking Technologies (ICC-CNT), (July 2018). https://doi.org/10.1109/icccnt.2018.8494045
13. Cyphers, B., Veeramachaneni, K.: AnonML: locally private machine learning over a network of peers' (2017)
14. Narayanan, A., Shmatikov, V.: 'Robust de-anonymization of large sparse datasets. In: Proceedings—IEEE Symposium on Security and Privacy, pp. 111–125 (2008)
15. Shokri, R., Stronati, M., Song, C., Shmatikov, V.: Membership inference attacks against machine learning models. In: 2017 IEEE Symposium on Security and Privacy (SP), IEEE, pp. 3–18 (2017)

An IoT Based Wearable Device for Healthcare Monitoring

J. Julian, R. Kavitha, and Y. Joy Rakesh

Abstract Nowadays IoT (Internet of Things) devices are popularly used to monitor humans remotely in the healthcare sector. There are many IoT devices that are being introduced to collect data from human beings in a different scenario. These devices are embedded with sensors and controllers in them to collect data. These devices help to support many applications like a simple counting step to an advanced rehabilitation for athletes. In this research work, a mini wearable device is designed with multiple sensors and a controller. The sensors sense the environment and the controller collects data from all the sensors and sends them to the cloud in order to do the analysis related to the application. The implemented wearable device is a pair of footwear, that consists of five force sensors, one gyroscope, and one accelerometer in each leg. This prototype is built using a Wi-Fi enabled controller to send the data remotely to the cloud. The collected data can be downloaded as xlsx file from the cloud and can be used for different analyses related to the applications.

Keywords Wearable sensors · IoT · Force sensor · Accelerometer · Gyroscope

1 Introduction

Healthcare has made many major breakthroughs in recent years with the help of science and technology. The advancement in IoT technologies supports the researchers in the healthcare sector to provide a better solution. IoT enabled devices are utilized to generate these datasets for different diseases, illness, injury, and other

J. Julian · R. Kavitha (✉) · Y. Joy Rakesh
Department of Computer Science, CHRIST (Deemed to be University), Bangalore, Karnataka, India
e-mail: kavitha.r@christuniversity.in

J. Julian
e-mail: julian.j@mca.christuniversity.in

Y. Joy Rakesh
e-mail: joy.rakesh@mca.christuniversity.in

S. C. Satapathy et al. (eds.), *Intelligent Data Engineering and Analytics*, Advances in Intelligent Systems and Computing 1177,
https://doi.org/10.1007/978-981-15-5679-1_50

physical and mental impairments. There is a great requirement for a large dataset for research in healthcare. IoT devices are designed to collect data that can be produced by human beings. There are two types of devices that are used to collect data. They are non-wearable and wearable devices. In order to collect data, a non-wearable device deals with different technologies like image processing, smartphones, and infrared thermography. Image processing is used to find the moving pattern of humans with the support of a camera. Smartphones are built with different sensors like gyroscope and accelerometer. With the support of these sensors, smartphones collect the data when humans are doing different activities. Infrared Thermography [1] is also used to collect data in the form of thermal images during human activities. The wearable devices are IoT enabled devices that are worn by human beings in order to collect data. Some of these devices are in the form of footwear, ankle bracelet, wrist band, and smartwatches. These wearable devices are embedded with a variety of sensors to collect different types of data. Some of the commonly used sensors are accelerometer and gyroscope. An electromechanical device accelerometer is used to measure the acceleration forces of human beings by dividing velocity or speed by time as the measurement of acceleration. The common way to represent this data is a 3D graph with the x-axis, y-axis, and z-axis. Gyroscope is used to measure orientation and angular velocity. It senses rotational motion and changes in orientation. It is also commonly represented in a 3D graph with the x-axis, y-axis, and z-axis. In this research, an IoT foot wearable device is designed using force sensor, accelerometer, and gyroscope. The collected data can be used to classify different activities in activity recognition [2]. It also supports to find the number of steps a pedestrian walked in pedestrian tracking systems. With the help of the collected data prediction of a fall can be detected or risk can be analyzed.

The rest of the paper is organized as follows. Section 2 discusses the related work. Detail design consideration of the proposed IoT wearable device is discussed in Sect. 3. Sensors used, Wi-fi enabled Controller, and IoT cloud platform is discussed in Sects. 3.1, 3.2, and 3.3, respectively. Section 4 provides an analysis of the basic understanding of human activity patterns. This section also gives the comparative study of the existing system and proposed system. Finally, the research work is concluded in Sect. 5.

2 Related Work

In the healthcare domain, there are different types of disease, illness, injury, and other physical impairments related to legs. In order to learn and make new discoveries, research needs a large dataset. In the human system, the organs like legs are involved in most of the human physical movements. Researchers have proposed and designed a wearable device using various types of sensors.

Chen et al. [3] has proposed a footwear solution with force sensors. Four sensors are placed in each footwear with 1-in. diameter sensors and circuit modules in each footwear and a base station. The circuit module has a PCB (printed circuit board), a battery, and a wireless module. The force signals are converted into electrical signals and transmitted to the base station. Low-power radio frequency is used to transmit the data with 100-Hz sampling frequency. The base station has an MCU (Microprogrammed control unit) which collects the data from the two-circuit module and together send it to the host computer using the serial port. Hong et al. [4] has proposed a wearable solution with three tri-axial accelerometers that are worn on waist, wrist, and thigh. The wireless communication is achieved with the Bluetooth module. Different features such as mean, entropy, and correlation are calculated from the collected data. The frequency for the collection of data is set with 256 sample windows. RFID module is used to identify different objects that are used to interact. The reader has a passive RFID tag that does not need a battery and it is very small, so it can be fixed in small objects. It works with a frequency of 13.56 MHz. Liu et al. [5] has proposed a wearable GRF (Ground Reaction Force) sensor system with five small tri-axial force sensors. The GRF and CoP (Center of Pressure) values are measured using the wearable sensor system. Each sensor's weight is 15 g and dimensions of the sensors are 20 mm × 20 mm × 5 mm. All the five tria-xial force sensors are mounted on an aluminum plate beneath the shoe. The total weight of the shoe with the sensors is about 300 g. A multi-channel data—logger is used to collect data from all the sensors. A battery of 300 mAh is used to power the whole sensors system. Shu et al. [6] has recently developed a soft pressure sensor using conductive textile fabric sensing elements. The sensors can measure the pressure ranging from 10 Pa to 800 kPa. The sensor is enclosed within silicon rubber to withstand dust and moisture without affecting the performance. Six sensors are placed within the insole. Polyimide film circuit which is a thin and flexible circuit is used for connection of all the sensors in the insole. An analog to digital converter is used with a Bluetooth module for wireless communication. A Li-ion battery is used for constant power supply of 3.3 V.

Vandewynckel et al. [7] has proposed a system with a shoe-mounted accelerometer and an alligator chip. The system consists of a standard battery, a tri-axial accelerometer, a USB Bluetooth dongle, and a PIC24 microcontroller. The tri-axial sensor is placed on the shoe is such a way that x-axis is parallel to the floor, the y-axis is perpendicular to the floor, and the z-axis is directed toward the inside of the foot. All the collected accelerometer data is transferred using the Bluetooth dongle to the server. The sampling rate at which the data is collected is 200 Hz. Hori et al. [8] has used a three-axis force sensor for measuring GRF distribution during straight walking. The sensor consists of a vertical force detector, a sheer force detector, a flexible cable, and a thermoplastic rubber. There are three pairs of Si-doped beams which are fabricated on the sensor using MEMS (Micro Electro Mechanical Systems). The total dimensions of the sensor are 25 × 25 × 7 mm and the weight of each sensor is 15 g. A circuit with the bridge, low-pass filters, Analog to digital converter, and amplification circuits are mounted to each sensor. Sixteen sensors are placed on each foot, each circuit manages 4 sensors. All the data is stored in the inbuilt memory and sent to a PC with serial communication. The serial communication had a baud rate of 921.6 kbps with a sampling frequency of 333 Hz. The whole circuit is powered by lithium batteries. All the circuits were placed in the backpack and connected by cables to the shoe. The whole system for each leg contained four sets of circuits, two batteries, and cables, and the total weight is 1100 g.

In this research work, in order to understand the physical movement of the human body, IoT enabled wearable device in the form of footwear is designed. This sensor embedded device collects the necessary data in the most convenient way and stores the collected data in the form of a file at the cloud for further analysis.

3 Proposed IoT Wearable Device

In order to build an IoT wearable device in the form of a footwear, leather sandals are used as a base for the whole device. Three types of sensors and a Wi-Fi enabled controller are used to design this footwear to collect and transmit data to the cloud.

The block diagram of the proposed IoT wearable device consists of three modules. The first module is a pair of sensor embedded footwear which includes five pressure sensors, a three-axial accelerometer, a three-axial gyroscope in each footwear. The second module consists of data acquisition and transmission which includes a Node MCU with a Wi-fi module for each footwear. This Node MCU extracts the sensed data from each sensor and transmits it to the IoT cloud platform. The third module is the cloud platform where the collected data can be downloaded in the tabular form. Figure 1 shows the Block Diagram of the proposed IoT wearable device.

3.1 Sensors Used

Inertial sensor, accelerometer, and gyroscope are used to measure the movement of human beings in a smart system. A component MPU 6050, which is based on MEMS technology, is used as an accelerometer and gyroscope sensor in this research. It has a six-axis IMU sensor, which generates six values as output that includes three values from the accelerometer and three from the gyroscope. It uses I2C (Inter-integrated Circuit) protocol for data communication and has a 1024 Byte FIFO buffer.

A general-purpose force sensor that measures the pushing and pulling forces of a leg is used in a smart monitoring system. Five FlexiForce A401 Sensors are placed in each footwear to understand human movement. Each sensor has two pins that act as a variable resistor. When the pressure is applied to the sensor, it gives low resistance and vice versa. Based on the pressure applied to the sensor, it increases or decreases the resistance. An ultra-thin FlexiForce A401 sensor has a 0.5" diameter circle sensing area with a flexible printed circuit. Since it comes in paper-thin size, it is very much suitable for placing between the sole and the footwear base. It can measure up to 111 N that is 0–25 lb. The force is calculated by finding the resistance of the sensor. The resistance is calculated by using the analog value from the sensor and the voltage supplied and the resistance of the parallel resistor. The force from

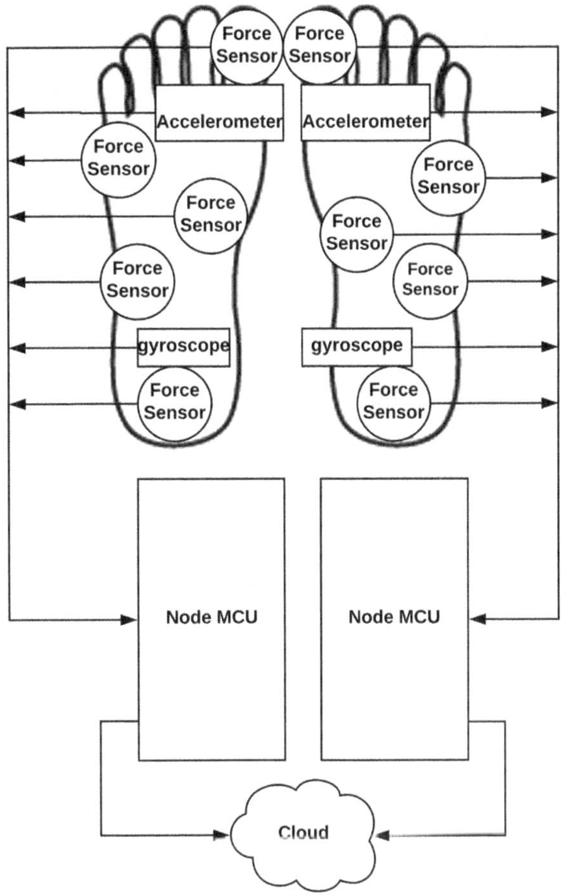

Fig. 1 Block diagram of proposed IoT wearable device

each sensor is collected by sending a 3-digit binary value as select signals to the multiplexer.

During physical movement, the pressure applied by the foot to the ground is not equally distributed. Since different pressure is applied in diverse parts of the foot, there is a challenge to find the correct location to place the force sensor on the footwear sole. In each footwear, one sensor is placed in the anterior region to find toe-off, three sensors are placed in the lateral region to understand left-right weight distribution, and one sensor is placed in the posterior region to recognize heel strike. Figure 2 depicts the position of force sensor on the sole and smart footwear, a sensor deployed IoT wearable device.

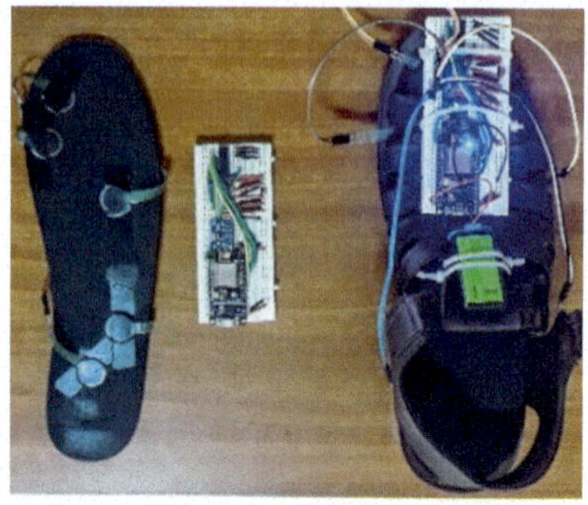

Fig. 2 Placement of force sensors on a sole and smart footwear

3.2 *Wi-Fi Enabled Controller*

An open-source IoT platform Node MCU 1.0 is used as the main controller in this footwear. Sensor embedded MPU 6050 connect to the Node MCU1.0 via two digital I/O pins to send the sensed data from both accelerometer and gyroscope. Each Flexi force sensor A401 needs a 3.3 O resistor in parallel with the same power supply and connect to the Node MCU 1.0 via an Analog I/O pin to send the force value to the controller. 74HC4051 analog Multiplexer and demultiplexer also used to multiplex the data from five FlexiForce sensors into a single Analog I/O pin. All five FlexiForce sensors are connected to the Multiplexer, and in turn, sensed data from five force sensor is sent to the Node MCU. The whole circuit is depicted in Fig. 3.

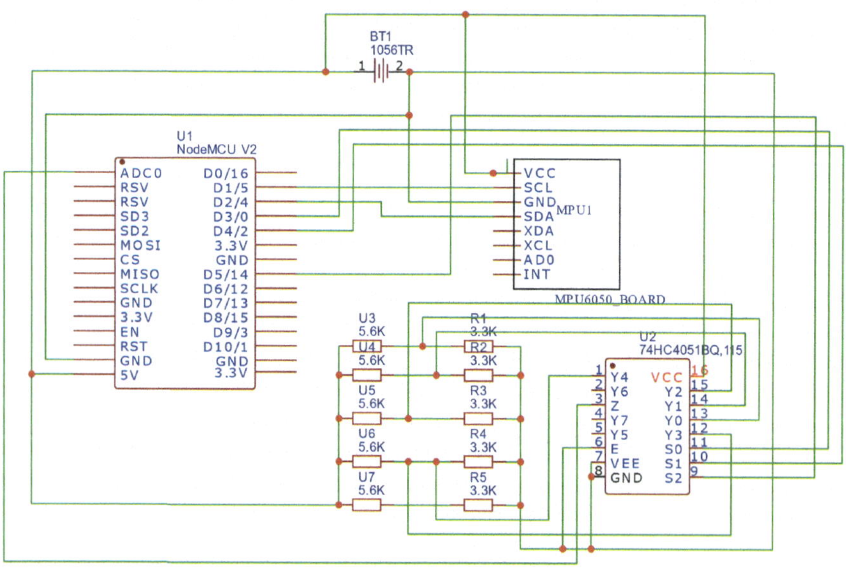

Fig. 3 Circuit diagram

A firmware ESP8266 a Wi-Fi SoC from Espressif Systems is being used in Node MCU to connect with the cloud platform. ESP-12E module-based hardware is used for Wi-Fi connection. It is the best choice because it has all the features of an Arduino UNO controller and an ESP8266 Wi-Fi module inbuilt in it. It has 80 KB of RAM memory and 4 MB of flash memory and operates on 80 MHz frequency. This controller needs a constant power supply of 3.3–5 V.

3.3 *IoT Cloud Platform*

In this research work, a Google spreadsheet is used as a cloud platform in the IoT wearable device. The google sheet REST API v4 is used to receive the collected data from the Wi-fi enabled controller. The sensed data is saved along with timestamps using the script editor in the spreadsheet. Force values from five force sensors, three values as X-axis, Y-axis, Z-axis from the accelerometer, and three values as X-axis, Y-axis, Z-axis from gyroscope from each footwear is saved in the spreadsheet. A python code with Jupyter Notebook is used to get the final dataset according to the timestamp.

The functionality of the proposed wearable device is represented as a flowchart in Fig. 4. When a subject wears this footwear and starts moving, the wearable device initializes, and all sensors start sensing and generate data related to the movement. In each footwear, the controller collects sensed data from accelerometer, gyroscope, and force sensor then it checks the connection with the cloud platform regularly. If the connection is established, the collected data is sent to the cloud platform otherwise it tries to reconnect. This whole process repeats in the IoT wearable device until data collection is over.

4 Result and Discussion

The collected dataset consists of twenty-three attributes which include, Timestamp, five force sensor values, three-axis values from accelerometer, and three axes values from the gyroscope, totally eleven attributes from each footwear. The attributes are named as DateTime, LAx, LAy, LAz, LGx, LGy, LGz, Lf1, Lf2, Lf3, Lf4, Lf5, RAx, RAy, RAz, RGx, RGy, RGz, Rf1, Rf2, Rf3, Rf4, Rf5. The accelerometer and gyroscope values are X, Y, Z-axis which can be positive or negative. The force values are in Newton ranging from 0 to 111.

As shown in Fig. 5, the sensed data from the wearable device is transmitted to the Google spreadsheets using a cloud platform. This dataset can be used in different applications such as gait pattern analysis, plantar pressure measurements, posture and activity recognition, energy expenditure estimation, biofeedback, fall risk assessment, fall detection applications, navigation, and pedestrian tracking systems, etc. [9] A pilot study was performed to verify the practicability of the proposed IoT

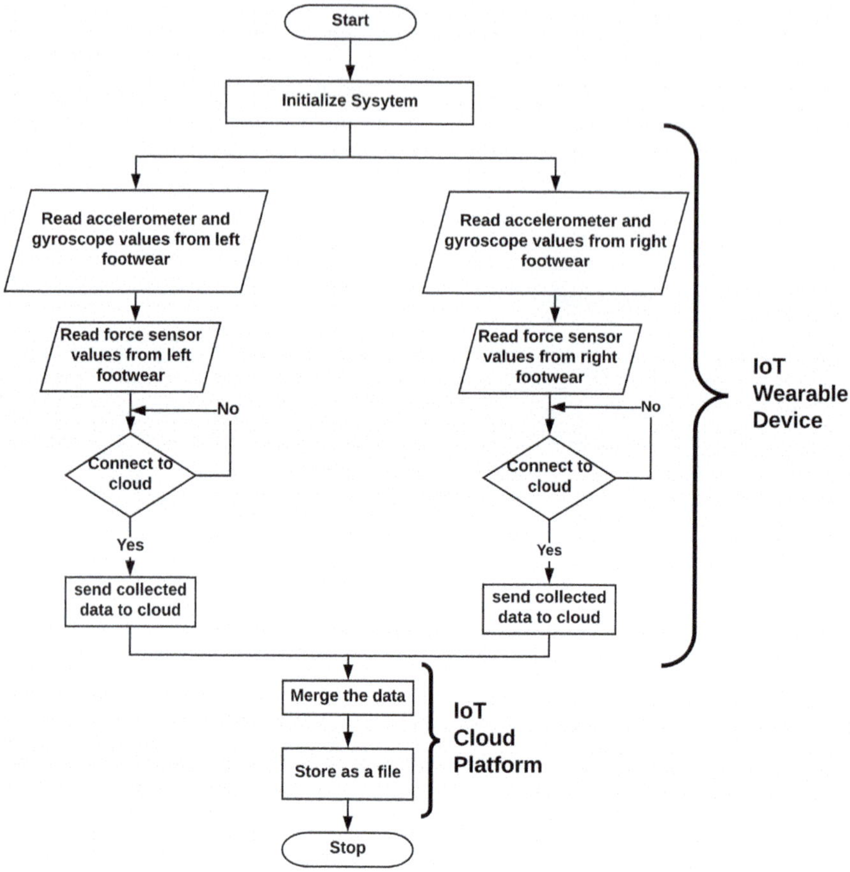

Fig. 4 Flowchart for data collection

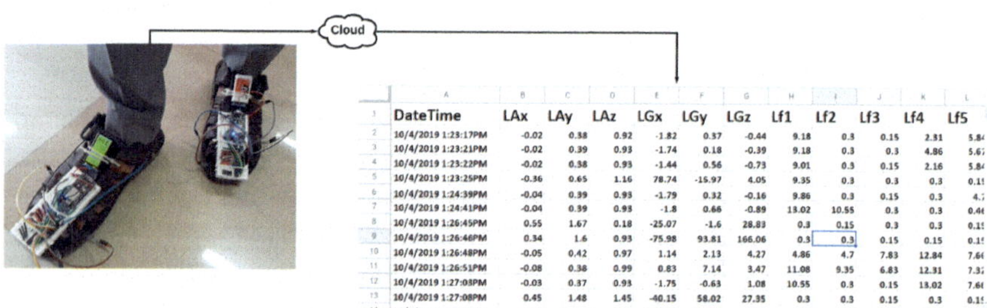

DateTime	LAx	LAy	LAz	LGx	LGy	LGz	Lf1	Lf2	Lf3	Lf4	Lf5
10/4/2019 1:23:17PM	-0.02	0.38	0.92	-1.82	0.37	-0.44	9.18	0.3	0.15	2.31	5.8
10/4/2019 1:23:21PM	-0.02	0.39	0.93	-1.74	0.18	-0.39	9.18	0.3	0.3	4.86	5.6
10/4/2019 1:23:22PM	-0.02	0.38	0.93	-1.44	0.56	-0.73	9.01	0.3	0.15	2.16	5.8
10/4/2019 1:23:25PM	-0.36	0.65	1.16	78.74	-15.97	4.05	9.35	0.3	0.3	0.3	0.1
10/4/2019 1:24:39PM	-0.04	0.39	0.93	-1.79	0.32	-0.16	9.86	0.3	0.15	0.3	4.
10/4/2019 1:24:41PM	-0.04	0.39	0.93	-1.8	0.66	-0.89	13.02	10.55	0.3	0.3	0.4
10/4/2019 1:26:45PM	0.55	1.67	0.18	-25.07	-1.6	28.83	0.3	0.15	0.3	0.3	0.1
10/4/2019 1:26:46PM	0.34	1.6	0.93	-75.98	93.81	166.06	0.3	0.3	0.15	0.15	0.1
10/4/2019 1:26:48PM	-0.05	0.42	0.97	1.14	2.13	4.27	4.86	4.7	7.83	12.84	7.6
10/4/2019 1:26:51PM	-0.08	0.38	0.99	0.83	7.14	3.47	11.08	9.35	6.83	12.31	7.3
10/4/2019 1:27:03PM	-0.03	0.37	0.93	-1.75	-0.63	1.08	10.55	0.3	0.15	13.02	7.6
10/4/2019 1:27:08PM	0.45	1.48	1.45	-40.15	58.02	27.35	0.3	0.3	0.15	0.3	0.1

Fig. 5 Data collection using proposed IoT device

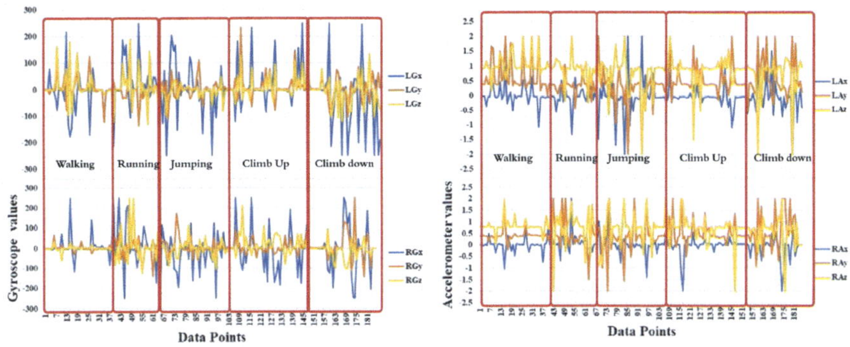

Fig. 6 Sensory data visualization of basic human activities

wearable device. In order to achieve this, data is collected from a human subject. The subject is instructed to do the basic activities like Walking, Running, Jumping, Climb Up (Stairs), Climb Down (Stairs) by wearing this device. Figure 6 shows the representation of data collected from the accelerometer and gyroscope. It clearly explains that the data collected from the accelerometer and gyroscope varies for each activity. The change in accelerometer values is less during slow activities such as walking than the fast activity running.

The accelerometer values show the change in acceleration in all three-axis based on the subject's physical movements. The change in acceleration towards X-axis represents the forward and backward movement. Accelerometer generates positive X values during the forward move and vice versa. The change in acceleration towards Y-axis represents the left and right movement. Accelerometer generates positive Y values during the left move and vice versa. The change in acceleration towards Z-axis represents the upward and downward movement. Accelerometer generates positive Z values during the upward move and vice versa. This implies that the acceleration is more toward the vertical direction.

The gyroscope values show the orientation or the angular velocity in each axis. The X-axis is perpendicular to the direction of the human motion. The change in the gyroscope X-axis value represents the angular velocity of the tilting motion of the foot during human motion. There is a unique change in the X-axis, Y-axis, and Z-axis values of both accelerometer and gyroscope based on human motion. This pilot study proves that using these unique values, it is possible to identify the different basic human activities related to the corresponding application. The comparative analysis of the existing system and the proposed system is shown in Table 1. Most of the existing systems are built with only one type of sensors such as force sensors, accelerometer, and gyroscope. The proposed IoT wearable device is designed with force sensors, accelerometers, and gyroscopes for collecting data which helps to understand the movement of an individual more accurately. With the earlier research wired or Bluetooth connection has its own limitations, as it needs a receiver to receive

Table 1 Comparative analysis of existing methods

Parameters	Chen et al. [3]	Hong et al. [4]	Vandewynckel et al. [7]	Proposed IoT wearable device
Sensors used	Four force sensors	Three tri-axial accelerometers	A tri-axial accelerometer	Force sensors, tri-axial accelerometer, and tri-axial gyroscope
Communication	Wired	Bluetooth	Bluetooth	Wi-fi
Storage	Local system	Local system	Local system	Cloud platform
Controller	Micro programmed control unit (MCU)	Free scale MMA7260Q	PIC24 microcontroller	Node MCU

the collected data. But Wi-fi connection does not need any such receivers, it only needs an internet connectivity.

5 Conclusion

In this research work, an efficient way for data collection using IoT based wearable devices has proposed with low-cost. This IoT based device was designed with five force sensors, accelerometer, and gyroscope, and are embedded in each footwear. Sensors are connected to a cloud via Wi-Fi enabled controller to store the sensed data. Collected data is stored in the cloud in the form of google sheets for further analysis. There are many applications like posture recognition, activity recognition, fall risk assessment, fall detection applications, pedestrian tracking systems, etc., can use this dataset to analyze the ground truth about the application with the help of machine learning algorithms. A user-friendly real-time remote monitoring system in the form of a mobile application is planned as future work. This application uses the collected data and will give healthcare support to the elders and patients by monitoring their regular physical activity.

References

1. Muro-de-la-Herrán, A., Garcia-Zapirain, B., Mendez-Zorrilla, A.: Gait analysis methods: an overview of wearable and non-wearable systems, highlighting clinical applications. Sensors **14**(2), 3362–3394 (2014)
2. Kavitha, R., Binu, S.: Ambient monitoring in smart home for independent living. Advances in Intelligent Systems and Computing, vol. 883. Springer, Singapore (2019)
3. Chen, B., Wang, X., Huang, Y., Wei, K., Wang, Q.: A foot-wearable interface for locomotion mode recognition based on discrete contact force distribution. Mechatronics **32**, 12–21 (2015)

4. Hong, Y.-J., Kim, I.-J., Ahn, S.C., Kim, H.-G.: Activity recognition using wearable sensors for elder care. In: 2008 Second International Conference on Future Generation Communication and Networking (2008)
5. Liu, T., Inoue, Y., Shibata, K.: A wearable ground reaction force sensor system and its application to the measurement of extrinsic gait variability. Sensors **10**(11), 10240–10255 (2010)
6. Shu, L., Hua, T., Wang, Y., Qiao Li, Q., Feng, D.D., Tao, X.: In-shoe plantar pressure measurement and analysis system based on fabric pressure sensing array. IEEE Trans. Inf. Technol. Biomed. **14**(3), 767–775 (2010)
7. Vandewynckel, J., Otis, M., Bouchard, B., Ménélas, B.-A.-J., Bouzouane, A.: Towards a real-time error detection within a smart home by using activity recognition with a shoe-mounted accelerometer. Procedia Comput. Sci. **19**, 516–523 (2013)
8. Hori, M., Nakai, A., Shimoyama, I.: Three-axis ground reaction force distribution during straight walking. Sensors **17**(10) (2017)
9. Hegde, N., Bries, M., Sazonov, E.: A comparative review of footwear-based wearable systems. Electronics **5**(4), 48 (2016)

Human Activity Recognition Using Wearable Sensors

Y. Joy Rakesh, R. Kavitha, and J. Julian

Abstract The advancement of the internet coined a new era for inventions. Internet of Things (IoT) is one such example. IoT is being applied in all sectors such as healthcare, automobile, retail industry etc. Out of these, Human Activity Recognition (HAR) has taken much attention in IoT applications. The prediction of human activity efficiently adds multiple advantages in many fields. This research paper proposes a HAR system using the wearable sensor. The performance of this system is analyzed using four publicly available datasets that are collected in a real-time environment. Five machine learning algorithms namely Decision tree (DT), Random Forest (RF), Logistics Regression (LR), K-Nearest Neighbor (kNN), and Support Vector Machine (SVM) are compared in terms of recognition of human activities. Out of this SVM responded well on all four datasets with the accuracy of 77%, 99%, 98%, and 99% respectively. With the support of four datasets, the obtained results proved that the performance of the proposed method is better for human activity recognition.

Keywords Activity recognition · Sensors · Machine leaning · Wearable computing · Classification

1 Introduction

Today internet has evolved largely because the internet has become accessible to everyone. Through the internet, the smart devices can be connected and communicated, the IoT has been evolved where the sensors are being embedded into the

Y. Joy Rakesh · R. Kavitha (✉) · J. Julian
Department of Computer Science, CHRIST (Deemed to Be University), Bangalore, Karnataka, India
e-mail: kavitha.r@christuniversity.in

Y. Joy Rakesh
e-mail: joy.rakesh@mca.christuniversity.in

J. Julian
e-mail: julian.j@mca.christuniversity.in

S. C. Satapathy et al. (eds.), *Intelligent Data Engineering and Analytics*, Advances in Intelligent Systems and Computing 1177,
https://doi.org/10.1007/978-981-15-5679-1_51

devices. Human Activity Recognition is a technique in which different daily activities are recognized by the system. Human activity recognition has many advantages where a different walking pattern relates to a particular human being. The first work on HAR was implemented in the late 90's [1]. The main purpose of HAR is to predict the basic activity pattern of human which can be further analyzed by doctors. HAR is possible in two ways: Image processing and sensor-based [9]. In the image processing method, the captured images are being analyzed using image processing algorithms. This method needs high-end computational devices to process image data. In a sensor-based approach, the sensors which produced data is being collected and stored. The collected data is then preprocessed and applied machine learning algorithm to predict the activities. In sensor-based human activity recognition, the data set is collected from either non-wearable or wearable sensors. In the non-wearable method, mobile phones are used as sensors. In wearable method sensors or sensor-embedded devices are used for data collection. The data sets used in this research are collected from wearable sensors where humans are wearing the sensors in their body. This method of data collection has more advantages than all the other methods since the actual humans are wearing the sensors during data collection. In the present world scenario, high accurate information is very useful for giving correct predictions in many applications. HAR can be very useful in assisting the patient by monitoring their regular activity. Suppose if the doctor has suggested few exercises to the diabetic patient that can be monitored regularly. Here HAR has an efficient application were in the activity of the patient that can be recorded and analyzed with the help of machine learning algorithms and provide a consolidated report to the caretaker.

The rest of the paper is structured as follows. The related review of the literature is discussed in Sect. 2. The architecture of the proposed design is discussed in Sect. 3. This section also gives detail information on five phases of the proposed system. Section 4 provides the performance analysis result of a proposed method on four data sets. Finally, Sect. 5 concludes the research work.

2 Literature Review

Human activity recognition is been in research activities for over a decade. HAR has many applications in the field of health care, home security, etc. Many researchers have proposed significant ideas in human activity recognition. The dataset and machine learning algorithms used are significant components in human activity recognition. This section summaries the work proposed by different researchers.

Kavitha et al. [8] proposed a model for an ambient monitoring system for elders. Here the researcher discusses necessary aspects for elders like smart home systems, activity recognition, etc. A new segmentation method called area-based segmentation is proposed using optimal change point detection. The performance of the proposed segmentation is analyzed using Naive Bayes, kNN, and SVM classifiers. Furthermore, this research work has a deep insight into human activity recognition. Bulbul et al. [9] proposes a model that uses machine learning algorithms to predict human activity using iterative model. The data set is collected using accelerometer and gyroscope sensors which are embedded in the smartwatch. The dataset is segmented based on 50 Hz of sampling rate and stored based on time series. The classification algorithms decision tree, support vector machine, k-nearest neighbors, Ensemble classification methods—Boosting, Bagging, and Stacking are used in this experiment. Out of these, the support vector machine predicted the sitting activity 99% accurately. AkramBayat et al. [2] proposed a model that identifies human activities using the accelerometer sensor in the user's cell phone. The various features are extracted from this data set and machine learning algorithms are used to predict the activities. Out of these multilayer perceptrons predicted the activities with the highest performance 89.48% accuracy. Zhuang et al. [3] proposed a sports-related activity recognition model where a different activity such as badminton and swimming are predicted. In this experiment smartwatch which consists of triaxial acceleration and the triaxial angular velocity sensor are used to collect the data. The machine learning algorithms CNN, k-NN, Naive Bayes, random forest, and support vector machine are used. In this experiment, SVM algorithm yields the highest accuracy.

3 Architecture

The structural flow of human activity recognition (HAR) system used in this research work is shown in Fig. 1. The workflow of the HAR system using wearable sensor data consists of five stages such as collecting data from wearable sensors, segmentation, feature extraction, model training, and activity recognition. Human activity can be recognized using sensor-embedded devices like mobile phones, wearable belts, shoes, etc. In the first stage of the HAR system, sensors play a vital role in

Fig. 1 Architecture diagram of HAR system

producing the data for human activity prediction. Mainly the sensors like accelerometer and gyroscope are used to recognize the human movement in terms of direction and rotation. Usually, wearable sensors produce a large volume of data. To avoid the complexities of handling these huge datasets, segmentation is introduced in the second stage. During the third stage, a set of features is extracted from each segment. In stage four, using these extracted features and identified machine learning algorithms the model is designed for activity recognition. Finally, in stage five human activities are recognized from the wearable sensor dataset.

3.1 Wearable Sensor Datasets

DATASET 1: This dataset was collected from fifteen participants performing seven different activities while wearing the chest-mounted accelerometer. The dataset is intended for Activity Recognition research purposes. This publicly available dataset [4] provides challenges for identification and authentication of people using motion patterns. Each file consists of six attributes namely sequential number, x acceleration, y acceleration, z acceleration, and activity label. The sampling frequency of the accelerometer is 52 Hz.

DATASET 2: This time-series dataset is produced by accelerometer and gyroscope sensors which are embedded in iPhone6. Twenty-four participants have performed six activities by keeping iPhone6 in their front packet. The dataset is generated from an accelerometer that is present inside the iPhone 6s. Datasets related to six activities namely downstairs, upstairs, walking, running, sitting, and standing are collected in the same environment. Twelve attributes related to accelerometer and gyroscope were recorded from each participant during data collection. The sampling frequency of this data collection is 50 Hz [5].

DATASET 3: This dataset was collected from three Colibri wireless IMUs (inertial measurement units) which are placed in the human hand, chest, and ankle. Each of the data-files contains 54 attributed namely timestamp, activity-ID, heart rate (bpm), seventeen attributes from IMU placed in the hand, seventeen attributes from IMU placed in the ankle, and seventeen attributes from IMU placed in the chest. The IMU sensory data contains 1 temperature (°C), 3D-acceleration data (ms^{-2}), 3D-acceleration data (ms^{-2}), 3D-gyroscope data, 3D-magnetometer data (μT) and orientation. Eighteen activities were collected from nine participants aged 27.22 ± 3.31 years. The sampling frequency of this data collection is 100 Hz [6].

DATASET 4: This dataset was collected from accelerometer and gyroscope to understand the human physical Activities. The subject was instructed to perform six activities namely Sitting, Standing, Walking, Running, Walking upstairs, Walking downstairs. Six attributed from Acceleration data x-axis, y-axis, z-axis, and Gyroscope data x-axis, y-axis, z-axis are recorded with sampling frequency 50 Hz [7].

Table 1 List of Features Extracted from Raw Sensor Data Stream

Feature name	Description
Count	Total number of values in array
Arithmetic mean	Arithmetic mean = $\sum (f_i * x_i)/f_i$
Standard deviation	Standard deviation = $\sqrt{[\Sigma\ (X_i - X_m)^2/(n-1)]}$
Min	Smallest value in array = Mini (S_i)
First quartile deviation	First quartile deviation = Q1 (s)
Second quartile deviation	Second quartile deviation = Q2 (s)
Third quartile deviation	Third quartile deviation = Q3 (s)
Max	Largest value in array = Max (S_i)
Kurtosis	Frequency signal kurtosis = $E\ [(s - s_i)^4]/(E\ (s - s_i)^2)^2$
Skewness	Frequency signal skewness = $E\lceil(s - s_i)^3/\sigma\rceil$
Median absolute deviation (MAD)	MAD = median $(\lvert s_i$ − median$(s_j)\rvert)$

3.2 *Segmentation*

The huge sensor data stream is divided into small fragments is called segmentation. Segmentation plays an important role in HAR by decreasing the complexity of the computation process. The data stream is divided into segments with no overlap. The size of the segment in each dataset is different since it is recorded in a different frequency. The human movement were recorded in DATASET-1, DATASET-2, DATASET-3, and DATASET-4 at 52 Hz, 50 Hz, 100 Hz, and 50 Hz respectively. I.e. DATASET-1 recorded 52 data points per second and so on.

3.3 *Feature Extraction*

Recognizing human activity from the inertial sensor data stream is generally lead by feature extraction stage. Frequency-domain features and time-domain features are two types of features popularly used in HAR. In this research, time-domain statistical features such as mean, median, quartiles, variance, kurtosis and skewness are extracted from each segment. The list features used in this research work which is extracted from the raw sensor data stream is listed in Table 1.

3.4 *Model Training*

Model training is an important process in HAR as it helps in the proper prediction of the activity performed by the subject. A good model is not when it performs well on the trained data, but it should give good accuracy when it is subjected to new data. The features extracted from the sensor data stream are used as an input to the machine learning algorithm for an activity classification as a model training. In this work, many machine learning algorithms were tried, out of that five algorithms namely Decision tree (DT), Random Forest (RF), Logistics Regression (LR), K-Nearest Neighbor (kNN), Support Vector Machine (SVM) are responded well.

Decision Tree: DT is a supervised machine learning algorithm. This efficient algorithm easy to implement, and it makes use of a divide and conquers method. DT is a graphical representation of the solutions which follows a sequence of IF ELSE statement where the next statement is based on previous statements results. The DT consists of nodes, links and leaf nodes. The nodes represent a predicted value and the links between nodes represent a decision made by the classifier the leaf nodes are expected outcome. The advantages of the DT algorithm are easy to understand, implement and generate the rules. The disadvantages are suffering for overfitting, weak to handle non-numeric data, need an additional approach pruning to handle large datasets.

Random Forest: RF is a machine learning algorithm that is derived based on the features of a decision tree. In RF the decisions are not made based on one decision tree, but all the 'K' decision tree predictions are considered to make predictions. This property of RF is called Ensemble. In RF the classifier creates a set of decision trees randomly selected from a training set. Then it further aggregates the votes from all the trees to decide the final result. Compared to DT, the RF algorithm works well with large datasets. Highly flexible in dealing with missing data. The disadvantage of RF algorithm is the process of building and testing the model is slower as it requires number of trees. Similar to bagging RF are more difficult to interpret.

k-Nearest Neighbors: kNN is a supervised machine learning algorithm that classifies data points using *k* of its already classified nearest neighbors. In this algorithm, the distance between the new data point and nearest classified data points decides the class of the new data point. kNN is suitable for a dataset with any distribution and it gives better results if the large dataset is used for training. The challenge of this algorithm is choosing the right *k* value.

Logistic Regression: LR is generally used for the dependent variable which predicts multi-class or binary class. The dependent variable can be "Yes" or "No". The independent variable can be either categorical or numerical variables. LR uses the probability score as predicted values. The regression models provide a simple and understandable algebraic equation to use. The regression models can match and beat the predictive power of other models. The disadvantage of the regression model is that the model cannot cover the poor and inadequate data quality. The regression models do not work with non-numeric and categorical variables.

Support Vector Machine: SVM is a supervised machine learning technique to build a linear classifier. The algorithm creates a hyperplane in a high dimension space that creates segments. The advantage of SVM is that it works well for a large number of features. SVM kernel places an important role in classification by operates using 'kernel trick'. This trick involves dealing with the relevant pair of data in feature space instead of using all data in feature space. The SVM works well when the number of features in the dataset is larger than the number of instances.

3.5 *Activity Recognition*

The final phase of the proposed HAR is activity recognition. Machine learning algorithms classify the segmented data stream based on the extracted features. In this research work, the basic activities walking, sitting, running, walking upstairs, and walking downstairs are used. The performance of the activity classification is verified using the accuracy measures precision, recall, F1-score, and accuracy.

4 Results and Discussion

Human Activity Recognition plays an important role such as monitoring a patient, fall detection, etc., in the field of healthcare. The HAR system helps to monitor a patient continuously by monitoring their regular activities such as walking, climbing stairs up and down. Suppose if the patient is suffering from a knee injury, the recovery process needs continuous monitoring. The HAR system provides a suitable solution by monitoring the patients and report to the authority which helps to take necessary action during an emergency. Most of the regular human activities are possible to represent using some sort of motion features. Same time, the feature values extracted from sensor data stream are different for distinct human activities. This idea assists to develop a HAR system. To recognize human activities from wearable sensor data, this research makes use of four datasets which are publicly available for research purpose. The proposed method is implemented using Python, an open source data analysis tool. Details of all four datasets are given in the earlier section. The sensory data visualization of a walking activity is depicted in Fig. 2.

There are multiple activities recorded in all four datasets. To understand the performance of the proposed HAR system, in this research common six activities namely walking, sitting, running, walking upstairs, and walking downstairs are considered from all data sets.

Based on the sampling frequency sensor data stream is divided into segments. A set of features are extracted from each segment and five different machine learning classifiers DT, RF, kNN, LR, and SVM are used to predict the activities. Figure 3 shows the visualization of the overall accuracy performance of all five classifiers on four datasets. In DATASET-1 the overall activity accuracy of DT, RF, kNN, LR, and SVM are 79%, 84%, 85%, 63%, and 77% respectively. This shows that kNN classifier recognizes the activities better with 85% accuracy. In DATASET- 2 the overall accuracy of DT, RF, kNN, LR, and SVM are 96%, 96%, 98%, 88%, and 99% respectively. In this dataset, the SVM recognized the activities with 99% accuracy which is best and highest. In DATASET-3 and DATASET-4 the overall activity accuracy of DT, RF, kNN, LR, and SVM are 91%, 92%, 98%, 78%, 98% and is 93%, 95%, 99%, 81%, and 99% respectively. In these two datasets, the classifiers' kNN and SVM model performed well and recognized the activities with the best and highest accuracy. Table 2 illustrates the best classifiers accuracy of all four datasets.

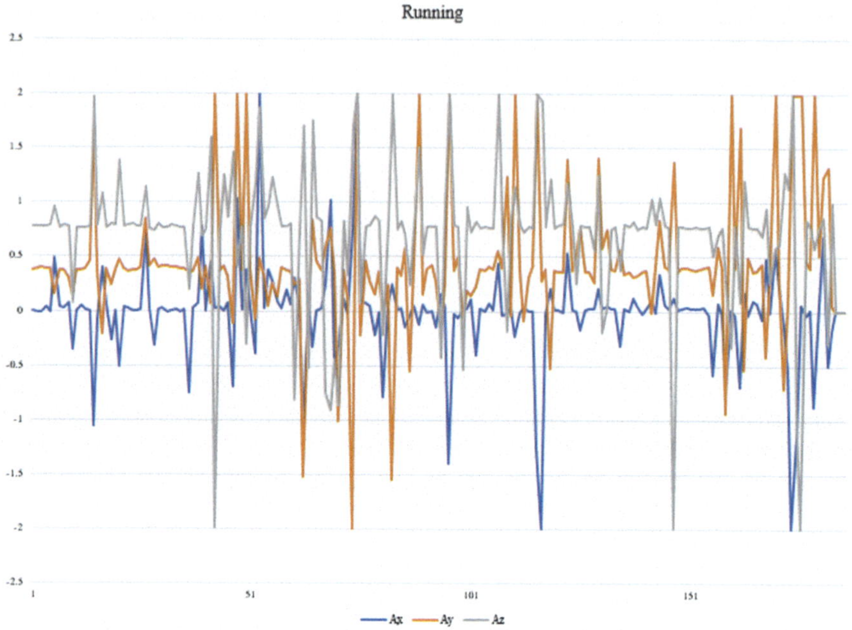

Fig. 2 Sensory data visualization of a dataset

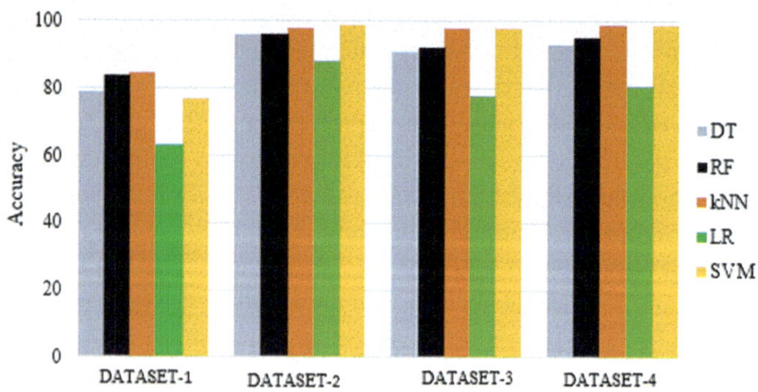

Fig. 3 Accuracy of all five classifiers on four datasets

The performance result of recognizing all five activities on all four datasets is analyzed. Out of this SVM classifier performed well in all datasets except the first one. The accuracy of individual activity recognition on all four data set using SVM is depicted in Fig. 4. SVM model classified all the activities accurately in DATASET-2, DATASET-4.

In order to assess the performance of the proposed method the accuracy measures Precision, Recall and F1-Score is calculated. The performance measures of five classifiers on four datasets are shown in Table 3. Here Precision refers to the posi-

Table 2 Four confusion matrix for best ML classification on four datasets

kNN	Walk	Run	Standing	Sitting	Up-stairs	Down-stairs	SVM	Walk	Run	Standing	Sitting	Up-stairs	Down-stairs
Walk	97	0	0	2	0	1	Walk	94	5	0	0	0	1
Run	27	60	3	0	5	5	Run	0	100	0	0	0	0
Standing	0	1	72	17	8	2	Standing	0	0	100	0	0	0
Sitting	0	3	9	87	0	1	Sitting	0	0	1	99	0	0
Up-stairs	1	9	21	11	55	3	Up-stairs	0	0	0	0	100	0
Down-stairs	0	0	16	12	8	64	Down-stairs	1	0	0	0	1	98
SVM	Walk	Run	Standing	Sitting	Up-stairs	Down-stairs	SVM	Walk	Run	Standing	Sitting	Up-stairs	Down-stairs
Walk	99	1	0	0	0	0	Walk	99	1	0	0	0	0
Run	0	99	1	0	0	0	Run	1	98	1	0	0	0
Standing	0	1	98	0	1	0	Standing	0	0	100	0	0	0
Sitting	0	0	3	97	0	0	Sitting	0	0	0	100	0	0
Up-stairs	0	0	0	0	98	2	Up-stairs	0	0	0	0	100	0
Down-stairs	1	0	0	0	2	97	Down-stairs	0	0	1	0	0	99

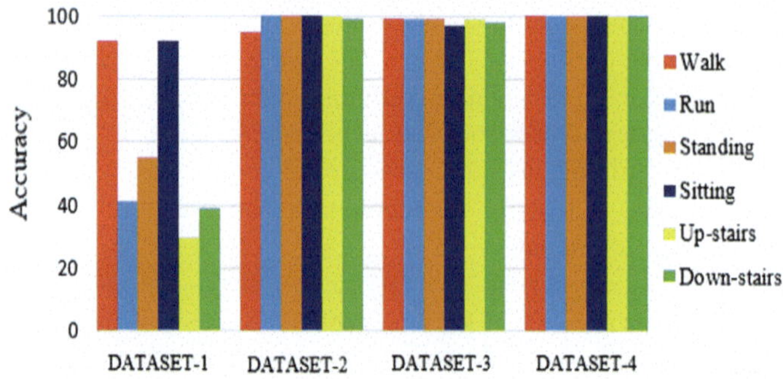

Fig. 4 Accuracy of individual activity recognition on four datasets using SVM

tively predicted activities with respect to a total number of activity instances classified positively. Recall defines the ratio of correctly classified activities. F1-Score measures the activity recognition with the combination of both precision and recall. The result shows strongly that our proposed method classifies the basic activities with better accuracy. This helps to understand human motion with respect to the environment. The result gives great confidence to use this proposed model in healthcare applications.

5 Conclusion

This research work presents a HAR system that can be used to recognize the human activities from wearable sensor data. In order to understand the performance of the proposed the method, four data sets are adopted. Sensor data stream were divided into segments and time-domain features were extracted from each segment. Five machine learning algorithms were used to classify the human activities using extracted features. The activity classification results are more accurate for the last three datasets compared to the result of the first dataset. The result proved, that the proposed method is suitable for human activity recognition. This work can be extended in many directions like creating hybrid classifiers which is combination of multiple classifiers for complex prediction and recognizing composite activities where one activity consists of multiple smaller activities.

Table 3 Performance measures of five classifiers on four datasets

Classifier	DATASET-1			DATASET-2			DATASET-3			DATASET-4		
	Precision	Recall	F1-Score	Precision	Recall	F1-Score	Precision	Recall	F1-Score	Precision	Recall	F1-Score
DT	0.8	0.8	0.8	0.97	0.97	0.97	0.91	0.91	0.91	0.94	0.94	0.94
RF	0.85	0.85	0.85	0.96	0.96	0.96	0.92	0.92	0.92	0.95	0.95	0.95
kNN	0.85	0.85	0.85	0.99	0.99	0.99	0.99	0.99	0.99	1	1	1
LR	0.63	0.63	0.63	0.88	0.88	0.88	0.8	0.8	0.8	0.82	0.82	0.82
SVM	0.78	0.78	0.78	0.99	0.99	0.99	0.99	0.99	0.99	1	1	1

References

1. Lara, O.D., Labrador, M.: A survey on human activity recognition using wearable sensors. IEEE Commun. Surv. Tutor. **15**(3), 1192–12099 (2013)
2. Bayat, A., Pomplun, M.: A study on human activity recognition using accelerometer data from smartphones. Procedia Comput. Sci. **34**, 450–457 (2014)
3. Zhuang, Z., Xue, Y.: Sport-related human activity detection and recognition using a smartwatch. Sensors **19**(22), 5001 (2019)
4. UCIrepository.https://archive.ics.uci.edu/ml/datasets/Activity+Recognition+from+Single+Chest-Mounted+Accelerometer. Last accessed 25 Nov 2019
5. GitHub. https://github.com/mmalekzadeh/motion-sense/tree/master/data. Last accessed 23 Nov 2019
6. UCIrepository.https://achive.ics.uci.edu/ml/dtasets/PAMAP2+Physical+Activity+Monitoring. Last accessed 26 Nov 2019
7. Kaggle. https://www.kaggle.com/uciml/human-activity-recognition-with-smartphones. Last accessed 25 Nov 2019
8. Kavitha, R., Binu, S.: Ambient monitoring in smart home for independent living. advanced computing and systems for security. Adv. Intell. Syst. Comput. **883** (2019)
9. Bulbul, E.: Human activity recognition using smartphones. In: 2nd International Sympo-sium on Multidisciplinary Studies and Innovative Technologies (ISMSIT) (2018)

Fingerspelling Identification for Chinese Sign Language via Wavelet Entropy and Kernel Support Vector Machine

Zhaosong Zhu, Miaoxian Zhang, and Xianwei Jiang

Abstract Sign language recognition is beneficial to help the hearing-impaired and the healthy communicate effectively and help hearing-impaired people integrate into society, making their study, work, and life more convenient, especially in speech therapy and rehabilitation. Fingerspelling identification plays an important role in sign language recognition, which has unique advantages in expressing abstract content, terminology, and specific words, and can also be utilized as the basis of learning gesture recognition based on Pinyin rules. We proposed a WE-kSVM approach, carrying out on 10-fold cross-validation, and achieved an overall accuracy of $88.76 \pm 0.59\%$. Maximum accuracy is 89.40% based on thirty categories. Here, Wavelet entropy technique can reduce the number of features and accelerate the training. Gaussian kernel (RBF) provided excellent classification performance. Meanwhile, 10-fold cross-validation prevented overfitting effectively. The experiment results indicate that our method is superior to the other five state-of-the-art approaches.

Keywords Sign language recognition · Fingerspelling identification · Wavelet entropy · Kernel support vector machine · 10-fold cross-validation

Z. Zhu · M. Zhang · X. Jiang
Nanjing Normal University of Special Education, Nanjing 210038, China
e-mail: zzs@njts.edu.cn

M. Zhang
e-mail: zmx@njts.edu.cn

M. Zhang
Zhou Enlai Government School of Management, Nankai University Tianjin, Tianjin, China

X. Jiang (✉)
Department of Informatics, University of Leicester, Leicester LE1 7RH, UK
e-mail: jxw@njts.edu.cn

S. C. Satapathy et al. (eds.), *Intelligent Data Engineering and Analytics*, Advances in Intelligent Systems and Computing 1177,
https://doi.org/10.1007/978-981-15-5679-1_52

1 Introduction

Sign Language (SL) refers to a complicated expression system consisting of hand shape, gesture, movement, body posture, and facial emotion, etc., which is a significant communication way between hearing-impaired and healthy people. Among all elements, hand shape and gesture are more important, which represent the meaning of sign language in most cases. Chinese Sign Language (CSL) is characterized by Chinese deaf and hearing-impaired, which can be classified into two categories: fingerspelling sign language and gesture sign language. Based on the universal sign language standard issued by the state in 2018, fingerspelling sign language includes 30 letters, that is, 26 single letters (A–Z) and 4 double letters (ZH, CH, SH, NG). It has its own characteristic: easy to learn and master, accurate to express, especially suitable for abstract concepts and terminology.

Sign Language Recognition (SLR) is regarded as a technique that can transform sign language information to other forms which are easy to understand and communicate, such as nature language, text, audio, video, etc. SLR can smooth the communication barrier between the hearing-impaired and the healthy. For instance, it can help to express hearing-impaired people's intentions to doctors in therapy examination. Based on the input method, we divided the SLR technologies into two: sensor-based and computer vision-based. The sensor-based SLR depends on wearable devices to gain the input data. Depth cameras are the main tool for collecting input data when computer vision-based SLR is employed. Compared to SLR based on sensor, SLR based on computer vision is closer to usage habits and more flexible to operate, which makes it more popular and cost-effective.

A large number of researchers have contributed to sign language recognition. Many classification methods and recognition algorithms including their variants are proposed. As a typical statistical analysis method, the Hidden Markov Model was employed by Cao [1], to identify Chinese sign language and achieve high accuracy. Combined additional two techniques: K-means and ant colony algorithm, HMM was shifted to recognize Taiwan sign language in paper [2], which gained an average accuracy of 91.3%. The template match is another method commonly utilized. Dynamic Time Warping (DTW) algorithm was introduced in [3], which proposed a threshold matrix for continuous Sign Language Recognition. Novel neural network technology provided strong competition. Based on the skin-color technique and convolutional neural networks, Wang et al. [4], identify gesture samples of various test environments and gained over 95% recognition rate. Jiang [5], recognized Chinese sign language fingerspelling via a 6-layer convolutional neural network with the leaky rectified linear unit (LReLU). Based on deep learning, some other researches on sign language recognition were introduced in the reference literature [6–8]. Additionally, Yang and Lee [9], proposed hierarchical conditional random fields (HCRF) method. The ANN classifier was trained to match words by Rao et al. [10]. Gray-Level Co-occurrence Matrix (GLCM) and Parameter-Optimized Medium Gaussian Support Vector Machine (MGSVM) method was employed to identify isolated Chinese sign language by Jiang [11].

Nevertheless, some shortcomings of these advanced methods should be focused on. The prerequisite for HMM to work smoothly depends on a complex initialization and a large number of computations, which is not friendly for human–computer interaction and not suitable for real-time implementation. DTW demands to construct a template first, which costs many calculation resources and a lot of time. Lacking identifiable gestures, it is utilized for static gesture recognition in most cases. Although neural network technology owns capability of self-learning and can obtain high accuracy, it requires a large amount of data sets and high training cost.

Thus, in this study, we proposed a suitable method for fingerspelling identification based on wavelet entropy and kernel support vector machine. Wavelet entropy was applied to reduce the number of features and accelerate the training. Gaussian kernel (RBF) was employed due to its effective performance, meanwhile, the experiment was carried out on 10-fold cross-validation to prevent overfitting.

2 Dataset

Our experiment materials are constructed by 510 self-built Chinese finger sign language samples, which were collected from 17 volunteers. As each volunteer provides 30 samples corresponding 30 categories in fingerspelling, thus totally 510 images were gained. We reprocessed these images using suitable software and normalized them to 256 × 256. Figure 1 showed the images of 30 categories.

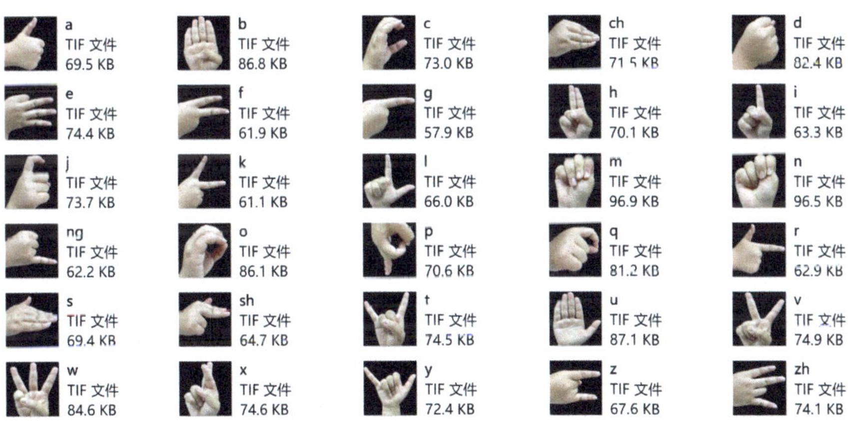

Fig. 1 30 categories of Chinese finger sign language

3 Methodology

3.1 Wavelet Entropy

Discrete wavelet transform (DWT) and entropy calculation are two important parts of wavelet entropy (WE), which is of benefit to analyze temporal features of the complicated signal. Choosing different coefficients, DWT decomposes images to preserve image information, which gives a hierarchical framework of information interpretation. The equations of DWT are represented as follows:

$$L(n) = \sum_{m} x(m) \times l(2n - m) \tag{1}$$

$$H(n) = \sum_{m} x(m) \times h(2n - m) \tag{2}$$

where $L(n)$ represents the approximation coefficient and $H(n)$ indicates the detail coefficient. m denotes a temporary variable. The low-pass filter is l and the high-pass filter is h.

Nevertheless, the DWT technique leads to excessive features, which brings burdens of computation and storage. To reduce features and improve performance, entropy is introduced, which can cut down the dimension of the dataset and maintain most variations.

Shannon entropy E_s is defined as follows, which is a random statistical measure and can be applied to characterize features.

$$E_s = -\sum p_j \log_2(p_j) \tag{3}$$

where j indicates gray-level of reconstructed coefficient and p_j denotes the probability when gray-level is j.

Taking a case of 2-level WE, the process can be divided into two steps (See Fig. 2). Firstly, based on 1-level DWT, the original 256 × 256 Chinese finger sign

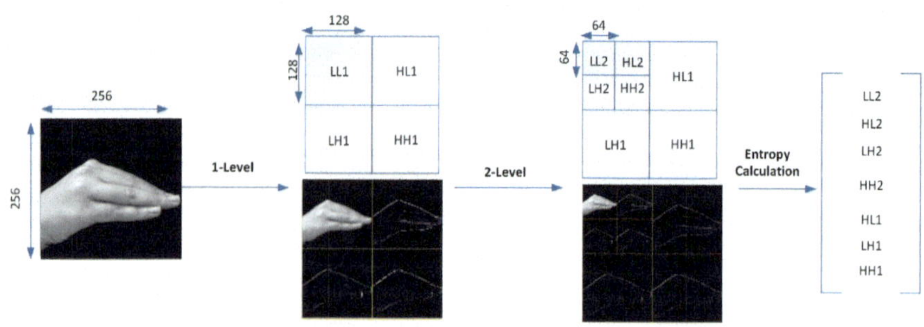

Fig. 2 Diagram of 2-level wavelet entropy

language image was decomposed into 4 subbands (LL1, HL1, LH1, HH1) with a size of 128 × 128. Then the LL1 subband will further carry out DWT and 4 smaller subbands (LL2, HL2, LH2, HH2) with a size of 64 × 64 were yield, which is the 2-level DWT. Here, L and H denote low-frequency coefficient and high-frequency coefficient, respectively. Secondly, we calculated the entropy on every subband and took these vectors as input. Thus, we can reduce a 256 × 256 fingerspelling image to 3 × n+1 vectors. Wavelet entropy can reduce the number of features and save computation time and storage memory.

3.2 SVM

As one of the most influential methods in supervised learning, support vector machines (SVM) can be used to deal with classification and regression. Based on a linear function $w^T x + b$, SVM only outputs categories, whose purpose is to find the best hyperplane to divide the samples in the N-dimensional space into two categories.

3.3 Kernel–SVM

Traditional linear support vector machines lack the capability of separating the practical data with complex distribution, thus, kernel trick is employed to SVMs as an important innovation. Since we can use the dot product between samples to represent many machine learning algorithms, the linear function of SVM can be redefined as follows:

$$w^T x + b = b + \sum_{i=1}^{n} \alpha_i x^T x^{(i)} \tag{4}$$

where α_i denotes coefficient vector, $x^{(i)}$ indicates the training sample. Meanwhile, as we can replace x with the output of the $\varphi(x)$ which is the eigenfunction, a kernel function is introduced to substitute the dot product. The formula is as follows:

$$k\left(x, x^{(i)}\right) = \varphi(x) \cdot \varphi\left(x^{(i)}\right) \tag{5}$$

where · means operation of the dot product. Thus, we can utilize the substitution function to predict.

$$F(x) = b + \sum_{i} \alpha_i k\left(x, x^{(i)}\right) \tag{6}$$

Here, the function $F(x)$ is nonlinear for x, which is completely equivalent to preprocessing all inputs with $\varphi(x)$ and then learning linear model in the new transformation space. From another perspective, it means that the classifier is a hyperplane in the high dimensional feature space, while nonlinear in the original input space.

There are two reasons why the kernel trick is so powerful. First of all, it guarantees the effective convergence of optimization techniques for learning nonlinear models. Here, φ is regarded as fixed and only α needs to be optimized. Secondly, the implementation of the kernel function $k\left(x, x^{(i)}\right)$ is much more efficient than constructing $\varphi(x)$ and then calculating the dot product.

In many cases, $\varphi(x)$ is difficult to calculate, but $k\left(x, x^{(i)}\right)$ is a nonlinear function of x and easy to gain, which shows the advantages of the kernel function. Among all of the kernel functions, the most commonly used is Gaussian kernel, also called as radial basis function (RBF), which can be defined as follows:

$$k_G\left(x, x^{(i)}\right) = N(x - x^{(i)}; 0; \sigma^2 I) \tag{7}$$

where N denotes standard normal density. RBF means its value decreased in the direction in which $x^{(i)}$ radiates outward from x. RBF also can be represented as the following formula in detail.

$$k_G\left(x, x^{(i)}\right) = \exp\left(-\gamma \left\| x - x^{(i)} \right\|^2\right) \tag{8}$$

Here, γ is the parameter that needs to be tuned.

AS another typical kernel function, the polynomial kernel is mentioned frequently. The kernels of the homogeneous polynomial (HPOL) and inhomogeneous polynomial (IPOL) with their formulas are expressed as follows, respectively.

$$k_{\mathrm{HPOL}}\left(x, x^{(i)}\right) = \left(x \cdot x^{(i)}\right)^{\mu} \tag{9}$$

$$k_{\mathrm{IPOL}}\left(x, x^{(i)}\right) = \left(x \cdot x^{(i)} + 1\right)^{\mu} \tag{10}$$

where μ indicates the adjusting parameter, which can fix the kernel according to practical data.

In general, kernel SVMs own several advantages: (1) need few parameters to tune; (2) employ convex quadratic optimization to train; (3) obtain remarkable success in many fields. Importantly, kernel SVMs provide unique and global solutions, preventing the convergence to local minima. In this paper, the Gaussian kernel was chosen due to its excellent performance.

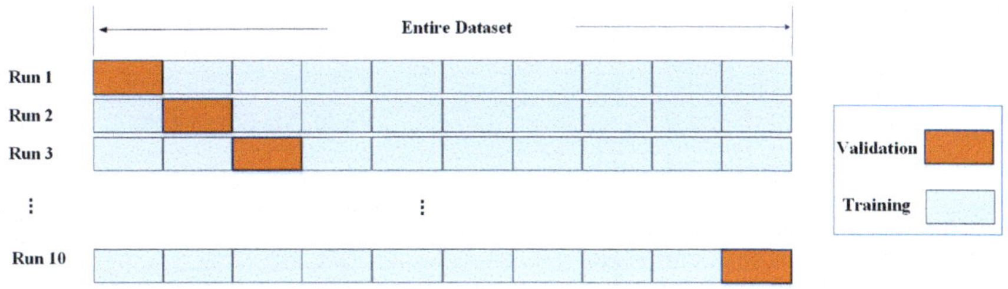

Fig. 3 illustration of 10-fold cross-validation

3.4 *10-Fold Cross-Validation*

K-fold cross-validation can make full use of all data when training and validating, which is becoming popular and required. The process of K-fold cross-validation is as following: the K-1 folds partitions are selected to train from the entire dataset and the remainder is left to be validated. There will be K iterations. According to experience, 10-fold cross-validation may achieve excellent performance. Figure 3 represents the implementation of 10-fold cross-validation.

Where the red partition denotes the fold of validation and blue partition denotes the K-1 folds for training in every run epoch. There are totally 10 runs to validate the whole dataset. K-fold cross-validation prevents overfitting and fulfills estimation out of sample, which can make classifiers more reliable and effective.

4 Experiment Results and Discussions

This experiment was carried out on a platform of the personal computer with Core i5 CPU and 8 GB memory, under the Windows 7 operating system. Overall accuracy (OA) is applied to evaluate the results, which indicates the ratio of the correct prediction over all test sets in the model to the total number.

4.1 *Statistical Results*

The statistical results of WE-kSVM method running 10 times under 10-fold cross-validation were demonstrated in Table 1. It can be observed that we achieved the value of means and standard deviation as $88.76 \pm 0.59\%$ and the maximum overall accuracy is 89.4%, which can be considered satisfactory and stable.

To pursue excellent performance, the different optimal decomposition level (n) was validated. The rank of level (n) was changed from 1 to 6 and the wavelet family was set to Haar. As can be seen from Fig. 4, the maximum overall accuracy of level

Table 1 Statistical results of WE-kSVM method

Run	Overall Accuracy (%)
1	87.89
2	**89.40 (Maximum OA)**
3	89.22
4	89.11
5	88.60
6	89.22
7	88.78
8	88.89
9	87.58
10	89.00
Mean ± SD	**88.76 ± 0.59**

Here, 89.40 indicates the maximum overall accuracy and 88.76 ± 0.59 indicates the value of means and standard deviation (Mean ± SD)

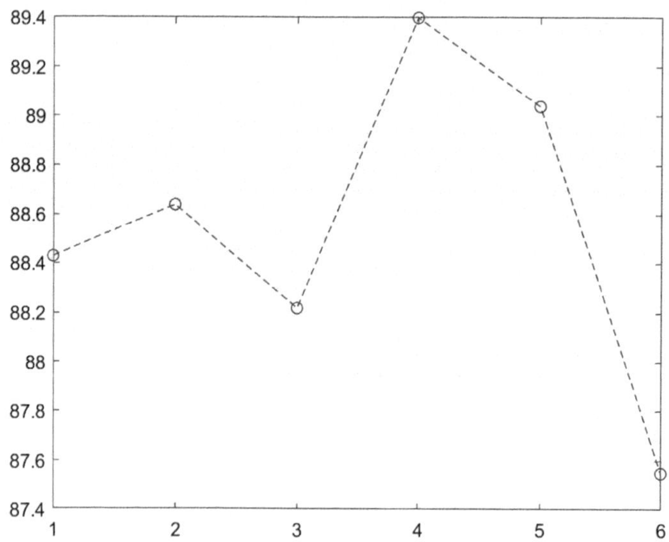

Fig. 4 Maximum overall accuracy with optimal decomposition level

(1), level (2), level (3), level (4), level (5), and level (6) are 88.43%, 88.64%, 88.22%, 89.40%, 89.04%, and 87.55%, respectively. As far as one certain run is concerned, the maximum OA reaches the highest at decomposition level (4).

Table 2 Comparison of the training algorithm

Training algorithm	Mean ± SD (%)
Linear SVM	84.18 ± 1.12
RBF kernel SVM	87.94 ± 0.75
WE-RBF kSVM	88.76 ± 0.59

4.2 Training Algorithm Comparison

Table 2 represents means and standard deviation using individual training algorithms, we compared three training algorithms: WE-RBF kSVM, RBF kernel SVM, Linear SVM, which gained Mean ± SD 88.76 ± 0.59%, 87.94 ± 0.75%, and 84.18 ± 1.12%, respectively. It can be found that the approach of WE-RBF kSVM obtained the best performance. As two advanced techniques (wavelet entropy and kernel SVM) were introduced, WE can improve the training speed of the classifier and RBF kernel SVM can avoid the convergence to local minima. Thus, the experiment results explained why WE-RBF kSVM is superior to Linear SVM about 4.5%.

4.3 Comparison to State-of-the-Art Approaches

In this study, our method WE-kSVM was compared with five state-of-the-art approaches: HMM [12], SVM-HMM [13], HCRF [9], GLCM-MGSVM [11], 6-layer CNN-LReLU [5]. The results are listed in Fig. 5, which represents that our approach is superior to HMM with OA 83.77%, SVM-HMM with OA 85.14%, GLCM-MGSVM with OA 85.3%, and 6-layer CNN-LReLU with OA 88.10 ± 1.48%, especially higher 10 points to HCRF with OA 78%. Three advanced techniques: wavelet entropy, kernel SVM and 10-fold cross-validation contributed to enhance the performance. Reducing feature number and speeding up training are the main advantages of WE, which remedied the shortcoming of SVM. Offering unique and global solutions and preventing the convergence to local minima, kernel SVMs improved the classification. Avoiding overfitting and accomplishing the estimation out of sample, 10-fold cross-validation made the contribution.

5 Conclusions

In this study, a novel Chinese finger sign language recognition method (WE-kSVM) was proposed, in which wavelet entropy was employed to extract and reduce the feature, kernel support vector machine using RBF kernel was applied to classify, and 10-fold cross-validation was implemented to avoid overfitting and accomplish the out of sample estimate. This approach achieved an overall accuracy of 88.76 ± 0.59%, which denotes its superiority in all six state-of-the-art approaches.

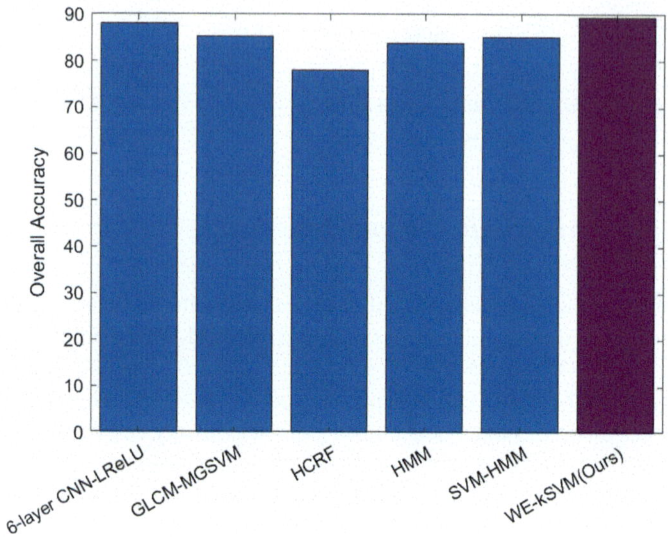

Fig. 5 Comparison plot of six state-of-the-art approaches

In the future, we shall do some contributions in the following areas. (1) we should realize the auto-preprocess of sign language images and cut the time taken for preprocessing. (2) trying to shift this method to other applications, such as Braille recognition, healthy and biomedical image identification, blind fever screening [14], clinical oncology [15], etc. (3) testing other feasible methods such as Principal Component Analysis PCA [16], Particle Swarm Optimization (PSO), Artificial Bee Colony algorithm(ABC) [17], and transfer learning [18, 19] in this theme.

Acknowledgements This work was supported from Jiangsu Overseas Visiting Scholar Program for University Prominent Young and Middle-aged Teachers and Presidents of China, The Natural Science Foundation of Jiangsu Higher Education Institutions of China (19KJA310002), The Surface Project of Natural Science Research in Colleges and Universities of Jiangsu China (16KJB520029, 16KJB520026), The Philosophy and Social Science Research Foundation Project of Universities of Jiangsu Province (2017SJB0668).

References

1. Cao, X.: Development of Wearable Sign Language Translator. University of Science and Technology of China, Hefei (2015)
2. Li, T.S., Kao, M., Kuo, P.: Recognition system for home-service-related sign language using entropy-based K-means algorithm and ABC-based HMM. IEEE Trans. Syst. Man Cybern. Syst. **46**(1), 150–162 (2016)
3. Jihai Zhang, W.Z., Li, H.: A threshold-based hmm-DTW approach for continuous sign language recognition. In: Proceedings of ACM International Conference on Internet Multimedia Computing and Service, p. 237 (2014)

4. Long Wang, H.L., Wang, B., et al.: Gesture recognition method based on skin color model and convolutional neural network. Comput. Eng. Appl. **53**(6), 209–214 (2017)
5. Jiang, X.: Chinese sign language fingerspelling recognition via six-layer convolutional neural network with leaky rectified linear units for therapy and rehabilitation. J. Med. Imaging Health Inform. **9**(9), 2031–2038 (2019)
6. Wu, D., Kindermans, P.J., et al.: Deep dynamic neural networks for multimodal gesture segmentation and recognition. IEEE Trans. Pattern Anal. Mach. Intell. **38**(8), 1583–1597 (2016)
7. Cui, L.H., Zhang, C.: Recurrent convolutional neural networks for continuous sign language recognition by staged optimization. In: IEEE Conference on Computer Vision and Pattern Recognition, pp. 1610–1618. IEEE (2017)
8. Huang, J.Z.W., Li, H et al.: Attention based 3D-CNNs for large-vocabulary sign language recognition. IEEE Trans. Circ. Syst. Video Technol. **1**, 1 (2018)
9. Yang, H.-D., Lee, S.-W.: Robust sign language recognition with hierarchical conditional random fields. In: 20th International Conference on Pattern Recognition, Istanbul, Turkey, pp. 2202–2205. IEEE (2010)
10. Rao, G.A., Kishore, P., Kumar, D.A., Sastry, A.: Neural network classifier for continuous sign language recognition with selfie video. Far East J. Electron. Commun. **17**(1), 49 (2017)
11. Jiang, X.: Isolated Chinese sign language recognition using gray-level co-occurrence matrix and parameter-optimized medium gaussian support vector machine. In: Frontiers in Intelligent Computing: Theory and Applications. Singapore, 2020, pp. 182–193: Springer Singapore
12. Kumar, P., Saini, R., Roy, P.P.: A position and rotation invariant framework for sign language recognition (SLR) using Kinect. Multimedia Tools Appl. **77**, 8823–8846 (2017)
13. Lee, G.C., Yeh, F., Hsiao, Y.: Kinect-based Taiwanese sign-language recognition system. Multimedia Tools Appl. **75**, 261–279 (2016)
14. Ng, E.Y.K., Kaw, G.J.L., Chang, W.M.: Analysis of IR thermal imager for mass blind fever screening. Microvasc. Res. Article **68**(2), 104–109 (2004). (in English)
15. Ng, E.Y.-K., Acharya, R.U.: Imaging as a diagnostic and therapeutic tool in clinical oncology. World J. Clin. Oncol. **2**(4), 169 (2011)
16. Artoni, A.D.F., Makeig, S.: Applying dimension reduction to EEG data by Principal Component Analysis reduces the quality of its subsequent Independent Component decomposition. NeuroImage **175**, 176–187 (2018)
17. Yang, J.: An adaptive encoding learning for artificial bee colony algorithms. J. Comput. Sci. **30**, 11–27 (2019)
18. Liu, J.: Detecting cerebral microbleeds with transfer learning. Mach. Vis. Appl. https://doi.org/10.1007/s00138-019-01029-5
19. Lu, S.: Pathological Brain Detection based on AlexNet and transfer learning. J. Comput. Sci. **30**, 41–47 (2019)

Clustering Diagnostic Codes: Exploratory Machine Learning Approach for Preventive Care of Chronic Diseases

K. N. Mohan Kumar, S. Sampath, Mohammed Imran, and N. Pradeep

Abstract High prevalence of chronic diseases along with poor health condition and the rising diagnosis and treatment costs necessitates concentration on prevention, early detection and disease management. In this paper correlation among the chronic diseases is examined with the help of diagnostic codes using unsupervised Machine Learning (ML) approaches. ML approaches pave the way to accomplish this objective. Healthcare data is categorized into clinical, Medi-claim, drugs and emergency information. In this work, Medi-claim data is used for exploring five types of chronic disorders such as Diabetes, Heart, Kidney, Liver and Cancer. Medi-claim data is acceptable because of its legitimacy, volume and demography qualities. Hierarchical Condition Category (HCC) and International Classification of Diseases (ICD) based coding of med-claim data are perfect with guaranteed informational index, this nature of ICD and HCC code urged us to work with Medi-claim records. The categorization of chronic and non-chronic diseases is built up through HCC codes utilizing different clustering techniques such as partitional, hierarchical and Fuzzy-K means clustering. The model is evaluated using various metrics such as Homogeneity, Completeness, V-measure, Adjusted Rand index, Adjusted Mutual Information. Among all the clustering techniques used K means and K means random have shown promising results. A compelling end on clustering of chronic diseases is made, remembering the clinical significance.

K. N. Mohan Kumar (✉) · S. Sampath
Adhichunchanagiri Institute of Technology, Chikkamagaluru, Karnataka, India
e-mail: mohan4183@gmail.com

S. Sampath
e-mail: 23.sampath@gmail.com

M. Imran
Ejyle Technology, Bangalore, Karnataka, India
e-mail: emraangi@gmail.com

N. Pradeep
Bapuji Institute of Engineering and Technology, Davanagere, Karnataka, India
e-mail: nmnpradeep@gmail.com

S. C. Satapathy et al. (eds.), *Intelligent Data Engineering and Analytics*, Advances in Intelligent Systems and Computing 1177,
https://doi.org/10.1007/978-981-15-5679-1_53

Keywords HCC · ICD · HIVE · Chronic condition · Clustering

1 Introduction

The recent advancements in computer technology has led to enormous developments in the healthcare sector. The advent of these technologies has provided another measurement to healthcare research. Much the same as how advanced mobile smartphones have made common man's life simpler, an endeavour to transform medical services progressively affordable by not settling on the nature of care that was there earlier. The best potential answer is to develop an intelligent healthcare decision support system. This has prompted different developments like keen rescue vehicle, smart emergency clinics and so on. These developments have served the member patients just like the healthcare specialists. Aside from all these, there are different problems to be taken care for quality healthcare, as to diminish the quantity of medical diagnosis, lessen healthcare costs for the member patient.

The recent developing pattern of chronic disorders, for example, Liver ailments, cardiac illness, diabetes has become an epidemic issue in the general public. These disorders include differential diagnosis and evaluation of various health parameters which prompts significant expense of social insurance for a member patient suffering from chronic illness. There is a need for efforts to decrease the medical tests for chronic disorders, hence lessening the total expense. The best solution is to utilize ML in developing algorithms for early detection of symptoms of chronic diseases. ML approaches are utilized to find relevant disease parameters in a gigantic dataset and extract useful information. Classification, Clustering and Association are the principal mechanisms in ML, having remarkable rules to solve contextual problems on clinical information [1, 2]. Successful utilization of all these procedures will mine out critical information helpful for preventive care of chronic diseases. Particularly, clustering is a type of ML algorithm that can infer conclusions from datasets that do not have labelled inputs [3–5].

The existing works have focussed only on ICD code to predict, classify and categorize the diseases and have not considered hierarchy of the disease occurrences. HCC codes have not been considered in existing work, without which, it would not be possible to find the chronic disease hierarchy [6]. In this paper, the objective is to explore the five chronic ailments, in particular Diabetes, Cancer, Heart, Liver, Kidney based on HCC codes utilizing Centers for Medicare and Medicaid Services (CMS) data [7]. CMS data comprises of separate files for carrier claims, inpatient, outpatient, beneficiary and prescription details. Clustering techniques are utilized to categorize member patients to be chronic and non-chronic based on the diagnostic codes. The clustering model is evaluated using various metrics, which is detailed in consecutive sections. At last, the importance and centrality of these codes were investigated [5].

The progression of this paper is as per the following: The related work is point by point in part II which features about existing information extraction, pre-handling and arrangement of medical records. Part III portrays the methods and materials followed in the analysis. Part IV gives the Experimental setup. Part V talks about the results and inferences. Part VI contains the summary and bearings for future improvement.

2 Related Work

Investigation of existing procedures would give better perceivability about the problem we are dealing with; this segment contemplates various methodologies utilized for categorization of chronic diseases.

Kaveh et al. [5] have developed a solution for the recommendation problem in insurance coverage. Recommendation of suitable coverage for the customers is challenging. They have mined Medi-claim data to dig out features associated with the customer to recommend them the best insurance coverage. They have used clustering mechanism in two stages, in the first stage, they have clustered the customers based on insurance coverage and in the second stage, they have used certain wrappers and filters to identify the required features. K means clustering is used for categorizing the customers and K nearest neighbour technique is used for classification. The Davies-Bouldin metric is used as an evaluation metric for clustering. Accuracy of 82.72% has been achieved in this work. Their suggestion is that the accuracy can be improved by using advanced clustering techniques.

Jiao et al. [8] have modelled a solution for categorization of patients from electronic medical records using clustering techniques. Chinese medical records are used in this work. They have constructed a dictionary on the medical domain to improve the accuracy of recognition in Chinese medical terminology. The features are extracted using latent Dirichlet technique and have used feature stitching to construct the feature vector. They have used certain indicators and constraints such as multi-dimensional attribute of the patient as supervised information, which guides the clustering technique. The model is evaluated using Normalized mutual information metric. The disadvantage is: the increase in constraints would adversely increase the accuracy but more constraints would incur more labour cost. Careful selection of constraint and data fusion is identified as future scope of research.

Birgit et al. [9] have done a commendable research on suicide using clustering technique. They have modelled a different approach other than the common classification of suicide as violent and non-violent using data-driven cluster analysis approach. All officially confirmed suicide data in Austria are assessed and used in this work. Age and gender is used as a feature for categorizing suicides. As a result of clustering, poisoning has evolved as a distinct method of suicide. The results of clustering have further divided into sub-clusters of violent suicides such as hanging, shooting, drowning. Their research has paved the way for advanced neurobiological research to develop potential suicide phenotypes.

Kun et al. [10] have devised a clustering technique to safeguard users privacy. In their work, they have used hybrid tables by establishing the relation between sensitivity attribute of the information. By the process of clustering, the records with minimum distance are selected and categorized. The validity of this work is proved using experiment and theory. The disadvantage of this work is that it does not consider the multi sensitive attributes and personalized privacy protection.

Chunlei et al. [11] have evaluated the feasibility of combining clustering and hash algorithm to find out the similarity in text free clinical records. They have used 1,063,893 text free clinical records of 32,682 unique patients. K-means clustering technique is used for categorizing the records. Fleiss' Kappa metric is used for reliability evaluation. 80–89% of precision is achieved in this experiment. They have opened the avenues for future work by stating that advanced clustering techniques to improve precision.

UshaRani et al. [12] have dealt with the classification of medical data using clustering of class-based records. They have used imputation approach and class labelling to overcome the difficulties of high dimension records. The classification is done on class-based records using clustering technique. Data sets such as Iris, Pima, Clevel and Ecoli, Wine and Wisconsin are used for the experiment. Techniques such as SVM, KNN and C4.5 are used for classification. They have recommended to extend this work using fuzzy measures.

Tiago et al. [13] have modelled a solution to detect suspicious medical claims from healthcare service providers. They have identified and dealt with difficulties faced in handling fake claims. Medi-claim records have been used for the experiment. They have used CRISP-DM and K-means clustering methods to ascertain the healthcare service providers. With their results, it is clear that this technique is successful in identifying outliers. Correlation matrix for cluster variables is used for model evaluation. Not all the attributes of claim records are used in the experiment, hence they recommend to use left out attributes of claim records to improve the efficiency of the model.

The review of the existing work has urged us to concentrate on utilizing ML approaches on CMS claim records [6], along with HCC codes. This will help to develop a productive model to set up the relationship among the chronic diseases and cluster them in terms of HCC codes. Mapping of HCC codes over ICD code would take out redundancy in data and makes it portrayal to focus on only required chronic ailments. This work focuses on the categorization of diagnostic codes to group them into chronic and non-chronic disease using different clustering techniques.

3 Methods and Techniques

In general, an unsupervised ML technique in the healthcare context is widely used to extract a pattern from medical records. Many applications on the healthcare domain use clustering method to categorize patient details from the medical record. In this work, we have used different clustering techniques on medi-claim data of CMS repository [6] to group the member patients. The different clustering techniques used in this work are, the popular K-means clustering, K-means clustering on

PCA, agglomerative clustering, K-median clustering and Fuzzy K-means clustering [4, 14, 15].

K-means method [5], partitions the set of data into subsets; the number of subsets is predefined by the number of clusters. The primary thought is to assign every vector to a given bunch of cluster centroids and after that, update the centroids given the recently settled task. This method is rehashed iteratively until a specific halting criterion is satisfied. The quantity of clusters to be discovered k is a necessary esteem input for the K-means technique. A set of k vectors are chosen from the medical record as introductory centroids, whose underlying qualities are considered haphazardly. The clustering procedure executes on an iterative calculation whose goal is to limit the general whole of separations between the centroids and objects, represented as halting criteria in a heuristic calculation. It is important to notice that one of the principal attributes of the algorithm is the separation considered between object, which is viewed as applicable in the kind of segmentation problem that should be illuminated.

K-medians clustering is a cluster analysis technique. It is another variant of K-means technique where as opposed to computing the mean for each cluster to decide its centroid, one rather considers the median value [8].

Principal Component Analysis (PCA) is a subspace algorithm. In this technique, the main concept is to find a subset of features F out of the universal feature set X, where $|F| \leq |X| = n$, and every feature $f \in F$ is extracted as an uncorrelated (symmetrical) linear mixture of universal set X. The resultant subset of features is identified as principal components of universal set X. In the process of determining the new subset of features, this technique focuses on minimizing the variance of data on every principal component [4, 14, 15].

The agglomerative clustering is the most well known sort of progressive clustering used to gather objects in groups dependent on their closeness. It's otherwise called AGNES (Agglomerative Nesting). The computation begins by regarding each object as a singleton bunch. Next, sets of groups are progressively converged until the sum total of what bunches has been converted into one major bunch containing all objects. The outcome is a tree-based portrayal of the objects, named dendrogram [8].

Fuzzy K-means clustering is the very same computation as K-means clustering. The main contrast in this technique is rather than assigning objects into only one cluster; it can have a type of fuzziness or cover between one or more clusters [13].

4 Experimental Setup

The fundamental step of any experiment is data acquisition. Our research is all about developing a model for preventive care of chronicle diseases using unsupervised ML. It is a great challenge to get medical records as it is sensitive and disclosure of medical records is a breach of privacy, hence medi-claim records (MCR) is used in our research. The medi-claim records are available in large quantity and proven. The main source of medi-claim records is from the CMS repository [7]. MCR includes

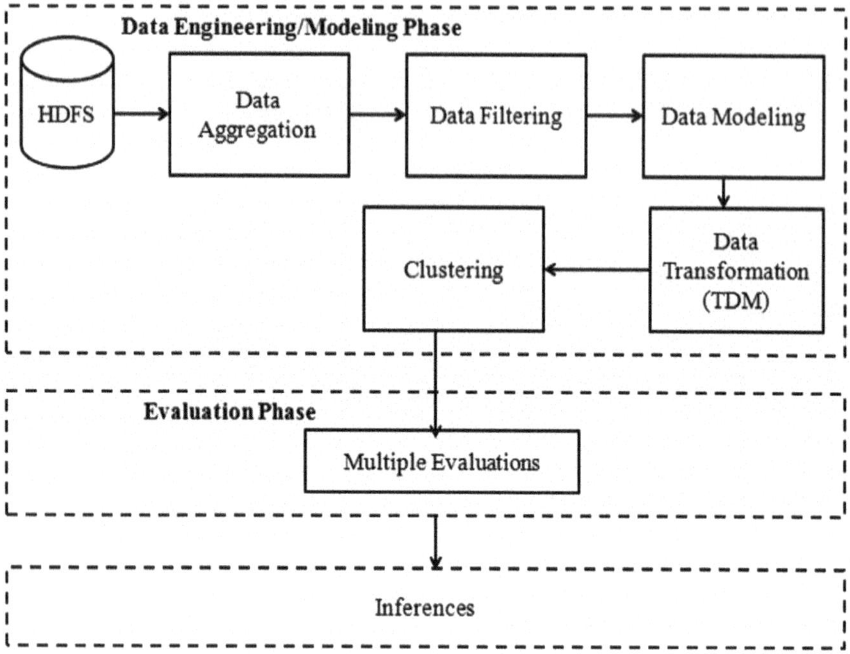

Fig. 1 Architecture of categorization of chronicle disease

beneficiary data describing demographic details of patient, inpatient, carrier claims, outpatient and prescription codes. The disease description is represented in term of ICD codes. We have not considered prescription codes in this work. The MCR is stored and processed using the Hadoop Distributed File System (HDFS) and Apache hive (Data warehouse). The integrated development environment used is python anaconda and pycharm. The process of the experiment is detailed in Fig. 1. Data aggregation step consolidates the records of the beneficiary, inpatient, carrier and outpatient files and represent the records in a structured format. Data filtering step will eliminate redundancy and selects the attributes containing ICD codes. Data modelling step divides the records into pre and post chronic. Data transformation step transforms the data into Term document matrix (TDM), in which row represents the patient records and columns represent the features. Clustering, evaluations and inferences are detailed in the below sections.

4.1 Data Modelling and Preprocessing

The essential test was the records in the gained information contained a large set of characteristics and a considerable lot of them are unimportant in our exploratory setting. Information modelling includes a cautious choice of attributes from the extracted information to fit the unique circumstance. Pre-handling of information includes the way towards changing crude information into the structure reasonable for ML methods. In the nutshell, big data technology (Hadoop) is used to store and process the information. HDFS is utilized for the storage of information and Apache hive is utilized for data preparation. With the use of HDFS and Apache hive, the information can be processed concurrently and the execution can be tuned by designing the quantity of mappers and reducers while SQL doesn't give this preferred position. Figure 1 subtleties model of the proposed methodology.

In data aggregation step, the records of the carrier, outpatient, and inpatient claims documents are moved into HDFS framework. Hive tables are made for each sort of claim files and beneficiary file [4, 15]. Progressively another table called super data (SD) is made to contain the consolidated information of all the three claim files referenced. The claim tables are populated with respect to the type of claim information. The essential key for every one of the tables is the beneficiary id. The SD table is populated with the selected ICD attributes from all the claim tables. The focus is to manage chronic ailments, thus ICD of non-chronic maladies ought to be eliminated [16]. In Data Filtering we extract the data of interest by selecting the attributes containing ICD codes. For this, ICD code in SD table is mapped with HCC code and resultant HCC codes are put away in a table called HCC data (HD) table. The records in HD table is modelled by isolating records into chronic and pre chronic records with reference to chronic diseases such as diabetes, heart, kidney, cancer and liver. The partitions are made with reference to the first occurrence of HCC code of respective diseases and the resultant records are stored in pre and post tables. At last, the pre and post chronic records are labelled into class 0 and a class 1 and placed in Final Data (FD) table. The class label is 0 for non-chronic record and 1 for chronic record. The records of the FD table are exported into final data Comma Separated Version (CSV) document.

The resultant data is not as demanded by ML approaches, thus the resultant records must be transformed into the form acceptable by legacy ML methods. The records in FD file is then transformed TDM, a sparse matrix containing columns representing HCC codes as features and rows as patient records. The existence HCC code is marked as 1 and its absence as 0 in patient records indicating whether the patient is suffering from that particular ailment or not. The chronic diseases that we are dealing with are diabetes, heart, kidney, cancer and liver disorders.

4.2 Clustering

As talked about in Sect. 2 above, numerous techniques have been proposed with the presentation of thoughts in clustering [10]. The essential thought of categorizing is to isolate an informational collection into a two classes (group) as indicated by the level of similarity with the goal that the information in a similar cluster have more grounded comparability than the information between groups [13]. Before starting with the process of clustering, the class labels are removes. The K-means, agglomerative, K-median and Fuzzy-K means isolates the tuples in the FD file into two identical classes as per the similarity of qualities, and independently sums them up. Categorization and comparability class division have comparable thoughts in information apportioning. In presenting clustering strategies, one of the keys is the means by which to gauge the comparability property estimations. In spite of the fact that there are numerous calculations to quantify the similarity, every technique has certain weaknesses. Those clustering estimations consider various information types, yet they don't all the more profoundly think about the correlation. Clustering doesn't consider various impacts of each property on the characterization. Clustering techniques have been applied to various chronic diseases records such as diabetes, heart, liver, kidney and cancer to categorize it chronic and non-chronic. The following section demonstrates the results of various clustering techniques [13].

5 Results and Discussion

The experiment is tried on different clustering methods to discover the ideal arrangement of records. After-effects of various clustering techniques and evaluation results are described in Tables 1, 2, 3, 4, 5. The results of the clustering techniques are evaluated using various metrics as mentioned below. The results are needed to be relooked and checked using different evaluation techniques to ensure that the model satisfies all the perspectives as we are experimenting on the life science data. Any mistake in clustering results would cost the life of the member patient, thus multiple evaluations are needed.

The performance evaluation metrics are as follows: M1-Homogeneity, M2-Completeness, M3-V-measure, M4-Adjusted Rand Index, M5-Adjusted Mutual Information. Homogeneity characterizes some instinctive measurement utilizing restrictive entropy examination and checks whether the elements in each cluster belongs to its own class. Completeness is a measure to indicate all members of a given class are assigned to the same group or not. The V-measure score is the harmonic mean among homogeneity and completeness. Adjusted Rand Index measures the similarity

Table 1 Results of clustering techniques on diabetes data

Diabetes										
Clustering techniques	Before PCA					After PCA				
	M1	M2	M3	M4	M5	M1	M2	M3	M4	M5
K-means	0.063	0.061	0.062	−0.079	0.060	1.0	1.0	1.0	1.0	1.0
Random	0.073	0.067	0.070	−0.069	0.066	1.0	1.0	1.0	1.0	1.0
Agglomerative	0.044	0.037	0.040	−0.034	0.036	0.260	0.228	0.243	0.378	0.227
K-median	0.016	0.035	0.022	−0.056	0.015	0.007	0.005	0.006	0.006	0.004
Fuzzy K-means	0.131	0.102	0.115	0.022	0.101	0.227	0.173	0.196	0.170	0.172

Table 2 Results of clustering techniques on cancer data

Cancer										
Clustering techniques	Before PCA					After PCA				
	M1	M2	M3	M4	M5	M1	M2	M3	M4	M5
K-means	0.574	0.611	0.592	0.628	0.574	1.0	1.0	1.0	1.0	1.0
Random	0.565	0.603	0.584	0.618	0.565	1.0	1.0	1.0	1.0	1.0
Agglomerative	0.226	0.259	0.242	0.277	0.226	0.278	0.280	0.279	0.319	0.277
K-median	0.225	0.249	0.236	0.285	0.224	0.003	0.004	0.004	0.006	0.002
Fuzzy K-means	0.550	0.555	0.552	0.661	0.550	0.074	0.074	0.074	0.092	0.073

Table 3 Results of clustering techniques on heart data

Heart										
Clustering techniques	Before PCA					After PCA				
	M1	M2	M3	M4	M5	M1	M2	M3	M4	M5
K-means	0.925	0.927	0.926	0.963	0.924	1.0	1.0	1.0	1.0	1.0
Random	0.925	0.927	0.926	0.963	0.924	1.0	1.0	1.0	1.0	1.0
Agglomerative	0.002	0.110	0.004	0.001	0.000	0.380	0.378	0.379	0.478	0.377
K-median	0.003	0.003	0.003	0.001	0.002	0.000	0.000	0.000	−0.002	−0.001
Fuzzy K-means	0.789	0.789	0.789	0.872	0.789	0.224	0.223	0.224	0.296	0.222

Table 4 Results of clustering techniques on liver data

Liver										
Clustering techniques	Before PCA					After PCA				
	M1	M2	M3	M4	M5	M1	M2	M3	M4	M5
K-means	0.717	0.721	0.719	0.791	0.717	1.0	1.0	1.0	1.0	1.0
Random	0.703	0.708	0.705	0.777	0.703	1.0	1.0	1.0	1.0	1.0
Agglomerative	0.282	0.287	0.284	0.353	0.281	0.424	0.443	0.433	0.456	0.423
K-median	0.403	0.410	0.406	0.493	0.402	0.971	0.971	0.971	0.987	0.971
Fuzzy K-means	0.470	0.474	0.472	0.570	0.469	0.777	0.777	0.777	0.860	0.776

of the two class, ignoring permutations and normalization. Adjusted Mutual Information is a measure of the agreement of the two class, ignoring permutations. The perfect score of the above discussed metrics is 1 indicating perfect categorization of records and 0 indicates improper categorization.

The clustering results for Diabetes data are detailed in Table 1. It is evident that none of the mentioned clustering techniques perform well on the diabetes TDM

Table 5 Results of clustering techniques on kidney data

Kidney										
Clustering techniques	Before PCA					After PCA				
	M1	M2	M3	M4	M5	M1	M2	M3	M4	M5
K-means	0.409	0.444	0.426	0.438	0.408	1.0	1.0	1.0	1.0	1.0
Random	0.370	0.411	0.390	0.387	0.369	1.0	1.0	1.0	1.0	1.0
Agglomerative	0.266	0.267	0.267	0.339	0.265	0.360	0.367	0.364	0.439	0.360
K-median	0.419	0.419	0.419	0.520	0.418	0.986	0.986	0.986	0.994	0.986
Fuzzy K-means	0.359	0.361	0.360	0.450	0.358	0.239	0.239	0.239	0.311	0.238

data. Our research continued to improve the model and for which we worked out to get subspace data with more correlation using PCA. It has been proved that we get better results on subspace data (PCA data) using K-means and K-means random. The results of clustering techniques such as Agglomerative, K-median and Fuzzy K-means have not improved on the subspace data (PCA) at all [15].

The clustering results for Cancer data are detailed in Table 2. It is evident that clustering techniques do not perform well on the cancer TDM data but K-means and K-means random show better results (>0.5). It has been evident that we get optimal results on subspace data (PCA data) using K-means and K-means random. The results of clustering techniques such as Agglomerative, K-median and Fuzzy K-means have not improved on the subspace data (PCA) at all [16].

The clustering results for Heart data are detailed in Table 3. It is clear from Table 3 that clustering techniques like K-means and K-means random produce good results (0.9 score) and Fuzzy K-means produce above average results (0.7) on heart TDM data but the rest of the techniques does not perform better at all. It is evident that we get optimal results on subspace data (PCA data) using K-means and K-means random (perfect score 1). The results of clustering techniques such as Agglomerative, K-median and Fuzzy K-means have not improved on the subspace data (PCA) at all [15, 16].

The clustering results for Liver data are detailed in Table 4. It is clear from Table 3 that clustering techniques like K-means and K-means random produce above average results (>0.7) and Fuzzy K-means and K-median produce below average results (0.4) on Liver TDM data but the rest of the techniques do not perform better at all. It has been evident that we get optimal results on subspace data (PCA data) using K-means and K-means random (perfect score 1).

The results of clustering techniques such as Agglomerative, K-median and Fuzzy K-means have little improved on the subspace data (PCA) with the score of 0.4, 0.9 and 0.7. The interesting part is K-median clustering has shown promising result of 0.97 on PCA data which is also acceptable [15, 17].

The clustering results for Kidney data are detailed in Table 5. It is evident that none of the mentioned clustering techniques perform well on the Kidney TDM data. It has been proved that we get the best results on subspace data (PCA data) using K-means and K-means random. The results of clustering techniques such as Agglomerative, and Fuzzy K-means have not improved on the subspace data (PCA) at all but the K-median technique shows a promising result of 0.98 on subspace PCA data [15, 18].

The plots Figs. 2, 3, 4, 5, 6, describes the clustering results of K-means technique on various diseases data graphically. The cross mark in each plot is the centre of the cluster. In each figure, the plots of the results of TDM data (left) and PCA subspace data (right) are adjacent to each other. From the figure, it is easy to compare and discriminate the improvement in the results of TDM data and PCA subspace data [15, 18].

The main agenda of this research work was to find an optimal clustering model for unlabelled health records to categorize or cluster the records into chronic and non-chronic. These clusters would indicate the status of the health condition of the member patients which helps the member patients to take preventive measures before the health condition worsens. The inferences from the clustering results are that the different techniques provide varied scores on TDM data and PCA subspace data contextually. It is observed that K-means and fuzzy K-means when applied on TDM data provide an average score of 0.5 on cancer, 0.9 and 0.7 on heart, 0.7 and 0.4 liver and 0.4 and 0.3 on kidney. The K-means and K-means random have provided

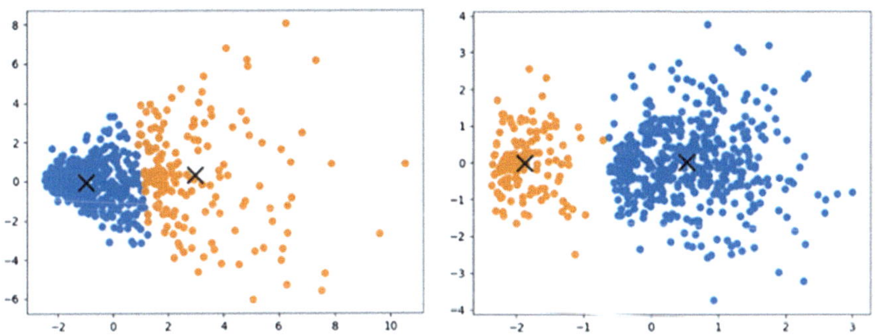

Fig. 2 Plot of clustering results of TDM and PCA diabetes data

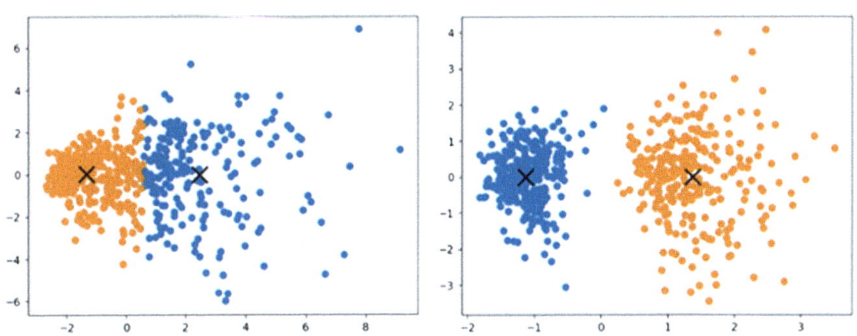

Fig. 3 Plot of clustering results of TDM and PCA cancer data

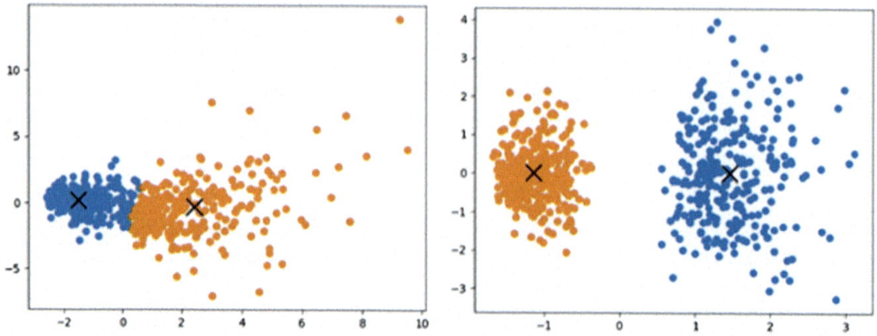

Fig. 4 Plot of clustering results of TDM and PCA heart data

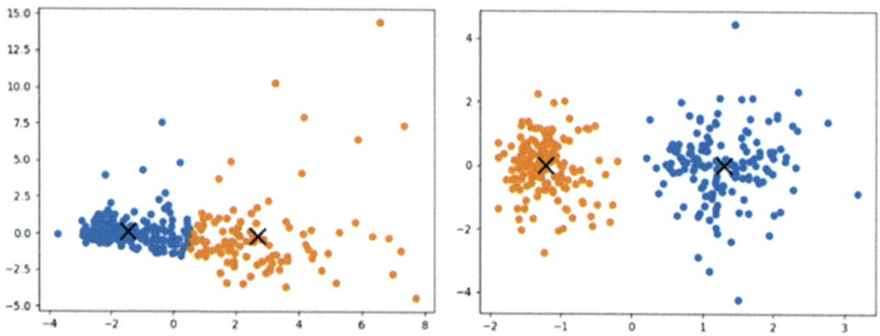

Fig. 5 Plot of clustering results of TDM and PCA liver data

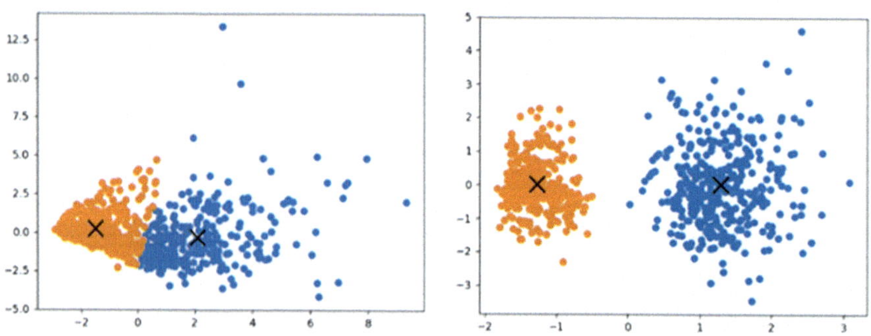

Fig. 6 Plot of clustering results of TDM and PCA kidney data

a perfect score of 1 on PCA subspace data. Agglomerative clustering technique has not shown any promising results. After the evaluation of the experimental results, it is clear that K-means and K-means random provide a perfect score of 1 on all the above discussed health records. Thus, K-means and K-means random are appropriate clustering model for categorization health records as chronic and non-chronic.

6 Conclusion

The focal point of our exploration work is to develop a model for preventive consideration of chronic illnesses. In this paper, we have investigated the Medi-Claim information particularly because of its validity. We have removed and rebuilt the information for various chronic diseases such as diabetes, heart, cancer, liver and kidney on account of the accessibility of huge. HCC codes were mapped with ICD and just HCC codes were held to build up the connection between's various levelled chronic disease. ML methods are then used to set up the relation among records to categorize the diseases as chronic and non-chronic. The experiment was driven by clustering techniques such as K-means, K-means random, agglomerative, K-median. The model was evaluated using Homogeneity, Completeness, V-measure, Adjusted Rand index and Adjusted Mutual Information metrics. The metric would indicate 1 as the best score for ideal cluster and any gradual decrease in score from 1 to 0 proportionally degrades the cluster. In this work, it has proved and observed that K-means and K-means random are appropriate clustering model for categorization health records as chronic and non-chronic. The clusters would indicate the status of the health condition of the member patients which helps the member patients to take preventive measures before the health condition worsens.

This work can be stretched out by incorporating extra features related to patients, for example, prescription codes, work type, habits and geographical location and so on. The med-claim data contains only the records of the old age patients and it would be better to develop a model that works on data of all age groups.

References

1. Jensen, P.B., Jensen L.J., Brunak, S.: Mining electronic health records: towards better research applications and clinical care. Nat. Rev. Genet. 395–405 (2012)
2. Bali, A., Bali, D., Iyer, N., Iyer, M.: Management of medical records: facts and figures for surgeons. J. Maxillofacial Oral Surg. **10**(3), 199–202 (2011)
3. https://medlineplus.gov/ency/patientinstructions/000602.html
4. Mohan Kumar, K.N., Sampath, S., Imran, M.: An overview on disease prediction for preventive care of health deterioration. IJEAT **8**(5S) (2019)
5. Khalili-Damghani, K., Abdi, F., Abolmakarem, S.: Solving customer insurance coverage recommendation problem using a two-stage clustering-classification model. Int. J. Manage. Sci. Eng. Manage. (2019)
6. Gupta, D., Khare, S., Aggarwal, A.: A method to predict diagnostic codes for chronic diseases using machine learning techniques. In: The Proceedings of IEEE International Conference on Computing, Communication and Automation (2016)
7. https://www.cms.gov/Research-Statistics-Data-and-Systems/Downloadable-Public-Use-Files/SynPUFs/DESample01
8. Zhang, J., Chang, D.: Semi-supervised patient similarity clustering algorithm based on electronic medical records. In: IEEE Healthcare Information Technology for the Extreme and Remote Environments, pp. 90705–90714 (2019)

9. Ludwiga, B., König, D., Kapustac, N.D., Blümlc, V., Dorffner, G., Vyssokib, B.: Clustering suicides: a data-driven, exploratory Machine Learning approach. In: Journal of the European Psychiatric Association, pp. 15–19. Elsevier (2019)
10. Wang, K., Zhao, W., Cui, J., Cui, Y., Hu, J.: A K-anonymous clustering algorithm based on the analytic hierarchy process. In: Journal of Visual Communication and Image Representation, pp. 76–83. Elsevier (2019)
11. Tanga, C., Plasekd, J.M., Xionge, Y., Batesa, D.W., Zhoua, L.: Clustering similar clinical documents in electronic health records. In: Proceedings of the International Conference on Data Science. Elsevier (2018)
12. Yelipe, U., Porika, S., Golla, M.: An efficient approach for imputation and classification of medical data values using class-based clustering of medical records. J. Comput. Electr. Eng. 487–504 (2018)
13. Tiago Hillermana, João Carlos F. Souzaa, Ana Carla B. Reisa, Rommel, "Applying clustering and AHP methods for evaluating suspect healthcare claims", Springer, 2017
14. Bai, Z.-J., Chan, R., Luk, F.: Principal component analysis for distributed data sets with updating. In: Proceedings of International workshop on Advanced Parallel Processing Technologies (APPT) (2005)
15. Qu, Y., Ostrouchovz, G., Samatovaz, N., Geıst, A.: Principal component analysis for dimension reduction in massive distributed data sets. In: Proceedings of IEEE International Conference on Data Mining (ICDM) (2002)
16. Gligorijevic, D., Stojanovic, J., Obradovic, Z.: Disease types discovery from a large database of inpatient records: a sepsis study. Expert Syst. Appl.**111**, 45–55 (2016)
17. Song, S., Warren, J., Riddle, P.: Developing high risk clusters for chronic disease events with classification association rule mining. In: The Proceedings of the Seventh Australasian Workshop on Health Informatics and Knowledge Management, vol. 153 (2014)
18. Jen, C.-H., Wangb, C.-C., Jiang, B.C., Chu, Y.-H., Chen, M.-S.: Application of classification techniques on development an early-warning system for chronic illnesses. Expert Syst. Appl. (2012)

NormCG: A Novel Deep Learning Model for Medical Entity Linking

Chen Tang, Weile Chen, Tao Wang, Chun Sun, JingChi Jiang, and Yi Guan

Abstract Medical Entity Linking is an essential task of medical natural language processing. It maps different ways of talking about the ontology to the standardized terminology. This paper presents NormCG, a novel deep learning model based on Convolutional Neural Network (CNN) and Gated Recurrent Unit (GRU) network, to rank candidate terminologies for each given entity in the free text. It considers the topical and global features of mentions in clinical narratives by extracting morphological and context information. NormCG models morphological features using the CNN model to compose phrase embeddings from word embeddings. The context features of mentions are represented by co-mention sequences within the same document. Considering both features, NormCG can give the probability of the candidate linked to the mention, by which we can choose the best candidate for the medical entity liking process. Our experimental results on the standard benchmark NCBI shows NormCG outperforms previous methods.

Keywords Deep learning · Medical entity linking · Gated recurrent unit · Convolutional neural network

C. Tang · W. Chen · T. Wang · J. Jiang · Y. Guan (✉)
Harbin Institute of Technology, Harbin, China
e-mail: guanyi@hit.edu.cn

C. Sun
Harbin Engineering University, Harbin, China

C. Tang · W. Chen · T. Wang · C. Sun · J. Jiang · Y. Guan
Shenzhen Research Institute of Big Data, 518000 Shenzhen, China

S. C. Satapathy et al. (eds.), *Intelligent Data Engineering and Analytics*, Advances in Intelligent Systems and Computing 1177,
https://doi.org/10.1007/978-981-15-5679-1_54

1 Introduction

Medical entity linking can be divided into two steps, candidate generation and candidate ranking. The step of candidate generation is mostly dependent on rule-based methods including phrase matching and Jaccard overlap. It aims to generate probable candidate entities from standard terminology database. The goal of candidate generation is to ensure the ontology of the given mention exits in its candidate set. While the step of candidate ranking aims to dig out potential connections between mentions and candidates and then rank the candidates for entity linking.

In recent years, deep learning methods have improved the performance of candidate ranking by a significant margin. Several promising solutions using machine learning have been proposed, such as the CNN-based [1], TaggerOne [2], and NormCo [3]. They tried to utilize semantic information by a single type of neural network. However, to our knowledge, none of them have fully utilized the context information by combining CNN [4] and GRU [5]. Besides, some models reply to external resources to raise linking accuracy, while in this way, the advantage of the model itself is hard to be discerned.

In this paper, we mainly focus on candidate ranking process and our candidate generation follows the work of the sieve based [6]. To solve the above issues of candidate ranking, we propose NormCG. NormCG refers to the idea of Li et al. [1] to get morphological features by CNN. The initialization of the word embeddings takes Zhang et al.'s work[1] [7]. To measure coherence and similarity of context, NormCG is based on NormCo [3] to exploit the co-mention sequence. Composing the merits of previous models, NormCG makes the following three contributions: (1) We build a new neural network to grab topical features and global features for linking. (2) Unlike the existing system, NormCG calculates the matching probability of the mention-candidate pair, rather than ranking candidates by the distance between mentions and candidates, as in this way, training is more efficient and effective. (3) We explore the capability of deep learning method in candidate ranking without external resources. The result of experiments is more objective.

The remaining of the paper is structured as follows: Sect. 2 introduces the methodology of medical entity liking. Section 3 focuses on displaying and explaining the experimental result. Section 4 is the conclusion and future work.

2 Methodology

Figure 1 is the schematic diagram of Medical Entity Linking process. Phrases with yellow underpainting denote mentions, which are extracted from clinical narratives through Named Entity Recognition (NER). The red phrase "WD" is now under the entity linking process. Its candidate entities are extracted from the database of

[1] https://github.com/ncbi-nlp/BioSentVec.

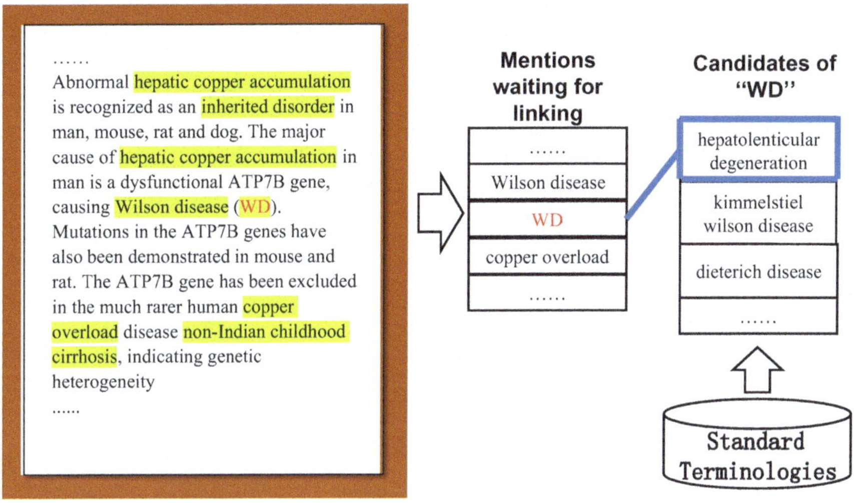

Fig. 1 Medical entity linking diagram

standard terminologies. Through candidate ranking, the terminology "hepatolenticular degeneration" is the best choice for linking process, thus "WD" is linked to "hepatolenticular degeneration".

2.1 Model Architecture

The diagram of Fig. 2 shows the overview of the architecture of NormCG. Let m_i be the mention waiting for linking and its corresponding candidate set C^i would be $\left\{c_0^i, c_1^i, \ldots, c_j^i\right\}$. The input of the model is the pair of the mention m_i and a candidate c_j^i. To better produce word embeddings for mentions and candidates, some preprocesses are taken. We refer to the work of Jennifer and Vincent [6] and Luo et al. [8] to filter phrases with several rules as Table 1 lists.

After preprocessing follows is the work of neural network. It is worth mentioning that each phrase uses the same CNN and each mention phrase sequence uses the same GRU. In other words, the CNN shares parameters when processing phrases, so does GRU. Because each mention has a different number of words, the first step is using CNN to convolute and max pool variously dimensional word embeddings into fixed-size phrase embeddings as Eq. 1 shows. The activation of the convolutional neural network block is ReLU [9]. We denote Φ as this process. By CNN, we can get the topical features of m_i and c_j^i.

$$\Phi(m \; or \; c) = \text{Max-pool}(Conv(m \; or \; c)) \tag{1}$$

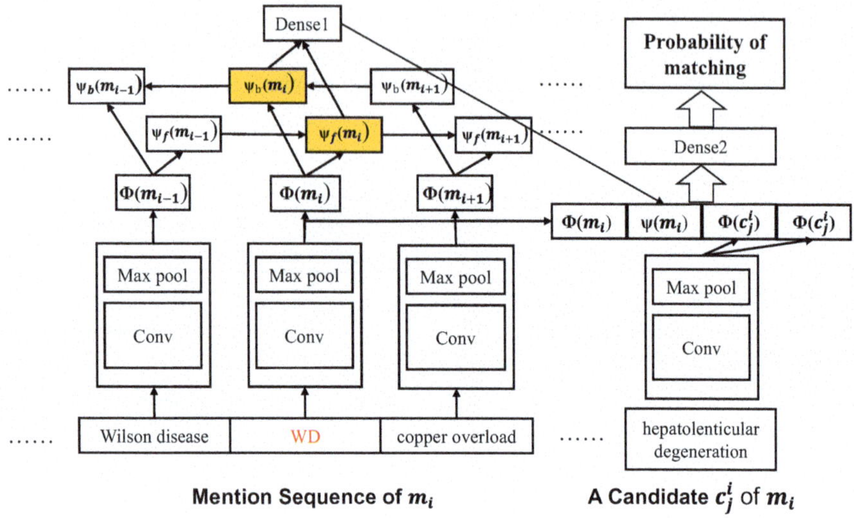

Fig. 2 The architecture of NormCG

Table 1 Rules for preprocessing

Rules	Examples
Remove digits	1, 2, 3…
Remove common words	"your", "his", "her", "their", "patient", "a", "an", "of", "any", "this", "these", "that", "those", "the", "be", "all", "feed"
Remove punctuations	,./;'[]\‘ ~ !@#$%^&*()- = _+{}:"? <>|

After obtaining phrase embeddings, the co-mention sequence of phrase embeddings $\{\ldots, \Phi(m_{i-1}), \Phi(m_i), \Phi(m_{i+1}), \ldots\}$ are sent to the bidirectional GRU model [5] and we can get forward sequence $\{\ldots, \psi_f(m_{i-1}), \psi_f(m_i), \psi_f(m_{i+1}), \ldots\}$ and backward sequence $\{\ldots, \psi_b(m_{i+1}), \psi_b(m_i), \psi_b(m_{i-1}), \ldots\}$ by Eqs. 2 and 3.

$$\begin{aligned} &\{\ldots, \psi_{\mathrm{f}}(m_{i-1}), \psi_{\mathrm{f}}(m_i), \psi_{\mathrm{f}}(m_{i+1}), \ldots\} \\ &\quad = \mathrm{GRU}(\ldots, \Phi(m_{i-1}), \Phi(m_i), \Phi(m_{i+1}), \ldots) \end{aligned} \tag{2}$$

and

$$\begin{aligned} &\{\ldots, \psi_{\mathrm{b}}(m_{i+1}), \psi_{\mathrm{b}}(m_i), \psi_{\mathrm{b}}(m_{i-1}), \ldots\} \\ &\quad = \mathrm{GRU}(\ldots, \Phi(m_{i+1}), \Phi(m_i), \Phi(m_{i-1}), \ldots) \end{aligned} \tag{3}$$

The global coherence of context can be represented as the concatenation of $\psi_b(m_i)$ and $\psi_f(m_i)$ as Eq. 4 shows. The concatenation is denoted by $\psi(m_i)$ and this is the global feature of the mention m_i.

$$\psi(m_i) = \text{concatenate}\,(\psi_b(m_i)\ \text{and}\ \psi_f(m_i)) \tag{4}$$

Finally, $\Phi(m_i)$, $\psi(m_i)$, and $\Phi(c^i_j)$ are all combined into the feature vector. To make the features from m_i and c^i_j with equal dimensions, we concatenate two $\Phi(c^i_j)$. After passing through the *Dense* Layer of neural network and the *Softmax* function, we can get the final matching probability $p\left(m_i, c^i_j\right)$ in Eq. 5. When all the corresponding probabilities of candidate set $\left\{p\left(m_i, c^i_0\right), p\left(m_i, c^i_1\right) \ldots, p\left(m_i, c^i_j\right)\right\}$ come out, we can rank candidates for linking according to it and the best linking candidate is obtained by Eq. 6.

$$p\left(m_i, c^i_j\right) = \text{Softmax}\left(\text{Dense}\left(\text{concatenate}\left(\Phi(m_i), \psi(m_i), \Phi\left(c^i_j\right), \Phi\left(c^i_j\right)\right)\right)\right) \tag{5}$$

$$\hat{c} = \text{argmax}\left\{p\left(m_i, c^i_0\right), p\left(m_i, c^i_1\right) \ldots, p\left(m_i, c^i_j\right)\right\} \tag{6}$$

2.2 *Training*

The Loss Function is defined in Eq. 7, where we use cross-entropy function to compute the loss between $p\left(m_i, c^i_j\right)$ and $\text{is\_match}\left(m_i, c^i_j\right)$. $\text{is\_match}\left(m_i, c^i_j\right)$ is a function to compare identifiers of m_i and c^i_j. If the identifier of m_i (manually annotated) is equal to the identifier of c^i_j (from MEDIC lexicon [10]), $\text{is\_match}\left(m_i, c^i_j\right)$ is assigned with 1, otherwise with 0.

$$L = \frac{\sum_{c^i_j \in C^i} \text{cross\_entropy}\left(p\left(m_i, c^i_j\right), \text{is\_match}\left(m_i, c^i_j\right)\right)}{\text{sizeof}\left(C^i\right)} \tag{7}$$

3 Experimental Results and Analysis

3.1 *Dataset*

The Dataset NCBI [11] contains 794 PubMed abstracts in total, in which 694 for training and development and 100 for test. It is the most popular dataset to evaluate the medical entity linking model. Each abstract was annotated manually, in which disorder mentions combined with their concept identifiers are listed in order. Through searching the identifier of the mention, we can find the ground-truth terminology in

Table 2 The statistics of NCBI dataset

	Documents	Mentions/identifiers	Unique mentions	Unique identifiers
Train	694	5932	1779	725
Test	100	960	404	202
Total	794	1654	1999	787

the MEDIC lexicon [10]. The statistics of NCBI are displayed in Table 2. Total unique mentions and identifiers mean the unique ones of all the data.

3.2 Training Details

In the candidate set of each mention, the ground truths are labeled as 1 and other candidates labeled as 0 by is_match function. The loss function we use is cross-entropy. Training epoch is set 50 but early stopping is also employed. The concrete neural network parameters are displayed in Table 3, which is chosen by experience and several trials.

Especially, GRU units equally include the units of GRU forward layer and backward layer. The pretrained word embeddings of Zhang et al.'s work [7] have choices of 200 and 400. Here, we use the 200-dimensional word2vec. Dropout [12] rate is the rate of inputs' values being set to zeros, usually used after layers of the neural network to avoid overfitting. Clip value is the bound to avoid gradient explosion. Shuffle means shuffling the input data.

Table 3 Parameters of NormCG

Parameter name	Value
Learning rate	1e−4
Batch size	200
Output size	2
GRU units	300
CNN units	400
CNN kernel size	3
CNN stride	1
Word embedding dimension	200
Dropout rate	0.2
Clip value	5
Is shuffle	True

3.3 Evaluation

After ranking the candidates set according to the corresponding probabilities, we select the highest rank candidates as the result of linking. The evaluation metric is Accuracy and Eq. 8 is the formula. To be concrete, each mention of the PubMed abstract is manually labeled with an identifier to represent the ground-truth terminology in the MEDIC lexicon. If the result we choose has the same identifier with the ground truth, then the result is true, otherwise it is false. The accuracy is represented as the ratio of the correctly linked mentions (TP) divided by the sum of mentions (TP + FP).

$$\text{Accuracy} = \text{TP}/(\text{TP} + \text{FP}) \tag{8}$$

3.4 Result and Analysis

The models above the dashed line are baseline models. As Table 4 records, NormCG outperforms existing baselines. Especially the CNN model (only use $\Phi(m_i)$, $\Phi\left(c_j^i\right)$) referring to the work of [1] and the GRU model (only use $\psi(m_i)$, $\Phi\left(c_j^i\right)$) referring to the work of [3] perform better than the previous models, respectively. It demonstrates the effectiveness and rationality of NormCG as well as the advantage of evaluating the probability by mention-candidate pairs.

Table 5 shows some typical examples of the prediction result, from which we can notice that the correct predictions often come along with very high probabilities. Among correct ones, some ground truths are dissimilar to the mentions, which illustrate the importance of digging out the connection between mentions and candidates from the context coherence aspect.

Table 4 The comparison of entity linking results in terms of accuracy

Model name	Accuracy (%)
Sieve-based	84.65
CNN-based	86.10
TaggerOne	87.7
NormCo	87.8
Only CNN model	88.75
Only GRU model	88.33
NormCG	89.79

Table 5 Typical examples of predictions

Mention	Ground truth	Prediction	Probability	Is_match
Copper toxicosis	Cirrhosis, familial	Copper storage disease	0.499	False
Hepatic copper accumulation	Liver diseases	Liver diseases	1.000	True
Inherited disorder	Genetic diseases	Genetic diseases	1.000	True
Wilson disease	Hepatolenticular degeneration	Hepatolenticular degeneration	0.995	True
CT	Cirrhosis, familial	Scan1	0.515	False

NormCG still suffers from dealing with the acronyms like "ct" and some out-of-vocabulary synonyms, which should resort to terminology dictionaries and rules. It seems that knowledge is still the key role of the neural network.

In the other hand, NormCG can consider the semantic meanings of mentions and candidates. Correct predictions may be beneficial from the similar mentions from the training dataset. Thus, we take "copper toxicosis", for example. Although the prediction is wrong, "copper storage disease" is close enough to the mention "copper toxicosis" in semantics to some extent.

In future work, combining the phrase with modifier tag may help it to make more accurate prediction.

4 Conclusion and Future Work

In this paper, we exploit the deep learning model to extract topical and global features of mentions and candidates. The experiment shows combining both features to boost the accuracy of model. Especially, the result of NormCG is not dependent on external data resources. In this way, it is more meaningful for this task itself, because the data resources are costly and hard to access in actual practice. In future work, more fresh research results in the open domain are going be taken into consideration when designing the neural network, such as large-scale pretrained models, attention mechanisms, meta-learning, and so on. We will try to make a more robust and effective neural network for practical use.

Acknowledgements This research was financially supported by the Open Research Fund from Shenzhen Research Institute of Big Data, under Grant No. 2019ORF01011. Meanwhile, Xiang Wan and Tsung-Hui Chang also gave us much help.

References

1. Li, H., Chen, Q., Tang, B., Wang, X., Xu, H., Wang, B., Huang, D.: CNN-based ranking for biomedical entity normalization. BMC Bioinform. **18**, (2017). https://doi.org/10.1186/s12859-017-1805-7
2. Leaman, R., Lu, Z.: TaggerOne: Joint named entity recognition and normalization with semi-Markov Models. Bioinformatics **32**, 2839–2846 (2016). https://doi.org/10.1093/bioinformatics/btw343
3. Wright, D., Katsis, Y., Mehta, R., Hsu, C.-N.: NormCo: deep disease normalization for biomedical knowledge base construction (2019)
4. Kim, Y.: Convolutional neural networks for sentence classification. EMNLP 2014–2014 Conf. Empir. Methods Nat. Lang. Process. Proc. Conf. 1746–1751 (2014). https://doi.org/10.3115/v1/d14-1181
5. Chung, J., Gulcehre, C., Cho, K., Bengio, Y.: Empirical evaluation of gated recurrent neural networks on sequence modeling. arXiv Prepr.: arXiv1412.3555. (2014)
6. D'Souza, J., Ng, V.: Sieve-based entity linking for the biomedical domain. In: ACL-IJCNLP 2015—53rd Annual Meeting of the Association for Computational Linguistics and the 7th International Joint Conference on Natural Language Processing of the Asian Federation of Natural Language Processing, Proceedings of the Conference, pp. 297–302 (2015)
7. Zhang, Y., Chen, Q., Yang, Z., Lin, H., Lu, Z.: BioWordVec, improving biomedical word embeddings with subword information and MeSH. Sci. Data. **6**, 52 (2019). https://doi.org/10.1038/s41597-019-0055-0
8. Luo, Y.F., Sun, W., Rumshisky, A.: MCN: A comprehensive corpus for medical concept normalization (2019)
9. Glorot, X., Bordes, A., Bengio, Y.: Deep sparse rectifier neural networks. J. Mach. Learn. Res. 315–323 (2011)
10. Davis, A.P., Wiegers, T.C., Rosenstein, M.C., Mattingly, C.J.: MEDIC: a practical disease vocabulary used at the comparative toxicogenomics database. Database (2012). https://doi.org/10.1093/database/bar065
11. Doğan, R.I., Leaman, R., Lu, Z.: NCBI disease corpus: A resource for disease name recognition and concept normalization. J. Biomed. Inform. **47**, 1–10 (2014). https://doi.org/10.1016/j.jbi.2013.12.006
12. Srivastava, N., Hinton, G., Krizhevsky, A., Sutskever, I., Salakhutdinov, R.: Dropout: A simple way to prevent neural networks from overfitting. J. Mach. Learn. Res. **15**, 1929–1958 (2014)

A Hybrid Model for Clinical Concept Normalization

Chen Tang, Weile Chen, Chun Sun, Tao Wang, Pengfei Li, Jingchi Jiang, and Yi Guan

Abstract Clinical Concept normalization is an essential task aiming to normalize recognized clinical concepts from clinical narratives. This paper presents a hybrid model combining traditional rule-based methods and a novel neural network to realize the normalization process. The experiment is set on MCN corpus and we generate candidates by modifying phrases and setting sieves to acquire terminologies from the UMLS system. Then the mentions are normalized through a flow of processes. Firstly, if mentions of testing data once appeared in the training data, the ontologies of mentions in training data are referred to. Secondly, we exert the exact matching between mentions and candidates after modifying phrases. Thirdly, a new deep neural network with an attention mechanism is designed for normalizing. If the above three methods failed to link the mention to its ontology, the first candidate in its candidate set will be recommended. The experimental results show our hybrid model performs better than both rule-based and deep learning methods and it also outperforms the baseline model of MCN corpus.

Keywords Clinical concept normalization · Attention mechanism · Deep learning · Hybrid model

---

C. Tang · W. Chen · T. Wang · P. Li · J. Jiang · Y. Guan (✉)
Harbin Institute of Technology, Harbin, China
e-mail: guanyi@hit.edu.cn

C. Sun
Harbin Engineering University, Harbin, China

C. Tang · W. Chen · C. Sun · T. Wang · P. Li · J. Jiang · Y. Guan
Shenzhen Research Institute of Big Data, 518000 Shenzhen, China

S. C. Satapathy et al. (eds.), *Intelligent Data Engineering and Analytics*, Advances in Intelligent Systems and Computing 1177,
https://doi.org/10.1007/978-981-15-5679-1_55

1 Introduction

Clinical Concept Normalization is a task to link clinical concepts in narratives to their corresponding terminologies. MCN corpus [1] is the newest corpus for medical concept normalization and our evaluation is based on it.

The first step of clinical concept normalization is to generate enough terminologies from controlled vocabularies to include probable ontologies as many as possible for linking mentions. This paper generates candidates based on [1] and [2]. We utilize sieves and modifying rules to acquire candidates from the UMLS [3]. UMLS is an authoritative system to index biomedical controlled vocabularies.

The next step is to find the most probable candidate for the searched mention. Many entity linking methods are tried in previous works. In particular, the sieve-based [2] and ensemble way [4] are rule-based methods. In recent years, deep learning started to become mainstream. Li [5] used Convolutional Neural Network (CNN [6]) and Nguy˜ên [7] used pairwise learning plus CNN to link medical concepts. NormCo [8] used Gated Recurrent Unit (GRU) [9]. However, to our knowledge, no one tried to combine traditional rule-based methods with deep learning, by which both traditional and deep learning methods can compensate each other. Thus, unlike existing systems, this paper presents a hybrid model for clinical concept normalization and its result exceeds the baseline on MCN corpus.

The prior contributions of our work can be concluded as follows: (1) A novel deep learning model employing attention mechanism [10] and Bert [11] is designed for clinical concept normalization. (2) Unlike existing methods, a hybrid model for clinical concept normalization is proposed. (3) Ways for candidate generation and phrase modifying are described in detail for the convenience of the following work.

2 Methodology

The whole process is described in Fig. 1. The recognized clinical concepts like "HIV positive" have many candidate ontologies in UMLS, such as "HIV Seropositivity", "test; HIV, positive, asymptomatic" and so on. The goal of the task is to find the actual ontology for the mention. The way to evaluate the result of the normalization

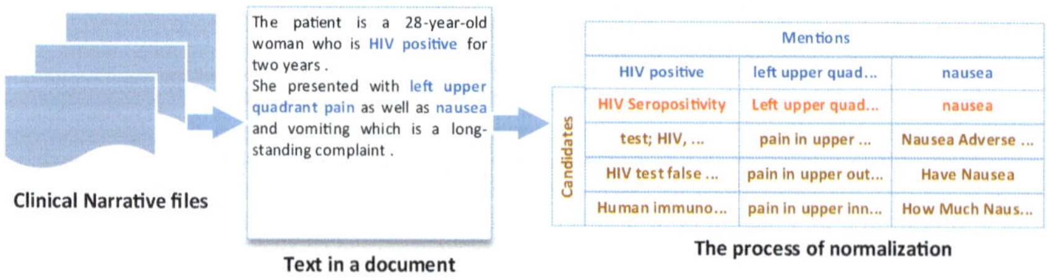

Fig. 1 The process of clinical concept normalization

is to compare the CUIs of mentions with that of linking results. For example, by comparing the CUIs, we can know the ground truth for linking "HIV positive" is "HIV Seropositivity". The following part 2.1 is to describe how to generate candidates for the mentions, and part 2.2 is to find the most probable candidates for linking.

2.1 Candidate Generation

Our candidate generation follows the work of Luo [1] and D'Souza [2]. Except for original 10 sieves, additional two rules are added to help generate candidates from UMLS. They can be concluded as follows:

1. Delete punctuations, digits, and common words. We define the common words as the set {"your", "his", "her", "their", "patient", "a", "an", "of", "any", "this", "these", "that", "those", "the", "be", "all", "feed"}. Those words appear due to the i2b2 phrase annotation policy and will incur some noise in clinical concept normalization.
2. Check spelling. Because clinical concepts are recognized from free text, some mentions are inevitably along with some spelling errors. For example, Terminology "Aztreonam" is wrongly spelled as "Aztrenam". We rectify them by aligning mentions with a medical dictionary.[1]

According to our error analysis, through incremental sieves and phrase modifying rules, our methods can largely decrease the ratio of mentions whose ground truth ontologies are not in their candidate sets.

2.2 Candidate Linking

Figure 2 shows the procedures for candidate linking. The deep learning model is based on NormCo, but it has some creative improvement. For better description, let us introduce the notation. First of all, let m_i be the mention waiting for normalization and let $\{c_0^i, c_1^i, \ldots\}$ be its corresponding candidate set. We use Bert-as-service[2] to transform mentions into phrase embeddings, because in recent years large-scale pretrained models like Bert boosted the performance in various Natural Language Processing (NLP) tasks. Then we can get $\{\ldots, \Phi(m_{i-1}), \Phi(m_i), \Phi(m_{i+1}), \cdots\}$ and $\{\ldots, \Phi(c_0^i), \Phi(c_1^i), \ldots\}$. Meanwhile, through GRU units, we can get the output sequence of $\{\ldots, \psi(m_{i-1}), \psi(m_i), \psi(m_{i+1}), \ldots\}$ as well and this process is the same as that of NormCo (Fig. 3).

Thereafter, the multi-head attention mechanism [10] is put onto the co-mention sequence $\{\ldots, \psi(m_{i-1}), \psi(m_i), \psi(m_{i+1}), \ldots\}$. Every $\psi(m_i)$ is multiplied by a

[1] https://github.com/glutanimate/wordlist-medicalterms-en.

[2] https://github.com/hanxiao/bert-as-service.

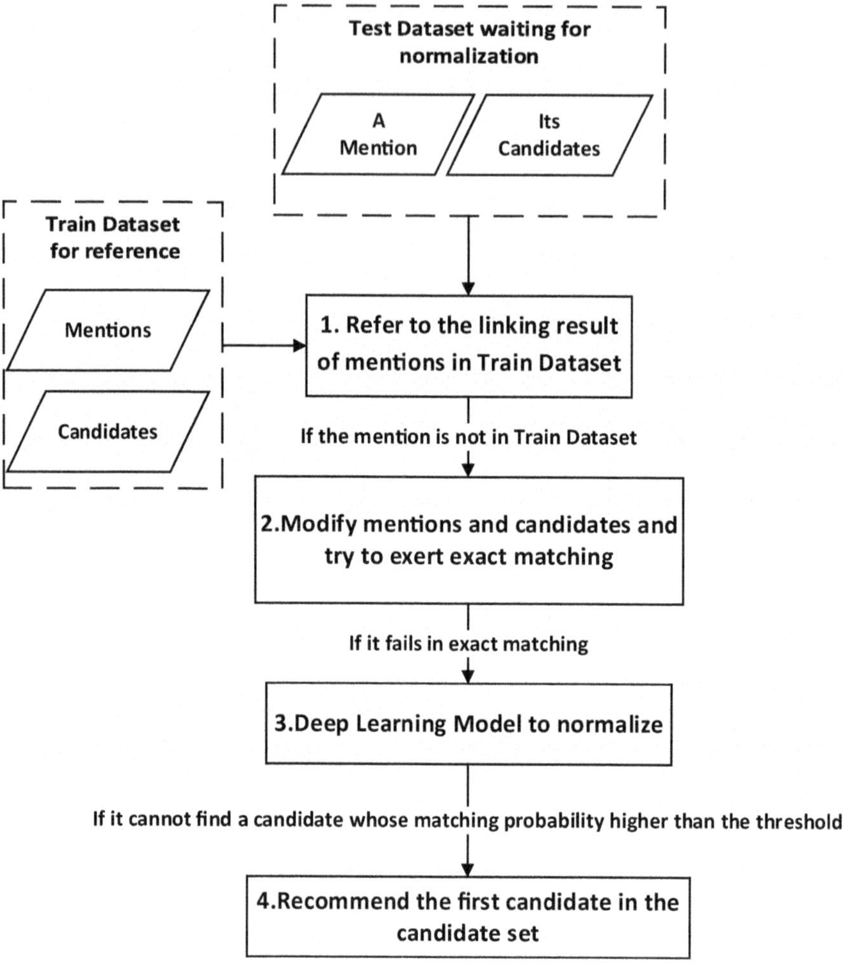

Fig. 2 The process of candidate linking

corresponding weight W_i. In Eq. 1, the input of attention mechanism is the concatenation of $\psi(m_i)$ and $\Phi(\mathrm{c}_\mathrm{j}^\mathrm{i})$. In Eq. 2, the next step is to project the input into a higher dimension matrix by multiplying W_1, where I denotes the length of Input and D is a hyperparameter of dimension. Through the projection, neural network can learn more complicated representation. Then by multiplying W_2, the matrix is projected to a lower dimension matrix with hyperparameter H as the multi-head of attention mechanism and we get M_i as the output.

$$\mathrm{Input} = \psi(\mathrm{m_i}) \| \Phi\left(\mathrm{c}_\mathrm{j}^\mathrm{i}\right) \tag{1}$$

$$M_i = \mathrm{Input}^{[1,I]} \times W\_1^{[I,D]} \times \mathrm{W\_2}^{[D,H]} \tag{2}$$

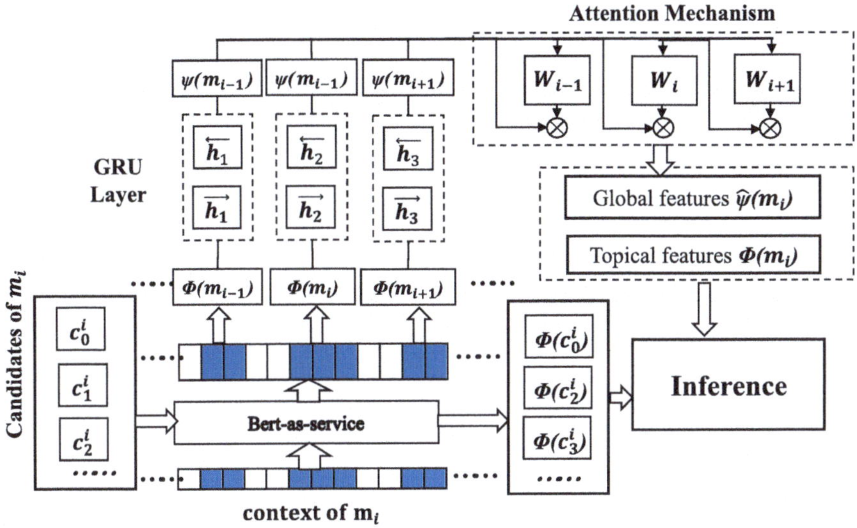

Fig. 3 The architecture of deep learning model

In Eq. 3, passing through the Softmax function, we can get the weights for the sequence $\{\ldots, \psi(m_{i-1}), \psi(m_i), \psi(m_{i+1}), \ldots\}$. Not forget to mention, the dimensions of M_i are $[1, H]$, where H, the multi-head of attention, means different weight sequences for multiplying.

$$\{\ldots, W_{i-1}, W_i, W_{i+1}, \ldots\} = \text{Softmax}(\{\ldots, M_{i-1}, M_i, M_{i+1}, \ldots\}) \tag{3}$$

Let us dub the matrix of $\{\ldots, \psi(m_{i-1}), \psi(m_i), \psi(m_{i+1}), \ldots\}$ as S and the matrix of $\{\ldots, W_{i-1}, W_i, W_{i+1}, \ldots\}$ as W. Then we get the global feature $\hat{\psi}(m_i)$ by Eq. 4, where Dense means fully connected layer.

$$\hat{\psi}(m_i) = \text{Dense}\left(W^T S\right) \tag{4}$$

Finally, $\Phi(m_i)$, $\psi(m_i)$ and $\Phi(c_j^i)$ are combined into the feature vector. To make the features from m_i and c_j^i with equal dimensions, we catenate two $\Phi(c_j^i)$. We get the matching probability between m_i and c_j^i by Eq. 5.

$$p\left(m_i, c_j^i\right) = \text{Softmax}\left(\text{Dense}\left(\text{concatenate}\left(\Phi(m_i), \psi(m_i), \Phi\left(c_j^i\right), \Phi\left(c_j^i\right)\right)\right)\right) \tag{5}$$

The final linking result is dependent on those corresponding probabilities of mention-candidate pairs by Eq. 6.

$$\hat{c} = \text{argmax}\left\{p\left(m_i, c_0^i\right), p\left(m_i, c_1^i\right) \ldots, p\left(m_i, c_j^i\right)\right\} \tag{6}$$

Table 1 The statistics of MCN Corpus

	Documents	Mentions/identifiers	Unique mentions	Unique identifiers
Train	50	6684	3438	2331
Test	50	6925	3733	2579
Total	100	13609	6032	3792

2.3 *Training*

In the hybrid model, only deep learning parts need to be trained. The Loss Function is defined in Eq. 7, where cross-entropy function is used to compute the loss between the matching probability and the result of matching ground truth with candidate.

$$L = \frac{\sum_{c_j^i \in C^i} \text{cross\_entropy}\left(p\left(m_i, c_j^i\right), \text{is\_match}\left(m_i, c_j^i\right)\right)}{\text{sizeof}\left(C^i\right)} \tag{7}$$

C^i stands for the candidate set of m_i.. If the CUI of m_i is equal to the CUI of c_j^i, is_match$\left(m_i, c_j^i\right)$ is assigned with 1, otherwise with 0.

3 Experimental Results

3.1 *Corpus*

The MCN corpus utilizes part of the i2b2 2010 dataset [12], in which clinical concepts are annotated as tests, treatments, and medical problems. The statistics of MCN Corpus are reported in Table 1.

3.2 *Evaluation*

We use accuracy as the evaluation metric to evaluate our hybrid model on MCN Corpus. The evaluation result is displayed in Table 2. "EM" stands for exact match and "wo-Com" stands for removing common word tokens. "GRU + Attention" stands for our deep learning model. MetaMap [13] is an official normalization tool released by UMLS.

Table 2 The comparison of clinical normalization results in terms of accuracy

Model name	Accuracy (%)
EM-UMLS-wo-Com	76.65
MetaMap-wo-Com	76.35
EM-train	60.45
GRU + attention	75.12
EM-train + UMLS	73.50
EM-train + UMLS + GRU + attention	77.94

4 Analysis

The models above the dashed line are the baseline models. The models listed below the dashed line are all preprocessed with phrase modifying methods given in Sect. 2.1. However, the accuracies of our regenerated models are lower than MCN [1]. EM-Train is only 60.45 compared with 76.27 in MCN. EM-Train + UMLS is 73.50 compared with 76.35 in MCN. The cause may be due to the poor performance of candidate generation or others. In this case, our hybrid model still performs well and the accuracy of 77.94% outperforms the baseline, which shows the advantage of combing multiple methods in normalization.

4.1 Merit Analysis

Through analysis of the normalization result, we find that the deep learning model performs better than rule-based methods on some mentions, especially those mentions largely different from their linking ground truths. However, in terms of the mentions which rule-based methods can correctly normalize, deep learning model fails to normalize part of them. For example, "resection" appeared in the train dataset and its ground truth is "Excision", which is also the ground truth of "resection" in the test dataset. It can be easily found by referring to the training dataset, but the neural network failed to normalize it. On the other hand, the neural network successfully normalizes the mention "DDI" to "Didanosine", for which rule-based methods are unable to. Thus, we combine different ways of normalization to make them work at the mention groups they perform best. In this way, the overall accuracy is higher than either way.

When it comes to the deep learning model, we exploit the large-scale pretrained model to make the representation learning more powerful and it performs better than the general word embeddings we use in our experiments.

Another novel point we make is the attention mechanism as Fig. 4 illustrates. Every mention from the context co-mention sequence is assigned with a weight to measure its importance for normalizing the mention "HIV seropositivity". The initial state of attention matrix is displayed in the left picture and through training,

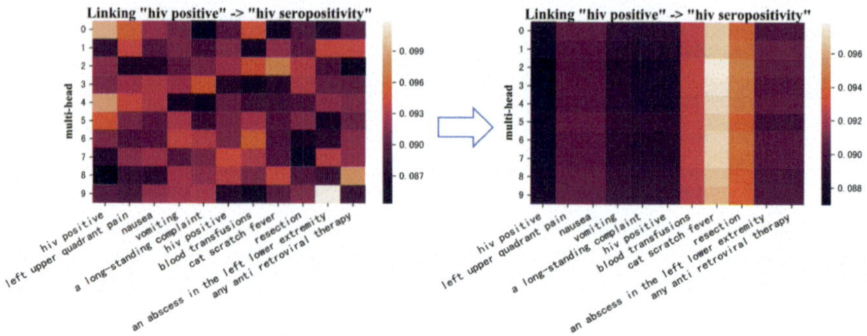

Fig. 4 Attention matrix for "HIV positive" before training and after training

we finally get the attention matrix displayed in the right picture. By allocating the importance index for the context, we can more efficiently and effectively utilize the context information.

4.2 Demerit Analysis

The demerits of the present model have been described in the above part. It can be concluded as follows:

(1) Not effective enough candidate generation and matching rules. The regeneration result of MCN shows the performance of EM-Train is a bit lower than expected. More research should be taken into the rule-based part.

(2) More improvement should be carried in the architecture of the deep learning model. The neural network still has some problems to be solved. Taking "Prinovil", for example, its corresponding ontology is "Prinivil", which is a simple spelling mistake. However, the neural network judges "Pharmacology" as the result of normalization with a very high probability. This means the neural network is not powerful enough to replace rule-based ways. After carrying deep research into the structure of the neural network, we may make it much more effective and robust on clinical normalization.

(3) More research should be done on how to better use the attention mechanism. Our new model is only one method to use attention mechanism and its effectiveness should be verified by more comparative experiments. The importance index system is good from the human's perspective but lacks a scientific evaluating system. Thus, the combining of attention mechanism and neural networks is not reliable at present, waiting for more research work.

5 Conclusion and Future Work

In this paper, firstly we exploit the effectiveness of the hybrid model in the Clinical Concept Normalization task. Secondly, we explore new ways to improve the existing deep learning models, such as large-scale pretrained models and attention mechanism. The performance of our hybrid model is more excellent than both the rule-based and deep learning methods. It is an effective and efficient way to combine the merits of both research methods. Besides, our regeneration result is lower than MCN and the accuracy of present models on MCN are relatively low compared with other similar shared tasks and corpora. That means our model still has a large improvement space. In future work, we will raise the effectiveness and efficiency of the candidate generation model. Besides, more front-edge methods should be taken into consideration such as multi-task learning and meta-learning.

Acknowledgements This research was financially supported by the Open Research Fund from Shenzhen Research Institute of Big Data, under Grant No. 2019ORF01011. Meanwhile, Xiang Wan and Tsung-Hui Chang also gave us much help.

References

1. Luo, Y.F., Sun, W., Rumshisky, A.: MCN: A comprehensive corpus for medical concept normalization. J. Biomed. Inform. **92**, (2019). https://doi.org/10.1016/j.jbi.2019.103132
2. D'Souza, J., Ng, V.: Sieve-based entity linking for the biomedical domain. In: ACL-IJCNLP 2015—53rd Annual Meeting of the Association for Computational Linguistics and the 7th International Joint Conference on Natural Language Processing of the Asian Federation of Natural Language Processing, Proceedings of the Conference, pp. 297–302 (2015)
3. Bodenreider, O.: The Unified Medical Language System (UMLS): Integrating biomedical terminology. Nucleic Acids Res. **32**, D267–D270 (2004). https://doi.org/10.1093/nar/gkh061
4. Rajani, N.F., Bornea, M., Barker, K.: Stacking with auxiliary features for entity linking in the medical domain. BioNLP. 39–47 (2017). https://doi.org/10.18653/v1/w17-2305
5. Li, H., Chen, Q., Tang, B., Wang, X., Xu, H., Wang, B., Huang, D.: CNN-based ranking for biomedical entity normalization. BMC Bioinform. **18**, (2017). https://doi.org/10.1186/s12859-017-1805-7
6. Kim, Y.: Convolutional neural networks for sentence classification. In: EMNLP 2014—2014 Conf. Empir. Methods Nat. Lang. Process. Proc. Conf. 1746–1751 (2014). https://doi.org/10.3115/v1/d14-1181
7. Nguy˜ên, T.N., Nguy˜ên, T.M., Đặng, T.H.: Disease named entity normalization using pairwise learning to rank and deep learning
8. Wright, D., Katsis, Y., Mehta, R., Hsu, C.-N.: NormCo: deep disease normalization for biomedical knowledge base construction (2019)
9. Chung, J., Gulcehre, C., Cho, K., Bengio, Y.: Empirical evaluation of gated recurrent neural networks on sequence modeling. arXiv Prepr.: arXiv1412.3555 (2014)
10. Vaswani, A., Shazeer, N., Parmar, N., Uszkoreit, J., Jones, L., Gomez, A.N., Kaiser, Ł., Polosukhin, I.: Attention is all you need. In: Advances in Neural Information Processing Systems, pp. 5999–6009 (2017)
11. Devlin, J., Chang, M.-W., Lee, K., Toutanova, K.: BERT: pre-training of deep bidirectional transformers for language understanding (2018)

12. Uzuner, Ö., South, B.R., Shen, S., DuVall, S.L.: 2010 i2b2/VA challenge on concepts, assertions, and relations in clinical text. J. Am. Med. Inform. Assoc. **18**, 552–556 (2011). https://doi.org/10.1136/amiajnl-2011-000203
13. Aronson, A.R.: Effective mapping of biomedical text to the UMLS Metathesaurus: the MetaMap program. In: Proceedings/AMIA Annual Symposium. AMIA Symposium, pp. 17–21. American Medical Informatics Association (2001)

Classification of Text Documents of an Electronic Archive Based on an Ontological Model

Anton Zarubin, Albina Koval, and Vadim Moshkin

Abstract A modern design organization has a significant electronic archive of documents in an unstructured form. Solving the problem of using the experience of previous projects to solve new problems can be based on the use of intelligent methods and algorithms for analyzing text documents of an organization in order to build a classification system for electronic archives. This work presents an ontological model of a text document as an electronic archive resource. The paper also presents an ontologically oriented classification algorithm for technical documents. In conclusion, the results of experiments confirming the effectiveness of models and algorithms in solving the problem of classifying a document archive are presented.

Keywords Electronic archive · Ontology · Classification · Graph

1 Introduction

Most of the documentation in large project organization is presented in textual unstructured form. This archive of textual documentation contains the experience and knowledge of a large number of highly qualified specialists.

The analysis of documents by predetermined details is difficult when increasing the volume of the electronic archive. Extensive knowledge of the subject area and skills in the field of semantic processing of a large amount of text documentation from decision-makers are required. As a result, often the important experience of

A. Zarubin · A. Koval
The Bonch-Bruevich Saint-Petersburg State University of Telecommunication, Saint-Petersburg, Russia
e-mail: azarubin@sut.ru

A. Koval
e-mail: akoval@sut.ru

V. Moshkin (✉)
Ulyanovsk State Technical University, Severny Venets Str., 32, Ulyanovsk 432027, Russia
e-mail: postforvadim@ya.ru

S. C. Satapathy et al. (eds.), *Intelligent Data Engineering and Analytics*, Advances in Intelligent Systems and Computing 1177,
https://doi.org/10.1007/978-981-15-5679-1_56

previous projects, recorded in electronic archives, remains unclaimed and the time taken to complete the decision-making process increases.

The solution to this problem can be based on the use of intelligent methods and algorithms for analyzing text documents of an organization.

Presentation of the contents of the archive in the form of classes containing text documents that are similar in topic reduces the search space and speeds up the search for the required documents by their content.

For the effective application of text analysis mining methods, it is not enough to consider a separate document as a set of terms from a limited subject area.

Taking into account the specifics of domain knowledge leads to the need to form an ontology of a special structure. This ontology includes a system of concepts of a subject domain, semantic relations between them, and interpretation functions. Thus, the electronic archive of the organization must possess the properties of an intellectual system.

In [1], the ant colony optimization algorithm is employed to classify the data mining and applied to the quick retrieval of intelligent archives big data.

In [2 and 3], the authors present an ontology designed to model the archival description of historical document collections. In [4], the authors present semantic web-related aspects of current digital library activities, and introduce their functionality; they show examples ranging from general architectural descriptions to detailed usages of specific ontologies. This paper [5] proposes a semantic search portal for cross-media cultural archives involving document, image, audio, and video.

Leading researchers in the field of ontological systems [6–8] note the relevance of studies based on the ontological approach. However, at present, there is a need for mathematical methods and algorithms to classify the contents of the electronic archive of text documents based on their content, taking into account the specifics of the subject area.

Therefore, the development of models, methods, and algorithms for constructing a system for classifying the contents of an electronic documentation archive based on a domain-specific ontology is relevant.

2 Basic Requirements for the Ontology of the Electronic Archive

2.1 *Structural and Analytical Model of Applied Ontology of the Electronic Archive*

Formally, the ontology of the electronic archive of text documentation is

$$O = C, W, R^D, F^D, \tag{1}$$

where C is a set of concepts of an intelligent archive;

W is a set of terms describing the concepts of the subject area;
R^D is a set of relations:
The set of concepts C is written as follows:

$$C = \left(C^{A_1} \cup C^{A_2} \cup \cdots \cup C^{A_k}\right) \cup C^E,$$

where $C^{A_i}, i = \overline{1, k}$—is the set of concepts of the ith subject area;
C^E is a set of concepts of the subject area, extracted from the text, which is already contained in the electronic archive of the organization.

Formally, a set of interpreting functions is

$$F^D = \left\{F^D_{WC}, F^D_{C^E C^A}\right\},$$

where $F^D_{WC} : \{W\} \to \{C\}$ is a function that associates a set of terms with a subset of the concepts of a subject domain, defined algorithmically;

$F^D_{C^E C^A} : \left\{C^E\right\} \to \left\{C^A\right\}$ is the interpretation function of a subset of concepts of electronic archive documents, which allows you to go to the level of concepts of certain subject areas.

2.2 *Formation of the Terminological Environment of the Concepts of Ontology of the Electronic Archive*

The terminological environment of an ontology is a set of terms (words) from technical documents of the electronic archive that are semantically close to this concept. The definition of the semantic distance between the concept and the terms of the document will be determined based on the idea of the analysis of semantic relations presented in [9], which consists of using the distance between words.

In [9], the relationship between a concept defined by an expert and at the same time a document term and a "regular" term located in one sentence should differ from the relationship between a concept and a term from two different paragraphs.

In addition, if an idea is repeated in several paragraphs, then it can be considered more important than if it were fixed in one paragraph. Formally, the semantic coefficient of the relationship (semantic distance) between the concept and the term is

$$S\left(c_i^{P(S)}, w_j\right) = \frac{\sum_{\text{occur}\left(c_i^{P(S)}, w_j\right)} \frac{1}{\exp(\text{sentence} \cdot (\text{paragraph}+1))}}{\text{num}\left(\text{occur}\left(c_i^{P(S)}, w_j\right)\right)} \times \frac{\text{num}\left(\text{paragraph} - \text{cooccur}\left(c_i^{P(S)}, w_j\right)\right)}{\text{num(total paragraph)}}$$

where c_i, w_j is the ith concept of ontology and jth term, respectively;

sentence is a distance as the number of sentences between a concept and a term;
paragraph is the distance as the number of paragraphs between the concept and the term;
$\text{num}(\text{occur}(c_i, w_j))$ is the number of matches c_i и w_j;
$\text{num}(\text{paragraph} - \text{cooccur}(c_i, w_j))$ is the number of paragraphs where c_i и w_j are together;
num(total paragraph) is—the number of paragraphs in the document.

It is necessary to determine the semantic distances between the selected concept and the terms of the document. Then you need to define a subset of terms that are closely and semantically related to the concept.

We use the λ-compactness hypothesis [10] in determining the terminological environment. This hypothesis is based on the concept of λ-distance, which takes into account the normalized distance d between the terms and the characteristic τ of the local density of terms in the vicinity of these elements.

If we determine the semantic distances between all pairs of terms, among which the terminological environment is determined, then we can construct a complete graph connecting all terms with all and find the longest edge. This edge is the diameter of the graph (D).

We select any two terms w_i and w_j and denote the length of the edge connecting them (semantic distance) $\alpha(w_i, w_j)$. The normalized distance between these terms is $d = \frac{\alpha}{D}$.

Find the edge among the edges adjacent to the edge (w_i, w_j). Its length is β_{min}. The ratio of the lengths of these adjacent segments is $\tau^* = \frac{\alpha}{\beta_{\text{min}}}$. Find the largest value of τ_{max} in the full graph to make this value normalized.

The quantity $\tau = \frac{\tau^*}{\tau_{\text{max}}}$ is the normalized characteristic of the local heterogeneity of the density of the set in the vicinity of the ontology terms w_i and w_j. The quantity $\lambda = f(\tau, d)$ is the λ-distance between the terms w_i and w_j. It is proposed to use the value $\lambda = \tau^2 \times d$ in [10] as a measure of this distance.

To determine the terminological environment of the concept of ontology at the design level it is necessary:

1. Define many terms related to the concept of ontology.
2. Define a variety of terms that are not included in the environment of this concept.
3. Determine the edge (w_i, w_j), along which the boundary separating these sets passes.

From here, formally, the final criterion (the λ-KRAB algorithm), showing the quality of separation of many terms, is

$$F = h^4 \tau^2 d \rightarrow \max,$$

where the criterion for the equidistance of these classes of terms is

$$h = 2 \cdot \frac{m^+}{m} \cdot \frac{m^-}{m},$$

where m^+ is the number of terms included in the terminological environment;

m^- is the number of other terms.

A subset of terms is defined using the λ-compactness hypothesis. This subset of terms is included in the terminological environment of the concept under consideration.

Formally, each terminological environment W^k of the concept c_k is

$$\left\{\left(w_1^k, f_1^k\right), \left(w_2^k, f_2^k\right), \ldots, \left(w_i^k, f_i^k\right), \ldots, \left(w_{l_k}^k, f_{l_k}^k\right)\right\},$$

where w_i^k is the ith term of the kth concept of ontology;

l_k is the total number of terms associated with the kth concept;

f_i^k is the normalized semantic weight of the ith term in the terminological environment of the kth concept (normalized semantic distance between the term and the concept within the framework of one terminological environment).

3 Ontological Model of a Text Document as an Electronic Archive Resource

The "connect_to" relationship defines a document's membership in a variety of domain concepts. The frequency of occurrence of the ith term in the document is used to determine the weight of the term f_i in the document. We will use the following rules to formulate estimates of terms:

High-frequency terms are not specific. However, high-frequency terms give a large number of matches when comparing query and document terms.

- Low-frequency terms do not make a big contribution to the search for the necessary documents, since rare terms give a small number of matches.
- The best condensation terms are terms that are not too rare and not too frequent [11, 12].

Formally, the dependence is

$$f_i = tf_i df_i = tf_i \cdot \log\left(\frac{N}{df(w_i)}\right),$$

where $tf_i df_i$ is the relative importance of the term w_i in the document;

tf_i is the normalized frequency of occurrence of the term w_i;

N is the number of all documents;

$df(w_i)$—the number of documents containing the term w_i.

3.1 Ontological Model of the Document

An ontological model of a document is a representation of the document corresponding to the state of the applied ontology of the electronic archive.

Formally, the document model is

$$oV_j^d = C^E, C^A,$$

where C^E, C^A is a subset of the concepts of the subject area of the jth document.

We write the presentation of the document from a formal position regarding the ontology (expression 1). Let the following conditions be satisfied for document d:

$$\begin{aligned} &\text{connect\_to}(d, C_1) = 1, \\ &\text{connect\_to}(d, C_2) = 1, \\ &\qquad\vdots \\ &\text{connect\_to}(d, C_k) = 1. \end{aligned}$$

This set of conditions means that document d is mapped into the concept space $C_1, C_2, \ldots, C_k$. Let w_i^d be the ith term of document d. Then the set of terms of document d will have the following form:

$$W^d = \{w_1^d, w_2^d, \ldots, w_l^d\},$$

where l is the number of terms in document d.

4 Classification of Ontological Representations of Electronic Archive Documents

The ontology-based document indexing algorithm was applied from [13].

It is necessary to define many concepts of ontology and a linguistic label to define classes. Electronic archive documents are divided into these classes. Linguistic label is used to determine the severity of concepts in the class.

The linguistic label defines the semantic representation for the interval of the degree of expression of the concept of ontology. For example, the linguistic label "High" may correspond to the value of the degree of expression of the concept from the interval from 0.7 to 1.0.

Thus, at the first step of the electronic archive classification algorithm, it is necessary to specify a set of G classes and determine their properties:

$$G = \{g_1, g_2, \ldots, g_i, \ldots, g_n\},$$

$$g_i = \{c_1, m, \ldots, c_k, m\},$$

$$m \in \left[\text{High, Middle, Low}\right],$$

$$\text{High} = [0.7 \ldots 1.0],$$

$$\text{Middle} = [0.5 \ldots 0.7),$$

$$\text{Low} = [0 \ldots 0.5),$$

where g_n is the nth class of documents (classification basis); c_k is the kth concept of ontology; m is the linguistic label.

At the second step, the degree of belonging of document d to each class g_i is calculated using the following expression:

$$s(g_i) = k - \sum_{i=1}^{k} (1 - \theta_k),$$

where k is the number of parameters of the class g_i;

θ_k is a sign of the correspondence of document d to the kth property of the class g_i, which is calculated using the following expression:

$$\theta_k = \begin{cases} 1, c_k \in d, \mu(c_k) \in m \\ 0 \end{cases}$$

Thus, document d corresponds to the θ_k attribute only if it contains concepts that characterize the given attribute and its degree of expression is included in the linguistic label interval.

5 Experiments

The test set consists of technical documents from the electronic archive of research and production association "MARS" (Ulyanovsk, Russia). This organization is developing software and hardware for warships. The test set consists of the following types of documents:

- Technical specifications;
- Reports on patent research;
- Testing programs and techniques;
- Manuals of the programmer, user, system administrator, etc.

Table 1 Assessment of the classification quality

Type of expert partition	Index type	Classification Time (s)	The value of the evaluation function
Documentation class	Ontological representation	53.2	0.63
	Classic index	810.6	0.80
Type of documentation	Ontological representation	900.2	0.45
	Classic index	15,548.4	0.46
Documentation section	Ontological representation	807.8	0.43
	Classic index	22,307.6	0.41
Subject of work	Ontological representation	807.8	0.48
	Classic index	22,307.6	0.21

The test set consists of 1037 design documents. The expert divided this sample into four grounds:

- According to the class of documentation (3 classes);
- By type of documentation (16 classes);
- In the documentation section (22 classes);
- On the subject of work (22 classes).

Based on the test set of documents, a set of indices was built, including ontological representations of documents, as well as classic indices containing pairs {term, frequency}. We used the criteria from [14] to evaluate the quality of the classification. Assessment of the quality of classification is presented in Table 1.

The classification of ontological representations is faster (up to 27 times) in comparison with the time of classification of classical indices. The quality of classification of ontological representations is slightly worse only when divided by the class of documentation.

The quality of classification of ontological representations is higher than that of classical indexes when dividing a multitude of documents by type of documentation, section of documentation, and subject of work.

6 Conclusions

Taking into account the specific knowledge of the subject area in which the organization operates, leads to the need for the formation of the ontology of a special structure, which includes a system of concepts of the subject area and semantic relations between them and interpretation functions.

As can be seen from the results of the experiments, the formation of the ontological representation of each individual document in the archive allows you to improve the quality of the automatic classification of documents.

In the future, we plan to introduce elements of fuzziness into the ontological representation of project documents.

Acknowledgements This paper has been approved within the framework of the federal target project "R&D for Priority Areas of the Russian Science-and-Technology Complex Development for 2014–2020", government contract No 05.604.21.0252 on the subject "The development and research of models, methods and algorithms for classifying large semistructured data based on hybridization of semantic-ontological analysis and machine learning".

References

1. Yong, W., Liming, L., Yongsheng, Q.: Improvement of big data retrieval algorithm in the intelligent archives management, pp. 487–491 (2015). https://doi.org/10.1109/icemi.2015.7494245
2. Pandolfo, L., Pulina, L., Zielinski, M.: Towards an ontology for describing archival resources (2017)
3. Pandolfo, L., Pulina, L., Adorni, G.: A framework for automatic population of ontology-based digital libraries. In: AI * IA 2016 Advances in Artificial Intelligence, pp. 406–417. Springer (2016)
4. Kruk, S.R., McDaniel, B.: Semantic Digital Libraries. Springer (2009)
5. Yan, Z., Scharffe, F., Ding, Y.: Semantic search on cross-media cultural archives. In: Wegrzyn-Wolska, K.M., Szczepaniak, P.S. (eds.) Advances in Intelligent Web Mastering. Advances in Soft Computing, vol. 43, pp. 375–380. Springer, Berlin, Heidelberg (2007)
6. Zagorulko, Y.A.: Semantic approach to the analysis of documents based on the ontology of the subject area. Zagorulko, Y.A, Kononenko, I.S., Sidorova, E.A., Electronics Resource. Access mode: http://www.dialog-21.ru/digests/dialog2006/materials/html/SidorovaE.htm
7. Gavrilova, T.A.: Knowledge Base of Intelligent Systems. St. Petersburg: Peter (2000)
8. Schneider, T., Hashemi, A., Bennett, M., Brady, M., Casanave, C., Graves, H., Grüninger, M., Guarino, N., Levenchuk, A., Lucier, E., Obrst, L., Ray, S., Sriram, R., Vizedom, A., West, M., Whetzel, T., Yim, P.: Ontology for big systems: the ontology summit 2012 communiqué. Appl. Ontol. **7**, 357–371 (2012). https://doi.org/10.3233/AO-2012-0111
9. Serrano-Guerrero, J., Olivas, J.A., de la Mata, J., Garces, P.: Physical and semantic relations to build ontologies for representing documents. In: En Liu, Y., Chen, G., Ying, M. (eds.) Fuzzy Logic, Soft Computing and Computational Intelligence (Eleventh International Fuzzy Systems Association World Congress IFSA), Beijing, China, vol. 1, pp. 503–508. Tsinghua University Press, Springer (2005)
10. Zagoruyko N.G.: Applied methods of data and knowledge analysis. Novosibirsk: IM SB RAS (1999)
11. Yarushkina, N. Moshkin, V., Filippov, A.: Development of a knowledge base based on context analysis of external information resources. In: DS-ITNT 2018. Proceedings of the International conference Information Technology and Nanotechnology, pp. 328–337. Samara, Russia (2018)
12. Namestnikov, A., Filippov, A., Avvakumova, V.: An ontology-based model of technical documentation fuzzy structuring. In: 2nd International Workshop on Soft Computing Applications and Knowledge Discovery. SCAKD (2016)

13. Filippov, A., Moshkin, V., Namestnikov, A., Guskov, G., Samokhvalov, M.: Approach to translation of RDF/OWL-ontology to the graphic knowledge base of intelligent systems. In: Proceedings of the II International Scientific and Practical Conference "Fuzzy Technologies in the Industry—FTI 2018", pp. 44–49. Ulyanovsk (2018)
14. Radionova, Y.A.: A method for constructing an evaluation function that determines the effectiveness of automatic clustering algorithms. Autom. Control Process **15**, 23–28 (2009)

Influence of Followers on Twitter Sentiments About Rare Disease Medications

Abhinav Choudhury, Shruti Kaushik, and Varun Dutt

Abstract Due to Twitter's popularity, companies have started incorporating this social platform into their product marketing. Followers play a pivotal role in the marketing process as they re-tweet recent tweets and spread recent product announcements and achievements of a company among other Twitter users. The main objective of this paper is to understand the influence of followers on the prevailing sentiments about products on Twitter. In this paper, we analyzed the sentiment of tweets of two medications for treating a rare disease. We collected tweets containing the company and medication names. We compared the sentiments of the tweets of the followers with the sentiment of the tweets of the non-followers using Hedonometer scores. We also analyzed differences in sentiments of followers and non-followers among positive and negative tweeted words. Our results indicated that there was no significant difference in the average sentiment of followers and non-followers. For positive words, the tweets of followers were significantly more positive than those of non-followers; however, no such effect was found in tweets containing negative words. We highlight the implications of our results for the product marketing of rare disease medications.

Keywords Sentiment analysis · Rare disease medications · Followers · Non-followers · Hedonometer scores · Twitter

---

A. Choudhury (✉) · S. Kaushik · V. Dutt
Applied Cognitive Science Laboratory, Indian Institute of Technology Mandi, Kamand 175005, India
e-mail: abhinav_choudhury@students.iitmandi.ac.in

S. Kaushik
e-mail: shruti_kaushik@students.iitmandi.ac.in

V. Dutt
e-mail: varun@iitmandi.ac.in

S. C. Satapathy et al. (eds.), *Intelligent Data Engineering and Analytics*, Advances in Intelligent Systems and Computing 1177,
https://doi.org/10.1007/978-981-15-5679-1_57

1 Introduction

Twitter is one of the most popular social media platforms, with an estimated 330 million monthly active users and nearly 500 million tweets per day [1]. With the global popularity of Twitter, companies have started incorporating this social platform as their primary source of product marketing. For example, more than 80% of Fortune 500 companies are active on Twitter, with high follower growth [1].

Followers may play a pivotal role in the marketing process on Twitter as they may re-tweet recent tweets of a company as well as mention recent product announcements of the company for other Twitter users. In fact, followers have been shown to feel more positively about the business than non-followers and they also help expand the company's reach and increase sales [2].

Sentiment analysis of tweets of followers and non-followers is likely to be an effective way of gauging public opinion about products [3]. For example, a company can retrieve timely feedback on a new product in the market by evaluating people's opinions on Twitter.

Although followers have been shown to possess positive sentiments compared to non-followers on Twitter [2], there is little research on sentiment analyses concerning followers and non-followers for medications for treating rare diseases. The World Health Organization (WHO) defines rare diseases as those which affect 0.65–1 out of every 1000 people [4]. This is a negligible number when compared to mainstream diseases. Thus, not many pharmaceutical companies invest in manufacturing rare disease medications. Therefore, it is very important that proper information about these drugs is easily available to patients and caregivers so that they can make informed decisions. An analysis of sentiments among followers and non-followers of rare disease medications on Twitter may help to understand how public perceptions about rare disease medications are shaped.

The primary objective of this paper is to understand the influence of followers on the prevailing sentiments about certain rare disease medications on Twitter. In this paper, we investigate Twitter data concerning two rare disease pharmaceutical companies, A and B, manufacturing rare disease medications, T and O, respectively.[1] We collected thousands of tweets containing the company and medication names. We then performed sentiment analysis on these tweets using Hedonometer analysis and evaluated the sentiments of the tweets of the followers compared to non-followers of companies. As the medications developed by the companies are used to treat rare diseases, we expect followers of the company to have a positive perception regarding the company and its products compared to non-followers.

In what follows, first, we describe the background and related studies in connection to Twitter sentiment analysis. Next, we describe the proposed methodologies for performing sentiment analysis. Section 4 compares the tweet sentiment of the followers with the non-followers and presents the results. Finally, we conclude by discussing the main implications of our results.

[1]The real names of companies and rare disease medications have not been disclosed due to a non-disclosure agreement on this project.

2 Related Work

Sentiment analysis is the process of determining and measuring the tone, attitude, opinion, and emotional state of responses [3]. More precisely, it is the concept of deciding whether a specific tweet is positive, negative, or neutral. Performing sentiment analysis on Twitter data can help companies obtain qualitative insights to understand how people are talking about their brand [2]. In their paper, reference [5] analyzed tweets containing the word "climate" collected between September 2008 and July 2014. They used the sentiment measurement tool called the Hedonometer [6, 7], to determine how collective sentiment varied in response to climate change news, events, and natural disasters. They found that natural disasters, climate bills, and oil-drilling contributed to a decrease in happiness while climate rallies, a book release, and green ideas contests contributed to an increase in happiness. Anger and Kittl [8] used sentiment analysis of Twitter to determine a grounded approach for measuring the social networking potential of individual Twitter users. Reference [9] used geospatial information related to tweets for estimating happiness in Italian cities.

Most customers that use a product may likely be followers of the company on Twitter. As such followers may play a pivotal role in the marketing process as they may likely re-tweet recent tweets of the company as well as mention recent product announcements.

In this paper, we analyze the sentiment of tweets of two companies A and B, which manufacture rare disease medications, T and O, respectively. We compare the sentiments of the tweets of the followers with the sentiment of the tweets of the non-followers to understand if the following provides any positive effect on the perception of users. We also analyze the effect of positive and negative words on the overall sentiment of the tweet for followers and non-followers.

3 Method

3.1 Data

In this paper, we focused on two pharmaceutical companies and the individual medications manufactured by them to treat the same rare disease. We collected all tweets mentioning company A or their Twitter handle (@a) and all tweets mentioning company B or their Twitter handle (@b). The tweets collected for company A were between January 2015 and November 2019 and those for company B were between September 2011 and November 2019. Next, we collected tweets mentioning the individual drugs manufactured by each company to treat the same rare disease; Company A manufactured medicine T; whereas, company B manufactured medicine O. The tweets collected for T were between April 2018 and November 2019 and those for medicine O were between November 2013 and November 2019.

3.2 Preprocessing of the Tweets

First, a filter was placed on each tweet so that only those tweets that contained the keywords (O, T, A, or B) or the Twitter handles of the respective company remained. Then, the tweets with insufficient information and the ones containing non-English words were omitted. After excluding the unrelated tweets, we separated the tweets based on followers and non-followers of each company. Comapny B and Comapny A had approximately 11,000 and 1300 followers respectively.

3.3 Sentiment Analysis

Next, we performed sentiment analysis on the tweets of the followers and non-followers using Hedonometer analysis. The Hedonometer is designed to calculate a happiness score for a large collection of text, based on the happiness of the individual words used in the text. The instrument uses sentiment scores collected by references [6, 7], where 10,222 of the most frequently used English words in four disparate corpora were given happiness ratings using Amazon's Mechanical Turk online marketplace. As people rated words as positive or negative, the scale is considered to be acceptable and accurate. The Hedonometer score of each tweet for both followers and non-followers was computed, and then this score was averaged over the total number of followers and non-followers for each company and each product. Furthermore, we collected all tweets containing positive words (Hedonometer score > 5) and negative words (Hedonometer score < 5) and calculated the average Hedonometer score of these tweets for both followers and non-followers. Lastly, we performed a t-test analysis [10] to determine if there was a significant difference between the average Hedonometer score of followers and non-followers.

4 Results

4.1 Sentiment Analysis

We carried out an independent sample t-test to compare the average sentiment (defined using the Hedonometer score) of tweets tweeted by the followers and non-followers of two companies and their products. First, we found that there was no significant difference in the mean Hedonometer score of followers of A compared to the non-followers of A (follower A HScore ($\mu = 5.42$, $\sigma = 0.47$) ~ non-follower A HScore ($\mu = 5.41$, $\sigma = 0.31$); $t(3850) = 0.39$, $p = 0.70$, $r = 0.010$). Similarly, we found that there was no significant difference in the Hedonometer score of tweets of the followers of B compared to the non-followers of B (follower B HScore ($\mu = 5.42, \sigma = 0.44$) ~ non-follower B HScore ($\mu = 5.41, \sigma = 0.34$); $t(38710) = 0.34, p = 0.73, s = 0.00$).

Next, we investigated the sentiments of tweets related to individual medicines manufactured by each company. In case of T, we did not find a significant difference in the Hedonometer score in tweets of non-followers of medication T compared to the tweets of followers of medication T (follower T HScore ($\mu = 5.32, \sigma = 0.30$) ~ non-follower T HScore ($\mu = 5.39, \sigma = 0.38$); $t(396) = -1.53, p = 0.13, r = 0.00$). Lastly, we investigated the tweet sentiment of followers and non-followers of company B's medication O. There was no significant difference in the average Hedonometer score of tweets tweeted by followers of medication O and the average Hedonometer score of tweets tweeted by non-followers of medication O (follower O HScore ($\mu = 5.40, \sigma = 0.37$) ~ non-follower O HScore ($\mu = 5.38, \sigma = 0.33$); $t(3389) = 1.68, p = 0.09, r = 0.02$).

4.2 Sentiment Analysis Using Positive and Negative Words

Next, we calculated the average Hedonometer score of the tweets containing only positive or negative words for both followers and non-followers of the two companies. As shown in Fig. 1, for positive words, the average Hedonometer scores of tweets of followers of A were greater compared to non-followers of A (follower A positive HScore ($\mu = 5.70, \sigma = 0.37$) > non-follower A positive HScore ($\mu = 5.20, \sigma = 1.57$); $t(189) = 2.09, p = 0.03, r = 0.15$). Next, we investigated the tweets containing negative words and found that there was no significant difference in the average Hedonometer score of tweets of followers and non-followers of company A (follower A negative HScore ($\mu = 4.76, \sigma = 1.11$) ~ non-follower A negative HScore ($\mu = 5.0, \sigma = 1.93$); $t(55) = -0.376, p = 0.70, r = 0.05$).

As shown in Fig. 2a, for positive words, we found that the average Hedonometer scores of tweets of followers of B were greater compared to non-followers of B (follower B positive HScore ($\mu = 5.63, \sigma = 0.54$) > non-follower B positive HScore ($\mu = 4.61, \sigma = 2.17$); $t(40{,}448) = 33.70, p = 0.00, r = 0.16$). Next, we investigated the tweets containing negative words and found that the average Hedonometer score of tweets of followers of B was greater compared to non-followers of B (follower B

Fig. 1 Average Hedonometer score of tweets for company A containing positive words

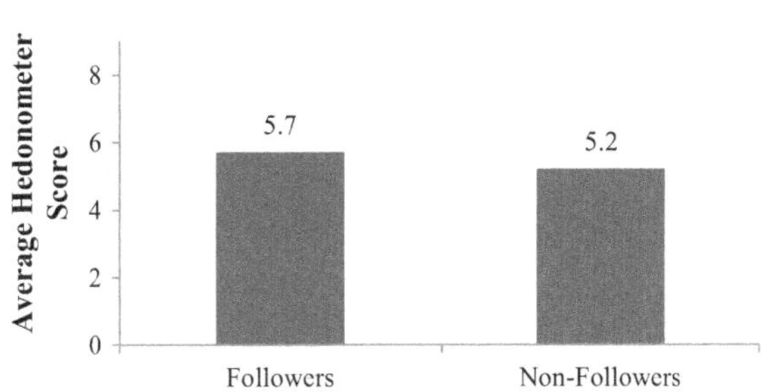

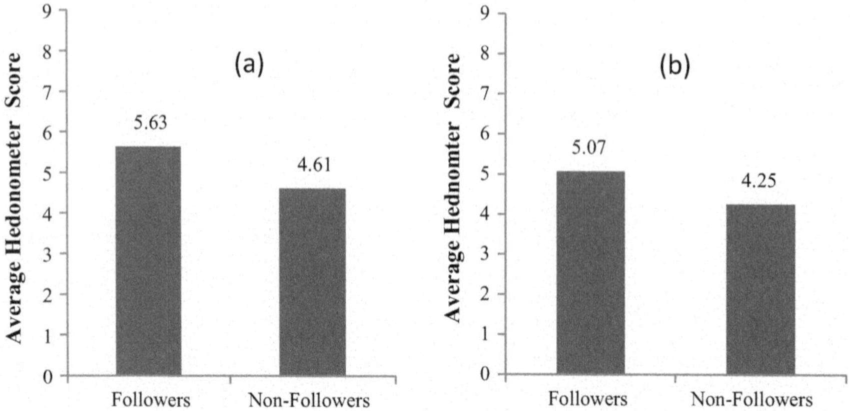

Fig. 2 Average Hedonometer score of tweets for company B containing. **a** Positive words, **b** Negative words

negative HScore ($\mu = 5.07, \sigma = 0.53$) > non-follower B negative HScore ($\mu = 4.25$, $\sigma = 2.23$); $t(14{,}220) = 14.46$, $p = 0.00$, $r = 0.12$) (see Fig. 2b).

Third, as shown in Fig. 3, for positive words, we found that the average Hedonometer score of tweets of followers of T was greater compared to non-followers of T (follower T positive HScore ($\mu = 5.62, \sigma = 0.4$) > non-follower T positive HScore ($\mu = 5.04$, $\sigma = 1.61$); $t(941) = 7.43$, $p = 0.00$, $r = 0.23$). Next, we investigated the tweets containing negative words and found that there was no significant difference in the average Hedonometer score of tweets of followers and non-followers of T (follower T negative HScore ($\mu = 5.19$, $\sigma = 0.58$) ~ non-follower T negative HScore ($\mu = 5.02$, $\sigma = 1.70$); $t(293) = 1.126$, $p = 0.26$, $r = 0.06$).

Lastly, as shown in Fig. 4, for positive words, we found that the average Hedonometer score of tweets of followers of O was greater compared to non-followers of O (follower O positive HScore ($\mu = 5.61$, $\sigma = 0.48$) > non-follower O positive HScore ($\mu = 5.19$, $\sigma = 1.62$); $t(6806) = 11.10$, $p = 0.00$, $r = 0.13$). Next, we investigated the tweets containing negative words and found that there was no significant difference in the average Hedonometer score of tweets of followers and

Fig. 3 Average Hedonometer score of tweets for T containing positive words

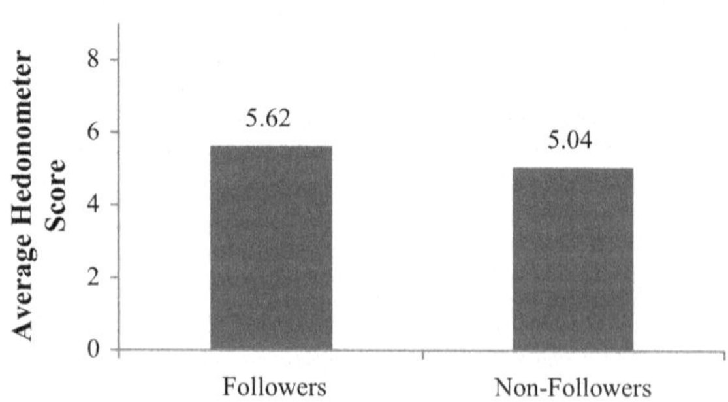

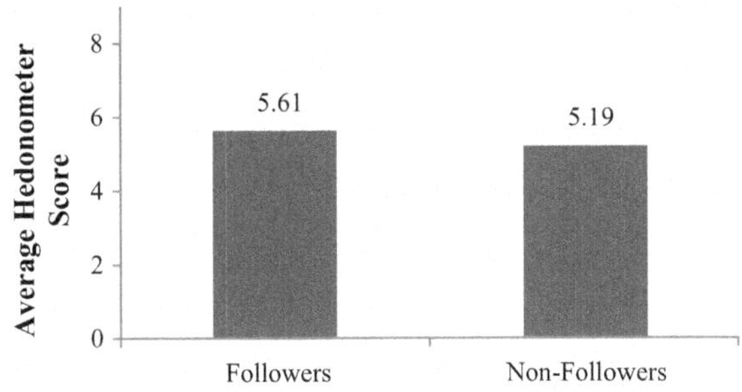

Fig. 4 Average Hedonometer score of tweets for O containing positive words

non-followers of O (follower O negative HScore ($\mu = 5.16, \sigma = 0.54$) ~ non-follower O negative HScore ($\mu = 5.27, \sigma = 1.42$), $t(1967) = -1.85$, $p = 0.06$, $r = 0.04$).

5 Discussion and Conclusion

Sentiment analysis is the concept of deciding whether a specific tweet is positive, negative, or neutral. Performing sentiment analysis on Twitter data can help companies obtain qualitative insights to understand how people are talking about their brand [2].

First, we found that there was no significant difference in the average Hedonometer score of tweets of followers compared to non-followers. The average Hedonometer score of tweets for both companies A and B as well their products T and O respectively were similar for both followers and non-followers of the company.

The Hedonometer is designed to calculate a happiness score for a large collection of text, based on the happiness of the individual words used in the text [6, 7]. So, the higher the score of the tweet, more positive is the perception of the user towards the product. For both company A and company B, the average Hedonometer score of tweets was around 5.4 for both followers and non-followers. While for the products T and O the average Hedonometer score of tweets was around 5.3 for both followers and Non-followers. As such both followers and non-followers possessed neutral sentiment towards the company and their products. This result is not as per our expectation as we expected followers to have a more positive sentiment towards the company and their products compared to non-followers. Since it is a rare disease medication, the followers will be mostly well-wishers and people suffering from that disease. As such they should have a more positive perception towards the company and their medication. The contrary result may be attributed to the fact that both companies may not be properly marketing the medication on Twitter. However, one other possibility could be that the treatment of rare disease conditions is such that it is difficult to increase followers' perception compared to the global (non-follower) perception.

Second, we found that tweets from followers containing positive words had significantly high average Hedonometer scores than the tweets of non-followers containing positive words. This result signifies that some followers do have a positive view towards the company; but, as a whole, the perception of the followers towards the companies and their products is mostly neutral. While for non-followers, the sentiment towards the company and its product was likely to be neutral.

Last, we found that tweets from followers containing negative words had similar average Hedonometer scores to the tweets of non-followers containing negative words. This result suggests that both followers and non-followers do not have a negative perception towards both companies.

Since followers are likely to be the main consumers of these medications it is paramount that they have proper knowledge regarding these medications. This analysis will likely help pharmaceutical companies assess the effectiveness of their marketing campaigns on Twitter and help them understand the perception of their users.

In the future, we intend to increase the number of companies and the products to understand if rare disease medications are generally being poorly marketed on Twitter, or the lack of positive outlook among followers was due to the nature of rare diseases. Furthermore, we will also investigate the effect of small world phenomena and current events on the sentiments of followers and non-followers.

Acknowledgements The project was supported from Grant (awards: #IITM/CONS/RxDSI/VD/33) to Varun Dutt.

References

1. Oberlo. https://www.oberlo.in/blog/Twitter-statistics. Last accessed 29 Nov 2019
2. Twitter, Customer Insights 2016: The value of a follower https://business.Twitter.com/en/basics/Twitter-followers.html. Last accessed 29 Nov 2019
3. Mehta, R., Mehta, D., Chheda, D., Shah, C., Chawan, P.M.: Sentiment analysis and influence tracking using Twitter. Int. J. Adv. Res. Comput. Sci. Electron. Eng. **1**(2), 72–79 (2012)
4. PharmaBioWorld. https://www.pharmabioworld.com/industry_orphan_drugs.html. Last accessed 20 Dec 2019
5. Cody, E.M., Reagan, A.J., Mitchell, L., Dodds, P.S., Danforth, C.M.: Climate change sentiment on Twitter: An unsolicited public opinion poll. PLoS one, **10**(8), (2015)
6. Dodds, P.S., Harris, K.D., Kloumann, I.M., Bliss, C.A., Danforth, C.M.: Temporal patterns of happiness and information in a global social network: Hedonometrics and Twitter. PLoS ONE **6**(12), (2011)
7. Mitchell, L., Frank, M.R., Harris, K.D., Dodds, P.S., Danforth, C.M.: The geography of happiness: connecting Twitter sentiment and expression, demographics, and objective characteristics of place. PLoS ONE **8**(5), (2013)
8. Anger, I., Kittl, C.: Measuring influence on Twitter. In: Proceedings of the 11th International Conference on Knowledge Management and Knowledge Technologies, p. 31. ACM (2011)
9. Allisio, L., Mussa, V., Bosco, C., Patti, V., Ruffo, G.F.: Felicittà: visualizing and estimating happiness in italian cities from geotagged tweets. In: 1st International Workshop on Emotion

and Sentiment in Social and Expressive Media: approaches and perspectives from AI, vol. 1096, pp. 95–106. CEUR Workshop Porceedings (2013)
10. Field, A.: Discovering Statistics Using IBM SPSS statistics. Sage publications, Thousand Oaks, California (2013)

Pulmonary Nodule Detection and False Acceptance Reduction: Review

Sheetal Pawar and Babasaheb Patil

Abstract Currently, lung cancer alone causes around 20% of all mortalities in cancer. In 2018 alone, 1.8 million deaths were accounted. Lung cancer does not show any prominent symptoms. The survival rate of affected persons can be enhanced by early detection of up to five years. Radiologists examine the chest CT scans slice wise for malignancy detection. The potential challenge in the prediction of the lung nodules is the variation in size, shape, the location of nodules, and the existence of nodules resembling objects. A manual detection is cumbersome and requires a high degree of skill and precision. Researchers throughout the world are currently working on developing a computer-aided approach to assist the radiologists. This study is intended to explore the different techniques for nodule detection, classification, and False Acceptance Reduction (FAR). This paper summarizes the advancements in feature-based and neural network based techniques along with their limitations and required further investigations. This study also highlights and compares a few evaluation metrics used for the assessment of the current methods.

Keywords Deep learning · Support vector machine · Pulmonary nodule

1 Introduction

Lung cancer is a widespread and malignant cancer. The life of victims highly depends on early diagnosis. The World Health Organization (WHO) reported that by 2030 up to 10 million patients will be victims of lung cancer [1]. A lung nodule is the result of abnormal cell division or cell death in lung tissue, or in the airways leading to the lungs. Pulmonary nodules resemble well or poorly defined round opacities.

S. Pawar (✉) · B. Patil
Walchand College of Engineering, Sangli, India
e-mail: pawarsheetal13@gmail.com

B. Patil
e-mail: babasaheb.patil@walchandsangli.ac.in

S. C. Satapathy et al. (eds.), *Intelligent Data Engineering and Analytics*, Advances in Intelligent Systems and Computing 1177,
https://doi.org/10.1007/978-981-15-5679-1_58

The occurrence of a spot on a lung does not necessarily specify cancer. It is a challenging work to identify the correct positions of the nodules from CT images and distinguish them as benign or malignant. Many noncancerous nodules (benign) are falsely accepted as malignant nodules (cancerous). The lung cancer detection can be broken down into three basic modules as lung segmentation, pulmonary nodule identification, and FAR. The key contributions are the following:

- It provides researchers an accumulated resource of accessing recent methods.
- It presents limitations of the current systems and necessary further improvements.

This paper comprises 4 sections. Section 2 focuses on the various preprocessing techniques employed for enhancement. Section 3 elaborates strategies used by recent methods for the detection of nodule, the classification of nodule, and FAR. Section 4 provides an overview of widely used metrics for the assessment of the current systems. Section 5 concludes the paper stating the current challenges and necessary future investigations.

2 Preprocessing Techniques

Noise affects the quality of an image severely. For automatic tumor segmentation and reducing misclassification, it is mandatory to preprocess the lung CT image. Talukdar et al. [2] employed an adaptive technique for binarization of the image. The Area of Interest (AOI) was segmented using a series of morphological operations. This method used a Canny edge detector. A Canny edge detector exhibits good localization property and is tolerant of the imaging artifacts like shading.

Rahman et al. preprocessed the input CT images using Gaussian Blur and Otsu's thresholding. The principle of border extraction was followed to locate AOI [3]. CT scan images are more prone to salt and pepper noise. To address this issue Nidhi Nadkarni et al. proposed the use of an ordered median filter. The AOI was extracted by following a series of morphological operations [4]. Suren Makaju et al. introduced a new method using median and Gaussian filtering followed by the watershed segmentation method [5].

Nastaran et al. [6] preferred lung segmentation using the convex hull algorithm and Euclidean Distance Transform. To evaluate the segmentation accuracy, similarity, and distance-based metrics were widely used. All preprocessing methods while segmenting mainly considered a gray level contrast between a lung region and surroundings. The efficacy of preprocessing highly relies on the selection of an optimum threshold. These methods are inappropriate for detection on the scans with severe pathologic background.

3 Nodule Detection, Classification and FAR

The task of nodule detection and classification can be divided into two broad categories based on the learning strategy employed for nodule detection and classification.

3.1 Feature-Based Learning

The feature-based learning involves the computation of complex features from the selected AOI. Talukdar et al. [2] used the area and the most primitive texture analysis tool in medical imaging; GLCM for feature extraction. Various statistical features like autocorrelation, entropy, difference entropy, dissimilarity, etc., were given as input to the SVM classifier. Nadkarni et al. [4] proposed the use of local binary pattern feature descriptor and SVM classifier with inputs extracted from lung AOI like area, perimeter, and eccentricity.

Nastaran et al. [6] suggested computation of tumor-related morphological, texture and density features to assess the risk of post-surgery recurrence. Sanjukta et al. [7] combined shape-based features, texture features, and intensity features as input to SVM classifier. SVM was trained using parameters, viz., area, perimeter, centroid, eccentricity, diameter and mean intensity. Bergtholdt et al. [8] introduced multi-threshold candidate detection. Total 35 features comprising geometrical, grayscale, and image features with adapting weighing schemes were identified as inputs to SVM. Choi et al. [9] experimented with various SVMs using Radial Basis Function, Polynomial function, and Minkowski distance function. They formulated a discriminating paradigm based on a set of geometric and texture features. Golosio et al. [10] proposed the use of geometrical features as input to Neural Network (NN) based classifier.

Almost all feature-based techniques used SVM. The progressive use of SVMs increases the number of support vectors and limits generalization. In feature-based learning, it is critical to identify a good discriminating feature space. Feature-based learning model provides limited representation. These features are not suitable to cope with the extensive changes in pulmonary nodules. These methods were associated with the computation of complex features and classification rules were stringent. Feature-based systems yield low accuracy and sensitivity.

3.2 NN-Based Learning

Deep learning techniques outperform the recent methods for detecting and classifying the objects in medical images. In the early days, Jyh-Shyan et al. [11] focused simultaneously on the detection of nodules and FAR. They developed a two-level

Convolutional Neural Network (CNN). CNN-I was trained directly on pre-scanned suspected nodule blocks. CNN-II was a dedicated unit designed for the nodule classification and reducing the number of false positives. Normalized disease index was computed from the activity distributions of output neurons from CNN-I. This proposed approach directly worked on suspected nodule regions. This system was tested on a small dataset with complex network architecture.

Penedo et al. [12] attempted the use of NN-based approach to detect the nodules on chest CTs. They proposed a two-stage NN. The first network identified possible nodule locations in low-resolution images. Curvature peaks of suspected nodule regions were provided as input to the second NN. The classification technique adopted in this method was much complex due to stringent rules regarding the compactness, circularity, and curvature of each point in suspected nodule areas.

Rahman et al. [3] put forth nodule detection and classification scheme for low configuration computers. They evaluated classifier frameworks like MobileNet, VGG Net, and Inception-v3. MobileNet achieved good accuracy and sensitivity as compared to VGG Net and Inception-v3. Liu et al. [13] investigated the cascading of CNN. Faster-Recurrent Convolutional Neural Net (F-RCNN) detected the nodule regions. A non-nodule filter network, followed by F-RCNN reduced the rate of false acceptance. The precision of this system was very low.

van Ginneken et al. [14] directly used the existing nodule detection system. This system offered the list of nodule locations and the likelihood of location to be a nodule. For every candidate location identified they computed a patch of 50×50 mm in different planes. These patches were trained using publically available OverFeat Network. Despite the large variations in nodule size, this system considered only fixed patch size.

Setio et al. [15] invented a novel FAR system using multi-view ConvNets. Lung nodules were detected using three existing detection algorithms. The outputs from multiple streams of 2D ConvNets were fused together. They tested different combinations of multi-view ConvNets and analyzed the prediction performance of each. This method failed to detect almost 18.7% of annotated nodules. The combination of multiple nodule detection algorithms increased the sensitivity but false-positive rate remain unaffected.

For automated pulmonary detection, Dou et al. [16] proposed a novel procedure by exploiting the properties of 3-D CNN. They proposed a multilevel encoding strategy for embedding contextual information. This architecture developed three 3-D CNN and encoded a specific level of surrounding information in each. The detection probabilities of each of these networks were combined together to project the result of the final classification. This methodology achieved the highest sensitivity at the cost of increased computational complexity and time. True nodules with very small sizes and in complex surroundings were classified with limited belief.

Moradi et al. [17] designed contextual multilevel 3-D CNN. They incorporated 4 different architectures with variable size receptive fields to process the nodules ranging from 3 mm to 3 cm. The knowledge gained from these networks was fused using meta-classifier. This architecture was tested only for slices with less than 2.5 mm. Cheng et al. [18] explored the deep learning approach for nodule detection using modified UNet. This architecture amalgamated ResNet with UNet and achieved the best segmentation results in terms of DSC. UNet accompanied by ResNet eliminated the issue of gradient disappearing as the network deepens. Instead of using a single-network classifier, this technique proposed a combination of Inception network, VGG Network and DenseNet for 3-D classification.

Xie et al. [19] proposed 2-D CNN architecture. This architecture modified F-RCNN based on region proposals and deconvolution layer. Dedicated modules were designed and trained for three different slice-sizes for detecting nodules. Dedicated boosting architecture was designed for reducing the number of false positives. Misclassified nodules were iteratively used to train the model. The proposed architecture worked effectively for the detection of the nodules greater than 10 mm.

Jiang et al. [20] formulated two new frameworks on the basis of multiple resolutions by inheriting the residual connections. These networks through residual connections facilitated the combination of features across multiple image resolutions. This approach utilized slice-wise segmentation and AOI-based training framework. It is observed that as network depth increases, the feature sparsity decreases in Incremental Multiple Resolution Residually Connected Network (MRRN). The proposed system demonstrated that Incremental MRRN achieves higher sensitivity as compared to Dense MRRN.

Motivated from the clinical practice followed by radiologists Zheng et al. [21] formulated a multi-view Maximum Intensity Projection (MIP) based CNN algorithm for lung nodule detection. This approach proved beneficial to discriminate nodules from vessels by imparting more representative information into 2-D CT images. This approach demonstrated well the use of multiple deep CNNs to reduce false positives. Liao et al. [22] introduced a 3-D deep NN to assess the spread of cancer. The model first detected all suspicious nodules using region proposals networks. Cancer probabilities of the top nodules identified based on prediction confidence were combined using OR gate to assess the malignancy. This method failed to detect the nodules that are less than 6 mm.

Almost all deep learning algorithms for nodule detection used a patch-based approach. Such patch-based systems hide the contextual information. There is a need to adhere to a robust approach while selecting the number of patches and patch size. Most of these algorithms outperformed in detecting and classifying the larger nodules but unable to detect the smaller nodules less than 6 mm. True nodules with very small size and in complex surroundings were classified with limited belief.

3-D CNNs can produce good spatial information. 3-D CNNs can extract more discriminative information with the aid of hierarchical network architecture. Extraction of information from 3-D patches and embedding contextual information may boost the sensitivity with least false acceptance, but 3D model occupies more memory and hence limits the batch size, model depth, and the running speed.

All these existing algorithms were tested on the standard databases like LIDC and tested on computational platforms, viz., NVIDIA GPUs like GTX Titan XP, TITAN Z, Google Cloud platform, GeForce GTX TITAN X.

4 Evaluation Metrics

All these existing algorithms were evaluated using performance measures like sensitivity (recall), specificity, precision, and accuracy. The index terms defined to compute these parameters were Genuinely Accepted (GA), Genuinely Rejected (GR), Falsely Accepted (FA), and Falsely Rejected (FR). Table 1 describes the significance of each index term.

- Sensitivity (Recall): It quantifies the proportion of genuinely malignant nodules that are identified as malignant.

$$\text{Sensitivity} = \frac{\text{GA}}{\text{GA} + \text{FR}} \tag{1}$$

- Specificity: It projects the proportion of genuinely benign nodules that are classified as benign.

$$\text{Specificity} = \frac{\text{GR}}{\text{GR} + \text{FA}} \tag{2}$$

- Precision: It is a fraction of genuinely positive instances among all the retrieved positive instances.

$$\text{Precision} = \frac{\text{GA}}{\text{GA} + \text{FA}} \tag{3}$$

- Accuracy: It measures the performance of the system in classifying malignant and benign nodules.

$$\text{Accuracy} = \frac{\text{GA} + \text{GR}}{\text{GA} + \text{GR} + \text{FA} + \text{FR}} \tag{4}$$

The sensitivity and corresponding false positives/scan of a few systems discussed above are depicted in Table 2.

Table 1 The significance of index terms for evaluation

Index term	Significance
GA	A malignant nodule identified as a malignant nodule
GR	A benign nodule identified as a benign nodule
FA	A benign nodule identified as a malignant nodule
FR	A malignant nodule identified as a benign nodule

Table 2 The comparison of sensitivity and false positives/scan (FP/scan)

Ref. no.	Author name	Sensitivity (%)	FP/scan
12	Penedo et al.	89	7
16	Dou et al.	92	8
17	Moradi et al.	81	2
18	Cheng et al.	71	2
19	Xie et al.	86	8

5 Conclusion

The survey of the literature reveals that most of the feature-based systems yield low accuracy and sensitivity. Deep learning algorithms shows promising results in classification and FAR, but with increased computational complexity. All the existing algorithms were tested on the standard databases. None of the studies had verified their frameworks on clinical datasets. To enhance the acceptability of CAD systems in clinical workflow, there is a need to develop more accurate, minimal network architecture for nodule detection and classification.

References

1. Organization, W.H.: Description of the Global Burden of NCDs, Their Risk Factors and Determinants. World-Health-Organization, Geneva Switzerland (2011)
2. Talukdar, J., Sarma, P.: An approach to classify lung nodules for detection of cancer cells. In: 5th International Conference on Advanced Computing and Communication Systems (ICACCS), pp. 189–192. Coimbatore, India (2019). https://doi.org/10.1109/icaccs.2019.8728332
3. Rahman, M.S., Shill, P.C., Homayra, Z.: A new method for lung nodule detection using deep neural networks for CT images. In: International Conference on Electrical, Computer and Communication Engineering (ECCE), pp. 1–6. Cox'sBazar, Bangladesh (2019). https://doi.org/10.1109/ecace.2019.8679439
4. Nadkarni, N.S., Borkar, S.: Detection of lung cancer in CT Images using image processing. In: 3rd International Conference on Trends in Electronics and Informatics (ICOEI), pp. 863–866. Tirunelveli, India (2019). https://doi.org/10.1109/icoei.2019.8862577
5. Makaju, S., Prasad, P.W.C., Alsadoona, A., Singhb, A.K., Elchouemi, A.: Lung cancer detection using CT scans. In: 6th International Conference on Smart Computing and Communications ICSCC. Procedia Computer Science 125, pp. 107–114. Sciencedirect, Kurukshetra, India (2017). https://doi.org/10.1016/j.procs.2017.12.016
6. Emaminejad, N., Qian, W., Guan, Y., Tan, M., Qiu, Y., Liu, H., Zheng, B.: Fusion of quantitative image and genomic biomarkers to improve prognosis assessment of early stage lung cancer patients. IEEE Trans. Biomed. Eng. **63**(5), 1034–1043 (2016). https://doi.org/10.1109/tbme.2015.2477688
7. Jena, S.R., George, T., Ponraj, N.: Texture analysis based feature extraction and classification of lung cancer. In: IEEE International Conference on Electrical, Computer and Communication Technologies (ICECCT), pp. 1–5. Coimbatore, India (2019). https://doi.org/10.1109/icecct.2019.8869369

8. Bergtholdt, M., Wiemker, R., Klinder.: Pulmonary nodule detection using a cascaded SVM classifier. In: Medical Imaging, International Society for Optics and Photonics, pp. 978513–978513. https://doi.org/10.1117/12.2216747
9. Choi, W.J., Choi, T.S.: Automated pulmonary nodule detection system in computed tomography images: a hierarchial block classification approach. Entropy **15**, 507–523. https://doi.org/10.3390/e15020507
10. Golosio, B., Masala, G.L., Piccioli, A., Oliva, P., Carpinelli, M., Cataldo, R., Ceello, P., De Carlo, F., Falaschi, F., Fantacci, M.E., Gargano, G., Kasae, P., Torsello, M.: A novel multi-threshold method for nodule detection in lung CT. Med. Phys. **36**, 3607–3618 (2009). https://doi.org/10.1118/1.3160107
11. Lin, J.-S., Lo, S.B., Hasegawa, A., Freedman, M.T., Mun, S.K.: Reduction of false positives in lung nodule detection using a two-level neural classification. IEEE Trans. Med. Imaging **15**(2), 206–217 (1996). https://doi.org/10.1109/42.491422
12. Penedo, M.G., Carreira, M.J., Mosquera, A., Cabello, D.: Computer-aided diagnosis: a neural-network-based approach to lung nodule detection. IEEE Trans. Med. Imaging **17**(6), 872–880 (1998). https://doi.org/10.1109/42.746620
13. Liu, C., Wang, B., Jiao, Q., Zhu, M.: Reducing false positives for lung nodule detection in chest X-rays using cascading CNN. In: 14th IEEE Conference on Industrial Electronics and Applications (ICIEA), pp. 1204–1207. Xi'an, China (2019). https://doi.org/10.1109/iciea.2019.8833699
14. van Ginneken, B., Setio, A.A.A., Jacobs, C., Ciompi, F.: Off-the shelf convolutional neural network features for pulmonary nodule detection in computed tomography scans. In: IEEE International Symposium on Biomedical Imaging, pp. 286–289 (2015). https://doi.org/10.1109/isbi.2015.7163869
15. Setio, A.A.A., Ciompi, F., Litjens, G., Gerke, P., Jacobs, C., van Riel, S.J., Wille, M.M.W., Naqibullah, M., Sánchez, C.I., van Ginneken, B.: Pulmonary nodule detection in CT images: false positive reduction using multi-view convolutional networks. IEEE Trans. Med. Imaging **35**(5), 1160–116. (2016). https://doi.org/10.1109/tmi.2016.2536809
16. Dou, Q., Chen, H., Yu, L., Qin, J., Heng, P.: Multilevel contextual 3-D CNNs for false positive reduction in pulmonary nodule detection. IEEE Trans. Biomed. Eng. **64**(7), 1558–1567 (2017). https://doi.org/10.1109/tbme.2016.2613502
17. Moradi, P., Jamzad, M.: Detecting lung cancer lesions in CT images using 3D convolutional neural networks. In: 4th International Conference on Pattern Recognition and Image Analysis (IPRIA), pp. 114–118. Tehran, Iran (2019). https://doi.org/10.1109/pria.2019.8785971
18. Cheng, H., Zhu, Y., Pan, H.: Modified U-Net block network for lung nodule detection. In: IEEE 8th Joint International Information Technology and Artificial Intelligence Conference (ITAIC), pp. 599–605. Chongqing, China (2019). https://doi.org/10.1109/itaic.2019.8785445
19. Xie, H., Yang, D., Sun, N., Chen, Z., Zhang, Y.: Automated pulmonary nodule detection in CT images using deep convolutional neural networks. Pattern Recogn. (2018). https://doi.org/10.1016/j.patcog.2018.07.031
20. Jiang, J., Hu, Y.-C., Liu, C.-J., Halpenny, D., Hellmann, M.D., Deasy, J.O.: Multiple resolution residually connected feature streams for automatic lung tumor segmentation from CT images. IEEE Trans. Med. Imaging **38**(1), 134–144 (2019). https://doi.org/10.1109/tmi.2018.2857800
21. Zheng, S. Guo, J., Cui, X., Veldhuis, R.N.J., Oudkerk, M., van Ooijen, P.M.A.: Automatic pulmonary nodule detection in CT scans using convolutional neural networks based on maximum intensity projection. IEEE Trans. Med. Imaging. https://doi.org/10.1109/tmi.2019.2935553
22. Liao, F., Liang, M., Li, Z., Hu, X., Song, S.: Evaluate the malignancy of pulmonary nodules using the 3-D deep leaky noisy-OR network. IEEE Trans. Neural Netw. Learn. Syst. **30**(11), 3484–3495 (2019). https://doi.org/10.1109/tnnls.2019.2892409

Leveraging Deep Learning Approaches for Patient Case Similarity Evaluation

Nachiket Naganure, Nayak U. Ashwin, and S. Sowmya Kamath

Abstract One of the fundamental problems in Health Informatics is evaluating the clinical similarity between two patients for treatment recommendation. Retrieving clinical records of existing patients who are potentially similar to a newly arrived patient could help a physician in faster diagnosis and recommending informed treatment options, especially in the case of areas where specialist medical care is scarce. In Western countries, patient records are extensively stored in the form of Electronic Health Records (EHR), which are created manually by human experts, which can take a lot of time and is a cost-intensive process. In developing countries like India, patient records are increasingly being stored in digital formats and often contain diverse, heterogeneous, unstructured reports of patients. These can be potentially utilized for designing patient similarity assessment and recommendation systems. In this paper, we propose a patient similarity evaluation framework built on two supervised learning models—Convolutional Neural Networks (CNN) and Gated Recurrent Unit (GRU). Our method creates an optimal patient representation for existing patients by aggregating reports collected over the duration of treatment, to overcome the loss of temporal information, for which a cohort of 16,723 patients across 8 disease categories was used. Both the models (CNN and GRU) learn by passing through the records of a patient chronologically and achieve an accuracy of 97.60 and 93.62%, respectively, on standard EHR dataset like MIMIC-III.

Keywords Patient similarity · Medical concept embedding · Deep learning · Electronic medical records

---

N. Naganure (✉) · N. U. Ashwin · S. S. Kamath
Department of Information Technology, National Institute of Technology Karnataka, Surathkal 575025, India
e-mail: nachiketun8@gmail.com

N. U. Ashwin
e-mail: ashwinnayak111@gmail.com

S. S. Kamath
e-mail: sowmyakamath@nitk.edu.in

S. C. Satapathy et al. (eds.), *Intelligent Data Engineering and Analytics*, Advances in Intelligent Systems and Computing 1177,
https://doi.org/10.1007/978-981-15-5679-1_59

1 Introduction

Over the past decade, Electronic Health Records (EHR) have been adopted in most hospitals in Western countries. EHRs contain vast information regarding the events, diagnoses reports, lab results, prescription data, etc., of patients. Leveraging this voluminous data to design precision medicine applications that can aid healthcare professionals toward informed decision-making is one of the fundamental goals of Health Informatics. Evaluating patient similarity is thus a crucial task, which can help pinpoint the ailments of newly admitted patients faster and more accurately, optimizing healthcare delivery time and costs. The degree of how similar a pair of patients are given their historical clinical information is termed as Patient Similarity, which has benefits not only in precision medicine but also in cohort analysis and other medical studies. In addition to this, learning patient similarities is the fundamental problem in medicine, which has been found to be useful in improving the quality of care, etc.

A significant challenge in this domain is, the historical data of patients can comprise multiple medical episodes or events, occurring over a period of time. Such medical events have to be effectively processed and represented optimally, while minimizing any loss of critical information. Most existing works on this problem have not addressed the issue of time sensitivity (temporal) and high sparsity. Information regarding time is important for designing health information management systems and as medical reports contain a lot of data and hence, dealing with high dimensionality is also of critical importance. Also, EHR data is often not easy to interpret due to coding styles, and if this problem is addressed well, the positive impact is immense.

Clinical Patient Similarity has since received a lot of research in the past decade. Chan et al. [1] devised a Support Vector Machine based model SimSVM which takes in fixed-size input of 14 features for predicting survival rate in cancer patients. SimSVM with RBF kernel gave better results than the linear or polynomial kernel. The drawback is that the process requires extensive curation of the data for engineering specific features for the model. Zhu et al. [2] presented a scheme that uses a convolutional neural network architecture that takes in patient reports represented by medical embeddings as the input and generates a score of patient similarity. Although it performs well due to the preservation of temporal data of patients, it fails to justify the direct summation of each word vector to get one report vector. Chao et al. [3] proposed a Recurrent Neural Network (RNN) architecture to make a personalized prediction of Parkinson's disease. This model learns the similarity between two longitudinal patient record sequences through dynamically matching temporal patterns inpatient sequences, but does not scale well to general EHR Data. A 15000 patient cohort with diabetes symptoms onset was used for performing risk analysis and similarity with locally Supervised Metric Learning (LSML) by Kenney et al. [7]. The same method in combination with custom interactive metric learning and feedback improved similarity scores is found in the work done by Jimeng Sun [8].

Based on our observations, we propose a solution based on word embeddings [9], for representing patients in terms of vectors and design suitable algorithms to derive similarity values between patients based on it. Unlike the usage of one-hot vectors which just indicate whether the person has a disease or not by representing it in forms of 0's and 1's, we use a fixed-length vector of size 300 which represents each patient. The given representation is generated by taking into consideration the context of the clinical terms in the EHR data. We use these 300 length vector representations to train learning architectures like CNN and GRU, for measuring the similarity between patient representations. We benchmarked our work on patient notes available in the MIMIC-III dataset, using standard metrics like precision, recall, and accuracy. The rest of the paper is organized as follows. In Sect. 2, we present a detailed discussion of the proposed methodology. Section 3 discusses the results of the experimental validation followed by conclusion and future work.

2 Proposed Methodology

2.1 Dataset and Pateint Cohort Selection

The first phase of the workflow is the patient cohort selection. For experimental validation of the proposed methodology, we used the MIMIC-III dataset [5], which is a publicly available EHR dataset, comprising around 40,000 deidentified critical care patients at various hospitals. This dataset includes basic patient information along with laboratory reports, nurse, and physician notes, ECG recordings, other health equipment recording, and more. Out of 26 tables, we used 5 tables namely PATIENTS, NOTEEVENTS, ADMISSIONS, DIAGNOSES_ICD, and D_ICD_DIAGNOSES. Each patient visits the hospital multiple times which results in multiple reports. A patient generally visits the hospital multiple times and for each hospital visit, the diagnosed diseases along with the priority of the disease are recorded. We choose 16,723 patients who have more than 25 reports but less than 400 reports.

2.2 Patient Report Representation

To represent the words in vector form, we used different pretrained word embeddings like Glove word embeddings. But these word embeddings were trained on Wikipedia Articles and cannot be used in our case, as we want to model clinical data contained in EHRs. Hence, we trained our own word embeddings on the TEXT column from the NOTEEVENTS table, which contains all the patient reports. Firstly, we cleaned the data by removing stop words and lower casing the text. We used Gensim [6] library to generate 300 length word embeddings using the Continuous Bag of Words

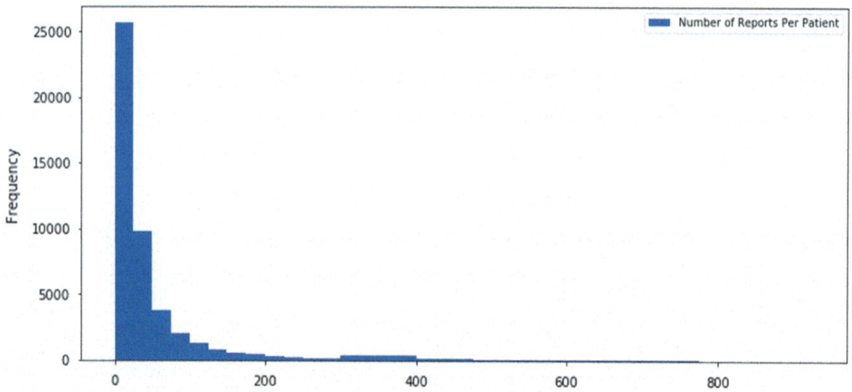

Fig. 1 Frequency distribution of number of reports per patient

algorithm. Then, we represent each word as 300 length vector in the report, after which, averaging of the word vectors present in the report is done, resulting in 300 length vector for one report. All the report vectors are stacked chronologically to generate a matrix of 300 rows by n (where n is the number of reports available for that patient) columns. From Fig. 1, we can infer that maximum patient density is within 0–400 reports category. To make the input embedding matrix of the same size, we add zero paddings to get the 400 columns. We decided to keep our lower limit as 25 so that we need not put more zero paddings with patients, who have a very less number of reports. Ultimately, an embedding matrix of size 300 rows and 400 columns is generated which will be used as input to the deep learning models for similarity evaluation.

2.3 *Patient Disease Representation*

The diseases in EHR are represented in the form of ICD9[1] codes, ranging from 0 to 1000. These codes are further categorized into 18 categories of which, we chose the eight most common disease categories, namely, diseases of the nervous system and sense organs (320–389), diseases of the circulatory system (390–459), diseases of the respiratory system (460–519), diseases of the digestive system (520–579), diseases of the genitourinary system (580–629), complications of pregnancy (630–679), diseases of the skin (680–709), and diseases of the musculoskeletal system (710–739). The priority of the disease is taken into account and the weighted average of the priority for each hospital visit is obtained. After this, we normalize the patient disease vector over all visits, which results in one 8 length vector for each patient. The *Patient Similarity Vector* captures the similarity between two patients across the 8 categories

[1]International Statistical Classification of Diseases and Related Health Problems, Revision 9.

previously used in the Patient Disease Vector. We perform two logical operations on the two vectors, v_1 and v_2 consecutively, to obtain the Patient Similarity Vector, which will be the input Y label for the deep learning models. The first operation is logical XOR and then logical NOT is performed as shown here, $S = \sim (v_1 \oplus v_2)$, where S is the Patient Similarity vector.

2.4 *Patient Report Similarity Prediction*

We propose two deep learning models—using Convolutional Neural Network (CNN) and the other, a Recurrent Neural Network (GRU), both of which are trained on the patient report matrix. Both the models take in two patient report matrices X and Y as input and output a vector of length 8.

2.4.1 CNN Model

Figure 2 depicts the CNN architecture used in the proposed approach. The convolution layers help in extracting important patterns from the input patient record matrix, i.e., medical concepts patterns that are present in the input record. In general, let $a_i \epsilon R^D$ be the 300 length report vector corresponding to the (ith)hospital event/visit.

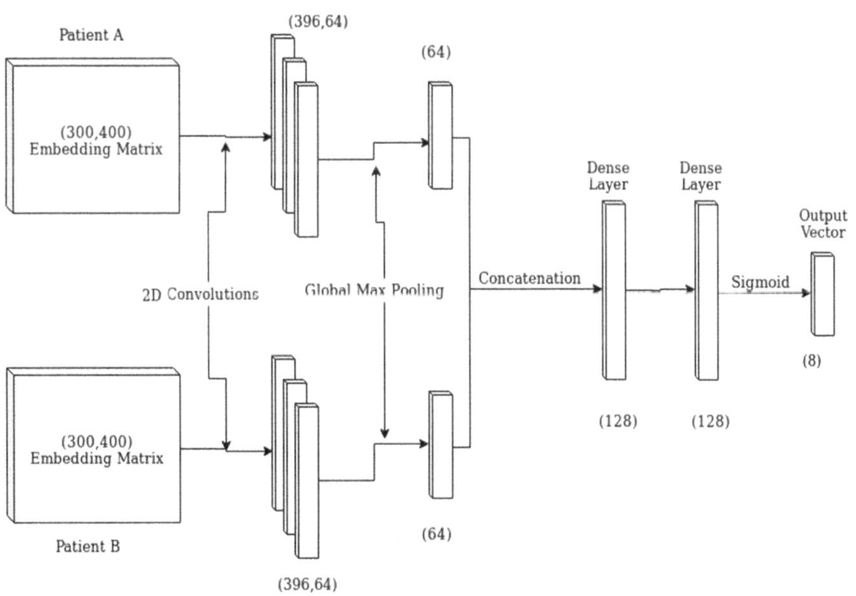

Fig. 2 CNN model architecture

A convolution operation has a filter $w \in R^{d \times h}$, which is applied to a window size of z hospital event features to produce a new feature. For example, a feature d_i is generated from a window of events $d_i = f(w.a_{i:i+z1}) + b$, where f is ReLU (Rectified Linear Unit) and b is the bias value. The output after performing convolution on the input of records is then passed through a pooling layer, which aggregates information received and reduces the representation dimension. Max Pooling is applied to each possible window of features in the embedding matrix (i.e., the patient report matrix), where $\{a_{1:z}, a_{2:z+1}, \ldots, a_{nz+1:n}\}$ is used to produce a feature map $d = [d_1, d_2, \ldots, d_{nz+1}]$, where d belongs to R^{nz+1}. Next, we apply Global Max Pooling over the feature maps and take the mean value $d = \text{mean}(\max(d))$. The primary goal here is to capture the most relevant feature which will play a decisive role in generating the score. The penultimate convolutional and pooling layers output two 64 length vectors which are flattened and concatenated to form a 128 length vector, and then passed to a fully connected sigmoid layer. Sigmoid layer computes the categorical or class-wise probability of occurrence of the diseases.

2.4.2 Gated Recurrent Unit (GRU)

The GRU model [4] with different input, output, and filter dimension sizes is depicted in Fig. 3. GRU has been particularly successful in language-related tasks because language is nothing but a sequence of words and the occurrence of each word is

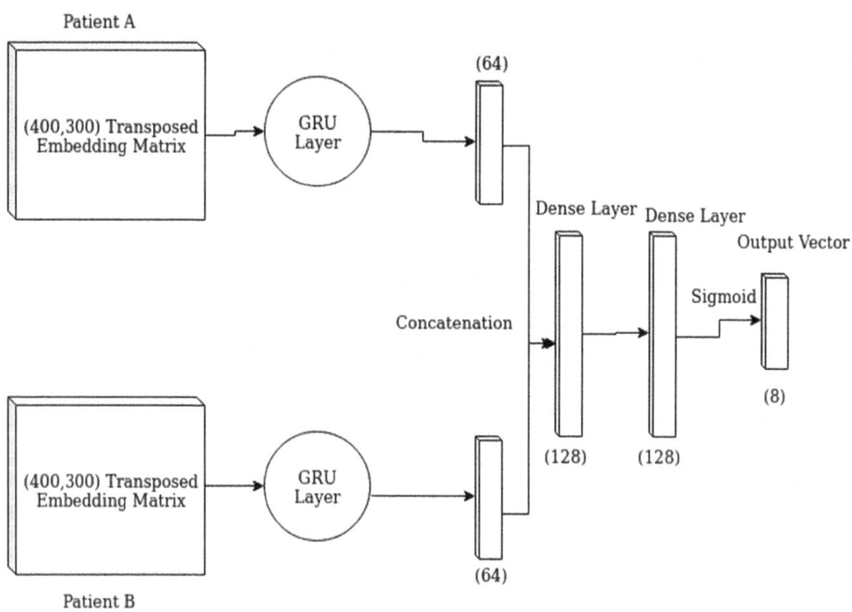

Fig. 3 GRU architecture

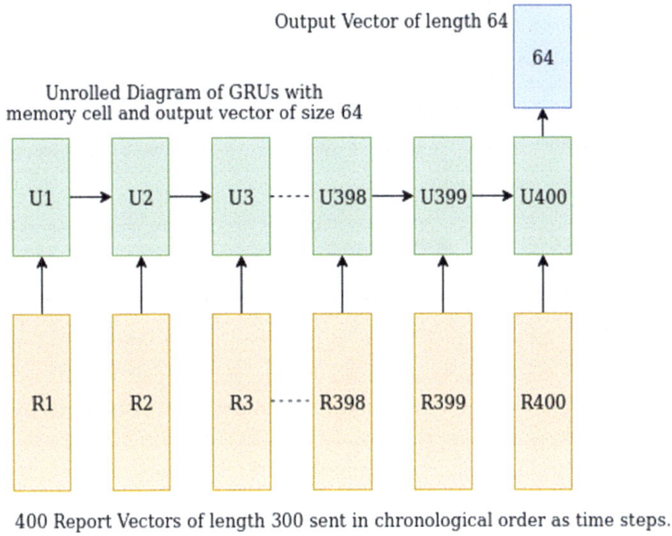

Fig. 4 GRU layer

dependent on previous words. In the context of EHR's, a patient's disease is linked to his/her past diseases. Therefore, GRU can be used in addressing this problem. The GRU has two gates, namely, update gate and reset gate. The update gate is used to help the model understand how much of the old information should be passed along to the future. The reset gate is used to find how much of the old information is needed to be forgotten.

The input to the GRU layer is a matrix generated using medical word embeddings for each patient report called the *embedding matrix*. We take a transpose of this matrix which is of shape 400 × 300, where 400 is the number of reports and 300 is the size of each report embedding vector. Figure 4 shows the GRU Layer, which has 400 units of GRU, with internal memory cell and output vector of size 64. Each report is passed to a GRU unit in a many-to-one sequence. We take the output vector of the last GRU unit from both patients GRU Layers and concatenate them. Now, this concatenated vector is passed through two dense layers and finally, we reach the output layer which has a sigmoid activation function with 8 output values.

3 Experimental Results and Discussion

The experimental validation was performed on patient records available in the standard open dataset, MIMIC-III. The patient cohort was generated as per the strategy outlined in Sects. 2.1 and 2.2. Following this, the representation generation process was applied to the clinical records of the patient cohort for deriving optimal patient

Table 1 MIMIC-III dataset metrics

Model	Training loss	Training accuracy (%)	Validation loss	Validation accuracy (%)	Test-accuracy (%)
CNN	0.046	97.80	0.040	98.00	97.60
GRU	0.090	96.55	0.0961	95.76	93.62
MLP	0.1731	91.30	0.1558	91.24	90.13

representations in the form of word embedding [8, 9] matrix. The next phase is to input the word embedding matrix to the deep learning models for patient similarity prediction. Since the architectures used require patient pair reports as input, we selected 16,723 patients and generated unique (Patient_ID1, Patient_ID2) pairs. We randomly shuffled these into approximately 100,000,000 patient pairs and selected 2,000,000 pairs for training and other random 100,000 pairs as our test data. For intuitive comparison, we also implemented a Multi-Layer Perceptron with 2 hidden layers each containing 512 nodes and 128 nodes, with a sigmoid output layer giving 8 length vector. The input to this model will be a stacked vector comprising of two flattened patient report matrices (each a vector of length 1200).

In Table 1, we tabulate the overall loss and accuracy for train and validation dataset generated by all models, while following tables present the class-wise accuracy, precision, F1-score, and recall performance for all the eight classes predicted by all models. There were a total of 18 disease classes present in the dataset, but a few categories like parasitic, neoplasms, etc., had very rare occurrences, hence, we decided to reduce the number of classes to 8. The GRU model achieved a 95.76% validation accuracy with loss of 0.09, whereas the CNN model achieved 98.00% validation accuracy with loss of 0.04. The loss used is binary cross-entropy. In our

Table 2 Categorical result metrics for CNN model

Disease category	Precision (%)	Recall (%)	Accuracy (%)	F1-score (%)
Nervous system and Sense organs	100.00	100.00	100.00	100.00
Circulatory system	98.40	97.29	98.02	99.17
Respiratory system	100.00	100.00	100.00	100.00
Digestive system	100.00	98.99	99.10	99.00
Genitourinary system	100.00	100.00	100.00	100.00
Pregnancy related	97.10	90.18	92.88	94.15
Skin diseases	100.00	100.00	100.00	100.00
Musculoskeletal system	95.53	91.90	93.33	92.29

Table 3 Categorical result metrics for GRU model

Disease category	Precision (%)	Recall (%)	Accuracy (%)	F1-score (%)
Circulatory system	99.65	99.32	99.86	99.23
Circulatory system	99.11	99.08	99.01	99.03
Respiratory system	95.13	94.72	94.87	94.17
Digestive system	99.70	99.06	99.33	99.02
Genitourinary system	100.00	100.00	100.00	100.00
Pregnancy related	88.07	90.75	86.88	89.34
Skin diseases	96.51	86.09	89.46	90.00
Musculoskeletal system	96.03	96.00	95.59	96.56

Table 4 Categorical result metrics for MLP model

Disease category	Precision (%)	Recall (%)	Accuracy (%)	F1-score (%)
Nervous system and sense organs	100.00	100.00	100.00	100.00
Circulatory system	97.65	93.12	94.76	95.33
Respiratory system	98.88	96.42	97.54	97.11
Digestive system	63.21	59.56	62.99	61.10
Genitourinary system	99.78	96.78	98.36	98.07
Pregnancy related	100.00	100.00	100.00	100.00
Skin diseases	96.51	86.09	89.46	90.00
Musculoskeletal system	97.23	92.53	93.89	94.56

case, our work is a multi-class, multi-label classification, hence, sigmoid activation function coupled with binary cross-entropy loss produced better results.

Tables 2,3, and 4 illustrate the category-wise precision, recall, accuracy, and F1 score for test dataset by all models (GRU, CNN, and MLP). Although GRUs are used to model dependencies over a longer length of text, in our case they fail to capture the dependencies over a very large number of patient reports. CNNs, on the other hand, manage to model better feature maps over these large reports. Any sentence or paragraph from the 400 reports can influence the similarity score, so we want our models to look at the whole input at a time rather than look at it sequentially. Hence, CNNs perform much better than GRUs and MLP in this particular task and are much faster in training and evaluation than GRUs.

4 Conclusion and Future Work

In this paper, a patient similarity evaluation framework built on two supervised learning models—Convolutional Neural Networks (CNN) and Gated Recurrent Unit (GRU) was presented. While processing large text data, generally meaningful features are not only difficult to obtain, but it is also difficult to justify their use. In our case, large-scale patient data is processed to meaningfully determine patient case similarity. Experimental results revealed that CNN-based model outperformed GRU-based model due to their ability to look at the entire input rather sequentially processing one report at a time as the GRUs do. As part of our future work, we intend to use state-of-the-art embedding models such as BERT, GPT-2 to enhance the quality of patient report representations further. Another aspect to address would be improving the representation of similarity and disease category vector to take into account the specific diseases instead of using generalized ICD9 categories.

Acknowledgements We gratefully acknowledge the use of facilities at the Department of Information Technology, NITK Surathkal, funded by the Government of India's DST-SERB Early Career Research Grant (ECR/2017/001056) to Sowmya Kamath S.

References

1. Chan, L.W.C., Chan, T., Cheng, L.F., Mak, W.S.: Machine learning of patient similarity: a case study on predicting survival in cancer patient after locoregional chemotherapy. In: IEEE International Conference on Bioinformatics and Biomedicine Workshop (2010)
2. Zhu, Z., Yin, C., Qian, B., Cheng, Y., Wei, J., Wang, F.: Measuring patient similarities via a deep architecture with medical concept embedding. In: IEEE 16th International Conference on Data Mining (2016)
3. Che, C., Xiao, C., Liang, J., Jin, B., Zhou, J., Wang, F.: An RNN architecture with dynamic temporal matching for personalized predictions of Parkinson's disease. In: Proceedings of the 2017 SIAM International Conference on Data Mining
4. Cho, K. et al.: Learning phrase representations using RNN encoder-decoder for statistical machine translation. arXiv:1406.1078
5. Pollard, T.J., Johnson, A.E.W.: The MIMIC-III clinical database (2016). http://dx.doi.org/10.13026/C2XW26
6. Gensim python language modelling library. https://pypi.org/project/gensim/
7. Ng, K. et al.: Personalized predictive modeling and risk factor identification using patient similarity. AMIA Summits on Translational Science Proceedings 2015, p. 132 (2015)
8. Sun, J. et al.: Supervised patient similarity measure of heterogeneous patient records. ACM Sigkdd Explor. Newslett. **14**(1), 16-24 (2012)
9. Mikolov, T., Sutskever, I., Chen, K., Corrado, G.S., Dean, J.: Distributed representations of words and phrases and their compositionality. In: Advance in Neural Information Processing Systems, pp. 3111–3119 (2013)

RUSDataBoost-IM: Improving Classification Performance in Imbalanced Data

Satyam Maheshwari, R. C. Jain, and R. S. Jadon

Abstract Real-world datasets often have skewed distribution; that is, one class may have much larger instances as compared to other classes. Rarely occurred events are very hard to judge due to their scarcity and casualty; however, the misinterpretation of rare events can lead to high costs. The traditional classifiers usually incline toward the standard class, and rare class is missed due to class imbalance. We propose RUSDataBoost-IM, a novel ensemble approach based on DataBoost-IM, which merges random undersampling (RUS) with the Boosting method. Our approach is as simple as DataBoost-IM but improves the precision of the base classifiers by using the random undersampling approach. Experiments on 29 highly imbalanced datasets show the superiority of our proposal over other ensemble methods. The experimental outcomes prove that our hybrid ensemble approach is proven to be a promising alternative on different benchmarks and can be applied to any imbalanced classification scenario.

Keywords Unbalanced datasets · Ensembles · Classification · Boosting · Bagging · Class distribution

S. Maheshwari (✉)
Department of CS/IT, S. A. T. I. (Engineering. College), Vidisha 464001, India
e-mail: satyam.vds@gmail.com

R. C. Jain
S. A. T. I. (Engineering. College), Vidisha 464001, India
e-mail: dr.jain.rc@gmail.com

R. S. Jadon
Department of MCA, M. I. T. S, Gwalior 474003, India
e-mail: rsjadon@gmail.com

S. C. Satapathy et al. (eds.), *Intelligent Data Engineering and Analytics*, Advances in Intelligent Systems and Computing 1177,
https://doi.org/10.1007/978-981-15-5679-1_60

1 Introduction

The imbalance class problem exists when the class allocation is not identical among the classes. In this, one class have a few number of samples (minority class) and the remaining samples belong to other class (majority class). The class that has fewer samples (rare class) is the most critical concept for the study. The presence of such type of data complexity appeared to be the topmost challenging issues in the data mining field [1]. This problem exists due to standard classification algorithms, which are inclined toward the common class and ignore the samples of minority class by considering them as noise. The problem of unbalanced datasets appears in various real-world applications such as fraudulent calls detection [2], medical diagnosis [3], new manufacturing plants [4], finding of oil spills from satellite pictures [5], text classification [6], and risk management [7].

In this work, our objective is to increase the performance of DataBoost-IM by improving the accuracy of base classifiers while preserving diversity. To perform the empirical comparison, we test 29 datasets from the KEEL source [8, 9]. We choose C4.5 [10] as an underlined classifier for our study and the *Area Under Curve* (AUC) [11, 12] for performance metrics. Our study shows that the RUSDataBoost-IM method outperforms over other approaches on this metric.

The remaining paper is arranged accordingly: Sect. 2 outlines the rare class problem, discusses the evaluation metrics, and discusses the various solutions to the rare class problem. In Sect. 3, we propose our Data generation Ensemble Undersampling-Boost (RUSDataBoost-IM) framework. Section 4 introduces the experimental settings. In Sect. 5, we make a comparison of our method with the rest of the approaches, and finally, we provide conclusion in Sect. 6.

2 The Problem of Unbalanced Data Classification

In this part, we discuss evaluation criteria to measure the classification performance in an unbalanced domain and analyze several ensemble methods that will be used for the analysis.

2.1 Evaluation Metrics

For binary classification, the confusion matrix (Table 1) shows the result of both classes.

The AUC-ROC value is calculated by the below formula:

Table 1 Confusion (error) matrix

	Predicted positive	Predicted negative
Positive class	"True positive"	"False negative"
Negative class	"False positive"	"True negative"

$$\mathrm{AUC} = \frac{1 + \mathrm{TPrate} - \mathrm{FPrate}}{2} \tag{1}$$

2.2 Taxonomy of Rare Class Problem

Data preprocessing is the most popular method to solve the imbalanced data problem. The goal of this approach is to balance the class allocation by modifying the prior allocation of prevalent and rare classes, either by increasing or decreasing the samples. The "Random Under-Sampling" (RUS) [13] and "Synthetic Minority Oversampling Technique" (SMOTE) [14] are popular data preprocessing techniques. The Algorithm-level Solutions select an appropriate inductive bias that reinforces the learning toward the rare class. Therefore, to develop an algorithm-level solution, one must acquire domain-specific knowledge. A cost-sensitive learning framework considers the variable misclassification cost of different classes [15]. This method attempts to reduce the overall cost of misclassification. The disadvantage of the cost-sensitive method is to identify misclassification costs, which are difficult to identify.

The role of ensemble-based classifier [16, 17] is to increase the efficiency of the individual model by combining different classifiers that exceed with every one of them.

Some popular boosting methods are AdaBoost [18], RUSBoost [19], and SMOTE-Boost [20], and some well known bagging methods are UnderBagging [21], Overbagging [22], and SMOTEBagging [22].

3 RUSDataBoost-IM: Random Undersampling with Boosting

In our work, we develop a novel hybrid RUSDataBoost-IM method that is constituted of DataBoost-IM ensemble and random undersampling (RUS).

3.1 Random Undersampling Method (RUS)

This technique attempts to adjust the class allocation by randomly removing common class instances. This method discards examples from the common class until the required class distribution is obtained. The major disadvantage of this approach is that, possibly, it can eliminate valuable data. However, this drawback is reduced when applying ensembles, since examples removed in one round can stay with others.

3.2 DataBoost-IM Method

Here, we consider the DataBoost-IM method, which is a popular boosting method known to increase the classification performance of weak learners. DataBoost-IM is an ensemble learning that integrates multiple learners to get an effective learner. DataBoost-IM improves the initial DataBoost method as follows: (i) We recognized hard instances from and produce artificial samples for various classes. (ii) We produce artificial instances with bias information toward the hard samples; we add further information for both the classes and hence avoid boosting overemphasizing the hard samples. (iii) The class allocation and the entire weights of various classes are well-balanced to update the training methods bias against the prevalent class by selecting a less number of typical samples among majority and minority classes. (iv) The total weights of both majorities, as well as minority classes in the recent learning set, are rebalanced to impose boosting methods to target not only the common instances but also the rare class samples.

3.3 Combining RUS and DataBoost-IM Method

We propose the RUSDataBoost-IM algorithm, which combines Random Undersampling (RUS) with DataBoost-IM. The RUS increase minority class identification with DataBoost-IM. RUS is effective to create ensembles since it supports high variety, which joined with base learners is converted into high accuracy ensembles. The complete procedure of RUSDataBoost-IM is explained in Algorithm 1.

Algorithm 1. RUSDataBoost–IM, RUS embedded in DataBoost-IM.

Input: Consider a series of n number of instances $\{(x_1, y_1).....(x_n, y_n)\}$ with labels $y_i \in Y = \{1,..,k\}$

Weak training method **WeakClassifier**, Number of repetitions T

Load D_1(i) = 1 / n for all value of *i*.

Do for t = 1, 2, …., T

1. Select hard samples from the initial data set of distinct classes.
2. Create artificial data to adjust the learning data of distinct classes.
3. Generate a training set $S_t^{'}$ with distribution $D_t^{'}$ using random undersampling.
4. Revise and adjust the total weights of distinct classes in the learning dataset.
5. Apply WeakClassifier and giving it a new learning data set $(S_t^{'})$, which consists of artificial data and re-balanced weights$(D_t^{'})$.
6. Generate a hypothesis h_t : X → Y.
7. Compute the error
   $h_t : e_t = \sum_{i:h_t(x_i) \neq y_i} D_t(i)$ if $e_t > 1/2$, then update T = t - 1 and terminate loop (where e_t is trained seed data).
8. Compute the weight update function:
   $\beta_t = e_t / (1 - e_t)$.
9. Revise distribution
   $D_t : D_{t+1}(i) = \frac{D_t(i)}{z_t} \times \left\{ \frac{b_t \leftarrow if\, h_t(x_i = y_i)}{1 \leftarrow otherwise}, \right\}$
   Here, z_t is a standarized value (preferred so that D_{t+1} will be a distribution).

Output: The final classifier:

$$h_{fin}(x) = argmax_{y \in Y} \sum_{t:h_t(x)=y} \log \frac{1}{\beta_t}$$

4 Experimental Framework

In Sect. 4.1, we briefly describe the methods that we have considered for the study and their setup specifications. In Sect. 4.2, a brief description of datasets is elaborated. In Sect. 4.3, we present the statistical tests performed for comparing results.

Table 2 The algorithms used for experimental purpose

Abbr.	Method	Short description
RUSC	RUS + C4.5	C4.5 applied to datasets previously preprocessed with RUS (random undersampling)
ADAB	AdaBoost	Adaptive Boosting and C4.5 as a base method
M1	AdaBoost. M1	Multi-Class extension of Adaptive Boosting combined with C4.5 as a primary method
RUS	RUSBoost	Random Under-Sampling with AdaBoost. M1 boosting with C4.5 as a base method
SBO	SMOTEBoost	SMOTE with AdaBoost. M2 boosting with C4.5 as a primary method
C2	Ada C2	AdaBoost with Costs and C4.5 as a base method
DB-IM	DataBoost-IM	Data generation combined with boosting and C4.5 as a base method

4.1 Algorithms and Parameters

Table 2 outlines the complete list of algorithms, the abbreviations that we consider in our study, and a brief overview of each.

4.2 Datasets

Table 3 presents a brief description of datasets: the number of data samples (#Ex.), the proportion of instances of every class, number of features (#Fea.), and the imbalanced ratio. The table is arranged based on IR. In experiments, the datasets are divided into fivefold cross-validation (5-fcv) procedure.

4.3 Analytical Tests for Comparing Algorithms

In this study, we perform hypothesis analysis to give analytical support for the analysis of results [23]. In our work, we use two types of tests:

1. Pair-wise tests: The Wilcoxon signed-rank [24] is pair-wise test (1×1) for comparing two methods.
2. Multiple tests: We consider the Iman-Davenport test [25] to find differences between two and more methods. Then, we use the Holm procedure [26] to compare the outcomes acquired by a control method across a set of methods (i.e., $1 \times n$ comparison). Finally, we apply the Shaffer test [27] to analyze which approach is different between n × n comparisons.

Table 3 Data set characteristics

Datasets	#Ex.	#Fea.	%Min.	%Maj.	IR
Wisconsin	683	9	35	65	1.86
Pima	768	8	34.84	66.16	1.87
Iris0	150	4	33.33	66.67	2.00
Yeast1	1484	8	28.91	71.09	2.46
Haberman	306	3	27.42	73.58	2.78
Vehicle2	846	18	28.37	71.63	2.88
Vehicle1	846	18	28.37	71.63	2.90
Vehicle3	846	18	28.37	71.63	2.99
Vehicle0	846	18	23.64	76.36	3.25
New-thyroid1	215	5	16.29	83.72	5.14
New-thyroid2	215	5	16.89	83.11	5.14
Segment0	2308	19	14.25	85.74	6.02
Yeast3	1484	8	10.99	89.02	8.10
Page-blocks0	5472	10	10.23	89.77	8.79
Yeast2vs4	514	8	9.92	90.08	9.08
Yeast05679vs4	528	8	9.66	90.34	9.35
Vowel0	988	13	9.01	90.99	9.98
Shuttle0vs4	1829	9	6.72	93.28	13.87
Yeast1vs7	459	7	6.54	93.28	14.30
Page-blocks13vs4	472	10	5.93	94.07	15.86
Abalone9vs18	731	8	5.75	94.25	16.40
Shuttle2vs4	129	9	4.65	95.35	20.50
Yeast1458vs7	693	8	4.33	95.67	22.10
Yeast2vs8	482	8	4.15	95.85	23.10
Yeast4	1484	8	3.43	96.57	28.10
Yeast1289vs7	947	8	3.17	96.83	30.57
Yeast5	1484	8	2.96	97.04	32.73
Yeast6	1484	8	2.49	97.51	41.40
Abalone19	4174	8	0.77	99.23	129.44

5 Empirical Comparison: RUSDataBoost-IM Versus Other Methods

In Sect. 5.1, we will examine whether the amending DataBoost-IM with random undersampling (Sect. 3.3) is appropriate or not. In Sect. 5.2, we will analyze our proposal against the existing ensemble method using various statistical tests. In Sect. 5.3, we discuss a summary of the results.

5.1 Amending DataBoost-IM with RUS: RUSDataBoost-IM

All algorithms results are evaluated using the AUC metric and presented in Table 4.

Table 4 AUC performance comparison of 29 imbalanced datasets using different methods. The best value in each dataset is spotted in boldface

Datasets	RUSC	C2	ADAB	M1	RUS	SBO	DB-IM	RUS DB-IM
Wisconsin	0.951	0.961	0.961	0.962	0.957	0.965	0.960	**0.968**
Pima	0.727	0.715	0.690	0.687	0.740	0.729	0.711	**0.793**
Iris 0	0.990	0.990	0.990	0.990	0.990	0.990	0.990	0.990
Yeast 1	0.713	0.660	0.652	0.655	0.705	0.713	0.700	**0.789**
Haberman	0.609	0.560	0.582	0.582	0.630	0.615	0.562	**0.675**
Vehicle 2	0.937	0.966	0.966	0.966	0.973	0.981	0.967	**0.988**
Vehicle 1	0.704	0.749	0.700	0.702	0.743	0.752	0.699	**0.775**
Vehicle 3	0.704	0.727	0.681	0.681	0.764	0.744	0.692	**0.791**
Vehicle 0	0.936	0.946	0.932	0.932	0.959	**0.976**	0.945	0.943
New-thyroid 1	0.916	0.961	0.932	0.932	0.932	0.989	0.957	**1.000**
New-thyroid 2	0.907	0.966	0.954	0.954	0.918	**0.972**	0.940	0.967
Segment 0	0.979	0.983	0.990	0.990	0.992	0.994	0.991	**0.994**
Yeast 3	0.902	0.907	0.841	0.835	0.916	0.888	0.898	**0.917**
Page-blocks 0	0.946	0.882	0.930	0.929	0.947	0.939	0.847	**0.950**
Yeast 2 versus 4	0.886	0.911	0.890	0.890	0.914	0.860	0.867	**0.921**
Yeast 05679 versus 4	0.786	0.761	0.645	0.645	0.845	0.782	0.737	**0.889**
Vowel 0	0.944	0.971	0.971	0.971	0.958	0.970	0.971	**0.971**
Shuttle c0 versus c4	**1.000**	1.000	1.000	1.000	**1.000**	1.000	1.000	**1.000**
Yeast 1 versus 7	0.660	0.705	0.619	0.619	**0.710**	0.687	0.664	0.698
Page-blocks 13 versus 4	0.944	0.998	0.998	0.998	0.972	0.972	0.998	**0.999**
Abalone 9 versus 18	0.689	0.717	0.692	**0.691**	0.791	0.753	0.680	0.690
Shuttle c2 versus c4	0.984	0.950	0.950	0.950	0.984	**1.000**	0.950	0.985
Yeast 1458 versus7	0.574	0.543	0.505	0.506	0.625	**0.599**	0.500	0.589
Yeast 2 versus 8	0.735	0.622	0.697	0.697	0.801	**0.818**	0.600	0.619
Yeast 4	0.819	0.720	0.617	0.619	**0.820**	0.704	0.580	0.697
Yeast 1289 versus 7	0.605	0.638	0.626	0.626	**0.738**	0.651	0.575	0.689
Yeast 5	0.938	0.883	0.826	0.826	0.960	0.947	0.929	**0.961**
Yeast 6	0.811	0.716	0.752	0.723	0.838	0.774	**0.913**	**0.913**
Abalone19	0.625	0.509	0.511	0.511	0.695	0.537	0.500	**0.699**
Average	0.825	0.814	0.797	0.795	0.856	0.838	0.804	**0.857**

5.2 RUSDataBoost-IM Versus state-of-the-art

Hereafter, we investigate whether the proposed approach gives more accurate ensembles than other methods. To do so, for two pairs of the algorithms we apply the Wilcoxon test, for comparing more than one method, we use the Iman-Davenport procedure, and after that, we use Holm test, if required. To show relationships between all methods, we perform the Shaffer test.

Initially, we consider the Wilcoxon procedure to find which one from each pair of algorithms is more appropriate (Table 5). This table demonstrates the ranks of each method, selection, or rejection of hypothesis with a significance value of $\alpha = 0.05$ and p-ratio, which shows differences between the two methods. The last column in the table indicates the setup we have chosen for the next stage, depending on the hypothesis being selected or rejected based on the ranks.

Now, we perform the Iman-Davenport procedure to find the average ranking of each method. Figure 1 depicts the average ranking of the methods calculated by the Iman and Davenport procedure. The p-ratio of the Iman and Davenport test is very low (−0), which shows that there exist some differences between these algorithms. Therefore, we apply the Holm post-hoc procedure (the results presented in Table 6).

The Holm procedure shows the optimum performance of RUSDB-IM with very unbalanced datasets. RUSDB-IM can statistically outperform except RUS and SBO. For this reason, we perform pair-wise comparisons applying the Wilcoxon test. In this manner, we can get a better understanding of their performance (in Table 7, we present the result).

Table 5 Wilcoxon tests to show the difference between each pair

Algorithm	R^+	R^-	Hypothesis ($\alpha = 0.05$)	p-ratio	Selected
RUSC versus C2	233.5	172.5	Not rejected	0.4789	RUSC
ADAB versus M1	220.5	185.5	Not rejected	0.6731	ADAB
RUS versus SBO	280.5	125.5	Not rejected	0.0748	RUS
DB-IM versus RUSDB-IM	7.50	427.5	Not rejected	1.0	RUSDB-IM

R^+ are ranks for RUSC, ADAB, RUS, DB-IM, and R^- for C2, M1, SBO, and RUSDB-IM

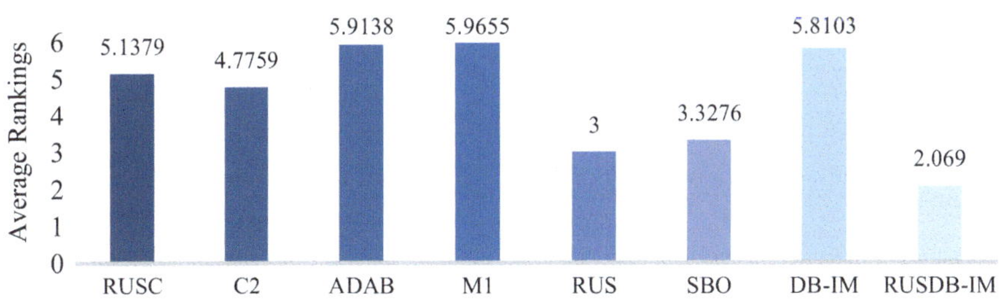

Fig. 1 The average ranking of state-of-the-art ensemble and proposed method RUSDB-IM

Table 6 Holm test results to compare RUSDB-IM with the rest of the methods

i	Method (Rank)	Z	p-ratio	Holm	Hypothesis ($\alpha = 0.05$)
7	M1 (5.9655)	6.057436	0	0.007143	**Rejected for RUSDB-IM**
6	ADAB (5.9138)	5.977027	0	0.008333	**Rejected for RUSDB-IM**
5	DB-IM (5.8103)	5.816211	0	0.01	**Rejected for RUSDB-IM**
4	RUSC (5.1379)	4.770901	0.000002	0.0125	**Rejected for RUSDB-IM**
3	C2 (4.7759)	4.208042	0.000026	0.016667	**Rejected for RUSDB-IM**
2	SBO (3.3276)	1.956605	0.050394	0.025	Not rejected
1	RUS (3.0)	1.447352	0.147798	0.05	Not rejected

Control method: RUSDB-IM, rank: 2.069

Table 7 Wilcoxon tests to analyze the proposed method RUSDB-IM with SBO and RUS

Algorithm	R^+	R^-	Hypothesis ($\alpha = 0.05$)	p-ratio	Selected
RUSDB-IM versus SBO	306.6	100.0	**Rejected for RUSDB-IM**	0.01843	RUSDB-IM
RUSDB-IM versus RUS	285.0	150.0	Not rejected	0.140384	RUSDB-IM

R^+ corresponds to RUSDB-IM and R^- to SBO and RUS

Following the outcome of the Wilcoxon test, we can say that RUSDB-IM outperforms than others; the test refuses the equivalence hypothesis of small p-ratio and is thus the ranks in support of RUSDB-IM, showing its supremacy.

In the end, we apply another post-hoc procedure to show the relationship between all approaches, i.e., $n \times n$ comparison. For this, we run the Shaffer test (Table 8). In the table, a (+) value shows that the proposal outperforms, while (−) indicates the opposite; and (=) indicates that compared algorithms are not distinct. The adjusted p-ratio also given in brackets. From the table, we can conclude that RUSDB-IM outperforms as compared to other methods.

Table 8 Shaffer tests to show the relationship between all algorithms

	RUSC	C2	ADAB	M1	RUS	SBO	DB-IM	RUS DB-IM
RUSC	×	=(2.867)	=(1.594)	=(1.586)	−(0.0142)	−(0.078)	=(1.775)	−(0)
C2	=(2.867)	×	=(0.845)	=(0.772)	−(0.086)	−(0.316)	=(1.077)	−(0.0005)
ADAB	=(1.594)	=(0.845)	×	=(2.867)	−(0.0001)	−(0.000)	=(2.867)	−(0)
M1	=(1.586)	=(0.772)	=(2.867)	×	−(0.0000)	−(0.000)	=(2.867)	−(0)
RUS	=(0.014)	+(0.086)	+(0.000)	+(0.0000)	×	=(2.867)	−(0.000)	=(1.330)
SBO	+(0.078)	+(0.316)	+(0.000)	+(0.0006)	=(2.8676)	×	−(0.001)	=(0.655)
DB-IM	=(1.775)	=(1.077)	=(2.867)	=(2.8676)	+(0.0002)	−(0.000)	×	−(0.002)
RUS DB-IM	+(0)	+(0.000)	+(0)	+(0)	=(1.3301)	=(0.655)	+(0.002)	×

5.3 Discussion: Summary of the Results

The outcomes obtained from the experimental study listed in the following section:

1. Table 4 represents that proposed RUSDataBoost-IM achieves the best average result in the AUC performance metric and outperforms than other state-of-the-art ensembles.
2. The analysis using the Iman-Davenport test shown in Fig. 1 depicts that the performance of RUSDataBoost-IM (RUSDB-IM) excels, followed by RUSBoost (RUS) and SMOTEBoost (SBO), respectively.
3. The results of the Holm test compiled in Table 6 demonstrate that RUSDM-IM outperforms all except RUS and SBO method. Due to this, we perform pair-wise comparisons using the Wilcoxon procedure.
4. Analysis using the Wilcoxon test compiled in Table 7 shows that RUSDB-IM outperforms than RUS and SBO, being the rank in support of RUSDB-IM.
5. The result of the Shaffer test presented in Table 8 shows significant differences among the methods. The proposed method outperforms as compared to RUSC, C2, ADAB, M1, DB-IM (marked as "+") and shows similarity for RUS and SBO (marked as "=") in the table. We show the superiority of RUSDB-IM in Table 7 by comparing it with RUS and SBO.

6 Conclusion

In this work, we merge the random undersampling (RUS) method with DataBoost-IM produce to get the best outcome between these two. DataBoost-IM is a combination of data generation and boosting methods. Furthermore, the advantage of random undersampling (RUS) is that it decreases the time required to develop a model. The major disadvantage of RUS is that it may lose some valuable data during the induction process, but this information loss may be overcome by embedding RUS into the DataBoost-IM method. RUS, combined with DataBoost-IM, works extremely well as compared to other methods. Therefore, RUSDataBoost-IM becomes an attractive solution for learning from unbalanced datasets due to its simplicity, speed, and excellent performance. In future work, we will continue to explore our methods using additional learners, performance metrics, and different application domains. Also, we will try to identify the effects of parameter selection on RUSDataBoost-IM (modifying the imbalanced ratio and the number of boosting iterations).

References

1. Yang, Q., Wu, X., Domingos, P., Elkan, C., Gehrke, J., Han, J., Heckerman, D., Keim, D., Liu, J., Madigan, D., Piatetsky-Shapiro, G., Raghavan, V.V, Rastogi, R., Stolfo, S.J., Tuzhilin, A., Wah, B.W.: Challenging problems in data mining research. Int. J. Info. Technol. Decis. Making **5**, 597–604 (2006)
2. Fawcett, T., Provost, F.J.: Adaptive fraud detection, Data Min. Knowl. Disc. **1**, 291–316 (1997)
3. Mazurowski, M.A., Habas, P.A., Zurada, J.M., Lo, J.Y., Baker, J.A., Tourassi, G.D.: Training neural network classifiers for medical decision making: The effects of imbalanced datasets on classification performance. Neural Networks **21**, 427–436 (2008). https://doi.org/10.1016/j.neunet.2007.12.031
4. Segal, R., Etzioni, O., Riddle, P., Healy, M., Newman, D., Pearson, C.: Representation design and brute-force induction in a boeing manufacturing domain. Appears Appl. Artif. Intell. **8**, 125–147 (1994)
5. Kubat, M., Holte, R.C., Matwin, S.: Machine Learning for the detection of oil spills in satellite radar images. Mach. Learn. **30**, 195–215 (1998). https://doi.org/10.1023/A:1007452223027
6. Cardie, C., Howe, N.: Improving minority class prediction using case-specific feature weights. In: proceeding of the Fourteenth International Conference on Machine Learning, pp. 57–65. Nashvile, TN (1997)
7. Ezawa, K., Singh, M., Norton, S.: Learning goal-oriented bayesian networks for telecommunications risk management. In: 13th International Conference on Machine Learning, pp. 139–147 (1996)
8. Alcalá-Fdez, J., Sánchez, L., García, S., del Jesus, M.J., Ventura, S., Garrell, J.M., Otero, J., Romero, C., Bacardit, J., Rivas, V.M., Fernández, J.C., Herrera, F.: KEEL: a software tool to assess evolutionary algorithms for data mining problems. Soft Comput. **13**, 307–318 (2009). https://doi.org/10.1007/s00500-008-0323-y
9. Alcalá-Fdez, J., Fernández, A., Luengo, J., Derrac, J., García, S., Sánchez, L., Herrera, F.: KEEL data-mining software tool: data set repository, integration of algorithms and experimental analysis framework. J. Multiple Valued Logic Soft Comput. **17**, 255–287 (2011). https://doi.org/10.1007/s00500-008-0323-y
10. Quinlan, J.R. C4.5: Programs for Machine Learning. Morgan Kaufmann Publishers (1993)
11. Bradley, A.P.: The use of the area under the ROC curve in the evaluation of machine learning algorithms. Pattern Recogn. **30**, 1145–1159 (1997). https://doi.org/10.1016/S0031-3203(96)00142-2
12. Huang, J., Ling, C.X.: Using AUC and accuracy in evaluating learning algorithms. IEEE Trans. Knowl. Data Eng. (2005). https://doi.org/10.1109/TKDE.2005.50
13. Batista, G.E.A.P.A., Prati, R.C., Monard, M.C.: A study of the behavior of several methods for balancing machine learning training data. ACM SIGKDD Explor. Newsl. **6**, 20–29 (2004). https://doi.org/10.1145/1007730.1007735
14. Chawla, N.V., Bowyer, K.W., Hall, L.O., Kegelmeyer, W.P.: SMOTE: synthetic minority over-sampling technique. J. Artif. Intell. Res. **16**, 321–357 (2002)
15. Margineantu, D.D.: Class probability estimation and cost-sensitive classification decisions. In: Lecture Notes in Computer Science (including subseries Lecture Notes in Artificial Intelligence and Lecture Notes in Bioinformatics) (2002)
16. Polikar, R.: Polikar, R.: Ensemble based systems in decision making. IEEE Circuit Syst. Mag. **6**, 21–45 (2006). https://doi.org/10.1109/MCAS.2006.1688199
17. Rokach, L.: Ensemble-based classifiers. Artif. Intell. Rev. (2010). https://doi.org/10.1007/s10462-009-9124-7
18. Freund, Y., Schapire, R.E.: A decision-theoretic generalization of on-line learning and an application to boosting. J. Comput. Syst. Sci. **55**, 119–139 (1997). https://doi.org/10.1006/jcss.1997.1504
19. Seiffert, C., Khoshgoftaar, T.M., Van Hulse, J., Napolitano, A.: RUSBoost: a hybrid approach to alleviating class imbalance. Syst. Hum. **40** (2010). https://doi.org/10.1109/TSMCA.2009.2029559

20. Chawla, N.V., Lazarevic, A., Hall, L.O., Bowyer, K.W.: SMOTEBoost: improving prediction of the minority class in boosting. In: Knowledge Discovery in Databases PKDD 2003. pp. 107–119. Springer, Berlin, Heidelberg (2003)
21. Barandela, R., Valdovinos, Sánchez, J.S.: New applications of ensembles of classifiers. Pattern Anal. Appl. **6**, 245–256 (2003). https://doi.org/10.1007/s10044-003-0192-z
22. Wang, S., Yao, X.: Diversity analysis on imbalanced data sets by using ensemble models. In: IEEE Symposium on Computational Intelligence and Data Mining. pp. 324–331. IEEE (2009)
23. García, S., Fernández, A., Luengo, J., Herrera, F.: A study of statistical techniques and performance measures for genetics-based machine learning: accuracy and interpretability. Soft. Comput. (2009). https://doi.org/10.1007/s00500-008-0392-y
24. Wilcoxon, F.: Individual comparisons by ranking methods. Biometrics Bull. (1945). https://doi.org/10.2307/3001968
25. Sheskin, D.J.: Handbook of Parametric and Nonparametric Statistical Procedures. (2004)
26. Holm, S.: A simple sequentially rejective multiple test procedure. Scand. J. Stat. (1979)
27. Shaffer, J.P.: Modified sequentially rejective multiple test procedures. J. Am. Stat. Assoc. (1986). https://doi.org/10.1080/01621459.1986.10478341

Performance Enhancement of Gene Mention Tagging by Using Deep Learning and Biomedical Named Entity Recognition

Ashutosh Kumar and Aakanksha Sharaff

Abstract Performance enhancement of biomedical named entity recognition tagging by applying a deep learning based framework is introduced with a combination of the word as well as character embedding. The input sentences first pass through the word and character level embedding, where word embedding is used to learn syntactic and semantic information and character level embedding handles those words which are out-of-vocabulary, then it passes through Bi-directional Long Short-Term Memory (BI-LSTM), where first it train the sentences in the forward direction and then again it trains the sentences in the backward direction, and finally it passes through CRF layer where the output comes in the form of gene mention tagging. The framework is tested over BioCreative II Gene mention task corpus. The deep learning framework combines with the conditional random field, and embedding techniques achieve 89.42% of *F*-score, which outperforms the various state-of-the-art techniques and top-ranked achieved system on BioCreative II Gene Mention competition.

Keywords Conditional random field (CRF) · Word embedding · Character level embedding · Biomedical named entity recognition (BIO-NER) · Long short-term Memory-CRF (LSTM-CRF)

1 Introduction

With the rapid increment in biological technology, biomedical literature is also growing, which creates a new opportunity in the area of biomedical text mining. Biomedical named entity recognition (BIO-NER) plays a vital role in the field of text mining in the biomedical domain. A different variation of BIO-NER comes to

A. Kumar · A. Sharaff (✉)
National Institute of Technology Raipur, Raipur, Chhattisgarh, India
e-mail: asharaff.cs@nitrr.ac.in

A. Kumar
e-mail: akumar.phd2018.cs@nitrr.ac.in

S. C. Satapathy et al. (eds.), *Intelligent Data Engineering and Analytics*, Advances in Intelligent Systems and Computing 1177,
https://doi.org/10.1007/978-981-15-5679-1_61

improve the performance, but ambiguities and irregularities make this task more challenging day by day. BIO-NER mainly consists of three approaches, a dictionary based (DB), rule based (RB), and machine learning based (MLB) approach. All these three approaches are different from each other. Manually constructed terms are required in the dictionary based approach. Hand-crafted features are characterized in a rule based approach, and a labeled corpus is required in machine learning based approach. Machine learning based technique mainly deals with the conditional random field (CRF), maximum entropy (ME), and support vector machine (SVM) [1, 2]. BIO-NER depends on the machine learning based technique generally focused on including important features which come from different classifiers although it is a better option for local as well as external features. Still, it is unable to achieve further improvement in the performance of the BIO-NER system. One of the suitable solutions to handle this problem is to combine multiple classifiers. The combination of multiple classifiers decreases the probability of local weakness, which is faced on single classifiers and gives an approximate, accurate recognition. The proposed framework adopts bi-direction LSTM along with CRF and introduced word level (WL) and character- evel (CL) embedding on it. BioCreative II Gene Mention task corpus [3] has been used to perform the proposed work. The results achieved by this work outperform the previous work done in the same corpus. The remaining paper is organized as: Sect. 2 describes the literature review done in the area of BIO-NER. Section 3 discusses the methodology and Sect. 4 defines various evaluation criteria. Section 5 discusses the result and analysis. Finally, the conclusion of the research paper along with the future work is mentioned in Sect. 6.

2 State of the Art

Most of the methodology has been proposed for the BioCreative II Gene Mention task corpus by combining the machine learning based classifiers, ensemble methods, consensus theory, and multiple classifiers. The combination of multiple classifiers performs better as compared to the single classifier. Two points are required to focus: first is accuracy and second is different classifiers' diversity. The random forest classifier performs better with respect to accuracy as compared to other classifiers. Currently, most of the researchers move toward ensemble-based learning techniques. The benefit of adopting ensemble methods is to processes the result by the single classifier as compared to the multiple classifiers. Li et al. [4] in the experiment combined the ensemble classifier technique with post-processing for improving the results. Ando [5] secured the first rank in the BioCreative II competition. Kuo et al. [6] combined likelihood scores of bidirectional MALLET parsing models to improve the performance of the existing model. Chang et al. [1] use MALLET and CRF++, they stated that using bidirectional parsing is not always beneficial. The forward parsing is always ahead of backward parsing. Recent BIO-NER research on the BioCreative II Gene Mention task has been focused on heuristic techniques.

3 Methodology

3.1 LSTM-CRF Networks

The combination of LSTM and CRF layers form the LSTM-CRF network is shown in Fig. 1. The purpose of the LSTM layer is to use past features and CRF is to use for sentence level tagging [7, 8]. This type of network used the past as well as future tags to predict the resent tag. The network shown in Fig. 1 first takes the input sentences then it passes through the LSTM network.

3.2 Bi-LSTM-CRF Networks

BI-LSTM-CRF architecture is similar to LSTM-CRF architecture, it is the blending of bi-directional LSTM and CRF [9]. The advantage of the BI-LSTM-CRF network over the LSTM-CRF network is it includes future input features to enhance the extra tagging accuracy (Fig. 2).

3.3 BI-LSTM-CRF with Embedding Networks

The proposed framework is shown in Fig. 3. This framework is designed for the BioCreative II Gene Mention corpus [10]. The framework adopts a deep learning based BIO-NER system along with a word or character embedding. First of all, it accepts an input sentence and converts it into the label form for each sentence. The character level and word level embedding take these inputs. Then it goes to the recent neural-based LSTM layers, where it trains the input in the forward and backward direction. Finally, it goes to the Conditional Random Field (CRF) layer for tagging. The detailed architecture of the proposed framework is shown in Fig. 3.

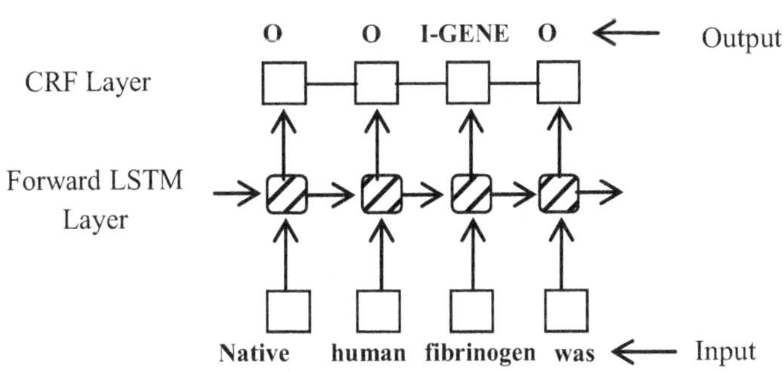

Fig. 1 Simple forward LSTM applied to the conditional random field

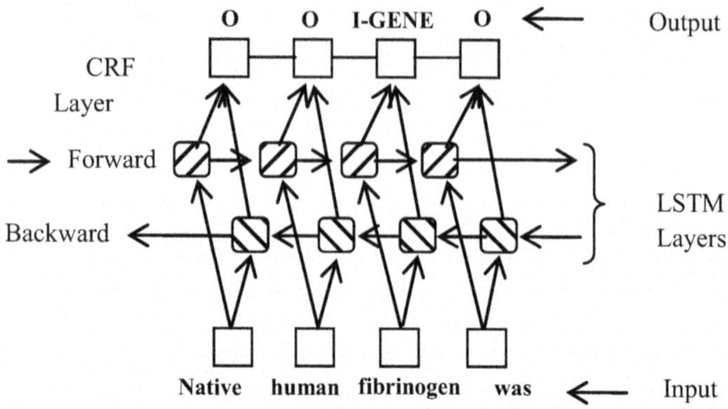

Fig. 2 The sentence is applied to the bi-directional LSTM model with CRF

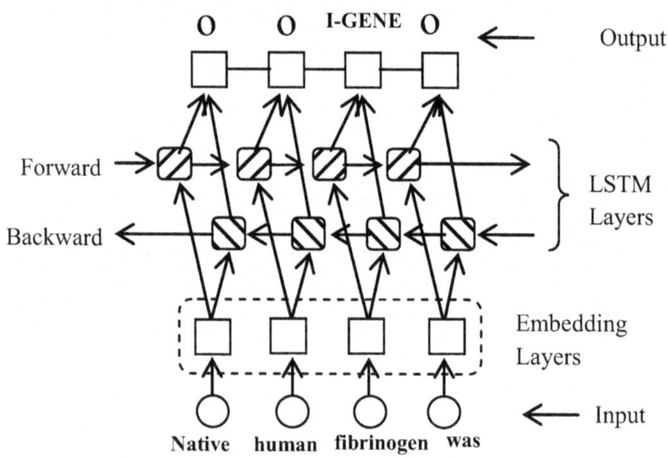

Fig. 3 Proposed a hybrid representation using the Bi-LSTM-based and embedding layer

3.4 Word Level (WL) Embedding

The WL embedding is the way to represent any document vocabulary. It takes care of the context of the word in the documents related to different type s of words in the vocabulary and helps to find out the relation among different words, also syntactic and semantic similarities, etc. Word embedding is also defined as the vector representation of any words. This experiment uses biomedical PubMed word2vec vocabulary.

Table 1 Description of BioCreative II Gene mentioning task corpus

Split dataset	*S*	*G*	*A*
Training set	15,000	18,265	14,499
Testing set	5000	6331	5061

• #: no of, *# Gene: G, # Sentences: S, # AltGene: A*

3.5 Character Level (CL) Embedding

The CL embedding is generally used for morphology as well as syntax and semantic information of the words. Unlike the WL embedding, CL embedding does not require a separate corpus to train the BIO-NER model.

4 Evaluation

4.1 Competitor System

The experiment compares with the methods proposed in the state of the art, and also compares with the performance of LSTM-CRF and as well as BI-LSTM-CRF.

4.2 Corpora Used

The proposed model was tested over BioCreative II Gene Mention corpus [10]. Detail description of the corpus is shown in Table 1 and this corpus contains 20 K sentences. To find out the entity boundaries, BIO tagging scheme is used, where ***B*** uses as a starting word of an entity. For inside the entity using ***I*** and ***O*** denotes to that words which do not represent any gene entity [11].

4.3 Text Preprocessing

Before inserting text into the model, preprocessing is required. First of all, sentences are split into the words and these words are tagged by the BIO tagging scheme then training is performed in preprocessed data.

4.4 Evaluation Measures

The efficiency of the framework is evaluated by Eqs. (1)–(3), which is used as a standard named entity recognition measures named as precision (P), recall (R), and F-measure score (F) [11].

$$\text{Precision}\,(P) = \frac{Z}{X} \tag{1}$$

$$\text{Recall}\,(R) = \frac{Z}{Y} \tag{2}$$

$$F\text{-Score}\,(F) = 2 * \frac{(P \cdot R)}{(P + R)} \tag{3}$$

where

- X represents the overall predicted entity in the sentence.
- Y represents the overall actual entity in the sentence.
- Z represents the overall correct entity.

4.5 Hyperparameter Setting

The proposed framework uses the Viterbi algorithm to train the model. This experiment uses the batch size of 25, that is, the size of the sentences is equal to or less than 25 and 8 epochs are used for learning training and validation testing, the dropout rate is taken as 0.5 with the softmax activation function. The word embedding dimension is considered as 20. Similarly, 20 dimensions are also used for character level embedding. The learning rate of the experiment is kept at 1.0.

4.6 Hardware/Software Requirement

This experiment is performed with the 8 GB of RAM and I-7 processor. Additional GPU enables the NVIDIA graphics processor of 1050 TI with Cuda enable and the experiment is performed in the Python environment.

5 Result and Analysis

In this section, the impact of the WL and CL embedding is observed in Table 2, which is performed over the BioCreative II Gene Mention task corpus.

Table 2 Comparison of the results with other supportive models

Models	Precision (%)	Recall (%)	*F*-Score (%)
LSTM-CRF's	83.85	82.70	83.27
Bi-LSTM-CRF's	85.05	86.57	86.03
Proposed	89.01	89.90	89.42

The effectiveness of using the word and character level embedding is clearly observed in the *F*-score column as shown in Table 2. It increases the *F*-score to 89.42%. To perform the experiment on the dataset, eight epochs are considered. Figure 3 shows as the number of epochs increases accuracy for both training and validation set increases and loss are gradually decreased as the number of epochs increased (Fig. 4).

Finally, Table 3 presents the comparisons of the proposed model with others to evaluate the same BioCreative II gene Mention corpus.

Ando et al. achieved the first rank in the BioCreative II Gene Mention task with the *F*-score of 87.21% [5]. Kuo et al. [6] secured the second rank of this task with

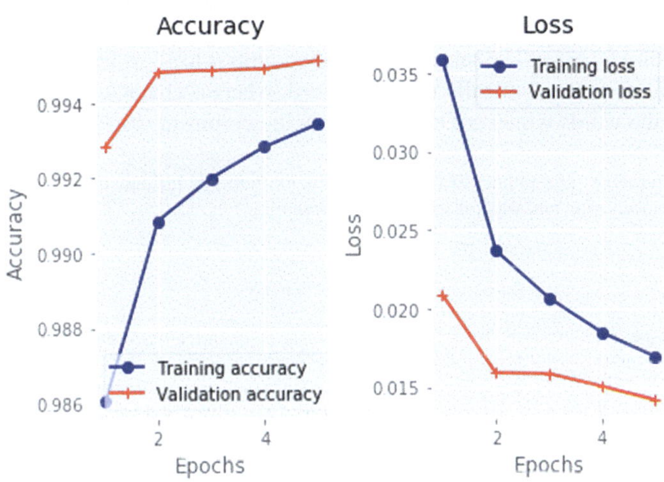

Fig. 4 Graphical representation for accuracy and loss while training

Table 3 Comparison of the proposed work with other related works in this area

Models	Precision (%)	Recall (%)	*F*-Score (%)
Proposed	89.01	89.90	89.42
Li et al.'s [4]	90.67	85.66	88.09
Chang et al.'s [1]	88.67	86.68	88.09
Ando's [5]	88.48	85.97	87.21
Kuo et al.'s [6]	89.30	84.49	86.83

86.83%. Chang et al. achieve an F-score of 87.67% [1] and Li et al. used shallow machine learning techniques and achieve the F-score of 88.09% [4].

Overall, the proposed model secured the highest F-score as a comparison to its competitor system. The F-score achieved by the proposed framework is +1.33% upgraded than the last highest score obtained in the BioCreative II Gene Mention task corpus.

6 Conclusion and Future Work

A deep learning based gene mention tagging framework is proposed in this paper. The framework contains, bidirectional LSTM and CRF layer with word and character level embedding. The experiment performed by the framework achieved 89.42% of F-measure on the BioCreative II Gene Mention corpus task, which is higher as compared with the other proposed methods and has obained top rank in performance in the BioCreative II Gene Mention task. The purpose of using the combination of bidirectional LSTM with CRF is to enhance the performance of the framework where CRF is responsible for higher tagging accuracy and LSTM is used to make past and future features. The role of word embeddings is to allow learning of the same representations for semantically the same word, whereas character level embedding deals with the words which are not covered in the corpus of vocabulary.

Acknowledgements The authors would like to thank the National Institute of Technology Raipur, India for the motivation and support for doing this research.

References

1. Chang, Y.M., Kuo, C.J., Huang, H.S., Lin, Y.S., Hsu, C.N.: Analysis and enhancement of conditional random fields gene mention taggers in BioCreative II challenge evaluation. LBM (Short Papers) **7**(1) (2007)
2. Wang, L., Sun, Y., Zhu, Z.: Knowledge points extraction of junior high school english exercises based on SVM method. In: Proceedings of the 2nd International Conference on E-Education, E-Business and E-Technology, pp. 43–47. ACM, New York, United State (2018)
3. Smith, L., Tanabe, L.K., nee Ando, R.J., Kuo, C.J., Chung, I.F., Hsu, C.N., Torii, M.: Overview of BioCreative II gene mention recognition. Genome Biol. **9**(2), S2 (2008)
4. Li, L., Sun, J., Huang, D.: Boosting performance of gene mention tagging system by classifiers ensemble. In: Proceedings of the 6th International Conference on Natural Language Processing and Knowledge Engineering (NLPKE-2010), pp. 1–4. IEEE, Beijing, China (2010)
5. Ando, R.K.: BioCreative II gene mention tagging system at IBM Wat-son. In: Proceedings of the Second BioCreative Challenge Evaluation Workshop, pp. 101–103. Centro Nacional de Investigaciones Oncologicas (CNIO) Madrid, Spain (2007)
6. Kuo, C.J., Chang, Y.M., Huang, H.S., Lin, K.T., Yang, B.H., Lin, Y.S., Chung, I.F.: Rich feature set, unification of bidirectional parsing and dictionary filtering for high F-score gene mention tagging. In: Proceedings of the Second BioCreative Challenge Evaluation Workshop, pp. 105–107. Centro Nacional de Investigaciones Oncologicas (CNIO) Madrid, Spain (2007)

7. Yang, X., Gao, Z., Li, Y., Pan, C., Yang, R., Gong, L., Yang, G.: Bidirectional LSTM-CRF for biomedical named entity recognition. In: 14th International Conference on Natural Computation, Fuzzy Systems and Knowledge Discovery (ICNC-FSKD), pp. 239–242. IEEE, Xi'an, China (2018)
8. Sutton, C., McCallum, A.: An Introduction to Conditional Random Fields. arXiv, p. arXiv preprint arXiv:1011.4088 (2010)
9. Laddha, A., Mukherjee, A.: Aspect Specific Opinion Expression Extraction using Attention based LSTM-CRF Network (2019). arXiv preprint arXiv:1902.02709
10. BioCreative Homepage. https://biocreative.bioinformatics.udel.edu. Last ccessed 01 Nov 2019
11. Huang, M.S., Lai, P.T., Tsai, R.T.H., Hsu, W. L.: Revised JNLPBA Corpus: A Revised Version of Biomedical NER Corpus for Relation Extraction Task (2019). arXiv preprint arXiv:1901.10219 (2019)

Mining of Cancerous Region from Brain MRI Slices with Otsu's Function and DRLS Segmentation

Manju Jain and C. S. Rai

Abstract Cancer is one of the deadly diseases and major causes of death in the human group. A significant number of methods are already proposed to detect and cure cancer in its early phase. Cancer cell growth in the brain section is known as the Brain-Cancer (BC) and the early identification and the cure may improve the life span of the patient. Magnetic-Resonance-Imaging (MRI) is normally considered to record the brain section and which can be considered for further checking. This article discusses a practice to extract and study the cancer segment from the MRIs of Flair, T1C, and T2 modalities. This paper implements Cuckoo-Search driven Otsu's thresholding to improve the MRI slice and Distance-Regularized-Level-Set (DRLS) to extort the infected segment for further evaluation. The aim of this article is to discuss the merit of the modalities, such as Flair, T1C, and T2 in the examination of the MRI of BC. The result confirms that Flair modality helps to achieve precise tumor and edema mining compared to other modalities. Further, the average accuracy in Flair is found to be >91%.

Keywords Brain-Cancer · Modalities · Processing · Inspection · Accuracy

1 Introduction

From the recent research works, it can be confirmed that major casualty in humans is due to cancer in internal organs [1–3]. Even though a notable amount of detecting and curing procedures are existing to detect cancer in humans, the casualty keeps on

---

M. Jain (✉)
Department of Computer Engineering, Directorate of Training and Technical Education, NCT of Delhi, Delhi 110083, India
e-mail: manju.jain69@gov.in

C. S. Rai
University School of Information Communication & Technology, Guru Gobind Inderprashta University, Delhi, New Delhi, India
e-mail: csrai@ipu.ac.in

S. C. Satapathy et al. (eds.), *Intelligent Data Engineering and Analytics*, Advances in Intelligent Systems and Computing 1177,
https://doi.org/10.1007/978-981-15-5679-1_62

increasing due to various reasons, such as heredity, lifestyle, irregular food habits, etc. [4]. Hence, scheduled as well as frequent health checkups are widely organized by the hospitals to discover the disease in interior organs in its early phase. If the disease is accurately identified, then the medical clinics will plan for a suitable treatment procedure to save the patients.

In clinics, most of the diseases are identified by recording the bio-signals as well as bio-images using a prescribed procedure under the guidance of a skilled doctor. Invasive and noninvasive bio-imaging techniques always provide essential information on the interior organ compared to the bio-signaling procedures. After recording the patient's information, the doctor will assist in the initial diagnosis and may request the expert's view to categorize the disease phase. In this article, the examination of Brain-Cancer (BC) based on the bio-imaging methodology called Magnetic-Resonance-Imaging (MRI) is adopted for the investigation.

MRI is quite famous in the medical field due to its noninvasive nature and wider modalities, in which every modality records a unique feature. MRIs are normally considered to record a variety of internal organs, and in this article, the brain section recorded with MRI is adopted for the study. Further, the MRI provides a three-dimensional (3D) picture, which can be analyzed in 3D or 2D form [5–8].

3D MRI requires more computation time contrast to 2D MRI; hence, this work considers the 2D MRI slices of Flair, T1C, and T2 for the examination. Initially, the BC section of MRRI is enhanced with the Otsu's and Cuckoo-Search (Otsu + CS) thresholding. In this action, the test image is detached into different levels, like background, brain tissue, and tumor section. Later, the DRLS techniques are then implemented to extract the tumor fragment from the threshold BC picture [8].

An added advantage of the proposed approach is the hybrid scheme which helps the operator to mine the tumor slice by a greater accuracy. In this effort, integration of the image processing procedure is implemented with the essential processing methodology, to extract the BC section from the input picture. Initially, a trilevel threshold is presented to develop the test picture and later the DRLS scheme is implemented to mine the tumor segment. Finally, a comparative evaluation is presented for the extorted section and the Ground-Truth (GT) to confirm the performance of the implemented technique by computing the essential values like the Accuracy (AC), Precision (PR), Sensitivity (SE), and Specificity (SP) parameters. The result of this article confirms that Flair modality images of the BC present better values contrast with other modality pictures. This verifies that the planned technique works well on the Flair modality picture than the T1C and T2 modalities and in future, it can be used to inspect the scientific class MRI to support the doctor who involves in the disease examination practice and the treatment planning for the patients to cure the BC.

2 Related Works

In recent years, a substantial quantity of brain picture assessment procedures is widely discussed and applied to provide possible support for the BC picture examination task.

The work by Rajinikanth and Satapathy [6] proposed a procedure to appraise the brain stroke pictures using a class of evaluation techniques [6]. The work by [9] discussed a hybrid technique to examine the magnetic-resonance-angiography picture using the soft-computing technique and from this they achieved an improved effect contrast to the accessible result [8]. The work of Revanth et al. [8] implemented an examination technique for Flair/DW modality approach [8]. The recent work of Satapathy and Rajinikanth [10] implemented Jaya algorithm based hybrid technique to assess the brain MRIs. These works confirm that brain MRI evaluation based on the hybrid technique offered superior outcomes compared to the traditional technique [10]. From these approaches, it can be established that heuristic algorithm based approach offers better results on MRI evaluation.

3 Methodology

The tactic executed in this article is presented in Fig. 1. Initially, the 2D slices of MRIs are extracted from the 3D database. The 2D slices of Flair, T1C, and T2 are

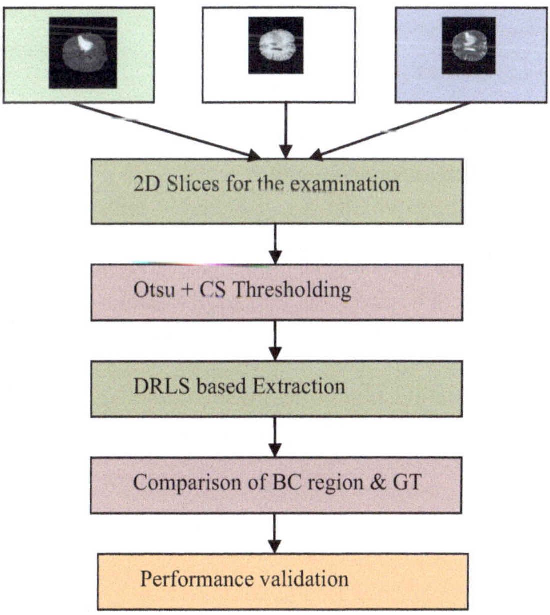

Fig. 1 Pictorial representation of the proposed technique

considered for the examination and the corresponding common GT pictures are then considered to verify the benefit of the proposed tool.

At first, the 2D slices of chosen modality are used for the examination. The Otsu + CS thresholding is initiated with trilevel thresholding as discussed in the earlier works [11–16]. This operation will improve the visibility of the BS region by identifying the best threshold and the enhanced picture is considered as the preprocessed image. Later, the DRLS segmentation practice is then applied to mine the BC segment from the enhanced MRI slice. A qualified appraisal between extracted BC section and GT is then executed to compute the essential values like; AC, PR, SE, and Sp are computed to confirm the merit of the tool. Based on the values of the above-mentioned parameters, the concert of the tool is established and the modality which suits this tool can be identified easily.

3.1 Otsu's + Cuckoo-Search (Otsu + CS) Thresholding

Otsu's thresholding is an essential image-enhancing procedure largely discussed to threshold the traditional and medical pictures. This work considers the Otsu's between-class practice to preprocess test picture to get better visibility and its explanation can be found in the following articles. The Cuckoo-Search (CS) algorithm is also a proven heuristic approach widely adopted for a class of medical image examination procedures [17]. The proposed approach is used to enhance the brain picture based on trilevel thresholding.

3.2 DRLS Segmentation

Segmentation procedures are widely adopted to extract the required sections from the raw and preprocessed medical pictures. In this work, Distance-Regularized-Level-Set (DRLS) technique is considered to mine the BC sector from the preprocessed Flair, T1C, and T2 modality pictures. This work is initiated as a bounding box with chosen axis coordinates and the box is permitted to shrivel toward the core of the BC segment based on the increased iteration. This work can be used to identify all the pixel groups related to the BC segment. Normally, in the MRI slice, the infected brain portion will have more visibility in the Flair modality than other brain tissues. Hence, identification of the infected brain section is easy when a grayscale picture is adopted [5].

3.3 Evaluation Based on the Study Among BC Segment and GT

To authenticate the dominance of implemented technique, a comparative assessment connecting the BC slice and the GT is implemented and the essential values like the Jaccard, Dice, sensitivity, specificity, accuracy, and precision are computed [18, 19]. The average of the performance measures of Flair, T1C, and T2 modalities are computed to justify the dominance of the proposed technique.

4 Results and Discussions

This section of the article presents the investigational result and its conversation obtained for the chosen test pictures. All the investigations are performed using the MATLAB software and the results are separately presented and discussed for each stage existing in the proposed image diagnosis tool.

Figure 2 depicts the sample test pictures of Flair, T1C, and T2 used for the examination. To simplify the assessment, in this article, only the 2D slices are considered as presented in Fig. 2. Figure 2a depicts the modality and the name of the 2D test

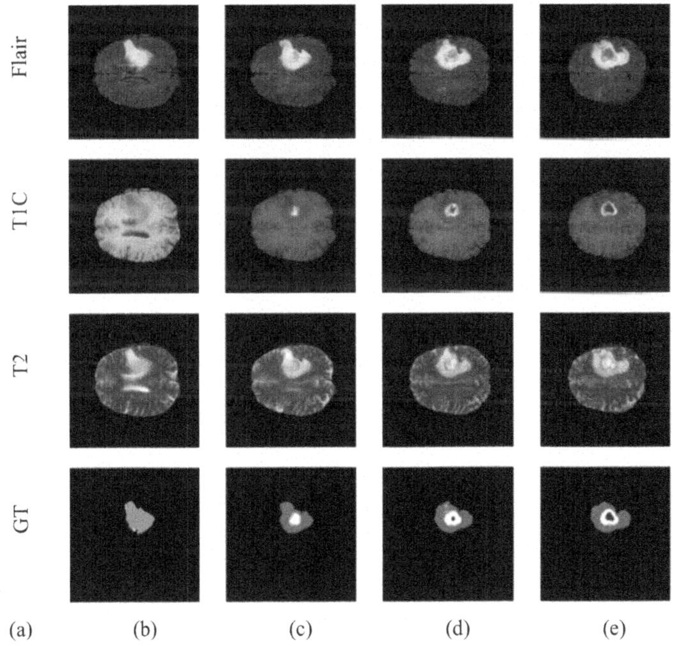

Fig. 2 Sample test pictures. **a** Modality, **b–e** 2D slices, such as 90, 95, 100, and 105

samples considered for the examination and Fig. 2b–e presents the various 2D slices of the MRIs and the associated GTs.

Initially, trilevel thresholding is applied for every image existing in Fig. 2 and the corresponding outcomes are recorded. Initially, the thresholding is executed with the Flair picture and the enhanced section is then recorded for the evaluation. Alike practice is executed for all other images cases and the matching outcome is tabulated as existing in Fig. 3. Figure 3a, b represents the threshold Flair picture and the extracted BC region. Similarly, Fig. 3c, d depicts the threshold T1C and the extracted BC core and Fig. 3e, f presents the results of T2 modality slices.

Figure 4 presents the execution of DRLS segmentation on the 2D BC slices. For the demonstration purpose, Flair modality MRI (slice90) is used for the estimation. Figure 4a presents the initialed bounding box with the DRLS. Later, the DRLS contour is allowed to regulate itself to recognize the entire possible BC region pixel group with an increase in the iteration value. When the DRLS identifies the entire achievable pixel groups, the DRLS modification stops as in Fig. 4b and the extracted section is presented in Fig. 4c. The extracted section is then compared with the related

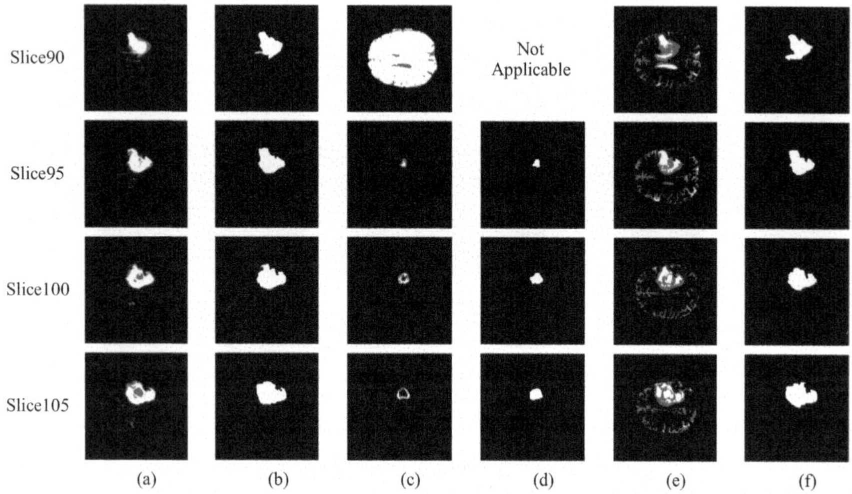

Fig. 3 Results achieved for Flair, T1C, and T2 modality pictures. **a**, **c**, and **e** Threshold pictures, **b**, **d** and **f** Extracted BC region

Fig. 4 Result of post-processing procedure. **a** Initial DRLS, **b** Converged DRLS, **c** Segmented BC region

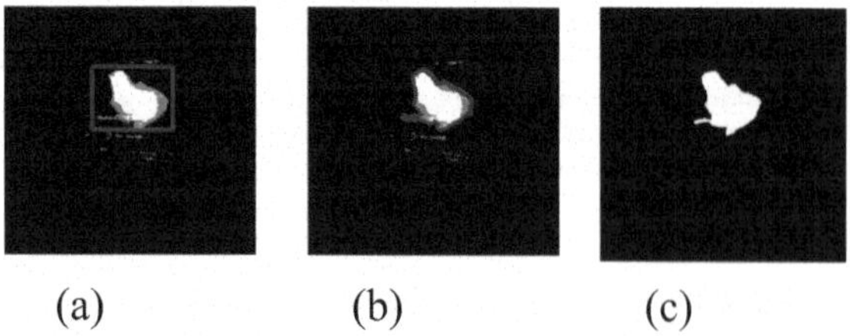

GT to find the possible image metrics to confirm the superiority of the proposed MRI assessment tool.

From this experiment, it can be noted that the present tool works fine on the Flair and the T2-modality picture. For the T1C case, it extracts only the tumor core and the image metrics attained with the proposed scheme for the T1C modality image is also very less compared to the Flair and the T2 MRI cases. The investigational result of this research substantiates that the hybrid practice will work well on the Flair and the T2-modality pictures. The enhanced exactness with the T1C can be attained by applying a probable image grouping practice as discussed in the related work [1, 20, 21].

5 Conclusion

The proposed work aims to develop an image examination proposal for the MRIs with modalities, like Flair, T1C, and T2. To exhibit the presentation of the projected tool, only 2D slices are considered for the examination. This work employs a preprocessing with Otsu + Cuckoo-Search (Otsu + CS) algorithm and the post-processing based on the DRLS. The initial part of the tool enhances the test picture and the later part will extract the BC region for further examination. The outcome attained with the proposed experimental work confirms that this tool works well on the Flair and T2 modality imagery and offers poor results on the T1C modality picture. This problem can be overcome by fusing the T1C with the Flair or the T2 modality MRI slices. Further, the average accuracy attained with the Flair modality is >91% and this substantiates that this tool works fine on the Flair modality case than the T1C and T2.

References

1. Rajinikanth, V., Satapathy, S.C., Fernandes, S.L., Nachiappan, S.: Entropy based segmentation of tumor from brain mr images—a study with teaching learning based optimization. Pattern Recognit. Lett. **94**, 87–94 (2016). https://doi.org/10.1016/j.patrec.2017.05.028
2. Jahmunah, V., et al.: Automated detection of schizophrenia using nonlinear signal processing methods. Artif. Intell. Med. **100**, 101698 (2019). https://doi.org/10.1016/j.artmed.2019.07.006
3. Acharya, U.R., et al.: Automated detection of Alzheimer's disease using brain MRI images—a study with various feature extraction techniques. J. Med. Syst. **43**, 302 (2019). https://doi.org/10.1007/s10916-019-1428-9
4. Chaki, J., Dey, N., Shi, F., Sherratt, R.M.: Pattern mining approaches used in sensor-based biometric recognition: a review. IEEE Sens. J. (2018). https://doi.org/10.1109/JSEN.2019.2894972
5. Rajinikanth, V., Fernandes, S.L., Bhushan, B., Sunder, N.R.: Segmentation and analysis of brain tumor using Tsallis entropy and regularised level set. Lecture Notes in Electrical Engineering, vol. 434, pp. 313–321 (2018)
6. Rajinikanth, V., Satapathy, S.C., Dey, N., Vijayarajan, R.: DWT-PCA image fusion technique to improve segmentation accuracy in brain tumor analysis, Lecture Notes in Electrical Engineering, vol. 471, pp. 453–462 (2018). https://doi.org/10.1007/978-981-10-7329-8_46

7. Rajinikanth, V., Satapathy, S.C.: Segmentation of ischemic stroke lesion in brain MRI based on social group optimization and Fuzzy-Tsallis entropy. Arab. J. Sci. Eng. **43**(8), 4365–4378 (2018). https://doi.org/10.1007/s13369-017-3053-6
8. Revanth K, Varsha shree TD, Raja NSM, Rajinikanth V.: Computational investigation of stroke lesion segmentation from flair/DW modality MRI. In: Fourth International Conference on Biosignals, Images and Instrumentation (ICBSII), pp. 206–212 (2018). https://doi.org/10.1109/icbsii.2018.8524617
9. Raja, N.S.M., et al.: Contrast enhanced medical MRI evaluation using Tsallis entropy and region growing segmentation. J. Ambient Intell. Hum. Comput. 1–12 (2018). https://doi.org/10.1007/s12652-018-0854-8
10. Satapathy, S.C., Rajinikanth, V.: Jaya algorithm guided procedure to segment tumor from brain MRI. J. of Optim. **2018**, 12 (2018). https://doi.org/10.1155/2018/3738049
11. Dey, N., Shi, F., Rajinikanth, V.: Leukocyte nuclei segmentation using entropy function and Chan-Vese approach. Inf. Technol. Intell. Transp. Syst. **314**, 255–264 (2019). https://doi.org/10.3233/978-1-61499-939-3-255
12. Moraru, L., et al.: Gaussian mixture model for texture characterization with application to brain DTI images. J. Adv. Res. **16**, 15–23 (2019). https://doi.org/10.1016/j.jare.2019.01.001
13. Li et al.: Dual feature selection and rebalancing strategy using metaheuristic optimization algorithms in X-ray image datasets. Multimed. Tools Appl. 1–21 (2019). https://doi.org/10.1007/s11042-019-7354-5
14. Bhateja, V., Nigam, M., Bhadauria, A.S., Arya, A., Yu-Dong Zhang, Y-D.: Human visual system based optimized mathematical morphology approach for enhancement of brain MR images, J. Ambient. Intell. Humaniz. Comput. 1–9 (2019). https://doi.org/10.1007/s12652-019-01386-z
15. Yang, X.S.: Nature-Inspired Metaheuristic Algorithms, 2nd edn. Luniver Press, Frome (2018)
16. Dey, N., Rajinikanth, V., Ashour, A.S., Tavares, J.M.R.S.: Social group optimization supported segmentation and evaluation of skin melanoma images. Symmetry **10**(2), 51 (2018). https://doi.org/10.3390/sym10020051
17. Nair. M.V. et al.: Investigation of breast melanoma using hybrid image-processing-tool. In. International Conference on Recent Trends in Advance Computing (ICRTAC), pp. 174–179. IEEE (2018). https://doi.org/10.1109/ICRTAC.2018.8679193
18. Menze, et al.: The multimodal brain tumor image segmentation benchmark (BRATS). IEEE Trans. Med. Imaging **34**(10), 1993–2024 (2015). https://doi.org/10.1109/TMI.2014.2377694
19. Manic S, Al Naimi IS, Hasoon FN, Rajinikanth V.: Jaya algorithm-assisted evaluation of tooth elements using digital bitewing radiography images. Comput. Tech. Dental Image Anal. 107–128 (2019). DOI: https://doi.org/10.4018/978-1-5225-6243-6.ch005
20. Lakshmi, B., Parthasarathy, S.: Human action recognition using median background and max pool convolution with nearest neighbor. Int. J. Ambient Comput. Intell. (IJACI) **10**(2), 34–47 (2019). https://doi.org/10.4018/IJACI.2019040103
21. Wang, R., Wang, G.: Web text categorization based on statistical merging algorithm in big data environment. Int. J. Ambient Comput. Intell. (IJACI) **10**(3), 17–32 (2019). https://doi.org/10.4018/IJACI.2019070102

An Automated Person Authentication System with Photo to Sketch Matching Technique

P. Resmi, R. Reshika, N. Sri Madhava Raja, S. Arunmozhi, and Vaddi Seshagiri Rao

Abstract Recently, considerable measures are taken to improve security in various locations. Most of the methods use the image-supported identification due to its accuracy and robustness. Usually, the picture assessment engages in capturing the raw image, identifying the vital information, extracting the features, and classification and authentication. This technique is used widely in human recognition and also in criminal science, recognition of the right individual is necessary for the verification necessities. In these applications, usually, the sketch drawn by an expert or a computer is matched alongside the digital photographs accessible in the criminal or the public database. Throughout this assessment, necessary facial features are mined from the photo and compared it with the original picture. This paper aims to evaluate the existing picture improvement and recognition procedures in the literature. After generating the necessary drawing for an individual, a relative examination against the digital picture of the person is executed and the image similarity measures are computed to authenticate photo with the sketch. This work proposes a methodology to evacuate the proposed technique using benchmark datasets and the result demonstrates a better result.

Keywords Person authentication · Sketch · Facial features · Detection · Accuracy

---

P. Resmi · R. Reshika · N. Sri Madhava Raja (✉)
Department of Electronics and Instrumentation Engineering, St. Joseph's College of Engineering, OMR, Chennai 600119, Tamil Nadu, India
e-mail: nsrimadhavaraja@stjosephs.ac.in

S. Arunmozhi
Department of Electronics and Communication Engineering, Manakula Vinayagar Institute of Technology, Madagadipet, Puducherry, India

V. S. Rao
Department of Mechanical Engineering, St. Joseph's College of Engineering, OMR, Chennai 600119, Tamil Nadu, India

S. C. Satapathy et al. (eds.), *Intelligent Data Engineering and Analytics*, Advances in Intelligent Systems and Computing 1177,
https://doi.org/10.1007/978-981-15-5679-1_63

1 Introduction

Rapid advancement in technology enhances human life and comfortless. Even though a number of security measures are taken, still there is a need for enhancing the security with recent methods; in order to have more comfortless. Person identification system is one of the security measures widely considered to identify the right/wrong person during the criminal identification and restrict the trespassing in highly secured areas [1–3].

The proposed work plays a chief role in crime investigation. When a crime happens, the inquiry team gathers all the essential evidence from the crime sight, which comprises the objects, fingerprint, and the eye witness/scrutiny camera picture of the suspect. If the scrutiny camera picture is not obtainable, then based on the intimation given by the witness, a specialist will draw the face-image of the suspect [4, 5]. This imaging practice is called the Composite-Sketch (Sketch), which plays a fundamental function in suspect recognition. An experienced composite artist may create the sketch of the suspect based on the information given by the victim/eyewitness. After getting the required sketch, the facial features are then checked against the imagery accessible in the criminal database maintained with the investigating division [6, 7]. This is an established practice in forensic/crime discipline and the correctness in suspect identification relies mainly on the comparison of the sketch with photo [8–10].

In the literature, a number of procedures are conferred to confirm the right/wrong person based on the existing sketches. The success of this procedure relies mainly on the procedures implemented to identify the person. The earlier work implemented a number of techniques in person identification system with better accuracy.

The main aim of the proposed work is to implement a procedure to enhance the outcome during the photo and sketch matching process. In this work, the procedures such as: (i) Firefly-Algorithm and Otsu-based image thresholding, (ii) Feature extraction based on Speeded-Up-Robust-Features (SURF) and Histogram of Oriented-Gradients (HOG), and (iii) Computation of image similarity measures and (iv) Authentication accuracy is implemented. During this work, the Photo-Sketch database of the Chinese University of Hong Kong (CUHK) dataset is considered for the examination [11]. This study was implemented with the MATLAB software and a cropped/resized image of size 250 $\times$ 300 pixels is considered for the examination.

This paper is arranged as follows; Sect. 2 explains the methodology, Sect. 3 presents the results and discussion and the conclusion is discussed in Sect. 4.

2 Methodology

This research presents a soft-computing based technique to authenticate a person based on the photo-sketch matching process and Fig. 1 shows the methodology executed.

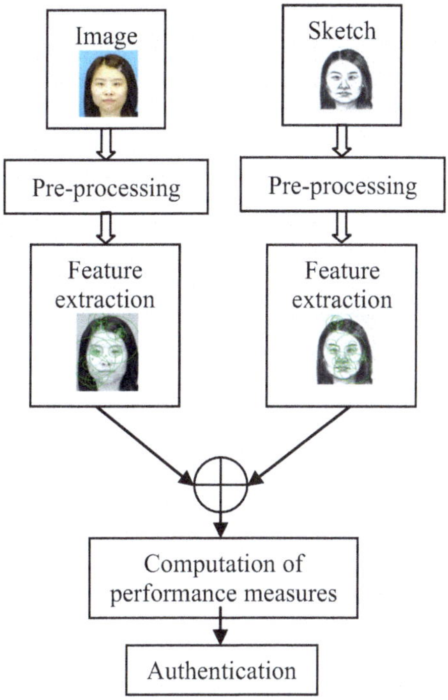

Fig. 1 Photo-sketch matching procedure of this work

Initially, the test images (Photo/Sketch) are collected from [11]. Later, these images are resized into 250 × 300 pixel-sized images to reduce the computation complexity. Then a four-level threshold is implemented to enhance the facial features of the test picture. Then, the features are mined with HOG and SURF for photo/sketch and then these features are compared. During the comparison, the essential image similarity measures are then computed and based on this computation, the identification of the photo with the sketch is authenticated.

2.1 *Photo/Sketch Database*

This work considers the benchmark CUHK dataset for the examination. This dataset consist of 88 numbers of RGB-scale photos and 88 numbers of the grayscale sketches. The image resizing is initially implemented to have unique sized images. The sample images of the dataset can be found in Fig. 2. This dataset is widely considered in the literature to examine the photo-sketch matching system.

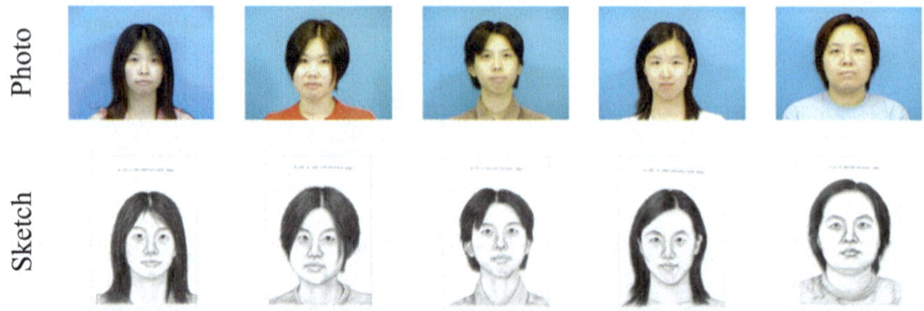

Fig. 2 Sample test pictures of CHUK database

2.2 *Image Enhancement*

In this work, the trial picture is then enhanced with the firefly algorithm and Otsu (FA + Otsu) based four-level threshold. This threshold enhances the picture (photo/sketch) by separating it into background, face separation from body, hair separation, and facial features. The essential details on (FA + Otsu) can be found in [12–17]. In this enhancement, the algorithm parameters are assigned as follows; no. of fireflies = 30, dimension = 4, stopping criteria = maximized Otsu's function and iteration = 2000. After the possible enhancement with the thresholding, the essential image features are then extracted using feature extraction process from the original photo and the thresholded photo and then compared against the features from the sketch. During this comparison, the image similarity features are computed and based on this value, the authentication accuracy is confirmed.

2.3 *Feature Extraction*

The facial features of original, thresholded, and sketch images are extracted using a chosen feature mining procedures. Feature extraction also plays a major role in machine-learning and deep-learning-based recognition systems [18–20]. In the literature, Speeded-Up-Robust-Features (SURF) and Histogram of Oriented-Gradients (HOG) features are widely considered in the face recognition and image classification tasks [21]. The SURF and HOG will mine all the essential information existing in the image based on the pixel and its density. In any automated system, the identification accuracy mainly relies on the considered database. The feature extraction process is considered to produce a feature vector of definite size (one dimension). The SURF and HOG features are combined together to form a single feature vector, which is then used to authenticate the person/photo based on the sketch. The other details on the feature extraction and authentication can be found in [18–20].

2.4 Similarity Measure and Validation

In the imagining literature, similarity measures are widely used to compare the processed picture with an unprocessed image. The measures, Root-Mean-Square-Error (RMSE), Peak-Signal-to-Noise Ratio (PSNR), Structural-Similarity-Index-Matrix (SSIM), Normalized-Absolute-Error (NAE), Normalized-Cross-Correlation (NCC), and Average-Difference (AD) are computed [22, 23]. During this evaluation, every image feature is compared against the features attained with the sketch and based on the rate of the similarity measures achieved with this comparison, the accuracy of the proposed system is validated based on the mean of the NAE [24–28].

3 Result and Discussion

This section of the paper presents the experimental outcomes and its deliberations. Initially, the feature extraction is implemented on the gray version of the test image, FA + Otsu threshold image, and the sketch. The outcome of the proposed technique is presented in Fig. 3. In this, Fig. 3a shows the original photo. Figure 3b, c represents the gray version and the SURF feature extraction process. Figure 3d, e presents the thresholded picture and the feature extraction, respectively. A similar result is also presented in Fig. 3f, g shows the outcome of the sketch picture.

This results reveals that the key features of the image texture are approximately similar in original, threshold, and sketch image in most of the cases and the HOG-based technique helps to extract the histogram oriented (pixel-level) information. A

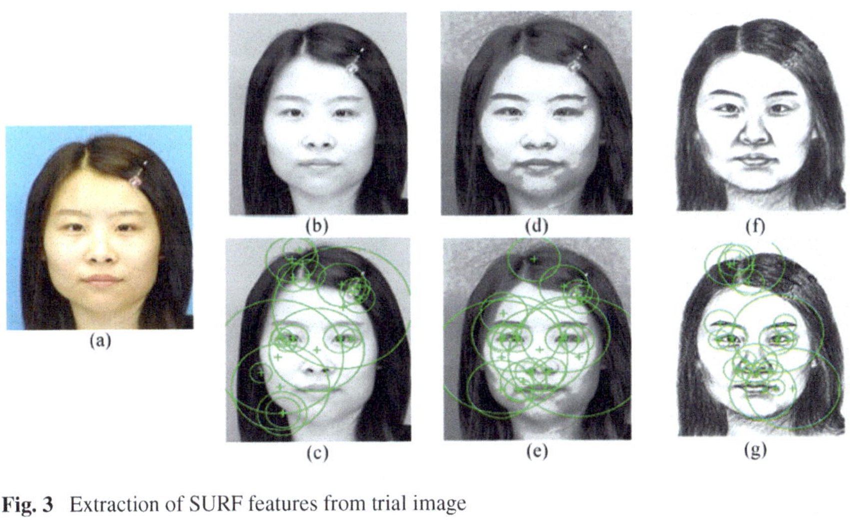

Fig. 3 Extraction of SURF features from trial image

Table 1 Average performance measure attained with the proposed work

Image	RMSE	PSNR	SSIM	NAE	NCC‘	AD
Threshold photo versus sketch	98.7306	24.69	0.9316	0.1085	0.9277	0.2679
Original photo versus sketch	104.5615	25.81	0.9274	0.1152	0.9216	0.2571

combination of these features will improve the detection accuracy during the photo-sketch matching process. Based on the accuracy, the performance of the proposed work is validated.

A similar procedure is realized for the HOG feature on all 88 pictures of the CUHK database test images and the combined SURF and HOG features are considered for the photo-sketch matching process. The proposed work is implemented on all the images and the performance measures and the accuracy are computed for the original image, threshold image, and the sketch and average values are shown in Table 1.

Finally, the accuracy attained with the photo-sketch matching is computed and the mean accuracy of the threshold photo versus sketch and photo versus sketch is considered to confirm the superiority of the proposed technique. From Fig. 4, it can be noted that the proposed work based on FA + Otsu thresholding and SURF and HOG feature helps to attain better accuracy compared to the conventional technique.

The future the proposed task can be extended using the following approaches:

(i) Implementing feature ensemble technique to enhance the detection.
(ii) Implementing entropy-based enhancement.
(iii) Analyzing the robustness of the proposed technique using the noise corrupted test pictures.

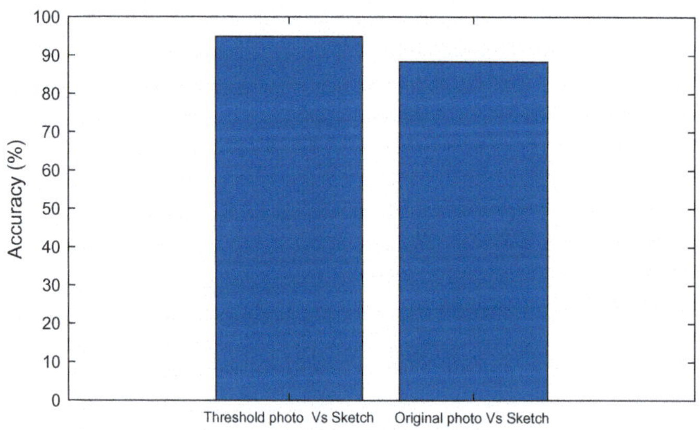

Fig. 4 Attained accuracy with the original and enhanced pictures

4 Conclusion

In security system and forensic science, person identification plays a vital role to solve a problem during the photo-sketch matching process. The proposed work implements a procedure; which implements the initial processing of photo with FA + Otsu thresholding and the SURF and HOG feature extraction to enhance the detection accuracy. After extracting the features of photo/sketch, a comparison is then implemented to identify the percentage of fit between the features of photo/sketch. During this evaluation, image similarity measures are computed and the outcome of this procedure confirms that the proposed work helps to achieve better accuracy (94.16%) compared to the traditional photo-sketch (89.38%) matching process. This result confirms that this technique works well on the CUHK database and in the future it can be tested using the real-time database existing along with its composite-sketch.

References

1. Li, SZ, Jain, AK: Handbook of Face Recognition, NY, USA. Springer (2011)
2. Zhao, W., Chellappa, R., Phillips, P.J., Rosenfeld, A.: Face recognition: a literature survey. ACM Comput. Surv. **35**(4), 399–458 (2003)
3. Medioni, G., Choi, J., Kuo, C.-H., Fidaleo, D.: Identifying non cooperative subjects at a distance using face images and inferred three dimensional face models. IEEE Trans. Syst. Man Cybern. A Syst. Hum. **39**(1), 12–24 (2009)
4. Vijayakumari, V.: Face recognition techniques: a survey. World J. Comput. Appl. Technol. **2**, 41–50 (2013)
5. Fernandes, S.L., Bala, G.J.: Self-similarity descriptor and local descriptor-based composite sketch matching. Adv. Intell. Syst. Comput. **436**, 643–649 (2016). https://doi.org/10.1007/978-981-10-0448-3_53
6. Fernandes, S.L., Bala, G.J.: A study on face recognition under facial expression variation and occlusion. Adv. Intell. Syst. Comput. **397**, 371–377 (2016). https://doi.org/10.1007/978-81-322-2671-0_35
7. Klum, C., Han, H., Klare, B., Jain, A.K.: The FaceSketchID system: matching facial composites to mugshots. IEEE Trans. Inf. Forensics Secur. (TIFS) **9**(12), 2248–2263 (2014)
8. Tang, X., Wang, X.: Face sketch recognition, IEEE Trans. Circuits Syst. Video Technol. (CSVT), Spec. Issue Image Video Based Biometrics **14**(1), 50–57 (2004)
9. Wang, X., Tang, X.: Face photo-sketch synthesis and recognition. IEEE Trans. Pattern Anal. Mach. Intell. (PAMI) **31** (2009)
10. Martinez, A.M., Benavente, R.: The AR Face Database. CVC Technical Report #24 (1998)
11. CUHK. http://mmlab.ie.cuhk.edu.hk/archive/facesketch.html
12. Lakshmi, B., Parthasarathy, S.: Human action recognition using median background and max pool convolution with nearest neighbor. Int. J. Ambient Comput. Intell. (IJACI) **10**(2), 34–47 (2019). https://doi.org/10.4018/IJACI.2019040103
13. Wang, R., Wang, G.: Web text categorization based on statistical merging algorithm in big data environment. Int. J. Ambient Comput. Intell. (IJACI) **10**(3), 17–32 (2019). https://doi.org/10.4018/IJACI.2019070102
14. Rajinikanth, V., Dey, N., Satapathy, S.C., Ashour, A.S.: An approach to examine magnetic resonance angiography based on Tsallis entropy and deformable snake model. Futur. Gener. Comput. Syst. **85**, 160–172 (2018). https://doi.org/10.1016/j.future.2018.03.025

15. Raja, N.S.M., Fernandes, S.L. Dey, N., Satapathy, S.C., Rajinikanth, V.: Contrast enhanced medical MRI evaluation using Tsallis entropy and region growing segmentation. J. Ambient. Intell. Humaniz. Comput. 1–12 (2018). https://doi.org/10.1007/s12652-018-0854-8
16. Raja, N.S.M., Rajinikanth, V., Latha, K.: Otsu based optimal multilevel image thresholding using firefly algorithm. Model Simul. Eng., 17. Article ID 794574 (2014)
17. Rajinikanth, V., Couceiro, M.S.: Optimal multilevel image threshold selection using a novel objective function. Inf. Syst. Des. Intell. Appl. Adv. Intell. Syst. Comput. **340**, 177–186 (2015)
18. Ali, et al.: Adam deep learning with SOM for human sentiment classification. Int. J. Ambient Comput. Int. (IJACI) **10**(3), 92–116 (2019). https://doi.org/10.4018/IJACI.2019070106
19. Acharya, U.R., et al.: Automated detection of Alzheimer's disease using brain MRI images—a study with various feature extraction techniques. J. Med. Syst. **43**, 302 (2019). https://doi.org/10.1007/s10916-019-1428-9
20. Jahmunah, V., et al.: Automated detection of schizophrenia using nonlinear signal processing methods. Artif. Intell. Med. **100**, 101698 (2019). https://doi.org/10.1016/j.artmed.2019.07.006
21. Mehmood, Z., et al.: Content-based image retrieval based on visual words fusion versus features fusion of local and global features. Arabian J. Sci. Eng. **43**(12), 7265–7284 (2018). https://doi.org/10.1007/s13369-018-3062-0
22. Fernandes, S.L., Rajinikanth, V., Kadry, S.: A hybrid framework to evaluate breast abnormality using infrared thermal images. IEEE Consum. Electron. Mag. **8**(5), 31–36 (2019). https://doi.org/10.1109/MCE.2019.2923926
23. Bhandary, A., et al.: Deep-learning framework to detect lung abnormality—a study with chest X-ray and lung CT scan images. Pattern Recogn. Lett. (2019). https://doi.org/10.1016/j.patrec.2019.11.013
24. Bhateja, V., Misra, M., Urooj, S.: Unsharp masking approaches for HVS based enhancement of mammographic masses: a comparative evaluation. Futur. Gener. Comput. Syst. **82**, 176–189 (2018)
25. Srivastava, A., Bhateja, V., Moin, A.: Combination of PCA and contourlets for multispectral image fusion. Adv. Intell. Syst. Comput. **469**, 577–585 (2017). https://doi.org/10.1007/978-981-10-1678-3_55
26. Rajinikanth, V., Fernandes, S.L., Bhushan, B., Sunder, N.R.: Segmentation and analysis of brain tumor using Tsallis entropy and regularised level set. Lecture Notes in Electrical Engineering, vol. 434, pp. 313–321 (2018
27. Fernandes, S.L., et al.: A reliable framework for accurate brain image examination and treatment planning based on early diagnosis support for clinicians. Neural Comput. Appl. 1–12 (2019). https://doi.org/10.1007/s00521-019-04369-5
28. Dey, N., et al.: Social-group-optimization based tumor evaluation tool for clinical brain mri of flair/diffusion-weighted modality. Biocybern. Biomed. Eng. **39**(3), 843–856 (2019). https://doi.org/10.1016/j.bbe.2019.07.005

Extraction of Leukocyte Section from Digital Microscopy Picture with Image Processing Method

R. Dellecta Jessy Rashmi, V. Rajinikanth, Hong Lin, and Suresh Chandra Satapathy

Abstract Appraisal of leukocyte sections is essential to predict the disease in humans. The increase in the leukocyte indirectly represents the disease in humans. This work proposes a methodology to extract the various leukocyte segments from the Microscopic Blood Smear (MBS) image with better accuracy. This work implemented an entropy assisted thresholding and morphology based segmentation to mine the segment. The threshold is executed with the Kapur's function with a Social Group Optimization (SGO). The mining of leukocyte is done with morphological extraction. The mined segment is then compared against the ground-truth (GT) and the performance metrics are then computed. Further, a confusion matrix is also constructed to confirm the performance of the proposed technique. In this analysis, the benchmark images of Leukocyte Images for Segmentation and Classification (LISC) is considered and the attained outcomes are authenticated with existing similar results in literature. The experimental outcome confirms that, proposed technique is efficient in extracting all the five leukocytes from the LISC catalog with improved image quality parameters. This confirms that, proposed system can be used to examine the clinical rank pictures in future.

Keywords Blood smear · Leukocyte · Kapur's threshold · Segmentation · Validation

---

R. Dellecta Jessy Rashmi · V. Rajinikanth (✉)
Department of Electronics and Instrumentation, St. Joseph's College of Engineering, Chennai 600119, Tamil Nadu, India
e-mail: rajinikanthv@stjosephs.ac.in

H. Lin
Department of Computer and Mathematical Sciences, University of Houston, Downtown, TX 77002, USA

S. C. Satapathy
School of Computer Engineering, Kalinga Institute of Industrial Technology (Deemed to Be University), Bhubaneswar 751024, Odisha, India

S. C. Satapathy et al. (eds.), *Intelligent Data Engineering and Analytics*, Advances in Intelligent Systems and Computing 1177,
https://doi.org/10.1007/978-981-15-5679-1_64

1 Introduction

Due to its clinical importance, a considerable number of image processing techniques are proposed and implemented by the researchers for a class of RGB and gray scaled pictures [1–5]. Assessment of clinical grade images is widely chosen, since the developed image evaluation system can be directly used for the clinical evaluation task. But, the availability of the clinical grade pictures is more difficult and all these images are to be recorded as per the ethical guidance and also require the ethical clearance. Hence, in most of the cases, the developed disease systems are initially tested using the benchmark images. If the outcome attained with these pictures is better, then the proposed tool can be suggested to appraise the medical grade pictures.

This work aims to propose an examination task on the leukocyte images of LISC dataset [6]. Evaluation of leukocyte is an essential task during the blood sample screening of the person suffering with the infectious disease. The screening involves in the counting of the leukocyte sections from the blood smear pictures and also its count. This process involves in; (i) Collection of the blood sample from the infected person, (ii) Creating the blood-smear slides with possible precautions, (iii) recording the blood slides using the digital microscopes, (iv) Judgment of these images manually or a dedicated computer program, (v) Detection of the abnormality and treatment planning to cure the infection [7, 8].

The proposed work aims to execute an imaging practice to assess the leukocyte images of LISC database with better accuracy. This work implements a thresholding and segmentation task to extract the leukocyte with an enhanced performance. The threshold is the preliminary technique to separate the segment to be evaluated. This work implements Kapur's entropy based three-level threshold monitored with a Social-Group-Optimization (SGO). The proposed work is implemented on the RGB sale picture, which enhanced the segment with better visibility. Afterwards, a morphology based extraction is executed to mine the leukocyte segment properly. The mined segment is then compared against the GT provided in LISC and the key performance values are then computed.

The earlier works with the SGO based imaging technique confirms that this algorithm helped to attain the better result with the dermoscopy [9] and brain tumor pictures [10–12]. The recent work on the LISC database also confirms that, the segmentation followed by a thresholding provides enhanced outcome. Hence, this work implemented a two-step examination procedure to extract the essential section from MBS pictures. Finally, the performance of the proposed approach is validated based on a comparative assessment with the ground-truth (GT) and computing the essential performance measures.

Remaining work is presented as follows; Sect. 2 explains the scheme, Sect. 3 reveals the results and deliberations and the conclusion is available in Sect. 4.

2 Methodology

The methodology implemented in this work is discussed in Fig. 1. The RGB scale test is collected from the LISC and the test image is then considered for the examination. The preliminary improvement is attained with the multi-thresholding with SGO and Kapur's function. Further, the enhanced section of MBS is then segmented using morphological extraction. Lastly an evaluation of GT and the blood-cell division is completed and the performance values are then computed.

Other discussion of this division describes the methodology considered in this work.

2.1 *Kapur's Thresholding*

Entropy based assessment is widely implemented in recent medical image evaluation tasks due to its better enhancement [13], Kapur's entropy is one of the established entropy measures widely implemented to examine the medical pictures. The essential details on the Kapur's threshold for the gray/RGB scaled pictures can be found in the recent works [14–16]. In this work, this entropy function is considered to enhance the MBS images of the LISC dataset based on a three-level threshold. This threshold separated the test image into, background, normal section and the blood-cell.

In this work, the essential thresholding is achieved with the help of the recent evolutionary technique called the SGO proposed in 2016 [17, 18]. The recent works confirmed the superiority of the SGO on medical image evaluation task. The SGO algorithm considered work and its essential detail can be found in the work of Dey et al. [9]. The initial SGO parameters are assigned as; no. of agents = 25, dimension = 3, iteration = 2000 and stopping criteria = maximized entropy.

2.2 *Morphology Based Segmentation*

The morphology based segmentation technique (MST) is previously discussed in the literature [19, 20]. The morphology based segmentation is closely related with the watershed approach. In MST, the essential information is extracted by identifying the identical pixel group in the threshold picture. This procedure examines the entire pixels of the picture based on the threshold value, and groups the pixel with a higher

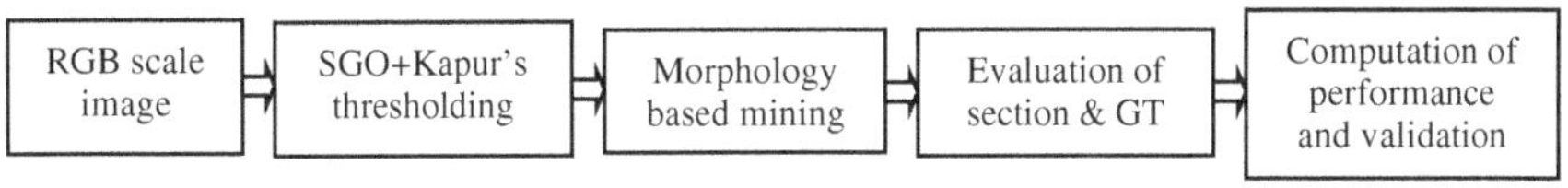

Fig. 1 Different stages of proposed method

value. After identifying the pixels, it groups the pixels and assigns a binary 1 and other pixel as binary 0. Later, the binary 1 will be the required section. The output of this section will be a binary image and later a morphological opening and closing function like in watershed, which further eliminates the additional small pixels groups existing in the binary picture.

2.3 *Image Performance Computation*

Capable estimation of the mined section with GT is done through pixel rank to discover the similarity among two sections. The essential events, like Jaccard, Dice, accuracy, precision, sensitivity, specificity and F1 score are computed [21, 22]. Further, the confusion matrix is also constructed using the true-positive (TP), true-negative (TN), false-positive (FP) and false-negative (FN) attained. The mathematical equations related to these measures can be found in [23–26].

3 Result and Discussion

This part of the paper describes the implementation of the proposed method and the attained results. The LISC benchmark images are considered for the evaluation and the experiment is executed with Matlab software.

The sample test images of the LISC database along with the GT is depicted in the picture. In this work, all the five class leukocytes are considered for the assessment with the proposed technique (Fig. 2).

The proposed work is then implemented on the considered trial pictures and the corresponding results are depicted in Fig. 3. Figure 3a shows the leukocyte class, Fig. 3b, c presents the trial picture and its GT. Figure 3d, e presents the outcomes of threshold and extraction processes respectively.

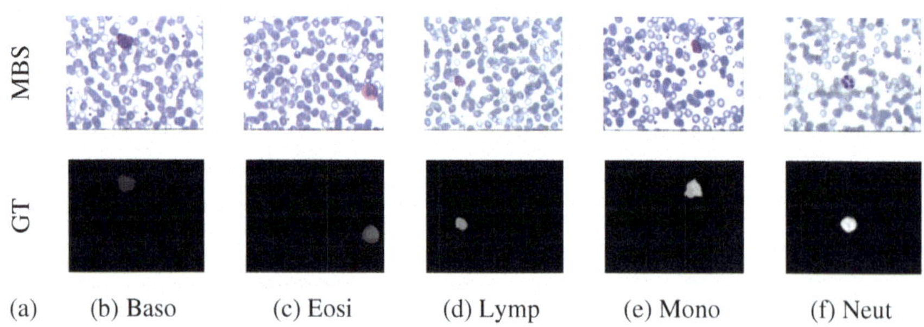

Fig. 2 Sample trial pictures of the LISC dataset

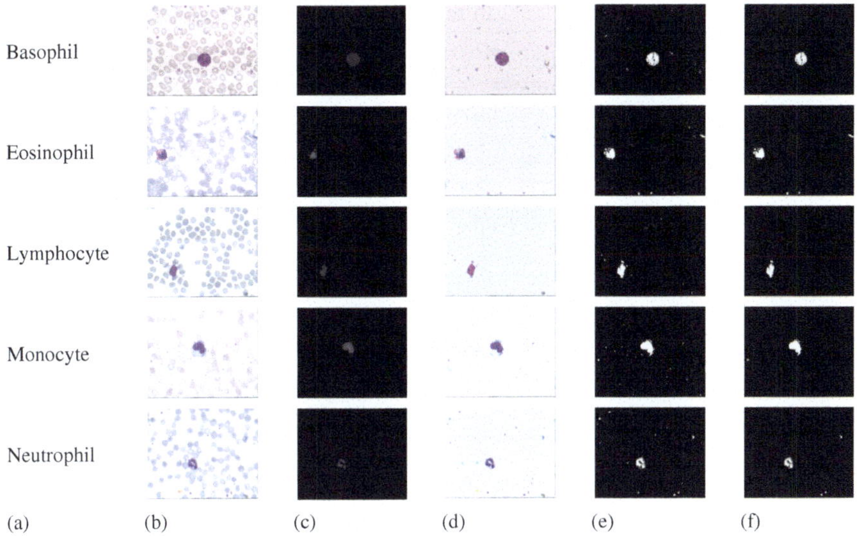

Fig. 3 Experimental outcome on sample trial images. **a** Pseudo name, **b** trial picture, **c** GT, **d** threshold image, **e**, **f** results attained with the segmentation

Later, the extracted section is mixed with the original test image and the corresponding outcomes are as in Fig. 4a, e. From these images, it is clear that, proposed method helps to extract the required region with better accuracy.

The mined section of MBS is then compared against the GT and the essential image performance values are then computed. The results are then assessed to identify the performance of the proposed method. Figure 5 presents the confusion matrix attained for a single image case. Similar result is obtained for all the images of the LISC database. Figure 6 shows the performance measure attained for the Basophil image case. Similar results are attained for other four image classes. From this result, it is confirmed that, proposed technique is efficient in extracting the required segment from the MBS and in future, its operation can be confirmed with the real clinical MBS.

Fig. 4 Extracted section from the test images. **a**–**e** extracted sections

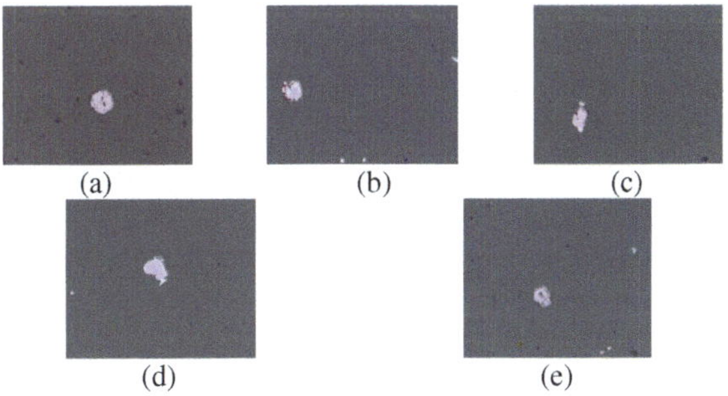

Actual Class

Extracted Class

T+ve= 4925	F+ve=1411	TPR=0.9992
F-ve= 4	T-ve=408380	FNR=0.0008
P=T+ve+F-ve=4929	N=F+ve+T-ve=409791	TNR=0.9966
Total pixels=414720		FPR=0.0034

Fig. 5 Confusion matrix for the chosen test picture

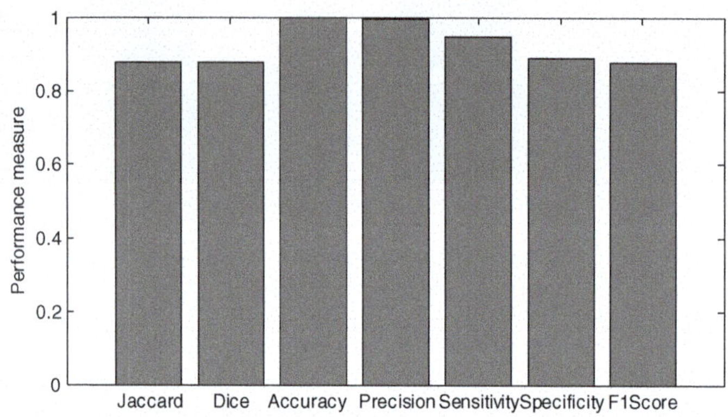

Fig. 6 Average performance appraisal of Basophil

4 Conclusion

The proposed procedure aims to implement a technique based on RGB threshold and morphological mining of leukocyte imagery. This work considered the benchmark images of LISC database and all the five class of leukocytes are considered for the assessment. The threshold enhances the leukocyte segment and then this segment is extracted using the morphological mining technique. The outcome of the proposed technique is then verified with a relative assessment on GT and the results confirm that, proposed technique helps to attain a better result on all the five class of leukocyte images. These result confirmed that, implemented method is best suitable to examine the Microscopic Blood Smear (MBS) images.

References

1. Raja, N.S.M., Fernandes, S.L. Dey, N., Satapathy, S.C., Rajinikanth, V.: Contrast enhanced medical MRI evaluation using Tsallis entropy and region growing segmentation. J. Ambient. Intell. Humaniz. Comput. 1–12 (2018). https://doi.org/10.1007/s12652-018-0854-8
2. Rajinikanth, V., Dey, N., Satapathy, S.C., Ashour, A.S.: An approach to examine magnetic resonance angiography based on Tsallis entropy and deformable snake model. Futur. Gener. Comput. Syst. **85**, 160–172 (2018). https://doi.org/10.1016/j.future.2018.03.025
3. Rezatofighi, S.H., Soltanian-Zadeh, H.: Automatic recognition of five types of white blood cells in peripheral blood. Comput. Med. Imaging Graph. **35**(4), 333–343 (2011)
4. Prinyakupt, J., Pluempitiwiriyawej, C.: Segmentation of white blood cells and comparison of cell morphology by linear and naïve Bayes classifiers. BioMed. Eng. OnLine **14**, 63 (2015). https://doi.org/10.1186/s12938-015-0037-1
5. Raja, N.S.M., et al.: Segmentation of breast thermal images using Kapur's entropy and hidden Markov random field. J. Med. Imag. Health Inf. **7**(8), 1825–1829 (2017). https://doi.org/10.1166/jmihi.2017.2267
6. LISC. http://users.cecs.anu.edu.au/~hrezatofighi/Data/Leukocyte%20Data.htm. Accessed on 11 Aug 2019
7. Dey, N., Shi, F., Rajinikanth, V.: Leukocyte nuclei segmentation using entropy function and Chan-Vese approach. Inf. Technol. Intell. Transp. Syst. **314**, 255–264 (2019). https://doi.org/10.3233/978-1-61499-939-3-255
8. Raja, N.S.M., Arunmozhi, S., Lin, H., Dey, N., Rajinikanth, V.: A study on segmentation of leukocyte image with Shannon's entropy. Histopathol. Image Anal. Med. Decis. Mak, 1–27 (2019). https://doi.org/10.4018/978-1-5225-6316-7.ch001
9. Dey, N., Rajinikanth, V., Ashour, A.S., Tavares, J.M.R.S.: Social group optimization supported segmentation and evaluation of skin melanoma images. Symmetry **10**(2), 51 (2018). https://doi.org/10.3390/sym10020051
10. Rajinikanth, V., Satapathy, S.C.: Segmentation of ischemic stroke lesion in brain MRI based on social group optimization and Fuzzy-Tsallis entropy. Arab. J. Sci. Eng. **43**(8), 4365–4378 (2018). https://doi.org/10.1007/s13369-017-3053-6
11. Rajinikanth, V., Fernandes, S.L., Bhushan, B., Sunder, N.R.: Segmentation and analysis of brain tumor using Tsallis entropy and regularised level set. Lecture Notes in Electrical Engineering, vol. 434, pp. 313–321 (2018)
12. Dey, N., et al.: Social-Group-Optimization based tumor evaluation tool for clinical brain MRI of Flair/diffusion-weighted modality. Biocybern. Biomed. Eng. **39**(3), 843–856 (2019). https://doi.org/10.1016/j.bbe.2019.07.005
13. Kapur, J.N., Sahoo, P.K., Wong, A.K.C.: A new method for gray-level picture thresholding using the entropy of the histogram. Comput. Vision. Graphics Image Process **29**, 273–285 (1985)
14. Rajinikanth, V., Satapathy, S.C., Fernandes, S.L., Nachiappan, S: Entropy based segmentation of tumor from brain MR images—a study with teaching learning based optimization. Pattern Recognit. Lett. **94**, 87–94 (2016). https://doi.org/10.1016/j.patrec.2017.05.028
15. Fernandes, S.L., Rajinikanth, V., Kadry, S.: A hybrid framework to evaluate breast abnormality using infrared thermal images. IEEE Consum. Electron. Mag. **8**(5), 31–36 (2019). https://doi.org/10.1109/MCE.2019.2923926
16. Rajinikanth, V., Dey, N., Satapathy, S.C., Kamalanand, K.: Inspection of crop-weed image database using Kapur's entropy and spider monkey optimization. Adv. Intell. Syst. Comput. **1048** (2019). https://doi.org/10.1007/978-981-15-0035-0_32
17. Satapathy, S., Naik, A.: Social group optimization (SGO): a new population evolutionary optimization technique. Complex Intell. Syst. **2**(3), 173–203 (2016)
18. Naik, A., Satapathy, S.C., Ashour, A.S., Dey, N.: Social group optimization for global optimization of multimodal functions and data clustering problems. Neural Comput. Appl. **30**(1), 271–287 (2016). https://doi.org/10.1007/s00521-016-2686-9

19. Bhateja, V., Nigam, M., Bhadauria, A.S., Arya, A., Yu-Dong Zhang, Y-D.: Human visual system based optimized mathematical morphology approach for enhancement of brain MR images, J. Ambient. Intell. Humaniz. Comput. 1–9 (2019). https://doi.org/10.1007/s12652-019-01386-z
20. Bhateja, V., Misra, M., Urooj, S.: Unsharp masking approaches for HVS based enhancement of mammographic masses: a comparative evaluation. Futur. Gener. Comput. Syst. **82**, 176–189 (2018)
21. Satapathy, S.C., El-Maleh, A., Bhateja, V.: Intelligent computing in multidisciplinary engineering applications. Arab. J. Sci. Eng. **43**(8), 3861–3862 (2018)
22. Wang, R., Wang, G.: Web text categorization based on statistical merging algorithm in big data environment. Int. J. Ambient Comput. Intell. (IJACI) **10**(3), 17–32 (2019). https://doi.org/10.4018/IJACI.2019070102
23. Ali, et al.: Adam deep learning with SOM for human sentiment classification. Int. J. Ambient Comput. Intell. (IJACI) **10**(3), 92–116 (2019). https://doi.org/10.4018/IJACI.2019070106
24. Yang, X., Jiang, X.: A hybrid active contour model based on new edge-stop functions for image segmentation. Int. J. Ambient Comput. Intell. (IJACI) **11**(1), 87–98 (2020). https://doi.org/10.4018/IJACI.2020010105
25. Acharya, U.R., et al.: Automated detection of Alzheimer's disease using brain MRI images– a study with various feature extraction techniques. J. Med. Syst. **43**, 302 (2019). https://doi.org/10.1007/s10916-019-1428-9
26. Jahmunah, V., et al.: Automated detection of schizophrenia using nonlinear signal processing methods. Artif. Intell. Med. **100**, 101698 (2019). https://doi.org/10.1016/j.artmed.2019.07.006

Brain MRI Examination with Varied Modality Fusion and Chan-Vese Segmentation

D. Abirami, N. Shalini, V. Rajinikanth, Hong Lin, and Vaddi Seshagiri Rao

Abstract Newly, varied techniques are implemented to improve and mine the apprehensive segment from the brain MRI obtained with diverse modalities. This work executes an image-fusion practice to combine the MRIs of dissimilar modalities, to improve the correctness in the assessment task. A Principle-Component-Analysis (PCA) supported fusion is employed to enhance tumor of brain MRI. This work considers the 2D slices of BRATS-2015 image for the assessment. Later, the segmentation based on Chan-Vese is executed and its performance is evaluated against DRLS. A comparative investigation among tumor section and ground-truth (GT) is implemented and vital picture-likeliness constraints are computed. The result of this work is compared against DWT-PCA procedure accessible in literature, and the experimental results of this technique offered better results, which substantiate the superiority of PCA practice.

Keywords Brain irregularity · MRI inspection · PCA fusion · Segmentation · Accuracy evaluation

---

D. Abirami · N. Shalini · V. Rajinikanth (✉)
Department of Electronics and Instrumentation Engineering, St. Joseph's College of Engineering, OMR, Chennai 600119, Tamil Nadu, India
e-mail: rajinikanthv@stjosephs.ac.in

H. Lin
Department of Computer and Mathematical Sciences, University of Houston, Downtown, TX 77002, USA

V. S. Rao
Department of Mechanical Engineering, St. Joseph's College of Engineering, OMR, Chennai 600119, Tamil Nadu, India

S. C. Satapathy et al. (eds.), *Intelligent Data Engineering and Analytics*, Advances in Intelligent Systems and Computing 1177,
https://doi.org/10.1007/978-981-15-5679-1_65

1 Introduction

Recently, a number of brain irregularity assessment procedures based on established and soft-computing methodologies are implemented to mine the necessary information from the brain imagery recorded with the Computed-Tomography (CT), Magnetic-Resonance-Imaging (MRI), and Magnetic-Resonance-Angiography (MRA) [1–3].

During the brain abnormality examination practice, principally the radiologist will record the whole brain part in the form of reconstructed three-dimensional MRI using chosen modalities, and also will supply the preopinion about the situation of the brain [4, 5]. Later, a clinical level assessment of the brain MRI is to be conducted to get the vital information like the nature of the disease, size, orientation, stage of the disease, etc. At the clinical level, usually the 2D brain MRI is examined by a capable doctor, and based on the study made by the doctor, the treatment planning procedure will be commenced. Due to its scientific importance, a considerable number of automated/semi-automated syndrome assessment procedures are proposed and implemented by the scientists to study the brain MRIs recorded with a group of modalities. Every modality will have its own merit, hence it is necessary to implement a unique tool to assess these MRIs. Most of the computerized tools are modality precise approaches, and may not supply enhanced outcomes for all the cases of MRIs.

Further, the correctness of the image inspection also provides a chief role in decision-making. Hence, a substantial quantity of methods are realized by the researchers to scrutinize the synthetic and real-time MRIs [6–9]. Hence, it is necessary to build up a tool, which will work efficiently on every modality of the MRI. The recent work of Rajinikanth et al. (2018), revealed that the mixing of two or more modality offers enhanced accuracy in MRI examination compared to the single modality case [10]. Their work also confirms that the included procedure is capable of offering the superior values of Picture-Likeliness-Constraints (PLC) during the comparative analysis among the Ground-Truth (GT) and the brain disease to be evaluated.

In this paper, a two-stage process together with; (i) Principle-Component-Analysis (PCA) based fusion and (ii) Chan-Vese segmentation is applied to the brain MRIs acquired from the benchmark BRATS-2015 database [11]. The outcome of this paper is validated against the DWT-PCA obtainable in [10]. Further, the segmentation correctness of the PCA is also confirmed with the well-known procedure called DRLS available in associated work [5]. During this work, skull-eliminated 2D brain MRIs of T1, T2, T1C, and Flair with a picture with dimension 216 $\times$ 160 pixels are used along with its GT. The result of this research substantiates that, the fused modalities will offer enhanced PLC compared to the MRI with solitary modality. In the future, the projected process can be tested on clinical MRIs.

2 Methodology

Different processing stages employed in this work are visibly depicted in Fig. 1. The 2D slices of MRI with dissimilar modalities are considered for assessment. The work of Rajinikanth et al. [10], confirmed that MRI assessment following fusing the 2D slices of dissimilar modalities will present improved outcomes compared with the MRI with solitary modality. Their work implemented the DWT-PCA for the fusion which offered an overall performance parameters with a value >91%.

In the proposed work, mostly used statistical-technique recognized as the Principal-Component-Analysis (PCA) is executed to fuse the 2D MRI slices, and also considered to mine the infected fragment. The PCA fusion work can be found in [12]. After integrating the selected modalities, like Flair + T2, Flair + T1C, and Flair + T1 + T1C, a segmentation technique called Chan-Vese is applied to mine the tumor segment from the fused image. The performance of the proposed method, which was employed to mine the tumor section, is then confirmed with a relative study with the GT. The chief aim of this paper is to progress the overall accuracy compared to DWT-PCA system.

2.1 PCA Fusion

In medical imaging literature, image-fusion is usually considered to improve the outcome of the analyzing events employed to assess the medical field image. The

Fig. 1 Implementation of the evaluation scheme

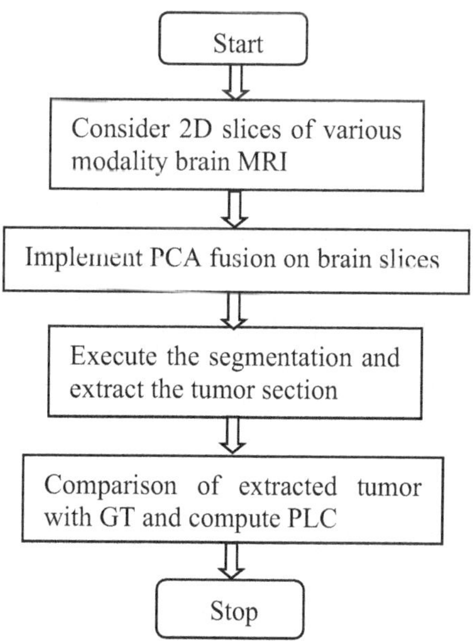

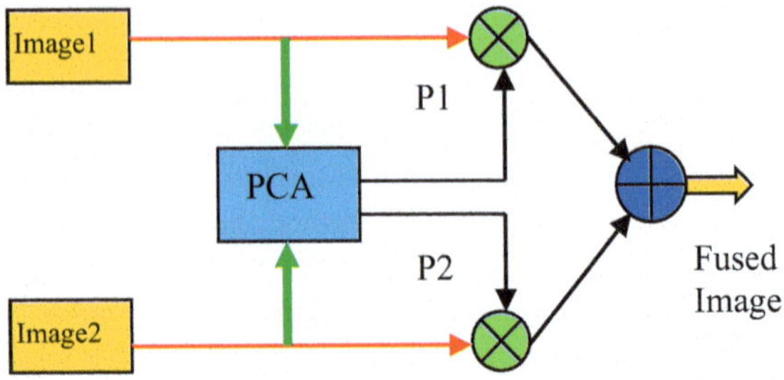

Fig. 2 PCA assisted fusion technique

fusion practice is usually considered to combine two or more identical aspect pictures without degrading its excellence, information, and the aspect [13, 14]. A number of procedures are proposed to fuse the brain pictures recorded with various modalities [12–15]. In this paper, the PCA based statistical procedure is considered to integrate the MRIs.

The working of the PCA is discussed below. The PCA conversion matrix is given as:

$$\text{Convariance matrix} \sum \sum_{i=1}^{N^2} (I_i - p)(I_i - p)^T \tag{1}$$

where I is ith spectral mark, p is mean spectral mark, and N^2 is the total spectral marks.

$$\text{Fuse Image} = (\text{Image 1}.\ P1) + (\text{Image 2}.\ P2) \tag{2}$$

The stages available in the PCA fusion are presented in Fig. 2. Equation (2) presents the action of fusion for two image case. For n-image case; this equation will be (Image 1. $P1$) + ··· + (Image n. Pn). The above method tender improved outcomes with minimal CPU time. More details regarding PCA fusion can be found in [12–14]. The result of the CV is then verified against the result by Level-Set (DRLS) segmentation.

2.2 *Chan-Vese Segmentation*

Chan-Vese is then straightly applied on the fused MRI and the infected fragment is extracted. The CV is a kind of Active-Contour (AC) procedure that works based on the energy minimization theory [16]. In CV, the curve is permitted to distinguish all the feasible alike pixels accessible in the picture. It is a semi-automated practice and necessitates an operator to start the elastic curve. When the iteration value rises, the flexible curve part will correct its orientation until the identical pixel group of the test picture is identified. Finally, the search stops after reaching the minimal energy value and displays the region which lies inside the contour. The details of CV can be found in [17–19].

2.3 *Picture-Likeliness-Constraints*

In the medical image examination, the merit of the proposed disease examination tool is validated by computing the PLC as discussed in the recent works. In this paper, the PLCs like the Jaccard (JI), Dice (DC), Sensitivity (SEN), and Specificity (SPE) are computed and its expressions are existing in [20–23].

3 Result and Discussion

This section highlights the sample images and their experimental results. Initially, the 2D MRI slices of the selected modalities and its connected GT are considered for the assessment. Table 1 shows the sample test pictures of BRATS2015 database. From this table, it is clear that the visibility of tumor in T1 modality is very poor, hence Flair, T1C, and T2 are considered for the examination. This work considers a total number of 25 slices from each modality case. The 2D slices of the MRIs are primarily integrated using PCA fusion and then the fused picture is examined with the proposed tool.

Figure 3 presents the fusion of Flair modality with T1C and the result of this process is accessible at the end. The correctness of the proposed technique depends only on the exactness in fusion. After receiving the required fused picture, then the preferred segmentation system is implemented to mine the cancer fragment.

Figure 4 describes different cases of the fused pictures in which the tumor accessibility in Flair + T1 is very poor, hence the abovesaid case is ignored throughout the proposed method. This figure also verifies the MRIs recorded with the Flair and T2,

Table 1 Sample test pictures of BRATS2015

Slice	Flair | T1C | T2 | T1 | GT
S1	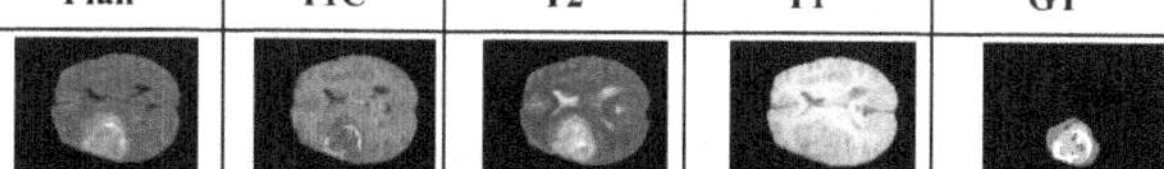
S2	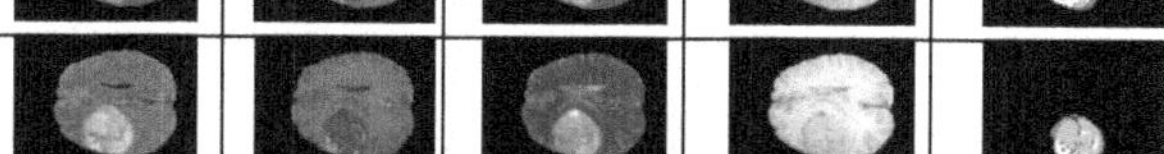
S3	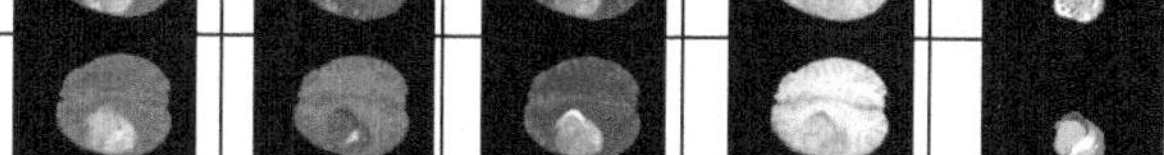
S4	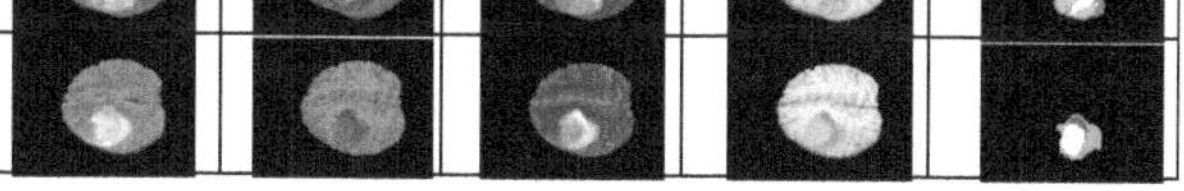

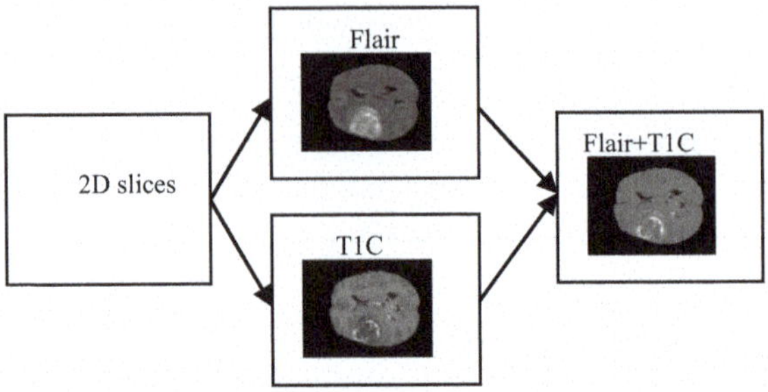

Fig. 3 Implemented fusion method for sample images

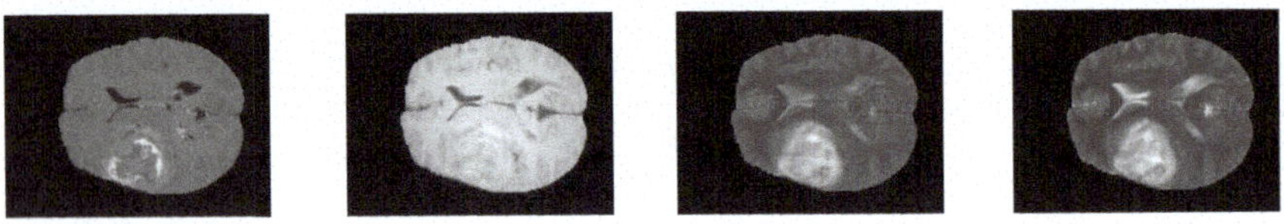

(a) Flair+T1C (b) Flair+T1 (c) Flair+T2 (d) Flair+T1C+T2

Fig. 4 Result of PCA fusion

which provides more information than other procedures. The fusion also enhances the visibility of the tumor section in Fig. 4a, c, d.

As shown in Fig. 5, primarily, the CV is implemented to mine the tumor fragment from the customary and fused pictures. From Fig. 6, it can be noted that the proposed process is very competent in extracting the cancer fragment without deviation. The extracted tumor fragment is then compared against GT, and the PLCs, such as JI, DC, SEN, and SPE are computed for these images to confirm the efficiency of the proposed tool. Further, the outcome of this procedure will offer better result than the DWT-PCA procedure discussed in [10].

As discussed earlier, the PLCs acquired with CV is better compared to DRLS [7]. Also, the run time taken by the DRLS is quite larger than CV, hence in this paper CV segmentation is adopted and its results are compared against the DWT-PCA to validate the dominance [10]. Comparable procedures are implemented for other test pictures adopted in this work and the average PLCs are tabulated as in Table 2.

Fig. 5 Implemented tumor mining procedures

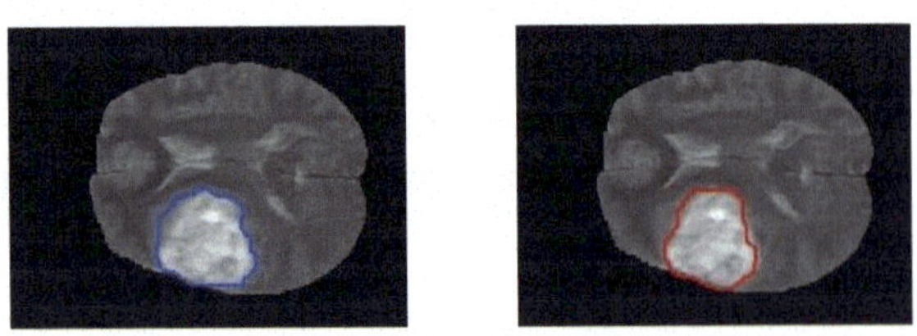

(a) Chan-Vese (b) DRLS

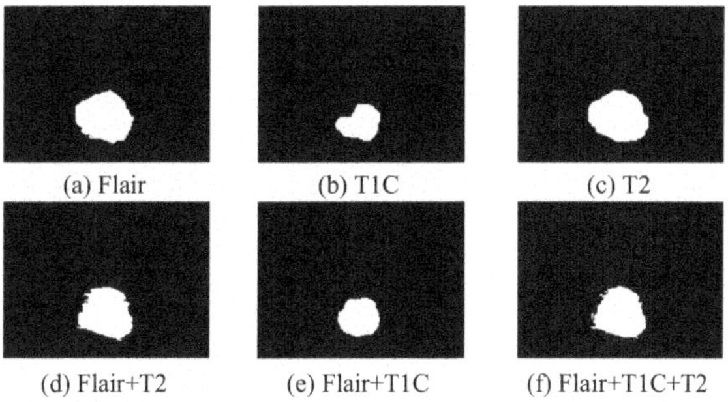

Fig. 6 Extracted tumor section from normal and fused pictures

Table 2 Image performance parameters

Method	Measure	Flair	T2	Flair + T2	Flair + T2 + T1C
Rajinikanth et al. [10]	JI	84.17	81.83	82.09	84.33
	DC	90.73	89.74	90.03	90.86
	SEN	99.84	99.36	99.40	99.93
	SPE	95.51	92.72	92.95	95.74
Level-Set (DRLS)	JI	83.96	80.48	81.57	83.84
	DC	89.58	91.24	85.28	88.52
	SEN	98.74	97.38	97.84	97.71
	SPE	96.12	94.31	95.37	96.48
Chan-Vese (CV)	JI	85.22	83.07	83.31	85.15
	DC	91.25	91.18	89.62	91.62
	SEN	98.51	97.36	98.85	98.57
	SPE	96.03	94.20	94.07	9610

This table confirms that the outcome obtained with CV is superior compared to DWT-PCA.

Figure 7 represents the graphical comparison of the result with DWT-PCA and PCA fusion. For each image case, the PLC is measured and compared against the existing method. This result also confirms that the outcome of PCA is comparatively better than the existing result (i.e., average PLC > 91%). This result confirms that the proposed approach offers better results on the implemented scheme compared to the existing approach. In the future, this tool can be further improved by considering the other fusion methods existing in the literature.

4 Conclusion

This work proposes a tactic to inspect the 2D brain image collected from different modalities. Primarily, a PCA based picture fusion process is employed to incorporate the brain MRIs recorded with different modalities. This test also authenticates that

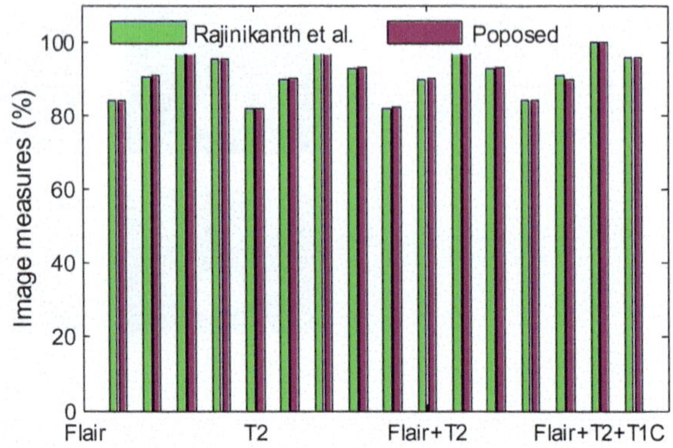

Fig. 7 Comparison of Chan-Vese with the existing approach [7]

Flair modality is to be adopted throughout the fusion to progress the result. Later the Chan-Vese and DRLS procedure is then considered to mine the infected fragment (tumor) from the brain MRI. The computation of the JI, DC, SEN, and SPE is also done to authenticate the performance of the DRLS and CV. The mean values of PLC also substantiate the advantage of this practice. The investigational study also validates that the proposed work provides improved results compared to DWT-PCA. In the future, this tool can be considered to assess the real medical pictures.

References

1. Bhateja, V., Nigam, M., Bhadauria, A.S., Arya, A., Yu-Dong Zhang, Y-D.: Human visual system based optimized mathematical morphology approach for enhancement of brain MR images. J. Ambient. Intell. Humaniz. Comput. 1–9 (2019). https://doi.org/10.1007/s12652-019
2. Menze, B., Reyes, M., Leemput, K.V., et al.: The multimodal brain tumor image segmentation benchmark (BRATS). IEEE Trans. Med. Imaging **34**(10), 1993–2024 (2015)
3. Rajinikanth, V., Satapathy, S.C., Fernandes, S.L., Nachiappan, S.: Entropy based segmentation of tumor from brain MR images—a study with teaching learning based optimization. Pattern Recognit. Lett. **94**, 87–94 (2016). https://doi.org/10.1016/j.patrec.2017.05.028
4. Fernandes, S.L., et al.: A reliable framework for accurate brain image examination and treatment planning based on early diagnosis support for clinicians. Neural Comput. Appl. 1–12 (2019). https://doi.org/10.1007/s00521-019-04369-5
5. Dey, N., et al.: Social-Group-Optimization based tumor evaluation tool for clinical brain MRI of flair/diffusion-weighted modality. Biocybern. Biomed. Eng. **39**(3), 843–856 (2019). https://doi.org/10.1016/j.bbe.2019.07.005
6. Jahmunah, V., et al.: Automated detection of schizophrenia using nonlinear signal processing methods. Artif. Int. Med. **100**, 101698 (2019). https://doi.org/10.1016/j.artmed.2019.07.006
7. Rajinikanth, V., Fernandes, S.L., Bhushan, B., Sunder, N.R.: Segmentation and analysis of brain tumor using Tsallis entropy and regularised level set. Lecture Notes in Electrical Engineering, vol. 434, pp. 313–321 (2018)
8. Rajinikanth, V., Dey, N., Satapathy, S.C., Ashour, A.S.: An approach to examine magnetic resonance angiography based on Tsallis entropy and deformable snake model. Futur. Gener. Comput. Syst. **85**, 160–172 (2018). https://doi.org/10.1016/j.future.2018.03.025

9. Raja, N.S.M., Fernandes, S.L. Dey, N., Satapathy, S.C., Rajinikanth, V.: Contrast enhanced medical MRI evaluation using Tsallis entropy and region growing segmentation. J. Ambient. Intell. Humaniz. Comput. 1–12 (2018). https://doi.org/10.1007/s12652-018-0854-8
10. Rajinikanth, V., Satapathy, S.C., Dey, N., Vijayarajan, R.: DWT-PCA image fusion technique to improve segmentation accuracy in brain tumor analysis. LNEE **471**, 453–462 (2018). https://doi.org/10.1007/978-981-10-7329-8_46
11. Farahani, K., Menze, B., Reyes, M.: Multimodal Brain Tumor Segmentation (BRATS 2013) (2013). http://martinos.org/qtim/miccai2013/
12. Vijayarajan, R., Muttan, S.: Local principal component averaging image fusion, Int. J. Imaging Robot. **13**(2), 94–103 (2014)
13. Vijayarajan, R., Muttan, S.: Fuzzy C-means clustering based principal component averaging fusion. Int. J. Fuzzy Syst. **16**(2), 153–159 (2014)
14. Vijayarajan, R., Muttan, S.: Iterative block level principal component averaging medical image fusion. Optik-Int. J. Light Electron Opt. **125**(17), 4751–4757 (2014)
15. Srivastava, A., Bhateja, V., Moin, A.: Combination of PCA and contourlets for multispectral image fusion. Adv. Intell. Syst. Comput. **469**, 577–585 (2017). https://doi.org/10.1007/978-981-10-1678-3_55
16. Chan, T.F., Vese, L.A.: Active contours without edges. IEEE Trans. Image Process **10**(2), 266–277 (2001)
17. Rajinikanth, V., Dey, N., Kumar, R., Panneerselvam, J., Raja, N.S.M.: Fetal head periphery extraction from ultrasound image using Jaya algorithm and Chan-Vese segmentation. Procedia Comput. Sci. **152**, 66–73 (2019). https://doi.org/10.1016/j.procs.2019.05.028
18. Manic, K.S., et al.: An approach to examine brain tumor based on Kapur's entropy and Chan-Vese algorithm. AISC **797**, 901–909 (2019). https://doi.org/10.1007/978-981-13-1165-9_81
19. Kowsalya, N., et al.: Skin-melanoma evaluation with Tsallis's thresholding and Chan-Vese approach. In. IEEE International Conference on System, Computation, Automation and Networking (ICSCA), pp.1–5. IEEE (2018). https://doi.org/10.1109/ICSCAN.2018.8541178
20. Rajinikanth, V., Satapathy, S.C.: Segmentation of ischemic stroke lesion in brain MRI based on social group optimization and Fuzzy-Tsallis entropy. Arab. J. Sci. Eng. **43**(8), 4365–4378 (2018). https://doi.org/10.1007/s13369-017-3053-6
21. Satapathy, S.C., Rajinikanth, V.: Jaya algorithm guided procedure to segment tumor from brain MRI. J. Optim. **2018**, 12 (2018). https://doi.org/10.1155/2018/3738049
22. Fernandes, S.L., Rajinikanth, V., Kadry, S.: A hybrid framework to evaluate breast abnormality using infrared thermal images. IEEE Consum. Electron. Mag. **8**(5), 31–36 (2019). https://doi.org/10.1109/MCE.2019.2923926
23. Bhandary, A., et al.: Deep-learning framework to detect lung abnormality–a study with chest X-ray and lung CT scan images. Pattern Recogn. Lett. (2019). https://doi.org/10.1016/j.patrec.2019.11.013

Examination of the Brain MRI Slices Corrupted with Induced Noise—A Study with SGO Algorithm

R. Pavidraa, R. Preethi, N. Sri Madhava Raja, P. Tamizharasi, and B. Parvatha Varthini

Abstract Brain irregularity is a chief disease in humans and conscientious for the loss of many lives. Due to its consequence, several image appraisal methods are proposed to study the abnormality in the brain. This research implements a two-stage practice to extort the cancer fragment from axial-view brain MRI slice stained with noise. Primarily, the selected 2D MRI slice is stained with the regular noises, and then the threshold and segmentation tactics are applied to mine the tumor sector. In this work, brain MRIs of Flair/T2 modalities are considered, and the thresholding is executed using Social Group Optimization (SGO) and Tsallis Entropy (SGO + TE) to improve its condition and later the Level Set Segmentation (LSS) is applied to extract the tumor. Lastly, the tumor area is compared against the Ground Truth (GT) to substantiate the benefit of the projected procedure. The outcome of this work verifies that, implemented work supplies improved result in original and noise corrupted MRI slices. The execution of this procedure is also quite straightforward compared with other procedures, so this work can be used to inspect the clinical MRIs in the future.

Keywords Brain MRI · Noise staining · Social group optimization · Thresholding · Level-set · Validation

R. Pavidraa · R. Preethi · N. Sri Madhava Raja (✉)
Department of Electronics and Instrumentation Engineering, St. Joseph's College of Engineering, Chennai 600119, Tamil Nadu, India
e-mail: nsrimadhavaraja@stjosephs.ac.in

P. Tamizharasi
Department of Electrical and Electronics Engineering, Adiparasakthi Engineering College, Melmaruvathur, Tamil Nadu, India

B. Parvatha Varthini
Department of Computer Science Engineering, St. Joseph's College of Engineering, Chennai 600119, Tamil Nadu, India

S. C. Satapathy et al. (eds.), *Intelligent Data Engineering and Analytics*, Advances in Intelligent Systems and Computing 1177,
https://doi.org/10.1007/978-981-15-5679-1_66

1 Introduction

Brain irregularity is the main danger for human beings and which causes a higher mortality rate every year. The brain connected sickness may raise the death rates and also the momentary or enduring disabilities in humans. Hence, a considerable quantity of assessment procedures is extensively adopted in hospitals to supervise and cure the brain associated syndrome [1–4].

Magnetic-Resonance-Imaging (MRI) is a prominent imaging tactic widely used in hospitals to trace the sickness existing in the interior body organs. MRI is reasonably prominent compared to other approaches [5–7]. The MRI can be scrutinized in 3D or 2D form [8–10].

In obtainable MRI modalities. Flair and T2 offer superior visibility on the abnormalities, like the tumor/stroke divisions compared to T1C, T1, and DW modalities [11]. Hence, Flair and T2 modality slices are considered in this investigation.

The complication of regular MRI is less and it can be examined with an experienced doctor, and also the existing picture appraisal schemes. In order to substantiate the robustness of the picture assessment events, it is necessary to authenticate its performance on the noise corrupted pictures. Hence, in this work, along with original 2D MRI slices, MRIs stained with the Salt and Pepper Noise (SPN) and Gaussian Noise (GN) are considered. For the examination, this paper considered MRIs slices of BRATS2015 database [12]. This dataset has a skull stripped slices of dimension 216×160 pixels.

This study implements a two-phase approach to inspect the brain MRI [13–16]. The preliminary phase implements a multi-threshold scheme based on SGO + TE to enhance the tumor section and then implements a Level-Set-Segmentation (LSS) to extort the tumor from the threshold picture. In order to recognize the optimal thresholds, a varied threshold with value 2–5 is implemented using SGO + TE.

The outcome of this threshold method authenticates that, for the original and noise stained pictures, the three-level threshold present improved visibility, and hence for other image cases, a trilevel threshold is applied. The performance of the proposed approach is then validated based on a relative study with the Ground Truth (GT) existing in the database.

The outcome of this study verifies that, for the original and SPN MRI pictures, the Picture Excellence Value (PEV) attained is better for both the Flair and T2 cases. But, for the GN cases, the results obtained are slightly poor. These results also confirm that the two-phase procedure based on SGO + TE and LSS approach works well on both the normal and noisy image cases with acceptable PEVs. In the future, this procedure can be applied in real-time medical MRI pictures.

2 Methodology

The soft-computing assisted MRI evaluation methods are largely used due to its better accuracy [13–15]. Different steps executed in this work are presented in Fig. 1.

Figure 1 depicts different stages involved in this system, firstly a chosen 2D MRI image is tainted with a chosen noise (noise amplitude = 0.1). Then the noise stained image is processed with SGO + TE. In this work, the threshold values are chosen from 2 to 5 and the optimal threshold is identified based on the allocated value, this is done using the SGO. The chief position of the SGO is to randomly vary the thresholds and calculate the entropy function. This procedure is sustained until the SGO discovers the maximal values of the entropy. This threshold procedure will help to improve the visibility of tumor section by integrating the analogous pixels of the chosen MRI. After the possible threshold task, the semi-automated segmentation process called the LSS is then implemented to mine the irregular division from the MRI. The LSS process requires the implementation of the Bounding Box (BB) by the operator around the tumor to be extracted. After the initialization, the BB is allowed to unite toward the tumor based on the augment in iteration. Finally, if the LSS identifies all the probable pixels, then it will extract the irregular fragment and displays the assessment for additional evaluation. Later a relative study with GT is implemented and the PEV is computed for different regions mined with LSS. Based on the PEV, the concluding judgment regarding the brain abnormality is to be taken. If the tumor size is less, it may be treated with possible drugs specified by the doctor, and if the tumor size is more, it can be removed with a suggested surgical procedure directed by an experienced doctor.

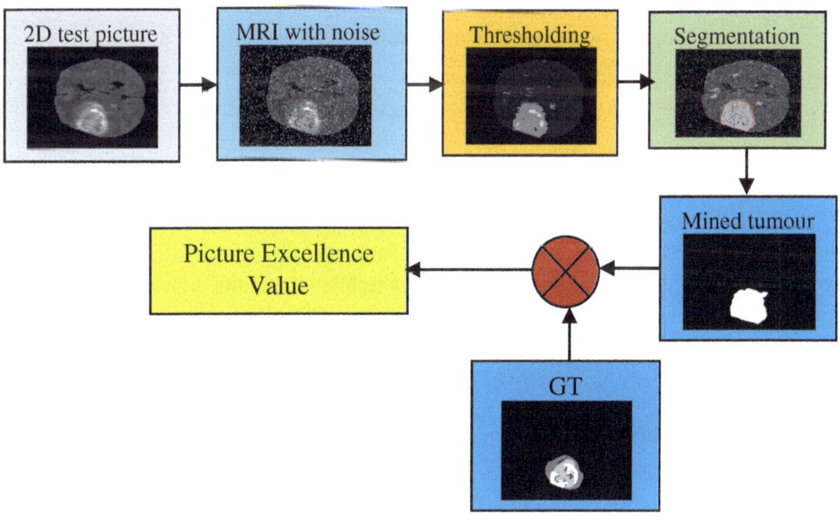

Fig. 1 Image examination technique implemented in this work

2.1 Brain MRI Database

The accessibility of the brain MRI information for research is extremely limited, and most of the data are protected with copyright/ethical principles. Hence, this work adopted the 2D slices of BRATS2015. The added benefit of this data is, it consists of the MRI of different modalities along with the GT image.

2.2 Thresholding

Threshold separates an image into various clusters and is a universal preprocessing approach extensively used to improve the conventional and medical pictures. Here, the threshold is implemented with Social Group Optimization and Tsallis Entropy (SGO + TE), and its details can be found in [5]. During this process, the multi-thresholding with the selected threshold values, like 2, 3, 4, and 5 is employed to enhance the MRI picture. In this work, SGO will regulate the threshold of the image until the TE reaches its maximal value. The detailed expression on the TE adopted in this paper can be found in [5, 17, 18].

2.3 Segmentation

The main role of segmentation is to mine the irregular division from the preprocessed image. This work implements the LSS discussed by Li et al. [19].

This works is based on the energy minimization procedure and it is expressed as

$$\Re_e(\varphi) \triangleq \int_{\Omega} e(|\nabla\varphi|)\mathrm{d}X \tag{1}$$

where, e = energy density value with $e = [0, \alpha] \rightarrow \Re$.

In LSS, an adoptable curve is authorized to recognize all the probable pixel values connected to the irregular division existing in the picture. After identifying all the likely pixel groups, it will mine the section which is within the converged curvature for the evaluation practice. This segment is the irregular part of the considered MRI.

2.4 Assessment

The merit of the picture appraisal tool is authenticated by computing the Picture Excellence Value (PEV) as discussed in [20–22]. In this paper, the PEVs like the Jaccard (JAC), Dice (DIC), Sensitivity (SEN), Specificity (SPE), Accuracy (ACC), and Precision (PRE). Based on the values of PEVs, the performance of the proposed technique is confirmed.

3 Result and Discussion

This division outlines the investigational results obtained from the current work. All the experiments are realized with Matlab software. Primarily, the parameters for SGO are assigned with the number of agents = 20, search dimension = threshold value, number of iterations = 2500, and stopping function = maximized TE.

In assessment, ten patient's MRI of Flair/T2 modalities are considered for the assessment. For each patient, around 30 numbers of 2D slices are selected for the assessment. Figures 2 and 3 describe the sample MRIs of different patients. For example, let us Initially, Fig. 2 is preferred for the examination, where Fig. 2a–c shows the preferred slices of patient 1, with images of Flair, T2 and GT, respectively. Each picture is obtained with a selected slice, and in GT the bright portion represents the tumor fragment and the gray represents edema. The projected process in this study is employed to mine tumor + edema division from the selected slices of the

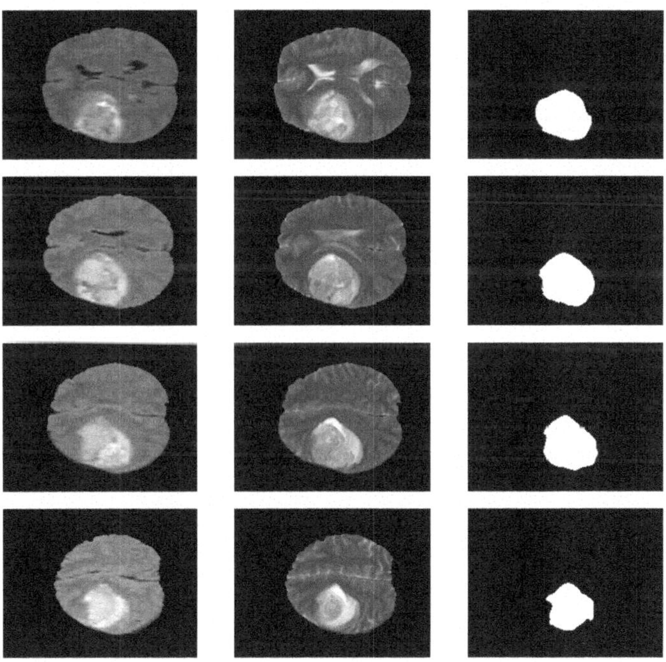

(a) Flair (b) T2 (c) GT

Fig. 2 Chosen test-pictures and GT of BRATS2015

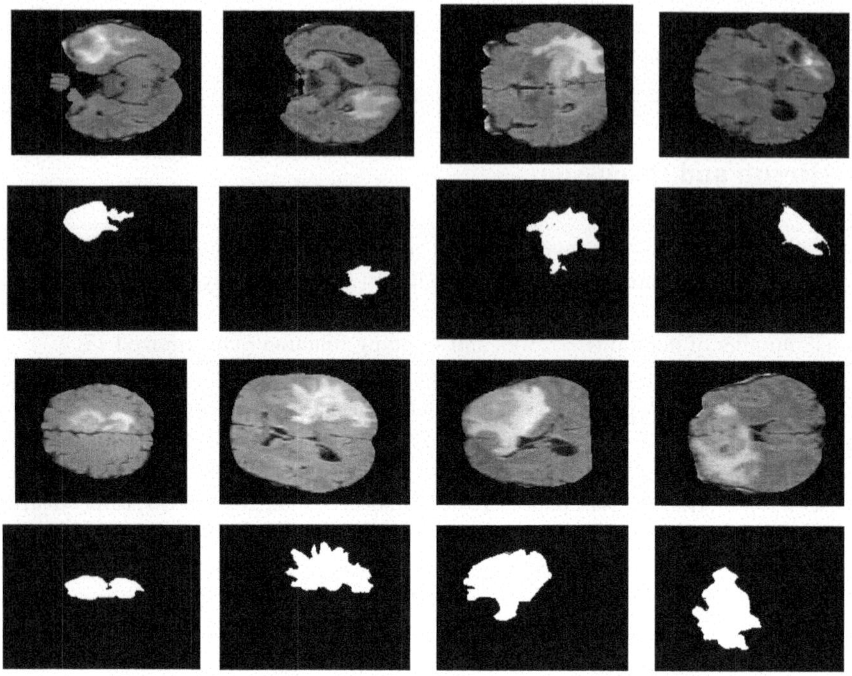

Fig. 3 Sample MRIs adopted in this study

MRI irrespective of its modality. Figure 3 also presents the MRI slices beside with GT. The image superiority of the selected MRI slice is then modified by injecting the SPN and GN, and the corresponding pictures are depicted in Fig. 4. Figure 4a presents the picture with SPN and Fig. 4b presents the picture with GN.

When the noise is added, it will corrupt the excellence of the image considerably based on the category and influence of the noise. From Fig. 4, it can be noted that the visibility and the excellence of the MRI is poor in noisy pictures contrast to original Flair-MRI. Further, the visibility of the SPN picture is poor than the GN picture. Later, the preprocessing based on the SGO + TE is implemented with various thresholds, and the result obtained for the original and noisy image cases is shown in Fig. 5.

In the above figure, Fig. 5a–d represent the threshold approach obtained for thresholds 2, 3, 4, and 5, respectively. From this figure, the observation is that bi-level thresholding offers a false product for SPN image. So, in all other image cases, the preprocessing is implemented by considering the three-level threshold (since 4 and

Fig. 4 Noise corrupted 2D MRI slices

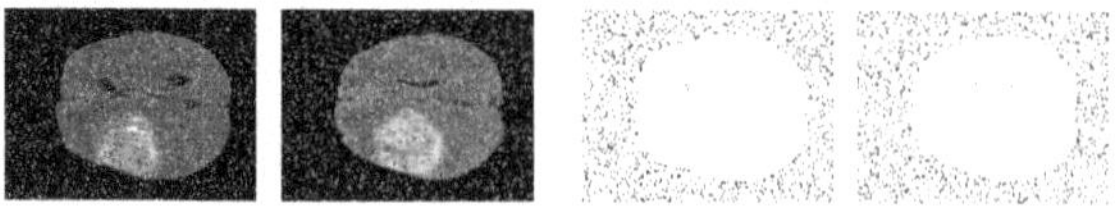

(a) SPN (b) GN

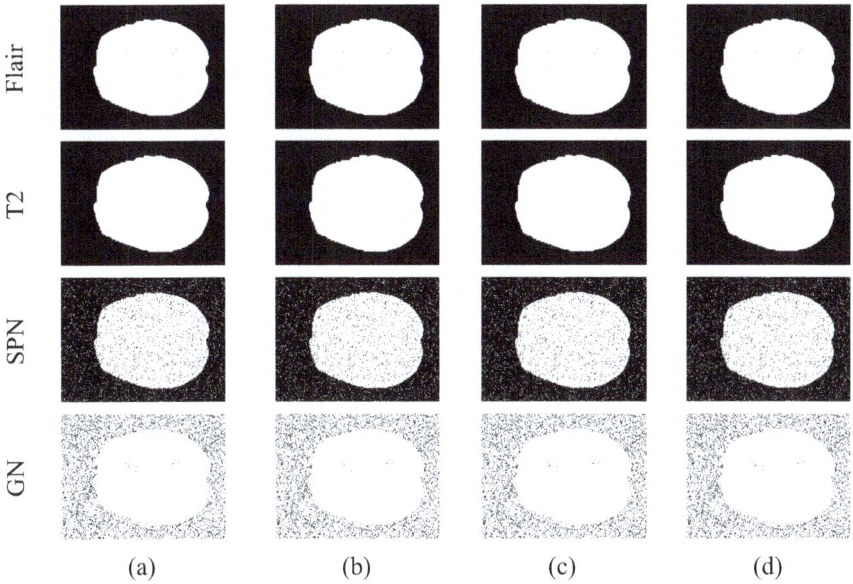

Fig. 5 Thresholded test picture for Thr = 2, 3, 4, 5

5 stage desires more investigational time compared to the bi and three-level thresholds). From this investigational work, it is established that three-level threshold is the best appropriate method to preprocess the selected brain MRI.

The mining of the irregular fragment with LSS is executed and show in Fig. 6. Figure 6a shows the preliminary bounding box assigned by the operator and Fig. 6b, d shows the converged LSS for the SPN and GN cases. The extracted brain abnormality is also presented in Fig. 6c, e for the noisy cases. Related results are obtained for other images are considered in this paper. The results obtained for the original image is better than the noisy image cases. Finally, a relative analysis between the extracted brain irregularity and the GT is done and the essential IQVs are computed for original and noisy images. The average values of IQVs obtained for the chosen images are presented in Tables 1 and 2. In this table, Flair (F), Flair with SPN (F_SPN), Flair with GN (F_GN), and T2 for original and noise cases are presented.

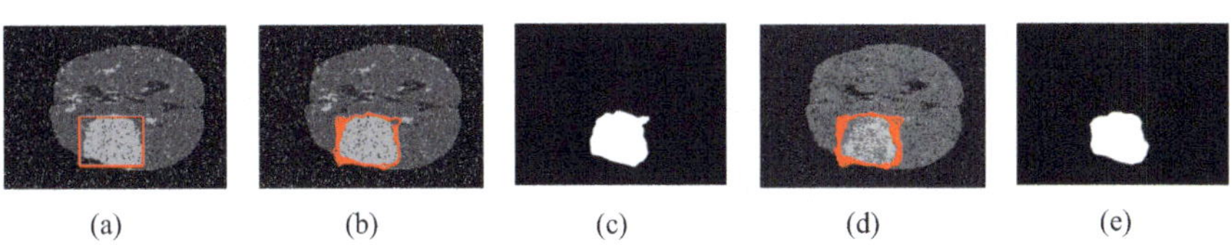

Fig. 6 Results obtained with LSS. **a** Sample test image, **b** SPN image, **c** extracted tumor, **d** GN image, **e** tumor

Table 1 Image performance values of Flair images

Image	JAC	DIC	SEN	SPE	ACC	PRE
Flair	0.8417	0.8502	0.9712	0.9297	0.9774	0.9707
F_SPN	0.8314	0.8311	0.9597	0.9243	0.9699	0.9604
F_GN	0.8327	0.8308	0.9618	0.9207	0.9687	0.9513

Table 2 PEVs for chosen images of BRATS2015

Image	JAC	DIC	SEN	SPE	ACC	PRE
Flair	0.8422	0.8666	0.9765	0.9516	0.9719	0.9448
F_SPN	0.8398	0.8364	0.9694	0.9433	0.9704	0.9398
F-GN	0.8353	0.8504	0.9621	0.9472	0.9628	0.9299
T2	0.8475	0.8616	0.9655	0.9408	0.9672	0.9408
T2_SPN	0.8297	0.8411	0.9500	0.9358	0.9605	0.9315
T2_GN	0.8316	0.8338	0.9482	0.9331	0.9561	0.9272

From these tables, it is clear that results obtained with the original and SPN class are improved than the GN class. Also, the Flair-MRI offers enhanced result compared to T2. The average PEVs are >85%, which confirms the superiority of the proposed image assessment scheme. In the future, this procedure can be considered to test real-time medical MRIs accessible with the skull division.

4 Conclusion

In this work, a robust two-stage picture assessment method is implemented to inspect the brain MRIs of BRATS2015 database. In this work, around 350 figures of 2D MRI slices of Flair/T2 are considered for the assessment. During this work, a combination of the SGO and TE based threshold and LSS extraction is implemented. To recognize the sturdiness of the projected system, the images are stained with the general noise signals with a noise power of 0.1. The extracted brain fragment is then compared with GT and essential PEVs are then computed. The results of this research confirm that the implemented procedure is very efficient for both the noise and conventional case of MRI and offers an average PEV of >85% for all the brain MRIs considered in this paper. In the future, this procedure can be considered to scrutinize the MRIs obtained from the hospitals.

References

1. Bhateja, V., Nigam, M., Bhadauria, A.S., Arya, A., Yu-Dong Zhang, Y-D.: Human visual system based optimized mathematical morphology approach for enhancement of brain MR images. J. Ambient. Intell. Humaniz. Comput. 1–9 (2019). https://doi.org/10.1007/s12652-019-01386-z
2. Satapathy, S.C., El-Maleh, A., Bhateja, V.: Intelligent computing in multidisciplinary engineering applications. Arab. J. Sci. Eng. **43**(8), 3861–3862 (2018)
3. Ali, et al.: Adam deep learning with SOM for human sentiment classification. Int. J. Ambient Comput. Intell. (IJACI) **10**(3), 92–116 (2019). https://doi.org/10.4018/IJACI.2019070106
4. Yang, X., Jiang, X.: A hybrid active contour model based on new edge-stop functions for image segmentation. Int. J. Ambient Comput. Intell. (IJACI) **11**(1), 87–98 (2020). https://doi.org/10.4018/IJACI.2020010105
5. Rajinikanth, V., Satapathy, S.C.: Segmentation of ischemic stroke lesion in brain MRI based on social group optimization and Fuzzy-Tsallis entropy. Arab. J. Sci. Eng. **43**(8), 4365–4378 (2018). https://doi.org/10.1007/s13369-017-3053-6
6. Satapathy, S.C., Rajinikanth, V.: Jaya algorithm guided procedure to segment tumor from brain MRI. J. Optim. **2018**, 12 (2018). https://doi.org/10.1155/2018/3738049
7. Fernandes, S.L., et al.: A reliable framework for accurate brain image examination and treatment planning based on early diagnosis support for clinicians. Neural Comput. Appl. 1–12 (2019). https://doi.org/10.1007/s00521-019-04369-5
8. Dey, N., et al.: Social-Group-Optimization based tumor evaluation tool for clinical brain MRI of Flair/diffusion-weighted modality. Biocybern. Biomed. Eng. **39**(3), 843–856 (2019). https://doi.org/10.1016/j.bbe.2019.07.005
9. Acharya, U.R., et al.: Automated detection of Alzheimer's disease using brain MRI images—a study with various feature extraction techniques. J. Med. Syst. **43**, 302 (2019). https://doi.org/10.1007/s10916-019-1428-9
10. Jahmunah, V., et al.: Automated detection of schizophrenia using nonlinear signal processing methods. Artif. Intell. Med. **100**, 101698 (2019). https://doi.org/10.1016/j.artmed.2019.07.006
11. Menze, B., Reyes, M., Leemput, K.V., et al.: The multimodal brain tumor image segmentation benchmark (BRATS). IEEE Trans. Med. Imaging **34**(10), 1993–2024 (2015)
12. Farahani, K., Menze, B., Reyes, M.: Multimodal Brain Tumor Segmentation (BRATS 2013). http://martinos.org/qtim/miccai2013/ (2013)
13. Rajinikanth, V., Fernandes, S.L., Bhushan, B., Sunder, N.R.: Segmentation and analysis of brain tumor using Tsallis entropy and regularised level set. Lecture Notes in Electrical Engineering, vol. 434, pp. 313–321 (2018)
14. Rajinikanth, V., Dey, N., Satapathy, S.C., Ashour, A.S.: An approach to examine magnetic resonance angiography based on Tsallis entropy and deformable snake model. Futur. Gener. Comput. Syst. **85**, 160–172 (2018). https://doi.org/10.1016/j.future.2018.03.025
15. Raja, N.S.M., Fernandes, S.L. Dey, N., Satapathy, S.C., Rajinikanth, V.: Contrast enhanced medical MRI evaluation using Tsallis entropy and region growing segmentation. J. Ambient. Intell. Humaniz. Comput. 1–12 (2018). https://doi.org/10.1007/s12652-018-0854-8
16. Rajinikanth, V., Satapathy, S.C., Dey, N., Vijayarajan, R.: DWT-PCA image fusion technique to improve segmentation accuracy in brain tumor analysis. LNEE **471**, 453–462 (2018). https://doi.org/10.1007/978-981-10-7329-8_46
17. Satapathy, S., Naik, A.: Social group optimization (SGO): a new population evolutionary optimization technique. Complex Intell. Syst. **2**(3), 173–203 (2016)
18. Naik, A., Satapathy, S.C., Ashour, A.S., Dey, N.: Social group optimization for global optimization of multimodal functions and data clustering problems. Neural Comput. Appl. (2016). https://doi.org/10.1007/s00521-016-2686-9
19. Li, C., Xu, C., Gui, C., Fox, M.D.: Distance regularized level set evolution and its application to image segmentation. IEEE Trans. Image Process. **19**(12), 3243–3254 (2010)
20. Fernandes, S.L., Rajinikanth, V., Kadry, S.: A hybrid framework to evaluate breast abnormality using infrared thermal images. IEEE Consum. Electron. Mag. **8**(5), 31–36 (2019). https://doi.org/10.1109/MCE.2019.2923926

21. Bhandary, A., et al.: Deep-learning framework to detect lung abnormality—a study with chest X-ray and lung CT scan images. Pattern Recogn. Lett. (2019). https://doi.org/10.1016/j.patrec.2019.11.013
22. Dey, N., Rajinikanth, V., Ashour, A.S., Tavares, J.M.R.S.: Social group optimization supported segmentation and evaluation of skin melanoma images. Symmetry **10**(2), 51 (2018). https://doi.org/10.3390/sym10020051

Segmentation and Assessment of Leukocytes Using Entropy-Based Procedure

S. Manasi, M. Ramyaa, N. Sri Madhava Raja, S. Arunmozhi, and Suresh Chandra Satapathy

Abstract Image evaluation plays a necessary dependability in medical discipline to support the sickness inspection. In this work, the assessment of White Blood Cell (WBC) is performed by using the Digital-Microscope-Pictures (DMP) of thin blood smears. In this study, a soft-computing procedure; Chaotic Firefly Algorithm and Kapur's Entropy (CFA + KE) focused bi-level threshold is considered to improve the test picture. Further, Distance Regularized Level Set (DRLS) Segmentation is considered to mine the WBC region. Later, the mined WBC region is then evaluated with the expert's Ground Truth (GT), and the well-known Picture-Quality-Parameters (PQP) is computed to substantiate the exactness of the proposed process. The experimental examination is implemented using the benchmark LISC images, and the results confirm that the proposed procedure helps to accomplish better values of PQP (average value >88%).

Keywords Blood cell · Digital microscopy · Firefly algorithm · Entropy · Level-set segmentation

S. Manasi · M. Ramyaa · N. Sri Madhava Raja (✉)
Department of Electronics and Instrumentation, St. Joseph's College of Engineering, Chennai 600119, Tamil Nadu, India
e-mail: nsrimadhavaraja@stjosephs.ac.in

S. Arunmozhi
Department of Electronics and Communication Engineering, Manakula Vinayagar Institute of Technology, Madagadipet, Puducherry, India

S. C. Satapathy
School of Computer Engineering, Kalinga Institute of Industrial Technology (Deemed to Be University), Bhubaneswar 751024, Odisha, India

S. C. Satapathy et al. (eds.), *Intelligent Data Engineering and Analytics*, Advances in Intelligent Systems and Computing 1177,
https://doi.org/10.1007/978-981-15-5679-1_67

1 Introduction

Recently, a number of image assessment procedures are implemented to examine a class of conventional and medical pictures based on the traditional and soft-computing procedures. In hospitals and clinics, the accurateness of the disease inspection procedure depends largely on the knowledge of the physicians [1–6]. Hence, in most of the imaging clinics, the available medical pictures are usually evaluated by a computer-assisted procedures to improve the examination accuracy.

In this work, the examination of White Blood Cell (WBC) is adopted for the study. Normally, the WBC examination involves in the following procedures; (i) collection of blood samples from the infected person, (ii) creating the thick/thin blood smears, (iii) staining the blood smear using the staining agents, and (iv) examination of the stained section using a microscope by an experienced technician (or) by a dedicated software tool. Based on the analysis, a report will be generated and sent to the doctor for further diagnosis and treatment scheduling [7, 8].

In recent years, the WBC sections are normally assessed using the semi-automated/automated image examination procedures due to its importance [9, 10]. The recent medical picture examination procedures also confirm that compared to the traditional procedure, the soft-computing guided procedures will help to attain better results with comparatively minimized evaluation time. Because of its merit, substantial soft-computing procedures are widely applied to examine MRI, microscopic pictures, thermal images, and retinal pictures [11–13]. These approaches also reveal the need for a two-phase procedure during the medical picture examination and its need.

Hence, in the proposed work, a two-step procedure which involves (i) Chaotic Firefly Algorithm (CFA) and (ii) Kapur's Entropy (KE) is utilized. This is performed for a bi-level thresholding and the Level-Set-Segmentation (DRLS) is implemented to scrutinize the WBC pictures. The thresholding will enhance the visibility of infection by improving its threshold value for a selected thresholding procedure. This is followed by a segmentation procedure that is implemented to mine the segment of interest for further examination. The chief task is to extract the stained section (WBC) for the microscopy picture. The CFA approach is already discussed in the work by Rajinikanth et al. [14]. In this CFA, the search process of the FA is monitored by the Ikeda-Map operator. This CFA will help to enhance the WBC picture by maximizing the KE function. Later, the WBC segment is extracted with the DRLS.

In this work, the WBC dataset called the Leukocyte-Images for-Segmentation-Classification (LISC) [15]. In this, all the pictures are in the form of RGB scale. The dimension of the image is 720 $\times$ 576 pixels. It is accessible along with its Ground Truth (GT). In this work, the implemented process primarily extracts the WBC from the considered sample picture. Later, the extracted WBC section is compared against the GT to compute the Image-Quality-Parameters (IQP), such as the Jaccard (JA), Dice (DI), Sensitivity (SE), Specificity (SP), Accuracy (AC), and Precision (PR) [16–20]. Initially, these values are individually computed for every picture of the

LISC dataset, later the average values of these parameters are considered to assess the dominance of the proposed method.

The experimental investigation of this paper is implemented using the Matlab-software, and the results are clearly presented and discussed for the WBC categories, such as Basophil (B), Lymphocyte (L), Eosinophil (E), Neutrophil (N), and Monocyte (M) cases. Finally, the average value of IQP of the LISC is considered to confirm the implemented procedure's dominance. The results attained from the present study confirms that, proposed entropy assited segmentation offers enhanced results on the considered test images.

2 Methodology

The imaging technique employed in this work is depicted in Fig. 1. Initially, the WBC image to be examined is preprocessed using CFA and KE based bi-level threshold. The threshold of the picture will enhance the stained WBC section and eliminates other blood-cell regions present in the test picture. Later, DRLS is then implemented to extract the WBC section with the help of operator initiated Bounding Box (BB). Finally, the extracted WBC is then compared with GT, and the IQP are then compared and validated.

2.1 *Thresholding*

This plays a major part in enhancing the images based on the operator's requirement. It can be implemented manually or along with a Heuristic Technique (HT). The manual procedure is time-consuming, and hence in recent days, thresholding based on a chosen method is normally done along with a HT. In this attempt, the thresholding of the WBC picture is performed with proposed CFA and KE techniques.

FA is one of the widely accepted soft-computing procedures and its prime functioning depends on luminosity shaped by firefly because of a reaction involving chemicals.

The proposed work, FA driven by the Chaotic Ikeda-map (IM) shown below is considered;

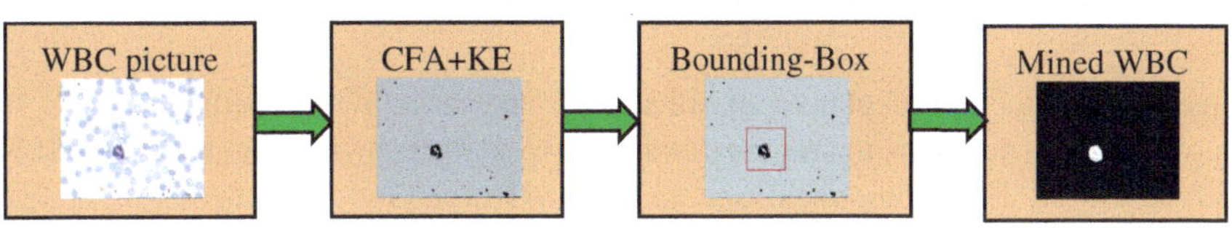

Fig. 1 Different process involved in WBC assessment

$$F_i^{t+1} = F_i^t \beta_0 e^{-\gamma d_{ij}^2}(F_j^t - F_i^t) + \alpha \cdot \text{sign}(\text{rand} - 1/2) \oplus \text{IM} \quad (1)$$

where, F_i^{t+1}—final position, F_i^t—initial point, $\beta_0 e^{-\gamma d_{ij}^2 (F_j^t - F_j^t)}$—attraction among fireflies, and the sign $\oplus$ signify entry wise multiplication. The related theory on the CFA can be found in [13, 14].

The optimal threshold for the picture is recognized using CFA by maximizing the KE value. The maximized KE is obtained using

$$J_{\max} = f_{\text{kapur}}(\text{Th}) = \sum_{j=1}^{k} H_j^C \quad (2)$$

The procedure involving KE is a popular procedure presented in [15]. In this attempt, the CFA parameters are assigned as follows; number of fireflies = 30, iteration = 1500, dimension = 2, and the termination function is obtained using $J_{\max}$.

2.2 Segmentation

DRLS method discussed by Li et al. [17], is employed in this work to extract WBC. This uses an energy minimization procedure which is presented as

$$\Re_e(\varphi) \triangleq \int_{\Omega} e(|\nabla\varphi|)\mathrm{d}X \quad (3)$$

where $e =$ energy density value with $e = [0, \alpha] \rightarrow \Re$

Here, an adoptable curve is approved to identify all the possible pixel values which are connected to the asymmetrical segment which are present in the sample. After obtaining the maximum possible pixel clusters, it extracts the section which is present inward of the converged curve which is used for the further evaluation process. This segmented region of the image is considered as abnormal in the WBC test picture [18].

2.3 Assessment and Validation

In general, segmentation plays a significant role in examining biomedical images. It is considered to delineate the region of interest for analysis. In this attempt, the DRLS with adjustable BB is employed to obtain the WBC section from the preprocessed image. Because of its advantages, the DRLS procedure is widely accomplished in the medical diagnosis task by the investigators.

The justification of the suggested system is performed by a relative study between the GT and the extracted WBC. In this procedure, various IQPs such as JA, SE, DI, AC, SP, and PR are obtained [19–22]. Based on these values, the outperformance of the implemented method is established.

3 Result and Discussion

This section of the paper demonstrates the outcome reached from this research-work. The microscopic blood smear picture is examined with the CFA + KE and DRLS. Initially, this technique is implemented on the RGB scale Basophil pictures shown in Fig. 2. Figure 2a denotes the pseudoname of pictures, Fig. 2b, c illustrate the test pictures and the GT. From Fig. 2b, it is clear that the examination of the RGB scale test picture is difficult compared to other pictures, since, in this picture, the entire picture frame is filled with the blood-cell sections. The stained section represents the WBC to be extracted and evaluated. The size of the test picture is large and the RGB image threshold may take a large threshold time. Hence, the RGB picture is converted into gray-scale during the threshold procedure.

Initially, the CFA and KE based bi-level thresholding is implemented for this picture, and the convergence of the CFA search toward the optimal value of the J_{max} (maximized entropy) is depicted in this Fig. 3. This figure represents the convergence attained for the pictures B1–B5. From this, for all the cases, it is clear that the optimized J_{max} lies between 12.4 and 12.55, and the search will converge before 1500 iterations. The outcome of this approach will offer the enhanced WBC which can be extracted by employing a suitable segmentation technique.

Figure 4 depicts the outcomes obtained from the proposed technique. Figure 4a presents the preprocessed picture, Fig. 4b depicts the initial BB initiated by the operator using the DRLS technique, and Fig. 4c presents the extracted WBC using the converged DRLS. Later, this WBC section is compared with the GT to compute the essential IQP.

Figure 5 presents the other WBC cases existing in the LISC. In this, a chosen picture can be identified with its pseudoname. Figure 5a, b present the sample picture and GT, respectively. Figure 5c represents the preprocessed picture. Later the DRLS is executed to extract the region.

The comparative investigation between the mined WBC segment and its corresponding GT is implemented and essential values of the IQP are computed. Table 1 presents the IQP of the sample test pictures shown in Fig. 2. Table 2 presents the average values of the IQP obtained for the whole LISC dataset with B, N, M, L, and E cell categories. Finally, the average values are computed and presented. During this analysis, the IQP are accomplished as follows; JA = 86.44%, DI = 90.49, SE = 95.33%, SP = 96.60%, AC = 97.60%, and 95.51%. This investigation also authenticates that this procedure offers superior results on the LISC dataset.

Fig. 2 Sample test pictures of Basophil

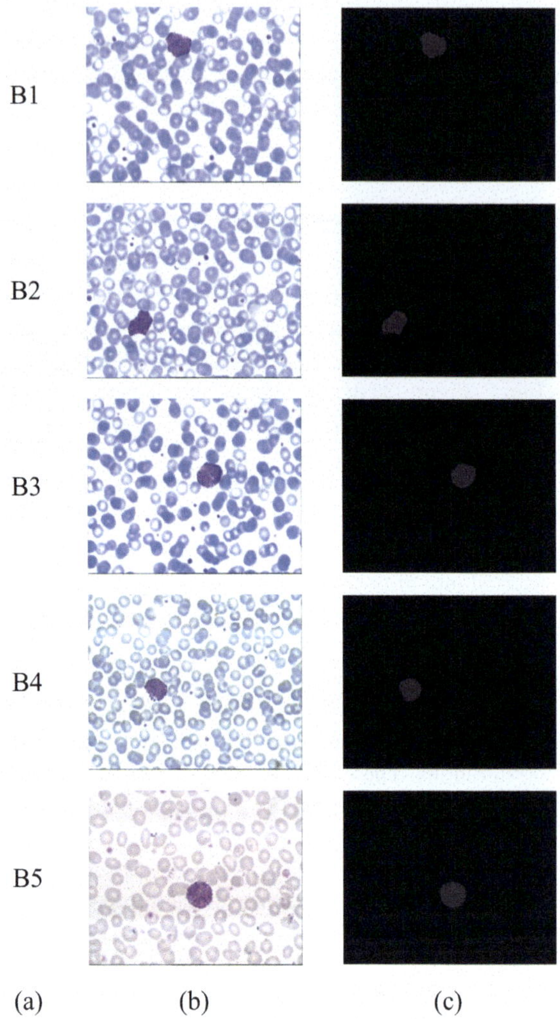

Fig. 3 Convergence of the CFA search with WBC picture

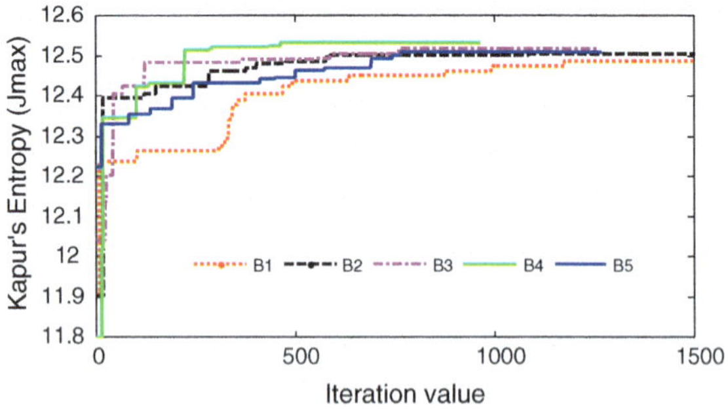

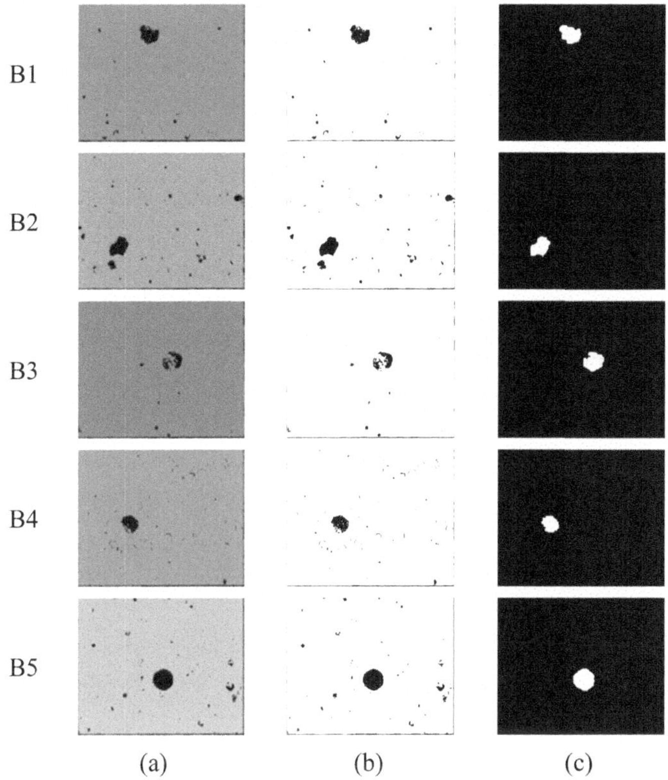

Fig. 4 Experimental outcome for Basophil pictures

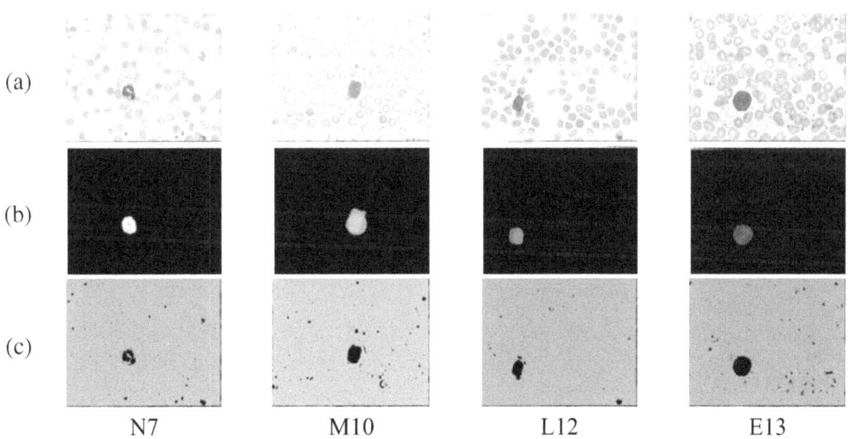

Fig. 5 Sample test images of LISC

Table 1 Comparison of Baso WBC with GT

Image	JA	DI	SE	SP	AC	PR
B1	0.8395	0.8827	0.9692	0.9726	0.9771	0.8917
B2	0.8197	0.8362	0.9274	0.9742	0.9693	0.9792
B3	0.8594	0.9173	0.9446	0.9835	0.9709	0.9517
B4	0.8873	0.9037	0.9583	0.9728	0.9866	0.9629
B5	0.8736	0.9105	0.9552	0.9617	0.9853	0.9758
Average	0.8559	0.8901	0.9509	0.9730	0.9778	0.9523

Table 2 Image-quality-parameter of LISC dataset

Image	JA	DI	SE	SP	AC	PR
B	0.8672	0.9027	0.9571	0.9668	0.9793	0.9382
N	0.8593	0.8762	0.9318	0.9537	0.9668	0.9443
M	0.8728	0.9274	0.9553	0.9683	0.9729	0.9538
L	0.8639	0.9115	0.9639	0.9774	0.9838	0.9962
E	0.8587	0.9068	0.9583	0.9638	0.9774	0.9429
Average	0.8644	0.9049	0.9533	0.9660	0.9760	0.9551

4 Conclusion

This work unified a two-step system in order to inspect the WBC of the microscopic blood smear pictures. Initially, the CFA and KE based thresholding is effected to augment the stained section of WBC picture. Further, the DRLS procedure is implemented to extract the stained segment from the preprocessed picture. This work considers the LISC database available with the GT section. The dominance of this approach is established based on a comparative study among the GT and WBC region. The experimental results reveal that the recommended procedure has more efficacy to extract the WBC segment and offered better values of IQP.

References

1. Bhandary, A., et al.: Deep-learning framework to detect lung abnormality—a study with chest X-ray and lung CT scan images. Pattern Recogn. Lett. (2019). https://doi.org/10.1016/j.patrec.2019.11.013
2. Fernandes, S.L., et al.: A reliable framework for accurate brain image examination and treatment planning based on early diagnosis support for clinicians. Neural Comput. Appl. 1–12 (2019). https://doi.org/10.1007/s00521-019-04369-5
3. Acharya, U.R., et al.: Automated detection of Alzheimer's disease using brain MRI images—a study with various feature extraction techniques. J. Med. Syst. **43**, 302 (2019). https://doi.org/10.1007/s10916-019-1428-9

4. Jahmunah, V., et al.: Automated detection of schizophrenia using nonlinear signal processing methods. Artif. Intell. Med. **100**, 101698 (2019). https://doi.org/10.1016/j.artmed.2019.07.006
5. Bhateja, V., Nigam, M., Bhadauria, A.S., Arya, A., Yu-Dong Zhang, Y-D.: Human visual system based optimized mathematical morphology approach for enhancement of brain MR images, J. Ambient. Intell. Humaniz. Comput. 1–9 (2019). https://doi.org/10.1007/s12652-019-01386-z
6. Satapathy, S.C., El-Maleh, A., Bhateja, V.: Intelligent computing in multidisciplinary engineering applications. Arab. J. Sci. Eng. **43**(8), 3861–3862 (2018)
7. Dey, N., Shi, F., Rajinikanth, V.: Leukocyte nuclei segmentation using entropy function and Chan-Vese approach. Inf. Technol. Intell. Transp. Syst. **314**, 255–264 (2019). https://doi.org/10.3233/978-1-61499-939-3-255
8. Raja, N.S.M., Arunmozhi, S., Lin, H., Dey, N., Rajinikanth, V.: A study on segmentation of leukocyte image with Shannon's entropy. Histopathol Image Anal. Med. Decis. Mak. 1–27 (2019). https://doi.org/10.4018/978-1-5225-6316-7.ch001
9. Rezatofighi, S.H., Soltanian-Zadeh, H.: Automatic recognition of five types of white blood cells in peripheral blood. Comput. Med. Imaging Graph. **35**(4), 333–343 (2011)
10. Prinyakupt, J., Pluempitiwiriyawej, C.: Segmentation of white blood cells and comparison of cell morphology by linear and naïve Bayes classifiers. BioMed Eng. OnLine **14**, 63 (2015). https://doi.org/10.1186/s12938-015-0037-1
11. Ali, et al.: Adam deep learning with SOM for human sentiment classification. Int. J. Ambient Comput. Intell. (IJACI) **10**(3), 92–116 (2019). https://doi.org/10.4018/IJACI.2019070106
12. Raja, N.S.M., Fernandes, S.L. Dey, N., Satapathy, S.C., Rajinikanth, V.: Contrast enhanced medical MRI evaluation using Tsallis entropy and region growing segmentation. J. Ambient. Intell. Humaniz. Comput. 1–12 (2018). https://doi.org/10.1007/s12652-018-0854-8
13. Dey, N., Rajinikanth, V., Ashour, A.S., Tavares, J.M.R.S.: Social group optimization supported segmentation and evaluation of skin melanoma images. Symmetry **10**(2), 51 (2018). https://doi.org/10.3390/sym10020051
14. Rajinikanth, V., Dey, N., Satapathy, S.C., Ashour, A.S.: An approach to examine magnetic resonance angiography based on Tsallis entropy and deformable snake model. Futur. Gener. Comput. Syst. **85**, 160–172 (2018). https://doi.org/10.1016/j.future.2018.03.025
15. LISC. http://users.cecs.anu.edu.au/~hrezatofighi/Data/Leukocyte%20Data.htm. Accessed on 18 October 2019
16. Rajinikanth, V., Satapathy, S.C.: Segmentation of ischemic stroke lesion in brain MRI based on social group optimization and Fuzzy-Tsallis entropy. Arab. J. Sci. Eng. **43**(8), 4365–4378 (2018). https://doi.org/10.1007/s13369-017-3053-6
17. Li, C., Xu, C., Gui, C., Fox, M.D.: Distance regularized level set evolution and its application to image segmentation. IEEE Trans. Image Process. **19**(12), 3243–3254 (2010)
18. Yang, X., Jiang, X.: A hybrid active contour model based on new edge-stop functions for image segmentation. Int. J. Ambient Comput. Intell. (IJACI) **11**(1), 87–98 (2020). https://doi.org/10.4018/IJACI.2020010105
19. Satapathy, S.C., Rajinikanth, V.: Jaya algorithm guided procedure to segment tumor from brain MRI. J. Optim. **2018**, 12 (2018). https://doi.org/10.1155/2018/3738049
20. Lakshmi, B., Parthasarathy, S.: Human action recognition using median background and max pool convolution with nearest neighbor. Int. J. Ambient Comput. Intell. (IJACI) **10**(2), 34–47 (2019). https://doi.org/10.4018/IJACI.2019040103
21. Wang, R., Wang, G.: Web text categorization based on statistical merging algorithm in big data environment. Int. J. Ambient Comput. Intell. (IJACI) **10**(3), 17–32 (2019). https://doi.org/10.4018/IJACI.2019070102
22. Rajinikanth, V., Fernandes, S.L., Bhushan, B., Sunder, N.R.: Segmentation and analysis of brain tumor using Tsallis entropy and regularised level set, Lecture Notes in Electrical Engineering, vol. 434, pp. 313–321 (2018)

Image Assisted Assessment of Cancer Segment from Dermoscopy Images

M. Santhosh, R. Rubin Silas Raj, V. Rajinikanth, and Suresh Chandra Satapathy

Abstract Previous research confirms the accessibility of an assortment of picture assessment schemes for the skin melanoma appraisal. Skin Melanoma (SM) is also one of the deadliest diseases, and the unnoticed SM may direct to casualty. In this study, the SM image assessment is based on the Bat Algorithm (BA) and Kapur's threshold. Active Contour Segmentation (ACS) is employed to mine and study the melanoma tainted skin fragment. In this work, the dermoscopy images of the benchmark data, like DermIS and Dermquest are considered for the inspection. Primarily, all the pictures are transformed into 256 × 256 pixels together with the ground truth (GT) slice, and these pictures are then used in the inspection. The efficacy of the projected procedure is then validated using a comparative examination among the mined skin segment and GT. The experimental result substantiates that this procedure helps to achieve a better Jaccard-Index and Dice value for the considered datasets; hence this procedure is appropriate to inspect the SM pictures.

Keywords Dermoscopy picture · Melanoma · Bat algorithm · Kapur's-function · Segmentation

1 Introduction

Numerous machine learning and deep learning procedures are widely applied to examine the medical pictures [1–5]. Since disease in the body parts are normally examined based on these pictures, hence numerous imaging schemes are suggested and realized by the researchers to examine these pictures [6–10]. The literature

M. Santhosh · R. Rubin Silas Raj · V. Rajinikanth (✉)
Department of Electronics and Instrumentation, St. Joseph's College of Engineering, Chennai 600119, Tamil Nadu, India
e-mail: v.rajinikanth@ieee.org

S. C. Satapathy
School of Computer Engineering, Kalinga Institute of Industrial Technology (Deemed to Be University), Bhubaneswar 751024, Odisha, India

S. C. Satapathy et al. (eds.), *Intelligent Data Engineering and Analytics*, Advances in Intelligent Systems and Computing 1177,
https://doi.org/10.1007/978-981-15-5679-1_68

reveals that cancer is a dangerous life-threatening illness in the human community, and is responsible for the increase in the mortality rate. Hence, numerous camps for alertness on precaution against cancer and its potential treatment procedures are conducted worldwide.

Skin Melanoma (SM) is also a deadliest cancer, originates in the skin sections. The untreated SM probably increases the threat and it allows the cancer-causing cells to spread to other body parts through bloodstream. Hence, it should be identified in its premature stage. In clinics, a routine visual examination by an experienced dermatologist is always suggested to identify the SM in its early stage, and if the chance of SM is suspected, then the dermatologist will suggest a computerized screening procedure in which a dermoscopy can be used to examine the skin section, and if any abnormalities are found, then the particular section is recorded as a digital image. Later, the recorded Digital Dermoscopy Picture (DDP) can be examined by an experienced doctor or the dedicated computerized image examination schemes. During this procedure, the thumb-rule called the ABCD rule is considered to examine the abnormal skin section [11–13].

Because of its consequences, a variety of routine and contemporary SM appraisal and localization events are comprehensively realized by the researchers [14–16]. Early works also authenticate that important forms of events are considered and engaged for the autonomous investigation of the oncogenic skin sections. The past research attempts, show the automated/partly-automated prognosis examination tools established with regular and heuristic techniques are engaged to evaluate the contaminated segment. If the information on skin slice is attainable, further it facilitates physicians to decide on advanced treating the patient. The treatment includes in (i) Therapeutic accomplishment to remove the skin irregularity with unique drugs, and (ii) Providing surgical action to eliminate the SM region.

In the proposed paper, the incorporation of the threshold and the segmentation is implemented to scrutinize the abnormal skin region. In this attempt, the standard SM datasets, DermIS, and Dermquest are employed for the examination. In order to have improved accuracy, the thresholding scheme is implemented based on the Bat Algorithm (BA) driven Kapur's thresholding and followed by segmentation executed with Active-Contour's adjustable snake procedure. After extracting the possible SM section, a comparative examination among the extracted SM and the Ground Truth (GT) is performed. Further, the familiar similarity parameters are computed. The experimental task is implemented with the Matlab software and the result confirms that the proposed procedure is substantial to extract the SM section and offers better results. In the future, it can be used to examine SM pictures obtained from the hospitals.

2 Methodology

The practice employed in this paper is graphically represented in Fig. 1.

Figure 1 presents the various stages existing in the proposed disease examination procedure. Initially, the test picture to be examined is preprocessed using the chosen image thresholding approach. Later, the SM section existing in the preprocessed procedure is mined by the adopted segmentation procedure. Finally, the extracted SM is related to its GT and the image similarity parameters (ISP) are calculated.

2.1 *Test Picture*

In this paper, the SM test images are obtained from DermIS and Dermquest [15]. In order to have the consistency, all the sample images and their corresponding GTs are resized into 256 × 256 sized pictures. These pictures are then considered for the examination.

2.2 *Preprocessing*

In this work, a two-stage procedure widely discussed in the research medical picture examination scheme is considered. In the preprocessing phase, a three-level thresholding procedure that uses Kapur's-Entropy (KE) is implemented. In order to monitor the performance of the KE, the well-established heuristic procedure called the Bat Algorithm (BA) is adopted [17–19].

The implementation of this procedure is as follows:

- Bat algorithm

The BA was originally conceived by emulating the trailing actions of microbats [13]. BA encloses successive arithmetical calculations in the form of equations

$$A_i^{(k+1)} = B_i^{(k)} + (A_i^{(k)} - G) \cdot H_i \quad (1)$$

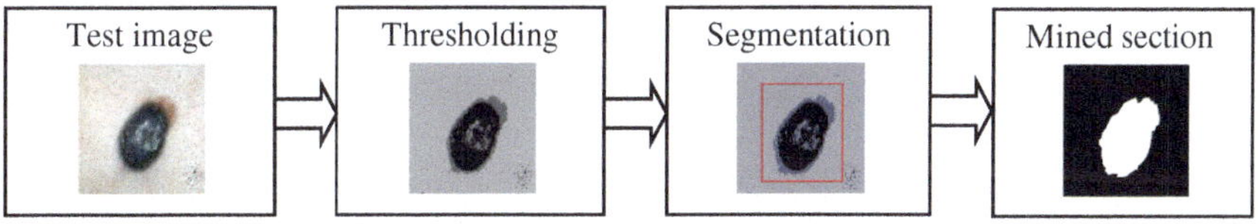

Fig. 1 Phases in the SM assessment process

$$B_i^{(k+1)} = B_i^{(k)} + A_i^{(k+1)} \tag{2}$$

$$H_i = H_{\min} + (H_{\max} - H_{\min}) \cdot \beta^1 \tag{3}$$

where $A_i^{(k+1)}$ and B_i^{k+1} represent the swiftness and position of bat, $H_{\min}$ is the minimum frequency and $H_{\max}$ is greatest.

The position update depends on

$$X_{\text{new}} = X_{\text{old}} + \varepsilon A^t \tag{4}$$

where, ε = random digit between $[-1, 1]$ and A is the strength of bats' produced sound during the course of the exploration in a new area. Other related details can be found in [17].

BA is responsible for identifying the optimal threshold values for the considered picture by maximizing the value of KE. The representation of maximized KE is provided by

$$J_{\max} = f_{\text{kapur}}(\text{Th}) = \sum_{j=1}^{k} H_j^C \tag{5}$$

KE is an extensively adopted technique. Its related outcomes can be widely found in [19]. In this attempt, the BA constraints assigned are; no. of bats = 20, exploration dimension = 3, iteration = 1000 and the termination function is $J_{\max}$.

2.3 *Post-processing and Validation*

In general, segmentation has a significant role in examining medical images. It is normally considered to obtain the region of attention for analysis. In this attempt, the Active Contour (AC)-based adaptable snake model is executed to mine the SM segment from a preprocessed image [20]. Because of its advantages, the AC method is widely executed in the medical inspection by the investigators.

The proposed scheme is validated by a comparison between the Ground truth and the mined SM. In this process, the Picture Similarity Parameters (PSP) like true + ve rate (TPR), false +ve rate (FPR), true –ve rate (TNR) and false –ve rate (FNR), Jaccard (JI), and Dice (DC) values are computed. The improved JI and DC approve the efficiency of the suggested procedure [21, 22].

3 Results and Discussions

The test pictures represented in Fig. 2 are employed for the inspection. In this, Fig. 2a–c represents the pseudo name, chosen test pictures, and the GT, respectively. Further, the tri-level thresholding which is based on BA and KE is implemented. The optimization-based search is allowed to accomplish until the BA determines the termination function J_{max}. The completion of this exploration delivers the optimized result. The obtained thresholded picture is depicted in Fig. 2d. The benefit of this preprocessing outline is that it will improve the prominence of the SM image. Later, the AC is employed to mine the SM section.

Figure 2 represents the sample SM images that are used for the examination. This image also reveals that every picture has a varied visibility and intensity, and the extraction of the SM section will be a challenging job for this picture. The literature reveals that RGB picture threshold is a time-consuming procedure, hence the gray-scale picture threshold is implemented in this paper for the chosen test pictures, which will enhance the visibility of SM based on the chosen thresholds.

Figure 3 presents the other sample test pictures adopted in this experimental investigation. This figure presents the 256 × 256 sized images of both the DermIS and Dermquest database [15].

Figure 4 represents the results acquired with the post-processing procedure. Figure 4a displays the bounding box initiated in the AC snake approach with the help of the operator. When the iteration value proceeds, the adaptable snake is allowed to adjust its orientation and it will find all the possible SM regions existing in the threshold picture, and the final converged box of the AC is presented in Fig. 4b. The extracted SM section for S1 picture is presented in Fig. 4c. Similar procedures are applied for other image cases and its outcome is clearly presented in Fig. 4d–i. Later, a predefined procedure is then implemented to compute the TPR, TNR, FPR, FNR, JI, and DC. This is based on a relative examination of the removed SM and the GT and the results are represented in Tables 1 and 2.

From Tables 1 and 2, it is clear that the picture similarity parameters obtained from the developed procedure is good. This confirms that the BA + KE and AC procedure can be considered to examine the SM pictures.

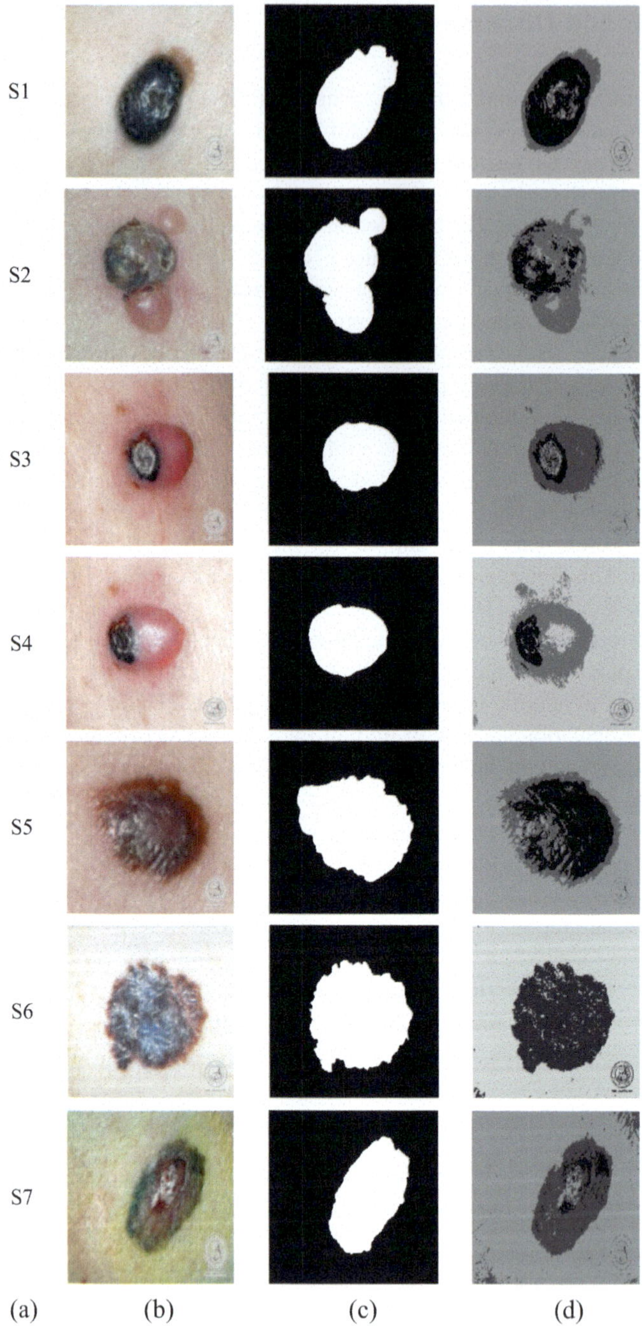

Fig. 2 Sample test images. **a** Picture, **b** GT, **c** preprocessed image

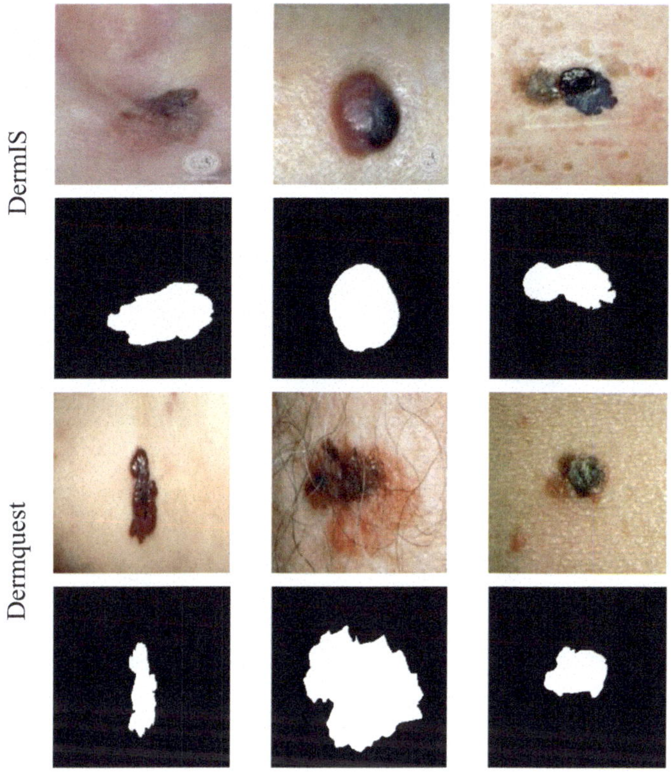

Fig. 3 Other sample test images existing in the dataset

4 Conclusion

In this attempt, an inspection technique for the skin melanoma with a two-stage process is incorporated. Initially, a preprocessing process that combines BA and KE is employed. This combination is used to improve the SM section from the selected picture for investigation on a tri-level thresholding outline. Further, AC segmentation is performed to remove the SM section. It mines the SM section in a better way from the image which is subjected to preprocessing. The validation of the study is performed by a relative valuation between the extracted SM and its GT is performed. This provides the vital picture similarity parameters. The results confirm that the suggested procedure offers a better outcome on the selected dataset (DermIS and Dermquest) and provides enhanced values of the JI and DC.

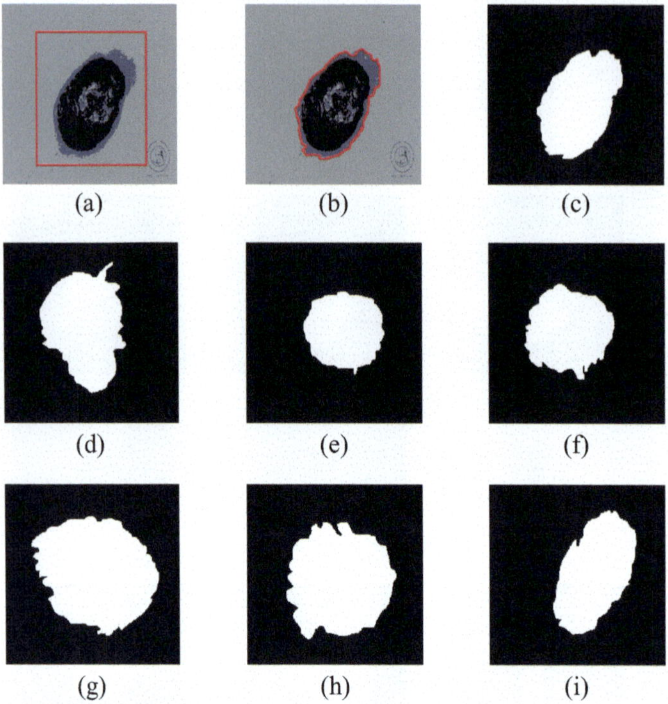

Fig. 4 Results obtained with the proposed technique

Table 1 Image similarity features for the selected test pictures

Image	TPR	FNR	TNR	FPR	JI	DC
S1	0.99	0.01	0.98	0.02	0.8316	0.8498
S2	0.99	0.01	0.97	0.03	0.8274	0.8397
S3	0.98	0.02	0.97	0.03	0.8193	0.8336
S4	0.99	0.01	0.98	0.02	0.8322	0.8503
S5	0.96	0.04	0.97	0.03	0.8287	0.8438
S6	0.98	0.02	0.97	0.03	0.8382	0.8552
S7	0.99	0.01	0.96	0.04	0.8417	0.8572

Table 2 Average similarity parameters

Dataset	TPR	FNR	TNR	FPR	JI	DC
DermIS	0.99	0.01	0.97	0.03	0.8418	0.8626
Dermquest	0.99	0.01	0.98	0.02	0.8532	0.8683

References

1. Bhandary, A., et al.: Deep-learning framework to detect lung abnormality—a study with chest X-ray and lung CT scan images. Pattern Recogn. Lett. (2019). https://doi.org/10.1016/j.patrec.2019.11.013
2. Fernandes, S.L., et al.: A reliable framework for accurate brain image examination and treatment planning based on early diagnosis support for clinicians. Neural Comput. Appl. 1–12 (2019). https://doi.org/10.1007/s00521-019-04369-5
3. Dey, N., et al.: Social-Group-Optimization based tumor evaluation tool for clinical brain MRI of Flair/diffusion-weighted modality. Biocybern. Biomed. Eng. **39**(3), 843–856 (2019). https://doi.org/10.1016/j.bbe.2019.07.005
4. Acharya, U.R., et al.: Automated detection of Alzheimer's disease using brain MRI images—a study with various feature extraction techniques. J. Med. Syst. **43**, 302 (2019). https://doi.org/10.1007/s10916-019-1428-9
5. Jahmunah, V., et al.: Automated detection of schizophrenia using nonlinear signal processing methods. Artif. Intell. Med. **100**, 101698 (2019). https://doi.org/10.1016/j.artmed.2019.07.006
6. Bhateja, V., Nigam, M., Bhadauria, A.S., Arya, A., Yu-Dong Zhang, Y-D.: Human visual system based optimized mathematical morphology approach for enhancement of brain MR images. J. Ambient. Intell. Humaniz. Comput. 1–9 (2019). https://doi.org/10.1007/s12652-019-01386-z
7. Satapathy, S.C., El-Maleh, A., Bhateja, V.: Intelligent computing in multidisciplinary engineering applications. Arab. J. Sci. Eng. **43**(8), 3861–3862 (2018)
8. Bhateja, V., Misra, M., Urooj, S.: Unsharp masking approaches for HVS based enhancement of mammographic masses: a comparative evaluation. Futur. Gener. Comput. Syst. **82**, 176–189 (2018)
9. Wang, R., Wang, G.: Web text categorization based on statistical merging algorithm in big data environment. Int. J. Ambient Comput. Intell. (IJACI) **10**(3), 17–32 (2019). https://doi.org/10.4018/IJACI.2019070102
10. Ali, et al.: Adam deep learning with SOM for human sentiment classification. Int. J. Ambient Comput. Intell. (IJACI) **10**(3), 92–116 (2019). https://doi.org/10.4018/IJACI.2019070106
11. Rajinikanth, V., Raja, N.S.M., Arunmozhi, S.: ABCD rule implementation for the skin melanoma assesment—a study. In: IEEE International Conference on System, Computation, Automation and Networking (ICSCAN), pp. 1–4. IEEE (2019). https://doi.org/10.1109/icscan.2019.8878860
12. Amelard, R., Glaister, J., Wong, A., Clausi, D.A.: Melanoma decision support using lighting-corrected intuitive feature models. In: Computer Vision Techniques for the Diagnosis of Skin Cancer, Series in BioEngineering, pp. 193–219 (2013)
13. Nachbar, F., Stolz, W., Merckle, T., et al.: The ABCD rule of dermatoscopy: High prospective value in the diagnosis of doubtful melanocytic skin lesions. J. Am. Acad. Dermatol. **30**, 551–559 (1994)
14. Dey, N., Rajinikanth, V., Ashour, A.S., Tavares, J.M R.S.: Social group optimization supported segmentation and evaluation of skin melanoma images. Symmetry **10**(2), 51 (2018). https://doi.org/10.3390/sym10020051
15. http://vip.uwaterloo.ca/demos/skin-cancer-detection
16. Amelard, R., Glaister, J.: Wong, A. and Clausi, D.A.: High-level intuitive features (HLIFs) for intuitive skin lesion description. IEEE Trans. Biomed. Eng. **62**(3), 820–831 (2015)
17. Satapathy, S.C., Raja, N.S.M., Rajinikanth, V., Ashour, A.S.: Dey, N: Multi-level image thresholding using Otsu and chaotic bat algorithm. Neural Comput. Appl. (2016). https://doi.org/10.1007/s00521-016-2645-5
18. Yang, X.S.: Nature-Inspired Metaheuristic Algorithms, 2nd edn. Luniver Press, Frome (2011)
19. Kapur, J.N., Sahoo, P.K., Wong, A.K.C.: A new method for gray-level picture thresholding using the entropy of the histogram. Comput. Vis. Graph. Image Process. **29**, 273–285 (1985)
20. Yang, X., Jiang, X.: A hybrid active contour model based on new edge-stop functions for image segmentation. Int. J. Ambient Comput. Intell. (IJACI) **11**(1), 87–98 (2020). https://doi.org/10.4018/IJACI.2020010105

21. Satapathy, S.C., Rajinikanth, V.: Jaya algorithm guided procedure to segment tumor from brain MRI. J. Optim. **2018**, 12 (2018). https://doi.org/10.1155/2018/3738049
22. Fernandes, S.L., Rajinikanth, V., Kadry, S.: A hybrid framework to evaluate breast abnormality using infrared thermal images. IEEE Consum. Electron. Mag. **8**(5), 31–36 (2019). https://doi.org/10.1109/MCE.2019.2923926

Examination of Optic Disc Sections of Fundus Retinal Images—A Study with Rim-One Database

S. Fuzail Ahmed Razeen, Emmanuel, V. Rajinikanth, P. Tamizharasi, and B. Parvatha Varthini

Abstract Eye is the most important sensory organ, which converts the available light signal into the respective images. The eye consists of various essential parts, in which, the infection/disease in the optic disc is to be cured as early as possible to prevent the loss in eyesight. In this work, an image processing technique to examine the disease in optic disc using the fusion of the thresholding and the segmentation procedure is presented. The implementation of the threshold is performed using the Modified Firefly Algorithm. It is based on Otsu's threshold, and the segmentation is employed using the Distance Regularized Level Set (DRLS) approach. This procedure is implemented and validated on the well-known Rim-One database with various levels of the Optic Disc (OD) pictures recorded with a Fundus Camera. The supremacy of the suggested approach is confirmed using a relative analysis between the mined OD with the Ground Truth (GT) existing in the Rim-One database. The outcomes of this work confirm that this suggested method offers better result on the considered dataset with better Picture Quality Parameters (PQP).

Keywords Fundus picture · Optic disc · Otsu · Firefly Algorithm · DRLS

S. Fuzail Ahmed Razeen · Emmanuel · V. Rajinikanth (✉)
Department of Electronics and Instrumentation Engineering, St. Joseph's College of Engineering, Chennai 600119, Tamil Nadu, India
e-mail: rajinikanthv@stjosephs.ac.in

P. Tamizharasi
Department of Electrical and Electronics Engineering, Adiparasakthi Engineering College, Melmaruvathur, Tamil Nadu, India

B. Parvatha Varthini
Department of Computer Science Engineering, St. Joseph's College of Engineering, Chennai 600119, Tamil Nadu, India

S. C. Satapathy et al. (eds.), *Intelligent Data Engineering and Analytics*, Advances in Intelligent Systems and Computing 1177,
https://doi.org/10.1007/978-981-15-5679-1_69

1 Introduction

Recently, medical picture evaluation attracted a larger research community because of its clinical significance. The recent approaches are dealing with the bio-signal examination and bio-image examination collected using dedicated sensors and imaging procedures. The bio-medical imaging procedure is significant to examine the vital internal organs and to identify the infected sections accurately to plan for the possible treatment procedures [1–5].

Because of its application and the disease prediction process, a variety of bio-image processing procedures are largely proposed and employed by the researchers [6–8]. Investigation of the gray picture is simple and requires less computational effort due to its simple pixel distribution process. But, the examination of the RGB picture always requires complex computations due to its Red, Green, and Blue pixels. Because of its complexity, a range of procedures are employed by the researchers to preprocess the RGB scale pictures to enhance the information during the evaluation.

The present research work aims to implement a Picture Evaluation Scheme (PES) to scrutinize the retinal optic disc (OD) picture. Normally, to identify the abnormality in the eye, an imaging procedure is to be followed. This procedure records the eye elements using the specialized camera, called the fundus camera, and the recorded picture is then inspected by the doctor or the dedicated software available in the eye clinics to identify the eye abnormality to plan for suitable treatment process [9, 10].

In this proposed attempt, the examination of the retinal OD is considered and the benchmark OD dataset called Rim-One [11, 12] is used to test the efficiency of the projected image processing procedure. Researchers have attempted a number of traditional and soft-computing assisted fundus picture examination procedures that are already proposed and instigated by many researchers [7, 8]. The recent works on OD examination confirm that a two-step procedure will always help to achieve better result compared to the single-step procedure. Hence, in the proposed paper, the combination of the RGB segmentation using multi-level threshold is integrated to enhance the outcome during the retinal OD examination. The RGB multi-threshold is implemented using the Modified Firefly Algorithm (MFA) [13, 14] based on Otsu's function, and the segmentation is executed using the DRLS segmentation [15].

The proposed experimental work is executed using the Matlab 7 software, and the performance of the proposed approach is validated by considering the Picture Quality Parameters (PQP), such as Accuracy (AC), Precision (PR), Sensitivity (SE), and Specificity (SP). The obtained values are computed by comparing the extracted OD section with the OD existing as the Ground Truth (GT) picture. This database consists of a total number of five GTs for every test picture. Finally, the average of these PQP is considered for the authentication of the performance of the proposed PES. In this work, the proposed imaging tool is very efficient and provides better values of PQP, such as SE (97.94%), SP (98.72%), AC (97.54), and PR (98.83%). In the future, this method can be used to examine other medical pictures recorded with various imaging techniques.

2 Methodology

In image examination literature, a number of methods are available to perform the multi-thresholding work and the segmentation work. In this paper, the various stages presented in Fig. 1 is adopted to examine the retinal OD of the Rim-One database.

2.1 *Rim-One Database*

Rim-One database is available in [11], in which a single retinal OD is available in edition-1 and two ODs are existing in edition-2. In this paper, OD available in edition-1 is considered for the examination and this dataset has ODs recorded with various categories, such as Deep, Early, Moderate, Normal, and OHT [12]. Each picture is associated with five numbers of the GTs, such as GT1, GT2, GT3, GT4, and GT5. Due to its clinical significance, this dataset is extensively considered by the investigators to estimate their image examination procedures [7, 8].

2.2 *Preprocessing*

Otsu's multi-thresholding was introduced in 1979, to enhance the grayscale pictures [16]. This procedure helps to obtain the optimum thresholds by maximizing the Between-Class-Variance (BCV) of image pixels. Because of its superiority, this method was extensively adopted to enhance RGB images [17].

In RGB case, let L denotes total greatness stages of range [0, 1, 2, …, $L-1$]. Then, the likelihood allocation A_i^C can be illustrated as

$$A_i^C = \frac{h_i^C}{N} \sum_{i=0}^{L-1} A_i^C = 1 \tag{1}$$

where i is a specific intensity stage of range $\{0 \le i \le L - 1\}$, for the color element $C = \{R, G, B\}$, N represents the whole pixels, and h_i^C is total pixels for the i intensity level in the element C. Comprehensive discussion concerning the Otsu's utility is existing in [18].

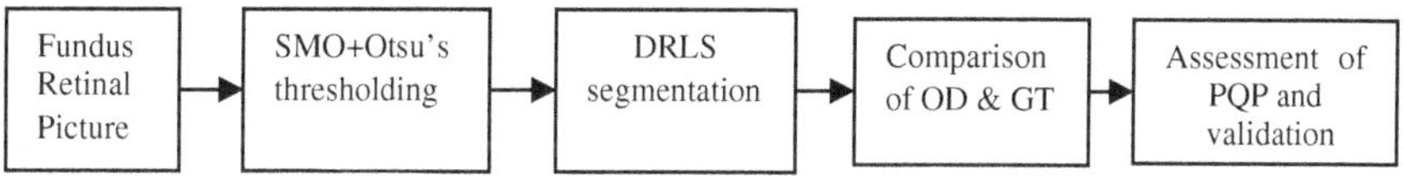

Fig. 1 Various steps involved in the proposed procedure

In the proposed paper, the identification of Otsu's J_{max} is done using the Brownian walk FA discussed in [17].

In this FA, the search process is guided by the BW strategy and its mathematical model is accessible below

$$X_i^{t+1} = X_i^t + \beta_0 e^{-\gamma d_{ij}^2}\left(X_j^t - X_j^t\right) + \alpha_1 \cdot \text{sign}\left(\text{rand} - 1/2\right) \oplus \text{BW} \tag{2}$$

The details on the FA considered in this paper are followed by [17]. The FA parameters are allocated in the following manner, total fireflies = 30, iteration number = 2000, search dimension = 3, and the stopping parameter = J_{max}.

The above procedure can be used as the preprocessing procedure, which will enhance the visibility of the OD based on Otsu's three-level thresholding procedure [7].

2.3 DRLS Segmentation

This work implements the DRLS segmentation of Li et al. [15]. DRLS follows energy minimization process which is expressed as

$$\Re_e(\varphi) \triangleq \int_{\Omega} e(|\nabla\varphi|)\mathrm{d}X \tag{3}$$

where e = energy density value with $e = [0, \alpha] \rightarrow \Re$.

In this work, an adoptable curve is allowed to identify all the possible pixel values related to the abnormal section existing in the image. After identifying all the possible pixel groups, it will extract the region which is inside the converged curve for the evaluation process.

2.4 Assessment

The merit of PES is confirmed by calculating the PQP as discussed in [4]. In this paper, the PQPs like SE, SP, AC, and PR are considered to evaluate the superiority of the proposed procedure [19, 20]. The required PQP values are computed by comparing the extracted OD against the five GTs available in the Rim-One. Later, the average values of these PQPs are recorded separately for every image group, and based on its value, the superiority is confirmed.

3 Results and Discussion

The investigational results obtained with the proposed PES are presented in this section. All these works are implemented with Matlab 7.

Figure 2 presents the sample test picture considered for the initial examination. It is from the Deep (D) case, and Fig. 2a signifies the representative name and Fig. 2b shows the test picture. Later, a tri-level threshold; which depends on MFA and Otsu is employed on these images and the preprocessed test pictures are shown in Fig. 2c. In this picture, the OD section is separated from the blood vessels and the background due to the preprocessing process. Later, the DRLS is then implemented to extract the OD segment. Figure 3 presents the sample test picture available in the Rim-One for other disease cases considered in this paper. In which Fig. 3a is the disease, Fig. 3b–d denotes the sample picture. Similar preprocessing and segmentation procedures are adopted for these pictures and the essential PQP are computed with a relative study using the GTs.

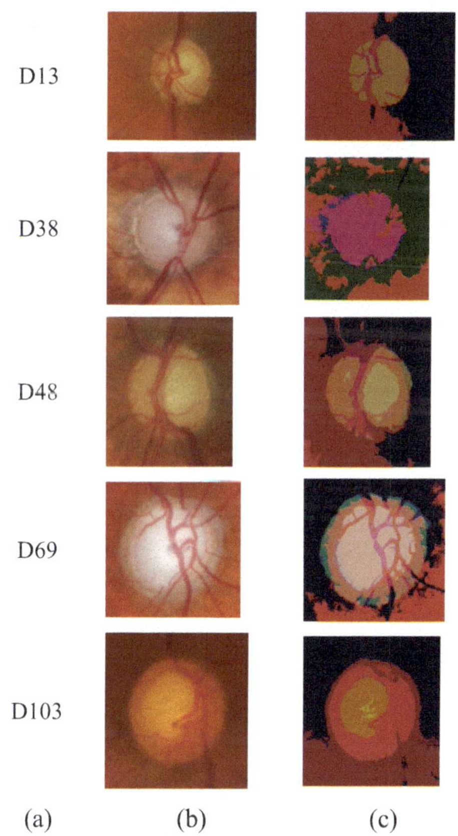

Fig. 2 Sample test picture of deep case

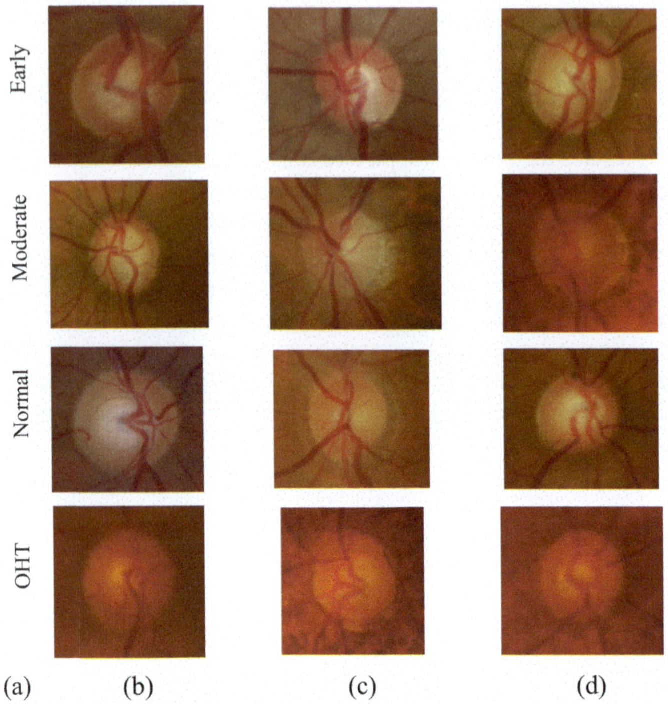

Fig. 3 Sample test pictures of various disease cases

Figure 4 provides the results acquired with the DRLS procedure. Initially, a bounding box is assigned on the OD section of preprocessed picture manually by choosing the X and Y-axis values. If the iteration is increased, the box is allowed to converge toward the OD, and finally, it extracts the OD section. Later, the OD section is then compared against every GT available from Fig. 4f–j and the required PQP are computed and the conforming values are tabulated in Table 1. Similar technique is then implemented for other test pictures and the corresponding average PQP for every disease case is presented in Table 2.

From the above table, it can be observed that the average PQP results obtained for Rim-One dataset are presented as follows; SE (97.94%), SP (98.72%), AC (97.54), and PR (98.83%). From this result, it is clear that these values are >97.50%, which confirms that, the projected method is more efficient for the examination of the fundus retinal pictures. In the future, this practice can also be used to examine other bio-medical pictures.

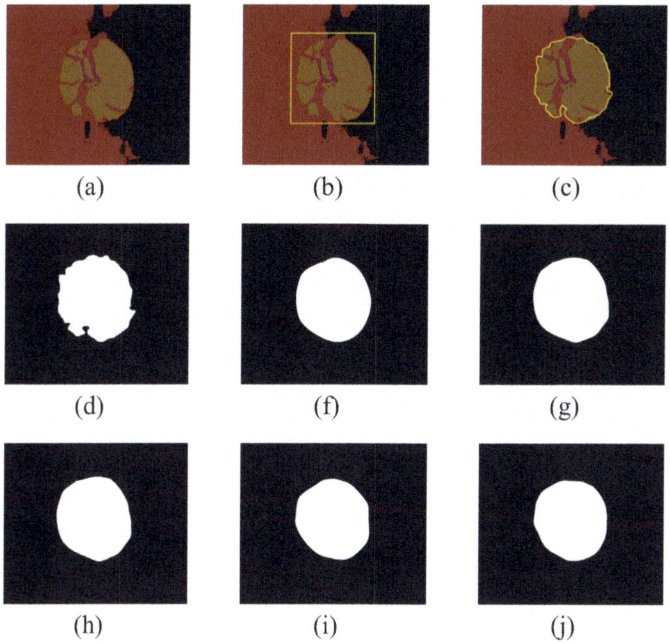

Fig. 4 Result obtained with the DRLS segmentation. **a** Preprocessed D13 picture, **b** initial bounding box **c** converged curve, **d** extracted OD, **f–j** GT1–GT5

Table 1 PQP values obtained for D13 picture

Image	SE	SP	AC	PR
GT1	0.9518	0.9715	0.9493	0.9619
GT2	0.9672	0.9743	0.9518	0.9727
GT3	0.9511	0.9685	0.9599	0.9638
GT4	0.9584	0.9690	0.9482	0.9682
GT5	0.9618	0.9704	0.9573	0.9558
Average	0.9581	0.9707	0.9533	0.9645

Table 2 PQP computed for the Rim-One database

Case	SE	SP	AC	PR
Deep	0.9816	0.9903	0.9783	0.9916
Early	0.9792	0.9884	0.9692	0.9879
Moderate	0.9814	0.9837	0.9800	0.9895
Normal	0.9773	0.9879	0.9724	0.9826
OHT	0.9774	0.9859	0.9771	0.9899
Average	0.9794	0.9872	0.9754	0.9883

4 Conclusion

This work implements a recent imaging method based on the MFA and Otsu based preprocessing and the DRLS based segmentation. The benchmark retinal OD picture database called the Rim-One is considered for the examination and the experimental work is implemented with Matlab 7. In this work, the Rim-One's images with the Deep, Early, Moderate, Normal, and OHT cases are separately examined and the PQP values such as SE, SP, AC, and PR are then computed based on the comparative inspection between the mined OD and the GTs existing in the dataset. The outcomes of this study approve that, this imaging system is efficient and offers an average PQP of >97.50%.

References

1. Dey, N., Rajinikanth, V., Ashour, A.S., Tavares, J.M.R.S.: Social group optimization supported segmentation and evaluation of skin melanoma images. Symmetry **10**(2), 51 (2018). https://doi.org/10.3390/sym10020051
2. Raja, N.S.M., Kavitha, N., Ramakrishnan, S.: Analysis of vasculature in human retinal images using particle swarm optimization based Tsallis multi-level thresholding and similarity measures. Lecture Notes in Computer Science, vol. 7677, pp. 380–387 (2012)
3. Vaishnavi, G.K., Jeevananthan, K., Begum, S.R., Kamalanand, K.: Geometrical analysis of schistosome egg images using distance regularized level set method for automated species identification. J. Bioinf. Intell. Control **3**(2), 147–152 (2014)
4. Rajinikanth, V., Raja, N.S.M., Kamalanand, K.: Firefly algorithm assisted segmentation of tumor from brain MRI using Tsallis function and markov random field. J. Control Eng. Appl. Inform. **19**(3), 97–106 (2017)
5. Rajinikanth, V., Fernandes, S.L., Bhushan, B., Sunder, N.R.: Segmentation and analysis of brain tumor using Tsallis entropy and regularised level set. Lecture Notes in Electrical Engineering, vol. 434, pp. 313–321 (2018)
6. Khan, M.W., Sharif, M., Yasmin, M., Fernandes, S.L.: A new approach of cup to disk ratio based glaucoma detection using fundus images. J. Integr. Des. Process Sci. **20**(1), 77–94 (2016). https://doi.org/10.3233/jid-2016-0004
7. Shree, T.D.V., Revanth, K., Raja, N.S.M., Rajinikanth, V.: A hybrid image processing approach to examine abnormality in retinal optic disc. Procedia Comput. Sci. 157–164 **125** (2018). https://doi.org/10.1016/j.procs.2017.12.022
8. Sudhan, G.H.H,, Aravind, R.G., Gowri, K., Rajinikanth, V.: Optic disc segmentation based on Otsu's thresholding and level set. In: International Conference on Computer Communication and Informatics (ICCCI), pp. 1–5. IEEE (2017). https://doi.org/10.1109/iccci.2017.8117688
9. Dey, N., Roy, A.B., Das, A., Chaudhuri, S.S.: Optical cup to disc ratio measurement for glaucoma diagnosis using Harris corner. In: Third International Conference on Computing Communication & Networking Technologies (ICCCNT). IEEE. https://doi.org/10.1109/icccnt.2012.6395971
10. Almazroa, R., Burman, K., Raahemifar, Lakshminarayanan, V.: Optic disc and optic cup segmentation methodologies for glaucoma image detection: a survey. J. Ophthalmol, 28. Article ID 180972. https://doi.org/10.1155/2015/180972
11. Fumero, F., Alayon, S., Sanchez, J.L., Sigut, J., Gonzalez-Hernandez, M.: RIM-ONE: an open retinal image database for optic nerve evaluation. In: 24th International Symposium on Computer-Based Medical Systems (CBMS), pp. 1–6, IEEE. https://doi.org/10.1109/cbms.2011.5999143

12. http://medimrg.webs.ull.es/research/retinal-imaging/rim-one/
13. Nair. M.V., et al.: Investigation of breast melanoma using hybrid image-processing-tool. In: International Conference on Recent Trends in Advance Computing (ICRTAC), pp. 174–179. IEEE (2018) https://doi.org/10.1109/ICRTAC.2018.8679193
14. Rajinikanth, V., Satapathy, S.C,, Dey, N., Vijayarajan, R.: DWT-PCA image fusion technique to improve segmentation accuracy in brain tumor analysis. Lecture Notes in Electrical Engineering, vol. 471, pp. 453–462 (2018). https://doi.org/10.1007/978-981-10-7329-8_46
15. Li, C., Xu, C., Gui, C., Fox, M.D.: Distance regularized level set evolution and its application to image segmentation. IEEE Trans. Image Process. **19**(12), 3243–3254 (2010). https://doi.org/10.1109/TIP.2010.2069690
16. Otsu, N.: A threshold selection method from gray-level histograms. IEEE Trans. Syst. Man Cybern. **9**(1), 62–66 (1979)
17. Rajinikanth, V., Couceiro, M.S.: RGB histogram based color image segmentation using firefly algorithm. Procedia Comput. Sci. **46**, 1449–1457 (2015)
18. Raja, N.S.M., Rajinikanth, V., Latha, K.: Otsu based optimal multilevel image thresholding using firefly algorithm. Model. Simul. Eng., 17 (2014), Article ID 794574, https://doi.org/10.1155/2014/794574
19. Rajinikanth, V., Satapathy, S.C., Fernandes, S.L., Nachiappan, S.: Entropy based segmentation of tumor from brain mr images—a study with teaching learning based optimization. Pattern Recogn. Lett. **94**, 87–95 (2017). https://doi.org/10.1016/j.patrec.2017.05.028
20. Grgic, S., Grgic, M., Mrak, M.: Reliability of objective picture quality measures. J. Electr. Eng. **55**(1–2), 3–10 (2004)

Inspection of 2D Brain MRI Slice Using Watershed Algorithm

D. Hariharan, S. Hemachandar, N. Sri Madhava Raja, Hong Lin, and K. Sundaravadivu

Abstract The abnormalities in the brain are frequently examined with signal/image supported methodologies. Image-based procedures will disclose more information concerning the functioning of brain contrast to signal assisted techniques. In this study, the examination of the Flair modality brain MRI is considered for testing. Compared to supplementary modalities, Flair MRI (FMRI) suggests improved visibility of the infected brain section; hence an automated procedure can be easily implemented for FMRI. This work implements an established automation procedure called the Marker Controlled Watershed Mining (MCWS) to extract the infected section from the chosen brain FMRI slice. The investigational exploration is implemented with the BRATS2015 database, and finally, the superiority of MCWS mining is confirmed by computing the picture similarity measures (PSM). This work confirms that MCWS practice offers an average PSM of >92% for the considered FMRIs.

Keywords Brain abnormality · Flair MRI · Watershed mining · Picture similarity measure · Validation

1 Introduction

Recently, therapeutic image evaluation exercise was mostly implemented due to its clinical significance and exactness in syndrome estimation and substantiation procedures. A large amount of imaging methodologies is existing to record the interior and external organs of the human body [1–4]. Magnetic Resonance Imaging (MRI) is a confirmed practice, broadly adopted in modern days to record the 3D/2D image

D. Hariharan · S. Hemachandar · N. Sri Madhava Raja (✉) · K. Sundaravadivu
Department of Electronics and Instrumentation, St. Joseph's College of Engineering, Chennai, Tamilnadu 600119, India
e-mail: nsrimadhavaraja@stjosephs.ac.in

H. Lin
Department of Computer and Mathematical Sciences, University of Houston, Downtown, TX 77002, USA

S. C. Satapathy et al. (eds.), *Intelligent Data Engineering and Analytics*, Advances in Intelligent Systems and Computing 1177,
https://doi.org/10.1007/978-981-15-5679-1_70

of vital organs, like the brain. MRI techniques are typically considered to document the working and disease existing in the human brain. Further, the MRI is active with various modalities, such as Flair, T1, T2, and T1C, etc [5–8].

Every MRI modality will have its own merit and demerits, in which the T1C will be recorded in association with a contrast agent, so the tumor core section will be more visible than the edema. In T1C, the contrast medium will improve the visibility of the most infected section than its associated edema, so it may offer a false prediction of the entire infection existing in the organ. The visibility existing in the T1-modality is poor compared to the alternatives. T2-modality will also offer better visibility compared to the T1 and T1C. The Flair modality MRI (FMRI) possibly improves the appearance of the tumor core and the edema; hence it is widely adopted in clinics to detect the infection/abnormality in human organs. In this paper, brain tumor evaluation based on FMRI is addressed. In the literature, a considerable number of conventional and soft-computing assisted tumor examination actions are proposed and implemented to examine various modality MRIs. Recently, a two-stage procedure based on the combination of thresholding and mining is also proposed to improve the accuracy of MRI evaluation [9, 10].

The researchers developed semi-automated and automated approaches to examine medical pictures. The work of this paper aims to implement an automated procedure to examine the infected section of the brain using FMRI. The availability of clinical MRI is limited to the users and it also requires the approval of the ethical board, which monitors the utilization of these biomedical datasets. So, in this paper, the benchmark MRI dataset called the Brain Tumor Image Mining Challenge (BRATS 2015) database [11] is adopted for the experimental study. BRATS 2015 consists of a three-dimensional (3D) skull stripped brain MRI recorded with T1, T1C, T2, and flair modalities, and all these pictures are combined with the expert given pictures called the Ground Truth (GT). For the assessment, the chosen 2D slices from the 3D picture are extracted using the ITK-SNAP tool [12].

Work of this paper implements an automated mining scheme based on the Watershed (WS) algorithm to mine the Region Of Interest (ROI) from the test pictures. The performance of WS is then confirmed by computing the standard Picture Similarity Measures (PSM) widely used during the ROI and GT assessment. In this study, a set of 50 pictures (10 slices from five patients) of dimension 216 × 160 pixels are considered and the experimental work is implemented using the Matlab software. The proposed tool directly implements the WS procedure to extract the infected section without any additional preprocessing procedures adopted in two-stage evaluation procedures [13, 14].

The experimental outcome confirms that the average PSM obtained in this study is >90%, which confirms that, the proposed tool is efficient in examining the FMRI. In the future, the proposed work can be extended to examine the brain MRIs recorded using T1, T1C, and T2 modalities. Further, it can also be tested by considering the real patient's images obtained from medical imaging centers.

2 Context

Clinical and benchmark brain MRI examination is largely addressed by the researchers due to the consequence of brain illness. Brain diseases like stroke and tumor may cause temporary or permanent disability in humans leading to mental stress and mortality. Previous works give the insight of single-stage and two-stage procedures implemented to examine the brain sickness. Rajinikanth et al. [9], implemented an entropy assisted two-stage methodology to examine the brain MRIs recorded using T1, T2, T1C, and flair modalities, and found that Shannon's based preprocessing offers better result compared to Kapur's and Tsallis entropy [9]. The recent work of Raja et al. (2018), implements Tsallis+level-set procedure to extract the tumor section from a class of MRIs, and this paper validates that the two-stage method offers better results on 2D MRI compared with the single-stage procedures [10]. Menze et al. [11] presented a review regarding various MRI assessment procedures available, and this work also developed a 3D brain MRI dataset called BRATS2015 [11]. The work of Rajinikanth et al. [9] implemented a comparative analysis on various semi- and fully automated mining procedures to extract the stroke lesion from brain MRI. This work also confirms the superiority of the two-stage method to evaluate illness with brain MRI [13]. The latest work of Dey et al. [6] implemented an evaluation technique and confirmed that soft-computing based assessment will offer better results during the assessment of the tumor section [6]. In this work, an automated, single-stage practice based on WS is implemented to extract the infected section from 2D FMRI slice. The outcome of this work confirms that this technique is computationally efficient and provides better PSM for the considered test pictures.

3 Methodology

The success of the medical image examination depends mainly on the procedures implemented to process test pictures. Figure 1 illustrates the different phases available in the proposed scheme.

3.1 FMRI Database

The test pictures considered in this study are obtained from BRATS 2015 database [11]. This consists of a 3D reconstructed brain picture, which includes a separate T1, T2, T1C, and flair modality cases with a picture size of 216 × 160 and the related GT picture. Al the test pictures are available without the skull section, which will help to reduce the complexity in MRI examination. The 2D slices from the 3D MRI are initially extracted using the ITK-SNAP tool of Yushkevich et al. [12]. During

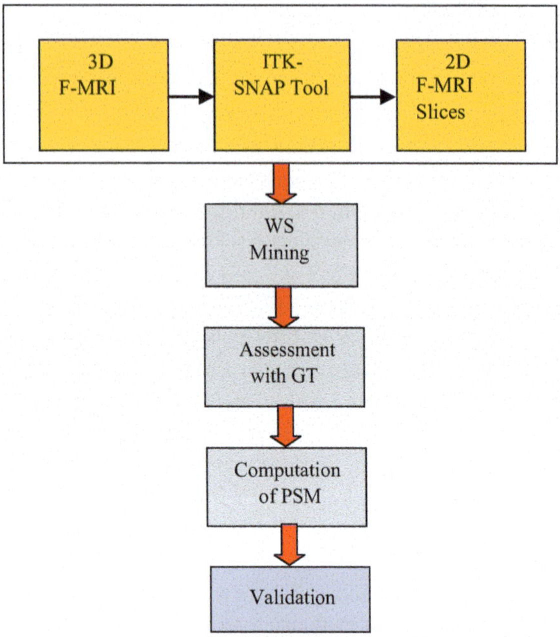

Fig. 1 Different stages in planned computerized tool

this process, five patient's MRIs are considered and ten 2D slices were chosen from each patient to test and validate the superiority of the proposed scheme.

3.2 Watershed Mining

The infected region of FMRI brain picture may be extracted using the semi- and fully-automated procedures. Semi-automated practice necessitates the help of a human operator to initiate the mining task [15]. Hence, automated practices are widely preferred during the assessment of medical pictures [16]. This work adopts the Marker Controlled Watershed Mining (MCWS) to mine the abnormal section in FMRI. MCWS involves edge recognition, formation of WS, adjustment of marker rate to spot the pixel groups, augmentation of recognized pixels, and extraction of the accepted region [17–19]. This MCWS helps to extract all the possible pixels from the FMRI. Similar procedure

3.3 Assessment of Extracted Section and Computing PSM

A relative study between the extracted tumor and GT is always recommended to judge the performance of the developed image assessment tool against the outcome by the experts. The previous works also confirm that appraisal against GT is necessary to compute the PSM, such as True Negative (TN), True Positive (TP), True Positive Rate (TPR), True Negative Rate (TNR), False Positive Rate (FPR), False Negative Rate (FNR). Further, the above values can be considered to obtain Jaccard (JA), Dice (DI), Sensitivity (SE), Specificity (SP), Accuracy (AC), and Precision (PR) as discussed in [20–22].

3.4 Validation

The result and the throughput of projected practice is authenticated based on the average value of PSM. The work of Rajinikanth et al. [9], authenticates that larger PSM values will designate the closeness of mined tumor against the GT [9]. So, if PSM is good, then the developed picture assessment toll can be considered to examine the brain FMRI.

4 Results and Discussion

This section of the paper demonstrates the outcome reached from this research work. The 2D FMRI is examined with the MCWS function and this approach is validated based on the PSM. Initially, this technique is implemented on the 2D test pictures shown in Fig. 2. Figure. 2a1 and b1 represent the chosen test pictures and Fig. 2a2 and b2 present the related GT. The brightest part in GT is the tumor and the gray part is the edema. The combination of tumor and edema sections is to be extracted from the FMRI using the MCWS.

Figure 3 presents the similar pictures for other patients existing in BRAT S2015. Figure 3a1–c1 represents the picture to be examined and Fig. 3a2–c2 shows the GTs. Later, the MCWS procedure is executed to mine the brain infection from the FMRI test image and the results are depicted in Fig. 4. Figure 4a presents recognized edges in FMRI test picture, Fig. 4b, c, and d shows the formation of WS, adjustment of the marker and identified tumor section, respectively. Figure 4e shows the enhanced tumor section and Fig. 4f depicts the extracted brain infection. A similar procedure

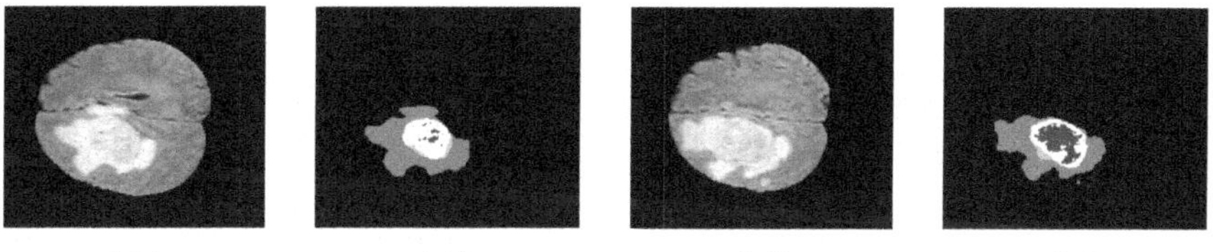

(a1) (a2) (b1) (b2)

Fig. 2 Sample test pictures

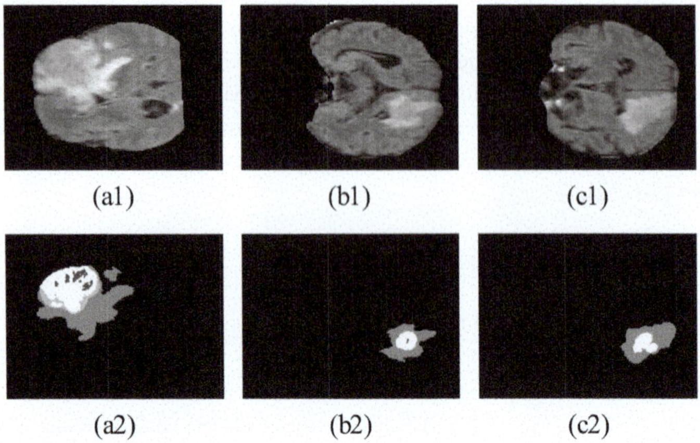

Fig. 3 Test-images and related GTs

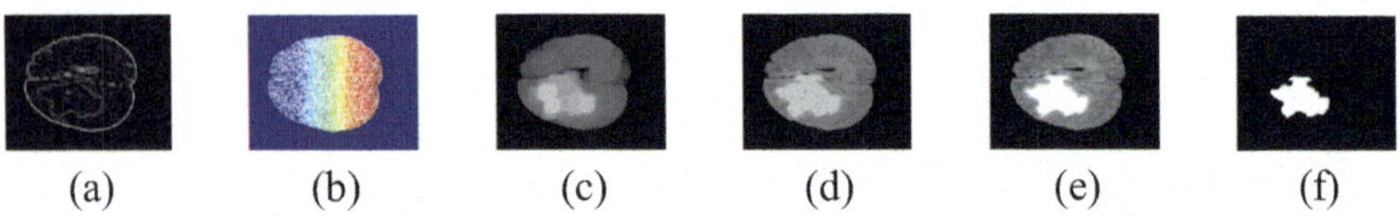

Fig. 4 Outcome of MCWS. **a** Edge recognition, **b** development of WS, **c** structure of marker, **d** recognition section, **e** enhanced section, **f** mined tumor

is executed on entire FMRIs (50 numbers) considered in this work and the sample results are presented in Fig. 5. Figure 5a3–e3 shows the mined tumor section for the test pictures existing in Fig. 2 and Fig. 3. After getting all the achievable infected division with the MCWS mining, a study to compare the pixel values of the tumor and GT is to be done to calculate the PSM, which helps to assess the superiority of MCWS.

In this procedure, the initiation of the adaptable curve sections is to be done by the operator. For most of the test image cases of this study, the AC and LS mining offered poor results, which can be improved by adopting a preprocessing section as discussed in [20]. Compared to the MCWS procedure, the accuracy of these procedures are less and the simulation time taken by the AC and LS is more. From this study, it can be confirmed that MCWS offers superior result with minimal run time.

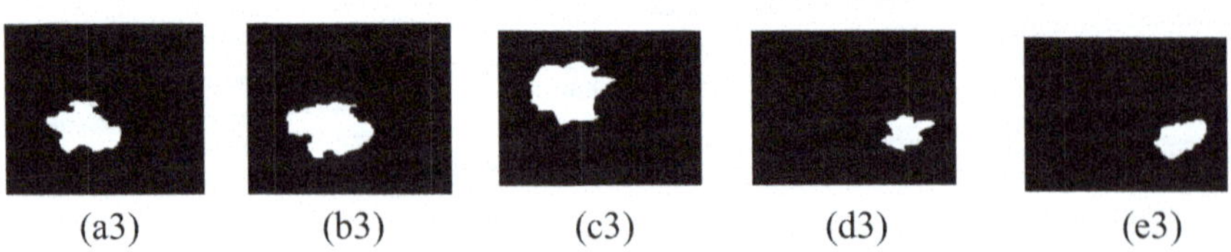

Fig. 5 Mined tumor section of the test pictures

Figure 6 shows the trial of investigation among tumor and GT. Through this appraisal, a pixel assisted evaluation of PSM is executed by adopting its TP and TN values. The outcome of the sample images (images of Fig. 2 and 3) are shown in Table 1 and 2. From Table 1, it is observable that, the sum of TPR and FNR is unity (i.e., TPR + FNR = 1) and similarly TNR + FPR = 1. Further, these tables also reveal that the TPR is equal to the SE and TNR is equal to SP. If these values are exactly unity (100%), then the MCWS is assumed as the best procedure to examine FMRI. For this picture, TPR = SE = 0.9737 (97.37%) and TNR = SP= 0.9890 (98.90%), which confirms that, single-stage procedure based on MCWS works well on the brain FMRIs.

This result is also depicted graphically as in Fig. 7. From this result, it is confirmed that proposed MCWS works well on the considered brain FMRIs. In the future, the MCWS approach can be validated against other approaches executed to examine brain MRIs. Further, the MCWS procedure can be extended to other modalities to confirm its superiority. This investigation also authenticates that the single-step procedure offers better throughput than two-step techniques.

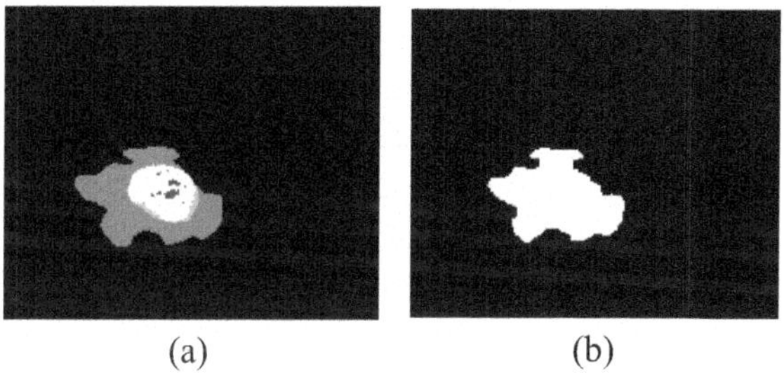

Fig. 6 Relative investigation among GT and tumor

Table 1 Picture performance measures

Image	TP	TN	TPR	TNR	FPR	FNR
1	2550	35009	0.9737	0.9890	0.0110	0.0263
2	2405	34496	0.6919	0.9991	0.0009	0.3081
3	3758	30410	0.9516	0.9934	0.0066	0.0484
4	4530	29645	0.9434	0.9962	0.0038	0.0566
5	3576	30721	0.9445	0.9983	0.0017	0.0555
6	3560	30357	0.9056	0.9966	0.0034	0.0944
7	1033	33402	0.9273	0.9987	0.0013	0.0727
8	1363	33069	0.9355	0.9990	0.0010	0.0645

Table 2 Similarity values between the GT and tumor

Image	JA	DI	SE	SP	AC	PR
1	0.8480	0.9178	0.9737	0.9890	0.9880	0.8679
2	0.6858	0.8136	0.6919	0.9991	0.9710	0.9873
3	0.9055	0.9504	0.9516	0.9934	0.9887	0.9492
4	0.9217	0.9592	0.9434	0.9962	0.9889	0.9757
5	0.9315	0.9645	0.9445	0.9983	0.9924	0.9854
6	0.8823	0.9375	0.9056	0.9966	0.9862	0.9716
7	0.8921	0.9429	0.9273	0.9987	0.9964	0.9591
8	0.9142	0.9552	0.9355	0.9990	0.9963	0.9757
Average	0.8726	0.9301	0.9092	0.9963	0.9885	0.9590

Alike procedure is employed for all the 50 test pictures and the average (avg) values are then computed. During this analysis, the PSM are achieved as follows; JA = 90.14%, DI = 93.85, SE = 92.73%, SP = 99.36%, AC = 98.2 and 96.15%

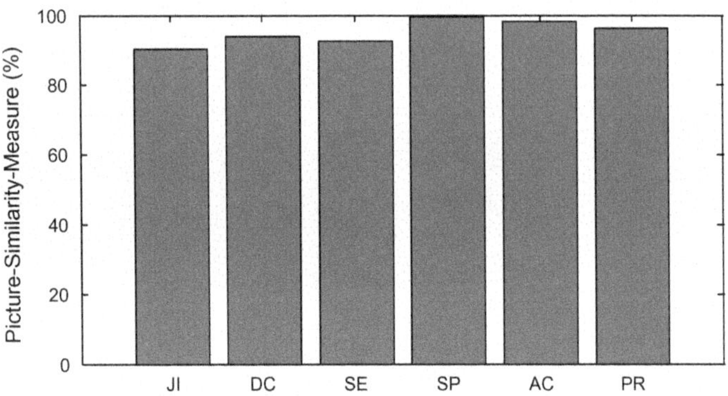

Fig. 7 Average picture similarity values for 50 test pictures

5 Conclusion

This work proposes a single-step methodology to examine the 2D brain FMRI slices. An MCWS algorithm is implemented to mine the infected section from the brain MRI using the MCWS procedure. This work considers the BRATS 2015, database available without the skull section. A sum of 50 pictures is examined in this work and the superiority of this approach is confirmed based on a relative study between the GT and tumor region. The experimental results reveal that the proposed procedure is efficient in extracting the suspicious section and offered an average PSM >90%. In the future, this technique is to be tested using other modality MRIs.

References

1. Bhateja, V., Nigam, M., Bhadauria, A.S., Arya, A., Yu-Dong Zhang, Y-D.: Human visual system based optimized mathematical morphology approach for enhancement of brain MR images. J. Ambient. Intell. Humaniz. Comput. 1–9 (2019). https://doi.org/10.1007/s12652-019-01386-z
2. Satapathy, S.C., El-Maleh, A., Bhateja, V.: Intelligent computing in multidisciplinary engineering applications. Arab. J. Sci. Eng. **43**(8), 3861–3862 (2018)
3. Wang, R., Wang, G.: Web text categorization based on statistical merging algorithm in big data environment. Int. J. Ambient Comput. Intell. (IJACI) **10**(3), 17–32 (2019). https://doi.org/10.4018/IJACI.2019070102
4. Ali, et al.: Adam deep learning with SOM for human sentiment classification. Int. J. Ambient Comput. Intell. (IJACI) **10**(3), 92–116 (2019). https://doi.org/10.4018/IJACI.2019070106
5. Fernandes, S.L., et al.: A reliable framework for accurate brain image examination and treatment planning based on early diagnosis support for clinicians. Neural Comput. Appl. 1–**12** (2019). https://doi.org/10.1007/s00521-019-04369-5
6. Dey, N., et al.: Social-group-optimization based tumor evaluation tool for clinical brain MRI of flair/diffusion-weighted modality. Biocybernetics Biomed. Eng. **39**(3), 843–856 (2019). https://doi.org/10.1016/j.bbe.2019.07.005
7. Acharya, U.R., et al.: Automated detection of Alzheimer's disease using brain MRI images—a study with various feature extraction techniques. J. Med. Syst. **43**, 302 (2019). https://doi.org/10.1007/s10916-019-1428-9
8. Jahmunah, V., et al.: Automated detection of schizophrenia using nonlinear signal processing methods. Artif. Intell. Med. **100**, 101698 (2019). https://doi.org/10.1016/j.artmed.2019.07.006
9. Rajinikanth, V., Dey, N., Satapathy, S.C., Ashour, A.S.: An approach to examine magnetic resonance angiography based on Tsallis entropy and deformable snake model. Futur. Gener. Comput. Syst. **85**, 160–172 (2018). https://doi.org/10.1016/j.future.2018.03.025
10. Raja, N.S.M., Fernandes, S.L., Dey, N., Satapathy, S.C., Rajinikanth, V.: Contrast enhanced medical MRI evaluation using Tsallis entropy and region growing segmentation. J. Ambient. Intell. Humaniz. Comput. 1–12 (2018). https://doi.org/10.1007/s12652-018-0854-8
11. Menze, et al.: The multimodal brain tumor image segmentation benchmark (BRATS). IEEE Trans. Med. Imaging **34**(10), 1993–2024 (2015). https://doi.org/10.1109/TMI.2014.2377694
12. Yushkevich, P.A., Gao, Y., Gerig, G.: ITK-SNAP: An interactive tool for semi-automatic segmentation of multi modality biomedical images. In: 38th Annual International Conference of the IEEE Engineering in Medicine and Biology Society (EMBC), IEEE, pp. 3342–3345 (2016). https://doi.org/10.1109/embc.2016.7591443
13. Rajinikanth, V., Satapathy, S.C.: Segmentation of ischemic stroke lesion in brain MRI based on social group optimization and Fuzzy-Tsallis entropy. Arab. J. Sci. Eng. **43**(8), 4365–4378 (2018). https://doi.org/10.1007/s13369-017-3053-6
14. Satapathy, S.C., Rajinikanth, V.: Jaya algorithm guided procedure to segment tumor from brain MRI. J. Optim. **2018**, 12 (2018). https://doi.org/10.1155/2018/3738049
15. Srivastava, A., Bhateja, V., Moin, A.: Combination of PCA and contourlets for multispectral image fusion. Adv. Intell. Syst. Comput. **469**, 577–585 (2017). https://doi.org/10.1007/978-981-10-1678-3_55
16. Lakshmi, B., Parthasarathy, S.: Human action recognition using median background and max pool convolution with nearest neighbor. Int. J. Ambient Comput. Intell. (IJACI) **10**(2), 34–47 (2019). https://doi.org/10.4018/IJACI.2019040103
17. Roerdink, J.B.T.M., Meijster, A.: The watershed transform: definitions, algorithms and parallelization strategies, Fundam. Informaticae **41**(1, 2), 187–228 (2000). https://doi.org/10.3233/fi-2000-411207
18. Fernandes, S.L., Rajinikanth, V., Kadry, S.: A hybrid framework to evaluate breast abnormality using infrared thermal images. IEEE Consum. Electron. Mag. **8**(5), 31–36 (2019). https://doi.org/10.1109/MCE.2019.2923926

19. Bhandary, A., et al.: Deep-learning framework to detect lung abnormality–a study with chest X-Ray and lung CT scan images. Pattern Recogn. Lett. (2019). https://doi.org/10.1016/j.patrec.2019.11.013
20. Dey, N., Rajinikanth, V., Ashour, A.S., Tavares, J.M.R.S.: Social group optimization supported segmentation and evaluation of skin melanoma images. Symmetry **10**(2), 51 (2018). https://doi.org/10.3390/sym10020051
21. Rajinikanth, V., Fernandes, S.L., Bhushan, B., Sunder, N.R.: Segmentation and analysis of brain tumor using Tsallis entropy and regularised level set. Lect. Notes Electr. Eng. **434**, 313–321 (2018)
22. Nair, M.V., et al.: Investigation of breast melanoma using hybrid image-processing-tool. In: International Conference on Recent Trends in Advance Computing (ICRTAC), IEEE, pp. 174–179 (2018). https://doi.org/10.1109/ICRTAC.2018.8679193

Extraction of Cancer Section from 2D Breast MRI Slice Using Brain Strom Optimization

R. Elanthirayan, K. Sakeenathul Kubra, V. Rajinikanth, N. Sri Madhava Raja, and Suresh Chandra Satapathy

Abstract The Magnetic Resonance Imaging (MRI) is a proven imaging modality in the medical domain to record the internal body organs for the disease diagnosis. This study considers the breast MRI slices to examine the breast tumor section. This work employs a hybrid imaging procedure based on Brain Storm Optimization Algorithm and Shannon's entropy (BSA + SE) thresholding, and a chosen segmentation procedure. In this work, the Active-Contour (AC) based segmentation is initially executed to mine the tumor section. The performance of the AC is then validated against the Chan-Vese (CV) and Level-Set (LS) techniques existing in the literature. In this technique, the Axial View 2D MRI slice along with the Ground Truth (GT) is considered for the assessment. After mining the tumor section, its dimension is compared against the GT, and the essential Image Performance Metrics (IPM) is then computed. Based on the values of IPMs, the performance of the segmentation technique is confirmed. The proposed experiment is executed on 150 2D breast MRI slices and the average accuracy of AC (> 93%) is better compared with the average accuracy of CV (> 92%) and LS (> 93%) technique. Further, the convergence speed of AC is superior to LS.

Keywords Breast tumor · MRI · Brain storm optimization · Shannon's threshold · Active-Contour · Verification

---

R. Elanthirayan · K. Sakeenathul Kubra
Department of Electrical and Electronics Engineering, St. Joseph's College of Engineering, Chennai 600119, Tamilnadu, India

V. Rajinikanth · N. Sri Madhava Raja (✉) · S. C. Satapathy
Department of Electronics and Instrumentation, St. Joseph's College of Engineering, Chennai 600119, Tamilnadu, India
e-mail: nsrimadhavaraja@stjosephs.ac.in

R. Elanthirayan · K. Sakeenathul Kubra · V. Rajinikanth · N. Sri Madhava Raja · S. C. Satapathy
School of Computer Engineering, Kalinga Institute of Industrial Technology (Deemed to Be University), Bhubaneswar 751024, Odisha, India

S. C. Satapathy et al. (eds.), *Intelligent Data Engineering and Analytics*, Advances in Intelligent Systems and Computing 1177,
https://doi.org/10.1007/978-981-15-5679-1_71

1 Introduction

In the women community, abnormality in breast arises in a variety of reasons and untreated irregularity may lead to death. Breast cancer is one of the threats, which needs major attention to prevent the death rate. In the year 2018, report of World Health Organization (WHO) confirms that 627,000 women died worldwide because of breast cancer (BC), and further, the WHO confirms approximately 15% of women's deaths are due to the BC [1].

If the BC is diagnosed in its premature phase, it can be easily cured with chemotherapy or a possible surgery [2, 3]. The identification of BC begins with a personal check followed by a clinical level screening. Due to its occurrence rate, recently, a substantial number of image modalities, such as mammogram, breast MRI, ultrasound imaging, and thermal imaging are considered to record the abnormal breast section for the clinical diagnosis [4–9]. Every modality has its own merits and demerits and the choice of the modality depends mainly on the disease severity and the doctor. In the proposed work, examination of BC is done using the breast MRI existing in [10, 11]. Other works executed with this dataset can be found in [12, 13]. This dataset consists of 3D breast MRI which can be viewed as axial, sagittal, and coronal views. The 2D breast slices of axial view are initially extracted using the ITK-SNAP [14], and the extracted 2D slice is then considered for the assessment. In this study, the breast MRI of four different patients is considered and from every patient, ten number of 2D breast MRI slices are collected for the examination.

The proposed research implements a semi-automated imaging tool based on the multi-level threshold and segmentation. After extracting the tumor part from the MRI, its pixels are then compared against the Ground Truth pixels available in the MRI database. This comparison provides a possible number of image similarity measures and based on this value, the performance of the proposed tool is assessed [15–18].

The evaluation capability of the proposed system is assessed based on a confusion matrix. The performance of the implemented segmentation procedure (Active-Contour) is also validated against other related techniques, such as Chan-Vese (CV) and Level-Set (LS) approaches and the overall result attained with study confirms that AC outperforms other approaches.

Remaining work is presented as follows; Sect. 2 explains the method, Sect. 3 reveals the results and discussions and the conclusion is obtainable in Sect. 4.

2 Methodology

The proposed hybrid imaging technique integrates the benefit of image enhancement and abnormal section extraction procedures. The overview of the system implemented in this work is presented Fig. 1.

Initially, the 3D breast MRI database is collected from [12, 13]. Assessment of MRI in its 3D form is quite complex and hence the original 3D MRI is to

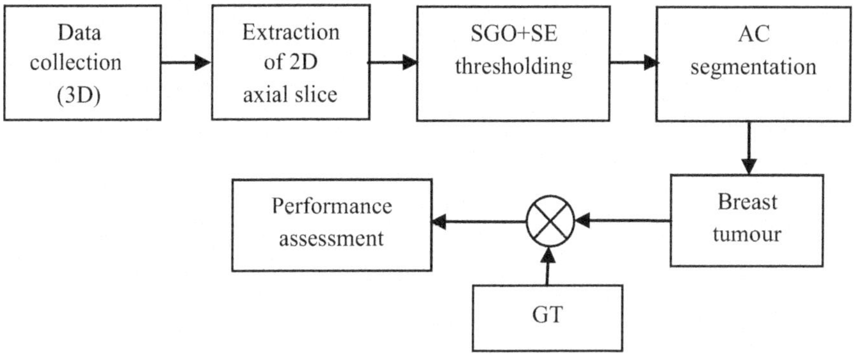

Fig. 1 Hybrid imaging technique to extract and evaluate the breast tumor

be separated into a number of 2D slices and then these slices can be examined easily with a selected image processing method. The threshold is implemented with SGO + SE and a three-level thresholding is implemented to separate the MRI into background, normal breast, and tumor segment. Later, a semi-automated practice called the Active-Contour (AC) segmentation is then implemented to mine the tumor segment. Finally, a comparison is implemented among the GT and tumor, and the performance of the proposed procedure is confirmed by computing the Image Performance Metrics (IPM). The performance of AC is also validated against the CV and LS segmentation methods.

2.1 Image Collection

The necessary breast MRI image is initially collected from the Cancer Imaging Achieve and the dataset and its download techniques can be accessed from [13]. The earlier work regarding this dataset is available in [12]. This data is in 3D form and to reduce the computation effort, this 3D database is converted into 2D axial MRI sliced with the ITK-Snap tool [14]. This procedure helps to attain RGB scaled MRI slices with dimension 288 × 288 pixels, and in this work about 50 MRI slices were collected and considered for the assessment. During the assessment, RGB to grayscale in implemented to further reduce the complexity in assessment, since the major task is to extract the tumor from the considered MRI slices.

2.2 Threshold Process

Entropy assisted estimation is extensively implemented in modern therapeutic picture evaluation tasks due to its better enhancement [19–23], Shannon's entropy is one of the confirmed procedure generally implemented to inspect the medical pictures. The necessary details on Shannon's threshold for the gray/RGB scaled pictures can be found in the latest work [21, 24]. In this work, this entropy utility is considered to improve the visibility of tumor in breast MRI slices based on a three-level threshold. This threshold separated the test image into, background, normal breast, and tumor.

In this work, the essential threshold is achieved with the help of the recent evolutionary technique called the Brain Storm Optimization Algorithm [25–27].

BSA was invented by Shi [25], by modeling the conclusion/discussion making practice in humans. This method works based on the problem-solving potential in humans.

This algorithm has the subsequent steps:

Step 1	Generate a brainstorming people's cluster with mixed surroundings
Step 2	Begin the algorithm to produce a selection of ideas
Step 3	Recognize and chose ideas to resolve the trouble
Step 4	Choose the suggestion with the improved probability
Step 5	Permit the leading person to accumulate several ideas as potential to resolve the specified mission
Step 6	Replicate the step till an improved resolution is attained for the selected problem

In this work, the BSA is considered to identify the optimal threshold by maximizing the objective values of SE. Other details on BSA can be found in [27]. The initial BSA parameters are assigned as; no. of agents = 30, dimension = 3, iteration = 1500, and stopping criteria = maximized entropy.

2.3 *Segmentation*

Image segmentation is a key technique used to mine the chosen part from a raw/processed picture. The AC is one of the semi-automated segmentation tactics used to extract a chosen part in medical pictures. In this, based on the region to be mined, a bounding box is initially assigned by the operator and the box is allowed to converge toward the section based on the increase in the iteration. After some time, the box will surround the pixel collection to be mined and the convergence stops. The extracted section is then converted into a visible region (binary 1) and the remaining part appears as dark (binary 0). This mined tumor section is then compared with the related GT to find the IPMs. Based on these values, the performance of AC is confirmed. The outcome of the Ac is then compared with the results attained with the CV and LS, and the mean performance measure is then computed to evaluate the outcome of AC, CV, and LS [3].

2.4 Image Performance Metrics

The merit of the proposed technique is assessed by comparing the mined segment with GT. The essential measures, like Jaccard, Dice, accuracy, precision, sensitivity, specificity, and F1score are computed. Further, the confusion matrix is also constructed with the true positive (TP), true negative (TN), false positive (FP), and false negative (FN). The arithmetical equations connected to these events can be found in [28–30].

3 Result and Discussion

This division of the study illustrates the realization of a hybrid imaging technique and its outcome. The benchmark breast MRI slices are used for the assessment, and the research is implemented using the MATLAB software.

The sample test images of dimension 288 × 288 pixels along with the GT are shown in Fig. 2. This figure confirms that the tumor segment is more visible compared to the normal section and background. The tumor is further enhanced with the thresholding and the enhanced segment is then mined with the segmentation practice.

Figure 3 depicts the outcome of the proposed methodology. Figure 3a presents the result of BSA + SE thresholding, Fig. 3b and c depicts the initial and the converged bounding box of AC, Fig. 3d depicts the extracted tumor section. The tumor section is a binary image with highly visible pixels (binary 1). This image is then compared against the GT and the essential IPMs are then computed and recorded for further analysis. A similar procedure is implemented for all the 5 numbers of the breast MRI slices, and for the discussion, only the result of five images is presented. Figure 4a–d depicts the segmentation results attained with other MRI slices shown in Fig. 2.

Assessment of tumor with the GT offers the essential tumor section and the confusion matrix attained for the tumor of Fig. 3d confirms that the proposed hybrid approach is better in evaluating the breast MRI slices. The better segmentation is

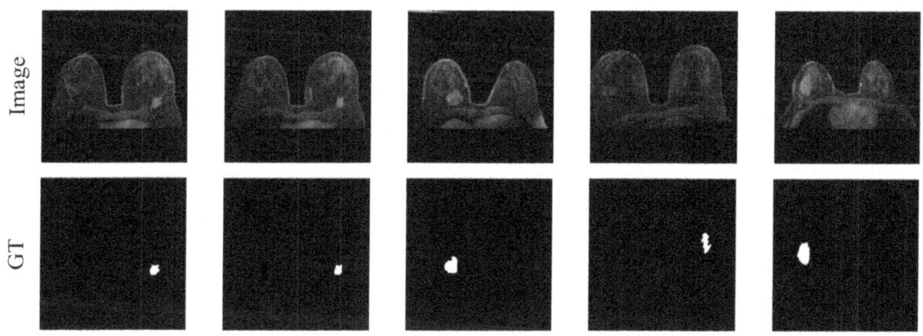

Fig. 2 Sample test images of breast MRI database

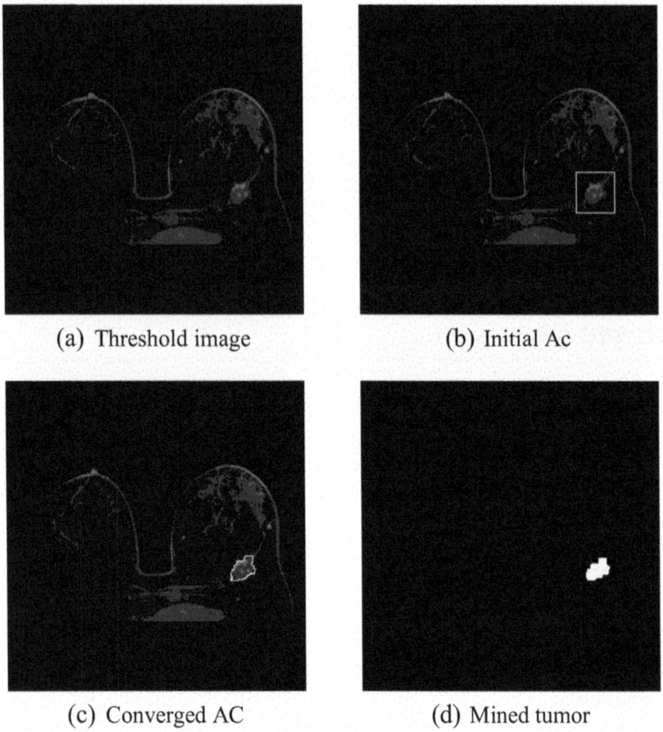

Fig. 3 Results attained in various stages of proposed procedure **a** threshold image **b** Initial Ac **c** Converged AC **d** Mined tumor

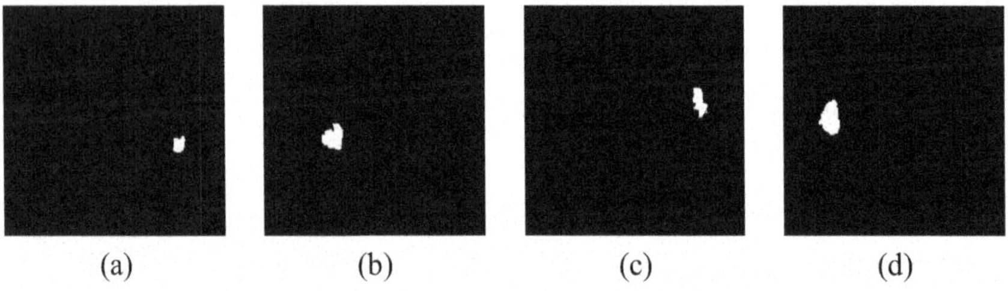

Fig. 4 Extracted tumor of other sample images presented in Fig. 2

evident with Confusion matrix as shown in table which is attained for the considered sample image (Fig. 5).

Figure 6 depicts the performance values attained with the AC, CV, and LS techniques. This measure confirms that the mean results attained with AC and LS is superior than CV. Compared to the Ac, the implementation of LS is fairly complicated and the convergence time of the LS is also larger than AC. In this work, the AC, CV, and LS are helped to attain mean accuracy of > 93%, > 92%, and >93%,

Target

Attained		
TP=242	FP=36	TPR=0.7539
FN=79	TN=82587	FNR=0.2461
P=TP+FN=321	N=FP+TN=82623	TNR=0.9996
Total pixels=P+N		FPR=0.0004

Fig. 5 Confusion matrix attained for a sample image

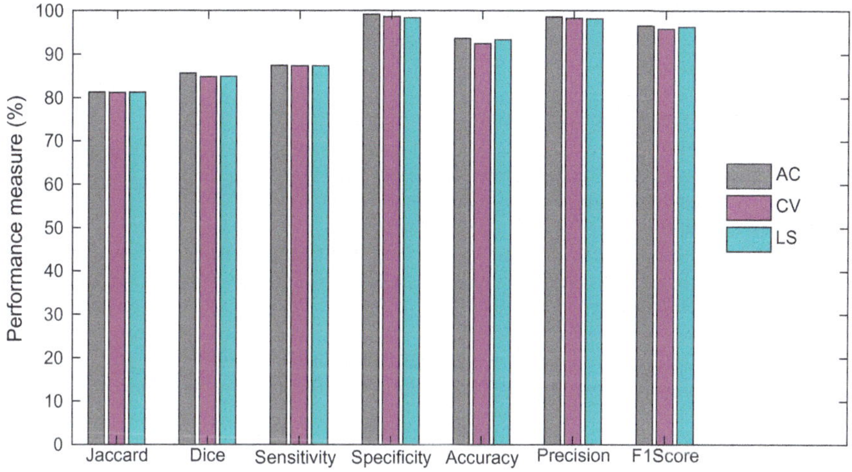

Fig. 6 Performance measures attained with AC, CV, and LS

respectively. In the future, the brain MRI can be assessed with other segmentation techniques existing in the literature.

4 Conclusion

This work aims to implement an image evaluation procedure to extract the tumor section from the 2D breast MRI. This procedure implemented a thresholding and segmentation approach to enhance and extract the cancer section from the MRI slice. The initial threshold is realized with BSA + SE, and the later extraction is achieved by employing the AC. The performance of the AC is also validated against the CV and LS techniques. After extracting the tumor, a relative assessment is done against the GT existing in the database. The comparison result of the tumor and GT confirms that this technique helps to achieve a better confusion matrix and the performance values. These results confirmed the superiority of the proposed technique to assess the breast MRIs.

References

1. https://www.who.int/cancer/prevention/diagnosis-screening/breast-cancer/en/
2. Raja, N.S.M., Rajinikanth, V., Fernandes, S.L., Satapathy, S.C.: Segmentation of breast thermal images using Kapur's entropy and hidden Markov random field. J. Med. Imaging Health Info. **7**(8), 1825–1829 (2017). https://doi.org/10.1166/jmihi.2017.2267
3. Fernandes, S.L., Rajinikanth, V., Kadry, S.: A hybrid framework to evaluate breast abnormality using infrared thermal images. IEEE Consum. Electron. Mag. **8**(5), 31–36 (2019). https://doi.org/10.1109/MCE.2019.2923926
4. Nair, M.V., Gnanaprakasam, C.N., Rakshana, R., Keerthana, N., Rajinikanth, V.: Investigation of breast melanoma using hybrid image-processing-tool. In: International Conference on Recent Trends in Advance Computing (ICRTAC), IEEE, 174–179 (2018). https://doi.org/10.1109/icrtac.2018.8679193
5. Rajinikanth, V., Raja, N.S.M., Satapathy, S.C., Dey, N., Devadhas, G.G.: Thermogram assisted detection and analysis of ductal carcinoma in situ (DCIS). In: International Conference on Intelligent Computing, Instrumentation and Control Technologies (ICICICT), IEEE 1641–1646 (2018). https://doi.org/10.1109/icicict1.2017.8342817
6. Raja, N.S.M., Sukanya, S.A., Nikita, Y.: Improved PSO based multi-level thresholding for cancer infected breast thermal images using Otsu. Procedia Comput. Sci. **48**, 524–529 (2015). https://doi.org/10.1016/j.procs.2015.04.130
7. Beers, A., Chang, K., Brown, J., Zhu, X., Sengupta, D., Willke, T.L., Gerstner, E., Rosen, B., Kalpathy-Cramer, J.: Anatomical DCE-MRI phantoms generated from glioma patient data. SPIE Medical Imaging, vol. 105732V. Houston: SPIE (2018). https://doi.org/10.1117/12.2294961
8. Vidya, K., Kurian, M.: Novel framework for breast cancer classification for retaining computational efficiency and precise diagnosis. Commun. Appl. Electron. **7**(15), 1–6 (2018)
9. Zhu, Y., Li, H., et al.: TU-CD-BRB-06: deciphering genomic underpinnings of quantitative MRI-based radiomic phenotypes of invasive breast carcinoma. Med. Phys. **42**(6), 3603 (2015). https://doi.org/10.1118/1.4925591
10. Meyer, C.R, Chenevert, T.L, Galbán, C.J., Johnson, T.D., Hamstra, D.A., Rehemtulla, A., Ross, B.D.: Data from RIDER_Breast_MRI. The Cancer Imaging Archive (2015). http://doi.org/10.7937/K9/TCIA.2015.H1SXNUXL
11. Clark, K., Vendt, B., Smith, K., Freymann, J., Kirby, J., Koppel, P., Moore, S., Phillips, S., Maffitt, D., Pringle, M., Tarbox, L., Prior, F.: The cancer imaging archive (TCIA): maintaining and operating a public information repository. J. Digit. Imaging **26**(6), 1045–1057 (2013)
12. Drukker, K., Li, H., Antropova, N., Edwards, A., Papaioannou, J., Giger, M.L.: Most-enhancing tumor volume by MRI radiomics predicts recurrence-free survival "early on" in neoadjuvant treatment of breast cancer. Cancer Imaging **18**(1) (2018). https://doi.org/10.1186/s40644-018-0145-9
13. Sutton, E.J., Huang, E.P., Drukker, K., Burnside, E.S., Li, H., Net, J.M., Rao, A., Whitman, G.J., Zuley, M., Ganott, M., Bonaccio, E., Giger, M.L., Morris, E.A.: Breast MRI radiomics: Comparison of computer- and human-extracted imaging phenotypes. Eur. Radiol. Exp. (2017). https://doi.org/10.1186/s41747-017-0025-2
14. Yushkevich, P.A., Gao, Y., Gerig, G.: ITK-SNAP: an interactive tool for semi-automatic segmentation of multi-modality biomedical images. In: 38th Annual International Conference of the IEEE Engineering in Medicine and Biology Society (EMBC), IEEE, 3342–3345 (2016). https://doi.org/10.1109/embc.2016.7591443

15. Bhateja, V., Misra, M., Urooj, S.: Unsharp masking approaches for HVS based enhancement of mammographic masses: a comparative evaluation. Futur. Gener. Comput. Syst. **82**, 176–189 (2018)
16. Wang, R., Wang, G.: Web text categorization based on statistical merging algorithm in big data environment. Int. J. Ambient Comput. Intell. (IJACI) **10**(3), 17–32 (2019). https://doi.org/10.4018/IJACI.2019070102
17. Ali, et al.: Adam deep learning with SOM for human sentiment classification. Int. J. Ambient Comput. Intell. (IJACI) **10**(3), 92–116 (2019). https://doi.org/10.4018/IJACI.2019070106
18. Yang, X., Jiang, X.: A hybrid active contour model based on new edge-stop functions for image segmentation. Int. J. Ambient Comput. Intell. (IJACI) **11**(1), 87–98 (2020). https://doi.org/10.4018/IJACI.2020010105
19. Rajinikanth, V., Dey, N., Satapathy, S.C., Ashour, A.S.: An approach to examine magnetic resonance angiography based on Tsallis entropy and deformable snake model. Futur. Gener. Comput. Syst. **85**, 160–172 (2018). https://doi.org/10.1016/j.future.2018.03.025
20. Raja, N.S.M., Fernandes, S.L.,. Dey, N., Satapathy, S.C., Rajinikanth, V.: Contrast enhanced medical MRI evaluation using Tsallis entropy and region growing segmentation. J. Ambient. Intell. Humaniz. Comput. 1–12 (2018). https://doi.org/10.1007/s12652-018-0854-8
21. Dey, N., Rajinikanth, V., Ashour, A.S., Tavares, J.M.R.S.: Social group optimization supported segmentation and evaluation of skin melanoma images. Symmetry **10**(2), 51 (2018). https://doi.org/10.3390/sym10020051
22. Rajinikanth, V., Satapathy, S.C.: Segmentation of ischemic stroke lesion in brain MRI based on social group optimization and Fuzzy-Tsallis entropy. Arab. J. Sci. Eng. **43**(8), 4365–4378 (2018). https://doi.org/10.1007/s13369-017-3053-6
23. Satapathy, S.C., Rajinikanth, V.: Jaya algorithm guided procedure to segment tumor from brain MRI. J. Optim. **2018**, 12 (2018). https://doi.org/10.1155/2018/3738049
24. Kannappan, P.L.: On Shannon's entropy, directed divergence and inaccuracy. Probab. Theory Rel. Fields **22**, 95–100 (1972). https://doi.org/10.1016/S0019-9958(73)90246-5
25. Shi, Y.: Brain storm optimization algorithm. Lect. Notes Comput. Sci. **6728**, 303–309 (2011). https://doi.org/10.1007/978-3-642-21515-5_36
26. Cheng, S., Shi, Y., Qin, Q., Zhang, Q., Bai, R.: Population diversity maintenance in brain storm optimization algorithm. J. Artif. Intell. Soft. Comput. Res. **4**(2), 83–97 (2014)
27. Cheng, S., Qin, Q., Chen, J., Shi, Y.: Brain storm optimization algorithm: a review. Artif. Intell. Rev. **46**(4), 445–458 (2016)
28. Dey, N., et al.: Social-group-optimization based tumor evaluation tool for clinical brain MRI of flair/diffusion-weighted modality. Biocybernetics Biomed. Eng. **39**(3), 843–856 (2019). https://doi.org/10.1016/j.bbe.2019.07.005
29. Acharya, U.R., et al.: Automated detection of Alzheimer's disease using brain MRI images–a study with various feature extraction techniques. J. Med. Syst. **43**, 302 (2019). https://doi.org/10.1007/s10916-019-1428-9
30. Bhandary, A., et al.: Deep-learning framework to detect lung abnormality–a study with chest X-ray and lung CT scan images. Pattern Recogn. Lett. (2019). https://doi.org/10.1016/j.patrec.2019.11.013

Air Quality Prediction Using Time Series Analysis

S. Hepziba Lizzie and B. Senthil Kumar

Abstract Thi paper aims at predicting the air quality of Chennai based on a Time Series Analysis which is one of the Machine Learning Algorithm. Here we consider the most important pollutants as Sulphur dioxide (SO_2), Nitrogen dioxide (NO_2), Respirable Suspended Particulate Matter (RSPM) as features causing pollution. These pollutant levels of these features can be predicted based on a daily to weekly basis. The technique used to predict these feature values is a statistical concept called Time series Analysis. This method is used to interpret collected and useful data recorded against over a period of time. Time Series Analysis involves developing models that best capture or describe an observed time series in order to understand the underlying causes. Time series forecasting involves taking models to fit on collected or historical data and using them to predict future observations..

Keywords Time series analysis · Auto correlation · Partial correlation · Estimation · Air quality index · Dataset · Machine learning

1 Introduction

Air pollution has become a major concern in all the cities of India. The reason behind this is the presence of many particulate matters in the air we breathe. With the emerging trends in technology, it is possible to predict the air quality of an area using the previously available records of pollution level in Chennai area. There is a common Prediction method used is Time Series Analysis (TSA). There are many methods in TSA which includes time domain, frequency domain and spectral analysis. All these techniques use the correlation, covariance and other statistical

S. Hepziba Lizzie · B. Senthil Kumar (✉)
Department of Electronics and Instrumentation Engineering, St. Joseph's College of Engineering, Chennai, India
e-mail: vlrsenthil@gmail.com

S. Hepziba Lizzie
e-mail: hepzibalizzie@gmail.com

S. C. Satapathy et al. (eds.), *Intelligent Data Engineering and Analytics*, Advances in Intelligent Systems and Computing 1177,
https://doi.org/10.1007/978-981-15-5679-1_72

elements to model and predict from the data set. The data set usually has many trend and seasonality components in them in context with random variables. Trend and seasonality can be described as the amount of dependency each data in the time series has with its previous data. In this paper, we consider a data set that has the recorded Respirable suspended particulate matter (rspm) data of Chennai for a period of 10 years from the year 2005–2015. In the work on Interrupted Time series analysis more robust approach has been demonstrated using Segmented Regression [1]. A segmented regression approach was analysed to be used to analyse an interrupted time series study by testing the effect of an intervention on the outcome of interest using an appropriately defined impact model. Reconstructing of missing Data for Dynamical systems has been considered which can lead to inefficient power estimation and increase in Standard Errors [2]. Seasonal trend analysis and ARIMA modelling of relative humidity and wind speed time series around [3], presents the modelling of the non-linear natural process for analysis and useful interpretation. Evaluation of Clustering Algorithms for the Prediction of Trends in Bus Travel [4], discusses on the proposed prediction methodology uses a PSF algorithm. An ARIMA modelling approach [5], discusses a specific case of modelling, examining the ADF test and residual analysis for the case discussed. Forecasting Australian Consumer Price Index (CPI) using ARIMA models. [6], discuss the modelling for annual CPI data of Australia, from 1960 to 2017. A Comparison of ARIMA and LSTM in Forecasting Time Series by Sima Siami-Namini in 2018 [7], describes a comparative study on LSTM models and ARIMA modelling by testing their various pros and cons. Time Series Based LSTM Model to Predict Air Pollutant's Concentration for Prominent Cities in India by [7], discusses the various pollutant modelling techniques.

2 Data Preprocessing

2.1 Dataset Preparation

The dataset is acquired from the website for data [8], set consists of several attributes like SO_2, NO_2, rspm, location are the attributes chosen. The data set needs to be processed form outliers and missing values. The data contained many missing values; this can be replaced by the mean values of the nearby value range. Also, the outliers had to be removed and the cleaned data is plotted as below (Fig. 1).

2.2 Handling Trend and Seasonality of Data

Time series data contains many components namely, error, seasonality and trend. The series can be decomposed by an additive or multiplicative model. An additive model is usually more appropriate when the seasonal or trend component is not

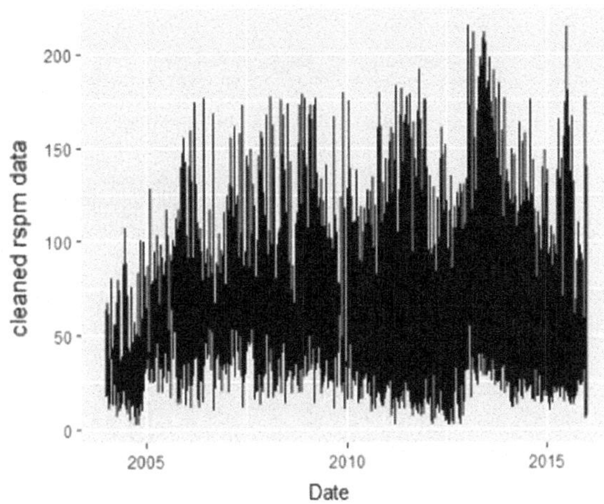

Fig. 1 Cleaned dataset of air quality index

proportional to the level of the series. If the seasonality component changes with the level or trend of the series, a simple overlay, or addition of components, won't be sufficient to reconstruct the series. In that case, a multiplicative model might be more appropriate.

$$Y = S_t * T_t * E_t \quad (1)$$

where S_t is the seasonal component, T_t is trend and cycle, and E_t is the remaining error and Y is the Chennai rspm series data. The trend components were separated for analysis (Fig. 2).

2.3 Checking the Stationary of Time Series

A series is said to be stationary when its mean, variance, and autocovariance are time-invariant. Fitting an ARIMA model requires the series to be stationary. The augmented Dickey–Fuller (ADF) test is a formal statistical test for checking the stationarity of a series. This test uses the method of using the null hypothesis in order to prove the stationarity. The null hypothesis assumes that the series is non-stationary. The ADF uses p-value <0.05 to prove the alternate hypothesis. On executing the ADF test over Chennai rspm data, p-value $= 0.01$. There is enough evidence to prove that the time series is stationary, and hence an ARIMA model can be built over it.

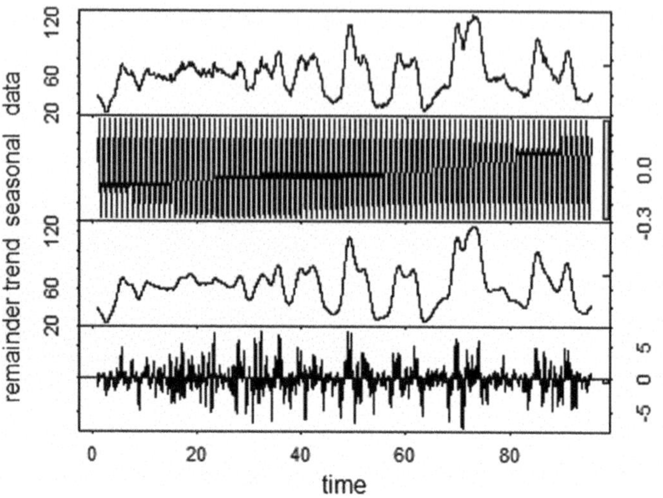

Fig. 2 Seperation of seasonality, trend in the time series data

3 Model Parameters

3.1 Building the Model

ARIMA (Autoregressive Integrated Moving average) model where it involves three main parameters (p, d, q). Each of these parameters represents the model asP is the parameter for AutoRegressive (AR). Model and q is the parameter representing Moving Average modelled is the differencing factor for slow processes involving integrator. MA (q) can be correlated noise structure in our data and goes beyond the traditional assumption where errors are not considered (Table 1).

The model order (5, 1, 7) has the lowest values for AIC and BIC which indicates that this model order is the best fit. However, in order to avoid over-fit issues, model order (5, 1, 0) is selected. The regression form of the chosen model order was selected. AIC and BIC are criteria based on which the best model is selected. The lower the values, better the fit. Akaike's information criterion (AIC) compares the quality of a set of statistical models to each other.

Table 1 Comparison of ARIMA model orders

Model order	Log—likelihood	AIC	BIC	σ^2 estimate
(3, 1, 3)	2280.06	−4546.12	−4498.63	0.02913
(5, 1, 7)	12708.06	−25390.12	−25503.3	0.00092
(5, 1, 0)	12278.81	−24545.61	−24505.3	0.00106

$$AIC = N * \log(RSS/N) + 2k(5) \quad (2)$$

with N being the number of observations and RSS the mentioned residual sum of squares. Bayesian information criterion (BIC) is another criterion for model selection that measures the trade-off between model fit and complexity of the model.

3.2 *Residual Analysis*

Residuals are estimates of experimental error obtained by subtracting the observed responses from the predicted responses. The other methods are the histogram plot with a normal distribution is bell-shaped denoting that the error is predictable around a mean value. The residuals also seem to be in normal distribution as the plot resembles the bell shape curve. Hence, based on the observations, we can see that the residuals are spread in the range (−0.5, 0.5). This is a good sign to indicate that the model is a good fit to the data and future predictions can be carried out (Fig. 3).

The data are plotted against a theoretical normal distribution in such a way that the points should form an approximate straight line. Departures from this straight line indicate departures from normality (Fig. 4).

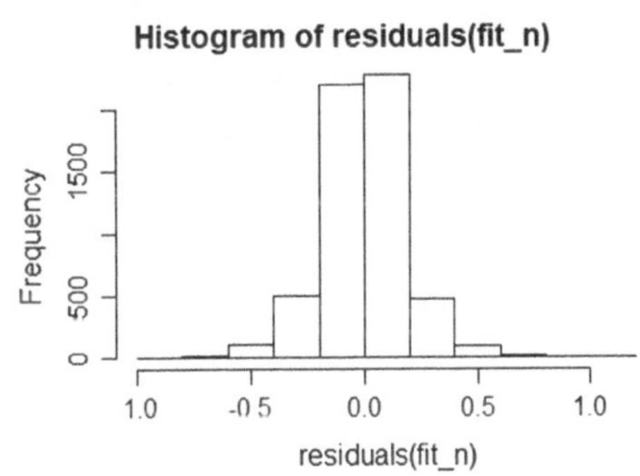

Fig. 3 Histogram of residuals

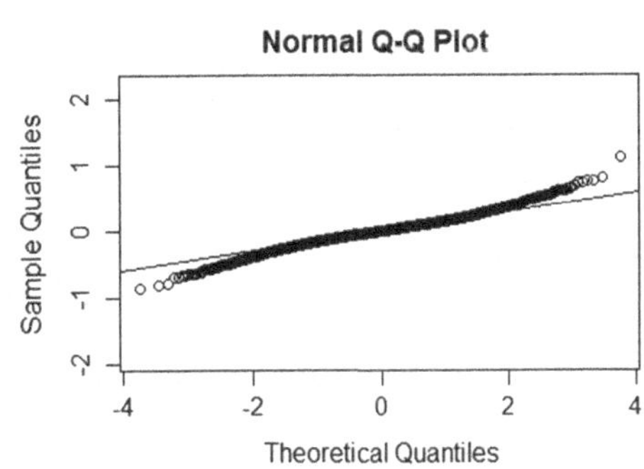

Fig. 4 Normal Q-Q plot

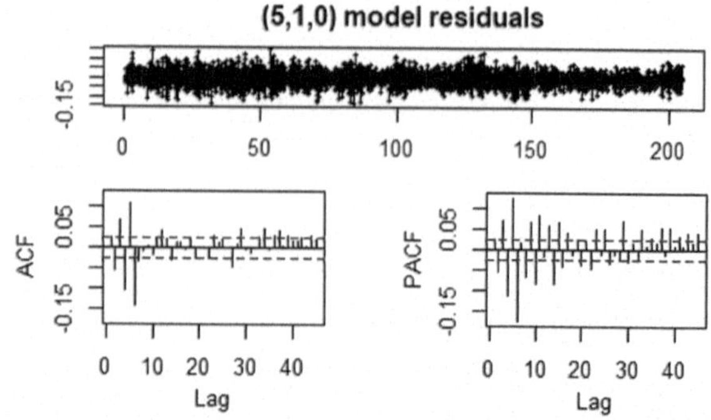

Fig. 5 Model residual with ACF and PACF

Here the q line fits exactly crosses the first and third quartile of the plot, proving the residuals to be under normal distribution. ACF and PACF for residuals: The residuals in the figure show that the residuals are in the range between(−1, 1). Hence, we can consider that they are in a normal distribution. ACF and PACF compare the present and the previously observed values to check the correlation in the data. Correlation is a measure of dependency observed in the data. Also, the Auto Correlation Function (ACF) shows that the correlation is significantly higher at lag 5. Partial Auto Correlation Function (PACF) also shows the same, and hence the fact is confirmed. Also, the ACF and PACF plot shows most of the values below the confidence interval, showing the lesser correlation. Generally, estimation algorithms treat the data as completely independent (Fig. 5).

4 Result and Conclusion

The time series plot for the forecast in an inbuilt library function in most programming that could be used directly to arrive at the conclusions. The result is the prediction for the next 50 days. Also, note that the predictions are done without considering the seasonality factor.

This forecast may be a naive model, but it illustrates the process of choosing an ARIMA model, and could also serve as a benchmark to grade against as more complex models are built. Hence, the seasonality factor has introduced a correlation to a certain amount in the dataset. This has reduced the flat line predictions as observed in Fig 6.

The validated model was compared with the predicted model with the last few existing values. Table 2 gives the exact figures of the plot. These values indicate that the values stick around the mean of the time series. We can see the deviation of the data, however, close to the values. These values could be further used to calculate the Air Quality index ratio and other standard parameters. Hence, the future RSPM values have been predicted effectively using time-based time series analysis (Fig. 7).

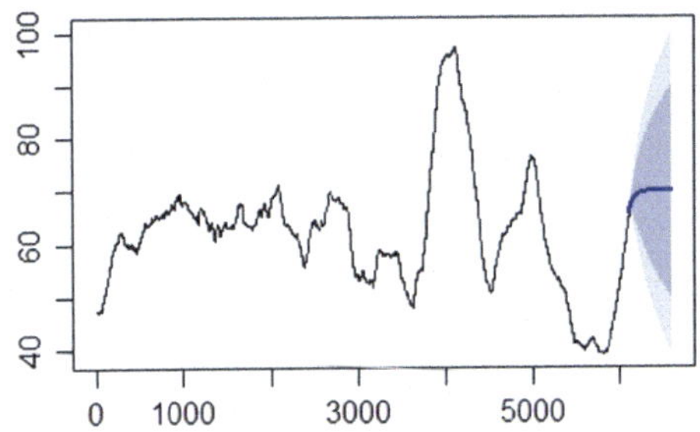

Fig. 6 ARIMA (5, 1, 0) forecast

Table 2 Actual values and fitted values of AIMA model

Actual values	Fitted values
90	64.99655
69	65.11362
77	65.20260
58	65.26123
50	65.33629
48	65.37744

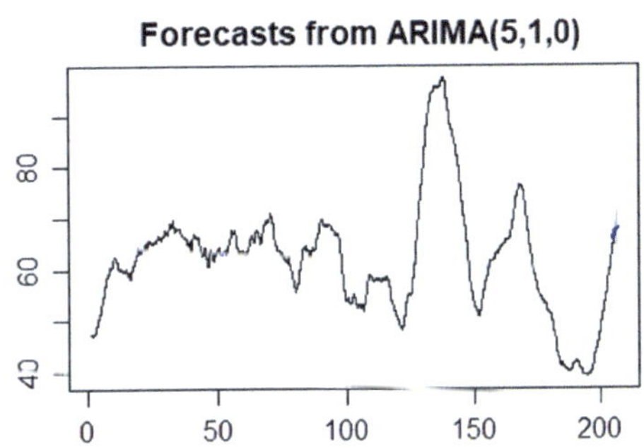

Fig. 7 Forecast with seasonality for model (5, 1, 0)

This concept of Time series analysis holds well for non-linear processes considered to be stationary in nature. A lot of statistical tools were explored and used for effective analysis and useful interpretation of the given data. The effectiveness of the ARIMA model was examined and found to be very effective even though it is a classical approach.

References

1. Bernal, J.L., Cummins, S., Gasparrini, A.: Interrupted time series regression for the evaluation of public health interventions: a tutorial. Int. J. Epidemiol. [Internet] dyw098 (2016). https://academic.oup.com/ije/article-lookup/doi/10.1093/ije/dyw098

2. Agarwal, P., Tangirala, A.K.: Reconstruction of missing data in multivariate processes with applications to causality analysis. Int. J. Adv. Eng. Sci. Appl. Math. [Internet] **9**, 196–213 (2017). http://link.springer.com/10.1007/s12572-017-0198-1
3. Eymen, A., Köylü, Ü.: Seasonal trend analysis and ARIMA modeling of relative humidity and wind speed time series around Yamula Dam. Meteorol. Atmos. Phys. [Internet] **131** 601–12 (2019). http://link.springer.com/10.1007/s00703-018-0591-8
4. Elsa Shaji, H,, Tangirala, A.K., Vanajakshi, L.: Evaluation of clustering algorithms for the prediction of trends in bus travel time. Transp. Res. Rec. J. Transp. Res. Board [Internet] **2672**, 242–252 (2018). http://journals.sagepub.com/doi/10.1177/0361198118791365
5. Nath, B., Dhakre, D., and, Bhattacharya, D.: Forecasting wheat production in India: An ARIMA modelling approach. J. Pharmacogn. Phytochem. **8**, 2158–2165 (2019)
6. Nyoni, T., Zimbabwe, H.: Forecasting Australian CPI using ARIMA models. Econ. Policy (2007)
7. Chaudhary, V., Deshbhratar, A., Kumar, V., Paul, D.: Time series based LSTM model to predict air pollutant' s concentration for prominent cities in India. Udm, vol. 18. (2018)
8. Bhargava, S. https://www.kaggle.com/shrutibhargava94/india-air-quality-data

A Comprehensive Survey on Down Syndrome Detection in Foetus Using Modern Technologies

I. M. Megha, S. Vyshnavi Kowshik, Sadia Ali, and Vindhya P. Malagi

Abstract Down syndrome is a genetic disorder which occurs due to the presence of an extra chromosome. This trisomy known as trisomy 21 leads to varying degrees of disability from physical disabilities like growth delays to mental disabilities. The paper focuses on exploring the different methodologies for the enhancement of ultrasound images and to extract the feature set from them; using machine learning algorithms to determine the probabilistic measure of the foetus being born with down syndrome. The proposed system aims to develop an end to end application, built upon the previous works in this field, utilising a dataset consisting of ultrasound scans of the first trimester, and hence would provide insights on the prominence of DS in India, and produces the probabilistic measure of the foetus being born with DS, to aid the doctor in prescribing further invasive test.

Keywords Down syndrome(DS) · Ultrasound images · First trimester

1 Introduction

Down syndrome also known as trisomy 21 is the most common aneuploidy in live born babies. It is an incurable genetic disorder and it causes retardation in cognitive and physical growth which can lead to developmental issues. Such individuals are

I. M. Megha (✉) · S. Vyshnavi Kowshik · S. Ali · V. P. Malagi
Department of Computer Science and Engineering, Dayananda Sagar College of Engineering, Bengaluru, Karnataka, India
e-mail: immeghai2@gmail.com

S. Vyshnavi Kowshik
e-mail: vyshnavikowshik@gmail.com

S. Ali
e-mail: alisadia303@gmail.com

V. P. Malagi
e-mail: vindhya-cs@dayanandasagar.edu

S. C. Satapathy et al. (eds.), *Intelligent Data Engineering and Analytics*, Advances in Intelligent Systems and Computing 1177,
https://doi.org/10.1007/978-981-15-5679-1_73

subject to some serious health complications. This syndrome has a different effect on all affected individuals.

The current protocol for screening Down syndrome includes maternal serum levels of PAPP-A, free beta-hCG and the ultrasound measurement of NT at the first trimester.

The most accurate tests for Down syndrome are Amniocentesis and Chorionic villi sampling.

Both these tests involve inserting a needle through the mother's abdomen owing to an increase in the risk of miscarriage.

The ultrasound markers for Down syndrome can be grouped into three categories.

High-level markers:

- Nuchal skinfold thickness. This marker was the first ultrasonographic marker to be associated with Down syndrome. Nuchal fold is the distance between the bone of the back of the head and overlying skin. A nuchal fold thickness greater than 5 mm at 15–18 weeks is considered as increased.
- Nasal Bone. Measurement of the nasal bone is feasible in the first trimester and can be obtained along with the nuchal translucency assessment. An ossified nasal bone can be seen as a solid and visible ridge within the top of the nose at 15–22 weeks. A non-ossified nasal bone between 15and 22 weeks increases the down syndrome risk factor.

Mid-level markers:

- Ratio between the length of the humerus and the average humerus length based on the baby's head size.

Low-Level markers:

- Ratio between the measured and expected size of the femur.
- Echogenic left ventricular foci.

2 Methodology

2.1 Development in the Medical Field

Down syndrome has been linked with many structural anomalies over the years. But with every new study in this regard, the significant markers for this syndrome have been deduced. Conducting tests for the presence of these markers can help in early diagnosis of the syndrome.

Following is a list of brief studies conducted in this relation to deduce vital features that indicate the presence of Down syndrome.

2.1.1 Consequences of Increased Thickness of Nuchal Translucency

In Nicolaides et al. [1], an ultrasound scan is used to examine the thickness of nuchal fold in the first trimester and its relation with chromosomal abnormalities. The study in [1], established two facts. The first fact confirmed that the presence of nuchal translucency of 3 mm or more increases the risk of chromosomal abnormality by more than 10 fold and the second one confirmed that the absence of translucency decreases the risk by 3 fold. A foetus with increased nuchal translucency is at a greater risk of having a chromosomal abnormality.

Cystic hygromas and nuchal oedema lead to increased nuchal translucency. The former can be associated with Turner's syndrome. The latter in contrast is indicative of the presence of trisomies like Down syndrome. Most foetuses under study which were detected to have chromosomal abnormalities were terminated in the first semester, this hindered detailed post-mortem evaluation of the foetuses as the pathology of the nuchal lesion could not be ascertained.

2.1.2 Nuchal Translucency, Nasal Bone Length and Maternal Age as Prime Factors in Down Syndrome

According to Otaño et al. [2], when maternal age and nuchal translucency at 11–14 weeks were used as screening parameters for Down syndrome, a detection rate of 77% and a false positive rate of 5% was achieved. This increased false positive rate became a problem and could lead to termination of a pregnancy. Combining the absence of nasal bone with the above two markers gave a much lower false positive rate. The detection rate improved to 85% and the false positive rate reduced to 1% (Fig. 1).

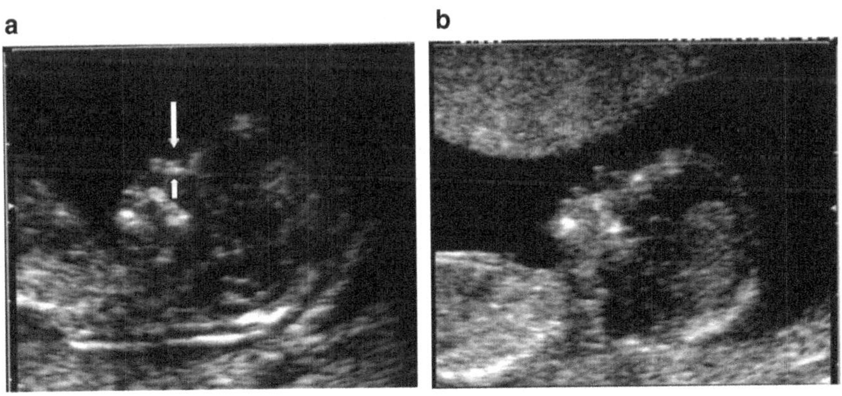

Fig. 1 Absence of foetal nasal bone and down syndrome. Sagittal view of the foetal face at 12 weeks of gestation showing **a** the skin of the nose (long arrow) and the ossified nasal bone (short arrow), and **b** the absence of nasal bone. (*Source* From to Otaño et al. [2])

2.1.3 Imaging Angles and Foetal Presentation

In Cho et al. [3], NFT which is measured between 18 and 20 weeks, is considered to be a soft marker for Down syndrome. The normal value of NFT is less than 6 mm. A value greater than this is indicative of a chromosomal anomaly like Down syndrome. NFT measurement shows some variability based on the imaging angles and foetal presentations. Foetuses in transverse presentation tend to have a greater value of NFT which is considered normal in comparison to foetuses in the breech presentation which tend to have lower NFT values. To measure NFT correctly the calliper positioning must be correct. Even with correct calliper positions, increased NFT might be measured which will lead to false positive results. These reasons must be kept in mind while taking nuchal fold measurements so as to reduce false positive cases and make the marker more relevant in the detection of Down syndrome.

2.1.4 Statistical Measure of Head Biometrics

In Sahli et al. [4], the authors have proposed a method through which head biometrics can be used for anomaly detection. The research conducted involved estimation of biometrics like HC, BPD, OFD. These metrics were selected based on the manual measurements made by an expert. The relation between these measurements and the occurrence of various head anomalies that are common was studied. The relation revealed that the statistical measure of the head biometrics for estimation of head anomalies shows high accuracy, and thus this correlation can be used to make deductions about head anomalies.

This study [4], of accurately measuring head biometrics can be extended to bone biometry. Accurate measurements of bone biometry can act as a supplementary marker for Down syndrome. Foetuses with Down syndrome have abnormal bone biometry and abnormal femur to biparietal diameter ratio. If these metrics are accurately measured then the classification of normal and abnormal values can be done more precisely.

2.2 Development in the Technological Front

Following is a list of studies that have been made in the field in this regard.

2.2.1 Automatic Edge Detection Scheme

Khashman et al. [5], has observed that the ultrasound images are very noisy and have many background unwanted intensities. To overcome this limitation, image recognition can be used wherein a model takes the raw noisy input, usually through edge detection techniques and recognises the various elements present in an image. The method suggested in [5], is called the Automatic Edge Detection Scheme (AEDS). The following features characterise this method. It reduces high computation, it's more user-friendly, and is adaptable to noise.

The method has two phases. In the first phase, the fast format of the Laplacian of the Gaussian edge detection is used. Multiscale processing in the edge detection systems is computation intensive. The computation can be reduced by using a properly scaled image as input. The second phase is the training of a neural network. The neural network consists of three layers. Input layer, hidden layer and output layer. Dimension adjustment for the images is done to reduce their size so that they can be fed to the input layer. A window of size 100 × 100 is used as input size. It is centred such that no useful information is lost. The output is digitally represented as a 0 or 1. The model achieved a 92% success rate on the test images and had a recognition time of 0.63 s.

2.2.2 Multi-Layer Perceptron with the Logistic Function

Wee et al. [6] study is divided into two parts. The first part involves finding the region of interest in an ultrasound scan and using it to train a network. The second part involves the automated measurement of NT using the BIFP computerised method.

This automated method can detect and recognise the foetal NT based on two-dimensional ultrasound images by using artificial neural network techniques (achieved an accuracy of 93.33%). ANN is used. Multi-layer perceptron with the logistic function. 150 input images of dimensions 30 × 50. 1500 input nodes, 100 hidden nodes, 1 output node. The backpropagation algorithm is used.

2.2.3 Neural Network—Backpropagation Algorithm

In Anjit et al. [7], a backpropagation neural network has been proposed for nasal bone detection. The approach in [7], uses a neural network based on a backpropagation algorithm. Still, images from the ultrasound scan are taken and preprocessed. Preprocessing involves smoothening the image using a median filter. These images are then segmented. Images with a properly developed nasal bone, as well as images without a nasal bone are used to train the model. The model learns by feature extraction and by minimising the error function through backpropagation. Feature extraction is done either in the spatial domain, where the actual pixels in the image are modified or it can be done in the transform domain where an orthogonal transform of the image is modified. After training, the model is tested on new unseen images of ultrasound scans with and without nasal bone. Testing reveals that the detection rate of nasal bone for the transform domain is higher than the spatial domain.

2.2.4 Assessment of Dermatoglyphic Indices Using Support Vector Machines

Wojtowicz et al. [8] assessment of dermatoglyphic indices of Down Syndrome in infants is done using an intelligent information system. Dermatoglyphics is the dermal ridge configurations on the digits, palms and soles. The dermatoglyphic prints are processed automatically for the medical diagnosis of genetic disorders. Unusual combinations of dermatoglyphic patterns in infants are an indication of the presence of genetic disorders.

Image processing and pattern recognition algorithms like SVMs (with RBF and triangular type kernels) is used for pattern classification of the hallucal area of sole and fingerprints. The method in [8], uses Principal Component Analysis, multi scale pyramid decomposition and the triangular kernel function in the SVM algorithm to detect pattern types that are common in Down syndrome, i.e. left and right loops of the palm print at an ATD angle which is calculated priorly. The method shows a classification accuracy of 95% for the left loop type and a classification accuracy of 96.7% for the right loop type.

2.2.5 Artificial Neural Network

In Soleimani et al. [9], the authors have proposed to develop an ANN to predict developmental disorder in foetuses. The input parameters for the network are prenatal and perinatal histories such as maternal age at delivery, parity, parental consanguinity, miscarriages, pregnancy complications, birth method, infant gender, neonatal and/or perinatal asphyxia, gestational age (completed weeks), birth weight, neonatal convulsion and/or postnatal epilepsy, hyperbilirubinemia leading to phototherapy or blood exchange transfusion. A three-layer MLP is used in [9]. The first (input) layer consists of 14 neurons (i.e. independent variables and factors). The second (hidden) layer is built up by examination (14 neurons) and the last (output) layer contains one neuron (i.e. outcome/dependent variable). The input variable selection of the ANN seems reasonable as most selected variables were reportedly associated with the developmental disorder: low birth weight, history of high risk pregnancy, prematurity, neonatal and/or postnatal convulsion, neonatal sepsis, RDS, icterus leading to phototherapy or exchange transfusion, parity (< 1 or > 3) and complications during pregnancy were determined as ordered important factors.

2.2.6 Discrete Wavelet Transform Filter and Hough Transform Algorithm

According to Sahli et al. [10], a fully automatic detection system of foetal head from ultrasound images has been proposed. The first stage is the Foetal Image Preprocessing. Here two filters are used for this purpose. Firstly, Anisotropic Diffusion Filter is a non-linear filtering method which focuses on variance reduction, mean conservation and boundary detection in the presence of speckle noise. Secondly, Discrete Wavelet Transform Filter recovers the discontinuous edge and non-stationary signal problems. This filter produces image enhancement and allows multi-resolution analysis. Furthermore, DWT employs two filters, one to decompose the noisy image and the other to process the image to obtain a denoised image. The last stage is the extraction of the Region of Interest from the ultrasound, in this case, the head dimensions. Here, after the denoising stage, head localization needs to be done. Hough Transform algorithm was used for this purpose, as it is excellent to find the geometrical edges, in this case, the somewhat circular shape of the foetal head. The number of pixels belonging to the circle is also estimated by HT. PNSR and NCC values of the DWT method gives a higher performance with improved image quality than ADF. For the circular form of the foetal head structure, HT technique is the most accurate as compared to expert detection based on ROI circular estimation. This method gives 97% detection accuracy which can be used for real ultrasound applications.

2.2.7 Artificial Neural Network: Feed-Forward and Feed-Backward

In Catic et al. [11], two neural network architectures have been proposed. The Artificial Neural Network takes in five input parameters. Maternal serum plasma protein-A, free beta-human chorionic gonadotropin, presence or absence of nasal bone, mother's age and foetal nuchal translucency thickness. The network has one output parameter which indicates whether the subject has one of the following prenatal syndromes or not. The syndromes are Patau syndrome, Edwards syndrome, Down syndrome, Turner syndrome, Klinefelter syndrome. The two proposed architectures are Feed-forward architecture, feedback architecture. In feed-forward architecture, the back-propagation algorithm is used. The output values are compared with the actual values and through backpropagation, the weights and biases are adjusted in every iteration. This architecture uses the Levenberg–Marquardt algorithm (LMA) for training. In the feedback architecture, the gradient descent algorithm is used for training. The error between the output value and the desired output is minimised by arriving at the global minimum in the gradient descent curve. A sensitivity of 92% has been achieved for feed-forward architecture and a sensitivity of 99% has been obtained for the feedback architecture.

2.2.8 OCR Algorithm, Adaptive Synthetic Sampling Approach and Naives Bayes Algorithm

According to Ramanathan et al. [12], the prediction is done based on the NT (Nuchal translucency), ultrasound scan data which shows the translucency, nasal bone size. In the data processing stage, the scanned data is converted to machine readable format by the application of the OCR algorithm. This method also handles the values of the missing attributes. In the next step, groups are formed based on the similarity in data, and labels are provided based on the available reference. Clustering methods are applied to derive the group labels. Firstly, K-means or K-medoids is applied. K-medoids gives better performance as it considers centre point for clustering. Secondly, Hierarchical clustering is applied which uses the Euclidean distance as the clustering criteria. Data augmentation is done using the Adaptive Synthetic Sampling Approach (ADASYN), as the data set might contain only some of the high-risk patients. Now, using the derived labelled data from the clustering of the dataset, classification methods are applied to them to find the probability measure of a foetus having trisomy 21. Firstly, Naive Bayes classification method is applied along with k-fold cross-validation algorithm to get the probability. Secondly, Random forest method is used which constructs many decision-tree based on the features in the feature set, this assigns the mean prediction value. Deep Learning method is applied to the feature set. Artificial Neural Network with ReLu and Softmax as the activation function is fed with the feature set data. We can conclude that the k-means performs well on the given data set and the Naive Bayes performs well, compared to other classification techniques.

2.2.9 Wavelet Analysis Tool

Sciortino et al. [13], proposes a new methodology to help the doctors to identify the nuchal region automatically and also to be able to measure the accurate thickness of the nuchal translucency measure. The proposed method [13], can detect the neck region from the US images automatically without any human intervention. Because of this, the inter and intra observer variability can be surpassed. To extract the feature set from the image, Wavelet analysis tool signal processing is applied. This comprises the Filter bank stage, which has a low pass and high pass filter in respective order. Next, Nuchal bounding box detection is applied. This consists of two steps: firstly, identification of the NT, Secondly, measurement of the NT. The identification is done by measuring the echogenic and anechogenic structures. After identifying the nuchal region, the delimit regions are focused and obtained. The result of this study provides the ability to properly select the thresholds; the best value parameters are identified by the quantitative analysis of the binary masks. The average error from the automatic detection is reduced as compared to the error from the human annotation. This methodology [13], based on the ground truth provided, the system has shown a true positive rate of 99.95% with respect to the detection of the nuchal translucency.

2.2.10 Fully Convolutional Network

Sinclair et al. [14], proposes an automated method utilising Fully Convolutional Networks (FCN) to determine the measurements of foetal head circumference (HC) and biparietal diameter (BPD). Data Preprocessing Steps followed are generating ellipse mask images from DICOM images. Retrieving unannotated frames from the video feed of the scan, and finally cropping and scaling the images so that they can be used for training the network in [14]. HC and BPD are measured from TV plane. An FCN pretrained with Image Net is used to estimate the head biometrics. The performance measure of this method is 98%.

2.2.11 Convolutional Neural Network with Resblocks

In Sobhaninia et al. [15], the authors have proposed a multi task CNN for segmentation of a US scan and estimating head circumference from it. It is based on semantic segmentation technique. Two modules have been proposed. The first one performs segmentation on 2D ultrasound images and feeds it to different layers. It is composed of Resblocks, which, in turn, are composed of convolution pooling layers. The second module is an Ellipse Tuner which consists of decoder blocks that perform upsampling of the feature maps and produce the final segmentation unit. To estimate ellipse parameters feature maps are used. This multi-layer network demonstrates better performance for measuring foetal biometrics than a single layer network.

3 Conclusion

In this literature survey, nuchal translucency and nasal bone length are deduced as the primary factor for determining the presence or absence of the possibility of Down syndrome in a foetus. Furthermore, the complexities in the extraction of the feature set from the ultrasound images are also discussed. Usage of image enhancement algorithms such as DWT are suggested. The authors intend to design a system with ultrasound images as the input, apply preprocessing algorithm on the images for enhancement, then extract the feature set from the image which includes NT, nasal bone length along with other head biometrics in their work. With the help of machine learning algorithms on the feature set, the probabilistic measure of foetus being born with down syndrome can be determined, which can aid the doctor to prescribe further invasive tests.

References

1. Nicolaides, K., Azar, G., Byrne, D., Mansur, C. Marks, K.: Fetal nuchal translucency: ultrasound screening for chromosomal defects in first trimester of pregnancy. BMJ Publishing Group **304**(6831), 867–869 (1992). https://doi.org/10.1136/bmj.304.6831.867
2. Otaño, L., Aiello, H., Igarzábal, L., Matayoshi, T. Gadow, E.: Association between first trimester absence of fetal nasal bone on ultrasound and down syndrome. Prenat. Diagn. **22**(10), 930–932 (2002). https://doi.org/10.1002/pd.431
3. Cho, J., Kim, K., Lee, Y., Toi, A.: Measurement of nuchal skin fold thickness in the second trimester: influence of imaging angle and fetal presentation. Ultrasound Obstet. Gyneco. **25**(3), 253–257 (2005). https://doi.org/10.1002/uog.1847

4. Sahli, H., Mouelhi, A., Hadada, F., Rachdi, R., Sayadi, M., Fnaiech, F.: Statistical analysis based on biometrie measures for fetal head anomaly characterization. In: 2018 IEEE 4th Middle East Conference on Biomedical Engineering (MECBME) (2018). https://doi.org/10.1109/mecbme.2018.8402440
5. Khashman, A., Curtis, K. (n.d.).: Automatic edge detection of foetal head and abdominal circumferences using neural network arbitration. In: ISIE '97 Proceeding of the IEEE International Symposium on Industrial Electronics (1997). https://doi.org/10.1109/isie.1997.648910
6. Wee, L.K., Min, T.Y., Arooj, A., Supriyanto, E.: Nuchal translucency marker detection based on artificial neural network and measurement via bidirectional iteration forward propagation. WSEAS Trans. Info. Sci. App (2010)
7. Anjit, T., Rishidas, S.: Identification of nasal bone for the early detection of down syndrome using back propagation neural network. In: 2011 International Conference on Communications and Signal Processing (2011). https://doi.org/10.1109/iccsp.2011.5739286
8. Wojtowicz, H., Wajs, W.: Medical decision support system for assessment of dermatoglyphic indices and diagnosis of down's syndrome. Intell. Decis. Technol., 69–78 (2012). https://doi.org/10.1007/978-3-642-29920-9_8
9. Soleimani, F., Teymouri, R., Biglarian, A.: Predicting developmental disorder in infants using an artificial neural network (2013)
10. Sahli, H., Ben Slama, A., Zaafouri, A., Sayadi, M., Rachdi, R.: Automated detection of current fetal head in ultrasound sequences. In: 2016 International Image Processing, Applications and Systems (IPAS) (2016). https://doi.org/10.1109/ipas.2016.7880142
11. Catic, A., Gurbeta, L., Kurtovic-Kozaric, A., Mehmedbasic, S., Badnjevic, A.: Application of neural networks for classification of patau, edwards, down, turner and klinefelter syndrome based on first trimester maternal serum screening data, ultrasonographic findings and patient demographics. BMC Med. Genomics **11**(1) (2018). https://doi.org/10.1186/s12920-018-0333-2
12. Ramanathan, S., Sangeetha, M., Talwai, S., Natarajan, S.: Probabilistic determination of down's syndrome using machine learning techniques. In: 2018 International Conference on Advances in Computing, Communications and Informatics (ICACCI) (2018). https://doi.org/10.1109/icacci.2018.8554392
13. Sciortino G, e. (2019). Automatic detection and measurement of nuchal translucency. PubMed—NCBI. [online] Ncbi.nlm.nih.gov. https://www.ncbi.nlm.nih.gov/pubmed/28126630. Retrieved 30 Nov. 2019. https://doi.org/10.1016/j.compbiomed.2017.01.008
14. Sinclair, M., Baumgartner, C., Matthew, J., Bai, W., Martinez, J., Li, Y., Smith, S., Knight, C., Kainz, B., Hajnal, J., King, A., Rueckert, D.: Human-level performance on automatic head biometrics in fetal ultrasound using fully convolutional neural networks. In: 2018 40th Annual International Conference of the IEEE Engineering in Medicine and Biology Society (EMBC) (2018). https://doi.org/10.1109/embc.2018.8512278
15. Sobhaninia, Z., Rafiei, S., Emami, A., Karimi, N., Najarian, K., Samavi, S., Soroushmehr, S.: Fetal ultrasound image segmentation for measuring biometric parameters using multi-task deep learning (2019). https://doi.org/10.1109/embc.2019.8856981

The manufacturer's authorised representative in the EU is Springer Nature Customer Service Centre GmbH, Europaplatz 3, 69115 Heidelberg, Germany. If you have any concerns regarding our products, please contact ProductSafety@springernature.com

Printed and bound by CPI Group (UK) Ltd, Croydon, CR0 4YY
30/09/2026
02225232-0016